Lecture Notes in Artificial Intelligence   10086

Subseries of Lecture Notes in Computer Science

More information about this series at http://www.springer.com/series/1244

Jinyan Li · Xue Li
Shuliang Wang · Jianxin Li
Quan Z. Sheng (Eds.)

Advanced Data Mining and Applications

12th International Conference, ADMA 2016
Gold Coast, QLD, Australia, December 12–15, 2016
Proceedings

 Springer

Editors
Jinyan Li
University of Technology Sydney
Ultimo, NSW
Australia

Xue Li
University of Queensland
Brisbane, QLD
Australia

Shuliang Wang
Beijing Institute of Technology
Beijing
China

Jianxin Li
University of Western Australia
Crawley, WA
Australia

Quan Z. Sheng
University of Adelaide
Adelaide, SA
Australia

ISSN 0302-9743   ISSN 1611-3349   (electronic)
Lecture Notes in Artificial Intelligence
ISBN 978-3-319-49585-9   ISBN 978-3-319-49586-6   (eBook)
DOI 10.1007/978-3-319-49586-6

Library of Congress Control Number: 2016957378

LNCS Sublibrary: SL7 – Artificial Intelligence

Printed on acid-free paper

This Springer imprint is published by Springer Nature
The registered company is Springer International Publishing AG
The registered company address is: Gewerbestrasse 11, 6330 Cham, Switzerland

Preface

The year 2016 marked the 12th anniversary of the International Conference on Advanced Data Mining and Applications (ADMA). This conference series has been organized and held in China since 2005. ADMA 2016 was the first time the conference was held outside China, taking place on Australia's Gold Coast during December 12–15, 2016. ADMA aims at bringing together both young researchers and senior experts from the world, and also at providing a leading international forum for the dissemination of original data mining findings and practical data mining experience. Over the years, ADMA has grown to become a flagship conference in the field of data mining and applications.

This volume comprises 57 thoroughly reviewed and carefully revised full-length papers that were all presented at the conference. The conference Program Committee received 105 high-quality manuscripts. Each of them was assigned to at least three Program Committee members for comments. All papers were rigorously reviewed and had at least three reviews. Some borderline papers were even reviewed by five members of the committee if the first three reviews were not adequate for making a rejection or acceptance decision. Several research teams showed strong interest in ADMA 2016, and submitted multiple high-quality papers of different topics. It was difficult for the committee chairs to exclude some of these papers so as to reserve some space for other teams to present their latest research achievements.

Although many papers were worthy of publication, only 18 spotlight research papers (acceptance rate of 17%, 15 pages each) and 39 regular research papers (13 pages each) could be accepted for presentation at the conference and publication in this volume. The selected papers covered a wide variety of important topics in the area of data mining, including parallel and distributed data mining algorithms, mining on data streams, graph mining, spatial data mining, multimedia data mining, Web mining, the Internet of Things, health informatics, and biomedical data mining. The conference program of ADMA 2016 was also complemented by four outstanding keynotes given by Yufei Tao, Pablo Moscato, Scott Johnston, and Gao Cong, as well as several research demonstrations. We would like to particularly thank the keynote speakers for presenting algorithms and ideas to solve data mining problems in the frontier areas of the field.

We thank the Program Committee members for their time and invaluable comments. The schedule of the reviewing process was extremely tight. The members' tremendous efforts to complete the review reports before the deadline are greatly appreciated. We read all of the review reports and found some of them to be truly excellent, as good as what is usually found in a survey—critical, digestive, and detailed. These comments were very helpful for us in selecting the papers. Thank you all and may the papers collected in this volume inspire the readers and open new windows of research.

We would like to express our gratitude to all individuals, institutions, and sponsors that supported ADMA 2016. This high-quality program would not have been possible without the expertise and dedication of our Program Committee members. We are grateful for the guidance of the general chairs (Michael Sheng and Osmar Zaiane), the tireless efforts of the demonstration chairs (Ji Zhang and Mingkui Tan), the publicity chair (Xiujuan Xu), the publication chair (Jianxin Li), the award chair (Zhifeng Bao), the Web chair (Wenjie Ruan), the special issue chair (Yongrui Qin), and the local organization chair (Weitong Chen). We would also like to acknowledge the support of the members of the conference Steering Committee. All of them helped make ADMA 2016 a success. Finally, we would like to thank all researchers, practitioners, and students who contributed with their work and participated in the conference. We hope that you find the papers in the proceedings interesting and stimulating.

December 2016   Jinyan Li
                                                        Xue Li
                                                   Shuliang Wang

Organization

General Co-chairs

Michael Sheng   University of Adelaide, Australia
Osmar Zaiane   University of Alberta, Canada

Program Co-chairs

Jinyan Li   University of Technology Sydney, Australia
Xue Li   University of Queensland, Australia
Shuliang Wang   Beijing Institute of Technology, China

Demo Co-chairs

Mingkui Tan   University of Adelaide, Australia
Ji Zhang   University of Southern Queensland, Australia

Proceedings Chair

Jianxin Li   University of Western Australia, Australia

Awards Committee Chair

Zhifeng Bao   RMIT, Australia

Publicity Chair

Xiujuan Xu   Dalian University of Technology, China

Special Issue Chair

Yongrui Qin   University of Huddersfield, UK

Sponsorship Co-chairs

Shichao Zhang   Guangxi Normal University, China
Xianxiu Si   China Academy of Telecom Research, China

Local Chair

Weitong Chen   University of Queensland, Australia

Web Chair

Wenjie Ruan   University of Adelaide, Australia

Steering Committee

Jie Cao	Nanjing University of Finance and Economics, China
Xue Li	University of Queensland, Australia (Chair)
Michael Sheng	University of Adelaide, Australia
Jie Tang	Tsinghua University, China
Kyu-Young Whang	Korea Advanced Institute of Science and Technology, Korea
Min Yao	Zhejiang University, China
Osmar Zaiane	University of Alberta, Canada
Chengqi Zhang	University of Technology Sydney, Australia
Shichao Zhang	Guangxi Normal University, China
Shuliang Wang	Beijing Institute of Technology, China

Program Committee

Ayan Acharya	University of Texas at Austin, USA
Djamal Benslimane	Lyon 1 University, France
Jin Chen	Michigan State University, USA
Liang Chen	RMIT University, Australia
Chenwei Deng	Beijing Institute of Technology, China
Xiangjun Dong	Qilu University of Technology, China
Xiu Susie Fang	University of Adelaide, Australia
Stefano Ferilli	University of Bari, Italy
Philippe Fournier-Viger	Harbin Institute of Technology Shenzhen Graduate School, China
Xiaoying Gao	Victoria University of Wellington, New Zealand
Sung Ho Ha	Kyungpook National University, South Korea
Jing He	Victoria University, Australia
Qing He	Institute of Computing Technology, CAS, China
Wei Hu	Nanjing University, China
Xiaohua Tony Hu	Drexel University, USA
Guangyan Huang	Deakin University, Australia
Akihiro Inokuchi	Kwansei Gakuin University, Japan
Daisuke Kawahara	Kyoto University, Japan
Gang Li	Deakin University, Australia
Haiquan Li	University of Arizona, USA
Jianxin Li	University of Western Australia
Jinyan Li	University of Technology Sydney, Australia
Xue Li	University of Queensland, Australia
Zhixin Li	Guangxi Normal University, China
Bo Liu	Guangdong University of Technology, China

Guohua Liu	Yanshan University, China
Yubao Liu	Sun Yat-Sen University, China
Xudong Luo	Sun Yat-Sen University, China
Jiangang Ma	Victoria University, Australia
Marco Maggini	University of Siena, Italy
Toshiro Minami	Kyushu Institute of Information Sciences and Kyushu University Library, Japan
Yasuhiko Morimoto	Hiroshima University, Japan
Pablo Moscato	University of Newcastle, Australia
Shirui Pan	University of Technology Sydney, Australia
Tao Peng	University of Illinois at Urbana-Champaign, USA
Shweta Purawat	San Diego Supercomputer Center, UCSD, USA
Tieyun Qian	Wuhan University, China
Kai Qin	RMIT University, Australia
Yongrui Qin	University of Huddersfield, UK
Wenjie Ruan	University of Adelaide, Australia
Michele Ruta	Politecnico di Bari, Italy
Arun Kumar Sangaiah	VIT University, India
Dharmendra Sharma	University of Canberra, Australia
Ali Shemshadi	University of Adelaide, Australia
Michael Sheng	University of Adelaide, Australia
Guojie Song	Peking University, Beijing, China
Eiji Uchino	Yamaguchi University, Japan
Li Wan	Google, USA
Hongzhi Wang	Harbin Institute of Technology, China
Qinsi Wang	Carnegie Mellon University, USA
Shuliang Wang	Beijing Institute of Technology, China
Xianzhi Wang	University of Adelaide, Australia
Zhihui Wang	Fudan University, China
Jia Wu	University of Technology, Sydney, Australia
Zhiang Wu	Nanjing University of Finance and Economics, China
Feng Xia	Dalian University of Technology, China
Zhipeng Xie	Fudan University, China
Guandong Xu	University of Technology Sydney, Australia
Xiujuan Xu	Dalian University of Technology, China
Bing Xue	Victoria University of Wellington, New Zealand
Zijiang Yang	York University, UK
Lina Yao	University of New South Wales, Australia
Min Yao	Zhejiang University, China
Dayong Ye	University of Wollongong, Australia
Chao Yu	University of Wollongong, Australia
Fusheng Yu	Beijing Normal University, China
Qi Yu	Rochester Institute of Technology, USA
Osmar R. Zaïane	University of Alberta, Canada
Chunxia Zhang	Beijing Institute of Technology, China

Guang Lan Zhang	Boston University, USA
Mengjie Zhang	Victoria University of Wellington, New Zealand
Shichao Zhang	Guangxi Normal University, China
Wei Emma Zhang	University of Adelaide, Australia
Xin Zhao	University of Queensland, Australia
Nenggan Zheng	Zhejiang University, China
Yong Zheng	Illinois Institute of Technology, USA
Mingyang Zhong	University of Queensland, Australia
Hill Zhu	Florida Atlantic University, USA
Xiaofeng Zhu	Guangxi Normal University, China
Yi Zhuang	Zhejiang Gongshang University, China

Additional Reviewers

Al-Sahaf, Harith
Bai, Xiaomei
Bekele, Teshome Megersa
Boyu, Zhang
Chen, Qi
Dam, Thu-Lan
Fan, Wei
Gao, Qian
Gong, Saisai
Gueniche, Ted
Hanane, Amirat
Keskin, Derin
Li, Xin
Liu, Shaowu
Lu, Guangquan
Nguyen, Bach
Nguyen, Thuong
Peng, Jiajie
Qian, Xinqi

Quang Huy, Duong
Rahim, Azizur
Siers, Michael
Sun, Zequn
Vu, Huy Quan
Wan, Yao
Wang, Biying
Wang, Dongjing
Wang, Hongtao
Wang, Qingxiang
Wang, Wei
Wu, Cheng Wei
Wu, Runze
Xu, Peipei
Yang, Shuiqiao
Yu, Shuo
Zhang, Boyu
Zhang, Yongshan
Zhao, Qi

Contents

Research Papers

Demo Papers

Spotlight Research Papers

Effective Monotone Knowledge Integration in Kernel Support Vector Machines

Christopher Bartley$^{(\boxtimes)}$, Wei Liu, and Mark Reynolds

School of Computer Science, University of Western Australia,
Perth, WA 6009, Australia
christopher.bartley@research.uwa.edu.au
http://web.csse.uwa.edu.au

Abstract. In many machine learning applications there exists prior knowledge that the response variable should be increasing (or decreasing) in one or more of the features. This is the knowledge of 'monotone' relationships. This paper presents two new techniques for incorporating monotone knowledge into non-linear kernel support vector machine classifiers. Incorporating monotone knowledge is useful because it can improve predictive performance, and satisfy user requirements. While this is relatively straight forward for linear margin classifiers, for kernel SVM it is more challenging to achieve efficiently. We apply the new techniques to real datasets and investigate the impact of monotonicity and sample size on predictive accuracy. The results show that the proposed techniques can significantly improve accuracy when the unconstrained model is not already fully monotone, which often occurs at smaller sample sizes. In contrast, existing techniques demonstrate a significantly lower capacity to increase monotonicity or achieve the resulting accuracy improvements.

1 Introduction

Prior domain knowledge has many forms, such as knowledge of class invariance in regions of the input space [15] (e.g. if $tumor > 4 \wedge lymphnodes > 5 \implies Recurrence$), class invariance under transformations of the input [12] (used mostly for image processing) and shape knowledge such as convexity [24] and monotonicity [2,13]. We focus in this paper on *monotone* knowledge. A monotone relationship between X and Y means that an increase in X should not lead to a decrease in Y. For example, a house with three bedrooms should not be cheaper than one with two bedrooms (all other factors constant).

As the base algorithm we use the flexible and powerful kernel Support Vector Machine (SVM). Although this is an older algorithm dating back to 1995 [3], its power and flexibility was recently (somewhat surprisingly) reasserted by the Radial Basis Function (RBF) kernel variant achieving second place in a comprehensive test of 179 classifiers on 121 datasets [8], despite competing against many newer algorithms. Thus we use the Radial Basis Function kernel in this

© Springer International Publishing AG 2016
J. Li et al. (Eds.): ADMA 2016, LNAI 10086, pp. 3–18, 2016.
DOI: 10.1007/978-3-319-49586-6_1

paper, although our technique is applicable to other Mercer kernels such as sigmoid or polynomial. By extending kernel SVM we seek to improve on one of the best non-parametric classifiers.

While a monotone SVM algorithm (MC-SVM) was proposed in 2014 by Li and Chen in [2,13], this paper proposes a new framework for *how* the monotone knowledge is integrated, called *partially monotone SVM* (PM-SVM). Our experiments suggest Li and Chen's techniques yield only limited increases in monotonicity, and as a result the accuracy improvements are also limited. In contrast our techniques substantially increase monotonicity, and hence also result in substantial accuracy increases. We hope to stimulate interest in this area and have provided a matlab implementation of the algorithm on github[1].

This paper is organised as follows. Section 2 outlines partially monotone classification and SVM. Section 3 is our core contribution of a measure for *partial monotonicity* and new techniques for integrating it into SVM. Section 4 describes the experiments and Sect. 5 reports the results, concluding in Sect. 6.

2 Background

2.1 Partially Monotone Classification

Monotone prior knowledge is easy to obtain from experts or domain knowledge for many types of problems, and is informative without being overly prescriptive. Monotonicity between input variable x_j and output y may be defined as:

> For an increase in variable x_j, variable y should not decrease (all other variables held constant).

Monotonicity can apply to problems where the model output is ordered, such as regression and ordinal classification. This paper considers *ordinal classification*, where the task is to assign objects to two or more classes when there is an *order* assigned to the classes, but the *distances* between classes are irrelevant. Examples of ordinal classification include credit ratings (AAA, AA, A, BBB, ...), where a better rating may be considered 'higher'. Similarly a cancer diagnosis of No/Yes may be considered ordered (in terms of the risk of being cancerous).

The classification literature on monotonicity [2,5–7,13,16,17,19–21,23] ubiquitously defines monotonicity in the context of the *dominance relation* $\succeq$ (a partial ordering), which is defined as:

$$\mathbf{x} \succeq \mathbf{x}' \Leftrightarrow x_j \geq x_j', \quad \mathbf{x}, \mathbf{x}' \in \Re^m, j = 1..m \tag{1}$$

Thus a function $f : \Re^m \to \Re$ is defined as *monotone* if and only if:

$$\mathbf{x} \succeq \mathbf{x}' \Rightarrow f(\mathbf{x}) \geq f(\mathbf{x}'), \ \forall \mathbf{x}, \mathbf{x}' \in \Re^m \tag{2}$$

This definition describes monotonicity in *all* features, and if some features are not necessarily monotone, they are sometimes removed prior to analysis

---

[1] https://github.com/chriswbartley/PMSVM.

(e.g. [6,10]). For this paper we cater for the more general situation of 'partial' monotonicity in *some* features. We define partial monotonicity similarly to Kamp et al. [9], making the *ceteris paribus* assumption (all other features being equal):

Definition 1 *Partial Dominance.* *Given monotone features $C \subseteq \{1,...,n\}$, the partial order $\preceq_C$ over $\mathbf{x}, \mathbf{x}' \in \Re^n$ is:*

$$\mathbf{x} \preceq_C \mathbf{x}' \Leftrightarrow \begin{cases} x_j \leq x'_j, \forall j \in C \\ x_j = x'_j, \forall j \in \{1,...,n\}\backslash C \end{cases} \tag{3}$$

Definition 2 *Partially Monotone Function.* *Function $f : \Re^n \to \Re$ is monotonic in $C \subseteq \{1,...,n\}$ if*

$$\mathbf{x} \preceq_C \mathbf{x}' \Rightarrow f(\mathbf{x}) \leq f(\mathbf{x}'), \forall \mathbf{x}, \mathbf{x}' \in \Re^n \tag{4}$$

This definition may also be considered as a multi-variate and deterministic form of the probabilistic definitions for *First Order Stochastic Dominance (FSD) Monotonicity* as proposed by Wellman as a *qualitative influence* (S^+) [25]. In this context the classifier, as a deterministic function, has the degenerate probability distribution and $cdf(f(\mathbf{x}))\ FSD\ cdf(f(\mathbf{x}')) \Leftrightarrow f(\mathbf{x}) \geq f(\mathbf{x}')$.

2.2 Monotone Support Vector Machines

The existing monotone SVM techniques are of the 'constrained optimisation' type and use hard discrete constraints on the solution space. The first such SVM was by Pelckmans et al. in 2005 [18] and incorporated inequality constraints on the SVM primal problem. However it was limited to ordered datasets, with all data-points monotone increasing in *all* features. In 2014 Chen and Li [2] reframed the Pelckmans et al. approach to allow arbitrary constraints, which meant the data points no longer needed to be monotone ascending. Li and Chen also extended the technique to Fuzzy MC-SVM [13].

The constrained SVM used in MC-SVM is as follows. The SVM classifier is $f(\mathbf{x}) = sign[\mathbf{w}^T\psi(\mathbf{x}) + b]$, where $\psi : \mathcal{X} \to \mathcal{H}$ is a transformation into a potentially higher or even infinite dimension space. The optimal model is then:

$$\min_{\mathbf{w},e} \frac{1}{2}\mathbf{w}^T\mathbf{w} + B\sum_{k=1}^{N} e_k \tag{5}$$

$$\text{subject to} \quad y_k(\mathbf{w}^T\psi(\mathbf{x}_k) + b) \geq 1 - e_k, \quad k = 1, 2, ..., N$$

$$e_k \geq 0, \quad k = 1, 2, ..., N$$

Observing that $\mathbf{w}^T\psi(\tilde{\mathbf{x}}) \geq \mathbf{w}^T\psi(\mathbf{x}) \implies f(\tilde{\mathbf{x}}) \geq f(\mathbf{x})$ the constrained SVM takes a set of data point pairs $MC = \{(\mathbf{x}_i, \tilde{\mathbf{x}}_i), i = 1..M\}$ and augments the SVM problem with the constraints:

$$\mathbf{w}^T\psi(\tilde{\mathbf{x}}_i) \geq \mathbf{w}^T\psi(\mathbf{x}_i), \quad i = 1, 2, ..., M \tag{6}$$

The Lagrangian can then be constructed to include the constraint inequalities using the Karush Kahn Tucker conditions [11] and introducing multipliers β_i in addition to the usual α_k (and ν_k):

$$\mathcal{L}(\mathbf{w}, b; \boldsymbol{\alpha}, \boldsymbol{\beta}, \boldsymbol{\nu}) = \frac{1}{2}\mathbf{w}^T\mathbf{w} + B\sum_{k=1}^{N} e_k - \sum_{k=1}^{N}\alpha_k(y_k[\mathbf{w}^T\boldsymbol{\psi}(\mathbf{x}_k) + b] - 1 + e_k)$$

$$- \sum_{i=1}^{M}\beta_i(\mathbf{w}^T\boldsymbol{\psi}(\tilde{\mathbf{x}}_i) - \mathbf{w}^T\boldsymbol{\psi}(\underset{\sim}{\mathbf{x}}_i)) - \sum_{k=1}^{N}\nu_k e_k, \quad (7)$$

$$\alpha_k, \nu_k \geq 0 \ \forall k = 1, ..., N,$$
$$\beta_i \geq 0 \ \forall i = 1, ..., M$$

The stationarity condition (setting partial derivatives to zero) reveals that $\mathbf{w} = \sum_{k=1}^{N}\alpha_k y_k \boldsymbol{\psi}(\mathbf{x}_k) + \sum_{i=1}^{M}\beta_i(\boldsymbol{\psi}(\tilde{\mathbf{x}}_i) - \boldsymbol{\psi}(\underset{\sim}{\mathbf{x}}_i))$ and $\nu_k = B - \alpha_k$, and the solution for $\boldsymbol{\alpha}$ and $\boldsymbol{\beta}$ is a quadratic problem (see [2] for full working):

$$\underset{\boldsymbol{\alpha}, \boldsymbol{\beta}}{max} - \frac{1}{2}[\boldsymbol{\alpha}^T, \boldsymbol{\beta}^T]\mathbf{G}\begin{bmatrix}\boldsymbol{\alpha} \\ \boldsymbol{\beta}\end{bmatrix} + \mathbf{1}^T\boldsymbol{\alpha} \qquad (8)$$

$$\text{subject to} \sum_{k=1}^{N}\alpha_k y_k = 0,$$
$$0 \leq \alpha_k \leq B, k = 1, 2, ..., N$$
$$\beta_i \geq 0, i = 1, 2, ..., M$$

where $\mathbf{G} = \begin{bmatrix}\mathbf{G}^{11} & \mathbf{G}^{12} \\ \mathbf{G}^{21} & \mathbf{G}^{22}\end{bmatrix}$

$\mathbf{G}_{k,l}^{11} = y_k y_l \boldsymbol{\psi}(\mathbf{x}_k)^T\boldsymbol{\psi}(\mathbf{x}_l), \ k, l = 1, ..., N$

$\mathbf{G}_{k,i}^{12} = \mathbf{G}_{i,k}^{21} = y_k(\boldsymbol{\psi}(\tilde{\mathbf{x}}_i) - \boldsymbol{\psi}(\underset{\sim}{\mathbf{x}}_i))^T\boldsymbol{\psi}(\mathbf{x}_k), \quad k = 1, .., N, i = 1, .., M$

$\mathbf{G}_{i,j}^{22} = (\boldsymbol{\psi}(\tilde{\mathbf{x}}_i) - \boldsymbol{\psi}(\underset{\sim}{\mathbf{x}}_i))^T(\boldsymbol{\psi}(\tilde{\mathbf{x}}_j) - \boldsymbol{\psi}(\underset{\sim}{\mathbf{x}}_j)), \quad i, j = 1, ..., M$

The unknown mapping ψ is typically not solved directly but instead restricted to come from a dot product space described by a kernel function $K(\mathbf{x}, \mathbf{x}')$. As originally shown by Nachman Aronszajn in 1950, positive definite kernel functions $K : \mathcal{X} \times \mathcal{X} \rightarrow \Re$, possessing both symmetry and having positive definite Gram matrices for all possible $\mathbf{x}_1, ...\mathbf{x}_m \in \mathcal{X}$, must necessarily have a corresponding space $\mathcal{H}$ (of dimension $n^{\mathcal{H}}$) and mapping $\psi : \mathcal{X} \rightarrow \mathcal{H}$ where $K(\mathbf{x}, \mathbf{x}') = \langle\boldsymbol{\psi}(\mathbf{x}), \boldsymbol{\psi}(\mathbf{x}')\rangle_{\mathcal{H}} = \boldsymbol{\psi}(\mathbf{x})^T\boldsymbol{\psi}(\mathbf{x}')$. As a result, if the solution can be represented in only dot product terms $\boldsymbol{\psi}(\mathbf{x})^T\boldsymbol{\psi}(\mathbf{x}')$ these can be replaced by

$K(\mathbf{x}, \mathbf{x}')$, as a result of the 'kernel trick'. The kernel used in this paper is the RBF (Gaussian) kernel (with parameter γ) $K_{rbf}(\mathbf{x}, \mathbf{x}') = exp(\frac{-||\mathbf{x}-\mathbf{x}'||^2}{\gamma})$. Other common kernels include polynomial, sigmoid and linear.

The quadratic problem (8) can then be solved by standard quadratic programming solvers such as MATLAB or R's quadprog. Problem convexity depends on $\mathbf{G}$ - if it is strictly positive definite, the solution is global and unique, and if positive semi-definite, the solution is only global. If it is indefinite, it may have no solution. Experimentally the addition of the constraints does sometimes cause $\mathbf{G}$ to be indefinite, with one or more negative eigenvalues. In this case as per [13] we used Tikhonov regularisation [22] to ensure $\mathbf{G}$ is positive definite, by adding $\sigma \mathbf{I}$ to $\mathbf{G}$, where $\sigma = 2|min(eigenvalues)|$.

In addition to this constrained SVM algorithm, MC-SVM requires a method for generating the constraint set, and two such algorithms are proposed in [2,13]. Both these techniques are 'conjunctive' in that each pair typically varies in *all* constrained features. In this paper we refer to these as:

1. **Randomised Conjunctive (CJ1)** [2]: M constraints are created by repeating the following M times: (a) select a random subset of m training data-points (m was unspecified and was set to 5 for our experiments); (b) set the *constrained* features of $\underset{\sim}{\mathbf{x}}_i$ to the minimum values in the subset, and $\tilde{\mathbf{x}}_i$ to the maximum values. (It was unclear what was used for *unconstrained* features. We used the values of one of the m data-points selected at random.)
2. **Uniform Partition Conjunctive (CJ2)** [13]: This technique generates M constraints by partitioning each constrained feature into M equal partitions between its minimum and maximum. (The values used for unconstrained features were unclear, so to influence the densest part of the input space we set these to the data-point closest (in Euclidean distance) to the centroid as determined by a 1-cluster normalised k-means analysis.)

We evaluate these MC-SVM techniques in Sect. 4. For reasons discussed in Sect. 3.1, we found that these techniques only increased monotonicity weakly, and as a result accuracy improvements were also low.

3 Partially Monotone Support Vector Machines

3.1 PM-SVM Technique

To address the limitations of the conjunctively constrained SVM by Li and Chen we propose the following procedure for PM-SVM:

1. Identify likely monotone features C from domain expertise or common sense;
2. Create constraints for the identified monotone features C, using one of the constraint generation techniques proposed in Algorithms 1 and 2;
3. Estimate optimal hyper-parameters (e.g. for SVM box constraint B and RBF kernel γ), for example by grid search and cross-validation;

4. Solve the *monotonicity constrained* C-SVM as per [2], using the training data and constraint set.

PM-SVM differs from the implementation in [2] by proposing a new constraint generation technique designed to more efficiently achieve monotonicity. We start by observing that the inherent problem with attempting to discretise dominance relations throughout $\Re^m$ in p monotone features is the sheer dimensionality of the space. The dimensionality is $(m - p) + p + p = (m + p)$, to allow for all monotone combinations of the p monotone feature values throughout the $\Re^{m-p}$ non-monotone feature space. The existing approaches CJ1 and CJ2 (Sect. 2.2) attempt to cover this space, using *conjunctive* constraints where *all* p monotone features increase between the first and second data point in each constraint. This is a critical issue, because the addition of M constraints to the SVM optimisation extends the matrix inversion by M, and thus has worst case complexity $O((N + M)^3)$, where N is the number of data-points and so increasing M can quickly make the problem practically infeasible.

However, the actual dimensionality of the space to be covered is much lower. In Definition 2, $\mathbf{x} \preceq_C \mathbf{x}'$ may be usefully considered to be made up of two cases: (a) strictly *univariate* changes (when exactly one feature $c_i \in C$ increases between $\mathbf{x}$ and $\mathbf{x}'$), and (b) strictly *conjunctive* changes (when more than one feature $\{c_i, c_j...\} \subseteq C$ increases). Then it is straightforward to show that univariate monotonicity in all $c_i \in C$ is equivalent to the conjunctive monotonicity in any $\{c_i, ...\} \subseteq C$ as per Theorem 1:

Theorem 1 (Equivalence of Univariate & Conjunctive Monotonicity).
Given monotone features $C \subseteq \{1, ..., m\}$ and function $f : \Re^m \to \Re$, the condition for conjunctive partial monotonicity in (4) is equivalent to:

$$\mathbf{x} \preceq_{c_i} \mathbf{x}' \Rightarrow f(\mathbf{x}) \le f(\mathbf{x}'), \ \forall \mathbf{x}, \mathbf{x}' \in \Re^m, c_i \in C \tag{9}$$

Proof:

(i) (4) $\implies$ (9): this follows since each $c \in C$ is simply a special case of (3), where $x_j = x'_j$ for $j \in (C \backslash c)$. In other words, univariate monotonicity in each monotone feature is simply a subset of conjunctive monotonicity.

(ii) (9) $\implies$ (4): Let function $f : \Re^m \to \Re$ comply with (9) for univariate monotonicity for all $c \in C$. For convenience and without loss of generality let $C = \{1, 2, ..., c_{max}\}, c_{max} \le m$. Let $\mathbf{x} = (x_1, x_2, ..., x_{c_{max}}, ..., x_{m-1}, x_m)$ and $\mathbf{x}' = (x_1 + \delta_1, x_2 + \delta_2, ..., x_{c_{max}} + \delta_{c_{max}}, ..., x_{m-1}, x_m)$, where $\delta_c \ge 0$. It is apparent that $\mathbf{x} \preceq_C \mathbf{x}', \forall \mathbf{x}, \mathbf{x}' \in \Re^m$ thus encompassing (4). We will now show that $f(\mathbf{x}') \ge f(\mathbf{x})$ using only (9):

$$
\begin{aligned}
f(\mathbf{x}') &= f((x_1 + \delta_1, x_2 + \delta_2, ..., x_{c_{max}} + \delta_{c_{max}}, ..., x_{m-1}, x_m)) \\
&= f((x_1, x_2 + \delta_2, ..., x_{c_{max}} + \delta_{c_{max}}, ..., x_{m-1}, x_m) + (\delta_1, 0, ..., 0, ..., 0, 0)) \\
&\ge f(x_1, x_2 + \delta_2, ..., x_{c_{max}} + \delta_{c_{max}}, ..., x_{m-1}, x_m) \\
&= f((x_1, x_2, ..., x_{c_{max}} + \delta_{c_{max}}, ..., x_{m-1}, x_m) + (0, \delta_2, ..., 0, ..., 0, 0)) \\
&\ge f(x_1, x_2, ..., x_{c_{max}} + \delta_{c_{max}}, ..., x_{m-1}, x_m) \\
&... \\
&\ge f(x_1, x_2, ..., x_{c_{max}}, ..., x_{m-1}, x_m) = f(\mathbf{x})
\end{aligned}
$$

Thus compliance with conjunctive monotonicity (4) in monotone features C for any $\mathbf{x}, \mathbf{x}' \in \Re^m, \mathbf{x} \preceq_C \mathbf{x}'$ is demonstrated for any function $f : \Re^m \to \Re$ complying with univariate monotonicity (9).

Thus it is sufficient to aim for *univariate* monotonicity in each monotone feature, with dimension $(m-1) + 1 + 1 = (m+1)$. This means that any strictly conjunctive constraints (where more than one feature increases) are redundant in the presence of univariate constraints. Hence we propose techniques for *univariate* constraint generation. MATLAB implementations are available on github.

The first technique (Algorithm 1) simply selects T random training points $\{\mathbf{x}_i \mid i = t_1..t_T\}$ for each constrained feature c_j and creates two constraints per point, a 'lower' one between the $(x_{i,1}, .. \min_i(x_{i,c_j}), ..x_{i,m})$ and $\mathbf{x}_i$, and a 'higher' one between $\mathbf{x}_i$ and $(x_{i,1}, .. \max_i(x_{i,c_j}), ..x_{i,m})$. We could simply set T to the number of training points for maximum coverage, but for the experimental data sets we found that monotonicity was adequate with $T = 25$.

Algorithm 1. Univariate Random Constraint Generation (UNR)

Input: Training data $\{\mathbf{x}_i = (x_{i,1}, ..x_{i,m}) \mid i = 1..N\}$, monotone features $C = \{c_j \mid j = 1..p\}$, number of base points per monotone feature $T, T < N$.
Output: Set of M constraint pairs $\{(\underline{\mathbf{x}}_k, \tilde{\mathbf{x}}_k) \mid k = 1..M\}$, $M \leq 2pT$
 1: Initialise set of constraints $\tau = \{\}$
 2: **for each** $c_j \in C$ **do**
 3:     Calculate minimum and maximum values for c_j $(\min_i(x_{i,c_j}), \max_i(x_{i,c_j}))$
 4:     Choose T points at random from training data $\{(\mathbf{x}_i, y_i) \mid i = t_1..t_T\}$.
 5:     Append T lower constraints to τ: $\{((x_{i,1}, .. \min_i(x_{i,c_j}), ..x_{i,m}), \mathbf{x}_i) \mid i = t_1..t_T\}$
 6:     Append T upper constraints to τ: $\{(\mathbf{x}_i, (x_{i,1}, .. \max_i(x_{i,c_j}), ..x_{i,m})) \mid i = t_1..t_T\}$
 7: **end for**
 8: Remove any duplicate constraints from τ (with identical $\underline{\mathbf{x}}_k$ and $\tilde{\mathbf{x}}_k$).
 9: Return τ

The second technique (Algorithm 2 - Adaptive Constraint Generation) uses the *unconstrained* SVM model to identify non-monotone regions and designs constraints to address them. Figure 1 illustrates what we term a 'Non-Monotone Region' (NMR) around a point. Note from this figure that given $f(\mathbf{x})$ is a binary function $\Re^m \Rightarrow \{1, 2\}$ and that $f(\mathbf{x}_{nmt}) = 2$ and feature i is monotone increasing, we only need to look for non-monotonicity in values of feature i that are higher than $x_{nmt,i}$. Evaluating $f(x)$ for higher values of feature i identifies a non-monotone hyperplane (when $f(x) = 1$), until at higher values we reach the 'extent' hyperplane (when $f(x) = 2$ again). To correct this non-monotone region, either the 'active NMR' (containing $\mathbf{x}_{nmt}$) or the 'passive NMR' (not-containing $\mathbf{x}_{nmt}$) need to fully reverse. We can encourage this correction by placing constraints between the active and passive NMRs.

Although a number of discrete constraints could be used to span these regions and ensure non-monotone 'islands' do not appear in the gaps, in practice we found that it sufficient to place a single constraint between the non-monotone

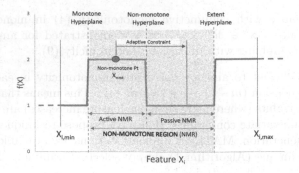

Fig. 1. Adaptive Constraint Generation. Around each point that is non-monotone in feature X_i is a 'Non-Monotone Region' (NMR) composed of an 'Active' NMR (containing the point) and a 'Passive' NMR (of the alternative class). An adaptive constraint is created between the non-monotone point and the midpoint of its 'Passive' NMR.

point and the mid-point of the passive NMR. We do note however that it is important to include $\mathbf{x}_{nmt}$ in the constraint, because the sparse SVM solution may select $\mathbf{x}_{nmt}$ as a support vector and indeed create a non-monotone island.

3.2 Measurement of Partial Monotonicity

The practicality of this approach depends on achieving satisfactory levels of monotonicity with reasonable numbers of constraints. To evaluate this we need a measure for the partial monotonicity of a classifier. The dominant measure in the classification literature is to conduct all possible pairwise comparisons of the data points and their predictions. The measure of (non)monotonicity is the proportion of (non)violations, and has various names such as Frequency Monotonicity Rate (FMR) [2,13], degree of monotonicity (NmDeg) [6,21], Non-monotonicity Index (NMHI) [16,17], fraction of monotone pairs [5,23].

Each of the data point comparisons ($N^2 - N$ excluding self-comparisons) is defined as 'comparable' if the points comply with the dominance relation (1):

$$\mathbf{x} \; comparable \, with \, \mathbf{x}' \Leftrightarrow (x_j \geq x'_j, \; \forall j = 1..m) \; OR \; (x_j \leq x'_j, \; \forall j = 1..m) \quad (10)$$

The 'incomparable' pairs (IP) are ignored. The comparable pairs (CP) are then assessed for compliance with monotonicity of the class predicted by the function or given by the data set (Eq. 2). *Comparable* pairs that do not comply with this equation (NM) are deemed non-monotone, and typically these are represented as a proportion of the total number of comparisons, for example the Non-Monotonicity Index (NMI), which is given by $NMI = NM/(N^2 + N)$.

This measure is useful for monotonicity in *all* features, but in practice cannot be used for our situation of *partial* monotonicity, because in addition to requiring that the constrained features are either all greater than (or less than) or equal, we must modify Eq. 10 to require *equality in all unconstrained features*. This dramatically reduces the number of comparable pairs, such that most of

Algorithm 2. Adaptive Constraint Generation (AD)

Input: Training data $\{\mathbf{x}_i = (x_{i,1}, .., x_{i,m}) \mid i = 1..N\}$, monotone features $C = \{c_j \mid j = 1..p\}$, trained classifier $f(\mathbf{x})$, continuous variable partitions $L > 0$
Output: Set of M constraint pairs $\{(\underline{\mathbf{x}}_k, \tilde{\mathbf{x}}_k) \mid k = 1..M\}$, $M \leq N$
1: Initialise empty set of constraints $\tau = \{\}$
2: **for each** $c_j \in C$ **do**
3:    Calculate minimum and maximum values for c_j $(\min_i(x_{i,c_j}), \max_i(x_{i,c_j}))$
4:    Set partition width $P = (\max_i(x_{i,c_j}) - \min_i(x_{i,c_j}))/L$
5:    Set feature grid vector $G = (\min_i(x_{i,c_j}) : P : \max_i(x_{i,c_j}))$
6:    **for each** $\mathbf{x}_i$ $(i = 1..N)$ **do**
7:       Set potential non-monotone dirn ψ_d as $>$ (above) if $f(\mathbf{x}_i) > 0$ else $<$ (below).
8:       Initialise $class = f(\mathbf{x}_i)$, $hyperplane = NULL$, $extent = NULL$
9:       **for each** $G_i \, \psi_d \, x_{i,c_j}$ **do**
10:          **if** $f((x_{i,1}, .. x_{i,c_j} = G_i, ..., x_{i,m})) \neq class$ **then**
11:             **if** $hyperplane \; IS \; NULL$ **then**
12:                $hyperplane = G_i, class = f((x_{i,1}, ..., x_{i,c_j} = G_i, .. x_{i,m}))$.
13:             **else**
14:                $extent = G_i$
15:             **end if**
16:          **end if**
17:          **if** $NOT \; hyperplane \; IS \; NULL$ **then**
18:             **if** $extent \; IS \; NULL$ **then** $extent = G_1$ or (G_{L+1}) as appropriate
19:             Set $midpt = (hyperplane + extent)/2$
20:             Append constraint to τ: $(\mathbf{x}_i, (x_{i,1}, .. x_{c_j} = midpt, .. x_{i,m}))$
21:          **end if**
22:       **end for**
23:    **end for**
24: **end for**
25: Return τ

---

the datasets in this paper had a comparability of 0.0 %. In practice even one continuous unconstrained feature will typically reduce comparability to zero. Even if equality is relaxed, comparability remains low, such as the Pima dataset [14], which only has categorical variables and still has comparability of only 4.6 %. This makes the NMI measure either impossible, or at least unreliable, in partially monotone situations.

We thus propose a new measure function monotonicity we call *Monotonicity Compliance* (MCC). Applying Theorem 1, we observe that if a function possesses univariate partial monotonicity in all constrained features $c_i \in C$ $(i = 1..p)$, it also possesses conjunctive monotonicity for any subset $C_s \subseteq C$. Thus we propose to simply measure univariate partial monotonicity for each $c_i \in C$ (MCC_{c_i}), and use the feature average $MCC_{featavg} = \frac{1}{p} \sum_{i=1}^{p} MCC_{c_i}$ as a measure of overall compliance. MCC is intuitively defined as the proportion of the input space where the requested monotonicity constraints are *not* violated, weighted by the joint probability distribution of the input space. This aims to provide a practical measure of how monotone a function is within the expected input space.

Definition 3 *Monotonicity Compliance (MCC).* *For function* $f(x)$:
$\Re^m \Rightarrow \Re$, $x \in X$ *with monotone constraints on features* $C = \{c_1, ..., c_p\} \subseteq$
$\{1, ..., m\}$, *and with joint probability density function* $P(x)$, *the Monotonicity*
Compliance of f *with respect to constrained feature* c_i *is*

$$MCC_{c_i}(f) = \int \cdots \int_X P(x_1, \ldots, x_m) m_{c_i}(f, x_1, \ldots, x_m) \, dx_1 \ldots dx_m \qquad (11)$$

where, for *increasing* monotonicity (for *decreasing* reverse as appropriate):

$$m_{c_i}(f, x) = \begin{cases} 1, & if \; m_{c_i}^!(f, x) \geq 0 \; and \; m_{c_i}^-(f, x) \geq 0 \\ \frac{1}{2}, & if \; (m_{c_i}^+(f, x), m_{c_i}^-(f, x)) \in \{(-1, +1), (+1, -1)\} \\ 0, & otherwise \end{cases} \qquad (12)$$

$$m_{c_i}^+(f, x) = \begin{cases} +1, & if \; \exists \; Q > 0 \; s.t. \; f(x) < f(x_1, \ldots x_{c_i} + Q, \ldots x_m) \; and \\ & \quad f(x) = f(x_1, \ldots, x_{c_i} + q, \ldots x_m) \; \forall \; 0 < q < Q \\ 0, & if \; f(x) = f(x_1, \ldots x_{c_i} + q, \ldots x_m) \; \forall \; q > 0 \\ -1, & otherwise \end{cases}$$
$$(13)$$

$$m_{c_i}^-(f, x) = \begin{cases} +1, & if \; \exists \; Q > 0 \; s.t. \; f(x) > f(x_1 \ldots, x_{c_i} - Q \ldots x_m) \; and \\ & \quad f(x) = f(x_1, \ldots, x_{c_i} - q, \ldots x_m) \; \forall \; 0 < q < Q \\ 0, & if \; f(x) = f(x_1, \ldots, x_{c_i} - q, \ldots x_m) \; \forall \; q > 0 \\ -1, & otherwise \end{cases}$$
$$(14)$$

Essentially the $m_{c_i}(\mathbf{x})$ function looks in the positive and negative directions
of x_{c_i} for the first change in f (if any). If in both of directions the function
f either does not change, or changes in the correct direction, $m_{c_i}(\mathbf{x})$ returns
1 (a monotone point). If the change in one direction is correct and the other
is incorrect, $m_{c_i}(\mathbf{x})$ returns $1/2$ (it is 'half' monotone). Otherwise, the point is
non-monotone and $m_{c_i}(\mathbf{x}) = 0$. In practice $P(\mathbf{x})$ is unknown, but for given data
set $\hat{X}$ of size N, $MCC_{c_i}(f)$ can be simply estimated by the plug-in estimate:

$$M\hat{C}C_{c_i}(f) = \frac{1}{N} \sum_{i=1}^{N} m_{c_i}(f, \mathbf{x}_i) \qquad (15)$$

MCC is analogous to the partial derivative based technique in [4], but for
non-continuous and non-differentiable functions. Although the partial derivative
of the real valued SVM function (prior to taking the sign) could have been
used, we preferred to evaluate class changes (i.e. *after* taking the sign). Partial
derivatives are a poor indicator of classifier non-monotonicity because a negative
slope does not necessarily result in a change in class (sign) over the input space.

4 Experiments and Datasets

Datasets. Nine datasets were used as described in Table 1, from the UCI Machine Learning Repository [14] and the KEEL Dataset Repository [1]. Rows with missing values were removed.

Table 1. Dataset summary table.

Dataset	Src	No. Rows	No. Feats	Output Class	Constrained Attributes
German Credit Ratings	UCI	1000	24	Good (-1)	**Increasing (4):** Loan Dur'n, Single Appl (vs Guarantor), Co-Appl (vs Guarantor), Renting
				Bad (+1)	**Decreasing (6):** Cheque Bal, Savings, Mths in Job, Age, Other Installment Plans, Owns House
Pima Indians Diabetes	UCI	394	8	Normal (-1)	**Increasing (8):** Num. Preg, Glucose Test, BP, Triceps Fold Thk, Serum Insulin, BMI, Diab Pedigree, Age.
				Diabetes (+1)	**Decreasing (0):** -
Cleveland Heart Disease	UCI	299	18	Normal (-1)	**Increasing (11):** Age, Typ Angina, Atyp Angina, Male, BP, Cholest, ECG=1, ECG=2, Exerc Ang, Exerc ST
				Disease (+1)	depn, Num Vessels Fluoro **Decreasing (1):** Max Hrt Rate
South African Heart Disease	KEEL	462	9	Normal (-1)	**Increasing (8):** Systolic BP, Cum Tobacco, Cholest, Adiposity, Family Hist, Type A Behav, Age, Obesity
				Disease (+1)	**Decreasing (0):** -
Ljubljana Breast Cancer	UCI	277	13	No Recurrence (-1)	**Increasing (3):** Tumor size, No. Inv Nodes, Degree of Malignancy.
				Recurrence (+1)	**Decreasing (2):** Nodes Encapsulated, Irradiation
Car Acceptability	UCI	390	6	Unacceptable (-1)	**Increasing (2):** Persons, Safety
				Acc (acc/gd/vgd) (+1)	**Decreasing (2):** Price, Maintenance Reqd
Auto Mileage	UCI	392	7	MPG<=28 (-1)	**Increasing (2):** Origin Japan, Year
				MPG>28 (+1)	**Decreasing (2):** Displacement, Weight
Haberman BC Survival	UCI	306	3	Died <5yrs (-1)	**Increasing (1):** Year
				Survived 5yrs (+1)	**Decreasing (2):** Age, Nodes
Wisconsin Breast Cancer	UCI	683	9	Benign (-1)	**Increasing (9):** Clump Thk, Uniform Size, Uniform Shape, Marg Adhes, Epit Size, Bare Nucl, Bland Chrom,
				Malignant (+1)	Norm Nucl, Mitos **Decreasing (0):** -

Experiment Design. For each dataset we conducted 50 experiments. The RBF kernel and the MATLAB quadprog solver was used on an Intel i7-4790 8 GB RAM PC. For each experiment 2/3 of available data points were randomly selected as the maximum training partition. This training partition was sequentially randomly sub-sampled down to 200, 100, and 50 data point samples to assess the effect of sample size. All sampling was stratified to retain class distribution. For each sample size, the remainder of the available data were used as the test partition. The same training/test partitions and CV partitions were used for all constraint techniques to ensure a fair comparison.

Four constraint techniques were evaluated: existing approaches CJ1 and CJ2, and the two proposed approaches UNR and AD. CJ2 was almost indistinguishable from unconstrained SVM and so the results were omitted for clarity. For UNR $T = 25$ was used, resulting in a maximum of $2pT$ constraints. For CJ1 $2pT$ constraints were used, to enable like-for-like comparison with UNR. For AD, the number of constraints varies depending on the non-monotone regions identified.

Each experiment proceeded as follows:

1. **Estimate optimal standard SVM hyper-parameters:** The RBF scale factor σ and SVM box constraint B were estimated using grid search on $C \in \{0.0001, 0.001, 0.01, 0.1, 1, 5, 10, 50, 100, 500, 1000, 5000, 10000, 50000, 100000\}, \sigma \in \{0.001, 0.01, 0.1, 1, 5, 10, 15, 25, 50, 100, 250, 500, 1000, 5000\}$. The optimal $\langle C, \sigma \rangle$ pair had the minimum MCR as determined by stratified 10-fold cross-validation.

2. **Fit SVM model:** The *unconstrained* SVM model is fitted to the whole training partition using the optimal hyper-parameters.
3. **For each constraint algorithm:**
   - **Create constraints:** Constraints were generated based on the *training* data as per the appropriate algorithm.
   - **Fit constrained SVM model:** The PM-SVM model was fitted to the training partition using the constraints and optimal SVM hyper-parameters.
   - **Estimate performance:** Accuracy and monotonicity metrics were both calculated on the *test* partition.

5   Results and Discussion

Classifier Partial Monotonicity. Figure 2 shows the impact of the constraint technique and sample size on classifier monotonicity ($MCC_{featavg}$). Firstly we note that the *unconstrained* car and WBCdiag models were almost perfectly monotone with no constraints! For the other datasets, it is clear that the two proposed approaches are more effective at inducing a monotone classifier than the existing CJ1 approach for all datasets. In fact CJ1 does not noticeably increase monotonicity except for SA Heart, Ljubjlana, Autompg and Haberman, where the increases were less than half those of the proposed techniques. It is most effective on Haberman, which is almost certainly due to that dataset's low dimensionality ($n = 3$ features), which makes it more likely to generate effective univariate constraints.

In contrast, the proposed random univariate UNR technique lifts monotonicity above 99 % for all data sets and sample sizes except Autompg and Haberman, where it is 97–98%. We emphasise that the monotonicity MCC measures were based on the *test* partition, so offer unbiased estimates. We find it somewhat surprising that so few discretised constraints (based on $T = 25$ random base points per constrained feature) can be so effective at inducing monotonicity throughout the input space. The Adaptive AD technique is generally similar to UNR, except for low sample sizes in Cleveland and SA Heart it is slightly less effective, and for higher sample sizes on Haberman it is more effective.

Impact of Partial Monotonicity on Accuracy. The performance metric Cohen's Kappa is summarised in Fig. 3. For the Car and WBCDiag datasets, the monotone models make negligable difference. This is because the unconstrained SVM is almost perfectly monotone ($MCC_{featavg}$ approaching 1, Fig. 2), so there is little room for improvement.

Table 2 shows the experiment-wise increase in Accuracy and Cohen's Kappa for $N = 50$, because monotonicity is most likely to improve accuracy at low sample sizes [9]. Apart from the Car and WBCDiag datasets (which are already virtually monotone without constraints), statistically significant accuracy and/or κ increases are seen for all remaining datasets. The existing CJ1 increases in κ and accuracy are typically a third or a half or less of the proposed approaches

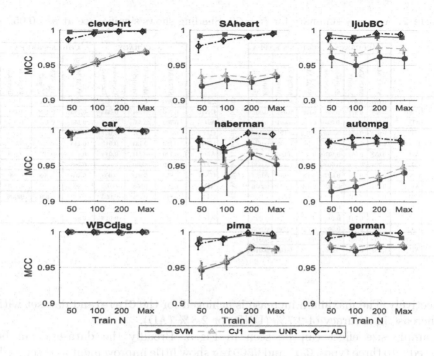

Fig. 2. MCC (feat avg) vs Sample Size

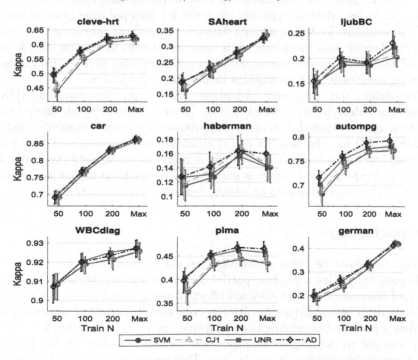

Fig. 3. Cohen's κ vs Sample Size. 90 % confidence interval shown.

Table 2. Accuracy summary for $N = 50$. Shading shows significance at $p = 0.05$.

		ACCURACY			COHEN'S KAPPA					ACCURACY			COHEN'S KAPPA		
		SVM Mean	MCSVM Mean*	Exptwise Increase	SVM Mean	MCSVM Mean*	Exptwise Increase			SVM Mean	MCSVM Mean*	Exptwise Increase	SVM Mean	MCSVM Mean*	Exptwise Increase
Car	CJ1	0.869	0.869	0.000	0.690	0.691	0.000	Ljub BC	CJ1	0.700	0.702	0.002	0.153	0.151	-0.002
	UNR	0.869	0.871	0.002	0.690	0.695	0.004		UNR	0.700	0.704	0.004	0.153	0.145	-0.008
	AD	0.869	0.869	0.000	0.690	0.691	0.001		AD	0.700	0.705	0.005	0.153	0.156	0.003
SAHeart	CJ1	0.652	0.656	0.005	0.162	0.171	0.009	Haberm	CJ1	0.720	0.727	0.007	0.115	0.119	0.004
	UNR	0.652	0.668	0.017	0.162	0.190	0.028		UNR	0.720	0.733	0.013	0.115	0.127	0.011
	AD	0.652	0.662	0.010	0.162	0.186	0.024		AD	0.720	0.734	0.014	0.115	0.127	0.012
Cleve	CJ1	0.724	0.728	0.004	0.440	0.448	0.009	WBCD	CJ1	0.959	0.959	0.000	0.908	0.909	0.001
	UNR	0.724	0.751	0.027	0.440	0.494	0.054		UNR	0.959	0.959	0.000	0.908	0.908	0.000
	AD	0.724	0.752	0.028	0.440	0.496	0.056		AD	0.959	0.959	-0.001	0.908	0.907	-0.001
Pima	CJ1	0.734	0.735	0.002	0.374	0.376	0.003	AutoMH	CJ1	0.875	0.878	0.003	0.682	0.688	0.006
	UNR	0.734	0.749	0.016	0.374	0.405	0.031		UNR	0.875	0.883	0.008	0.682	0.702	0.020
	AD	0.734	0.747	0.013	0.374	0.398	0.025		AD	0.875	0.889	0.014	0.682	0.716	0.034
German	CJ1	0.698	0.700	0.002	0.184	0.189	0.005								
	UNR	0.698	0.707	0.009	0.184	0.202	0.018								
	AD	0.698	0.702	0.005	0.184	0.199	0.015								

respectively. The maximum increase is achieved for the Cleveland dataset with an increase in accuracy of 2.7 % (UNR) or 2.8 % (AD).

Sample size effect can be seen in Fig. 3. Broadly the datasets can be grouped into three types. Car and WBCDiag show little improvement as discussed. Cleveland, SA Heart and German datasets show diminishing returns as the sample size increases, which is typically expected for incorporating domain knowledge. Interestingly, for the remaining datasets (Ljubjlana, Autompg, Haberman, Pima) the benefit from monotonicity is maintained with increasing sample size, even when the maximum training data is used, suggesting perhaps that noise is obscuring the monotone relationships rather than sample size.

Comparing the proposed Adaptive (AD) and Randomised (UNR) approaches, although AD induced slightly lower monotonicity at smaller samples sizes, we note that the accuracy improvements remained similar or better. In addition AD has the benefits of (a) eliminating the question of how many constraints to use, and (b) a *much* lower number of constraints (typically 10–20% in these experiments), which is particularly beneficial at higher sample sizes because of the polynomial cost of adding constraints ($O(N + M)^3$).

6 Conclusions

We have presented two new constraint generation techniques for incorporating domain knowledge regarding partial monotonicity into kernel SVM classifiers, and demonstrated that they are effective using a new measure for partial monotonicity. The techniques involve the creation of discretised constraint sets by either a random (Algorithm 1) or adaptive approach (Algorithm 2).

In contrast to existing techniques, monotonicity is substantially increased with a reasonable number of constraints. This results in increases in accuracy in seven of the nine datasets, particularly at lower sample sizes. The lack of increased accuracy in the remaining two datasets was also shown to be because the unconstrained model was already almost perfectly monotone and there was little room for improvement.

Although discretised constraint approaches are not ideal in that they do not guarantee *global* monotonicity in the same way that model specification would, we believe the inelegance of our approach is worthwhile and necessary to extend such a powerful family nonlinear nonparametric classifiers. We note that the accuracy increases are particularly significant because RBF kernel C-SVM is already one of the most powerful classification algorithms [8]. Future exciting research avenues are extensions to multi-class, comparison with other monotone classifiers, and development of a more sophisticated monotone feature selection.

References

1. Alcalá, J., Fernández, A., Luengo, J., Derrac, J., García, S., Sánchez, L., Herrera, F.: Keel data-mining software tool: data set repository, integration of algorithms and experimental analysis framework. J. Multiple-Valued Logic Soft Comput. **17**, 255–287 (2010)
2. Chen, C.C., Li, S.T.: Credit rating with a monotonicity-constrained support vector machine model. Expert Syst. Appl. **41**(16), 7235–7247 (2014)
3. Cortes, C., Vapnik, V.: Support-vector networks. Mach. Learn. **20**(3), 273–297 (1995)
4. Daniels, H., Kamp, B.: Application of MLP networks to bond rating and house pricing. Neural Comput. Appl. **8**(3), 226–234 (1999)
5. Duivesteijn, W., Feelders, A.: Nearest neighbour classification with monotonicity constraints. In: Daelemans, W., Goethals, B., Morik, K. (eds.) ECML PKDD 2008. LNCS (LNAI), vol. 5211, pp. 301–316. Springer, Heidelberg (2008). doi:10.1007/978-3-540-87479-9_38
6. Feelders, A., Pardoel, M.: Pruning for monotone classification trees. In: R. Berthold, M., Lenz, H.-J., Bradley, E., Kruse, R., Borgelt, C. (eds.) IDA 2003. LNCS, vol. 2810, pp. 1–12. Springer, Heidelberg (2003). doi:10.1007/978-3-540-45231-7_1
7. Feelders, A.J.: Prior knowledge in economic applications of data mining. In: Zighed, D.A., Komorowski, J., Żytkow, J. (eds.) PKDD 2000. LNCS (LNAI), vol. 1910, pp. 395–400. Springer, Heidelberg (2000). doi:10.1007/3-540-45372-5_42
8. Fernández-Delgado, M., Cernadas, E., Barro, S., Amorim, D.: Do we need hundreds of classifiers to solve real world classification problems? J. Mach. Learn. Res. **15**(1), 3133–3181 (2014)
9. Kamp, R., Feelders, A., Barile, N.: Isotonic classification trees. In: Adams, N.M., Robardet, C., Siebes, A., Boulicaut, J.-F. (eds.) IDA 2009. LNCS, vol. 5772, pp. 405–416. Springer, Heidelberg (2009). doi:10.1007/978-3-642-03915-7_35
10. Kotlowski, W.: Statistical approach to ordinal classification with monotonicity constraints. Ph.D. thesis, Pozna Univ of Techn Inst of Computing Science (2008)
11. Kuhn, H.W., Tucker, A.W.: Nonlinear programming. In: Proceedings of the Second Berkeley Symposium on Mathematical Statistics and Probability. pp. 481–492. University of California Press, Berkeley, Calif. (1951)
12. Lauer, F., Bloch, G.: Incorporating prior knowledge in support vector machines for classification: a review. Neurocomputing **71**(7), 1578–1594 (2008)

18   C. Bartley et al.

13. Li, S.T., Chen, C.C.: A regularized monotonic fuzzy support vector machine for data mining with prior knowledge. IEEE Trans. Fuzzy Syst. **PP**(99) (2014)
14. Lichman, M.: UCI machine learning repository (2013). http://archive.ics.uci.edu/ml
15. Mangasarian, O.L., Wild, E.W.: Nonlinear knowledge-based classification. IEEE Trans. Neural Netw. **19**(10), 1826–1832 (2008)
16. Marsala, C., Petturiti, D.: Rank discrimination measures for enforcing monotonicity in decision tree induction. Inf. Sci. **291**, 143–171 (2015)
17. Milstein, I., David, A.B., Potharst, R.: Generating noisy monotone ordinal datasets. Artif. Intell. Res. **3**(1), p30 (2013)
18. Pelckmans, K., Espinoza, M., De Brabanter, J., Suykens, J.A., De Moor, B.: Primal-dual monotone kernel regression. Neural Proc. Letters **22**(2), 171–182 (2005)
19. Potharst, R., Ben-David, A., van Wezel, M.: Two algorithms for generating structured and unstructured monotone ordinal data sets. Eng. App. Art. Intell. **22**(4), 491–496 (2009)
20. Potharst, R., Bioch, J.C.: Decision trees for ordinal classification. Intell. Data Anal. **4**(2), 97–111 (2000)
21. Potharst, R., Feelders, A.J.: Classification trees for problems with monotonicity constraints. ACM SIGKDD Explor. Newsletter **4**(1), 1–10 (2002)
22. Tikhonov, A., Arsenin, V.: Solutions of ill-posed problems. Scripta series in mathematics, Winston (1977)
23. Velikova, M., Daniels, H.: Decision trees for monotone price models. CMS **1**(3-4), 231–244 (2004)
24. Wang, Y., Ni, H.: Multivariate convex support vector regression with semidefinite programming. Knowl.-Based Syst. **30**, 87–94 (2012)
25. Wellman, M.P.: Fundamental concepts of qualitative probabilistic networks. Artif. Intell. **44**(3), 257–303 (1990)

Textual Cues for Online Depression in Community and Personal Settings

Thin Nguyen[✉], Svetha Venkatesh, and Dinh Phung

Deakin University, Geelong, Australia
{thin.nguyen,svetha.venkatesh,dinh.phung}@deakin.edu.au

Abstract. Depression is often associated with poor social skills. The Internet allows individuals who are depressed to connect with others via online communities, helping them to address the social skill deficit. While the difficulty of collecting data in traditional studies raises a bar for investigating the cues of depression, the user-generated media left by depression sufferers on social media enable us to learn more about depression signs. Previous studies examined the traces left in the posts of online depression communities in comparison with other online communities. This work further investigates if the content that members of the depression community contribute to the community blogs different from what they make in their own personal blogs? The answer to this question would help to improve the performance of online depression screening for different blogging settings. The content made in the two settings were compared in three textual features: affective information, topics, and language styles. Machine learning and statistical methods were used to discriminate the blog content. All three features were found to be significantly different between depression Community and Personal blogs. Noticeably, topic and language style features, either separately or jointly used, show strong indicative power in prediction of depression blogs in personal or community settings, illustrating the potential of using content-based multi-cues for early screening of online depression communities and individuals.

Keywords: Computer mediated communication · Weblog · Social media analysis · Mental health · Textual cues · Affective norms · Language styles · Topics

1 Introduction

In their lifetime, 12.1 % and 4.1 % of people have suicide ideation and attempts, respectively [15]. Psychiatry disorders were found in 90 % of suicide victims and the relative risk of suicide among those with the conditions was increased more than 9-fold [3]. Of the disorders, depression is among the major risk factor. Knowing what are the signs of depression would help detect early the sufferers and ultimately help alleviate the suicide crisis.

The signals could be non-verbal activities, such as "lack of eye contact", "head drooped, looking at ground", and "mouth turned down", which were found

© Springer International Publishing AG 2016
J. Li et al. (Eds.): ADMA 2016, LNAI 10086, pp. 19–34, 2016.
DOI: 10.1007/978-3-319-49586-6_2

as salient cues for depression [22]. Other markers of depression were vocal characteristics [11]. Social skills were also found as an indicator of depression [7]. Another aspect, language styles, such as the use of sadness or swearing words, was identified as depression cues for personal diaries and online blogs [19]. Most of these studies were conducted in small scale, such as for the study on the linguistic cues [19], the text was collected from fifty-seven participants.

On the other hand, the Internet and social media and networking systems built on it offers great sources to investigate causes and cues for depression in large scale. The new media has provided an excellent venue for individuals to express their ideas and like-minded people to share stories, including depression sufferers. This unintentionally leaves a huge data which are often difficult to be collected in traditional studies.

In this study, we collect data from *depression.livejournal.com*, the largest Live Journal community interested in "depression". Investigations into online communities have often been at the community setting, such as in social capital and mood patterns [13,18]. However, it is still questionable whether blogging by members of depression communities in personal blogs is different from community blogs. For example, *fillers* or *swearing* are expected to be used more in personal than in community pages, probably suggesting different weights for the cues when predicting depression in the two settings.

This paper examines if the content that members of the depression community contribute to the community blogs (*Community*) different from what they make in their own personal blogs (*Personal*)? In this study, the content is seen in three aspects: the topics discussed, the language styles expressed, and the affective information conveyed. These features were used as the base to detect online community [12], especially they have been found to be strong predictors of autism, and differentiate mental health communities from other online communities [14]. The features can be considered as potential text-based cues for depression.

We present an analysis of a large scale cohort of data authored by more than 4,000 members of the depression community. The content made by members of the depression community in their own personal blogs is also examined in comparison with what they make in the community blog. The way to collect data unobtrusively from online sources, such as online depression communities and individuals in this study, provides a valuable alternative approach to research in mental health, avoiding the issue of privacy as in traditional data collection for clinical studies.

The main contributions of this work are: (1) to introduce a relatively comprehensive view of the content made in depression personal and community blogs, regarding three aspects: sentiment information, language style, and topics of interest; (2) to propose an efficient approach to select features and do regression simultaneously, providing a set of powerful predictors of depression blogs in personal and community settings; and (3) to provide statistically reliable empirical evidence in a data-driven approach compared with small-scale questionnaire-based method in psychology. The result shows the potential of the new media in screening and monitoring of online depression blogging in both individual and community context. In addition, the same framework could be employed for at-risk individuals and communities. In a broad sense this work demonstrates the application of machine learning in medical practice and research.

The remainder of this paper is organized as follows. Section 2 describes the methodology. Section 3 presents the difference of blogging in personal and community settings. Section 4 shows the performance of the classification of Community vs. Personal blogs. Section 5 discusses the limitation of the work and Sect. 6 concludes the paper.

2 Method

2.1 Datasets

In this paper Live Journal data was chosen since it allows people to create or join communities of interest, along with their own personal pages. In particular, to examine blogging within online depression communities, data from the largest community in Live Journal interested in "depression" – depression.livejournal.com was crawled. The community is described in its profile as "a safe and open community for those experiencing depression".

The community was founded in December 2000 and as of September 2015, it has more than 7,000 members and 40 thousand posts. This is the *Community* cohort in this study. We then construct a control dataset. The posts made in personal blogs of members of the depression community were also crawled. This is the *Personal* cohort. Only those members who have made posts in both cohorts were taken into the experiments, resulting in a corpus of 25,012 community posts and 104,033 personal posts made by 4,439 users. To create a balanced dataset, the same number of posts (the smaller number of posts, 25,012, which appeared in the Community category) for Personal category was randomly selected into the study, resulting in a corpus of 50,024 posts.

2.2 Feature Sets

To characterised the posts made in community or personal blogs, three types of features are extracted: (1) Topics: what are discussed in the posts? (2) Language styles: how the posts are expressed? and (3) Expressed emotion: the affective information is conveyed in the content.

- *Affective information:* For the affective aspect, we used ANEW lexicon [2] to extract the sentiment conveyed in the content. Words in this lexicon are rated in term of valence, arousal, and dominance. The valence of words is on a scale of one, *very unpleasant*, to nine, *very pleasant*. The arousal is measured in the same scale, one for *least active* and nine for *most active*. The dominance is also in the same scale, ranging from *submissive* to *dominant*.

– *LIWC features:* We examined the proportions of words in psycho-linguistic categories as defined in the LIWC package [17]: linguistic, social, affective, cognitive, perceptual, biological, relativity, personal concerns and spoken.[1]
– *Topics:* to extract topics, latent Dirichlet allocation (LDA) [1] was used as a Bayesian probabilistic modeling framework. LDA extracts the probabilities p(vocabulary | topic) - that is, words in a topic, and then assigns a topic to each word in a document. For the inference part, we implemented Gibbs inference detailed in [10]. We set the number of topics to 50, run the Gibbs for 5000 samples and use the last Gibbs sample to interpret the results.

2.3   Statistical Testing

Statistical tests were conducted to examine the difference between Community and Personal blogs in the use of each feature. For each of 50 topics, 68 linguistic categories, and three ANEW sentiment scores, two following hypotheses were tested:

– H_1: $mean_{comm} = mean_{pers}$: the null hypothesis that, for the tested feature, the data made by the two examined populations, Community vs. Personal blogs, are samples from normal distributions with equal means. This test is conducted using the two-sample t-test, a parametric test.
– H_2: $median_{comm} = median_{pers}$: the null hypothesis that, for the tested feature, the data made by the two examined populations, Community vs. Personal blogs, are samples from continuous distributions with equal medians. This test is conducted using the Wilcoxon rank sum tests, a non-parametric test.

The relative difference in the use of a feature is determined as

$$diff = \frac{mean_{comm} - mean_{pers}}{(mean_{comm} + mean_{pers})/2} * 100 \tag{1}$$

If $diff > 0$, the feature is used more in Community than in Personal, and vice versa.

2.4   Classification

Lasso as the Classifier and Feature Selector. Denote by $\mathcal{B}$ a corpus of N posts made in community or personal blogs. A document $d \in \mathcal{B}$ is denoted as $\mathbf{x}^{(d)} = [\ldots, x_i^{(d)}, \ldots]$, a vector of features. The feature sets experimented in this work are topics, extracted through topic modeling (LDA) and language styles (LIWC). When topics are the features, $x_i^{(d)}$ is the probability of topic i in document d. If LIWC processes are the features, $x_i^{(d)}$ represents the quantity of the process i in document d. Our experimental design examines the effect of

---

[1] http://www.liwc.net/descriptiontable1.php, retrieved Sept. 2015, cached: http://bit.ly/1PPbeSv.

these two feature sets in classifying a blog post into one of two target classes. Given a document $d \in \mathcal{B}$, we predict if the document belongs to a Community or Personal blog based on the textual features $\mathbf{x}^{(d)}$.

We are interested in not only which sets of features perform well in the classification but also which features in the sets are strongly predictive of depression. For this purpose, the least absolute shrinkage and selection operator (Lasso) [9], a regularized regression, is chosen. Lasso does logistic regression and selects features simultaneously, enabling an evaluation on both the prediction performance and the importance of each feature in the classification. Particularly, in prediction of community posts, Lasso assigns positive and negative weights to features associated with community and personal posts, respectively. To the features irrelevant to the prediction, Lasso assigns zero weight. Thus, by examining its weights, we can learn the importance of each feature in the prediction.

The regularization parameter (λ) in the regression model is chosen such that it is the largest number and the accuracy is still within one standard error of the optimum (1se rule). This way prevents over-fitting since not too many features are included in the model while the accuracy of classification is still assured.

For each run, we use five-fold cross-validation, that is, one held-out fold is used for testing and other four folds for training. Accuracy was used to evaluate the performance of the classifications.

Other Classifiers. For comparison with the classification performed by Lasso, classifiers from other paradigms were also included:

– Naive Bayes (NB): one of the probabilistic methods that construct the conditional probability distributions of underlying features given a class label. The classification on unseen cases is then done by comparing the class likelihood.
– Support vector machines (SVM): a non-probabilistic binary classifier that finds the separating plane between two classes with maximal margins.
– Logistic regression (LR): a non-regularized logistic regression model, as opposed to the regularized one as Lasso.

These classifiers will perform the binary classifications of *Community* versus *Personal* posts, using LIWC, topics, and a combination of them as the feature sets. The accuracy is used to compare with that of Lasso on the same classifications.

Table 1. The mean (±std) of sentiment scores for community and personal posts.

Sentiment score	Community	Personal	h_1	h_2
Valence	5.29 ± 0.64	5.66 ± 0.62	1	1
Arousal	4.29 ± 0.34	4.23 ± 0.39	1	1
Dominance	5.33 ± 0.44	5.54 ± 0.42	1	1

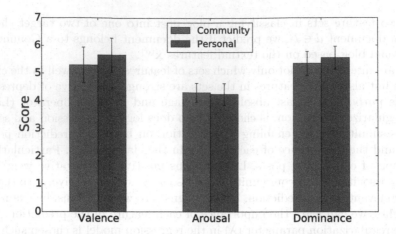

Fig. 1. The mean of sentiment scores for posts made in community and personal blogs.

3 Depression in Community and Personal Settings

In this section, the differences between *Community* and *Personal* blogs made by members of the depression community, with respect to the three feature sets, are discussed.

3.1 Affective Information

Table 1 shows the mean of sentiment scores for *Community* and *Personal* posts, accompanied with the result of t-test and rank sum test. If the result is one then the test rejects the null hypothesis of equal mean or median at the 5 % significance level. Otherwise the tests fail to reject the null hypothesis. As shown in the table, both tests rejected the null hypotheses.

Figure 1 shows that the mean valence of the community posts was lower than that of the personal posts. This means that the posts made in the community blogs have more unpleasant words and/or fewer pleasant words than those in the personal blogs.

Similarly, the average dominance score of community posts was lower than those of personal posts. It is likely that as a member of a depression community, people use more submissive and/or less dominant words, in comparison with what they write in their own pages.

On the other hand, the average arousal score of *Community* posts was slightly higher than that in *Personal* posts. This probably implies that members tend to use more high active words and/or less low active words in their posts.

3.2 Psycho-Linguistic Features

Table 2 shows the mean of LIWC linguistic features for the posts made in community and personal blogs, as well as the result of the statistical tests. For a

Table 2. The mean (±std) of LIWC features for posts made in community and personal blogs. Except for *wc* and *wps* are for the number of words for each post and number of words for each sentence in a post, respectively, the values for other features are the percentage of the LIWC categories in a post.

LIWC	Community	Personal	h_1	h_2	LIWC	Community	Personal	h_1	h_2
wc	225 ± 260	238 ± 456	1	1	anger	1.38 ± 1.9	1.09 ± 2.2	1	1
wps	15.2 ± 19.5	14.8 ± 24.3	0	1	sad	1.44 ± 1.8	0.60 ± 1.4	1	1
dic	90.88 ± 6.5	82.38 ± 14.3	1	1	cogmech	19.56 ± 5.3	15.57 ± 6.5	1	1
sixltr	12.98 ± 5.2	14.36 ± 7.2	1	1	insight	3.28 ± 2.2	2.12 ± 2	1	1
funct	60.86 ± 7.2	52.86 ± 13.1	1	1	cause	1.80 ± 1.6	1.39 ± 1.7	1	1
pronoun	20.86 ± 5.0	16.56 ± 6.8	1	1	discrep	2.28 ± 1.9	1.69 ± 2.0	1	1
ppron	14.29 ± 4.5	11.34 ± 5.6	1	1	tentat	3.43 ± 2.6	2.64 ± 2.5	1	1
i	11.13 ± 4.6	7.60 ± 5.0	1	1	certain	1.73 ± 1.6	1.36 ± 1.7	1	1
we	0.28 ± 0.8	0.49 ± 1.2	1	1	inhib	0.53 ± 1.0	0.57 ± 1.5	1	1
you	1.10 ± 2.2	1.65 ± 3.1	1	1	incl	4.32 ± 2.4	4.08 ± 2.9	1	1
shehe	1.20 ± 2.1	1.12 ± 2.1	1	1	excl	3.88 ± 2.4	2.84 ± 2.5	1	1
they	0.58 ± 1.1	0.49 ± 1.0	1	1	percept	2.34 ± 2.1	2.45 ± 2.8	1	1
ipron	6.57 ± 3.2	5.22 ± 3.4	1	1	see	0.54 ± 1.0	0.92 ± 1.7	1	1
article	4.10 ± 2.3	4.85 ± 3.2	1	1	hear	0.54 ± 1.0	0.62 ± 1.4	1	1
verb	18.21 ± 4.6	14.93 ± 6.0	1	1	feel	1.17 ± 1.4	0.74 ± 1.4	1	1
auxverb	10.94 ± 3.6	9.01 ± 4.5	1	1	bio	2.95 ± 2.7	2.86 ± 3.5	1	1
past	3.39 ± 2.7	3.25 ± 3.1	1	1	body	0.82 ± 1.3	0.97 ± 1.9	1	0
present	12.55 ± 4.6	9.56 ± 5.2	1	1	health	1.41 ± 1.8	0.76 ± 1.6	1	1
future	0.94 ± 1.1	0.99 ± 1.4	1	1	sexual	0.49 ± 1.1	0.70 ± 1.7	1	1
adverb	6.62 ± 3.0	5.43 ± 3.6	1	1	ingest	0.34 ± 1.0	0.57 ± 1.7	1	1
preps	11.63 ± 3.5	11.02 ± 4.4	1	1	relativ	12.83 ± 4.8	13.19 ± 6.2	1	1
conj	7.04 ± 2.8	5.94 ± 3.4	1	1	motion	1.64 ± 1.5	1.91 ± 2.2	1	1
negate	2.82 ± 2.1	1.91 ± 2.0	1	1	space	4.97 ± 2.7	5.16 ± 3.4	1	1
quant	2.92 ± 2.0	2.56 ± 2.2	1	1	time	5.84 ± 3.3	5.77 ± 4.2	0	1
number	0.65 ± 1.0	0.73 ± 1.4	1	1	work	1.17 ± 1.7	1.68 ± 2.7	1	1
swear	0.52 ± 1.2	0.57 ± 1.6	1	1	achieve	1.35 ± 1.4	1.32 ± 1.8	1	1
social	8.14 ± 5.2	8.04 ± 5.8	0	1	leisure	0.64 ± 1.2	1.45 ± 2.4	1	1
family	0.40 ± 0.9	0.33 ± 1.0	1	1	home	0.34 ± 0.7	0.49 ± 1.3	1	1
friend	0.36 ± 0.8	0.31 ± 1.0	1	1	money	0.29 ± 0.8	0.56 ± 1.5	1	1
humans	0.71 ± 1.1	0.72 ± 1.4	0	1	relig	0.22 ± 0.8	0.37 ± 1.3	1	1
affect	7.54 ± 3.8	6.57 ± 4.6	1	1	death	0.37 ± 1.0	0.23 ± 1.0	1	1
posemo	3.04 ± 2.5	3.82 ± 3.5	1	1	assent	0.25 ± 0.7	0.63 ± 1.5	1	1
negemo	4.39 ± 3.2	2.68 ± 3.2	1	1	nonfl	0.18 ± 0.5	0.23 ± 0.7	1	1
anx	0.67 ± 1.1	0.36 ± 0.9	1	1	filler	0.02 ± 0.2	0.02 ± 0.4	0	1

Table 3. The mean (±std) of topic distribution for posts made in community and personal blogs.

Topic	Community	Personal	h_1	h_2	Topic	Community	Personal	h_1	h_2
T1	3.24 ± 3.3	1.56 ± 1.1	1	1	T26	2.14 ± 1.7	1.92 ± 1.7	1	1
T2	1.42 ± 0.8	1.93 ± 2.7	1	1	T27	2.41 ± 2.0	2.07 ± 1.7	1	1
T3	1.55 ± 0.8	1.81 ± 2.1	1	1	T28	1.75 ± 1.3	2.19 ± 2.5	1	1
T4	2.31 ± 2.3	1.76 ± 1.5	1	1	T29	1.55 ± 1.7	1.69 ± 2.2	1	1
T5	1.30 ± 0.8	1.67 ± 3.2	1	1	T30	1.73 ± 1.2	2.08 ± 2.7	1	1
T6	2.25 + 1.7	1.86 ± 1.5	1	1	T31	1.81 ± 1.9	2.40 ± 4.0	1	1
T7	2.11 ± 2.1	2.24 ± 2.6	1	1	T32	1.85 ± 1.6	1.83 ± 1.6	0	1
T8	2.44 ± 2.1	1.75 ± 1.3	1	1	T33	2.09 ± 1.7	2.06 ± 1.9	0	1
T9	2.08 ± 1.9	2.03 ± 1.9	1	0	T34	1.60 ± 1.2	2.18 ± 2.7	1	1
T10	2.08 ± 1.8	1.90 ± 1.6	1	1	T35	2.28 ± 3.3	2.05 ± 2.0	1	1
T11	1.83 ± 1.6	2.08 ± 2.4	1	1	T36	2.26 ± 1.8	1.90 ± 1.5	1	1
T12	2.54 ± 2.2	1.79 ± 1.4	1	1	T37	1.75 ± 1.3	1.87 ± 1.9	1	1
T13	2.49 ± 2.2	1.85 ± 1.5	1	1	T38	2.13 ± 1.6	2.00 ± 1.9	1	1
T14	1.55 ± 0.9	2.06 ± 2.2	1	1	T39	1.90 ± 1.6	2.18 ± 2.3	1	1
T15	1.81 ± 1.4	2.30 ± 4.3	1	1	T40	2.45 ± 2.1	1.76 ± 1.3	1	1
T16	1.68 ± 1.2	1.89 ± 1.9	1	1	T41	1.86 ± 1.7	2.16 ± 2.3	1	1
T17	1.68 ± 1.4	2.31 ± 2.9	1	1	T42	1.86 ± 1.5	2.16 ± 2.2	1	1
T18	2.21 ± 1.6	2.01 ± 1.5	1	1	T43	1.79 ± 1.5	2.15 ± 2.4	1	1
T19	1.63 ± 1.2	2.29 ± 2.8	1	1	T44	1.72 ± 1.2	1.87 ± 2.1	1	1
T20	2.49 ± 2.7	2.05 ± 1.9	1	1	T45	1.63 ± 1.0	2.25 ± 3.6	1	1
T21	2.49 ± 2.0	1.74 ± 1.3	1	1	T46	1.52 ± 0.9	2.27 ± 3.2	1	1
T22	1.58 ± 1.6	2.08 ± 3.2	1	1	T47	2.09 ± 1.5	1.89 ± 1.4	1	1
T23	2.30 ± 2.1	1.90 ± 1.5	1	1	T48	1.61 ± 1.1	1.81 ± 1.9	1	1
T24	2.19 ± 2.3	2.10 ± 2.4	1	1	T49	2.84 ± 3.5	1.73 ± 1.7	1	1
T25	1.74 ± 1.2	2.37 ± 2.4	1	1	T50	2.41 ± 3.0	2.20 ± 2.6	1	1

majority of LIWC features, t-test and/or rank sum test rejected the null hypothesis of no difference between Community and Personal cohorts. There were only six of 68 LIWC features (wps, social, humans, body, time, and filler) whose the null hypothesis were failed to be rejected by either t-test or rank-test or both, at the 0.05 level.

Figure 2 shows the relative difference in the use of LIWC features by Community and Personal cohorts. On the *affective* processes, it is observed that posts in the depression community blogs contained more *negative* (including *anxiety*, *sadness*, and *anger*) emotion and less *positive* words than those in the depression personal blogs. Community blogs also have more *affective* and *negate* words, compared with Personal blogs. In addition, they have more *pronoun* in the posts

Table 4. The prominent topics in favor of Community (number in red) and Personal (number in blue) blogs.

Topic	Word cloud	Topic	Word cloud
T1	**depression** anxiety therapist mental therapy disorder	T46	**movie** movies watch watching character film season episode series scene star watched story
T49	**doctor meds** hospital medication pills anti appointment doctors weeks psychiatrist	T19	**hair** black wear blue white red clothes
T21	**cry stop crying** anymore horrible upset wrong head tears cried sit hurts reason away	T45	**favorite** current color hair sex phone school movie crush food yep worst kissed
T12	**self** thoughts feelings mind emotions emotional anger negative control	T17	**sun water** rain snow fire sky feet tree ground weather
T40	**deal mood lately worse** normal past problem worry anymore weeks control completely stress reason constantly	T25	**birthday friday** party saturday sunday monday thursday house wednesday tuesday

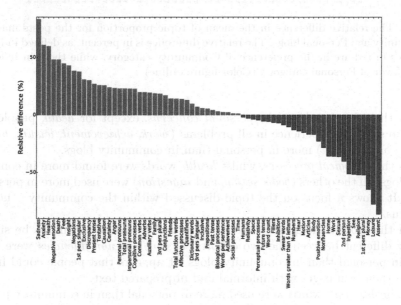

Fig. 2. The relative difference between Community and Personal blogs in the mean of LIWC features in the posts. The relative difference is in percent, as defined in Eq. 1. Features in red are in the preference of Community category, while those in blue are in the favor of Personal category. (Color figure online)

than do their Personal counterparts, except for *first personal plural* and *second personal*. This is partly in line with previous findings that depression people tend to use more pronouns [20].

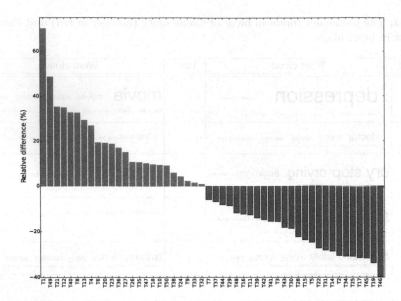

Fig. 3. The relative difference in the mean of topic proportion for the posts made in Community and Personal blogs. The relative difference is in percent, as defined in Eq. 1. Features in red are in the preference of Community category, while those in blue are in the favor of Personal category. (Color figure online)

On the other hand, on the *personal concerns*, except for *death*, people are likely to share all experience in all problems (*work, achievement, leisure, home, money,* and *religion*) more in personal than in community blogs.

On the *biological processes*, whilst *health* words were found more in community blogs, all the others (*body, sexual,* and *ingestion*) were used more in personal blogs. It shows a focus on the topic discussed within the community – health concerns.

On the *spoken* categories, except *fillers* which was not found to be significantly different between the two cohorts, *assent* and *non-fluencies* were used more in personal than in community blogs. It appears that people could freely post in their own pages with informal and unprepared text.

Similarly, *swear* words were used more in personal than in community pages. It may be because posting to the communities is often gone through a moderation process. So, a post with inappropriate words could be rejected to be posted to the community pages.

3.3   Topical Representation

Table 3 shows the mean of the proportion of topics for community and personal blogs, as well as the result of the statistical tests.[2] For 47 of 50 topics, t-test

---

[2] All 50 topics learned from the corpus by LDA are placed at http://bit.ly/1KEgjpM.

Table 5. Performance, in terms of the predictive accuracy (percentage of correct predictions), of different classifiers on different feature sets in the binary classifications of Community versus Personal posts.

Feature	SVM	NB	LR	Lasso
LIWC	59.6	68	77.5	**78.9**
Topic	77.7	69.7	**77.9**	77.8
Joint features	59.2	74.6	80.3	**80.5**

and/or rank sum test rejected the null hypothesis of no difference between community and personal cohorts.

Figure 3 shows the difference in the distribution of topics for posts made by community and personal cohorts. The topics with the highest difference were presented in Table 4. It is obvious to see in the table that topic 1 with depression themes ("depression", "anxiety", "therapist") was used much more in community than in personal blogs. Likewise, topic 49 with "doctor", "medicine", "hospital" was found in the preference of community posts, with the second highest relative difference, in comparison with personal posts. This partly confirms the finding in the language section that community posts were more focused on health concerns than were personal posts.

However, it is not straightforward to interpret the meaning of the other three topics in the favor of the community cohort. Nevertheless, "emotional and submissive" sense could be observed across the three topics. They also consisted of many affective words, confirming that community posts contained more *affective* words than did personal posts.

On the other hand, personal cohort favored more on concrete concepts, from "movie" or fashion ("hair") to weather ("sun", "rain") or "birthday", than did Community cohort. Most of these topics contain words in the LIWC *"leisure"* feature, which was found to be a favor language category for personal posts.

4   Classification

4.1   Performance

The Lasso model [9] is used for the classification. Using the coefficients derived from the Lasso method, we implemented a pair-wise classifier of *Community* versus *Personal* posts, using three feature sets: LIWC, topics, and a combination of them. The accuracy of this classifier in different feature sets are shown in Table 5, accompanied with that of SVM, Naive Bayes, and Logistic Regression. Lasso outperformed other classifiers when LIWC and a combination of LIWC and topics were used as the features, and was second to LR (77.8 % and 77.9 %), when topics were the features.

Classifying based on the derived topics outperformed the LIWC linguistic feature analysis in three of the four classifiers. Noticeably, a fusion of the features gained the best performance in all the classifications, except for the case when SVM was the classifier. This shows the potential of using multi-cues for mental health prediction.

4.2 Linguistic Features as the Predictors

Figure 4a shows the model using language style cues as features to predict community versus personal posts. Several features found significantly different between community vs. personal cohorts (see Sect. 3.2) were chosen into the prediction model. For example, for the *affective* category, *negative*, *sadness*, and *anxiety* were assigned large positive coefficients, whilst *positive* emotion had negative coefficient. The positive predictors also included *negation*, which was found in the preference of community blogs, confirming the finding in Sect. 3.2.

Of *personal concerns* chosen into the model, only *death* was assigned positive coefficient, whilst the rest (*work, leisure, home, money*, and *religion*) had negative coefficients. This is consistent with the findings presented in Sect. 3.2.

Similarly, on the *biological processes*, *health* was a positive predictor while *body, sexual*, and *ingestion* were negative ones.

As expected, *assent* and *swear* words were also negative predictors in the model. Noticeably, *assent* had the largest negative coefficient, becoming the most predictive feature of personal posts.

4.3 Topics as the Predictors

Figure 4b shows the classification model to predict community versus personal posts using topics as features. Table 6 shows the word cloud of topics whose the weights is largest in the model. It confirms that the top five topics in the preference of community blogs, which were "emotional and submissive" and focused on health concerns, as shown in Table 4, were among the top six positive predictors in the model.

On the other hand, fashion and leisure topics, which were in the preference of personal blogs, were assigned negative coefficients. The topic on sex was also a negative predictor, which is in line with the finding on LIWC as the features that *sexual*, a *biological process*, was a negative predictor. Similarly, the topic of "eat", whose words belong to *ingestion*, another *biological process*, was also assigned a negative coefficient in the prediction model.

5 Limitation and Further Research

In the current work, the mental health status of the individuals in the online communities was not validated. Instead, labeling was "by affiliation" [5]. As such, it cannot be stated that all the individuals whose communications were analysed were experiencing depression. Future studies would benefit from attempting to validate this either by direct contact with the individuals or by analyzing the conversations for admission of diagnosis [4–6,8]. If an admission is identified, other

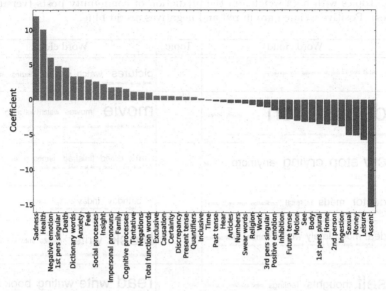

(a) Lasso model coefficients for prediction of community (versus personal) posts using LIWC features as the predictors.

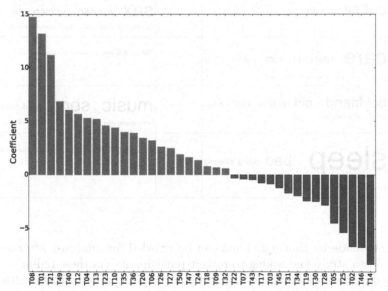

(b) Lasso model coefficients for prediction of community (versus personal) posts using topics as the predictors.

Fig. 4. Prediction models of community (versus personal) posts using topics and language styles as features. Features in red are positive predictors of community posts whilst the blues are the negatives. Individual coefficients that were not significant have been omitted, as have topics and LIWC features with no significant coefficients. (Color figure online)

Table 6. Topics with high weights in the prediction of community posts (versus personal ones). Positive weights are in red and negatives are in blue.

Topic	Word cloud	Topic	Word cloud
T8	end live kill suicide anymore death dead living	T14	pictures watch picture amazing video yourwintr funny camera mike photos big pics youtube ryan watching
T1	**depression** anxiety therapist mental therapy disorder	T46	**movie** movies watch watching character film season episode series scene star watched story
T21	**cry stop crying** anymore horrible upset wrong head tears cried	T2	mark david finished james peter scott george michael robert bob jane king frank various richard read world
T49	doctor meds hospital medication pills anti appointment doctors weeks psychiatrist	T25	birthday friday party saturday sunday monday thursday house wednesday tuesday
T40	deal mood lately worse normal past problem worry anymore weeks control completely stress	T5	**block** writer oacute automatically tweets ntilde eacute shipped loudwitter iacute asma aacute
T12	**self** thoughts feelings mind emotions emotional anger negative control	T28	**read** write writing book reading story books written wrote paper art stories
T4	**cut pain** cutting stop blood body hurt deep	T30	**sex** woman women men gay sexual type female male test personality children straight marriage partner
T13	**care** hurt anymore trust matter deserve fault away lie real truth cares wrong	T19	hair black wear blue white red clothes wearing shirt dress pants
T23	**boyfriend** girl relationship months girlfriend girls past close broke date break	T34	**music song** band listening songs listen playing rock heard concert sound hear lyrics sing singing
T10	**sleep** bed wake sleeping woke asleep stay late fall slept hour	T31	eat weight eating food fat

statements made by that individual can be crawled for analysis. Alternatively, establishing a study and asking to recruit participants via these online communities would allow standardized diagnostic measurements of mental health to be administered. By having a more complete understanding of the mental health of the individuals in these communities, we are able to improve the profiling of their discussion and linguistic features so that a clearer match between online behavior and diagnosis can be made.

An example of this approach was conducted in [16], where depression scores of Twitter participants were collected and LIWC features were used as the predictors of the scores. Similarly, the depression ground-truth for Twitter users was collected individually in [21] (as opposed to "affiliation labeling" in this work), where tweeting behaviors were employed as the predictors.

6 Conclusion

We have investigated online depression blogs in community and personal settings. Affect information, language styles, and topics were extracted from the posts made in depression community blogs, as well as those made by members of the depression community in their own personal blogs. Statistical and machine learning techniques were used to discriminate the content in the two settings. A majority of features in all aspects were found to be significantly different between community and personal blogging by members of the community. Also, language styles and topics were found to have strong indicative powers in prediction of depression in personal and community settings. High performance in the classifications illustrates the effectiveness of the content-based cues as the discriminators. The result shows the potential of the new media in early screening and monitoring of online depression individuals and communities.

References

1. Blei, D.M., Ng, A.Y., Jordan, M.I.: Latent Dirichlet allocation. J. Mach. Learn. Res. **3**, 993–1022 (2003)
2. Bradley, M.M., Lang, P.J.: Affective norms for English words (ANEW): instruction manual and affective ratings. Technical report, University of Florida (1999)
3. Bridge, J.A., Goldstein, T.R., Brent, D.A.: Adolescent suicide and suicidal behavior. J. Child Psychol. Psychiatry **47**(3–4), 372–394 (2006)
4. Coppersmith, G., Dredze, M., Harman, C.: Quantifying mental health signals in Twitter. In: Proceedings of the Workshop on Computational Linguistics and Clinical Psychology: From Linguistic Signal to Clinical Reality, pp. 51–60 (2014)
5. Coppersmith, G., Dredze, M., Harman, C., Hollingshead, K.: From ADHD to SAD: analyzing the language of mental health on Twitter through self-reported diagnoses. In: Proceedings of the Workshop on Computational Linguistics and Clinical Psychology: From Linguistic Signal to Clinical Reality, pp. 1–10 (2015)
6. Coppersmith, G., Harman, C., Dredze, M.: Measuring post traumatic stress disorder in Twitter. In: International AAAI Conference on Weblogs and Social Media, pp. 579–582 (2014)
7. Cruwys, T., Haslam, S.A., Dingle, G.A., Haslam, C., Jetten, J.: Depression and social identity: an integrative review. Pers. Soc. Psychol. Rev. **18**(3), 215–238 (2014)
8. De Choudhury, M., Counts, S., Horvitz, E.: Major life changes and behavioral markers in social media: Case of childbirth. In: Proceedings of the 2013 Conference on Computer Supported Cooperative Work, pp. 1431–1442 (2013)
9. Friedman, J., Hastie, T., Tibshirani, R.: Regularization paths for generalized linear models via coordinate descent. J. Stat. Softw. **33**(1), 1–22 (2010)
10. Griffiths, T.L., Steyvers, M.: Finding scientific topics. Proc. Natl. Acad. Sci. **101**(90001), 5228–5235 (2004)

11. Mundt, J.C., Vogel, A.P., Feltner, D.E., Lenderking, W.R.: Vocal acoustic biomarkers of depression severity and treatment response. Biol. Psychiatry **72**(7), 580–587 (2012)
12. Nguyen, T., Phung, D., Adams, B., Venkatesh, S.: A sentiment-aware approach to community formation in social media. In: Proceedings of the International AAAI Conference on Weblogs and Social Media, pp. 527–530 (2012)
13. Nguyen, T.: Mood patterns and affective lexicon access in weblogs. In: Proceedings of the ACL Student Research Workshop, pp. 43–48 (2010)
14. Nguyen, T., Phung, D., Venkatesh, S.: Analysis of psycholinguistic processes and topics in online autism communities. In: Proceedings of the IEEE International Conference on Multimedia and Expo, pp. 1–6 (2013)
15. Nock, M.K., Green, J.G., Hwang, I., McLaughlin, K.A., Sampson, N.A., Zaslavsky, A.M., Kessler, R.C.: Prevalence, correlates, and treatment of lifetime suicidal behavior among adolescents: Results from the national comorbidity survey replication adolescent supplement. JAMA Psychiatry **70**(3), 300–310 (2013)
16. Park, M., Cha, C., Cha, M.: Depressive moods of users portrayed in Twitter. In: Proceedings of the ACM SIGKDD Workshop on Healthcare Informatics, pp. 1–8 (2012)
17. Pennebaker, J.W., Francis, M.E., Booth, R.J.: Linguistic Inquiry and Word Count (LIWC) [Computer software]. LIWC Inc. (2007)
18. Phung, D., Gupta, S., Nguyen, T., Venkatesh, S.: Connectivity, online social capital, and mood: a Bayesian nonparametric analysis. IEEE Trans. Multimedia **15**(6), 1316–1325 (2013)
19. Rodriguez, A.J., Holleran, S.E., Mehl, M.R.: Reading between the lines: the lay assessment of subclinical depression from written self-descriptions. J. Pers. **78**(2), 575–598 (2010)
20. Tausczik, Y.R., Pennebaker, J.W.: The psychological meaning of words: LIWC and computerized text analysis methods. J. Lang. Soc. Psychol. **29**(1), 24–54 (2010)
21. Tsugawa, S., Kikuchi, Y., Kishino, F., Nakajima, K., Itoh, Y., Ohsaki, H.: Recognizing depression from Twitter activity. In: Proceedings of the ACM Conference on Human Factors in Computing Systems, pp. 3187–3196 (2015)
22. Waxer, P.H.: Nonverbal cues for depth of depression: set versus no set. J. Consult. Clin. Psychol. **44**(3), 493 (1976)

Confidence-Weighted Bipartite Ranking

Majdi Khalid[✉], Indrakshi Ray, and Hamidreza Chitsaz

Colorado State University, Fort Collins, USA
majdi.khaled@gmail.com, indrakshi.ray@colostate.edu,
chitsaz@chitsazlab.org
http://www.colostate.edu

Abstract. Bipartite ranking is a fundamental machine learning and data mining problem. It commonly concerns the maximization of the AUC metric. Recently, a number of studies have proposed online bipartite ranking algorithms to learn from massive streams of class-imbalanced data. These methods suggest both linear and kernel-based bipartite ranking algorithms based on first and second-order online learning. Unlike kernelized ranker, linear ranker is more scalable learning algorithm. The existing linear online bipartite ranking algorithms lack either handling non-separable data or constructing adaptive large margin. These limitations yield unreliable bipartite ranking performance. In this work, we propose a linear online confidence-weighted bipartite ranking algorithm (CBR) that adopts soft confidence-weighted learning. The proposed algorithm leverages the same properties of soft confidence-weighted learning in a framework for bipartite ranking. We also develop a diagonal variation of the proposed confidence-weighted bipartite ranking algorithm to deal with high-dimensional data by maintaining only the diagonal elements of the covariance matrix. We empirically evaluate the effectiveness of the proposed algorithms on several benchmark and high-dimensional datasets. The experimental results validate the reliability of the proposed algorithms. The results also show that our algorithms outperform or are at least comparable to the competing online AUC maximization methods.

Keywords: Online ranking · Imbalanced learning · AUC maximization

1 Introduction

Bipartite ranking is a fundamental machine learning and data mining problem because of its wide range of applications such as recommender systems, information retrieval, and bioinformatics [1, 25, 29]. Bipartite ranking has also been shown to be an appropriate learning algorithm for imbalanced data [6]. The aim of the bipartite ranking algorithm is to maximize the Area Under the Curve (AUC) by learning a function that scores positive instances higher than negative instances. Therefore, the optimization problem of such a ranking model is formulated as the minimization of a pairwise loss function. This ranking problem can be solved by applying a binary classifier to pairs of positive and negative

© Springer International Publishing AG 2016
J. Li et al. (Eds.): ADMA 2016, LNAI 10086, pp. 35–49, 2016.
DOI: 10.1007/978-3-319-49586-6_3

instances, where the classification function learns to classify a pair as positive or negative based on the first instance in the pair. The key problem of this approach is the high complexity of a learning algorithm that grows quadratically or subquadratically [19] with respect to the number of instances.

Recently, significant efforts have been devoted to developing scalable bipartite ranking algorithms to optimize AUC in both batch and online settings [12,15,17,24,37]. Online learning approach is an appealing statistical method because of its scalability and effectivity. The online bipartite ranking algorithms can be classified on the basis of the learned function into linear and nonlinear ranking models. While they are advantageous over linear online ranker in modeling nonlinearity of the data, kernelized online ranking algorithms require a kernel computation for each new training instance. Further, the decision function of the nonlinear kernel ranker depends on support vectors to construct the kernel and to make a decision.

Online bipartite ranking algorithms can also be grouped into two different schemes. The first scheme maintains random instances from each class label in a finite buffer [20,37]. Once a new instance is received, the buffer is updated based on stream oblivious policies such as Reservoir Sampling (RS) and First-In-First-Out (FIFO). Then the ranker function is updated based on a pair of instances, the new instance and each opposite instance stored in the corresponding buffer. These online algorithms are able to deal with non-separable data, but they are based on simple first-order learning. The second approach maintains the first and second statistics [15], and is able to adapt the ranker to the importance of the features [12]. However, these algorithms assume the data are linearly separable.

Moreover, these algorithms make no attempt to exploit the confidence information, which has shown to be very effective in ameliorating the classification performance [8,13,35]. Confidence-weighted (CW) learning takes the advantage of the underlying structure between features by modeling the classifier (i.e., the weight vector) as a Gaussian distribution parameterized by a mean vector and covariance matrix [13]. This model captures the notion of confidence for each weight coordinate via the covariance matrix. A large diagonal value corresponding to the i-th feature in the covariance matrix results in less confidence in its weight (i.e., its mean). Therefore, an aggressive update is performed on the less confident weight coordinates. This is analogous to the adaptive subgradient method [14] that involves the geometric structure of the data seen so far in regularizing the weights of sparse features (i.e., less occurring features) as they are deemed more informative than dense features. The confidence-weighted algorithm [13] has also been improved by introducing the adaptive regularization (AROW) that deals with inseparable data [10]. The soft confidence-weighted (SCW) algorithm improves upon AROW by maintaining an adaptive margin [35].

In this paper, we propose a novel framework that solves a linear online bipartite ranking using the soft confidence-weighted algorithm [35]. The proposed confidence-weighted bipartite ranking algorithm (CBR) entertains the fast training and testing phases of linear bipartite ranking. It also enjoys the capability of the soft confidence-weighted algorithm in learning confidence-weighted model, handling linearly inseparable data, and constructing adaptive large margin. The proposed framework follows an online bipartite ranking scheme that maintains a finite buffer for each class label while updating the buffer by one of the stream oblivious policies such as Reservoir Sampling (RS) and First-In-First-Out (FIFO) [32]. We also provide a diagonal variation (CBR-diag) of the proposed algorithm to handle high-dimensional datasets.

The remainder of the paper is organized as follows. In Sect. 2, we briefly review closely related work. We present the confidence-weighted bipartite ranking (CBR) and its diagonal variation (CBR-diag) in Sect. 3. The experimental results are presented in Sect. 4. Section 5 concludes the paper and presents some future work.

2 Related Work

The proposed bipartite ranking algorithm is closely related to the online learning and bipartite ranking algorithms. What follows is a brief review of recent studies related to these topics.

Online Learning. The proliferation of big data and massive streams of data emphasize the importance of online learning algorithms. Online learning algorithms have shown a comparable classification performance compared to batch learning, while being more scalable. Some online learning algorithms, such as the Perceptron algorithm [30], the Passive-Aggressive (PA) [7], and the online gradient descent [38], update the model based on a first-order learning approach. These methods do not take into account the underlying structure of the data during learning. This limitation is addressed by exploring second-order information to exploit the underlying structure between features in ameliorating learning performance [4,10,13,14,27,35]. Moreover, kernelized online learning methods have been proposed to deal with nonlinearly distributed data [3,11,28].

Bipartite Ranking. Bipartite ranking learns a real-valued function that induces an order on the data in which the positive instances precede negative instances. The common measure used to evaluate the success of the bipartite ranking algorithm is the AUC [16]. Hence the minimization of the bipartite ranking loss function is equivalent to the maximization of the AUC metric. The AUC presents the probability that a model will rank a randomly drawn positive instance higher than a random negative instance. In batch setting, a considerable amount of studies have investigated the optimization of linear and nonlinear kernel ranking function [5,18,22,23]. More scalable methods are based on stochastic or online learning [31,33,34]. However, these methods are not specifically designed to optimize the AUC metric. Recently, a few studies have focused on the optimization of the AUC metric in online setting. The first approach adopted a framework that maintained a buffer with limited capacity to store random instances to deal with the pairwise loss function [20,37]. The other methods maintained only the first and second statistics for each received instance and optimized the AUC in one pass over the training data [12,15]. The work [12] exploited the second-order technique [14] to make the ranker aware of the importance of less frequently occurring features, hence updating their weights with a higher learning rate.

Our proposed method follows the online framework that maintains fixed-sized buffers to store instances from each class label. Further, it exploits the online second-order method [35] to learn a robust bipartite ranking function. This distinguishes the proposed method from [20,37] employing first-order online learning. Also, the proposed method is different from [12,15] by learning a confidence-weighted ranker capable of dealing with non-separable data and learning an adaptive large margin. The most similar approaches to our method are [33,34]. However, these are not designed to directly maximize the AUC. They also use classical first-order and second-order online learning whereas we use the soft variation of confidence-weighted learning that has shown a robust performance in the classification task [35].

3   Online Confidence-Weighted Bipartite Ranking

3.1   Problem Setting

We consider a linear online bipartite ranking function that learns on imbalanced data to optimize the AUC metric [16]. Let $\mathcal{S} = \{x_i^+ \cup x_j^- \in \mathcal{R}^d | i = \{1,\ldots,n\}, j = \{1,\ldots,m\}\}$ denotes the input space of dimension d generated from unknown distribution $\mathcal{D}$, where x_i^+ is the i-th positive instance and x_j^- is the j-th negative instance. The n and m denote the number of positive and negative instances, respectively. The linear bipartite ranking function $f : \mathcal{S} \to \mathcal{R}$ is a real valued function that maximizes the AUC metric by minimizing the following loss function:

$$\mathcal{L}(f;\mathcal{S}) = \frac{1}{nm} \sum_{i=1}^{n} \sum_{j=1}^{m} I(f(x_i^+) \leq f(x_j^-)),$$

where $f(x) = w^T x$ and $I(\cdot)$ is the indicator function that outputs 1 if the condition is held, and 0 otherwise. It is common to replace the indicator function with a convex surrogate function,

$$\mathcal{L}(f;\mathcal{S}) = \frac{1}{nm} \sum_{i=1}^{n} \sum_{j=1}^{m} \ell(f(x_i^+) - f(x_j^-)),$$

where $\ell(z) = max(0, 1-z)$ is the hinge loss function. Therefore, the objective function we seek to minimize with respect to the linear function is defined as follows:

$$\frac{1}{2}||w||^2 + C \sum_{i=1}^{n} \sum_{j=1}^{m} max(0, 1 - w(x_i^+ - x_j^-)), \tag{1}$$

where $||w||$ is the euclidean norm to regularize the ranker, and C is the hyper-parameter to penalize the sum of errors. It is easy to see that the optimization problem (1) will grow quadratically with respect to the number of training instances. Following the approach suggested by [37] to deal with the complexity of the pairwise loss function, we reformulate the objective function (1) as a sum of two losses for a pair of instances,

$$\frac{1}{2}||w||^2 + C \sum_{t=1}^{T} I_{(y_t=+1)}g_t^+(w) + I_{(y_t=-1)}g_t^-(w), \tag{2}$$

where $T = n + m$, and $g_t(w)$ is defined as follows

$$g_t^+(w) = \sum_{t'=1}^{t-1} I_{(y_{t'}=-1)}max(0, 1 - w(x_t - x_{t'})), \tag{3}$$

$$g_t^-(w) = \sum_{t'=1}^{t-1} I_{(y_{t'}=+1)}max(0, 1 - w(x_{t'} - x_t)). \tag{4}$$

Instead of maintaining all the received instances to compute the gradients $\nabla g_t(w)$, we store random instances from each class in the corresponding buffer. Therefore, two buffers B_+ and B_- with predefined capacity are maintained for positive and negative classes, respectively. The buffers are updated using a stream oblivious policy. The current stored instances in a buffer are used to update the classifier as in Eq. (2) whenever a new instance from the opposite class label is received.

The framework of the online confidence-weighted bipartite ranking is shown in Algorithm 1. The two main components of this framework are UpdateBuffer and UpdateRanker, which are explained in the following subsections.

3.2   Update Buffer

One effective approach to deal with pairwise learning algorithms is to maintain a buffer with a fixed capacity. This raises the problem of updating the buffer to store the most informative instances. In our online Bipartite ranking framework, we investigate the following two stream oblivious policies to update the buffer:

Reservoir Sampling **(RS)**: Reservoir Sampling is a common oblivious policy to deal with streaming data [32]. In this approach, the new instance (x_t, y_t) is added to the corresponding buffer if its capacity is not reached, $|B_{y_t}^t| < M_{y_t}$. If the buffer is at capacity, it will be updated with probability $\frac{M_{y_t}}{|B_{y_t}^t|}$ by randomly replacing one instance in $B_{y_t}^t$ with x_t.

First-In-First-Out **(FIFO)**: This simple strategy replaces the oldest instance with the new instance if the corresponding buffer reaches its capacity. Otherwise, the new instance is simply added to the buffer.

Algorithm 1. A Framework for Confidence-Weighted Bipartite Ranking (CBR)

  Input:
  - the penalty parameter C
  - the capacity of the buffers M_+ and M_-
  - η parameter
  - $a_i = 1$ for $i \in 1, \ldots, d$

  Initialize: $\mu_1 = \{0, \ldots, 0\}^d$, $B_+ = B_- = \emptyset$
  $\qquad\qquad \Sigma_1 = diag(a)$ or $G_1 = a$
  for t = 1, ..., T **do**
  $\quad$ Receive a training instance (x_t, y_t)
  $\quad$ **if** $y_t = +1$ **then**
  $\qquad B_-^{t+1} = B_-^t$
  $\qquad C_t = C\ max(1, |B_-^t|/M_-)$
  $\qquad B_+^{t+1} = \text{UpdateBuffer}(x_t, B_+^t, M_+)$
  $\qquad [\mu_{t+1}, \Sigma_{t+1}] = \text{UpdateRanker}(\mu_t, \Sigma_t, x_t, y_t, C_t, B_-^{t+1}, \eta)$ or
  $\qquad [\mu_{t+1}, G_{t+1}] = \text{UpdateRanker}(\mu_t, G_t, x_t, y_t, C_t, B_-^{t+1}, \eta)$ (CBR-diag)
  $\quad$ **else**
  $\qquad B_+^{t+1} = B_+^t$
  $\qquad C_t = C\ max(1, |B_+^t|/M_+)$
  $\qquad B_-^{t+1} = \text{UpdateBuffer}(x_t, B_-^t, M_-)$
  $\qquad [\mu_{t+1}, \Sigma_{t+1}] = \text{UpdateRanker}(\mu_t, \Sigma_t, x_t, y_t, C_t, B_+^{t+1}, \eta)$ or
  $\qquad [\mu_{t+1}, G_{t+1}] = \text{UpdateRanker}(\mu_t, G_t, x_t, y_t, C_t, B_+^{t+1}, \eta)$ (CBR-diag)
  $\quad$ **end if**
  end for

3.3 Update Ranker

Inspired by the robust performance of second-order learning algorithms, we apply the soft confidence-weighted learning approach [35] to updated the bipartite ranking function. Therefore, our confidence-weighted bipartite ranking model (CBR) is formulated as a ranker with a Gaussian distribution parameterized by mean vector $\mu \in \mathcal{R}^d$ and covariance matrix $\Sigma \in \mathcal{R}^{d \times d}$. The mean vector μ represents the model of the bipartite ranking function, while the covariance matrix captures the confidence in the model. The ranker is more confident about the model value μ_p as its diagonal value $\Sigma_{p,p}$ is smaller. The model distribution is updated once the new instance is received while being close to the old model distribution. This optimization problem is performed by minimizing the Kullback-Leibler divergence between the new and the old distributions of the model. The online confidence-weighted bipartite ranking (CBR) is formulated as follows:

$$(\mu_{t+1}, \Sigma_{t+1}) = \underset{\mu, \Sigma}{\operatorname{argmin}} D_{KL}(\mathcal{N}(\mu, \Sigma) || \mathcal{N}(\mu_t, \Sigma_t)) \tag{5}$$

$$+ C\ell^{\phi}(\mathcal{N}(\mu, \Sigma); (z, y_t)),$$

where $z = (x_t - x)$, C is the the penalty hyperparamter, $\phi = \Phi^{-1}(\eta)$, and Φ is the normal cumulative distribution function. The loss function $\ell^{\phi}(\cdot)$ is defined as:

$$\ell^{\phi}(\mathcal{N}(\mu, \Sigma); (z, y_t)) = max(0, \phi\sqrt{z^T \Sigma z} - y_t \mu \cdot z).$$

The solution of (5) is given by the following proposition.

Proposition 1. *The optimization problem (5) has a closed-form solution as follows:*

$$\mu_{t+1} = \mu_t + \alpha_t y_t \Sigma_t z,$$

$$\Sigma_{t+1} = \Sigma_t - \beta_t \Sigma_t z^T z \Sigma_t.$$

The coefficients α and β are defined as follows:

$$\alpha_t = min\{C, max\{0, \frac{1}{v_t \zeta}(-m_t \psi + \sqrt{m_t^2 \frac{\phi^4}{4} + v_t \phi^2 \zeta})\}\},$$

$$\beta_t = \frac{\alpha_t \phi}{\sqrt{u_t} + v_t \alpha_t \phi}. \text{ where } u_t = \frac{1}{4}(-\alpha_t v_t \phi + \sqrt{\alpha_t^2 v_t^2 \phi^2 + 4v_t})^2,$$

$$v_t = z^T \Sigma_t z, \ m_t = y_t(\mu_t \cdot z), \ \phi = \Phi^{-1}(\eta), \ \psi = 1 + \frac{\phi^2}{2}, \ \zeta = 1 + \phi^2, \text{ and}$$
$$z = x_t - x.$$

The proposition 1 is analogous to the one derived in [35].

Though modeling the full covariance matrix lends the CW algorithms a powerful capability in learning [9,26,35], it raises potential concerns with high-dimensional data. The covariance matrix grows quadratically with respect to the data dimension. This makes the CBR algorithm impractical with high-dimensional data due to high computational and memory requirements.

We remedy this deficiency by a diagonalization technique [9,14]. Therefore, we present a diagonal confidence-weighted bipartite ranking (CBR-diag) that models the ranker as a mean vector $\mu \in \mathcal{R}^d$ and diagonal matrix $\hat{\Sigma} \in \mathcal{R}^{d \times d}$. Let G denotes $diag(\hat{\Sigma})$, and the optimization problem of CBR-diag is formulated as follows:

$$(\mu_{t+1}, G_{t+1}) = \underset{\mu, G}{\operatorname{argmin}} D_{KL}(\mathcal{N}(\mu, G) || \mathcal{N}(\mu_t, G_t)) \tag{6}$$

$$+ C\ell^{\phi}(\mathcal{N}(\mu, G); (z, y_t)).$$

Proposition 2. *The optimization problem (6) has a closed-form solution as follows:*

$$\mu_{t+1} = \mu_t + \frac{\alpha_t y_t z}{G_t},$$

$$G_{t+1} = G_t + \beta_t z^2.$$

The coefficients α and β are defined as follows

$$\alpha_t = min\{C, max\{0, \frac{1}{v_t \zeta}(-m_t \psi + \sqrt{m_t^2 \frac{\phi^4}{4} + v_t \phi^2 \zeta})\}\},$$

$$\beta_t = \frac{\alpha_t \phi}{\sqrt{u_t} + v_t \alpha_t \phi},$$

where $u_t = \frac{1}{4}(-\alpha_t v_t \phi + \sqrt{\alpha_t^2 v_t^2 \phi^2 + 4v_t})^2$, $v_t = \sum_{i=1}^d \frac{z_i^2}{G_i + C}$, $m_t = y_t(\mu_t \cdot z)$, $\phi = \Phi^{-1}(\eta)$, $\psi = 1 + \frac{\phi^2}{2}$, $\zeta = 1 + \phi^2$, and $z = x_t - x$.

The Propositions 1 and 2 can be proved similarly to the proof in [35]. The steps of updating the online confidence-weighted bipartite ranking with full covariance matrix or with the diagonal elements are summarized in Algorithm 2.

Algorithm 2. Update Ranker
___

Input:

- μ_t : current mean vector
- Σ_t or G_t : current covariance matrix or diagonal elements
- (x_t, y_t) : a training instance
- B : the buffer storing instances from the opposite class label
- C_t : class-specific weighting parameter
- η : the predefined probability

Output: updated ranker:

- μ_{t+1}
- Σ_{t+1} or G_{t+1}

Initialize: $\mu^1 = \mu_t$, $(\Sigma^1 = \Sigma_t$ or $G^1 = G_t)$, $i = 1$
for $x \in B$ **do**
    Update the ranker (μ^i, Σ^i) with $z = x_t - x$ and y_t by
    $(\mu^{i+1}, \Sigma^{i+1}) = \operatorname*{argmin}_{\mu, \Sigma} D_{KL}(\mathcal{N}(\mu, \Sigma) || \mathcal{N}(\mu^i, \Sigma^i)) + C\ell^\phi(\mathcal{N}(\mu, \Sigma); (z, y_t))$
    or
    Update the ranker (μ^i, G^i) with $z = x_t - x$ and y_t by
    $(\mu^{i+1}, G^{i+1}) = \operatorname*{argmin}_{\mu, G} D_{KL}(\mathcal{N}(\mu, G) || \mathcal{N}(\mu^i, G^i)) + C\ell^\phi(\mathcal{N}(\mu, G); (z, y_t))$
    $i = i + 1$
end for
Return $\mu_{t+1} = \mu^{|B|+1}$
        $\Sigma_{t+1} = \Sigma^{|B|+1}$ or $G_{t+1} = G^{|B|+1}$

---

4   Experimental Results

In this section, we conduct extensive experiments on several real world datasets in order to demonstrate the effectiveness of the proposed algorithms. We also compare the performance of our methods with existing online learning algorithms in terms of AUC and classification accuracy at the optimal operating point of the ROC curve (OPTROC). The running time comparison is also presented.

4.1   Real World Datasets

We conduct extensive experiments on various benchmark and high-dimensional datasets. All datasets can be downloaded from LibSVM[1] and the machine learning repository UCI[2] except the Reuters[3] dataset that is used in [2]. If the data are provided as training and test sets, we combine them together in one set. For cod-rna data, only the training and validation sets are grouped together. For rcv1 and news20, we only use their training sets in our experiments. The multi-class datasets are transformed randomly into class-imbalanced binary datasets. Tables 1 and 2 show the characteristics of the benchmark and the high-dimensional datasets, respectively.

Table 1. Benchmark datasets

Data	#inst	#feat	Data	#inst	#feat	Data	#inst	#feat
glass	214	10	cod-rna	331,152	8	australian	690	14
ionosphere	351	34	spambase	4,601	57	diabetes	768	8
german	1,000	24	covtype	581,012	54	acoustic	78,823	50
svmguide4	612	10	magic04	19,020	11	vehicle	846	18
svmguide3	1284	21	heart	270	13	segment	2,310	19

Table 2. High-dimensional datasets

Data	#inst	#feat
farm-ads	4,143	54,877
rcv1	15,564	47,236
sector	9,619	55,197
real-sim	72,309	20,958
news20	15,937	62,061
Reuters	8,293	18,933

---

[1] https://www.csie.ntu.edu.tw/~cjlin/libsvmtools/.
[2] https://archive.ics.uci.edu/ml/.
[3] http://www.cad.zju.edu.cn/home/dengcai/Data/TextData.html.

4.2 Compared Methods and Model Selection

Online Uni-Exp [21]: An online pointwise ranking algorithm that optimizes the weighted univariate exponential loss. The learning rate is tuned by 2-fold cross validation on the training set by searching in $2^{[-10:10]}$.

OPAUC [15]: An online learning algorithm that optimizes the AUC in one-pass through square loss function. The learning rate is tuned by 2-fold cross validation by searching in $2^{[-10:10]}$, and the regularization hyperparameter is set to a small value 0.0001.

OPAUCr [15]: A variation of OPAUC that approximates the covariance matrices using low-rank matrices. The model selection step is carried out similarly to OPAUC, while the value of rank τ is set to 50 as suggested in [15].

OAM$_{seq}$ [37]: The online AUC maximization (OAM) is the state-of-the-art first-order learning method. We implement the algorithm with the Reservoir Sampling as a buffer updating scheme. The size of the positive and negative buffers is fixed at 50. The penalty hyperparameter C is tuned by 2-fold cross validation on the training set by searching in $2^{[-10:10]}$.

AdaOAM [12]: This is a second-order AUC maximization method that adapts the classifier to the importance of features. The smooth hyperparameter δ is set to 0.5, and the regularization hyperparameter is set to 0.0001. The learning rate is tuned by 2-fold cross validation on the training set by searching in $2^{[-10:10]}$.

CBR$_{RS}$ and **CBR$_{FIFO}$**: The proposed confidence-weighted bipartite ranking algorithms with the Reservoir Sampling and First-In-First-Out buffer updating policies, respectively. The size of the positive and negative buffers is fixed at 50. The hyperparameter η is set to 0.7, and the penalty hyperparameter C is tuned by 2-fold cross validation by searching in $2^{[-10:10]}$.

CBR-diag$_{FIFO}$: The proposed diagonal variation of confidence-weighted bipartite ranking that uses the First-In-First-Out policy to update the buffer. The buffers are set to 50, and the hyperparameters are tuned similarly to CBR$_{FIFO}$.

For a fair comparison, the datasets are scaled similarly in all experiments. We randomly divide each dataset into 5 folds, where 4 folds are used for training and one fold is used as a test set. For benchmark datasets, we randomly choose 8000 instances if the data exceeds this size. For high-dimensional datasets, we limit the sample size of the data to 2000 due to the high dimensionality of the data. The results on the benchmark and the high-dimensional datasets are averaged over 10 and 5 runs, respectively. A random permutation is performed on the datasets with each run. All experiments are conducted with Matlab 15 on a workstation computer with $8 \times 2.6\,G$ CPU and 32 GB memory.

4.3   Results on Benchmark Datasets

The comparison in terms of AUC is shown in Table 3, while the comparison in terms of classification accuracy at OPTROC is shown in Table 4. The running time (in milliseconds) comparison is illustrated in Fig. 1.

The results show the robust performance of the proposed methods CBR_{RS} and CBR_{FIFO} in terms of AUC and classification accuracy compared to other first and second-order online learning algorithms. We can observe that the improvement of the second-order methods such as OPAUC and AdaOAM over first-order method OAM_{seq} is not reliable, while our CBR algorithms often outperform the OAM_{seq}. Also, the proposed methods are faster than OAM_{seq}, while they incur more running time compared to AdaOAM except with spambase, covtype, and acoustic datasets. The pointwise method online Uni-Exp maintains fastest running time, but at the expense of the AUC and classification accuracy. We also notice that the performance of CBR_{FIFO} is often slightly better than CBR_{RS} in terms of AUC, classification accuracy, and running time.

Table 3. Comparison of AUC performance on benchmark datasets

Data	CBR_{RS}	CBR_{FIFO}	Online Uni-Exp	OPAUC	OAM_{seq}	AdaOA
glass	**0.825** ± 0.043	0.823 ± 0.046	0.714 ± 0.075	0.798 ± 0.061	0.805 ± 0.047	0.794 ± 0.061
ionosphere	0.950 ± 0.027	**0.951** ± 0.028	0.913 ± 0.018	0.943 ± 0.026	0.946 ± 0.025	0.943 ± 0.029
german	**0.782** ± 0.024	0.780 ± 0.019	0.702 ± 0.032	0.736 ± 0.034	0.731 ± 0.028	0.770 ± 0.024
svmguide4	0.969 ± 0.013	**0.974** ± 0.013	0.609 ± 0.096	0.733 ± 0.056	0.771 ± 0.063	0.761 ± 0.053
svmguide3	0.755 ± 0.022	**0.764** ± 0.036	0.701 ± 0.025	0.737 ± 0.029	0.705 ± 0.033	0.738 ± 0.033
cod-rna	0.983 ± 0.000	**0.984** ± 0.000	0.928 ± 0.000	0.927 ± 0.001	0.951 ± 0.025	0.927 ± 0.000
spambase	0.941 ± 0.006	**0.942** ± 0.006	0.866 ± 0.016	0.849 ± 0.020	0.897 ± 0.043	0.862 ± 0.011
covtype	0.816 ± 0.003	**0.835** ± 0.001	0.705 ± 0.033	0.711 ± 0.041	0.737 ± 0.023	0.770 ± 0.010
magic04	0.799 ± 0.006	**0.801** ± 0.006	0.759 ± 0.006	0.748 ± 0.033	0.757 ± 0.015	0.773 ± 0.006
heart	0.908 ± 0.019	**0.909** ± 0.021	0.733 ± 0.039	0.788 ± 0.054	0.806 ± 0.059	0.799 ± 0.079
australian	0.883 ± 0.028	**0.889** ± 0.019	0.710 ± 0.130	0.735 ± 0.138	0.765 ± 0.107	0.801 ± 0.037
diabetes	0.700 ± 0.021	**0.707** ± 0.033	0.633 ± 0.036	0.667 ± 0.041	0.648 ± 0.040	0.675 ± 0.034
acoustic	0. 879 ± 0.006	**0.892** ± 0.003	0.876 ± 0.003	0.878 ± 0.003	0.863 ± 0.011	0.882 ± 0.003
vehicle	**0.846** ± 0.031	**0.846** ± 0.034	0.711 ± 0.053	0.764 ± 0.073	0.761 ± 0.078	0.792 ± 0.049
segment	0.900 ± 0.013	**0.903** ± 0.008	0.689 ± 0.061	0.828 ± 0.024	0.812 ± 0.035	0.855 ± 0.008

4.4   Results on High-Dimensional Datasets

We study the performance of the proposed CBR-diag$_{FIFO}$ and compare it with online Uni-Exp, OPAUCr, and OAM_{seq} that avoid constructing the full covariance matrix. Table 5 compares our method and the other online algorithms in terms of AUC, while Table 6 shows the classification accuracy at OPTROC. Figure 2 displays the running time (in milliseconds) comparison.

The results show that the proposed method CBR-diag$_{FIFO}$ yields a better performance on both measures. We observe that the CBR-diag$_{FIFO}$ presents a competitive running time compared to its counterpart OAM_{seq} as shown in Fig. 2. We can also see that the CBR-diag$_{FIFO}$ takes more running time compared to the OPAUCr. However, the CBR-diag$_{FIFO}$ achieves better AUC and

Table 4. Comparison of classification accuracy at OPTROC on benchmark datasets

Data	CBR$_{RS}$	CBR$_{FIFO}$	Online Uni-Exp	OPAUC	OAM$_{seq}$	AdaOAM
glass	**0.813 ± 0.044**	0.811 ± 0.049	0.732 ± 0.060	0.795 ± 0.046	0.788 ± 0.040	0.783 ± 0.047
ionosphere	**0.946 ± 0.028**	**0.946 ± 0.022**	0.902 ± 0.028	0.936 ± 0.018	0.943 ± 0.017	0.938 ± 0.018
german	0.780 ± 0.022	**0.787 ± 0.019**	0.741 ± 0.027	0.754 ± 0.022	0.751 ± 0.028	0.770 ± 0.030
svmguide4	0.951 ± 0.014	**0.956 ± 0.012**	0.829 ± 0.021	0.843 ± 0.024	0.839 ± 0.022	0.848 ± 0.020
svmguide3	0.784 ± 0.015	**0.793 ± 0.016**	0.784 ± 0.019	0.777 ± 0.024	0.780 ± 0.020	0.777 ± 0.024
cod-rna	0.948 ± 0.002	**0.949 ± 0.000**	0.887 ± 0.001	0.887 ± 0.001	0.910 ± 0.019	0.887 ± 0.001
spambase	**0.899 ± 0.009**	0.898 ± 0.009	0.818 ± 0.019	0.795 ± 0.022	0.849 ± 0.053	0.809 ± 0.014
covtype	0.746 ± 0.005	**0.766 ± 0.003**	0.672 ± 0.018	0.674 ± 0.021	0.685 ± 0.016	0.709 ± 0.008
magic04	0.769 ± 0.011	**0.771 ± 0.006**	0.734 ± 0.007	0.731 ± 0.015	0.736 ± 0.013	0.752 ± 0.008
heart	**0.883 ± 0.032**	0.875 ± 0.026	0.716 ± 0.021	0.753 ± 0.038	0.777 ± 0.043	0.772 ± 0.053
australian	0.841 ± 0.023	**0.842 ± 0.022**	0.711 ± 0.056	0.725 ± 0.070	0.742 ± 0.064	0.768 ± 0.036
diabetes	**0.714 ± 0.029**	0.705 ± 0.032	0.683 ± 0.037	0.692 ± 0.040	0.694 ± 0.044	0.689 ± 0.040
acoustic	0. 844 ± 0.005	**0.850 ± 0.003**	0.840 ± 0.005	0.839 ± 0.002	0.832 ± 0.005	0.841 ± 0.003
vehicle	**0.816 ± 0.018**	0.814 ± 0.018	0.764 ± 0.027	0.797 ± 0.014	0.790 ± 0.029	0.805 ± 0.021
segment	**0.838 ± 0.015**	0.836 ± 0.008	0.691 ± 0.031	0.768 ± 0.027	0.755 ± 0.024	0.796 ± 0.014

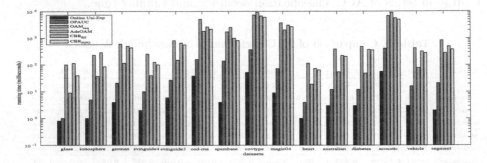

Fig. 1. Running time (in milliseconds) of CBR and the other online learning algorithms on the benchmark datasets. The y-axis is displayed in log- scale.

classification accuracy compared to the OPAUCr. The online Uni-Exp algorithm requires the least running time, but it presents lower AUC and classification accuracy compared to our method.

Table 5. Comparison of AUC on high-dimensional datasets

Data	CBR-diag$_{FIFO}$	Online Uni-Exp	OPAUCr	OAM$_{seq}$
farm-ads	**0.961 ± 0.004**	0.942 ± 0.006	0.951 ± 0.004	0.952 ± 0.005
rcv1	**0.950 ± 0.007**	0.927 ± 0.015	0.914 ± 0.016	0.945 ± 0.008
sector	**0.927 ± 0.009**	0.846 ± 0.019	0.908 ± 0.013	0.857 ± 0.008
real-sim	**0.982 ± 0.001**	0.969 ± 0.003	0.975 ± 0.002	0.977 ± 0.001
news20	**0.956 ± 0.003**	0.939 ± 0.005	0.942 ± 0.006	0.944 ± 0.005
Reuters	**0.993 ± 0.001**	0.985 ± 0.003	0.988 ± 0.002	0.989 ± 0.003

Table 6. Comparison of classification accuracy at OPTROC on high-dimensional datasets

Data	CBR-diag$_{FIFO}$	Online Uni-Exp	OPAUCr	OAM$_{seq}$
farm-ads	**0.897** ± 0.007	0.872 ± 0.012	0.885 ± 0.008	0.882 ± 0.007
rcv1	**0.971** ± 0.001	0.967 ± 0.002	0.966 ± 0.003	0.970 ± 0.001
sector	**0.850** ± 0.012	0.772 ± 0.011	0.831 ± 0.015	0.776 ± 0.008
real-sim	**0.939** ± 0.003	0.913 ± 0.005	0.926 ± 0.002	0.929 ± 0.001
news20	**0.918** ± 0.005	0.895 ± 0.005	0.902 ± 0.009	0.907 ± 0.006
Reuters	**0.971** ± 0.004	0.953 ± 0.006	0.961 ± 0.006	0.961 ± 0.006

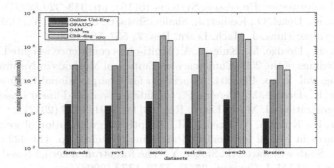

Fig. 2. Running time (in milliseconds) of CBR-diag$_{FIFO}$ algorithm and the other online learning algorithms on the high-dimensional datasets. The y-axis is dis- played in log-scale.

5 Conclusions and Future Work

In this paper, we proposed a linear online soft confidence-weighted bipartite ranking algorithm that maximizes the AUC metric via optimizing a pairwise loss function. The complexity of the pairwise loss function is mitigated in our algorithm by employing a finite buffer that is updated using one of the stream oblivious policies. We also develop a diagonal variation of the proposed confidence-weighted bipartite ranking algorithm to deal with high-dimensional data by maintaining only the diagonal elements of the covariance matrix instead of the full covariance matrix. The experimental results on several benchmark and high-dimensional datasets show that our algorithms yield a robust performance. The results also show that the proposed algorithms outperform the first and second-order AUC maximization methods on most of the datasets. As future work, we plan to conduct a theoretical analysis of the proposed method. We also aim to investigate the use of online feature selection [36] within our proposed framework to effectively handle high-dimensional data.

References

1. Agarwal, S.: A study of the bipartite ranking problem in machine learning (2005)
2. Cai, D., He, X., Han, J.: Locally consistent concept factorization for document clustering. IEEE Trans. Knowl. Data Eng. **23**(6), 902–913 (2011)
3. Cavallanti, G., Cesa-Bianchi, N., Gentile, C.: Tracking the best hyperplane with a simple budget perceptron. Mach. Learn. **69**(2–3), 143–167 (2007)
4. Cesa-Bianchi, N., Conconi, A., Gentile, C.: A second-order perceptron algorithm. SIAM J. Comput. **34**(3), 640–668 (2005)
5. Chapelle, O., Keerthi, S.S.: Efficient algorithms for ranking with svms. Inf. Retrieval **13**(3), 201–215 (2010)
6. Cortes, C., Mohri, M.: Auc optimization vs. error rate minimization. In: Advances in Neural Information Processing Systems 16(16), pp. 313–320 (2004)
7. Crammer, K., Dekel, O., Keshet, J., Shalev-Shwartz, S., Singer, Y.: Online passive-aggressive algorithms. J. Mach. Learn. Res. **7**, 551–585 (2006)
8. Crammer, K., Dredze, M., Kulesza, A.: Multi-class confidence weighted algorithms. In: Proceedings of the 2009 Conference on Empirical Methods in Natural Language Processing, vol. 2, pp. 496–504. Association for Computational Linguistics (2009)
9. Crammer, K., Dredze, M., Pereira, F.: Confidence-weighted linear classification for text categorization. J. Mach. Learn. Res. **13**(1), 1891–1926 (2012)
10. Crammer, K., Kulesza, A., Dredze, M.: Adaptive regularization of weight vectors. In: Advances in Neural Information Processing Systems, pp. 414–422 (2009)
11. Dekel, O., Shalev-Shwartz, S., Singer, Y.: The forgetron: a kernel-based perceptron on a budget. SIAM J. Comput. **37**(5), 1342–1372 (2008)
12. Ding, Y., Zhao, P., Hoi, S.C., Ong, Y.S.: An adaptive gradient method for online auc maximization. In: AAAI, pp. 2568–2574 (2015)
13. Dredze, M., Crammer, K., Pereira, F.: Confidence-weighted linear classification. In: Proceedings of the 25th International Conference on Machine Learning, pp. 264–271. ACM (2008)
14. Duchi, J., Hazan, E., Singer, Y.: Adaptive subgradient methods for online learning and stochastic optimization. J. Mach. Learn. Res. **12**, 2121–2159 (2011)
15. Gao, W., Jin, R., Zhu, S., Zhou, Z.H.: One-pass auc optimization. In: ICML (3), pp. 906–914 (2013)
16. Hanley, J.A., McNeil, B.J.: The meaning and use of the area under a receiver operating characteristic (roc) curve. Radiology **143**(1), 29–36 (1982)
17. Hu, J., Yang, H., King, I., Lyu, M.R., So, A.M.C.: Kernelized online imbalanced learning with fixed budgets. In: AAAI, pp. 2666–2672 (2015)
18. Joachims, T.: A support vector method for multivariate performance measures. In: Proceedings of the 22nd International Conference on Machine Learning, pp. 377–384. ACM (2005)
19. Joachims, T.: Training linear svms in linear time. In: Proceedings of the 12th ACM SIGKDD International Conference on Knowledge Discovery and Data Mining, pp. 217–226. ACM (2006)
20. Kar, P., Sriperumbudur, B.K., Jain, P., Karnick, H.: On the generalization ability of online learning algorithms for pairwise loss functions. In: ICML (3), pp. 441–449 (2013)
21. Kotlowski, W., Dembczynski, K.J., Huellermeier, E.: Bipartite ranking through minimization of univariate loss. In: Proceedings of the 28th International Conference on Machine Learning (ICML 2011), pp. 1113–1120 (2011)

22. Kuo, T.M., Lee, C.P., Lin, C.J.: Large-scale kernel ranksvm. In: SDM, pp. 812–820. SIAM (2014)
23. Lee, C.P., Lin, C.B.: Large-scale linear ranksvm. Neural Comput. **26**(4), 781–817 (2014)
24. Li, N., Jin, R., Zhou, Z.H.: Top rank optimization in linear time. In: Advances in Neural Information Processing Systems, pp. 1502–1510 (2014)
25. Liu, T.Y.: Learning to rank for information retrieval. Found. Trends Inf. Retrieval **3**(3), 225–331 (2009)
26. Ma, J., Kulesza, A., Dredze, M., Crammer, K., Saul, L.K., Pereira, F.: Exploiting feature covariance in high-dimensional online learning. In: International Conference on Artificial Intelligence and Statistics, pp. 493–500 (2010)
27. Orabona, F., Crammer, K.: New adaptive algorithms for online classification. In: Advances in Neural Information Processing Systems, pp. 1840–1848 (2010)
28. Orabona, F., Keshet, J., Caputo, B.: Bounded kernel-based online learning. J. Mach. Learn. Res. **10**, 2643–2666 (2009)
29. Rendle, S., Balby Marinho, L., Nanopoulos, A., Schmidt-Thieme, L.: Learning optimal ranking with tensor factorization for tag recommendation. In: Proceedings of the 15th ACM SIGKDD International Conference on Knowledge Discovery and Data Mining, pp. 727–736. ACM (2009)
30. Rosenblatt, F.: The perceptron: a probabilistic model for information storage and organization in the brain. Psychol. Rev. **65**(6), 386 (1958)
31. Sculley, D.: Large scale learning to rank. In: NIPS Workshop on Advances in Ranking, pp. 1–6 (2009)
32. Vitter, J.S.: Random sampling with a reservoir. ACM Trans. Math. Softw. (TOMS) **11**(1), 37–57 (1985)
33. Wan, J., Wu, P., Hoi, S.C., Zhao, P., Gao, X., Wang, D., Zhang, Y., Li, J.: Online learning to rank for content-based image retrieval. In: Proceedings of the Twenty-Fourth International Joint Conference on Artificial Intelligence, IJCAI, pp. 2284–2290 (2015)
34. Wang, J., Wan, J., Zhang, Y., Hoi, S.C.: Solar: Scalable online learning algorithms for ranking
35. Wang, J., Zhao, P., Hoi, S.C.: Exact soft confidence-weighted learning. In: Proceedings of the 29th International Conference on Machine Learning (ICML-12), pp. 121–128 (2012)
36. Wang, J., Zhao, P., Hoi, S.C., Jin, R.: Online feature selection and its applications. IEEE Trans. Knowl. Data Eng. **26**(3), 698–710 (2014)
37. Zhao, P., Jin, R., Yang, T., Hoi, S.C.: Online auc maximization. In: Proceedings of the 28th International Conference on Machine Learning (ICML 2011), pp. 233–240 (2011)
38. Zinkevich, M.: Online convex programming and generalized infinitesimal gradient ascent (2003)

Mining Distinguishing Customer Focus Sets for Online Shopping Decision Support

Lu Liu[1], Lei Duan[1,2(✉)], Hao Yang[1], Jyrki Nummenmaa[3,4],
Guozhu Dong[5], and Pan Qin[1]

[1] School of Computer Science, Sichuan University, Chengdu, China
liulu1221@stu.scu.edu.cn, leiduan@scu.edu.cn,
{hyang.cn,panqin.cn}@outlook.com
[2] West China School of Public Health, Sichuan University, Chengdu, China
[3] School of Information Sciences, University of Tampere, Tampere, Finland
jyrki.nummenmaa@uta.fi
[4] Sino-Finnish Centre, Tongji University, Shanghai, China
[5] Department of Computer Science and Engineering,
Wright State University, Dayton, USA
guozhu.dong@wright.edu

Abstract. With the development of e-commerce, online shopping becomes increasingly popular. Very often, online shopping customers read reviews written by other customers to compare similar items. However, the number of customer reviews is typically too large to look through in a reasonable amount of time. To extract information that can be used for online shopping decision support, this paper investigates a novel data mining problem of mining distinguishing customer focus sets from customer reviews. We demonstrate that this problem has many applications, and at the same time, is challenging. We present dFocus-Miner, a mining method with various techniques that makes the mined results interpretable and user-friendly. Our experimental results on real world data sets verify the effectiveness and efficiency of our method.

Keywords: Distinguishing customer focus · Shopping decision support · Data mining

1 Introduction

As e-commerce provides a convenient and flexible way to purchase consumer products at prices that are often cheaper than purchases from traditional stores, more and more people now prefer online shopping to in-store shopping. Different from traditional in-store shopping where customers seldom exchange their opinions about the items they bought, most e-commerce web sites provide a platform

This work was supported in part by NSFC 61572332, the Fundamental Research Funds for the Central Universities 2016SCU04A22, and the China Postdoctoral Science Foundation 2014M552371, 2016T90850.

© Springer International Publishing AG 2016
J. Li et al. (Eds.): ADMA 2016, LNAI 10086, pp. 50–64, 2016.
DOI: 10.1007/978-3-319-49586-6_4

for customers to post their reviews on those items. As Ghose and Ipeirotis [1] pointed out, the opinions expressed in online reviews are important factors that impact the sale of the items being reviewed. Typically, customer reviews can

- provide more user-oriented descriptions of various items. For example, instead of presenting the exact weight of a laptop in kilograms, a customer may state that "*it is convenient to travel with it*".
- affect future customers' shopping decisions. For example, products with good word-of-mouth tend to have bigger sales.

As a result, most online shopping websites encourage customers to share their assessments on the shopping items they bought. Table 1 lists the number of customer reviews on *Apple iPhone6* and *Samsung Note4*, the two most popular smart phones, at Amazon and Best Buy (two famous online shopping websites in North America) as well as at JD, Tmall and Suning (three famous online shopping websites in China).

Table 1. Number of customer reviews on Apple iPhone6 and Samsung Note4

Product	Number of customer reviews				
	Amazon	Best Buy	JD	Tmall	Suning
iPhone6	1620	1466	370050	32551	51837
Note4	1118	992	48903	1150	3653

From Table 1, we can see that (1) customer reviews are available at many popular online shopping websites, and (2) the number of customer reviews for a hot product is huge. Clearly, it is time consuming for customers to look through hundreds or thousands of reviews. In addition, for two similar shopping items, such as two smart phones or two hotels at the same scenic spot, the customer reviews can be quite similar. For example, the most frequent customer reviews on both *Apple iPhone6* and *Samsung Note4* talk about "*ringtones large*", "*nice look*", "*quick response*", etc.

Intuitively, customers are more interested in the reviews describing distinguishing aspects, rather than the common aspects, on the items under comparison. For example, "*big screen*" is a comment that frequently occurs in the reviews of *Note4* but infrequently occurs in those of *iPhone6*. This comment can be influential in helping customers who like big screen to select *Note4*.

Based on the above, online shopping websites should provide more distinguishing information when comparing reviews on different items to help customers make shopping decisions. We note that existing recommendation methods are lacking on addressing the above important needs.

This leads us to a novel data mining problem. We say that a customer description of a certain aspect of a given shopping item is a *customer focus* on the item; examples include "*big screen*", "*nice look*". Given two shopping items, a *distinguishing customer focus set* is a set of customer focuses that frequently occur in the reviews of one of the shopping item but infrequently occur in the reviews of the other shopping item. For example, *"big screen"*, *"android system"* are customer focuses that frequently occur in the reviews of *Note4* but infrequently occur in the reviews of *iPhone6*; we call such a phrase set a *distinguishing customer focus set*. Mining distinguishing customer focus sets is an interesting problem with many useful applications. On one hand, it can provide more information to support customers' shopping decision making. On the other hand, it is possible for product manufacturers to use the distinguishing strong points and weak points of their products, mined from customers reviews of their products compared against reviews of their rivals' products.

To tackle the problem of mining distinguishing customer focus sets, we need to address several technical challenges. First, we need to have a comprehensive yet complete way to represent customer focuses. Naturally, a customer focus should not only reflect the opinions of customers but should also be easy for customers to understand. In addition, considering a variety of expressions can arise for different products, different time, and different customers, and the total set of customer focuses should not be predefined.

Second, we need to have an effective way to evaluate the similarities among candidate customer focuses, so that redundant customer focuses can be removed, the mining results can be more concise, and the algorithm can be more efficient.

Third, we also need to find effective techniques to efficiently discover top distinguishing customer focus sets.

This paper makes the following main contributions to mining distinguishing customer focus sets for shopping decision support: (1) introducing a novel data mining problem of top-k distinguishing customer focus set mining; (2) designing an effective yet flexible method for distinguishing customer focus set mining; (3) conducting extensive experiments using the online reviews from various types of e-commerce websites, to evaluate our distinguishing customer focus set mining method, and to demonstrate that the proposed method can find interesting customer focus sets and it is effective and efficient.

The rest of the paper is organized as follows. We formulate the problem of distinguishing customer focus set mining in Sect. 2, and review related work in Sect. 3. In Sect. 4, we discuss the critical techniques of our method (called dFocus-Miner). We report a systematic empirical study in Sect. 5, and conclude the paper in Sect. 6.

2   Problem Definition

We start with some preliminaries. Given a shopping item q, let $\mathcal{R}(q)$ be the set of customer reviews of q in a website. A *customer focus* is a description of a certain aspect of a given shopping item from the users' perspective. For example, in the customer review *"At the business center of the hotel we were treated extremely coldly and arrogantly"*, the customer focus is about the "service" of the hotel's business center; we consider *"treated extremely coldly and arrogantly"* as its customer focus.

Given a customer review $r \in \mathcal{R}(q)$ and a customer focus f, if f is "semantically" contained by r, then we say that r *supports* f, denoted by $f \dot{\sqsubseteq} r$. For example, the review "*We enjoyed the nightly turn down service as well as an exceptionally clean room and bathroom*" supports the customer focus "neat room" since "*neat*" is semantically similar with "*clean*".

A *customer focus set* fs is a set of customer focuses. The *support* of fs on a shopping item q, denoted by $Sup(fs, q)$, is defined by Eq. 1.

$$Sup(fs, q) = \frac{|\{r \in \mathcal{R}(q) \mid \forall f \in fs, f \dot{\sqsubseteq} r\}|}{|\mathcal{R}(q)|} \qquad (1)$$

Definition 1. *Given a customer focus set fs and two shopping items q_1 and q_2, the* contrast score *of fs from q_2 to q_1, denoted by $cScore(fs, q_1, q_2)$, is*

$$cScore(fs, q_1, q_2) = Sup(fs, q_1) - Sup(fs, q_2) \qquad (2)$$

To select top-k distinguishing customer focus sets, we now define a total order on all customer focus sets.

Definition 2. *Given two shopping items, q_1 and q_2, and two customer focus sets, fs and fs', we say fs has a higher precedence than fs' (or fs precedes fs') if:*

1. $cScore(fs, q_1, q_2) > cScore(fs', q_1, q_2)$, or
2. $cScore(fs, q_1, q_2) = cScore(fs', q_1, q_2)$, but $Sup(fs, q_1) > Sup(fs', q_1)$,
3. $cScore(fs, q_1, q_2) = cScore(fs', q_1, q_2)$ and $Sup(fs, q_1) = Sup(fs', q_1)$, but fs is lexically smaller than fs'.

Definition 3. *Given an integer k, a query item q_1, and a reference query item q_2, the problem of mining top-k distinguishing customer focus sets is to find the customer focus sets with top-k precedence from $\mathcal{R}(q_1)$ and $\mathcal{R}(q_2)$.*

3 Related Work

Our study is related to previous studies on recommender systems, opinion mining, and contrast mining.

3.1 Recommendation

Most studies on recommendation employ model-based collaborative filtering techniques for ranking [2,3]. Recently, location-based social networks are widely used in POI recommendation [4], which is related to user behavior study [5], event and activity recommendations [6] and online retail store placement [7]. There are several studies exploring customer reviews for prediction [8,9]. Both rating information and user preference can be integrated into review-based recommender systems [8]. Other studies include [10,11].

Different from studies on recommendation, the target of distinguishing customer focus set mining is providing more information for shopping decision support rather than recommending shopping items. Thus, methods for recommendation cannot be applied to our problem directly.

3.2 Opinion Mining

Opinion mining (also called sentiment analysis) tries to identify customers' opinions by applying NLP techniques to extract opinions related noun words and adjectives words. Technically, several approaches, including association-rule based [12] and model based [13], have been proposed to extract opinion words and opinion targets. The mining of user opinions is also useful to improve the performance of recommender systems [9,14].

A customer focus reflects the customer's assessments on certain aspects of a shopping item. Conceptually, the definition of customer focus is similar to that of customer opinion. However, the target of this study is discovering distinguishing customer focus sets that frequently occur in the set of reviews for one item but infrequently occur in the set of reviews of another item. Thus, our problem is significantly different from the problem of opinion mining.

To the best of our knowledge, the work by Wang *et al.* [15] which summarizes product reviews by selecting the most representative review sentences, is the most related previous studies to our paper. Generally, not so good to use [15] as a substantive in English language. But others do so too. There are two essential differences between [15] and this study. First, [15] extracts representative sentences from reviews, while our method extracts customer focuses (quality phrases) from reviews. Second, the target of [15] is to summarize customer reviews, while the target of our method is comparing two sets of customer reviews.

3.3 Contrast Mining

Contrast mining discovers patterns and models that manifest significant differences between data sets. Dong and Bailey [16] presented a comprehensive review of contrast mining, together with a series of real-life applications. Several types of contrast patterns have been proposed, such as emerging pattern [17], contrast set [18], and distinguishing sequential pattern [19].

To the best of knowledge, there is no previous work on mining contrast patterns from customer reviews. It is possible to treat each review as a sequence in which each word is an element, so that distinguishing sequential patterns can be discovered. However, it is hard for people to understand such patterns, since the previous methods [19,20] don't consider the comprehensibility of the discovered patterns. Thus, it is necessary to design a novel algorithm for the problem studied in this paper.

4   Design of dFocus-Miner

In this section, we present our method, dFocus-Miner, for mining top-k distinguishing customer focus sets from $\mathcal{R}(q_1)$ and $\mathcal{R}(q_2)$. The framework of dFocus-Miner includes three main steps: (i) generating candidate customer focuses (Sect. 4.1), (ii) extracting customer focuses from the candidates (Sect. 4.2), and

(iii) evaluating the contrast scores of each customer focus set (Sect. 4.3). Below we discuss the most important techniques used in each step of dFocus-Miner, emphasizing the new ideas that make the results interpretable and user-friendly.

4.1 Candidate Customer Focus Generation

As stated in Sect. 2, in our work a customer focus is a user comment on a certain aspect of an item. Intuitively, a customer focus should be: (i) representative, i.e., reflects the opinions of large number of customers; (ii) non-predefined, i.e., can be dynamically formed using unsupervised learning from customer reviews; (iii) comprehensible, i.e., it is easy for customers to understand.

Observation 1. *Customer reviews on a given shopping item contain phrases that reflect users' concerns with respect to some aspects of the shopping item.*

Example 1. Given a user review *"My wife loves this phone. Very easy to use. She loves the big screen size"*. We can see that the user concerns include: *"easy to use"* and *"big screen size"*.

By Observation 1, we extract "qualified" phrases from the customer reviews as the *candidate customer focuses*. We employ *Quality Phrases Miner*, the algorithm proposed by Liu *et al.* [21], to generate candidate customer focuses from the set of customer reviews. Previous studies, such as [21,22], verified that:

- *Quality Phrase Miner* is *efficient, effective* and *scalable* for mining quality phrases.
- *Quality Phrase Miner* extracts phrases by considering the requirements on *frequency, concordance* and *completeness*.

Clearly, the performance of *Quality Phrase Miner* satisfies our requirements on generating candidate customer focuses. Given two query items q_1 and q_2, the size of $\mathcal{R}(q_1)$ may be unbalanced to that of $\mathcal{R}(q_2)$. Thus, we apply *Quality Phrase Miner* to separately extract quality phrases from $\mathcal{R}(q_1)$ and $\mathcal{R}(q_2)$.

4.2 Customer Focus Selection

Table 2 contains example quality phrases of reviews of a hotel in New York city. Clearly, the semantics of some phrases are similar. For example, *"short walk"*, *"minutes walk"* and *"walking distance"* are related to the hotel location. Such similar semantic phrases may cause a redundancy problem. To solve this problem, we use K-means clustering algorithm to remove the semantically redundant focuses. The similarity measure used in clustering is based on the path similarity of individual words, which is a well-known semantic distance defined for WordNet [23].

Given two words w and w', we denote the path similarity between w and w' by $PathSim(w, w')$. As defined in [23], the range of $PathSim(w, w')$ is $[0.0, 1.0]$. Larger values of $PathSim(w, w')$ indicate higher degree of similarity between w

Table 2. Partial list of quality phrases of two products

Cutomer reviews	Quality phrases
r_1	short walk, comfy bed, studio suite
r_2	minutes walk, fully equipped, staff is nice
r_3	bedroom suite, grand central station
r_4	good location, walking distance

and w'. In practice, the phrases are rather short, and based on our experimental analysis, we propose the following similarity measure for candidate customer focuses (phrases). Given two candidate customer focuses f and f', the similarity between f and f', denoted by $Sim(f, f')$, is defined by Eq. 3.

$$Sim(f, f') = \frac{\sum_{w \in f, w' \in f'} PathSim(w, w')}{|f| \times |f'|} \tag{3}$$

where $w \in f$ means f contains w, and $|f|$ is the number of words in f.

After clustering, we use the centers of clusters as *customer focuses*. For the sake of clarity, given a phrase f, we denote the center of the cluster that f belongs to by $C(f)$.

Once the customer focuses are available, we use them to label the reviews. Specifically, for a given customer review r, we use a set of customer focuses, which are the closest to the review semantically, to represent it. The set of labels of r, denoted by $\mathcal{L}(r)$, is

$$\mathcal{L}(r) = \{C(f) \mid f \text{ is a phrase in } r\}. \tag{4}$$

4.3 Mining Top-k Distinguishing Customer Focus Sets

Given the representative customer focuses of two different products, we want to find k focus sets with biggest $cScore$ values. Intuitively, we can apply an emerging pattern (EP) mining method for this task. Recall that an item-set is an EP if its support in one class is greater than α while the support in the other class is less than β [16]. Both α and β are parameters whose values are set by users.

To the best of our knowledge, DPMiner [24] is the most efficient method to mine emerging patterns. Clearly, it is difficult for a user to set up proper α and β parameters. To solve this problem, we designed a top-k distinguishing frequent pattern mining method based on the above emerging pattern mining method.

dFocus-Miner discovers the focus sets with top-k $cScore$. Specifically, let FS be the set of all candidate focus-sets, $Sup_1(fs)$ is the number of reviews for the query item that contain focus set fs, and $Sup_2(fs)$ the number of reviews for the query reference item that contain focus set fs. Then the mining result of dFocus-Miner is

$$\{fs \in FS \mid fs \text{ with top } k \text{ precedence.}\}$$

Algorithm 1. dFocus-Miner(q_1, q_2, k)

Input: q_1, q_2: two shopping items to be compared; k: an integer
Output: $\mathcal{F}$: the set of top-k distinguishing customer focus sets
 1: $P_1 \leftarrow$ quality phrases in $\mathcal{R}(q_1)$; $P_2 \leftarrow$ quality phrases in $\mathcal{R}(q_2)$;
 2: $\mathcal{P} \leftarrow P_1 \cup P_2$;
 3: $F \leftarrow$ representative customer focuses in $(\mathcal{P})$;
 4: $L_1 \leftarrow \emptyset$; $L_2 \leftarrow \emptyset$;
 5: **for** each review $r \in \mathcal{R}(q_1)$ **do**
 6:     $L_1 \leftarrow L_1 \cup \mathcal{L}(r)$; // $\mathcal{L}(r)$ is the set of customer focuses labels of r (Equation 4)
 7: **end for**
 8: **for** each review $r \in \mathcal{R}(q_2)$ **do**
 9:     $L_2 \leftarrow L_2 \cup \mathcal{L}(r)$;
10: **end for**
11: $\mathcal{F} \leftarrow$ topkEPMiner(L_1, L_2, k);
12: **return** $\mathcal{F}$;

Due to the limited space, we only describe some important techniques of mining top-k distinguishing customer focus sets. Two pruning rules are used to improve the efficiency.

Pruning Rule 1. *For focus f, if there is no review $r \in \mathcal{R}(q_1)$ satisfying $f \sqsubseteq r$, then any focus set fs containing f can be pruned.*

If focus f only occurs in reference query item $\mathcal{R}(q_2)$, then any focus set fs containing f cannot be in top-k distinguishing customer focus sets.

Pruning Rule 2. *Let $cScore_k$ be the k-th largest cScore found. If focus f satisfies $Sup(f, \mathcal{R}(q_1)) < cScore_k$, then any focus set fs containing f can be pruned.*

If the support of a focus set is less than $cScore_k$, none of its super-sets can satisfy the definition of top-k distinguishing customer focus sets.

The whole framework of our Distinguishing Customer Focus Set Miner (dFocus-Miner) is presented in Algorithm 1.

5 Empirical Evaluation

This section presents the results of our experimental studies; we have studied the effectiveness and efficiency of our proposed method (dFocus-Miner). All experiments were conducted on a PC with an Intel Core i7-4770 3.40 GHz CPU, and 16GB main memory. dFocus-Miner was implemented using Java and Python.

5.1 Case Study for Effectiveness Evaluation

We use three real-world data sets crawled from different web sites containing reviews of hotels, movies and products, as summarized in Table 3. In Table 3, the items labeled '+' are query items, while the ones labeled '−' are query

Table 3. Statistics of data sets

Category	Item	#customer reviews
Hotel	Lotte New York Palace (+)	822
	Dumont NYC (−)	359
Movie	Life of Pi (+)	747
	The Revenant (−)	1091
Laptop	Dell Inspiron i7559-763BLK (+)	690
	ASUS K501UX (−)	496

reference items. For every two sets of customer reviews from the same website, we apply dFocus-Miner to the reviews to find top-10 distinguishing customer focus sets (Table 4). In addition, for the sake of comparison, we list top-10 frequent customer focuses of the two query items in the same website, denoted by F_{10}^+ and F_{10}^-, respectively, in Table 5.

Table 4. Top-10 distinguishing customer focus sets

Hotel	Movie	Laptop
blocks from the empire state	**ang lee**	**screen bleed**
barking dog restaurant	human spirit	**coms battery**
chrysler building	human spirit, ang lee	gb ram
studio suite	**richard parker**	single screw
residential area	**suraj sharma**	screen bleed, brand new
murray hill	crouching tiger	stick of ram
minutes walk	suraj sharma, ang lee	sleep mode
east river	suraj sharma, human spirit	sleep mode, ultra settings
queen beds	richard parker, ang lee	ultra settings, brand new
times square	richard parker, human spirit	sata bay, dell website

TripAdvisor. First, we consider mining distinguishing customer focus sets from reviews on hotels provided by *TripAdvisor* (http://www.tripadvisor.com), which provides over 200 million supposedly unbiased traveler reviews of hotels, restaurants and vacation rentals. We compare *Lotte New York Palace Hotel* against *Dumont NYC*, as they were ranked 97 and 98 of 470 hotels, respectively, in New York city. Their recent prices are the same – $449 per night. Both of their average ratings from customers are the same – 4.5 out of 5.

By Table 4, we observe that the interesting focuses mined by our method are meaningful from the user view point. For example, "*studio suite*" and "*queen*

Table 5. Top-10 frequent customer focuses

	Hotel	Movie	Laptop
F_{10}^+	empire state	life of pi	brand new
	new york	ang lee	ultra settings
	barking dog restaurant	crouching tiger	sleep mode
		richard parker	stick of ram
	times square	suraj sharma	battery life
	murray hill	pacific ocean	discrete graphics
	residential area	visual poetry	number pad
	corner suite	irfan khan	gb ram
	chrysler building	yann martel	screen bleed
	better than	named richard parker	display driver
	grand central		
F_{10}^-	new york	leonardo	brand new
	control panel	dicaprio's	light weight
	front desk	robert redford	battery life
	patrick's cathedral	hugh glass	stick of ram
	grand central	bear grylls	sleep mode
	like royalty	g rard depardieu	number pad
	screen tv	long takes	discrete graphics
	times square	using only natural light	k lx
	better than	cinematic experience	drive bay
	even though	alejandro gonz	keyboard backlight
		zoo owner	

beds" are the meaningful distinguishing focuses about *Lotte New York Palace*. From the website, we can find that this hotel provides many types of suites that are popular. Thus, these two customer focuses are interpretable descriptions which are different from hotel introduction. Another discovered focus "*blocks from the empire state*" is also an interpretable quality phrase, since the hotel's location is given by a landmark building instead of a detailed address.

From Table 5, it is interesting to see that the phrases in top-10 distinguishing customer focus sets of *Lotte New York Palace Hotel* against *Dumont NYC* are different from the top-10 frequent customer focuses of *Lotte New York Palace Hotel*. For example, "new york" is frequent in the reviews of both *Lotte New York Palace Hotel* against *Dumont NYC*, thus it is not a result of dFocus-Miner. While we can see that "empire state" is a distinguishing characteristic of *Lotte New York Palace Hotel*. Clearly, dFocus-Miner can provide more information for users.

IMDb. Second, we consider mining distinguishing customer focus sets from movie reviews provided by *IMDb* (http://www.imdb.com), which is the world's most popular and authoritative source for movie information. We compare *Life* *of Pi* against *The Revenant,* both of which are Oscar winning movies, and have similar ratings of 8 and 8.1 in *IMDb.*

From Table 4, which lists the top-10 distinguishing customer focus sets identified by dFocus-Miner, we find that the top-10 distinguishing customer focus sets are mostly about the director (Ang Lee), the leading actor (Suraj Sharma) and the leading role (Richard Parker) of *Life of Pi.* Intuitively, the famous director and actors are main factors to attract people to watch a movie. Our proposed method can find these main factors for customers. Please note that the distinguishing customer focus sets in Table 5 (F_{10}^{+}) are very dissimilar to the distinguishing customer focus sets in Table 4. For example, "life of pi", a top frequent customer focus, is not in a top distinguishing customer focus set. The reason is that dFocus-Miner clusters the candidate customer focuses, and uses the centers as customer focuses to label each review. For the cluster, whose center is "life of pi", there are several candidate customer focuses belonging to the reviews for *The Revenant.* Intuitively, the distinguishing customer focus sets listed in Table 4 offer more information highly related to the query-items.

Amazon. Last, we apply dFocus-Miner to the reviews on electronic products provided by *Amazon* (http://www.amazon.com), which is a famous online shopping website. We compare a *DELL* laptop against to an *ASUS* laptop, since they have similar configurations in CPU, RAM, hard disc, and screen size, and the same price $799.99. The product information for these products can make the choice difficult for buyers who are not professional in computer. dFocus-Miner can help them to find distinguishing comments in comprehensible focuses.

As listed in Table 4, dFocus-Miner finds some interesting distinguishing focus sets for the *DELL* laptop. For example, *"screen bleed"* means that LCD products can get light leaking from the back. By this customer focus, it is better for users to check the screen if they bought a *DELL* laptop. Also many users mentioned *"coms battery"*, which indicates a quality issue. Moreover, from Table 5, we can see that these two laptops share several characteristics from users' views, such as "brand new", "stick of ram", and "battery life". However, dFocus-Miner can find distinguishing users' concerns of the query item.

Effect of k. We investigate the effect of mining results of dFocus-Miner (i.e., distinguishing customer focus sets) when the value of k (the number of top distinguishing customer focus sets) changes from 10 to 50 by step of 10. Figure 1 illustrates the average of *cScore* with respect to k. It is clear that the value of *cScore* decreases with the increase of k. As stated in Sects. 4.1 and 4.2, dFocus-Miner extracts "qualified" phrases from the customer reviews, and clusters them to get customer focuses. Figure 2 illustrates the average length of customer focus, i.e., the average number of words in a customer focus. It is interesting to see that with the increase of k, the average length changes slightly. The reason is that the length of a qualified phrases extracted by *Quality Phrase Miner* is typically between 2 and 3. Please recall that the mining results of dFocus-Miner are top distinguishing customer focus sets. From Fig. 3, we can see that the average size

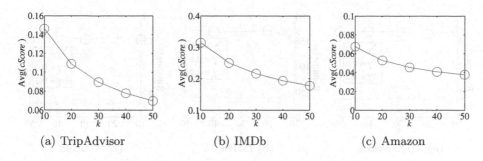

Fig. 1. Average of *cScore* of top-*k* distinguishing customer focus sets

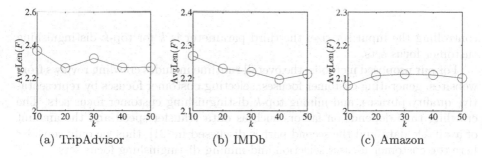

Fig. 2. Average length of customer focuses in top-*k* distinguishing customer focus sets

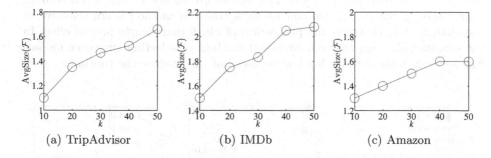

Fig. 3. Average size of of top-*k* distinguishing customer focus sets

of customer focus sets increase when the k changes from 10 to 50, because when the value of k increases, more customer focus sets can be found by dFocus-Miner. This is consistent with the intuition.

5.2   Efficiency Evaluation

We characterize the dFocus-Miner executions with three parameters: γ is the ratio of K-means clustering numbers – if we have m customer focuses, then $K = \gamma \cdot m$; μ describes the proportion of the review data set being used thus

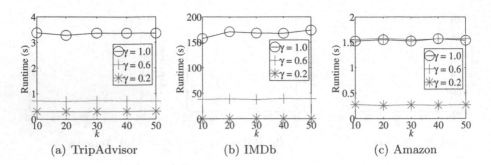

Fig. 4. Runtime of dFocus-Miner w.r.t. the ratio of K-means clustering numbers

controlling the input set size; the third parameter is k for top-k distinguishing customer focus sets.

For our proposed methods, the overall runtime includes crawling reviews from websites, generating customer focuses, selecting customer focuses by representative quality phrases, and mining top-k distinguishing customer focus sets. The crawling time depends on factors such as data transfer speed and the amount of available data and the second part is discussed in [21], thus we only consider here the customer focuses selection and mining distinguishing focus sets.

We first use the three data sets described in Table 3 to evaluate how the runtime changes when k changes from 10 to 50 by step of 10. Here, we use the full review sets, i.e., $\mu = 1.0$. The results are shown in Fig. 4. Three lines representing μ values 0.2, 0.6, and 1.0 show that our method is not sensitive to the change of k. However, the proportion of clustering clearly has an effect in the runtime, showing that our proposed method of selecting customer focuses can not only remove the redundant focuses but also reduce the runtime.

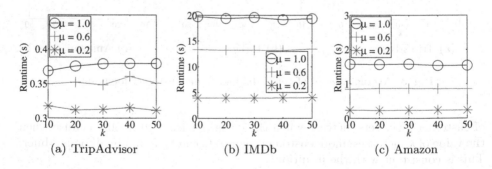

Fig. 5. Runtime of dFocus-Miner w.r.t. the number of reviews

From Fig. 4, we see that the number of clusters will influence the runtime of our proposed method, i.e., K is decided by the number of customer focuses. However the number of customer focuses m is decided by the number of reviews.

We randomly chose 20 %, 60 % and 100 % of the reviews for different test runs. Figure 5 shows the results. We can see that the runtime increases linearly with the input size. Both figures prove that our pruning rules of our novel method in mining top-k distinguishing customer focus sets are efficient.

6   Conclusions

In this paper, we proposed and studied a new problem of mining distinguishing customer focus sets which has wide applications, and which is useful for online shopping decision support. We designed an algorithm called dFocus-Miner to find such customer focuses. Our experiments on real world data sets verify the effectiveness and efficiency of dFocus-Miner.

As future work, we plan to introduce a domain-independent phrase collection method to reduce the number of candidate customer focuses, and try some other methods to generate customer focus besides the clustering method used in this work. It is also interesting to introduce some natural language processing methods to refine the reviews. Moreover, we will explore some distributed computing techniques to improve the performance of dFocus-Miner. In addition, we will explore some domain-driven methods for quality phrase mining to improve the effectiveness of dFocus-Miner in specific domains.

References

1. Ghose, A., Ipeirotis, P.G.: Designing novel review ranking systems: predicting the usefulness and impact of reviews. In: Proceedings of the 9th International Conference on Electronic Commerce: The Wireless World of Electronic Commerce, pp. 303–310 (2007)
2. Hu, Y., Koren, Y., Volinsky, C.: Collaborative filtering for implicit feedback datasets. In: Proceedings of the 8th IEEE International Conference on Data Mining, ICDM, pp. 263–272 (2008)
3. Koren, Y., Bell, R.M., Volinsky, C.: Matrix factorization techniques for recommender systems. IEEE Comput. **42**(8), 30–37 (2009)
4. Li, X., Xu, G., Chen, E., Li, L.: MARS: a multi-aspect recommender system for point-of-interest. In: Proceedings of the 31st IEEE International Conference on Data Engineering, ICDE, pp. 1436–1439 (2015)
5. Zheng, V.W., Cao, B., Zheng, Y., Xie, X., Yang, Q.: Collaborative filtering meets mobile recommendation: a user-centered approach. In: Proceedings of the 24th AAAI Conference on Artificial Intelligence, AAAI (2010)
6. Zhang, W., Wang, J., Feng, W.: Combining latent factor model with location features for event-based group recommendation. In: Proceedings of the 19th ACM International Conference on Knowledge Discovery and Data Mining, KDD, pp. 910–918 (2013)
7. Karamshuk, D., Noulas, A., Scellato, S., Nicosia, V., Mascolo, C.: Geo-spotting: mining online location-based services for optimal retail store placement. In: Proceedings of the 19th ACM International Conference on Knowledge Discovery and Data Mining, KDD, pp. 793–801 (2013)

8. Mukherjee, S., Basu, G., Joshi, S.: Incorporating author preference in sentiment rating prediction of reviews. In: Proceedings of the 22nd International World Wide Web Conference, WWW, pp. 47–48 (2013)
9. Wang, H., Lu, Y., Zhai, C.: Latent aspect rating analysis on review text data: a rating regression approach. In: Proceedings of the 16th ACM International Conference on Knowledge Discovery and Data Mining, KDD, pp. 783–792 (2010)
10. Zhang, F., Zheng, K., Yuan, N.J., Xie, X., Chen, E., Zhou, X.: A novelty-seeking based dining recommender system. In: Proceedings of the 24th International Conference on World Wide Web, WWW, pp. 1362–1372 (2015)
11. Li, X., Xu, G., Chen, E., Li, L.: Learning user preferences across multiple aspects for merchant recommendation. In: Proceedings of the 15th IEEE International Conference on Data Mining, ICDM, pp. 865–870 (2015)
12. Hu, M., Liu, B.: Mining opinion features in customer reviews. In: Proceedings of the 19th AAAI Conference on Artificial Intelligence, 16th Conference on Innovative Applications of Artificial Intelligence, AAAI, pp. 755–760 (2004)
13. Zhao, Q., Wang, H., Lv, P., Zhang, C.: A bootstrapping based refinement framework for mining opinion words and targets. In: Proceedings of the 23rd ACM International Conference on Conference on Information and Knowledge Management, CIKM, pp. 1995–1998 (2014)
14. McAuley, J.J., Leskovec, J.: Hidden factors and hidden topics: understanding rating dimensions with review text. In: Proceedings of the 7th ACM Conference on Recommender Systems, pp. 165–172 (2013)
15. Wang, D., Zhu, S., Li, T.: Sumview: a web-based engine for summarizing product reviews and customer opinions. Expert Syst. Appl. **40**(1), 27–33 (2013)
16. Dong, G., Bailey, J. (eds.): Contrast Data Mining: Concepts, Algorithms, and Applications. CRC Press (2012)
17. Dong, G., Li, J.: Efficient mining of emerging patterns: discovering trends and differences. In: Proceedings of the 5th ACM International Conference on Knowledge Discovery and Data Mining, KDD, pp. 43–52 (1999)
18. Bay, S.D., Pazzani, M.J.: Detecting group differences: mining contrast sets. Data Min. Knowl. Disc. **5**(3), 213–246 (2001)
19. Ji, X., Bailey, J., Dong, G.: Mining minimal distinguishing subsequence patterns with gap constraints. Knowl. Inf. Syst. **11**(3), 259–286 (2007)
20. Yang, H., Duan, L., Dong, G., Nummenmaa, J., Tang, C., Li, X.: Mining itemset-based distinguishing sequential patterns with gap constraint. In: Proceedings of the 20th International Conference on Database Systems for Advanced Applications, DASFAA, pp. 39–54 (2015)
21. Liu, J., Shang, J., Wang, C., Ren, X., Han, J.: Mining quality phrases from massive text corpora. In: Proceedings of the 36th ACM International Conference on Management of Data, SIGMOD, pp. 1729–1744 (2015)
22. El-Kishky, A., Song, Y., Wang, C., Voss, C.R., Han, J.: Scalable topical phrase mining from text corpora. PVLDB **8**(3), 305–316 (2014)
23. Miller, G.A.: Wordnet: a lexical database for english. Commun. ACM **38**(11), 39–41 (1995)
24. Li, J., Liu, G., Wong, L.: Mining statistically important equivalence classes and delta-discriminative emerging patterns. In: Proceedings of the 13th ACM International Conference on Knowledge Discovery and Data Mining, KDD, pp. 430–439 (2007)

Community Detection in Networks with Less Significant Community Structure

Ba-Dung Le$^{(\boxtimes)}$, Hung Nguyen, and Hong Shen

School of Computer Science, The University of Adelaide, Adelaide 5005, Australia
{badung.le,hung.nguyen,hong.shen}@adelaide.edu.au

Abstract. Label propagation is a low complexity approach to community detection in complex networks. Research has extended the basic label propagation algorithm (LPA) in multiple directions including maximizing the modularity, a well-known quality function to evaluate the goodness of a community division, of the detected communities. Current state-of-the-art modularity-specialized label propagation algorithm (LPAm+) maximizes modularity using a two-stage iterative procedure: the first stage is to assign labels to nodes using label propagation, the second stage merges smaller communities to further improve modularity. LPAm+ has been shown able to achieve excellent performance on networks with significant community structure where the network modularity is above a certain threshold. However, we show in this paper that for networks with less significant community structure, LPAm+ tends to get trapped in local optimal solutions that are far from optimal. The main reason comes from the fact that the first stage of LPAm+ often misplaces node labels and severely hinders the merging operation in the second stage. We overcome the drawback of LPAm+ by correcting the node labels after the first stage. We apply a label propagation procedure inspired by the meta-heuristic Record-to-Record Travel algorithm that reassigns node labels to improve modularity before merging communities. Experimental results show that the proposed algorithm, named meta-LPAm+, outperforms LPAm+ in terms of modularity on networks with less significant community structure while retaining almost the same performance on networks with significant community structure.

Keywords: Community detection · Label propagation · LPAm · meta-LPAm · LPAm+ · meta-LPAm+

1 Introduction

Complex networks often represent the network structure of complex systems in the real-world [24]. Communities in complex networks are informally defined as groups of nodes where interactions between nodes in the same group are more frequent than interactions between nodes in different groups [25]. Revealing community structure in complex networks is important to understand the organization of the networks [11].

© Springer International Publishing AG 2016
J. Li et al. (Eds.): ADMA 2016, LNAI 10086, pp. 65–80, 2016.
DOI: 10.1007/978-3-319-49586-6_5

Community detection has recently become an active research topic in various disciplines such as computer science, physics and sociology [7, 26]. Among the existing community detection techniques, the label propagation algorithm (LPA) [27] is known as a time-efficient method for detecting communities. Extensive research has improved LPA in different aspects including network modularity [1], real-time detection [17], overlapping community detection [10], and robustness [30]. The modularity-specialized label propagation algorithm (LPAm) [1] is an enhanced version of LPA that improves quality of the detected community structure by propagating labels of nodes to maximize network modularity, a well-known quality function to evaluate the goodness of a community division [22]. Advanced modularity-specialized label propagation (LPAm+) [18] is a further improvement of LPAm that iteratively combines LPAm with merging pairs of communities to increase network modularity.

LPAm+ achieves excellent performance on networks with significant community structure where the network modularity is above a certain threshold [18], which is practically set at 0.3 [20]. Networks with lower modularity, such as the dependency networks of software packages in [32], often have a less significant community structure, emerging from random connections between network entities. The performance of LPAm+ on networks with less significant community structure, which is the focus of this paper, is generally not well understood. Detecting less significant community structure is important, for it would help to predict the evolution of communities early.

We first show that LPAm+ tends to get trapped in poor local maxima on networks with less significant community structure. The poor local optimal solution often contains misplaced nodes which are placed in the same community with nodes belonging to different communities in the global optimal solution. We address this issue in LPAm+ by employing a meta-heuristic based label propagation procedure before merging communities to adjust the misplaced nodes. Our main contribution is a novel way of using the meta-heuristic Record-to-Record Travel [6] to improve LPAm+ but keeping the overall complexity low. Experimental results show that the proposed algorithm, named meta-LPAm+, performs better, in term of modularity, than LPAm+ on networks with less significant community structure while retaining almost the same performance on networks with significant community structure.

2   Background: LPA, LPAm and LPAm+

2.1   LPA

LPA [27] works by initially assigning a unique label to every node in the network. The algorithm then iteratively updates the labels of nodes in a random sequential order. At each iteration, each node updates its label by the label that the maximum number of its neighbours hold. If a node has many candidate labels to update, one of the labels is chosen randomly. The label updating rule for a node u in an undirected and unweighted network of n nodes can be mathematically specified as

$$l_u^{new} = \underset{l_u'}{\operatorname{argmax}} \sum_{v=1}^{n} A_{uv} \delta(l_u', l_v),$$

where l_u^{new} is the new label to be assigned to node u, l_u' is a label of the neighbours of u, v is a node in the network, l_v is the label of node v, A_{uv} is the element of the adjacency matrix of the network representing the connection between node u and node v, and δ is the Kronecker delta function.

The label updating process stops when node labels remain unchanged after an iteration. Communities are identified as groups of nodes holding the same labels. To ensure the convergence of the algorithm, node labels are updated asynchronously which means that a node updates its label based on the labels in the previous iteration of some of its neighbours and the labels in the current iteration of the other neighbours. LPA has a near linear time complexity of $O(m)$, where m is the number of edges in the network [27]. The objective function of LPA can be understood as finding a division of the network that maximizes the number of edges falling within communities [1]. The trivial solution of LPA is to assign every node in the network the same label. This illustrates a potential drawback of LPA that the detected community structure does not necessarily have any meaningful interpretation.

2.2   LPAm

LPAm [1] modifies the label updating rule in LPA to drive the solution toward the most improvement of network modularity - a well-known quality measurement of community structure [23]. Modularity measures the fraction of edges within communities minus the expected fraction of edges falling inside the same community division on a random network with the same node degrees [23]. The modularity of a network is defined as [23]

$$Q = \frac{1}{2m} \sum_{g} \sum_{i,j \in g} \left(A_{ij} - \frac{k_i k_j}{2m} \right),$$

where g is a community in the network, k_i and k_j are the degrees of node i and node j respectively, and m is the number of edges in the network.

Starting from an initial community division, LPAm performs a label propagation step at each iteration. The label propagation step updates the label of every node by a label of its neighbours that leads to the maximal increase in network modularity. Let l_u be the current label of a node u, and l_u' be the new label for node u after updating. Let g_u and g_u' denote the communities of nodes holding label l_u and l_u' respectively. The gain in modularity when updating label l_u' for node u is

$$\Delta_Q = \frac{1}{2m}\left(\sum_{i,j\in g'_u+u}\left(A_{ij}-\frac{k_ik_j}{2m}\right)+\sum_{i,j\in g_u-u}\left(A_{ij}-\frac{k_ik_j}{2m}\right)\right)$$

$$-\frac{1}{2m}\left(\sum_{i,j\in g'_u}\left(A_{ij}-\frac{k_ik_j}{2m}\right)+\sum_{i,j\in g_u}\left(A_{ij}-\frac{k_ik_j}{2m}\right)\right), \qquad (1)$$

where g_u-u denotes the community formed by removing node u from its community, and g'_u+u denotes the community formed by adding node u into community g'_u.

We can rewrite Eq. 1 as

$$\Delta_Q = \frac{1}{2m}\left(\sum_{i,j\in g'_u+u}\left(A_{ij}-\frac{k_ik_j}{2m}\right)-\sum_{i,j\in g'_u}\left(A_{ij}-\frac{k_ik_j}{2m}\right)\right)$$

$$-\frac{1}{2m}\left(\sum_{i,j\in g_u}\left(A_{ij}-\frac{k_ik_j}{2m}\right)-\sum_{i,j\in g_u-u}\left(A_{ij}-\frac{k_ik_j}{2m}\right)\right)$$

or equivalently,

$$\Delta_Q = \frac{1}{m}\sum_{j\in g'_u}\left(A_{uj}-\frac{k_uk_j}{2m}\right)-\frac{1}{m}\sum_{j\in g_u-u}\left(A_{uj}-\frac{k_uk_j}{2m}\right) \qquad (2)$$

Since the second term of Eq. 2 remains unchanged for every choice of label l'_u, choosing label l'_u that maximizes Δ_Q is equivalent to choosing label l'_u that maximizes the sum

$$\sum_{j\in g'_u}\left(A_{uj}-\frac{k_uk_j}{2m}\right) \qquad (3)$$

Therefore, the label updating rule of LPAm can be expressed as

$$l_u^{new} = \underset{l'_u}{\mathrm{argmax}}\sum_{v=1}^{n}\left(A_{uv}-\frac{k_uk_v}{2m}\right)\delta(l'_u,l_v) \qquad (4)$$

As LPAm is a greedy heuristic for finding an optimal community division regarding network modularity, the algorithm can therefore easily get trapped in local maxima of the modularity space [18].

2.3   LPAm+

LPAm+ [18] drives LPAm out of local maxima by iteratively combining LPAm with merging pairs of communities that improves network modularity the most. LPAm+ utilizes the multistep greedy algorithm (MSG) in [29] to merge multiple pairs of communities simultaneously at a time [18]. After each round of LPAm, two communities having labels l_1 and l_2 are merged if

$$(\Delta Q_{l_1\to l_2} > 0)\wedge[!\exists l:(\Delta Q_{l_1\to l} > \Delta Q_{l_1\to l_2})\vee(\Delta Q_{l_2\to l} > \Delta Q_{l_1\to l_2})]$$

with $\Delta Q_{l_1 \to l_2}$ denoting the gain in modularity when community having label l_1 is merged with community having label l_2. The algorithm stops when no pair of communities can be merged to increase network modularity.

LPAm+ achieves excellent performance on networks with significant community structure where the network modularity is above 0.3 [18]. The performance of LPAm+ for networks with less significant community structure has not been widely explored in the literature. We study the performance of LPAm+ on networks with modularity below 0.3 in the rest of this section.

Figure 1a illustrates a toy network with two intuitively divided communities (one community with nodes in white colour and the other community with nodes in dark colour) and the modularity Q is 0.2117. Applying LPAm+ on this toy network, the first round of LPAm in LPAm+ results in a local maxima with modularity $Q = 0.1097$ (Fig. 1b), for the initial community division of nodes in their own community and the sequence of node orders $\{6, 0, 9, 5, 7, 11, 10, 1, 8, 2, 3, 4\}$ to update the node labels. The first round of merging communities in LPAm+ further improves the modularity by merging community labeled 'a' and community labeled 'd', and merging community labeled 'c' and community labeled 'e' with the new modularity $Q = 0.1607$ (Fig. 1c). Note that this modularity value is still significantly below the modularity value of 0.2117 achieving in Fig. 1a. LPAm+ gets trapped in the local maxima $Q = 0.1607$ mainly due to the initially misplaced nodes 0, 3, 6 and 9, in Fig. 1b. These nodes are assigned the same community with the nodes which actually belong to different communities in the global optimal community solution by the first round of LPAm. LPAm+ does not have any mechanism to correct this misplacement of node labels and cannot adjust the labels in any other round of LPAm or merging communities.

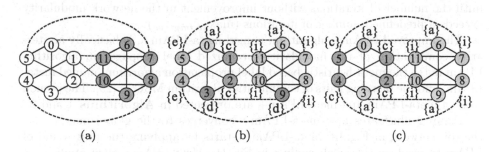

(a)   (b)   (c)

Fig. 1. A toy network consists of two intuitively divided communities with modularity $Q = 0.2117$ (a). The first round of LPAm in LPAm+ results in a local maxima with modularity $Q = 0.1097$ (b). The first round of merging communities in LPAm+ merges community labeled 'a' and community labeled 'd', and merges community labeled 'c' and community labeled 'e' with the new modularity $Q = 0.1607$ (c). LPAm+ gets trapped in this local maxima because the misplaced nodes 0, 3, 6 and 9 cannot be adjusted by any other round of LPAm or merging communities.

3   Meta-LPAm+

We propose a method to overcome the drawback of LPAm+ identified in the previous section by employing a meta-heuristic based label propagation procedure before merging communities. Our meta-heuristic based label propagation procedure searches for a better local maxima than that given by LPAm to correct the misplaced nodes in the intermediate solutions given by LPAm. Meta-heuristics [3] can find a better optimal solution than simple heuristics by accepting a worse solution temporarily to explore more thoroughly the solution space.

The proposed meta-heuristic based label propagation procedure is inspired by the Record-to-Record Travel (RRT) algorithm [6] for a balance between performance and computational complexity. The RRT algorithm is a variant of Simulated Annealing (SA) algorithm [14] with a different mechanism to accept a worse solution. RRT accepts a worse solution if the difference between the accepted solution and the best found solution is less than a specific threshold. RRT is reported to outperform SA with much lower running time for some optimization problems [6]. To the best of our knowledge, RRT has never been applied to the task of community detection.

To escape local maxima, we first performs a label propagation step that accepts a decrease in network modularity. In the label updating step, the label of every node is updated sequentially by a label of its neighbours if the decrease in network modularity is less than a predetermined threshold (DEVIATION). The decrease in network modularity is calculated as the difference between the current modularity and the highest network modularity found. After accepting a worse solution, the meta-heuristic based label propagation procedure performs a round of LPAm to quickly improve network modularity. The process is repeated until the number of iterations without improvement in the network modularity exceeds a predefined number of iterations ($\max_{no\_improve}$).

The improved LPAm+ algorithm, named meta-LPAm+, is basically an iterative combination of the modularity-specialized label propagation algorithm LPAm in [1], the meta-heuristic based label propagation algorithm, named meta-LPAm, and the merging pairs of communities algorithm MSG in [29]. The pseudo code of meta-LPAm and meta-LPAm+ are presented in **Algorithms 1** and **2**.

Figure 2 illustrates how meta-LPAm+ converges to the global maxima for the toy network in Fig. 1a. Meta-LPAm+ starts by applying the first round of LPAm to produce the local maxima in Fig. 1b. Meta-LPAm+ then applies the first round of meta-LPAm with the input parameters DEVIATION = 0.01 and $\max_{no\_improve}$ = 50 to improve this local maxima by the following steps. First, the first round of meta-LPAm performs a label propagation step that assigns label 'e' for node 1 and label 'i' for node 2 with modularity decreased by 0.0083 and the modularity is 0.1014. The next round of LPAm embedded in meta-LPAm then reaches to a better local maxima with modularity $Q = 0.1811$. The first round of merging communities in meta-LPAm+ merges community labeled 'a' and community labeled 'i' to increase modularity to 0.2117. The algorithm stops at the global maximal solution because the network modularity cannot be improved by any further round of LPAm, meta-LPAm or merging communities.

Algorithm 1. meta-LPAm

Input: An initial community division S, the threshold DEVIATION and the maximum number of iterations without improvement in modularity $max_{no\_improve}$
Output: The best found community division RECORD
1: Set RECORD = S and $n_{no\_improve}$ = 0
2: While $n_{no\_improve}$ < $max_{no\_improve}$
3:      Update the label of each node sequentially by a label of its neighbours if
4:      the new network modularity $Q_{S'} \geq Q_{RECORD} - DEVIATION$
5:      Maximize network modularity by LPAm
6:      Set $n_{no\_improve} = n_{no\_improve} + 1$
7:      If $Q_S > Q_{RECORD}$ Then
8:          Set RECORD = S and $n_{no\_improve}$ = 0
9:      End if
10: End While
/* S' is the new community division to assign to S if the node label is updated */

Algorithm 2. meta-LPAm+

Input: The threshold DEVIATION and the maximum number of iterations without improvement in modularity $max_{no\_improve}$ for the embedded meta-LPAm algorithm
Output: The best found community division
1: Assign every node into its own community
2: Maximize network modularity by LPAm
3: Find a better local maxima by meta-LPAm
4: While ∃ a pair of communities (l_1, l_2) with $\Delta Q_{l_1, l_2} > 0$
5:     Merging pairs of communities by MSG
6:     Maximize network modularity by LPAm
7:     Find a better local maxima by meta-LPAm
8: End While
/* $\Delta Q_{l_1, l_2}$ is the change in network modularity when merging community labeled 'l1' and community labeled 'l2' */

As with many other meta-heuristics, the values of the input parameters in meta-LPAm+, DEVIATION and $max_{no\_improve}$, determines the trade-off between performance and complexity of the algorithm. Increasing DEVIATION and $max_{no\_improve}$ allows the algorithm to explore larger solution space but requires more time to converge. For example, while LPAm+ finds the global optimal solution on the toy network in Fig. 1a in about 50 times out of 100 runs with randomized initial solutions, the number of times that meta-LPAm+ finds the global maxima depends on the input parameter settings. Meta-LPAm+ finds the global maxima in about 70 times over 100 runs for DEVIATION = 0.01 and $max_{no\_improve}$ = 10. However, for DEVIATION = 0.1 and $max_{no\_improve}$ = 10, meta-LPAm+ finds the global maxima in almost all of the 100 runs of the algorithm. The input parameter $max_{no\_improve}$ has less effect on the results of meta-LPAm+ providing that the parameter is set to be large enough for the algorithm to explore the neighborhood modularity space. After conducting a large number of experiments for many values of DEVIATION and $max_{no\_improve}$, the best

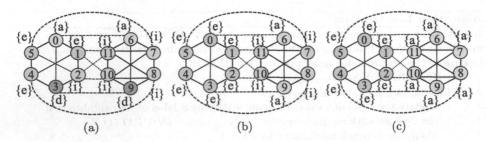

Fig. 2. A pathway to converge to global maxima by meta-LPAm+ for the toy network in Fig. 1a. Giving the community solution in Fig. 1b as the local maxima returned by the first round of LPAm in meta-LPAm+, the first round of meta-LPAm in meta-LPAm+ improves this local maxima by: first performing a label propagation step that decreases modularity by 0.0083 and the modularity is 0.1014 (a), and then performing a round of LPAm to reach to another local maxima with modularity $Q = 0.1811$ (b). The first round of merging communities in meta-LPAm+ results in the global maxima with modularity $Q = 0.2117$ (c).

combination was found to be 0.01 and 50 respectively for large-scale networks with more than a thousand of nodes.

4   Experimental Results

To demonstrate the effectiveness of our modification on LPAm+, we compared the performance of LPAm, LPAm+ and meta-LPAm+ on both synthetic networks and real-world networks. We used the standard Lancichinetti-Fortunato-Radicchi (LFR) benchmark [16] to generate the artificial networks for two different network sizes $N = 1000$ and $N = 5000$. The mixing parameter μ ranges from 0.1 to 0.8 to vary the modularity of the generated networks. The other network parameters including the average degree of nodes, the maximum degree of nodes, the minimum community size and the maximum community size are set to 20, 50, 20 and 100 respectively. The power-law exponents for the node degree sequence and the community size sequence are set to -2. As for real-world networks, we use the commonly studied networks including the Zachary's karate club network [31], the Lusseau's dolphins' network [19], the Political Books network [15], the American college football network [8], the Jazz musicians network [9], the C.elegans metabolic network [13], the University email network [12] and the Condmat2003 network [21]. For the network sizes, one can refer to [18].

Tables 1 and 2 show the average of network modularity, denoted by Q_{avg}, the maximum network modularity, denoted by Q_{max}, and the running time (in second), denoted by t, of LPAm, LPAm+ and meta-LPAm+ on the LFR benchmark networks. Table 3 shows the performance of LPAm, LPAm+ and meta-LPAm+ on the real-world networks. The experiments were performed on a desktop PC with Intel ® Core™ i7 @ 3.40 GHz CPU and 8 GB RAM. Each data point is the average over 100 graph realizations.

Table 1. Comparison on the performance of LPAm, LPAm+ and meta-LPAm+ on the LFR benchmark networks with N=1000.

Networks	LPAm			LPAm+			meta-LPAm+		
	Q_{avg}	Q_{max}	t(s)	Q_{avg}	Q_{max}	t(s)	Q_{avg}	Q_{max}	t(s)
$\mu = 0.1$	0.844	0.860	0.005	0.844	0.860	0.005	0.844	0.860	0.171
$\mu = 0.2$	0.746	0.762	0.005	0.747	0.762	0.007	0.747	0.762	0.204
$\mu = 0.3$	0.645	0.663	0.010	0.647	0.663	0.009	0.647	0.663	0.255
$\mu = 0.4$	0.542	0.559	0.012	0.547	0.559	0.014	0.547	0.561	0.310
$\mu = 0.5$	0.441	0.462	0.013	0.449	0.462	0.018	0.449	0.462	0.478
$\mu = 0.6$	0.336	0.360	0.023	0.346	0.364	0.033	0.348	0.364	0.677
$\mu = 0.7$	0.224	0.241	0.034	0.237	0.258	0.062	0.250	0.265	3.325
$\mu = 0.8$	0.190	0.204	0.034	0.205	0.212	0.071	0.219	0.226	2.698

Table 2. Comparison on the performance of LPAm, LPAm+ and meta-LPAm+ on the LFR benchmark networks with N = 5000.

Network	LPAm			LPAm+			meta-LPAm+		
	Q_{avg}	Q_{max}	t(s)	Q_{avg}	Q_{max}	t(s)	Q_{avg}	Q_{max}	t(s)
$\mu = 0.1$	0.888	0.890	0.037	0.889	0.890	0.054	0.889	0.890	2.337
$\mu = 0.2$	0.788	0.791	0.044	0.790	0.791	0.071	0.790	0.791	2.985
$\mu = 0.3$	0.687	0.691	0.054	0.690	0.692	0.088	0.690	0.692	3.731
$\mu = 0.4$	0.584	0.591	0.071	0.591	0.593	0.111	0.591	0.593	4.449
$\mu = 0.5$	0.480	0.489	0.087	0.493	0.495	0.151	0.493	0.495	5.303
$\mu = 0.6$	0.375	0.385	0.111	0.394	0.396	0.204	0.394	0.396	6.610
$\mu = 0.7$	0.273	0.281	0.201	0.293	0.298	0.453	0.294	0.298	14.250
$\mu = 0.8$	0.201	0.207	0.229	0.211	0.215	0.633	0.228	0.232	33.388

As can be seen from Tables 1 and 2, meta-LPAm+ can find solutions with higher modularity than LPAm+ on the LFR benchmark networks with μ from 0.7 and above, corresponding to networks with modularity below 0.3. The improvement in the average of network modularity by meta-LPAm+ on LPAm+ is as high as the improvement of LPAm+ on LPAm, upto 6.5%, on the low modularity networks, with acceptable time in a few seconds. Meta-LPAm+ and LPAm+ result in almost the same network modularity on the LFR benchmark networks with $\mu < 0.7$, where the network modularity is above 0.3, and the real-world networks (Table 3).

The results confirm that by employing the meta-heuristic based label propagation procedure before merging communities, meta-LPAm+ can adjust misplaced nodes in community solutions leading to the increase in network modularity on networks with less significant community structure. Meta-LPAm+ and LPAm+ achieve almost the same modularity on networks with significant

Table 3. Comparison on the performance of LPAm, LPAm+ and meta-LPAm+ on the real-world networks.

Network	LPAm			LPAm+			meta-LPAm+		
	Q_{avg}	Q_{max}	t(s)	Q_{avg}	Q_{max}	t(s)	Q_{avg}	Q_{max}	t(s)
Karate	0.351	0.399	na	0.418	0.420	na	0.418	0.420	0.005
Dolphins	0.495	0.502	na	0.522	0.528	na	0.526	0.529	0.012
Political books	0.494	0.516	na	0.527	0.527	na	0.527	0.527	0.033
Football	0.582	0.603	na	0.604	0.605	na	0.604	0.605	0.030
Jazz	0.435	0.445	0.003	0.444	0.445	0.005	0.445	0.445	0.075
C. elegans	0.378	0.404	0.005	0.443	0.451	0.016	0.444	0.451	0.630
Email	0.493	0.537	0.011	0.576	0.581	0.035	0.578	0.582	1.897
PGP	0.703	0.718	0.071	0.883	0.885	0.980	0.883	0.885	67.701

community structure since LPAm is less likely to be trapped in local maxima with many misplaced nodes on these networks. This property of LPAm has been observed in previous works [1,18]. The meta-heuristic based label propagation procedure is therefore more or less unnecessary on networks with significant community structure. However, it is worth applying meta-LPAm+ on these networks without prior knowledge of the community structure significance because the meta-heuristic based label propagation procedure generally requires an extra running time within only a few seconds. The total running time of meta-LPAm+ on the tested networks of about ten thousands of nodes is often less than a minute which is adequate for practical applications.

The computational complexity of meta-LPAm+ is mainly from the complexity of the three components embedded in the algorithm: (1) the modularity specialized label propagation algorithm LPAm, (2) the meta-heuristic based label propagation algorithm meta-LPAm, and (3) the merging communities algorithm MSG.

One round of LPAm in meta-LPAm+ has a computational complexity of $rO(m)$, with r being the average number of label propagation steps required for the round of LPAm to converge [1]. One round of meta-LPAm in meta-LPAm+ takes a computational cost of $l(O(m)+rO(m))$, with l being the average number of times that the round of meta-LPAm accepts a worse solution. The combination of LPAm and meta-LPAm in meta-LPAm+ thus has a computational complexity of $sO(m)$ with $s = r + l(1 + r)$ being the average number of label propagation steps needed for the round of LPAm and meta-LPAm to stop iterations. The computational cost of one round of MSG in meta-LPAm+ is $O(m \log N)$ [29]. Therefore, the total computational cost of meta-LPAm+ is

$$s_1 O(m) + h(O(m \log N) + s_2 O(m))$$

where s_1 is the average number of label propagation steps of the round of LPAm and meta-LPAm before the while loop in meta-LPAm+, s_2 is the average number

Table 4. The values of s_1, s_2 and h when applying meta-LPAm+ on the LFR benchmark networks.

LFR benchmark networks	N = 1000			N = 5000		
	s_1	s_2	h	s_1	s_2	h
$\mu = 0.1$	154.60	0	0	157.60	152.0	1.60
$\mu = 0.2$	156.10	0	0	158.60	152.0	2.10
$\mu = 0.3$	156.70	45.60	0.30	159.10	152.0	2.20
$\mu = 0.4$	159.20	76.0	0.50	166.30	152.0	2.50
$\mu = 0.5$	162.90	106.40	0.70	170.900	152.0	2.90
$\mu = 0.6$	212.60	107.95	0.80	262.20	152.0	3.40
$\mu = 0.7$	1134.60	492.0	1.20	498.60	267.44	3.80
$\mu = 0.8$	1605.90	185.0	0.40	2859.50	444.50	0.70

Table 5. The values of s_1, s_2 and h when applying meta-LPAm+ on the LFR benchmark networks of different sizes for $\mu = 0.7$ and $\mu = 0.8$.

LFR benchmark networks	$\mu = 0.7$			$\mu = 0.8$		
	s_1	s_2	h	s_1	s_2	h
N = 1000	1208.32	395.67	1.05	1535.29	114.35	0.18
N = 2000	998.69	412.10	2.58	2069.66	98.35	0.11
N = 3000	775.20	324.07	3.04	2313.99	280.67	0.34
N = 4000	654.94	280.86	3.42	2694.37	364.51	0.52
N = 5000	543.48	266.60	3.74	3317.09	464.60	0.77
N = 6000	503.91	241.29	4.09	3599.37	650.20	1.23
N = 7000	503.76	230.22	4.24	3674.05	828.16	1.54
N = 8000	512.91	224.34	4.39	4106.22	874.32	1.99
N = 9000	503.21	219.22	4.68	4479.50	773.13	2.20
N = 10000	499.88	218.94	4.84	4810.25	793.28	2.69

of label propagation steps of the round of LPAm and meta-LPAm in the while loop, and h is the average number of iterations for the while loop.

Table 4 shows the values of s_1, s_2 and h when applying meta-LPAm+ on the LFR benchmark networks. According to the tables, the values of s_1 and s_2 are almost stable for $\mu \leq 0.7$, where the network modularity is high above 0.3, whereas the values of s_1 and s_2 are much larger for μ from 0.7 and above, where the network modularity is low below 0.3. This evidently indicates that LPAm results in community solutions with more misplaced nodes on the lower modularity networks leading to the subsequent increase in the number of iterations needed for the round of LPAm and meta-LPAm to converge.

Table 5 reports the values of s_1, s_2, and h when applying meta-LPAm+ on the LFR benchmark networks of different sizes for $\mu = 0.7$ and $\mu = 0.8$, where

Table 6. Comparison on the performance of meta-LPAm+, FastGreedy, MSG-VM and the Louvain method on the LFR benchmark networks with N = 1000.

Network	meta-LPAm+		Fast greedy		MSG-VM		Louvain	
	Q_{avg}	Q_{max}	Q_{avg}	Q_{max}	Q_{avg}	Q_{max}	Q_{avg}	Q_{max}
$\mu = 0.1$	0.844	0.860	0.833	0.845	0.831	0.846	0.844	0.860
$\mu = 0.2$	0.747	0.762	0.707	0.726	0.732	0.748	0.747	0.762
$\mu = 0.3$	0.647	0.663	0.584	0.602	0.631	0.646	0.647	0.663
$\mu = 0.4$	0.547	0.561	0.467	0.490	0.530	0.543	0.547	0.561
$\mu = 0.5$	0.449	0.462	0.356	0.381	0.431	0.445	0.449	0.462
$\mu = 0.6$	0.348	0.364	0.260	0.277	0.331	0.346	0.347	0.364
$\mu = 0.7$	0.250	0.265	0.202	0.218	0.222	0.238	0.231	0.250
$\mu = 0.8$	0.219	0.226	0.189	0.196	0.195	0.203	0.198	0.208

Table 7. Comparison on the performance of meta-LPAm+, FastGreedy, MSG-VM and the Louvain method on the LFR benchmark networks with N = 5000.

Network	meta-LPAm+		Fast greedy		MSG-VM		Louvain	
	Q_{avg}	Q_{max}	Q_{avg}	Q_{max}	Q_{avg}	Q_{max}	Q_{avg}	Q_{max}
$\mu = 0.1$	0.889	0.890	0.875	0.879	0.877	0.879	0.889	0.890
$\mu = 0.2$	0.790	0.791	0.745	0.757	0.776	0.780	0.790	0.791
$\mu = 0.3$	0.690	0.692	0.626	0.644	0.676	0.680	0.690	0.692
$\mu = 0.4$	0.591	0.593	0.514	0.530	0.577	0.581	0.591	0.593
$\mu = 0.5$	0.493	0.495	0.403	0.418	0.478	0.481	0.493	0.495
$\mu = 0.6$	0.394	0.396	0.297	0.312	0.379	0.383	0.394	0.396
$\mu = 0.7$	0.294	0.298	0.220	0.233	0.268	0.277	0.291	0.297
$\mu = 0.8$	0.228	0.232	0.188	0.197	0.192	0.197	0.206	0.211

s_1 and s_2 are expected to be maximum on the networks of the same size with different μ. As can be seen from Table 5, s_1 and s_2 are essentially upper-bounded by N and a small fraction (about $\frac{1}{10}$) of N respectively. Therefore, we can safely set $s_1 = O(N)$ and $s_2 = O(N)$. The values of h are almost bounded by a small constant and therefore h can be estimated to be $\log(n)$, the depth of the dendrogram describing the hierarchical decomposition of a network with balanced hierarchical community structure into communities [18]. Therefore, the total computational complexity of meta-LPAm+ is

$$O(mN) + \log N(O(m\log N) + O(mN))$$

or equivalently, $O(mN\log N)$.

We also compare the performance of meta-LPAm+ with some existing modularity optimization algorithms. These algorithms use a single or a combination of node label propagation and merging communities to maximize network modularity. The compared algorithms include FastGreedy [4], which repeatedly

Table 8. Comparison on the performance of meta-LPAm+, FastGreedy, MSG-VM and the Louvain method on the real-world networks.

Network	meta-LPAm+		Fast greedy		MSG-VM		Louvain	
	Q_{avg}	Q_{max}	Q_{avg}	Q_{max}	Q_{avg}	Q_{max}	Q_{avg}	Q_{max}
Karate	0.415	0.420	0.381	0.381	0.401	0.420	0.419	0.419
Dolphins	0.525	0.529	0.495	0.495	0.522	0.522	0.519	0.519
Political books	0.527	0.527	0.502	0.502	0.520	0.521	0.520	0.520
Football	0.604	0.605	0.550	0.550	0.596	0.596	0.605	0.605
Jazz	0.445	0.445	0.439	0.439	0.444	0.445	0.443	0.443
C. elegans	0.445	0.453	0.404	0.404	0.439	0.440	0.441	0.441
Email	0.578	0.582	0.500	0.500	0.565	0.556	0.543	0.543
PGP	0.883	0.885	0.853	0.583	0.875	0.875	0.883	0.883

merges communities in pairs, MSG-VM [29], which iteratively combines merging multiple pairs of communities at a time and node label propagation, and the Louvain method [2], which is basically a multilevel node label propagation with each node at a level representing a community found in the previous level. Our implementation of MSG-VM follows [29] closely. The implementation of Fastgreedy and the Louvain method can be found in the public network software library igraph [5]. The source code of meta-LPAm+ can be downloaded at https://github.com/badungle/meta-LPAm_plus.

Tables 6 and 7 show the average network modularity and the maximum network modularity found by meta-LPAm+, FastGreedy, MSG-VM and the Louvain method on the LFR benchmark networks with N = 1000 and N = 5000 respectively. Table 8 shows the performance of meta-LPAm+ and the compared algorithms on the real-world networks. As can be seen from Tables 6 and 7, meta-LPAm+ detects communities with higher modularity than FastGreedy and MSG-VM on all the benchmark networks. Meta-LPAm+ results in almost the same modularity with the Louvain method on the LFR benchmark networks with $\mu < 0.7$, where the network modularity is high above 0.3. However, meta-LPAm+ performs better than the Louvain method on the LFR benchmark networks with $\mu \geq 0.7$, where the network modularity is below 0.3. As for the real-world networks, meta-LPAm+ performs better than FastGreedy and MSG-VM on all the real-world networks. Meta-LPAm+ performs notably better than the Louvain method on the Dolphins network, the Political books network and the Email network while the network modularity found by meta-LPAm+ and the Louvain method are almost the same on the other real-world networks. It seems that after performing the first round of node label propagation, meta-LPAm+ and the Louvain method produce more misplaced nodes in the Dolphins network, the Political books network and the Email network than in the other networks. However, meta-LPAm+ is able to further adjust the misplaced nodes to improve network modularity at node level while the Louvain method can only further improve network modularity at community level.

5   Conclusions

The advanced modularity-specialized label propagation algorithm (LPAm+) addresses the local maxima issue of LPAm, a time-efficient modularity optimization method to detect communities in complex networks, by iteratively combining LPAm with community merging.

In this paper, we first demonstrate that LPAm+ achieves excellent performance on networks with significant community structure but can easily get trapped in poor local maxima with many misplaced nodes on networks with less significant community structure. We then present an improved modularity-specialized label propagation, named meta-LPAm+, employing a meta-heuristic based label propagation procedure before merging communities to correct misplaced nodes in community solutions. The meta-heuristic based label propagation procedure is inspired by the Record-to-Record Travel algorithm for a balance between performance and complexity. The experimental results show that meta-LPAm+ finds community divisions with higher modularity than LPAm+ on networks with less significant community structure while resulting in almost the same community detection solutions with LPAm+ on networks with significant community structure. Meta-LPAm+ also outperforms the other modularity optimization methods, including FastGreedy, MSG-VM and the Louvain method, on networks with less significant community structure.

Meta-LPAm+, as a basic optimization method, can be applied to optimize other quality metrics of community structure such as the map equation [28]. Extending meta-LPAm+ to detect hierarchical community structures and overlapping communities in complex networks could be interesting future directions for research in community detection.

Acknowledgments. The authors would like to thank the maintainers and contributors of the igraph packages used in this research.

References

1. Barber, M.J., Clark, J.W.: Detecting network communities by propagating labels under constraints. Phys. Rev. E **80**(2), 026129 (2009)
2. Blondel, V.D., Guillaume, J.L., Lambiotte, R., Lefebvre, E.: Fast unfolding of communities in large networks. J. Stat. Mech. Theory Exp. **2008**(10), P10008 (2008)
3. Blum, C., Roli, A.: Metaheuristics in combinatorial optimization: overview and conceptual comparison. ACM Comput. Surv. (CSUR) **35**(3), 268–308 (2003)
4. Clauset, A., Newman, M.E.J., Moore, C.: Finding community structure in very large networks. Phys. Rev. E **70**(6), 066111 (2004)
5. Csardi, G., Nepusz, T.: The igraph software package for complex network research. InterJournal Complex Syst. **1695**(5), 1–9 (2006)
6. Dueck, G.: New optimization heuristics: the great deluge algorithm and the record-to-record travel. J. Comput. Phys. **104**(1), 86–92 (1993)
7. Fortunato, S.: Community detection in graphs. Phys. Rep. **486**(3), 75–174 (2010)

8. Girvan, M., Newman, M.E.J.: Community structure in social and biological networks. Proc. Natl Acad. Sci. **99**(12), 7821–7826 (2002)
9. Gleiser, P.M., Danon, L.: Community structure in jazz. Adv. Complex Syst. **6**(4), 565–573 (2003)
10. Gregory, S.: Finding overlapping communities in networks by label propagation. New J. Phys. **12**(10), 103018 (2010)
11. Guimera, R., Amaral, L.A.N.: Cartography of complex networks: modules and universal roles. J. Stat. Mech. Theory Exp. **2005**(2), P02001 (2005)
12. Guimera, R., Danon, L., Diaz-Guilera, A., Giralt, F., Arenas, A.: Self-similar community structure in a network of human interactions. Phys. Rev. E **68**(6), 065103 (2003)
13. Jeong, H., Tombor, B., Albert, R., Oltvai, Z.N., Barabsi, A.L.: The large-scale organization of metabolic networks. Nature **407**(6804), 651–654 (2000)
14. Kirkpatrick, S., Gelatt, C.D., Vecchi, M.P., et al.: Optimization by simmulated annealing. Science **220**(4598), 671–680 (1983)
15. Krebs, V.: A network of co-purchased books about us politics sold by the online bookseller amazon.com (2008). http://www.orgnet.com/
16. Lancichinetti, A., Fortunato, S., Radicchi, F.: Benchmark graphs for testing community detection algorithms. Phys. Rev. E **78**(4), 046110 (2008)
17. Leung, I.X., Hui, P., Lio, P., Crowcroft, J.: Towards real-time community detection in large networks. Phys. Rev. E **79**(6), 066107 (2009)
18. Liu, X., Murata, T.: Advanced modularity-specialized label propagation algorithm for detecting communities in networks. Physica A Stat. Mech. Appl. **389**(7), 1493–1500 (2010)
19. Lusseau, D., Schneider, K., Boisseau, O.J., Haase, P., Slooten, E., Dawson, S.M.: The bottlenose dolphin community of doubtful sound features a large proportion of long-lasting associations. Behav. Ecol. Sociobiol. **54**(4), 396–405 (2003)
20. Newman, M.E.J.: The structure of scientific collaboration networks. Proc. Natl Acad. Sci. **98**(2), 404–409 (2001)
21. Newman, M.E.J.: Fast algorithm for detecting community structure in networks. Phys. Rev. E **69**(6), 066133 (2004)
22. Newman, M.E.J.: Modularity and community structure in networks. Proc. Natl Acad. Sci. **103**(23), 8577–8582 (2006)
23. Newman, M.E.J., Girvan, M.: Finding and evaluating community structure in networks. Phys. Rev. E **69**(2), 026113 (2004)
24. Newman, M.E.: The structure and function of complex networks. SIAM Rev. **45**(2), 167–256 (2003)
25. Newman, M.E., Girvan, M.: Mixing patterns and community structure in networks. In: Pastor-Satorras, R., Rubi, M., Diaz-Guilera, A. (eds.) Statistical Mechanics of Complex Networks, pp. 66–87. Springer, Heidelberg (2003)
26. Porter, M.A., Onnela, J.P., Mucha, P.J.: Communities in networks. Not. AMS **56**(9), 1082–1097 (2009)
27. Raghavan, U.N., Albert, R., Kumara, S.: Near linear time algorithm to detect community structures in large-scale networks. Phys. Rev. E **76**(3), 036106 (2007)
28. Rosvall, M., Axelsson, D., Bergstrom, C.T.: The map equation. Eur. Phys. J. Spec. Top. **178**(1), 13–23 (2009)
29. Schuetz, P., Caflisch, A.: Efficient modularity optimization by multistep greedy algorithm and vertex mover refinement. Phys. Rev. E **77**(4), 046112 (2008)

30. Šubelj, L., Bajec, M.: Robust network community detection using balanced propagation. Eur. Phys. J. B Condens. Matter Complex Syst. **81**(3), 353–362 (2011)
31. Zachary, W.W.: An information flow model for conflict and fission in small groups. J. Anthropol. Res. **33**, 452–473 (1977)
32. Zanetti, M.S., Schweitzer, F.: A network perspective on software modularity. In: ARCS Workshops (ARCS) 2012, pp. 1–8. IEEE (2012)

Prediction-Based, Prioritized Market-Share Insight Extraction

Renato Keshet, Alina Maor, and George Kour[✉]

Hewlett Packard Labs, Guthwirth Park, Technion, 32000 Haifa, Israel
{renato.keshet,george.kour}@hpe.com, alina.maor@gmail.com
http://www.labs.hpe.com/

Abstract. We present an approach for Business Intelligence (BI), where market share changes are tracked, evaluated, and prioritized dynamically and interactively. Out of all the hundreds or thousands of possible combinations of sub-markets and players, the system brings to the user those combinations where the most significant changes have happened, grouped into related insights. Time-series prediction and user interaction enable the system to learn what "significant" means to the user, and adapt the results accordingly. The proposed approach captures key insights that are missed by current top-down aggregative BI systems, and that are hard to be spotted by humans (e.g., Cisco's US market disruption in 2010).

1 Introduction

Business Intelligence (BI) is about providing decision makers with the critical information for running their businesses. Traditional BI systems typically provide reports, which list main summaries and business insights, and/or dashboards, which show aggregated indicators about the business "health." These summaries and aggregations are key for informed decisions.

There are often hundreds or thousands of ways one can aggregate events to look for significant indicators. For instance, if a sales table contains columns such as product, country, region, category, and vendor, then one could aggregate products by countries, or categories by regions, or regions by products AND categories, etc., all combinations possible. In this paper, we refer to each such aggregations as "players" in "markets"; e.g., if one checks the aggregation of a product "milk" in "Canada", we call milk the "player" and Canada the "market". Notice that "Canada" could also be a player in the "milk" market.

What analysts usually do is to analyze players in a pre-selected, small number of markets. Each widget in a dashboard is typically a graph of the performances of players in a particular market. The pre-selected players and markets are typically high-level (e.g., products "world-wide", in "Americas", or in "Europe", etc.; or regions for "Meat", or for "Milk", etc.). At each point in time, the user visually inspects one or a few markets, and if they visually see something interesting (strange, anomalous) there, they dig down into more specific markets, until they discover the root cause.

© Springer International Publishing AG 2016
J. Li et al. (Eds.): ADMA 2016, LNAI 10086, pp. 81–94, 2016.
DOI: 10.1007/978-3-319-49586-6_6

This approach is known as *Online Analytics Processing* (OLAP) [1]. It is semi-manual in the sense that the aggregations are done automatically by the OLAP engine, but the choice of players and markets are made by the analyst. The space of all possible attribute combinations is called *OLAP cube*, and it can be explored with the aid of operations such as: "Slice", "dice", "drill-down", "roll-up", and "pivot" [2]. BI vendors usually provide various ranges of dashboarding, reporting and OLAP capabilities.

One of the difficulties with the traditional OLAP is that its top-down, semi-manual nature often leads to analysts missing critical insights, because key events may be diluted or canceled out when aggregated. As a consequence, these events "fly under the radar" and are missed. This usually leads to costly, inaccurate decisions. Another difficulty is that the analyst needs to manually explore the different markets, by actively slicing and dicing, and thus navigating the maze of markets and sub-markets. It is not feasible to manually check all possible players in all possible markets events for information.

Some works propose the use of analytics to help the user perform OLAP exploration. In [3,4], bottom-up approaches provide visual clues to the analyst the help them choose where to drill down. However, the solution still requires active guidance from the user.

What we propose is an approach and a computer system for mining market-share[1] data, and providing the major changes/trends that are relevant to the user. Our analysis engine performs automatically all possible slicing and dicing of the OLAP cube, analyze the market-share behavior in each of its "faces", and returns a *prioritized* list of anomalous players in specific markets. The main challenge is in computing a "significance score" for each player in each market, that can be used to return *relevant* insights to the analyst. The process is completely automatic, so the user does not have to explore the OLAP cube, as the engine digs deep into its faces, and mines the important information. The system is nevertheless interactive, enabling the user to choose the attributes for analysis, and tuning relevance weights for chosen markets and/or players.

Differently from previous solutions, our approach analyzes player "shares" (rather than "values") in each market, and perform the analysis based on time-series predictors. The use of market-shares is advantageous because (1) it is often what interests the analyst, and (2) neutralizes player oscillations in oscillating markets. The time-series analysis enables us to provide the user with significant, unpredicted player *changes* period over period. Analyzing *market-share* series require non-standard modeling, as their samples are not normally distributed.

Our paper is organized as follows. In Sect. 2 time-series predictors are briefly reviewed, and in Sect. 3 our solution is described. In Sect. 4, we dip dive into the analytics of the prediction-based "surprise factor" that is used in the computation of the "significance score". Section 5 is devoted to the experiments performed and their results, and Sect. 6 provides the conclusion.

---

[1] The share of a player in a market is the ratio between the volume of that player in that market and the total volume of that market.

2 Review of Time-Series Predictors

One can find many approaches for sequence prediction in the literature. In financial data analysis, the simplest predictor of a certain value is the *zero-order hold predictor*, in which the prediction of a sample is equal to the previous one.

More advanced, but still fairly simple techniques, which work especially well for short time series, are *linear regression*, *geometric*, and *exponential* predictors. In all those cases, the forecast is a combination of weighted past samples. The weight coefficients themselves are either fixed (e.g., the predicted value is twice the last seen value minus the value seen a time-unit before), or learned from the time series past. With linear regression, one assumes that the series has a context-related, fixed slope, which may then be learned empirically. With geometric prediction, the weight coefficients are based on average growth rate, whereas in the exponential predictor case the prediction is a fixed linear combination of the previous sample and the previous prediction.

The exponential predictor can be seen as a simplified version of the *Autoregressive* (AR) predictor, which constructs the forecast of the next values as a combination of few past values (context), learning the length of the context and the weights that should be given to each context samples from the entire sequence of the past values.

In turn, the AR model is a component of a more generalized model: the *Auto-Regressive Integrated Moving Average* [6,7] (ARIMA). The ARIMA(p, d, q) model constructs the forecast of the next value as a linear combination of the last p input values, last q prediction errors, operating on d recurrent differences between the series values. The model topology (the p, q, and d parameters) is selected for each input series according to a given model selection criterion, the most common of which being the *Akaike Information Criterion* (see [8]).

ARIMA-based schemes are considered fairly robust and strong for financial analysis, particularly for the quarterly forecasts [9,10]. However, ARIMA may be ineffective for short sequences, due to the training time and additional pre-processing effort, the length of the minimal required training context, the variety of optimization techniques, and mismatched criteria for selecting the best model.

3 Our Solution

Our solution is based on two elements: (1) A prediction-based analytic engine that mines the data, and (2) an interactive interface that empowers the user to determine what is relevant. We have integrated those in a interactive market share analytic tool, which checks all market events, in all existing markets, detects the most significant market share changes, prioritizes those, groups related insights together, and returns the corresponding visual graphs.

3.1 Analytics Engine

Consider a table of sales; each row is a deal, and each column is an attribute associated to that deal. Suppose at least one of those attributes is a "time period" where the deal occurred (e.g., week, month, quarter, etc.) – let A_t be that attribute, and $T = \{t_1, t_2, \ldots\}$ be the different data values for A_t. At least one additional attribute is a "deal value" which is used to measure market shares (e.g., revenue, number of units, etc.) – let A_v be that attribute. Let **A** be the collection of all the attributes that are neither time period or value; we call those the "market attributes". If $A \in \mathbf{A}$ is a market attribute, then Val(A) is the set of all values that attributes has in the data. E.g., if A is "country", then "USA" could be an element of Val(A).

Let a "market type" $M = \{A_1, A_2, \ldots\}$ be any subset of **A**, and a "market" m be any instance $\{A_1 = a_1, A_2 = a_2, \ldots\}$, where $a_i \in$ Val(A_i), $\forall i$. Let P be "player type in market M" if $M \subset P \subseteq \mathbf{A}$, and a player p in market m is of the format $\{A_1 = a_1, A_2 = a_2, \ldots\}$, where $m \subset p$. E.g., $m = \{$"country" $=$ "USA"$\}$ and $p = \{$"country" $=$ "USA", "OS" $=$ "Windows"$\}$. In practice, we say that p is the player "Windows" in the market "USA". Notice that sometimes "USA" can be a market, sometimes a player in some market, all combinations being possible. Now, at a time $t \in T$, the **volume** of market m, denoted $V_t(m)$, is the sum of all the values in column A_v for the deals in that market. Similarly for defining the **player volume**.

We call the ratio $z_t(p|m) = V_t(p)/V_t(m)$ the **market share** of p in m at time t.

For every possible market m, every possible player p within m, and every time t, we compute a "significance" score $\rho_t(p, m)$ to the associated market share change. This significance score is of the form:

$$\rho_t(p, m) = S_t(p|m) \cdot R_t(p|m) \cdot W_{p,m}, \tag{1}$$

where:

- $S_t(p|m)$ is a "surprise factor". This indicates how the engine "is surprised" by the occurrence of the market share $z_t(p|m)$, given all the knowledge it has about previous values of that player's share in that market, and other players' in the same market. For instance, if a value $z_t(p|m)$ is significantly different from all the values in $\{z_{t'}(p|m)\}_{t'<t}$, then the surprise factor for $z_t(p|m)$ is high. We discuss the computation of this factor in detail in Sect. 4.
- $R_t(p|m)$ is a "relevance factor". This factor reflects the relevance of each player/market to a particular vertical or use-case. For instance, some specific markets could be of more interest than others for a particular use-case. The premise regarding the relevance factor is that, differently from the surprise factor discuss above, it cannot be determined from the data alone; it is based on domain expertise that is external to the dataset.
- $W_{p,m}$ is a user-defined weight. The engine supports user feedback, either by directly setting a weight for a particular player in a particular market, or by providing hints (such as "like"/"dislike" tags for individual output insights) that is transformed into specific weights (not in the scope of this article). The weights $W_{p,m}$ are originally set to 1 for all p and m, and are modified as the user interacts.

In this article, we focus on sales transactions, and therefore we use the following relevance factor function:

$$R_t(p|m) = V_t(m). \tag{2}$$

We made this choice because, usually, the larger the market volume, the more significant it is to the user. However, we could envision a more elaborate function, involving player trend strength [5], market activity, and other elements (outside the scope of this article).

In our implementation, we do not compute for all the possible markets and players, but only those markets that have up to two elements, and players that have one additional element. However, for stronger hardware or if one is willing to wait longer, there is no need to set the above limits.

After they are computed, the significance scores are then used to sort the share occurrences for all the players in all the markets. The output of the engine is the top N insights. For instance, the top insight could be that Vendor X share in the market Country = USA and OS = Windows is surprising.

3.2   Display and Interactivity

We display the top market share changes (those with highest significant scores) in a GUI (Fig. 1). We consider each such change an "insight," which we bring to the attention of the user. The GUI clusters similar insights to the user to compose a "story." Moreover, graphical information (a pie chart or a multi-value line chart) is presented to the user for each different insight, to help understand that market.

The system is interactive in a number of ways; it enables the user to choose the time periods to compare, the value attribute to consider, and the attributes to use in the analysis (the more attributes the longer it takes to compute). In addition, the user can filter the results by specifying "terms" (attribute values of interest). The user can also create custom filters/categories and save them for future use (the upper boxes in the Fig. 1). The user can provide feedback for each insight ("like"/"dislike"), and the system adapts the score function weights $W_{p,m}$ accordingly.

4   Surprise Factor

As mentioned in Sect. 3.1, "surprise factor" is one of the factors used for computing the "significance score," which in turn is used to score the insights. The surprise factor is purely data-dependent, i.e., it is learned from the past data (as opposed to the "relevance factor", which is dependent on external domain knowledge).

The surprise factor assumes the existence of a given prediction procedure. This predictor takes some context (past samples) of a given series of market shares, and predicts the value of the next sample. How to generate the predictor

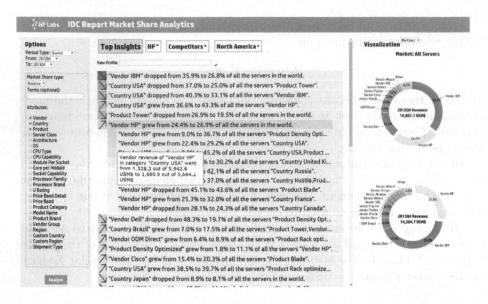

Fig. 1. The proposed Market Share analytics tool, applied to the IDC WW Server Tracker data. At the upper left, the user chooses the period to consider, and the list in the center displays the top insights for that period. Upon clicking on an insight, related events are displayed. We think of each insight as a headline in a news outlet, and the related events convey the story. To the right, graphical information related to the story is provided. To the bottom left, the user can interact with the results, by choosing attributes and setting filters. To the top, the user can create categories of interest.

is outside of the scope of this article. We assume that the predictor is given, and use it to compute a surprise factor.

In this article, we investigate the usage of two predictors: the zero-order holder and ARIMA (see Sect. 2). However, the following approach is suitable for any particular predictor (in particular, any of those described in Sect. 2).

Let $\Pi(z|D)$ be a predictor for a market share z, given the data D, and let $f(z|\pi)$ be the distribution of z given a particular prediction value $\pi = \Pi(z|D)$. We shall assume that $f(z|\pi)$ has the following format:

$$f(z|\pi) = C \cdot \exp\left(-F_\pi(z)\right), \tag{3}$$

where $F_\pi(\cdot)$ is a convex parametric function with minimum at $z = \pi$, and C is a normalization constant. Under this assumption, for the true value z we define the "surprise factor" as follows:

$$\rho_\pi(z) = -\log \frac{f(z|\pi)}{f(\pi|\pi)} = F_\pi(z) - F_\pi(\pi). \tag{4}$$

Notice that, if $z = \pi$ then $\rho_\pi(z) = 0$.

In order to establish the surprise factor, what is needed is (1) to verify if $f(z|\pi)$ can indeed be approximated by (3), and (2) estimate the parameters of F_π accordingly.

For illustration, if $f(z|\pi)$ can be approximated by a Gaussian distribution with mean π, then $F_\pi(z) = (z - \pi)^2/(2\sigma_\pi^2)$, where σ_π is the standard deviation of $f(z|\pi)$. In this case, $\rho_\pi(z)$ is the Mahalanobis distance [11] between z and $f(z|\pi)$.

Algorithm 1 summarizes the general approach.

Data: Table of categorical columns, aggregate column A, period array T, a
      prediction function Π, a set of weights $\{W_{p,m}\}$, N
Result: top N market-share change insights at current time t_c
read the input data table;
let M be all possible markets (all up-to-two attribute combinations);
let P_m be all the possible players in market $m \in M$;
for *all market* $m \in M$ **do**
    for *all player* $p \in P_m$ **do**
        for *all time* $t \in T$ **do**
            | let $z_t(p|m)$ be the market share of player p in market m at time t;
        end
    end
    for *all player* $p \in P_m$ **do**
        compute prediction $\pi = \Pi[z_{t_c}(p|m)]$;
        estimate $F_\pi(z|\pi)$ as a function of Π and the collection $\{z_t(p'|m)\}_{t,p'}$
        (see Algorithm 2);
        compute the surprise $S_{t_c}(p|m) = F_\pi(z_{t_c}(p|m)) - F_\pi(\pi)$;
        compute the relevance $R_{t_c}(p|m)$;
        compute the significance score $\rho_{t_c}(p, m)$ according to (1);
    end
end
return the insights (p, m) ordered by their significance score;

Algorithm 1. Prediction-based insight discovery approach.

4.1 Logit Transform

In practice, since z is a market share, and thus exists only inside the finite interval $[0, 1]$, we do not expect $f(z|\pi)$ to be normal. We explore a particular possibility in the remainder of this section.

Let us consider here the input data to be of a financial nature. One typical way of modeling financial data is by means of a multiplicative model. For instance, demand is viewed as the product of a deterministic and a random variables [12,13], where the random variable is an "uncertainty" quantity, and the deterministic variable is a prediction.

Following this multiplicative assumption, let us model the volume $V(p)$ of a player p given the prediction π as a random variable Y of the form:

$$Y = \exp(X), \tag{5}$$

where X is normally distributed. Thus, deviations in X translate into multiplicative uncertainty in Y.

The player size is not our focus, though, but its share in a given market m, i.e., $z(p|m) = V(p)/V(m)$. Following the model (5), the market share $z(p|m)$ given π is then a random variable Z of the form:

$$Z = \frac{Y}{Y + C} = \frac{\exp(X)}{\exp(X) + C}, \tag{6}$$

where C is the sum of the sizes of all players in market m except for p, and X is normally distributed as before. We consider C to be a constant for the analysis of the particular player p in the market m.

One can express X in terms of Z:

$$X = \log\left(\frac{C \cdot Z}{1 - Z}\right) = \log(C) + L(Z), \tag{7}$$

where:

$$L(Z) = \log\left(\frac{Z}{1 - Z}\right). \tag{8}$$

The function $L(\cdot)$ is the inverse logistic transform, called "logit", and is usually used as a link function for several distributions in generalized linear models [14].

What (7) implies is that, if one applies the logit transform (8) to realizations of a market share predicted to be equal to π, then the results should be approximately normally distributed (assuming the multiplicative model holds). In other words, $f\left(L(z)|\pi\right)$ is a Gaussian distribution.

As noted above, a normally distributed $f(z|\pi)$ corresponds to a surprise factor $\rho_\pi(\cdot)$ being the Mahalanobis distance. In the logit case, this is true up to a logit transformation, yielding the following surprise factor:

$$S_t(p|m) = [L(z_t(p|m)) - L(\pi)]^2 / (2\sigma_\pi^2). \tag{9}$$

Algorithm 2 describes the logit-based algorithm for computing the surprise factor, assuming the multiplicative-model.

5   Experiments

In order to evaluate the proposed approach, we performed a number of experiments. In Sect. 5.1, we describe experiments on synthetic data, which we generated, and in Sect. 5.2, we go over our experiments with real-world data, specifically with the IDC World-Wide Server Tracker [15], from Q1-2015.

Data: a player p, the collection of market-shares $\{z_t(p'|m)\}_{t,p'}$, predictor Π
Result: estimation of $F_\pi(p|m)$ for player p
for *all player p'* **do**
    for *all time $t \in T$* **do**
        let $L_t(p'|m) = L[z_t(p'|m)]$, according to the "logit" transform (8);
        let $\Pi[L_t(p'|m)]$ be the prediction of $L_t(p'|m)$;
        let $e_t(p'|m) = L_t(p'|m) - \Pi[L_t(p'|m)]$;
    end
end
use $\{e_t(p|m)\}_{t,p}$ in order to estimate the mean μ_π and the variance σ_π for the
Gaussian distribution $f\left(L_t(p|m)|\pi\right)$;
return $F_\pi(\cdot) = (\cdot - \mu_\pi)^2/(2\sigma_\pi^2)$;

Algorithm 2. Estimation of the "surprise function" F_π, assuming the multiplicative-model.

5.1 Synthetic Data

We generated synthetic data using a multiplicative model to simulate tables of sales transactions in a supermarket. Each row of the table relates to a sale transaction with the following fields: Month, product, country, customer type, channel, and value. We considered six (for the zero-order predictor case) or thirty (in the ARIMA case) different values for "month". For the first month, 10,000 rows were drawn, using the distributions given in Table 1 (each field drawn independently from the others). All the 10,000 rows received value equal to 1.0.

Table 1. Distributions used for generating the synthetic data.

Product		Country		Customer Type		Channel	
Bread	0.3	USA	0.3	Frequent	0.5	In store	0.8
Care	0.25	China	0.3	Unfrequent	0.2	Online	0.2
Dairy	0.2	Mexico	0.18	Sporadic	0.3		
Produce	0.15	UK	0.12				
Meat	0.1	Canada	0.1				

For the second to the penultimate months, we copied the same 10,000 transactions from the first month, but added multiplicative noise to the transaction values. More specifically, we replaced the value 1.0 for each row with the value v given by:

$$v = n^{2r-1}, \tag{10}$$

where r is a random number uniformly drawn from the $[0,1)$ interval, and n is a noise parameter. In our experiments, we tried all the values $\{1.1, 1.5, 2.0, 2.5, 3.0, 3.5\}$ for the noise n.

Finally, for the last month, we repeated the same procedure as for the previous four months, but introduced a change in a particular market, for a particular player. For each of the targeted (player,market)s, we multiplied the transaction value by a factor. For the factor, we tried all the values {0.7, 0.75, 0.8, 0.85, 0.9, 1.1, 1.15, 1.2, 1.25, 1.3}.

For illustration, we present here the results for the following two targeted modifications:

1. "Country: USA" in the market "Product: Care",
2. "Country: Mexico" in the market "Product: Dairy" AND "Customer Type: Unfrequent".

Tables 2 and 3 show the results of running these synthetic datasets (for each combination of noise and factor) with zero-order hold and ARIMA as predictors, respectively. The numbers provided in each of the tables are the position of the first returned insight that matches the player and market values for that targeted modification. For instance, suppose that for noise = 1.1, factor = 0.7, and target modification #1, the algorithm returns "Product: Care" in market "Country: USA" (switched on purpose) as the top insight (highest surprise factor), then we set the appropriate cell in the appropriate table to 0. If it is the fifth insight, then we set it to 4, and so on.

In the ARIMA experiments, the p parameter varied between 1 and 2, and the q parameter from zero to three, while d remained zero (thus, effectively ARMA).

The results shown in Tables 2 and 3 are typical; other similar targeted modifications yield similar results. The reader can notice that targeted modifications in a deeper market (i.e., a market defined by two values – such as "Product:

Table 2. Results for zero-order hold: Positions of matched targeted modification in the top insight list, for target modifications (a) "Country: USA" in the market "Product: Care", and (b) "Country: Mexico" in the market "Product: Dairy" AND "Customer Type: Unfrequent". The lower the number, the better; zero means top result.

Matched positions for target modification #1

| Factor | \multicolumn{6}{c}{Noise} |
	1.1	1.5	2.0	2.5	3	3.5
0.7	0	0	0	0	0	0
0.75	0	0	0	0	0	0
0.8	0	0	0	0	0	0
0.85	0	0	0	0	0	0
0.9	0	0	0	2	4	59
1.1	0	0	0	2	11	2
1.15	0	0	0	0	0	0
1.2	0	0	0	0	1	0
1.25	0	0	0	0	0	2
1.3	0	0	0	0	0	0

(a)

Matched positions for target modification #2

| Factor | \multicolumn{6}{c}{Noise} |
	1.1	1.5	2.0	2.5	3	3.5
0.7	0	0	0	0	0	0
0.75	0	0	0	0	0	27
0.8	0	0	0	0	0	37
0.85	0	0	0	51	335	46
0.9	0	11	0	15	8	80
1.1	0	3	63	0	83	57
1.15	0	0	0	15	4	0
1.2	0	0	0	17	81	286
1.25	0	0	7	55	44	7
1.3	1	0	0	42	0	-

(b)

Table 3. Results for ARIMA: Positions of matched targeted modification in the top insight list, for target modifications (a) "Country: USA" in the market "Product: Care", and (b) "Country: Mexico" in the market "Product: Dairy" AND "Customer Type: Unfrequent". The lower the number, the better; zero means top result.

Matched positions for target modification #1

Factor	1.1	1.5	2.0	2.5	3.0	3.5
0.70	0	0	0	1	0	0
0.75	0	0	0	0	0	0
0.8	1	0	0	0	0	0
0.85	0	0	0	1	0	0
0.9	0	0	0	0	0	0
1.1	0	0	1	0	0	0
1.15	0	0	0	0	0	0
1.2	1	0	0	1	0	0
1.25	0	0	0	0	0	0
1.3	1	0	0	0	0	0

(a)

Matched positions for target modification #2

Factor	1.1	1.5	2.0	2.5	3.0	3.5
0.7	0	0	0	0	0	0
0.75	0	0	0	0	0	0
0.8	0	0	0	0	0	0
0.85	0	0	0	0	0	0
0.9	0	0	0	0	8	0
1.1	0	0	0	0	0	4
1.15	0	0	0	0	8	0
1.2	0	0	0	0	0	0
1.25	0	0	0	0	0	0
1.3	0	0	0	0	0	0

(b)

Dairy" AND "Customer Type: Unfrequent" – rather than one) have less robust results. For the shallower modification (#1), only very large levels of noise (2.5 or above in the zero-order hold case) confused the algorithm in such a way that the targeted modification is not detected as the first insight. Even when the noise is very high, the confusion happened only on small modifications (factors 0.9 and 1.1), in which case the modification was still detected among the top insights in most cases. For deeper modifications (#2), the algorithm correctly detected the correct insight for not-so-high levels of noise (up to 2.0 in the zero-order hold case) and not-so-small modifications.

Notice that these experiments on synthetic data are purely academic; in real life, it is not clear what noise is, as market shares are dynamic, so typically one cannot be sure if a change is "true" or due to "noise".

5.2   Real-World Data

More interesting than those with synthetic data are the experiments with real-life data. The main dataset we worked with is the *IDC WW Quarterly Server Tracker* [15]. It consists of a table of hundreds of thousands of server sale transactions, involving server vendors since 2003 until today (the version we worked with is up to quarter 1 of 2015). Each sale record (row) contains several attributes, besides the time stamp and the revenue value: Vendor, region, country, product, shipment type, processor, CPU capabilities, and several other fields. In the vendor column there are several dozens of values, as in the product column.

Our first step was to check whether our assumptions hold, that is, that the data can be approximated by a multiplicative model. To that end, let us collect

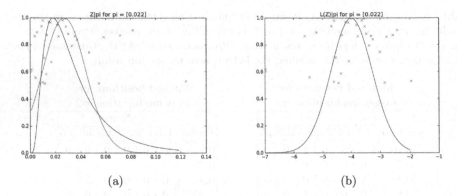

Fig. 2. Statistical characteristics of the IDC WW Server Tracker data, for a specific market ("Country: USA"), and a specific predictor value ($\pi = 0.022$). In (a) one can see that $f(z|\pi)$ is not well approximated by a Gaussian distribution (red curve), whereas in (b) one sees that the logit-transformed data $f(L(z)|\pi)$ is approximately normal (green curve). In (a), the green curve depicts the inversely transformed distribution. (Color figure online)

market-share data $\{r\}$ for a specific market, and check whether $f(r|\pi)$ is normally distributed. For illustration, consider the market "Country: USA", and all the "Vendor" players in that market. Suppose that $\pi = 0.022$. For every player in every time period t, we collect the pair (r, s), where s is the distance of p's market-share prediction to π, and r is the actual p's market share. The value of s is equal to 1 when the prediction equals π, and decreases towards 0 according to a bell-curve distance from π with $\sigma = 0.015$ (value empirically chosen). Figure 2(a) shows the collection of points (r, s), and its s-weighted Gaussian approximation (in red). It turns out that $\sum s$ for points below and above the average are 15.4 and 9.76, respectively, which indicates that this series is definitely not symmetrical, and therefore the normal distribution (red line) is not a good approximation.

We repeat the above procedure, this time in the logit domain. I.e., we collect the pairs $(L(r), s)$ for the same values of r and s as above. Figure 2(b) shows the transformed collection, and its approximation by a Gaussian distribution (green line). This time, the sum of similarities for points below and above the mean (11.7 and 13.6, respectively) are much closer to each other. Even more significantly, the percentage of points within one standard deviation from the mean is 69.9 %, which is very close to what is expected in a Gaussian distribution (68 %), as opposed to 82 % in the previous case (data before the logit transformation). This suggests that the Gaussian approximation in the logit domain is justified, and that the multiplicative model can be assumed. The green line in Fig. 2(a) depicts the logit-Gaussian curve of Fig. 2(b) transformed back to the original, raw domain. Notice that the green peak in Fig. 2(a) is closer to the predictor value (0.022) than the red peak, which indicates that the Gaussian approximation for the raw data is also biased.

Now that we checked that the data is consistent with our assumptions, the next step is to evaluate the output results. We have built a system that enables the user to obtain top insights from the IDC report. The predictor we used here was the zero-order hold. Figure 1 shows an instance of what the user can see, in this case for the particular period between 2012 and 2013 fourth quarters. The tracker table contains about one million deals, and the tool takes about ten seconds to compute the results between two quarters for three selected attributes.

Evaluating this output is harder than in the synthetic case, as typically there is no ground truth for one to compare to. It is not feasible to ask an expert to sort hundreds or thousands of market-share changes in order to use it as ground truth; and even if one does that, it is likely that two experts would rank differently, thus the algorithm may not be wrong if it sorts also differently. The approach we took was to correlate some of the results with real life events.

One test we performed was to check the year of 2010, when Cisco entered the US market, and gained significant market share. BI systems were slow to detect that trend due to the relative low volumes involved. Our tool showed Cisco's growth in the Blade market in the US as one of the top insights (#16 out of 22), and when we searched for term "Blade", it showed that same insight as #1 on the list. The top insights the system returned for that year were: 1. Oracle's overall market share from 0 % to 6.4 %, and 2. Sun Microsystems's overall market share dropped from 6.3 % to 0 %, which means that the system successfully "discovered" the purchase of Sun by Oracle in that year. A similar check for 2014 showed a significant drop for IBM in product category x86 (insight #1), at the same time that a similar increase was detected for Cisco in that same category (insight #5), which correlates well with the fact that Cisco indeed purchased the x86 sever line from IBM in 2014.

6   Conclusion

We presented a tool for analyzing market shares, and returning the top insights. The tool is based on prediction of market shares for all possible players in all possible sub-markets, and presenting those with largest "significance scores".

The significance score consists of the product of a "surprise factor", a "relevance factor", and user-provided weights. The surprise factor is computed from a function of the deviation from a time-series prediction, where the surprise function is estimated by fitting the data to a logit-transformed Gaussian distribution.

We have built an interactive tool, which enables an user to explore their data using the insight engine. We have evaluated it with synthetic data, as well as with a sales transaction table. The real-life results indicated good correlation between the insights obtained and significant market events.

References

1. "Overview of Online Analytical Processing" in a Microsoft support site. https://support.office.com/en-us/article/Overview-of-Online-Analytical-Processing-OLAP-15d2cdde-f70b-4277-b009-ed732b75fdd6
2. OLAP cube in the English Wikipedia. https://en.wikipedia.org/wiki/OLAP_cube
3. Sarawagi, S., Agrawal, R., Megiddo, N.: Discovery-driven exploration of OLAP data cubes. In: Schek, H.-J., Alonso, G., Saltor, F., Ramos, I. (eds.) EDBT 1998. LNCS, vol. 1377, pp. 168–182. Springer, Heidelberg (1998). doi:10.1007/BFb0100984
4. Du, J., Spence, I., McGuffin, M.J.: Visual guidance in the exploration of large databases. In: Proceedings of the 2010 Conference of the Center for Advanced Studies on Collaborative Research, pp. 128–138. IBM Corp., Riverton, NJ, USA (2010)
5. Hamed, K.H., Rao, A.R.: A modifieed Mann-Kandell trend test for autocorrelated data. J. Hydrol. **204**, 182–196 (1998)
6. http://people.duke.edu/~rnau/411arim.htm
7. Box, G.E.P., Jenkins, G.M.: Time Series Analysis: Forecasting and Control. Holden Day, San Francisco (1976). Revised ed.
8. Petrov, B.N., Csaki, F.: Information theory and an extension of the maximum likelihood principle. In: 2nd International Symposium on Information Theory, Akademia Kiado, pp. 267–281 (1973)
9. Jochen, M., Setzer, T.: On the robustness of ARIMA-based benchmarks for corporate financial planning quality. In: 47th Hawaii International Conference on System Sciences (HICSS), pp. 1221–1229 (2014)
10. Lorek, K.S.: Trends in statistically based quarterly cash-flow prediction models. Acc. Forum **38**(2), 145–151 (2014). Elsevier
11. See "Mahalanobis Distance" in the English Wikipedia
12. Talluri, K.T., van Ryzin, G.J.: The Theory and Practice of Revenue Management. Springer, New York (2004). ISBN-13: 978-0387243764
13. Wilson, J.G., MacDonald, L., Anderson, C.: A comparison of different demand models for joint inventory-pricing decisions. J. Revenue Pricing Manage. **10**(6), 528 (2011)
14. Logit function in the English Wikipedia. https://en.wikipedia.org/wiki/Logit
15. IDC Server Tracker. http://www.idc.com/tracker/showproductinfo.jsp

Interrelationships of Service Orchestrations

Victor W. Chu[1,2][(✉)], Raymond K. Wong[1,2], Fang Chen[2], and Chi-Hung Chi[3]

[1] University of New South Wales, Sydney, Australia
wchu@cse.unsw.edu.au
[2] Data61, CSIRO, Sydney, Australia
[3] Data61, CSIRO, Hobart, Australia

Abstract. Despite topic models have been successfully used to reveal hidden orchestration patterns from service logs, the potential uses of their interrelationships have yet to be explored. In particular, the popularity of an orchestration pattern is a leading indicator of other orchestrations in many situations. Indeed, the research in capturing relationships by induced networks has been active in some areas, such as in spatial problems. In this paper, we propose a structure discovery process to reveal relationship networks among service orchestrations. In practice, more robust business logic can be formulated by having a good understanding of these relationships that leads to efficiency gains. Our proposed interrelationship discovery process is performed by a set of optimizations with adaptive regularization. These features make our proposed solution efficient and self-adjusted to the dynamics in service environments. The results from our extensive experiments on service consumption logs confirm the effectiveness of our proposed solution.

Keywords: Service orchestration · Interrelationship · Bayesian networks

1 Introduction

Although web service orchestrations are commonly referred to as the integrations of web services, they can be interpreted as the implementations of business processes by combining multiple service components. In this context, service orchestrations produce end products for businesses which are constructed by the interactions among internal and external services [23]. The definition of interactions includes both business logic and task execution orders, where their components can span multiple applications and organizations. Due to the popularity of using service orchestrations to implement business products, a clear view of how they are consumed is critical for ongoing improvements.

Asynchronous web service invocations is a popular technique to provide reliability and scalability in service orchestrations [36], e.g., Ajax [12]. However, the capability to invoke services concurrently and to execute them in distributed environments have made it difficult for people to identify their embedded business processes, such as abstract processes [23]. Due to the randomness in

© Springer International Publishing AG 2016
J. Li et al. (Eds.): ADMA 2016, LNAI 10086, pp. 95–110, 2016.
DOI: 10.1007/978-3-319-49586-6_7

log entry generations, our visions are obscured by out-of-order and fragmented service logs. Despite topic models have been successfully used to reveal hidden orchestration patterns from service logs [7], their interrelationships have yet to be explored. In particular, the popularity of an orchestration pattern is a leading indicator of other orchestrations in many situations [33,34]. In practice, a good understanding of these relationships will lead to more robust business processes and efficiency gains. Until now, related work only focuses on the explicit representation of the relationships among partner services in modeling and enacts service orchestrations [16].

The research in network-based approach in relationship discovery has been active in many areas, such as in [6,32]. Lately, similar approach has been introduced to service computing, e.g., [5]. This solution captured the quality of service relationships among services by a network model for service recommendations. While there were some previous investigations in how to capture time-dependent relationships by Bayesian networks, e.g., [22,27], Song et al. [30] proposed time-varying dynamic Bayesian networks (TVDBNs) to model non-stationary relationships. Different from other conventional network discovery methods, such as Markov chain Monte Carlo (MCMC) sampling adopted in [28], TVDBNs takes a more scalable approach by reducing a structure discovery problem into a set of optimizations.

Regularization is commonly used to prevent overfitting in network discovery yet a constant level is often used. However, Schmidt et al. [29] reported that the degree of inter-vertex connectivity do vary in many real-world networks. In addition, the exploration into time-varying relationships also makes sample data sparse in each epoch [30], that intensifies data imbalance problem [15]. Inspired by similar concern in spatial analysis, e.g., [6], we propose an adaptive regularization solution in our interrelationship model to address the problems. It is achieved by exploiting the piecewise linear property of regularization paths for regularization parameter selections. Approximate homotopy is used for regularization path constructions making the selections tractable. These features make our proposed model to be a scalable-and-adaptive method to discover interrelationships among service orchestrations.

The rest of the paper is organized as follows. Section 2 reviews related work. Section 3 presents our proposed interrelationship discovery method for service orchestrations. Section 4 presents our experiments. Section 5 concludes this paper.

2   Related Work

The research focus in service orchestration pattern discovery has been diverse. Examples of recent investigations are as follows: (i) web service composition re-engineering by analyzing disparities between discovered model and initial design [11], (ii) to enable and improve automated process integration by analyzing service patterns [18], (iii) to reveal unexpected and interesting service composition patterns [38], and (iv) to discover the transactional behaviors of composite services for reliability improvements [1], etc. Out of the above examples, only Zheng and Bouguettaya [38] highlighted a practical issue of combinatorial explosion. They tried to develop discovery algorithms that can scale well with the number of services. In general, deterministic approaches have been mostly taken. Though probabilistic topic models have been recently introduced to service computing to reveal hidden orchestration patterns from service logs, such as in [7].

Over the last decade, latent Dirichlet allocation (LDA) [3] has been the state-of-the-art unsupervised topic model in text mining. Despite LDA is good at identifying latent topics, it does not capture any relationship between topics and side information which is essential for predictions. McAuliffe and Blei [21] introduced supervised latent Dirichlet allocation (SLDA). SLDA utilizes side information, such as ratings or labels associated with documents, to discover more predictive low dimensional topical representations. A maximum-likelihood procedure is used for parameter estimation, which uses variational approximations to handle intractable posterior expectations. On the other hand, there are several partially supervised topic models appeared in recent literature, such as labeled LDA [25] and partially labeled Dirichlet allocation [26]. More recently, Zhu et al. [39] proposed maximum entropy discrimination latent Dirichlet allocation (MedLDA) model which integrates the mechanism behind the max-margin prediction models with the mechanism behind topic models.

Different from other related work, we investigate service orchestrations from a different perspective by looking at how to capture their interrelationships by Bayesian networks. Dynamic Bayesian networks [22] have been used to model sequences of variables. They are regarded as a method to overcome the limitations of expressive power in hidden Markov model and Kalman filter. However, dynamic Bayesian networks are in fact time-invariant models, where the structure of a network is fixed but still capable to model dynamic systems [30]. Non-stationary dynamic Bayesian networks were investigated by Robinson and Hartemink [27,28] and Wang et al. [32] in recent time to model time-varying network structures, in which Markov chain Monte Carlo (MCMC) sampling and particle filtering were used for structure discovery, respectively. However, Song et al. [30] pointed out that such an approach is unlikely to be scalable, and it is also prone to overfitting. In parallel, Grzegorczyk and Husmeier [13,14] developed an alternative approach. However, their assumption of a fixed network structure is deemed to be too restrictive [8]. Song et al. proposed TVDBNs to overcome those weaknesses [30].

3   Interrelationships of Service Orchestrations

In this section, we first extend topic modeling for service orchestration discovery, including both unsupervised and supervised options. Secondly, we present our proposed model for the discovery of interrelationships among service orchestrations.

3.1 Service Orchestration Discovery by Topic Modeling

Latent Dirichlet allocation (LDA) [3] is a popular topic model for text mining where each topic is a multinomial distribution over a vocabulary. It is an unsupervised model that infers latent topics solely based on observed documents. The generative process behind LDA assumes that the words in each document are generated by a two-stage process:

(i) randomly choose a distribution over topics, and
(ii) for each word in a document, randomly choose a topic from the distribution in step (i) and then randomly choose a word from a distribution over a vocabulary.

In service orchestration discovery, we substitute our notions of words, documents and corpora by web services, service log entries and service logs respectively. In our model, service orchestrations are represented by topics. We assume that service log entries in a collection are sharing the same set of orchestrations, but each service log entry records the consumptions of those orchestrations in different proportion [2]. Topic learning in LDA is achieved by computing the posterior distribution of latent variables based on service log entries. Although the calculation is intractable, several approximation methods exist for practical uses [3].

The number of web services in a service log entry is relatively small in comparison with the total number of web services that leads to sparsity problem. Similar problem has been identified in text mining. For example, Zhao *et al.* [37] suggested that LDA may have difficulty in handling short messages. There are several proposals in recent literature trying to alleviate the difficulty from different directions, such as labeled LDA (LLDA) [24], partially labeled Dirichlet allocation (PLDA) [26], etc. One of the latest methods is biterm topic model (BTM) that approaches short message problem by modeling the generation of co-occurrence patterns in the whole dataset [35]. A biterm represents an appearance of unordered word-pair in a short message. Short message is equivalent to a short service log entry as recommended by [7]. BTM is used to model the generative process of web service biterm pattern rather than on a single web service basis. The patterns are mined from the whole service log rather than at service log entry level to overcome data sparsity problem. The time complexity of BTM is about $(\bar{l}-1)/2$ times of LDA, where $\bar{l}$ is the average length of service log entry in a service log [35]. The generative processes of LDA and BTM are summarized in Algorithms 1 and 2 for the ease of comparison of their mechanisms. Note that Dirichlet($\cdot$) and Multinomial($\cdot$) represent Dirichet and multinomial distributions respectively, where $V \sim D(\mathbf{p})$ is a short notation to denote random variable V has a distribution D with parameter(s) $\mathbf{p}$.

Topic modeling with preferred labels is a more sophisticated approach. For example, LLDA and PLDA consider additional side information to discover more predictive, low-dimensional topical representations. LLDA assumes that there is a set of discrete labels, where each of them is characterized by a multinomial distribution over a common set of services. Each service log entry uses only a

---

Algorithm 1. Generative process in LDA [3]

1: **for** each service log entry **e** in a service log $\mathcal{L}$ **do**
2:     Choose $N \sim \text{Poisson}(\xi)$
3:     Choose a orchestration distribution $\Theta \sim \text{Dirichlet}(\alpha)$
4:     **for** each of the N services s **do**
5:         Choose a orchestration $z \sim \text{Multinomial}(\Theta)$
6:         Choose a service s from $p(s|z, \beta)$, a multinomial probability conditioned on orchestration z
7:     **end for**
8: **end for**

Algorithm 2. Generative process in BTM [35]

1: **for** each orchestration $z \in \mathcal{Z}$ **do**
2:     Choose a orchestration-specific service distribution $\phi_t \sim \text{Dirichlet}(\beta)$
3: **end for**
4: Choose a orchestration distribution $\Theta \sim \text{Dirichlet}(\alpha)$ for the whole service log
5: **for** each biterm b in the biterm set $\mathcal{B}$ **do**
6:     Choose a orchestration assignment $t \sim \text{Multinomial}(\Theta)$
7:     Choose two services $s_i, s_j \sim \text{Multinomial}(\phi_t)$
8: **end for**

---

subset of labels and it could have a preference on some labels over others as represented by a multinomial distribution over labels. Each service in a service log entry is drawn from a service distribution that is associated with the labels. The service is drawn in proportion both to how much the a label is preferred and to how much that label prefers the service. As an alternative, PLDA is used in the situation of possible unknown background orchestrations. Similar to LLDA, PLDA also assumes that there is a set of labels. While each of the labels is assigned with a number of orchestrations, each orchestration only takes part in exactly one label. Different from LLDA, a special label is defined, which represents a global background orchestration class that exists in a service log.

Although LLDA and PLDA can be used to associate discretized side information with orchestrations, their prediction power is constrained by the lost of information in the discretization process. McAuliffe and Blei [21] introduced supervised latent Dirichlet allocation (SLDA) to better address prediction problems for continuous data. Different from the aforementioned models, i.e., LDA, LLDA, PLDA and BTM, SLDA accommodates a variety of response types, such as unconstrained real values, ordered or unordered class labels, non-negative integers, etc. In our investigations, the side information can be execution time, a measure of quality of service, location, etc. SLDA jointly model service log entries and their responses to find latent topics that will best predict the response variables. Under SLDA model, each service log entry and its response are considered together in a generative process. It is summarized algorithmically in Algorithm 3. Note that GLM(·) represents generalized linear model (GLM) [19].

Algorithm 3. Generative algorithm of SLDA [21]
___
1: Draw orchestration proportions $\theta|\alpha \sim \text{Dirichlet}(\alpha)$
2: **for** each service **do**
3:     Draw orchestration assignment $z_n|\theta \sim \text{Multinomial}(\theta)$
4:     Draw service $s_n|z_n, \beta_{1:K} \sim \text{Multinomial}(\beta_{z_n})$
5: **end for**
6: Draw response variable $r|z_{1:N}, \eta, \delta \sim \text{GLM}(\bar{z}, \eta, \delta)$, where $\bar{z} := (1/N)\sum_{n=1}^{N} z_n$
___

In SLDA, maximum-likelihood is used for parameter estimation, which is based on variational approximations to handle intractable posterior expectations.

3.2 Proposed Model for Interrelationship Discovery

We propose to model the interrelationships of service orchestrations by Bayesian networks. Similar to TVDBNs, our model takes a more scalable approach by reducing a structure discovery problem into a set of optimizations. Let $\mathbf{Q}^t = (Q_1^t, \ldots, Q_S^t)^\top \in \mathbb{R}^S$ represents a random vector of measures from S services at time-step t, a dynamic process of such time dependent condition can be modeled by a first-order Markovian transition model $p(\mathbf{Q}^t|\mathbf{Q}^{t-1})$ which defines the probability distribution of variables at time-step t (current time) given those at time-step $t-1$ (previous time). The probability of observing a scenario from these services over a period $t \in \{1 \ldots T\}$ can be expressed by:

$$p(\mathbf{Q}^1, \ldots, \mathbf{Q}^T) = p(\mathbf{Q}^1) \prod_{t=2}^{T} p(\mathbf{Q}^t|\mathbf{Q}^{t-1}). \qquad (1)$$

Assume that the structure of the networks is specified by a set of relationships $\mathbf{Q}_{\pi_i}^{t-1} = \{Q_j^{t-1} : Q_j^{t-1} \text{ regulates } Q_i^t\}$, where $i, j \in \{1 \ldots S\}$ and $\pi_i \subseteq \{1 \ldots S\}$. We factorize the transition model $p(\mathbf{Q}^t|\mathbf{Q}^{t-1})$ over individual services. Equation (1) is then rewritten to:

$$p(\mathbf{Q}^1, \ldots, \mathbf{Q}^T) = p(\mathbf{Q}^1) \prod_{t=2}^{T} \prod_{i=1}^{s} p(Q_i^t|\mathbf{Q}_{\pi_i}^{t-1}). \qquad (2)$$

Let graph $\mathcal{G}^t = (\mathcal{V}, \mathcal{E}^t)$ represents the conditional dependence between random vectors $\mathbf{Q}^{t-1}$ and $\mathbf{Q}^t$, where the vectors represent feature values from different services at time-step $t-1$ and at time-step t respectively. Each vertex in $\mathcal{V}$ corresponds to a sequence of variables $Q_i^1, \ldots, Q_i^T$, and the edge set $\mathcal{E}^t \subseteq \mathcal{V} \times \mathcal{V}$ contains directed edges from components of $\mathbf{Q}^{t-1}$ to components of $\mathbf{Q}^t$. The time dependent transition model $p^t = (\mathbf{Q}^t|\mathbf{Q}^{t-1})$ is expressed by an auto-regressive form $\mathbf{Q}^t = \mathbf{A}^t\mathbf{Q}^{t-1} + \epsilon$, where $\mathbf{A}^t \in \mathbb{R}^{S \times S}$ is a matrix of coefficients relating the variables at time-step $t-1$ from all services to the variables of the services in the next time-step t, and $\epsilon \sim \text{Normal}(\mathbf{0}, \sigma^2\mathbf{I})$ is an error term. The service time-varying structure is represented by non-zero entries (connections), and zeros (disconnections) in the estimated matrices $\hat{\mathbf{A}}^t$ at time-step t. In this model, the estimation of the strength of dependencies is accomplished by minimizing a set of squared loss functions with regularization, one for each vertex at each time-step $\tau \in \{1 \ldots T\}$.

In this model, sample scarcity is addressed by information sharing across time. To estimate a network structure, we first decompose the estimation along time (t) and service feature vector $(\mathbf{q})$. The neighborhood for each service is estimated separately by using linear regression and then these neighborhoods are joined together to form the overall networks. As a result, the estimation problem is effectively reduced to a set of optimizations with one for each vertex $i \in \{1 \ldots |\mathcal{V}|\}$ for time-steps $\tau \in \{1 \ldots T\}$:

$$\hat{\mathbf{A}}_{i.}^{\tau} = \underset{\mathbf{A}_{i.}^{\tau} \in \mathbb{R}^{1 \times S}}{\operatorname{argmin}} \frac{1}{T} \sum_{t=1}^{T} w^{\tau}(t)(q_i^t - \mathbf{A}_{i.}^{\tau} \mathbf{q}^{t-1})^2 + \lambda \|\mathbf{A}_{i.}^{\tau}\|_1 \qquad (3)$$

where λ is a network-wide regularization parameter which controls the sparsity of the networks, and $w^{\tau}(t)$ is the weighting of an observation from time-step t defined as $w^{\tau}(t) = \frac{K_h(t-\tau)}{\sum_{t=1}^{T} K_h(t-\tau)}$ in which $K_h(\cdot)$ is a symmetric and non-negative kernel function and h is the kernel bandwidth. The end product is a set of $\hat{\mathbf{A}}_{i.}^{\tau}$ (one per service) which can be combined to obtain $\hat{\mathbf{A}}^t$, where the strength and direction of the connections between services i and j from previous time (time-step $t-1$) to current time (t) are represented by $\hat{\mathbf{A}}_{ij}^t$. The non-zero and zero entries in the matrices $\hat{\mathbf{A}}^t$ represent the time-varying connections and disconnections between services over the time period:

$$\hat{\mathcal{E}}_{\mathcal{G}}^t = \{(i,j) \in \mathcal{V} \times \mathcal{V} | \hat{\mathbf{A}}_{ij}^t \neq 0\}. \qquad (4)$$

However, the assumption of all vertices only require the same overfitting control in TVDBNs is insufficient to address the dynamics of service environments. Schmidt et al. [29] suggested that the degree of inter-vertex connectivities could vary in many networks and the level of regularization should not be a network-wide constant. To address this problem, we exploit the piecewise linear property of regularization paths to select the best fitted regularization parameter for each vertex [6]. Approximate homotopy method is used for regularization path constructions making the selection tractable. In our proposed model for the inter-relationships of service orchestrations, namely TVDBN+, we only consider the values of λ that would trigger structural changes. Due to the continuous piecewise property, the changes only happen at the junctions (kinks) between the piecewise linear segments in a regularization path.

Although the number of linear segments is exponential in the number of variables under the worst-case scenario [20], an ε-approximate regularization path $\tilde{\mathcal{P}} \equiv \left\{ \tilde{\boldsymbol{\theta}}(\lambda) : \lambda > 0 \right\}$ can always be found by approximate homotopy method [20]. This method generates $\tilde{\mathcal{P}}$ with at most $\mathcal{O}(1/\sqrt{\varepsilon})$ linear segments, where $\varepsilon \in [0, 1]$ is the required precision. Equation (3) is transformed into the lasso [31] by pushing $w^{\tau}(t)$ into the square loss function:

$$\operatorname*{argmin}_{\mathbf{A}_{i\cdot}^{\tau} \in \mathbb{R}^{1 \times K}} \frac{1}{2T} \sum_{t=1}^{T} \left(\mathring{q}_i^t - \mathbf{A}_{i\cdot}^{\tau} \mathring{\mathbf{q}}^{\,t-1} \right)^2 + \lambda_i^{\tau} \|\mathbf{A}_{i\cdot}^{\tau}\|_1 \qquad (5)$$

where $\mathring{q}_i^t = \sqrt{w^{\tau}(t)} q_i^t$ and $\mathring{\mathbf{q}}^{\,t-1} = \sqrt{w^{\tau}(t)}\mathbf{q}^{t-1}$. We define a regularization path $\mathcal{P}$ as a set of all solutions for $\lambda_i^{\tau} > 0$:

$$\mathcal{P} = \left\{ \hat{\mathbf{A}}_{i\cdot}^{\tau}(\lambda_i^{\tau}) : \lambda_i^{\tau} > 0 \right\}. \qquad (6)$$

According to the optimality and uniqueness conditions of the lasso solution [10], $\hat{\mathbf{A}}_{i\cdot}^{\tau}(\lambda_i^{\tau})$ is a solution of Eq. (5) if and only if for all $j \in \{1, \ldots, K\}$,

$$\left(\mathring{\mathbf{q}}_j^{\,t-1} \right)^{\top} \left(\mathring{q}_i^t - \hat{\mathbf{A}}_{i\cdot}^{\tau}(\lambda_i^{\tau})\mathring{\mathbf{q}}^{\,t-1} \right) = \lambda_i^{\tau} sign(\hat{\mathbf{A}}_{i\cdot}^{\tau})(\lambda_i^{\tau})$$

$$\text{if } \mathbf{A}_{i\cdot}^{\tau}(\lambda_i^{\tau}) \neq 0,$$

$$\left| \left(\mathring{\mathbf{q}}_j^{\,t-1} \right)^{\top} \left(\mathring{q}_i^t - \hat{\mathbf{A}}_{i\cdot}^{\tau}(\lambda_i^{\tau})\mathring{\mathbf{q}}^{\,t-1} \right) \right| \leq \lambda_i^{\tau} \text{otherwise.} \qquad (7)$$

Let

$$J = \left\{ j \in \{1, \ldots, |\mathcal{V}|\} : \left| \left(\mathring{\mathbf{q}}_j^{\,t-1} \right)^{\top} \left(\mathring{q}_i^t - \hat{\mathbf{A}}_{i\cdot}^{\tau}(\lambda_i^{\tau})\mathring{\mathbf{q}}^{\,t-1} \right) \right| = \lambda_i^{\tau} \right\},$$

$\mathcal{P}$ is well defined, unique and continuous piecewise linear if matrix $\left[\mathring{\mathbf{q}}_j^{\,t-1} \right]_{j \in J}$ is full-rank [9].

Focusing on vertex i at time step τ, we simplify our notations by using $\mathbf{y}$, $\mathbf{X}$, $\boldsymbol{\theta}$ and λ to represent $\mathring{q}_i^t$, $\mathring{\mathbf{q}}^{\,t-1}$, $\hat{\mathbf{A}}_{i\cdot}^{\tau}$ and λ_i^{τ} respectively. In our structure discovery process, we only consider the values of λ that would trigger structural changes as governed by Eq. (4). Due to the continuous piecewise property, these changes only happen at kinks between the piecewise linear segments in $\mathcal{P}$. Although the number of linear segments is exponential in the number of variables under the worst-case scenario, an ε-approximate regularization path $\tilde{\mathcal{P}} \equiv \{\tilde{\boldsymbol{\theta}}(\lambda) : \lambda > 0\}$ can always be found by an approximate homotopy method [20]. The method generates $\tilde{\mathcal{P}}$ with at most $\mathcal{O}(1/\sqrt{\varepsilon})$ linear segments, where $\varepsilon \in [0, 1]$ is the required precision. Let λ_{∞} be the λ at the beginning of the path and λ_{min} be the last (minimum) value accepted. It terminates in at most $\left\lceil \frac{\log(\lambda_{\infty}/\lambda_{min})}{\gamma\sqrt{\varepsilon}} \right\rceil$ iterations, where $\lambda_{\infty} = \|(\mathbf{X}^{\top}\mathbf{y})\|_{\infty}$ and $\gamma = 1 + \varepsilon/2 - \sqrt{\varepsilon}/2$. The algorithm of our proposed TVDBN+ structure discovery method is summarized in Algorithm 4.

Moreover, a more detailed explanation of the approximate homotopy method is given as follow. $\tilde{\mathcal{P}}$ is guaranteed to be optimal up to a relative ε-duality gap $\delta_{\lambda}(\tilde{\boldsymbol{\theta}}, \boldsymbol{\vartheta}) = f_{\lambda}(\tilde{\boldsymbol{\theta}}) - g_{\lambda}(\boldsymbol{\vartheta})$, such that $0 \leq f_{\lambda}(\tilde{\boldsymbol{\theta}}) - f_{\lambda}(\boldsymbol{\theta}^{\star}(\lambda)) \leq \delta_{\lambda}(\tilde{\boldsymbol{\theta}}, \boldsymbol{\vartheta})$ [4], where $f_{\lambda}(\cdot)$ and $g_{\lambda}(\cdot)$ are the objective functions of the prime problem defined in Eq. (5) and its dual respectively, and $\boldsymbol{\vartheta}$ is a variable in the dual problem. To obtain an approximated solution, an optimal condition $OPT_{\lambda}(\varepsilon_1, \varepsilon_2)$ similar to Eq. (7) is defined by introducing perturbations $\varepsilon_1 \geq 0$ and $\varepsilon_2 \geq -\varepsilon_1$. $\tilde{\boldsymbol{\theta}}$ satisfies $OPT_{\lambda}(\varepsilon_1, \varepsilon_2)$ if an only if $\forall j \in \{1, \ldots, K\}$,

$$\lambda(1 - \varepsilon_2) \leq \mathbf{x}^{j\top}(\mathbf{y} - \mathbf{X}\tilde{\boldsymbol{\theta}})sign(\tilde{\boldsymbol{\theta}}_j) \leq \lambda(1 + \varepsilon_1) \text{ if } \tilde{\boldsymbol{\theta}}_j \neq 0,$$

$$|\mathbf{x}^{j\top}(\mathbf{y} - \mathbf{X}\tilde{\boldsymbol{\theta}})| \leq \lambda(1 + \varepsilon_1), \text{ otherwise.} \qquad (8)$$

Algorithm 4. TVDBN+ structure discovery algorithm

1: **for** $\tau \in \{1, \ldots, T\}$ **do**
2:     $\hat{\mathbf{A}}^\tau \leftarrow [\,]$; $w^\tau(t) \leftarrow \frac{\mathcal{K}_h(t-\tau)}{\sum_{t=1}^{T} \mathcal{K}_h(t-\tau)}$
3:     **for** $i \in \{1, \ldots, K\}$ **do**
4:         $\mathring{q}_i^t \leftarrow \sqrt{w^\tau(t)} s_i^t$; $\mathring{\mathbf{q}}^{t-1} \leftarrow \sqrt{w^\tau(t)} \mathbf{q}^{t-1}$
5:         $\tilde{\mathcal{P}} \leftarrow$ ApproxmateHomotopy$(\mathring{q}_i^t, \mathring{\mathbf{q}}^{t-1}, \varepsilon, \lambda_{min})$
6:         $\lambda \leftarrow$ MinCrossValidationError$(\mathring{q}_i^t, \mathring{\mathbf{q}}^{t-1}, \tilde{\mathcal{P}})$
7:         $\hat{\mathbf{A}}^\tau \leftarrow \left[\hat{\mathbf{A}}^\tau; \tilde{\mathcal{P}}(\lambda) \right]$
8:     **end for**
9: **end for**
10: **return** $\hat{\mathbf{A}}^{1:T}$

For a $\tilde{\boldsymbol{\theta}}$ which satisfies $OPT_\lambda(\varepsilon_1, \varepsilon_2)$, it can be shown that

$$\delta_\lambda(\tilde{\boldsymbol{\theta}}, \vartheta) \leq max \left(\frac{\varepsilon_1^2}{(1+\varepsilon_1)^2}, \frac{\varepsilon_1 + \varepsilon_2}{1 + \varepsilon_1} \right) f_\lambda(\tilde{\boldsymbol{\theta}}). \tag{9}$$

The number of segments in $\tilde{\mathcal{P}}$ is at most $\mathcal{O}(1/\sqrt{\varepsilon})$ based on Eqs. (8) and (9). $\varepsilon/2$ is assigned to both ε_1 and ε_2 where $\tilde{\boldsymbol{\theta}}(\lambda)$ remains as an ε-approximation for all λ with a step size $\xi \geq \lambda\gamma\sqrt{\varepsilon}$ between kinks. For $\xi < \lambda\gamma\sqrt{\varepsilon}$, a solution for $\lambda' = \lambda(1 - \gamma\sqrt{\varepsilon})$ such that $OPT_{\lambda'}(\varepsilon/2, \varepsilon/2)$ is searched by first-order method by using the current $\tilde{\boldsymbol{\theta}}(\lambda)$ as warm start. Based on $\tilde{\mathcal{P}}$, λ selection is conducted by finding a value from kinks that generates a model with the lowest cross validation error, i.e., the function MinCrossValidationError in Algorithm 4.

4 Experiments

The following experiments are conducted by using a HP EliteBook 8470p with Intel® Core™ i5-3320M CPU 2.60 GHz × 4 and 16GB RAM with Ubuntu 12.04.1 LTS, Python 2.7.3, GNU Octave 3.2.4, Matlab 2012b, and OpenJDK Java 7 Runtime.

4.1 Service Orchestration Discovery by Topic Modeling

In this experiment, we first apply unsupervised LDA and BTM models to a publicly available web log dataset – MSNBC.com anonymous web log dataset. This dataset records the page visit of MSNBC.com on September 28, 1999 (PST). Visits are recorded in time order at the URL category level. The number of services of this dataset is 17. After executing LDA and BTM on this dataset with 25 orchestrations, we match the orchestration from the two models by Hungarian method [17]. A low average per topic Jensen-Shannon divergence (JSD) value (0.272) between the two orchestration-set from LDA and BTM confirms that a close matching is obtained. Relative to LDA, the orchestration-set from BTM provides additional service details with an additional time complexity of

$(\bar{l} - 1)/2 = 4.3$ times of LDA. Based on the orchestrations identified by LDA, we randomly assign an integer value to each of them. The labels of each service log entry in MSNBC.com are then calculated by a weighted sum of the integer values according to their orchestrations distributions. We then apply supervised SLDA model to the labeled dataset. Similar orchestrations are identified from supervised model with less variations.

Secondly, we apply LDA and BTM models to twenty synthetic service logs. The service logs are created by twenty randomly generated $5,000 \times 500$ binary matrices with increasing sparsity value. The number of services of this dataset is 500. After applying LDA and BTM on this dataset with 25 orchestrations, we execute orchestration matching and also calculate the total JSD from each optimal allocation. The autoJSD of LDA and BTM from the role-set of each log are also calculated. Note that autoJSD is a measure of divergence within a orchestration set [7]. This experiment shows that the increase in divergence between the topic-sets between BTM and LDA accelerated when data sparsity is above 90 % and autoJSD's converge to the same value around 95 % sparsity. Ten randomly generated $5,000 \times 5,000$ binary matrices with increasing sparsity are translated into service logs. The number of services is further increased to 5,000 to achieve even higher sparsity. BTM is found more robust to sparse service log than LDA.

4.2 Proposed Model for Interrelationship Discovery

In this experiment, we verify the effectiveness of our proposed model for interrelationships discovery. Once service orchestrations are discovered by topic modeling in Sect. 4.1, a matrix of log-orchestration-distribution is generated with each row vector represents a probability distribution of orchestrations in service logs. By counting the daily number of orchestration used from service log from the matrix in time order, time-series that represent the popularity of orchestration are derived. Due to the limited size of the MSNBC.com dataset, we use the quality of service of wsdream.net dataset to simulate the popularity of orchestrations. The dataset consists of performance values of more than 4,500 web services invoked from 142 geographically dispersed workstations in 64 time intervals. We obtain an estimated time-dependent orchestration interrelationship structure by using TVDBNs. The relationships are captured in matrix $\hat{\mathbf{A}}^t$.

Two set of structures are retrieved by using regularization factors $\lambda = 0.4$ and 0.8, respectively. By comparing the two set of matrices from time-step 32 (in the middle of the series), there is no visible feature difference can be found between the two matrices; and therefore, a lower $\lambda = 0.4$ is used in the subsequent experiments. The prominent-bands of the matrix are selected for further analysis. Prominent-bands are groups of services which exhibit similar relations with other services. Below is a brief description of the identified bands:

- Bands H1 – H3 are the horizontal bands, where the services on these bands at current time (t) are connected to most of the services at previous time $(t-1)$.

Table 1. Prominent-bands of services and their providers

Bands	Web services	Provider URLs	
H1	1063 – 1087	http://lcdnl.co.uk	⊢⊣
		http://www.zimalimited.co.uk	⊢▷
H2	1389 – 1433	http://www.confcooperative.it	◁⊣
		http://www.agrotrace.coop	⊢⊣
		http://www.serviziocivile.coop	⊢⊣
H3	2359 – 2383	http://www.napier.ac.uk	◁⊣
V1	867 – 895	http://lipm-bioinfo.toulouse.inra.fr	⊢⊣
		http://www.hametbenoit.info	⊢⊣
V2	2162 – 2351	http://phoebus.cs.man.ac.uk	⊢⊣
V3	2668 – 2711	http://www.webservicex.net	⊢⊣

– Bands V1 – V3 are the vertical bands, where the services on these bands at previous time $(t-1)$ tend to disconnect with most of the services at current time (t). However, please note that somehow band V2 is connected to bands H1, H2 and H3. This feature will be further investigated.

Associated services of the bands are summarized in Table 1. To verify the accuracy of the bands in fitting to their service providers, we further inspect the left and right services next to the service ranges. For the ranges with both left and right services different from the service provider within the range, we mark a symbol of ⊢⊣ next to their provider URLs, whereas the ones continue with the same provider to the left only is labeled with ◁⊣ and to the right only is labeled with ⊢▷. Based on the results, most of the bands are with ⊢⊣, some with one sided continuity, and none of them with continuity to both sides outside their ranges. Extracts of the relationship structures between the first 500 services at time-steps $t-1$ and services from 1063 to 1087 (band H1) at time-steps $t = 10$ and 60 are inspected. Although the structure variations are not obvious, a detailed visual inspection confirms that there is a time-varying relationships among different time-steps.

We next apply our proposed TVDBN+ to improve the accuracy of the inter-relationship networks. The regularization parameters (λ's) selected for web services 1063 to 1087 (band H1) at time-step 32 are reported in Fig. 1, while the elapse times of this TVDBN+ structure discovery process is shown in Fig. 2. The run time of regularization path construction is more or less constant, and the overall execution time is heavily determined by the path length as expected. By using TVDBN+, we achieve on average 24% reduction in cross validation error (improvement in accuracy) in band H1. Similar significant improvements in accuracy are also obtained from other prominent-bands.

At a high level, our experiment results indicate that the services at band H1, H2 and H3 are connected with the performance of most of the services at

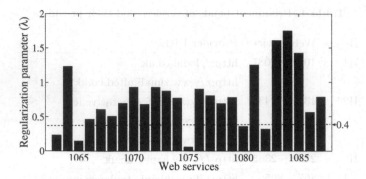

Fig. 1. Regularization parameters (λ's) proposed by TVDBN+ for web services 1063 to 1087 (band H1) at time-step $t = 32$, where the dotted line marked the position of $\lambda = 0.4$ that we used previously

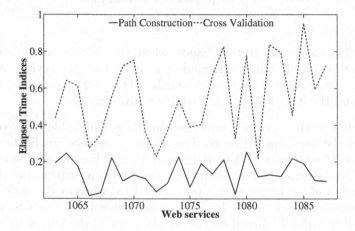

Fig. 2. Elapsed times of TVDBN+ structure discovery for web services 1063 to 1087 (band H1) at time-step $t = 32$

previous time $(t - 1)$, except for the services that fall into bands V1 and V3. However, although the services performance in band V2 at previous time $(t-1)$ are disconnected with most of the services at current time (t), they are somehow connected to the services in bands H1, H2 and H3. It gives an indication that the service providers in bands H1, H2 and to a lesser extent in band H3 are likely utilizing the services in band V2. However, they are unlikely to be hosted by a closely connected infrastructure. On the other hand, the services in bands H1, H2 and H3 seem to be highly sensitive to the overall performance of the Web in general. As indicated by the symbols ⊢⊣, ◁⊣ and ⊢▷ in Table 1, 8 out of 10 are having a clear cut of service providers at their boundaries. It provides evidence to the reliability of our proposed model.

4.3 Performance Analysis

After verifying the accuracy of TVDBN+, we next compare the network discovery run time complexity of TVDBN+ with two recently proposed time-varying dynamic Bayesian networks: (i) TVDBNs by Song et al. [30] which is without adaptive regularization, and (ii) TVDBNw by Wang et al. [32] which is based on particle filtering. The execution elapse times of TVDBN+ and TVDBNs with 2 to 50 vertices are firstly recorded. The number of possible network configurations of TVDBNw at a time step is found to be exponential that is quickly exploded. An indicative curve for TVDBNw is drawn on Fig. 3 for comparison. By rescaling the results by their respective maximum values, they are found tracking closely to each other as shown in Fig. 3. The additional complexity of automated overfitting control in TVDBN+ is found offset by approximate homotopy method effectively as shown in Fig. 2. These results provide additional evidences that our proposed TVDBN+ offer good trade-off between speed and accuracy for the discovery of interrelationships among service compositions.

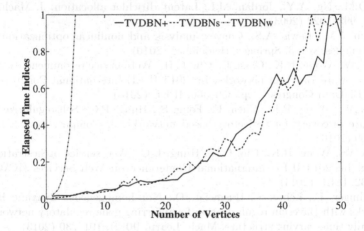

Fig. 3. Run time complexity (please note that the TVDBNw curve is indicative only)

5 Conclusions

The capability to invoke services concurrently and to execute them in distributed environments have made it difficult for people to identify their embedded business processes. Despite topic models have been successfully used to reveal hidden orchestration patterns from service logs in recent time, the potential uses of their interrelationships have yet to be explored. In particular, the popularity of an orchestration pattern is a leading indicator of other orchestrations in many situations. In service relationship modeling, recent work has so far only supporting the explicit representation of the relationships between partner services in modeling and enacting service orchestrations. However, a good understanding of the relationships among service orchestrations will provide better understanding of the dynamics of service environments and hence will lead to more robust business processes. Indeed, the research in capturing relationships by induced networks has been active in some areas. In this paper, we extend orchestration discovery with supervised models and propose a structure discovery process to reveal the interrelationships among service orchestrations. Our discovery process is achieved by a set of optimizations with adaptive regularization. These features make our proposed model efficient and adaptive to the dynamics in service environments. The results from our extensive experiments on service consumption logs confirm the effectiveness of our proposed solution.

References

1. Bhiri, S., Gaaloul, W., Godart, C.: Mining and improving composite web services recovery mechanisms. Int. J. Web Serv. Res. (IJWSR) **5**(2), 23–48 (2008)
2. Blei, D.M.: Probabilistic topic models. Commun. ACM **55**(4), 77–84 (2012)
3. Blei, D.M., Ng, A.Y., Jordan, M.I.: Latent dirichlet allocation. J. Mach. Learn. Res. **3**, 993–1022 (2003)
4. Borwein, J.M., Lewis, A.S.: Convex analysis and nonlinear optimization: theory and examples, vol. 3. Springer, Heidelberg (2010)
5. Chu, V.W., Wong, R.K., Chen, F., Chi, C.H.: Web service recommendations based on time-aware Bayesian networks. In: 2015 IEEE International Congress on Big Data (BigData Congress), pp. 359–366. IEEE (2015)
6. Chu, V.W., Wong, R.K., Chen, F., Fong, S., Hung, P.C.: Self-regularized causal structure discovery for trajectory-based networks. J. Comput. Syst. Sci. **82**(4), 594–609 (2016)
7. Chu, V.W., Wong, R.K., Chi, C.H., Hung, P.C.: Web service orchestration topic mining. In: 2014 IEEE International Conference on Web Services (ICWS), pp. 225–232. IEEE (2014)
8. Dondelinger, F., Lèbre, S., Husmeier, D.: Non-homogeneous dynamic Bayesian networks with Bayesian regularization for inferring gene regulatory networks with gradually time-varying structure. Mach. Learn. **90**(2), 191–230 (2013)
9. Efron, B., Hastie, T., Johnstone, I., Tibshirani, R., et al.: Least angle regression. Ann. Stat. **32**(2), 407–499 (2004)
10. Fuchs, J.J.: Recovery of exact sparse representations in the presence of bounded noise. Inf. Theor. IEEE Trans. **51**(10), 3601–3608 (2005)
11. Gaaloul, W., Baïna, K., Godart, C.: Log-based mining techniques applied to web service composition reengineering. Serv. Oriented Comput. Appl. **2**(2–3), 93–110 (2008)
12. Garrett, J.J.: Ajax: A new approach to web applications, 7(3) (2007). http://adaptivepath.com/ideas/essays/archives/000385.php
13. Grzegorczyk, M., Husmeier, D.: Non-stationary continuous dynamic Bayesian networks. In: NIPS, pp. 682–690 (2009)
14. Grzegorczyk, M., Husmeier, D.: Non-homogeneous dynamic Bayesian networks for continuous data. Mach. Learn. **83**(3), 355–419 (2011)
15. He, H., Garcia, E.A.: Learning from imbalanced data. Knowl. Data Eng. IEEE Trans. **21**(9), 1263–1284 (2009)

16. Kapuruge, M., Han, J., Colman, A.: Representing service-relationships as first class entities in service orchestrations. In: Wang, X.S., Cruz, I., Delis, A., Huang, G. (eds.) Web Information Systems Engineering - WISE 2012. LNCS, vol. 7651, pp. 257–270. Springer, Heidelberg (2012)
17. Kuhn, H.W.: The Hungarian method for the assignment problem. Naval Res. Logistics (NRL) **52**(1), 7–21 (2005)
18. Liang, Q.A., Chung, J.Y., Miller, S., Ouyang, Y.: Service pattern discovery of web service mining in web service registry-repository. In: IEEE International Conference on e-Business Engineering, 2006. ICEBE 2006, pp. 286–293. IEEE (2006)
19. Madsen, H., Thyregod, P.: Introduction to General and Generalized Linear Models. CRC Press, Boca Raton (2010)
20. Mairal, J., Yu, B.: Complexity analysis of the lasso regularization path. In: ICML (2012)
21. Mcauliffe, J.D., Blei, D.M.: Supervised topic models. In: NIPS, pp. 121–128 (2008)
22. Murphy, K.: Dynamic Bayesian Networks: Representation, Inference and Learning. Ph.D. thesis, UC Berkeley, Computer Science Division (2002)
23. Peltz, C.: Web services orchestration and choreography. Computer **36**(10), 46–52 (2003)
24. Ramage, D., Dumais, S.T., Liebling, D.J.: Characterizing microblogs with topic models. In: ICWSM (2010)
25. Ramage, D., Hall, D., Nallapati, R., Manning, C.D.: Labeled LDA: A supervised topic model for credit attribution in multi-labeled corpora. In: Proceedings of the 2009 Conference on Empirical Methods in Natural Language Processing. vol. 1, pp. 248–256. Association for Computational Linguistics (2009)
26. Ramage, D., Manning, C.D., Dumais, S.: Partially labeled topic models for interpretable text mining. In: Proceedings of the 17th ACM SIGKDD International Conference on Knowledge Discovery and Data Mining, pp. 457–465. ACM (2011)
27. Robinson, J.W., Hartemink, A.J.: Non-stationary dynamic Bayesian networks. In: NIPS, pp. 1369–1376 (2008)
28. Robinson, J.W., Hartemink, A.J.: Learning non-stationary dynamic Bayesian networks. J. Mach. Learn. Res. **11**, 3647–3680 (2010)
29. Schmidt, M., Niculescu-Mizil, A., Murphy, K., et al.: Learning graphical model structure using L1-regularization paths. AAAI **7**, 1278–1283 (2007)
30. Song, L., Kolar, M., Xing, E.P.: Time-varying dynamic Bayesian networks. In: NIPS, pp. 1732–1740 (2009)
31. Tibshirani, R.: Regression shrinkage and selection via the lasso. J. Royal Stat. Soc. Ser. B (Methodological) **58**, 267–288 (1996)
32. Wang, Z., Kuruoglu, E.E., Yang, X., Xu, Y., Huang, T.S.: Time varying dynamic Bayesian network for nonstationary events modeling and online inference. Signal Proc. IEEE Trans. **59**(4), 1553–1568 (2011)
33. Wetzstein, B., Danylevych, O., Leymann, F., Bitsaki, M., Nikolaou, C., van den Heuvel, W.J., Papazoglou, M.: Towards monitoring of key performance indicators across partners in service networks. In: Workshop on Service Monitoring, Adaptation and Beyond, p. 7 (2009)
34. Wetzstein, B., Karastoyanova, D., Leymann, F.: Towards management of sla-aware business processes based on key performance indicators. In: 9th Workshop on Business Process Modeling, Development and Support (BPMDS 2008) (2008)
35. Yan, X., Guo, J., Lan, Y., Cheng, X.: A biterm topic model for short texts. In: Proceedings of the 22nd International Conference on World Wide Web, pp. 1445–1456. International World Wide Web Conferences Steering Committee (2013)

36. Zdun, U., Voelter, M., Kircher, M.: Design and implementation of an asynchronous invocation framework for web services. In: Jeckle, M., Zhang, L.-J. (eds.) ICWS-Europe 2003. LNCS, vol. 2853, pp. 64–78. Springer, Heidelberg (2003). doi:10. 1007/978-3-540-39872-1_6
37. Zhao, W.X., Jiang, J., Weng, J., He, J., Lim, E.-P., Yan, H., Li, X.: Comparing twitter and traditional media using topic models. In: Clough, P., Foley, C., Gurrin, C., Jones, G.J.F., Kraaij, W., Lee, H., Mudoch, V. (eds.) ECIR 2011. LNCS, vol. 6611, pp. 338–349. Springer, Heidelberg (2011). doi:10.1007/978-3-642-20161-5_34
38. Zheng, G., Bouguettaya, A.: Service mining on the web. Services Comput. IEEE Trans. 2(1), 65–78 (2009)
39. Zhu, J., Ahmed, A., Xing, E.P.: Medlda: maximum margin supervised topic models. J. Mach. Learn. Res. 13(1), 2237–2278 (2012)

Outlier Detection on Mixed-Type Data: An Energy-Based Approach

Kien Do$^{(\boxtimes)}$, Truyen Tran, Dinh Phung, and Svetha Venkatesh

Centre for Pattern Recognition and Data Analytics,
Deakin University, Geelong, Australia
dkdo@deakin.edu.au

Abstract. Outlier detection amounts to finding data points that differ
significantly from the norm. Classic outlier detection methods are largely
designed for single data type such as continuous or discrete. However,
real world data is increasingly heterogeneous, where a data point can
have both discrete and continuous attributes. Handling mixed-type data
in a disciplined way remains a great challenge. In this paper, we propose a new unsupervised outlier detection method for mixed-type data
based on Mixed-variate Restricted Boltzmann Machine (Mv.RBM). The
Mv.RBM is a principled probabilistic method that models data density.
We propose to use *free-energy* derived from Mv.RBM as outlier score
to detect outliers as those data points lying in low density regions. The
method is fast to learn and compute, is scalable to massive datasets. At
the same time, the outlier score is identical to data negative log-density
up-to an additive constant. We evaluate the proposed method on synthetic and real-world datasets and demonstrate that (a) a proper handling mixed-types is necessary in outlier detection, and (b) free-energy
of Mv.RBM is a powerful and efficient outlier scoring method, which is
highly competitive against state-of-the-arts.

1 Introduction

Outliers are those deviating significantly from the norm. Outlier detection has
broad applications in many fields such as security [8,16,26], healthcare [32], and
insurance [14]. A common assumption is that outliers lie in the low density
regions [6]. Methods implementing this assumption differ in how the notion of
density is defined. For example, in nearest neighbor methods (k-NN) [2], large
distance between a point to its nearest neighbors indicates isolation or a low
density region around this point. Gaussian mixture models (GMM), on the other
hand, estimate the density directly through a parametric family of clusters [21].

A real-world challenge rarely addressed in outlier detection is mixed-type
data, where each data attribute can be any type such as continuous, binary,
count or nominal. Most existing methods, however, assume homogeneous data
types. Gaussian mixture models, for instance, require data to be continuous
and normally distributed. One approach to mixed-type data is to reuse existing methods. For example, we can transform multiple types into a single type

© Springer International Publishing AG 2016
J. Li et al. (Eds.): ADMA 2016, LNAI 10086, pp. 111–125, 2016.
DOI: 10.1007/978-3-319-49586-6_8

– the process known as coding in the literature. A typical practice of coding nominal data is to use a set of binary variables with exactly one active element. But it leads to information loss because the derived binary variables are considered independent in subsequent analysis. Another drawback of coding is that numerical methods such as GMM and PCA ignore the binary nature of the derived variables. Another way of model reusing is to modify existing methods to accommodate multiple types. However, the modification is often heuristic. Distance-based methods would define type-specific distances, then combine these distances into a single measure. Because type-specific distances differ in scale and semantics, finding a suitable combination is non-trivial.

A disciplined approach to mixed-type outlier detection demands three criteria to be met: (i) *capturing correlation structure between types*, (ii) *measuring deviation from the norm*, and (iii) *efficient to compute* [18]. To this end, we propose a new approach that models multiple types directly and at the same time, provides a fast mechanism for identifying low density regions. To be more precise, we adapt and extend a recent method called Mixed-variate Restricted Boltzmann Machine (Mv.RBM) [31]. Mv.RBM is a generalization of the classic RBM – originally designed for binary data, and now a building block for many deep learning architectures [3,12]. Mv.RBM has been applied for representing *regularities* in survey analysis [31], multimedia [23] and healthcare [22], but not for outlier detection, which searches for *irregularities*. Mv.RBM captures the correlation structure between types through factoring – data types are assumed to be independent given a generating mechanism.

In this work, we extend the Mv.RBM to cover counts, which are then modeled as Poisson distribution [27]. We then propose to use *free-energy* as outlier score to rank mixed-type instances. Note that *free-energy* is notion rarely seen in outlier detection. In RBM, free-energy equals the negative log-density up to an additive constant, and thus offering a principled way for density-based outlier detection. Importantly, estimation of Mv.RBM is very efficient, and scalable to massive datasets. Likewise, free-energy is computed easily through a single matrix projection. Thus Mv.RBM coupled with free-energy meets all the three criteria outlined above for outlier detection. We validate the proposed approach through an extensive set of synthetic and real experiments against well-known baselines, which include the classic single-type methods (PCA, GMM and one-class SVM), as well as state-of-the-art mixed-type methods (ODMAD [15], Beta mixture model (BMM) [4] and GLM-t [19]). The experiments demonstrate that (a) a proper handling mixed-types is necessary in outlier detection, and (b) free-energy of Mv.RBM is a powerful and efficient outlier scoring method, being highly competitive against state-of-the-arts.

In summary, we claim the following contributions:

– Introduction of a new outlier detection method for mixed-type data. The method is based on the concept of free-energy derived from a recent method known as Mixed-variate Restricted Boltzmann Machine (Mv.RBM). The method is theoretically motivated and efficient.
– Extension of Mv.RBM to handle counts as Poisson distribution.

- A comprehensive evaluation on synthetic and real mixed-type datasets, demonstrating the effectiveness of the proposed method against classic and state-of-the-art rivals.

2 Related Work

Outliers, also known as anomalies or novelties, are those thought to be generated from a mechanism different from the majority. Outlier detection is to recognize data points with unusual characteristics, or in other word, instances that do not follow any regular patterns. When there is very little or no information about outliers provided, which is common in real world data, the regular patterns need to be discovered from normal data itself. This is called *unsupervised* anomaly detection. A variant known as *semi-supervised* is when the training data is composed of just normal data [6].

Single Type Outlier Detection. A wide range of unsupervised methods have been proposed, for example, distance-based (e.g., k-NN [2]), density-based (e.g., LOF [5], LOCI [25]), cluster-based (e.g., Gaussian mixture model or GMM), projection-based (e.g., PCA) and max margin (One-class SVM). Distance-based and density-based methods model the local behaviors around each data point at a high level of granularity while cluster-based methods group similar data points together into clusters [1]. Projection-based methods, on the other hand, find a data projection that is sensitive to outliers. A comprehensive review of these methods were conducted by Chandola *et al.* [6].

Mixed-Type Outlier Detection. Although pervasive in real-world domains, mixed-type data is rarely addressed in the literature. When data is mixed (e.g., continuous and discrete), measuring distance between two data points or estimating data density can be highly challenging. A nave solution is to transform mixed-types into a single type, e.g., by coding nominal variables into 0/1 or discretizing continuous variables. This practice can significantly distort the true underlying data distribution and result in poor performance [15]. In order to handle mixed-type data directly, several methods have been proposed. LOADED [10] uses frequent pattern mining to define the score of each data point in the nominal attribute space and link it with a precomputed correlation matrix for each item set in continuous attribute space. Since there are a large number of item sets generated, this method suffers from high memory cost. RELOAD [24] is a memory-efficient version of LOADED, which employs a set of Nave Bayes classifiers with continuous attributes as inputs to predict abnormality of nominal attributes instead of aggregating over a large number of item sets.

Koufakou *et al.* [15] propose a method named ODMAD to detect outliers in sparse data with both nominal and continuous attributes. Their method first computes the anomaly score for nominal attributes using the same algorithm as LOADED. Points detected as outliers at this step are set aside and the remaining are examined over continuous attribute space with cosine similarity as a measurement. In [4], separate scores over nominal data space and numerical data space are calculated for each data point. The list of two dimensional score vectors of data was then modeled by a mixture of bivariate beta distributions. Similar to other cluster-based methods, abnormal objects could be detected as having a small probability of belonging to any components. Although the idea of beta modeling is interesting, the calculation of scores is still very simple, which is k-NN distance for continuous attributes and sum of item frequencies for nominal attributes.

The work of [33] adopts a different approach called Pattern-based Outlier Detection (POD). A pattern is a subspace formed by a particular nominal fields and all continuous fields. A logistic classifier is trained for each subspace pattern, in which continuous and nominal attributes are explanatory and response variables, respectively. The probability returned by the classifier measures the degree to which an instance deviates from a specific pattern. This is called Categorical Outlier Factor (COF). The collection of COFs and k-NN distance form the final anomaly score for a data example. Given a nominal attribute, POD models the functional relationship between continuous variables. The dependency between nominal attributes, however, is not actually captured. Moreover, when data only contains nominal attributes, the classifier cannot be created.

For all the methods mentioned above, their common drawback is that they are only able to capture correlation between a set of nominal and numerical attributes but not pair-wise correlations. The most recent work of Lu *et al.* [19] overcomes the mentioned drawback and models the data distribution. They design a Generalized Linear Model framework accompanied with a latent variable for correlation capturing and an another latent variable following Student-t distribution as an error buffer. The main advantage of this method is that it provides a strong statistical foundation for modeling distribution of different data types. However, the inference for detecting outliers is inexact and expensive to compute.

Restricted Boltzmann Machine (RBM) is a probabilistic model of binary data, formulated as a bipartite Markov random field. This special structure allows efficient inference and learning [11]. More recently, it was used as a building block for Deep Belief Networks [12], the work that started the current revolution of deep learning [17]. Recently RBM has been used for single-type outlier detection [9].

3   Mixed-Type Outlier Detection

In this section, we present a new density-based method for mixed-type outlier detection. Given a data instance x we estimate the density $P(x)$ then detect if the instance is an outlier using a threshold on the density:

$$- \log P(x) \geq \beta \tag{1}$$

for some predefined threshold β. Here $- \log P(x)$ serves as the outlier scoring function.

3.1 Density Estimation for Mixed Data

Estimating $P(\boldsymbol{x})$ is non-trivial in mixed-type data since we need to model corre-lation structures within-type and between-types. A direct correlation between-types demands a careful specification for each type-pair. For example, for two variables of different types x_1 and x_2, we need to specify either $P(x_1, x_2) = P(x_1)P(x_2 \mid x_1)$ or $P(x_1, x_2) = P(x_2)P(x_1 \mid x_2)$. With this strategy, the num-ber of pairs grows quadratically with the number of types. Most existing methods follow this approach and they are designed for a specific pair such as binary and Gaussian [7]. They neither scale to large-scale problems nor support arbitrary types such as binary, continuous, nominal, and count.

Mixed-variate Restricted Boltzmann Machine (Mv.RBM) is a recent method that supports arbitrary types simultaneously [31]. It bypasses the problems with detailed specifications and quadratic complexity by using latent binary variables. Correlation between types is not modeled directly but is factored into indirect correlation with latent variables. As such we need only to model the correlation between a type and the latent binary. This scales linearly with the number of types.

Mv.RBM was primarily designed for data representation which transforms mixed data into a homogeneous representation, which serves as input for the next analysis stage. Our adaptation, on the other hand, proposes to use Mv.RBM as outlier detector directly, without going through the representation stage.

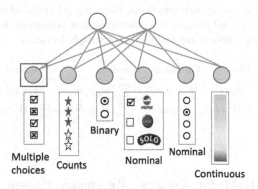

Fig. 1. Mix-variate Restricted Boltzmann machines for mixed-type data. Filled circles denote visible inputs, empty circles denote hidden units. Multiple choices are modeled as multiple binaries, denoted by a filled circle in a clear box.

3.2 Mixed-Variate Restricted Boltzmann Machines

We first review Mv.RBM for a mixture of binary, Gaussian and nominal types, then extend to cover counts. See Fig. 1 for a graphical illustration. Mv.RBM is an extension of RBM for multiple data types. An RBM is a probabilistic neural network that models binary data in an unsupervised manner. More formally, let $x \in \{0,1\}^N$ be a binary input vector, and $h \in \{0,1\}^K$ be a binary hidden vector, RBM defines the joint distribution as follows:

$$P(x, h) \propto \exp(-E(x, h))$$

where $E(x, h)$ is energy function of the following form:

$$E(x, h) = -\left(\sum_i a_i x_i + \sum_k b_k h_k + \sum_{ik} W_{ik} x_i h_k \right) \quad (2)$$

Here (a, b, W) are model parameters.

For subsequent development, we rewrite the energy function as:

$$E(x, h) = \sum_i E_i(x_i) + \sum_k \left(-b_k + \sum_i G_{ik}(x_i) \right) h_k \quad (3)$$

where $E_i(x_i) = -a_i x_i$ and $G_{ik}(x_i) = -W_{ik} x_i$.

Mv.RBM extends RBM by redefining the energy function to fit multiple data types. The energy function of Mv.RBM differs from that of RBM by the using multiple type-specific energy sub-functions $E_i(x_i)$ and $G_{ik}(x_i)$ as listed[1] in Table 1. The energy decomposition in Eq. (3) remains unchanged.

Table 1. Type-specific energy sub-functions. Here $\delta(x_i, c)$ is the identity function, that is, $\delta(x_i, c) = 1$ if $x_i = c$, and $\delta(x_i, c) = 0$ otherwise. For Gaussian, we assume data has unit variance. Multiple choices are modeled as multiple binaries.

Func.	Binary	Gaussian	Nominal	Count
$E_i(x_i)$	$-a_i x_i$	$\frac{x_i^2}{2} - a_i x_i$	$-\sum_c a_{ic}\delta(x_i, c)$	$\log x_i! - a_i x_i$
$G_{ik}(x_i)$	$-W_{ik} x_i$	$-W_{ik} x_i$	$-\sum_c W_{ikc}\delta(x_i, c)$	$-W_{ik} x_i$

Extending Mv.RBM for Counts. We employ Poisson distributions for counts [27]. The sub-energy sub-functions are defined as:

$$E_i(x_i) = \log x_i! - a_i x_i; \qquad G_{ik}(x_i) = -W_{ik} x_i \quad (4)$$

Note that count modeling was not introduced in the original Mv.RBM work.

Learning. Model estimation in RBM and Mv.RBM amounts to maximize data likelihood with respect to model parameters. It is typically done by n-step Contrastive Divergence (CD-n), which is an approximate but fast method. In particular, for each parameter update, CD-n maintains a very short Mote Carlo

---

[1] The original Mv.RBM also covers rank, but we do not consider in this paper.

Markov chain (MCMC), starting from the data, runs for n steps, then collects the samples to approximate data statistics. The MCMC is efficient because of the factorizations in Eq. (5), that is, we can sample all hidden variables in parallel through $\hat{h} \sim P(h \mid x)$ and all visible variables in parallel through $\hat{x} \sim P(x \mid h)$. For example, for Gaussian inputs, the parameters are updated as follows:

$$b_k \leftarrow b_k + \eta \left(\bar{h}_{k|x} - \bar{h}_{k|\hat{x}}\right)$$
$$a_i \leftarrow a_i + \eta (x_i - \hat{x}_i)$$
$$W_{ik} \leftarrow W_{ik} + \eta \left(x_i \bar{h}_{k|x} - \hat{x}_i \bar{h}_{k|\hat{x}}\right)$$

where $\bar{h}_{k|x} = P(h_k = 1 \mid x)$ and $\eta > 0$ is the learning rate. This learning procedure scales linearly with n and data size.

Mv.RBM as a Mixture Model of Exponential Size. In Mv.RBM, types are not correlated directly but through the common hidden layer. The posterior $P(h \mid x)$ and data generative process $P(x \mid h)$ in Mv.RBM are factorized as:

$$P(h \mid x) = \prod_k P(h_k \mid x); \qquad P(x \mid h) = \prod_i P(x_i \mid h) \qquad (5)$$

Here types are conditionally independent given h, but since h are hidden, types are dependent as in $P(x) = \sum_h P(x, h)$.

The posterior has the same form across types – the activation probability $P(h_k = 1 \mid x)$ is sigmoid $(b_k - \sum_i G_{ik}(x_i))$. On the other hand, the generative process is type-specific. For example, for binary data, the activation probability $P(x_i = 1 \mid h)$ is sigmoid $(a_i + \sum_k W_{ik}h_k)$; and for Gaussian data, the conditional density $P(x_i \mid h)$ is $\mathcal{N}(a_i + \sum_k W_{ik}h_k; 1)$.

Since h is discrete, Mv.RBM can be considered as a mixture model of 2^K components that shared the same parameter. This suggests that Mv.RBM can be used for outlier detection in the same way that GMM does.

3.3   Outlier Detection on Mixed-Type Data

Recall that for outlier detection as in Eq. (1) we need the marginal distribution $P(x) = \sum_h P(x, h)$, which is:

$$P(x) \propto \sum_h \exp(-E(x, h)) = \exp(-F(x))$$

where $F(x) = -\log \sum_h \exp(-E(x, h))$ is known as *free-energy*. Notice that the free-energy equals the negative log-density up to an additive constant:

$$F(x) = -\log P(x) + \text{constant}$$

Thus *we can use the free-energy as the outlier score* to rank data instances, following the detection rule in Eq. (1).

Computing Free-Energy. Although estimating the free-energy amounts to summing over 2^K configurations of the hidden layer, we can still compute the summation efficiently, thanks to the decomposition of the energy function in Eq. (3). We can rewrite the free-energy as follows:

$$F(\boldsymbol{x}) = \sum_i E_i(x_i) - \sum_k \log \left(1 + \exp \left(b_k - \sum_i G_{ik}(x_i) \right) \right) \qquad (6)$$

This free-energy can be computed in linear time.

Controlling Model Expressiveness. A major challenge of unsupervised outlier detection is the phenomenon of *swamping effect*, where an inlier is misclassified as outlier, possibly due a large number of true outliers in the data [28]. When data models are highly expressive – such as large RBMs and Mv.RBMs – outliers are included by the models as if they have patterns themselves, even if these patterns are weak and differ significantly from the regularities of the inliers. One way to control the model expressiveness is to limit the number of hidden layers K (hence the number of mixing components 2^K). Another way is to apply early stopping – learning stops before convergence has occurred.

Summary. To sum up, Mv.RBM, coupled with free-energy, offers a disciplined approach to mixed-type outlier detection that meet three desirable criteria: (i) *capturing correlation structure between types*, (ii) *measuring deviation from the norm*, and (iii) *efficient to compute*.

4   Experiments

We present experiments on synthetic and real-world data. For comparison, we implement well-known single-type outlier detection methods including Gaussian mixture model (GMM), Probabilistic Principal Component Analysis (PPCA) [29] and one-class SVM (OCSVM) [20]. The number of components of PPCA model is set so that the discarded energy is the same as the anomaly rate in training data. For OCSVM, we use radial basis kernel with $\nu = 0.7$. GMM and PPCA are probabilistic, and thus data log-likelihood can be computed for outlier detection.

Since all of these single-type methods assume numerical data, we code nominal types using dummy binaries. For example, a A in the nominal set $\{A, B, C\}$ is coded as (1,0,0) and B as (0,1,0). This coding causes some nominal information loss, since the coding does not ensure that only one value is allowed in nominal variables. For all methods, the detection threshold is based on the α percentile of the training outlier scores. Whenever possible, we also include results from other recent mixed-type papers, ODMAD [15], Beta mixture model (BMM) [4] and GLM-t [19]. We followed the same mechanism they used to generate outliers.

4.1   Synthetic Data

We first evaluate the behaviors of Mv.RBM on synthetic data with controllable complexity. We simulate mixed-type data using a generalized Thurstonian theory, where Gaussians serve as underlying latent variables for observed discrete values. Readers are referred to [30] for a complete account of the theory. For this study, the underlying data is generated from a GMM of 3 mixing components with equal mixing probability. Each component is a multivariate Gaussian distributions of 15 dimensions with random mean and positive-definite covariance matrix. From each distribution, we simulate 1,000 samples, creating a data set size 3,000. To generate outliers, we randomly pick 5 % of data, and add uniform noise to each dimension, i.e., $x_i \leftarrow x_i + e_i$ where $e_i \sim \mathcal{U}$. For visualization, we use t-SNE to reduce the dimensionality to 2 and plot the data in Fig. 2.

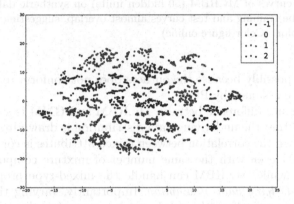

Fig. 2. Synthetic data with 3 normal clusters (cluster IDs 0,1,2) and 1 set of scattered outliers (ID: -1, colored in red). Best viewed in color. (Color figure online)

Out of 15 variables, 3 are kept as Gaussian and the rest are used to create mixed-type variables. More specifically, 3 variables are transformed into binaries using random thresholds, i.e., $\bar{x}_i = \delta(x_i \geq \theta_i)$. The other 9 variables are used to generate 3 nominal variables of size 3 using the rule: $\tilde{x}_i = \arg\max(x_{i1}, x_{i2}, x_{i3})$.

Models are trained on 70 % data and tested on the remaining 30 %. This testing scheme is to validate the generalizability of models on unseen data. The learning curves of Mv.RBM are plotted in Fig. 3. With the learning rate of 0.05, learning converges after 10 epochs. No overfitting occurs.

The decision threshold β in Eq. (1) is set at 5 percentile of the training set. Figure 4 plots the outlier detection performance of Mv.RBM (in F-score) on test data as a function of model size (number of hidden units). To account for random initialization, we run Mv.RBM 10 times and average the F-scores. It is apparent that the performance of Mv.RBM is competitive against that of GMM. The best F-score achieved by GMM is only about 0.35, lower than the worst F-score by Mv.RBM, which is 0.50. The PCA performs poorly, with

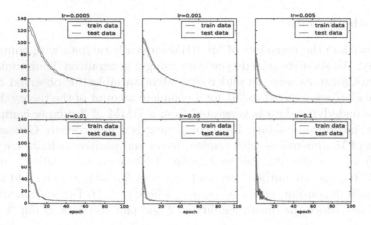

Fig. 3. Learning curves of Mv.RBM (50 hidden units) on synthetic data for different learning rates. The training and test curves almost overlap, suggesting no overfitting. Best viewed in color. (Color figure online)

F-score of 0.11, possibly because the outliers does not conform to the notion of residual subspace assumed by PCA.

The performance difference between Mv.RBM and GMM is significant considering the fact that the underlying data distribution is drawn from a GMM. It suggests that when the correlation between mixed attributes is complex like this case, using GMM even with the same number of mixture components cannot learn well. Meanwhile, Mv.RBM can handle the mixed-type properly, *without knowing the underlying data assumption*. Importantly, varying the number of hidden units does not affect the result much, suggesting the stability of the model and it can free users from carefully crafting this hyper-parameter.

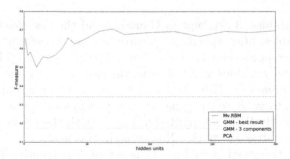

Fig. 4. Performance of Mv.RBM in F-score on synthetic data as a function of number of hidden units. Horizontal lines are performance measures of PCA (in green) and GMM (best result in red; and with 3 components in yellow). Best viewed in color. (Color figure online)

4.2 Real Data

For real-world applications, we use a wide range of mixed-type datasets. From the UCI repository[2], we select 7 datasets which were previously used as benchmarks for mixed-type anomaly detection [4,10,19]. Data statistics are reported in Table 2. We generate outliers by either using rare classes whenever possible, or by randomly injecting a small proportion of anomalies, as follows:

- **Using rare classes**: For the KDD99 *10 percent* dataset (KDD99-10), intrusions (outliers) account for 70% of all data, and thus it is not possible to use full data because outliers will be treated as normal in unsupervised learning. Thus, we consider all normal instances from the original data as inliers, which accounts for 90% of the new data. The remaining 10% outliers are randomly selected from the original intrusions.
- **Outliers injection**: For the other datasets, we treat data points as normal objects and generate outliers based on a contamination procedure described in [4,18]. Outliers are created by randomly selecting 10% of instances and modifying their default values. For numerical attributes (Gaussian, Poisson), values are shifted by 2.0 to 3.0 times standard deviation. For discrete attributes (binary, categorical), the values are switched to alternatives.

Numerical attributes are standardized to zero means and unit variance. For evaluation, we randomly select 30% data for testing, and 70% data for training. Note that since learning is unsupervised, outliers must also be detected in the training set since there are no ground-truths. The outliers in the test set is to test the generalizability of the models to unseen data.

Table 2. Characteristics of mixed-type datasets. The proportion of outliers are 10%.

Dataset	No. Instances		No. Attributes				
	Train	Test	Bin.	Gauss.	Nominal	Poisson	Total
KDD99-10	75,669	32,417	4	15	3	19	41
Australian Credit	533	266	3	6	5	0	14
German Credit	770	330	2	7	11	0	20
Heart	208	89	3	6	4	0	13
Thoracic Surgery	362	155	10	3	3	0	16
Auto MPG	303	128	0	5	3	0	8
Contraceptive	1136	484	3	0	4	1	8

---

[2] https://archive.ics.uci.edu/ml/datasets.html.

Models Setup. The number of hidden units in Mv.RBM is set to $K = 2$ for the KDD99-10 dataset, and to $K = 5$ for other datasets. The parameters of Mv. RBM are updated using stochastic gradient descent, that is, update occurs after every mini-batch of data points. For small datasets, the mini-batch size is equal to the size of the entire datasets while for KDD99-10, the mini-batch size is set to 500. The learning rate is set to 0.01 for all small datasets, and to 0.001 for KDD99-10. Small datasets are trained using momentum of 0.8. For KDD99-10, we use Adam [13], with $\beta_1 = 0.85$ and $\beta_2 = 0.995$. For small datasets, the number of mixture components in GMM is chosen using grid search in the range from 1 to 30 with a step size of 5. For KDD99-10, the number of mixture components is set to 4.

Results. Figure 5(a) shows a histogram of free-energies computed using Eq. (6) on the KDD99-10 dataset. The inliers/outliers are well-separated into the low/high energy regions, respectively. This is also reflected in an Area Under the ROC Curve (AUC) of 0.914 (see Fig. 5(b)).

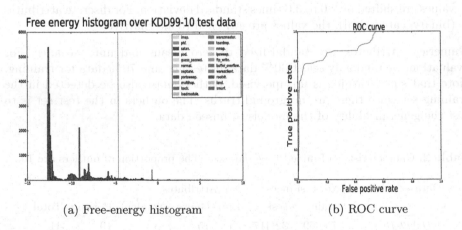

(a) Free-energy histogram   (b) ROC curve

Fig. 5. Outlier detection on the KDD99-10 dataset. (a) Histogram of free-energies. The vertical line separates data classified as inliers (left) from those classified as outliers (right). The color of majority (light blue) is inlier. Best viewed in color. (b) ROC curve (AUC = 0.914). (Color figure online)

The detection performance in term of F-score on test data is reported in Tables 3. The mean of all single type scores is 0.66, of all competing mixed-type scores is 0.77, and of Mv.RBM scores is 0.91. These demonstrate that (a) a proper handling of mixed-types is required, and (b) Mv.RBM is highly competitive against other mixed-type methods for outlier detection. Point (a) can also be strengthened by looking deeper: On average, the best competing mixed-type method (BMM) is better than the best single-type method (OCSVM). For point (b), the gaps between Mv.RBM and other methods are significant:

Table 3. Outlier detection F-score.

Dataset	Single type			Mixed-type			
	GMM	OCSVM	PPCA	BMM	ODMAD	GLM-t	Mv.RBM
KDD99-10	0.42	0.54	0.55	–	–	–	**0.71**
Australian Credit	0.74	0.84	0.38	**0.972**	0.942	–	0.90
German Credit	0.86	0.86	0.02	0.934	0.810	–	**0.95**
Heart	0.89	0.76	0.64	0.872	0.630	0.72	**0.94**
Thoracic Surgery	0.71	0.71	0.70	**0.939**	0.879	–	0.90
Auto MPG	**1.00**	**1.00**	0.67	0.625	0.575	0.64	**1.00**
Contraceptive	0.62	0.84	0.02	0.673	0.523	–	**0.91**
Average	*0.75*	*0.79*	*0.43*	*0.84*	*0.73*	*0.68*	*0.91*

On average, Mv.RBM is better than the best competing method – the BMM (mixed-type) – by 8.3 %, and better than the worst method – the PPCA (single type), by 111.6 %. On the largest dataset – the KDD99-10 – Mv.RBM exhibits a significant improvement of 29.1 % over the best single type method (PPCA).

5 Discussion

This paper has introduced a new method for mixed-type outlier detection based on an energy-based model known as Mixed-variate Restricted Boltzmann Machine (Mv.RBM). Mv.RBM avoids direct modeling of correlation between types by using binary latent variables, and in effect, model the correlation between each type and the binary type. We derive free-energy, which equals the negative log of density up-to a constant, and use it as the outlier score. Overall, the method is highly scalable. Our experiments on mixed-type datasets of various types and characteristics demonstrate that the proposed method is competitive against the well-known baselines designed for single types, and recent models designed for mixed-types. These results (a) support the hypothesis that in mixed-data, proper modeling of types should be in place for outlier detection, and (b) show Mv.RBM is a powerful density-based outlier detection method.

Mv.RBM opens several future directions. First, Mv.RBM transforms multiple types into a single type through its hidden posteriors. Existing single-type outlier detectors can be readily employed. Second, Mv.RBM can serve as a building block for deep architecture, such as Deep Belief Networks and Deep Boltzmann Machine. It would be interesting to see how deep networks perform in non-prediction settings such as outlier detection.

Acknowledgments. This work is partially supported by the Telstra-Deakin Centre of Excellence in Big Data and Machine Learning.

124   K. Do et al.

References

1. Aggarwal, C.C.: Outlier Analysis. Data Mining. Springer, Heidelberg (2015)
2. Angiulli, F., Pizzuti, C.: Fast outlier detection in high dimensional spaces. In: Elomaa, T., Mannila, H., Toivonen, H. (eds.) PKDD 2002. LNCS, vol. 2431, pp. 15–27. Springer, Heidelberg (2002). doi:10.1007/3-540-45681-3_2
3. Bengio, Y., Courville, A., Vincent, P.: Representation learning: A review and new perspectives. IEEE Trans. Pattern Anal. Mach. Intell. 35(8), 1798–1828 (2013)
4. Bouguessa, M.: A practical outlier detection approach for mixed-attribute data. Expert Syst. Appl. 42(22), 8637–8649 (2015)
5. Breunig, M.M., Kriegel, H.-P., Ng, R.T., Sander, J.: Lof: identifying density-based local outliers. In: ACM Sigmod Record, vol. 29, pp. 93–104. ACM (2000)
6. Chandola, V., Banerjee, A., Kumar, V.: Anomaly detection: A survey. ACM Comput. Surv. (CSUR) 41(3), 15 (2009)
7. De Leon, A.R., Chough, K.C.: Analysis of Mixed Data: Methods & Applications. CRC Press (2013)
8. Diehl, C.P., Hampshire, J.B.: Real-time object classification and novelty detection for collaborative video surveillance. In: Proceedings of the 2002 International Joint Conference on Neural Networks, 2002. IJCNN 2002, vol. 3, pp. 2620–2625. IEEE (2002)
9. Fiore, U., Palmieri, F., Castiglione, A., De Santis, A.: Network anomaly detection with the restricted Boltzmann machine. Neurocomputing 122, 13–23 (2013)
10. Ghoting, A., Otey, M.E., Parthasarathy, S.: Loaded: Link-based outlier and anomaly detection in evolving data sets. In: ICDM, pp. 387–390 (2004)
11. Hinton, G.E.: Training products of experts by minimizing contrastive divergence. Neural Comput. 14, 1771–1800 (2002)
12. Hinton, G.E., Salakhutdinov, R.R.: Reducing the dimensionality of data with neural networks. Science 313(5786), 504–507 (2006)
13. Kingma, D., Ba, J., Adam: A method for stochastic optimization. arXiv preprint (2014). arXiv:1412.6980
14. Konijn, R.M., Kowalczyk, W.: Finding fraud in health insurance data with two-layer outlier detection approach. In: Cuzzocrea, A., Dayal, U. (eds.) DaWaK 2011. LNCS, vol. 6862, pp. 394–405. Springer, Heidelberg (2011). doi:10.1007/978-3-642-23544-3_30
15. Koufakou, A., Georgiopoulos, M., Anagnostopoulos, G.C.: Detecting outliers in high-dimensional datasets with mixed attributes. In: DMIN, pp. 427–433. Citeseer (2008)
16. Kruegel, C., Vigna, G.: Anomaly detection of web-based attacks. In: Proceedings of the 10th ACM Conference on Computer and Communications Security, pp. 251–261. ACM (2003)
17. LeCun, Y., Bengio, Y., Hinton, G.: Deep learning. Nature 521(7553), 436–444 (2015)
18. Yen-Cheng, L., Chen, F., Wang, Y., Chang-Tien, L.: Discovering anomalies on mixed-type data using a generalized student-t based approach. IEEE Trans. Knowl. Data Eng. 28, 858–872 (2016). doi:10.1109/TKDE.2016.2583429
19. Lu, Y.-C., Chen, F., Wang, Y., Lu, C.-T.: Discovering anomalies on mixed-type data using a generalized student-t based approach (2016)
20. Manevitz, L.M., Yousef, M.: One-class SVMs for document classification. J. Mach. Learn. Res. 2, 139–154 (2001)

21. McLachlan, G.J., Basford, K.E.: Mixture models. inference, applications to clustering. Statistics: Textbooks and Monographs, New York: Dekker, 1988, 1 (1988)
22. Nguyen, T.D., Tran, T., Phung, D., Venkatesh, S.: Latent patient profile modelling and applications with mixed-variate restricted boltzmann machine. In: Pei, J., Tseng, V.S., Cao, L., Motoda, H., Xu, G. (eds.) PAKDD 2013. LNCS (LNAI), vol. 7818, pp. 123–135. Springer, Heidelberg (2013). doi:10.1007/978-3-642-37453-1_11
23. Nguyen, T.D., Tran, T., Phung, D., Venkatesh, S.: Learning sparse latent representation and distance metric for image retrieval. In: Proceedings of IEEE International Conference on Multimedia & Expo, California, USA, July 15–19 2013
24. Otey, M.E.: Srinivasan Parthasarathy, and Amol Ghoting. Fast lightweight outlier detection in mixed-attribute data. Techincal Report, OSU-CISRC-6/05-TR43 (2005)
25. Papadimitriou, S., Kitagawa, H., Gibbons, P.B., Faloutsos, C.: Loci: Fast outlier detection using the local correlation integral. In: 19th International Conference on Data Engineering, Proceedings, pp. 315–326. IEEE (2003)
26. Portnoy, L., Eskin, E., Stolfo, S.: Intrusion detection with unlabeled data using clustering. In: Proceedings of ACM CSS Workshop on Data Mining Applied to Security (DMSA-2001. Citeseer (2001)
27. Salakhutdinov, R., Hinton, G.: Semantic hashing. Int. J. Approximate Reasoning **50**(7), 969–978 (2009)
28. Serfling, R., Wang, S.: General foundations for studying masking and swamping robustness of outlier identifiers. Stat. Methodol. **20**, 79–90 (2014)
29. Tipping, M.E., Bishop, C.M.: Probabilistic principal component analysis. J. Royal Stat. Soc. Ser. B **61**(3), 611–622 (1999)
30. Tran, T., Phung, D., Venkatesh, S., Machines, T.B.: Learning from Multiple Inequalities. In: International Conference on Machine Learning (ICML), Atlanta, USA, June 16–21 2013
31. Tran, T., Phung, D.Q., Venkatesh, S.: Mixed-variate restricted Boltzmann machines. In: Proceedings of 3rd Asian Conference on Machine Learning (ACML), Taoyuan, Taiwan (2011)
32. Tran, T., Phung, D., Luo, W., Harvey, R., Berk, M., Venkatesh, S.: An integrated framework for suicide risk prediction. In: Proceedings of the 19th ACM SIGKDD International Conference on Knowledge Discovery and Data Mining, pp. 1410–1418. ACM (2013)
33. Zhang, K., Jin, H.: An effective pattern based outlier detection approach for mixed attribute data. In: Li, J. (ed.) AI 2010. LNCS (LNAI), vol. 6464, pp. 122–131. Springer, Heidelberg (2010). doi:10.1007/978-3-642-17432-2_13

Low-Rank Feature Reduction and Sample Selection for Multi-output Regression

Shichao Zhang[1,2]([✉]), Lifeng Yang[1,2], Yonggang Li[1,2], Yan Luo[1,2], and Xiaofeng Zhu[1,2]

[1] Guangxi Key Lab of Multi-source Information Mining and Security, Guangxi Normal University, Guilin 541004, Guangxi, People's Republic of China
zhangsc@mailbox.gxnu.edu.cn
[2] College of CS & IT, Guangxi Normal University, Guilin 541004, Guangxi, People's Republic of China

Abstract. There are always varieties of inherent relational structures in the observations, which is crucial to perform multi-output regression task for high-dimensional data. Therefore, this paper proposes a new multi-output regression method, simultaneously taking into account three kinds of relational structures, *i.e.*, the relationships between output and output, feature and output, sample and sample. Specially, the paper seeks the correlation of output variables by using a low-rank constraint, finds the correlation between features and outputs by imposing an $\ell_{2,1}$-norm regularization on coefficient matrix to conduct feature selection, and discovers the correlation of samples by designing the $\ell_{2,1}$-norm on the loss function to conduct sample selection. Furthermore, an effective iterative optimization algorithm is proposed to settle the convex objective function but not smooth problem. Finally, experimental results on many real datasets showed the proposed method outperforms all comparison algorithms in aspect of aCC and aRMSE.

Keywords: Multi-output regression · Feature reduction · Low-rank constraint · Sample selection · Orthogonal subspace learning

1 Introduction

Multi-output regression [4], aiming at learning a mapping from the feature variables to the multiple real-value output variables, is an extreme crucial research topic in the fields of data mining and statistics, and so on. In real life, multi-output regression is encountered frequently in lots of applications, such as electrical discharge machining domain predict gap control and flow control by utilizing mean values and deviations of many considered machining parameters [17]; stock analysis domain predict simultaneously many indicators of a stock via constructing a bridge from the related economic variables and already known prices [21]; the airline ticket price system [22] predicts the next day price of multiple airlines according to the different records of flights, including the past prices, the number of stations will pass by, and departure date; and so on.

© Springer International Publishing AG 2016
J. Li et al. (Eds.): ADMA 2016, LNAI 10086, pp. 126–141, 2016.
DOI: 10.1007/978-3-319-49586-6_9

In realistic application, due to the advanced computer science and storage technology, the data utilized for achieving the multi-output regression always own high-dimensional features and have a big scale of quantity [28, 32]. However, dealing with these high dimensional data often encounter the problem of curse of dimensionality, and the interference of noise and outliers [25, 33, 42]. Therefore, These problems always result in a big challenge of utilizing a large number of high-dimensional data to perform multi-output regression task effectively. To this end, we desire to conduct feature selection to eliminate redundant features and noise, as well as conduct sample selection to remove the outliers. In the meantime, we search for the inherent correlation structures in these data to improve the multi-output regression predictive capacity.

Traditional linear regression always can effectively settle the single-output regression problem, but it usually deal with the multiple-output regression via performing the single-regression to predict each output variable separately. Therefore, the traditional linear regression neglects the advantage of utilizing likely correlation among multiple outputs [20, 27].

In order to uncover the relationship among multiple outputs for improving the multi-output regression model, Aderson et al. [1], proposed an effective method of using a low-rank constraint on the rank of coefficient matrix. And researchers [2] proposed a method leveraging the trace-norm regularizer to mine the low-rank structure existing in multiple outputs. However, these methods did not explicitly determine the rank of the coefficient matrix. Thus, Izenwan, et al. [16], proposed an reduced-rank regression (RRR) method, which implied that the coefficient matrix can be represented by a product of two matrices owning a selected rank, thus select explicitly a rank of the coefficient matrix and can reduce the number of estimate parameters to improve the efficiency of estimation. However, as these high-dimensional data own a large number of features, many redundant features and noise in data might have unrelated information for prediction and even may lead to adding more computational cost. Then, this method simply takes advantage of the low-rank constraint to seek the correlation of multi-outputs, but without taking into account reducing of the quantity of feature and the removing of unnecessary noise.

In recent years, sparse learning has been widely researched in many fields and always can gain a rather trustful effectiveness [26, 34, 36, 37], because it has the strong power of making matrix sparsity to capture the important features and eliminate redundant information in high-dimensional data. Thus, some researchers [24] added the effect of sparse learning on the traditional linear regression to seek the correlation of features, i.e., conducting feature selection. Although these methods dealt well with the removing of noise and redundant features, did not catch the correlation of multiple outputs. Therefore, some researchers [5] proposed a new method combining the process of sparse learning with a low-rank constraint, which not only can take advantage of low-rank constraint to find the correlation of multiple outputs, but also find out the correlation between features and outputs to improve the predictive capacity of regression model. However, many outliers or corrupted samples are usually involved in realistic dataset [40], and may have a negative effect on the training of multi-output regression model.

To this end, we pursue to get a better multi-output regression by considering fully three kinds of inherent correlation structures in high-dimensional data, interpretation as Fig. 1. Then we propose a new method called Low-rank Feature Reduction and Sample Selection for multiple-output, (shorted for LFR_SS). The rationale of proposed multi-output regression model is simultaneously performing feature selection, sample selection and low-rank regression in a framework. Specially, the correlation structure of multiple output variables can be searched by employing the low-rank constraint on the rank of coefficient matrix. To remove the useless features, we also conduct feature selection by imposing an $\ell_{2,1}$-norm regularization to penalize all coefficients in each row of coefficient matrix. Furthermore, in order to catch the correlation of samples and remove the outliers in dataset, we also combine the $\ell_{2,1}$-norm with the loss function to conduct sample selection.

The major contributions of our proposed LFR_SS method are described as below:

- This work imposes an $\ell_{2,1}$-norm regularization on the coefficient matrix to get row-sparsity of the coefficient matrix. In this way, the multi-output regression model can effectively make use of the correlation structure of feature variables to eliminate the redundant features and noise [35,39] to improve the predictive capacity of multi-output regression, as well as can address the curse of dimensionality problem.
- In the proposed model, we first combine the $\ell_{2,1}$-norm regularization with loss function to find the correlation of samples. Therefore, we can remove effectively the interference of the outliers and utilize the samples which have more information to train the multi-output regression model. However, present existing regression methods always neglect this idea.
- This proposed model takes advantage of the low-rank constraint to explicitly determine the rank of coefficient matrix. Furthermore, it can effectively search for the correlation structure of multiple outputs, and reduce the number of parameters to improve the predictive accuracy of the multi-output regression model.

The remainder of this paper is organized as follow: we briefly summarize the preliminary in Sect. 2. And we introduce the innovative multi-output regression model for high-dimensional data, follow by the algorithm to get the optimal solution of objective function and then the proving of the convergence in Sect. 3. Extensive experiments are performed to demonstrate effectiveness of the multi-output regression in Sect. 4. Finally, Sect. 5 concludes this paper.

2   Preliminary

In this section, we briefly summarize the notations used in this paper in Table 1. we denote the Frobenius norm, ℓ_1-norm, ℓ_2-norm, $\ell_{2,1}$-norm of a matrix $\mathbf{X}$ as

Table 1. Notations used in this paper.

Notation	Description
Matrix	Boldface uppercase letter
Vector	Boldface lowercase letter
Scalar	Normal italic letter
$\mathbf{x}^i$	i-th row of a matrix
$\mathbf{x}_j$	j-th column of a matrix
$\mathbf{X}^T$	Transpose operator of a matrix
$tr(\mathbf{X})$	Trace operator of a matrix
$\mathbf{X}^{-1}$	Inverse of a matrix

$$\|\mathbf{X}\|_F = \sqrt{\sum_i \|\mathbf{x}^i\|_2^2} = \sqrt{\sum_j \|\mathbf{x}_j\|_2^2}, \|\mathbf{X}\|_1 = \sum_i \sum_j |x_{ij}|, \|\mathbf{X}\|_2 = \sum_i \sum_j |x_{ij}|^2,$$

$$\|\mathbf{X}\|_{2,1} = \sum_i \|\mathbf{X}_i\|_2^2 = \sum_i \sqrt{\sum_j x_{ij}^2}, \text{ respectively.}$$

3 Method

In this section, we utilize these three kinds of correlation structures inhered in data, as interpretation in Fig. 1, to construct our designed multi-output regression model. Then, we propose an iterative algorithm to optimize the objective function to get the optimal solution. At last, we demonstrate the convergence of objective function.

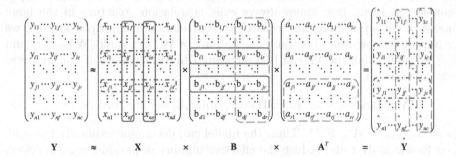

Fig. 1. An illustration of the three kinds of inhered information in the data. The red dotted rectangles, the blue solid rectangles, and the green dash rectangles, imply the relationship of sample-sample, the relationship of feature-output, the relationship of output-output, respectively. Here each row and column denotes a sample and a feature respectively in X. (Color figure online)

3.1   LFR_SS Algorithm

Let $\mathbf{D} = \{(\mathbf{X}^1, \mathbf{Y}^1), ..., (\mathbf{X}^n, \mathbf{Y}^n)\} = (\mathbf{X}, \mathbf{Y}) \in R^{n \times (d+c)}$ denotes the training dataset, where $(\mathbf{X}^i, \mathbf{Y}^i)(i = 1, ..., n)$ denotes a training sample, n is the number of samples, $\mathbf{X}^i \in R^{1 \times d}$ and $\mathbf{Y}^i \in R^{1 \times c}$ denote d dimensional feature variables and c outputs of the i-th sample, respectively. Let $\mathbf{X} = [\mathbf{X}^1, ..., \mathbf{X}^i, ..., \mathbf{X}^n]^T \in R^{n \times d}$ and $\mathbf{Y} = [\mathbf{Y}^1, ..., \mathbf{Y}^i, ..., \mathbf{Y}^n]^T \in R^{n \times c}$ be the feature matrix and output matrix, respectively. The multi-output regression aims at finding a mapping to predict the outputs $\mathbf{Y}$ from the feature matrix $\mathbf{X}$ in dataset $\mathbf{D}$. As known, the linear regression model tries to search a linear relation, that is,

$$\mathbf{Y}_{ij} = \mathbf{X}^i \mathbf{W}_j + e_{ij}, \qquad (i = 1, ..., n \text{ and } j = 1, ..., c)$$

Where $\mathbf{W}$ and e_{ij} denote the regression coefficient matrix and the error term respectively. The above regression model can be described by matrix notation as below, for there exist n samples in dataset,

$$\mathbf{Y} = \mathbf{XW} + \mathbf{E}$$

Where $\mathbf{E} \in R^{n \times c}$ denotes the error matrix. And we desire the predictive multi-output $\tilde{\mathbf{Y}} = \mathbf{XW}$ getting closest to the ground truth outputs $\mathbf{Y}$, $i.e.$, $\mathbf{E}$ gets minimal value. The least square loss function is useful to address this problem, $i.e.$,

$$\min_{\mathbf{W}} \quad ||\mathbf{Y} - \mathbf{XW}||_F^2 \tag{1}$$

It can be noted from the solution of Eq. (1), $i.e.$, $\mathbf{W} = (\mathbf{X}^T\mathbf{X})^{-1}\mathbf{X}^T\mathbf{Y}$, that this method just transfer a multi-output regression problem to a single-output regression problem. But, there always exist correlation structure in the large amount of outputs. Therefore, to catch the correlation of multiple outputs, we can impose the low-rank constraint on the rank of $\mathbf{W}$, $i.e.$, rank $(\mathbf{W}) = r \leq \min (d, c)$. In additional, the $\mathbf{W}$ can be represented by the product of two matrices with the rank r, $i.e.$,

$$\mathbf{W} = \mathbf{BA}^T \quad s.t., \quad rank(\mathbf{B}) = r, rank(\mathbf{A}) = r, r \leq min(d, c) \tag{2}$$

where $\mathbf{B} \in R^{d \times r}$, $\mathbf{A} \in R^{c \times r}$. Thus, the model can determine explicitly the rank of coefficient matrix and reduce the effective number of parameters. Therefore, the Eq. (1) can be transformed to the following problem,

$$\min_{\mathbf{A}, \mathbf{B}} \quad ||\mathbf{Y} - \mathbf{XBA}^T||_F^2 \quad s.t., \quad rank(\mathbf{B}) = r, rank(\mathbf{A}) = r, r \leq min(d, c) \tag{3}$$

To reduce the features of data, feature selection is always utilized to select the discriminating features from high-dimensional data. Consequently, the computational cost of various analysis is reduced for high-dimensional data, and simultaneously the noise in data are eliminated. Therefore, feature selection is usually used to obtain a better model performance in practice [8,13,14,38].

Recent years, sparse-based feature selection approaches [29,30,43] attract increasing attentions of researchers. Such methods can always find out the correlation of features and indicate that high-dimensional data can be represented by several low dimensional subspaces. Thus, researchers [9,11] tried to imposing the $\ell_{2,1}$-norm regularization on the coefficient matrix to conduct feature selection effectively, as a result, such sparse-based feature selection methods always obtain a reliable performance of regression model. The $\ell_{2,1}$-norm is defined as below

$$||\mathbf{W}||_{2,1} = \sum_i \left(\sum_j |w_{ij}|^2 \right)^{1/2} \tag{4}$$

where $\mathbf{W} \in R^{d \times c}$ denotes a matrix. The $\ell_{2,1}$-norm can effectively shrink some rows of the matrix to be zeros, and the nonzero rows correspond to the selected features, so that effectively complete the task of feature selection.

Eliminating an useless feature corresponds to setting an entire row of the matrix as zeros. Therefore, we design an $\ell_{2,1}$-norm regularization on the coefficient matrix $\mathbf{B}$ to penalty its row to perform feature selection and make use of an orthogonal constraint, *i.e.*, $\mathbf{A}^T \mathbf{A} = \mathbf{I}$, to keep the low-rank structure of data.

$$\min_{\mathbf{A},\mathbf{B}} \ ||\mathbf{Y} - \mathbf{XBA}^T||_F^2 + \alpha||\mathbf{B}||_{2,1} \quad s.t., \ \ \mathbf{A}^T\mathbf{A} = \mathbf{I} \tag{5}$$

where α is a positive tuning parameter. Moreover, the orthogonal constraint $\mathbf{A}^T \mathbf{A} = \mathbf{I}$ is used to conduct a subspace learning to enhance the performance of the feature selection process.

However, there are many outlier samples in big scale of real high-dimensional data, which might have an bad effect on the training process of regression model. Thus it is very necessary to remove the samples with little information in dataset. To this end, imposing the $\ell_{2,1}$-norm on loss function to effectively take advantage of the correlation of samples [18]. Final, we obtain the following optimization objective function,

$$\min_{\mathbf{A},\mathbf{B}} \ ||\mathbf{Y} - \mathbf{XBA}^T||_{2,1} + \alpha||\mathbf{B}||_{2,1} \quad s.t., \ \ \mathbf{A}^T\mathbf{A} = \mathbf{I} \tag{6}$$

We impose the $\ell_{2,1}$-norm on the matrix $\tilde{\mathbf{Y}} = \mathbf{Y} - \mathbf{XBA}^T$ to obtain its row-sparse and then complete the process of sample selection, *i.e.*, all element values in the corresponding row of the matrix $\tilde{\mathbf{Y}}$ are 0, if the sample is useless to train the regression model and should be removed.

According to above discussion, we can conclude that the objective function considers simultaneously the advantage of low-rank regression, feature selection, sample selection and subspace learning in a framework to improve the predictive performance of model. Specifically, we utilize the low rank constraint, *i.e.*, rank $(\mathbf{B}) = r$, rank $(\mathbf{A}) = r, r \leq \min(d,c)$, to limit the rank of $\mathbf{A}$ and $\mathbf{B}$, and directly leading to the rank of coefficient matrix can be explicitly selected. Moreover, the low-rank constraint is also leveraged to search the correlation between features and multi-outputs to construct a better mapping from feature matrix to multiple outputs. And an $\ell_{2,1}$-norm regularization on the matrix $\mathbf{B}$ is designed to select the features with high correlation by considering the correlation of features, and remove the redundancy or irrelevant features. Furthermore, the multi-output regression model can find out the correlation of samples by combining the $\ell_{2,1}$-norm regularization with the loss function, and remove the outlier samples to train a robust multi-output regression model. Therefore, the proposed model not only can take advantage of the correlation between features and outputs by low-rank constraint, but also can select the representative features and samples by the sparse regression model. Additional, the distribution of low-rank structures may be different from each other after conducting feature selection, thus the subspace learning is used to keep the structure of multi-output by rotating the output matrix.

3.2 Optimization

The objective function, *i.e.*, Eq. (6), is a convex but non-smooth function for involving in the $\ell_{2,1}$-norm regularization. Consequently, to obtain the global optimize solution of the objective function, we propose an effectively iterative algorithm. In detail, we iteratively conduct the below two steps until the predefined conditions is satisfied.

(1) Fixed $\mathbf{A}$ and update $\mathbf{B}$

As the constraint $\mathbf{A}^T\mathbf{A} = \mathbf{I}$, there exist an orthogonal matrix $(\mathbf{A}, \mathbf{A}^\perp)$, and $\mathbf{A}^\perp$ is a matrix with orthogonal column. Then the optimization of objective function becomes the following problem,

$$\min_{\mathbf{B}} ||\mathbf{Y} - \mathbf{XBA}^T||_{2,1} + \alpha||\mathbf{B}||_{2,1}$$

$$= \min_{\mathbf{B}} || \left(\mathbf{Y} - \mathbf{XBA}^T\right) \left(\mathbf{A}, \mathbf{A}^\perp\right) ||_{2,1} + \alpha||\mathbf{B}||_{2,1}$$

$$= \min_{\mathbf{B}} ||\mathbf{YA} - \mathbf{XB}||_{2,1} + ||\mathbf{YA}^\perp||_{2,1} + \alpha||\mathbf{B}||_{2,1} \qquad (7)$$

After we fix the $\mathbf{A}$, the above optimization problem can be reduced to

$$\min_{\mathbf{B}} \ \ ||\mathbf{YA} - \mathbf{XB}||_{2,1} + \alpha||\mathbf{B}||_{2,1} \qquad (8)$$

Note that, Eq. (8) can be easily rewritten as below

$$\min_{\mathbf{B}} \ \ tr\left[(\mathbf{YA} - \mathbf{XB}^T)\mathbf{N}\left(\mathbf{YA} - \mathbf{XB}\right)\right] + \alpha tr\left(\mathbf{B}^T\mathbf{QB}\right) \qquad (9)$$

where $\mathbf{N} \in R^{n \times n}$ and $\mathbf{Q} \in R^{d \times d}$ are diagonal matrices, and their diagonal elements are $\mathbf{N}_{ii} = \frac{1}{2||(\mathbf{YA}-\mathbf{XB})^i||_2}$ $(i = 1, ..., n)$ and $\mathbf{Q}_{jj} = \frac{1}{2||\mathbf{B}^j||_2}$ $(j = 1, ..., d)$, respectively. By setting the derivative of Eq. (9) *w.r.t.* $\mathbf{B}$ to be zero, then we can have

$$\mathbf{B} = \left(\mathbf{X}^T\mathbf{NX} + \lambda\mathbf{Q}\right)^{-1}\mathbf{X}^T\mathbf{NYA} \qquad (10)$$

(2) Fixed **B** and update **A**

When the matrix **B** is fixed, the optimization problem of Eq. (6) can be reduced to as below

$$\min_{\mathbf{A}} \ \|\mathbf{Y} - \tilde{\mathbf{X}}\mathbf{A}^T\|_{2,1}, \quad s.t., \ \mathbf{A}^T\mathbf{A} = \mathbf{I} \tag{11}$$

where $\tilde{\mathbf{X}} = \mathbf{XB} \in R^{n \times r}$. It can be known that the Eq. (11) is an orthogonal Procrustes problem [15], so that the optimal solution of **A** is $\mathbf{UV}^T$, where $\mathbf{U} \in R^{c \times r}$ and $\mathbf{V} \in R^{r \times r}$ are obtained from the singular value decomposition of $\mathbf{Y}^T\tilde{\mathbf{X}} = \mathbf{UDV}^T$, and $\mathbf{D} \in R^{r \times r}$ is a diagonal matrix.

The above discussion leads to the below Algorithm 1 [19,31].

Algorithm 1. The pseudo code of solving objective function

    Input: X $\in R^{n \times d}$, **Y** $\in R^{n \times c}$, α, r;
    Output: A $\in R^{c \times r}$, **B** $\in R^{d \times r}$;
1 Initialize the iterative number t=0 ;
2 Initialize **A**(0) as a random diagonal matrix;
3 **while** *the value of objective function not converged* **do**
4   Update **B**$(t + 1)$ via Eq.(10);
5   Update **A**$(t + 1)$ via Eq.(11);
6   Calculate the diagonal matrix **N**$(t + 1)$ as $\mathbf{N}_{ii} = \frac{1}{2\|(\mathbf{YA}-\mathbf{XB})^i\|_2}$ $(i = 1, ..., n)$;
7   Calculate the diagonal matrix **Q**$(t + 1)$ as $\mathbf{Q}_{jj} = \frac{1}{2\|\mathbf{B}^j\|_2}$ $(j = 1, ..., d)$;
8   Calculate the value of objective function;
9      $t = t+1$;
10 **end**

3.3 Proving of the Convergence

It can be demonstrated that the value of objective function monotonically decrease in each iteration. Note that, the objective function can be reduced to

$$\min_{\mathbf{B}} \ tr\left[(\mathbf{YA} - \mathbf{XB})^T\mathbf{N}(\mathbf{YA} - \mathbf{XB})\right] + \alpha tr(\mathbf{B}^T\mathbf{QB})$$

Therefore, we have

$$tr\left[(\mathbf{YA}_{(t+1)} - \mathbf{XB}_{(t+1)})^T\mathbf{N}_{(t+1)}(\mathbf{YA}_{(t+1)} - \mathbf{XB}_{(t+1)})\right] + \alpha tr(\mathbf{B}_{(t+1)}^T\mathbf{Q}_{(t+1)}\mathbf{B}_{(t+1)})$$
$$\leq tr\left[(\mathbf{YA}_t - \mathbf{XB}_t)^T\mathbf{N}_t(\mathbf{YA}_t - \mathbf{XB}_t)\right] + \alpha tr(\mathbf{B}_t^T\mathbf{Q}_t\mathbf{B}_t)$$

$$\Rightarrow \sum_{i=1}^{n} \frac{||\mathbf{Y}^i \mathbf{A}_{(t+1)} - \mathbf{X}^i \mathbf{B}_{(t+1)}||_2^2}{2||\mathbf{Y}^i \mathbf{A}_t - \mathbf{X}^i \mathbf{B}_t||_2} + \sum_{i=1}^{d} \frac{||\mathbf{B}_{(t+1)}^i||_2^2}{2||\mathbf{B}_t^i||_2} \leq \sum_{i=1}^{n} \frac{||\mathbf{Y}^i \mathbf{A}_t - \mathbf{X}^i \mathbf{B}_t||_2^2}{2||\mathbf{Y}^i \mathbf{A}_t - \mathbf{X}^i \mathbf{B}_t||_2} + \sum_{i=1}^{d} \frac{||\mathbf{B}_t^i||_2^2}{2||\mathbf{B}_t^i||_2}$$

$$\Rightarrow \sum_{i=1}^{n} ||\mathbf{Y}^i \mathbf{A}_{(t+1)} - \mathbf{X}^i \mathbf{B}_{(t+1)}||_2 - \sum_{i=1}^{n} ||\mathbf{Y}^i \mathbf{A}_{(t+1)} - \mathbf{X}^i \mathbf{B}_{(t+1)}||_2 + \alpha \sum_{i=1}^{d} ||\mathbf{B}_{(t+1)}^i||_2$$

$$+ \sum_{i=1}^{n} \frac{||\mathbf{Y}^i \mathbf{A}_{(t+1)} - \mathbf{X}^i \mathbf{B}_{(t+1)}||_2^2}{2||\mathbf{Y}^i \mathbf{A}_t - \mathbf{X}^i \mathbf{B}_t||_2} - \alpha \sum_{i=1}^{d} ||\mathbf{B}_{(t+1)}^i||_2 + \alpha \sum_{i=1}^{d} \frac{||\mathbf{B}_{(t+1)}^i||_2^2}{2||\mathbf{B}_t^i||_2}$$

$$\leq \sum_{i=1}^{n} ||\mathbf{Y}^i \mathbf{A}_t - \mathbf{X}^i \mathbf{B}_t||_2 - \sum_{i=1}^{n} ||\mathbf{Y}^i \mathbf{A}_t - \mathbf{X}^i \mathbf{B}_t||_2 + \sum_{i=1}^{n} \frac{||\mathbf{Y}^i \mathbf{A}_t - \mathbf{X}^i \mathbf{B}_t||_2^2}{2||\mathbf{Y}^i \mathbf{A}_t - \mathbf{X}^i \mathbf{B}_t||_2}$$

$$+ \alpha \sum_{i=1}^{d} ||\mathbf{B}_t^i||_2 - \alpha \sum_{i=1}^{d} ||\mathbf{D}_t^i||_2 + \alpha \sum_{i=1}^{d} \frac{||\mathbf{B}_t^i||_2^2}{2||\mathbf{B}_t^i||_2}$$

$$\Rightarrow \sum_{i=1}^{n} ||\mathbf{Y}^i \mathbf{A}_{(t+1)} - \mathbf{X}^i \mathbf{B}_{(t+1)}||_2 + \alpha \sum_{i=1}^{d} ||\mathbf{B}_{(t+1)}^i||_2$$

$$- \left(\sum_{i=1}^{n} ||\mathbf{Y}^i \mathbf{A}_{(t+1)} - \mathbf{X}^i \mathbf{B}_{(t+1)}||_2 - \sum_{i=1}^{n} \frac{||\mathbf{Y}^i \mathbf{A}_{(t+1)} - \mathbf{X}^i \mathbf{B}_{(t+1)}||_2^2}{2||\mathbf{Y}^i \mathbf{A}_t - \mathbf{X}^i \mathbf{B}_t||_2} \right)$$

$$- \alpha \left(\sum_{i=1}^{d} ||\mathbf{B}_{(t+1)}^i||_2 - \sum_{i=1}^{d} \frac{||\mathbf{B}_{(t+1)}^i||_2^2}{2||\mathbf{B}_t^i||_2} \right)$$

$$\leq \sum_{i=1}^{n} ||\mathbf{Y}^i \mathbf{A}_t - \mathbf{X}^i \mathbf{B}_t||_2 + \alpha \sum_{i=1}^{d} ||\mathbf{B}_t^i||_2$$

$$- \left(\sum_{i=1}^{n} ||\mathbf{Y}^i \mathbf{A}_t - \mathbf{X}^i \mathbf{B}_t||_2 - \sum_{i=1}^{n} \frac{||\mathbf{Y}^i \mathbf{A}_t - \mathbf{X}^i \mathbf{B}_t||_2^2}{2||\mathbf{Y}^i \mathbf{A}_t - \mathbf{X}^i \mathbf{B}_t||_2} \right)$$

$$- \alpha \left(\sum_{i=1}^{d} ||\mathbf{B}_t^i||_2 - \sum_{i=1}^{d} \frac{||\mathbf{B}_t^i||_2^2}{2||\mathbf{B}_t^i||_2} \right)$$

The literature [7] had showed that for any nonzero vectors have

$$\sum_{i=1} ||\mathbf{B}_{(t+1)}^i||_2 - \sum_{i=1} \frac{||\mathbf{B}_{(t+1)}^i||_2^2}{2||\mathbf{B}_{(t+1)}^i||_2} \leq \sum_{i=1} ||\mathbf{B}_t^i||_2 - \sum_{i=1} \frac{||\mathbf{B}_t^i||_2^2}{2||\mathbf{B}_t^i||_2}$$

$$\sum_{i=1} ||\mathbf{Y}^i \mathbf{A}_{(t+1)} - \mathbf{X}^i \mathbf{B}_{(t+1)}||_2 - \sum_{i=1} \frac{||\mathbf{Y}^i \mathbf{A}_{(t+1)} - \mathbf{X}^i \mathbf{B}_{(t+1)}||_2^2}{2||\mathbf{Y}^i \mathbf{A}_{(t+1)} - \mathbf{X}^i \mathbf{B}_{(t+1)}||_2} \leq$$

$$\sum_{i=1} ||\mathbf{Y}^i \mathbf{A}_t - \mathbf{X}^i \mathbf{B}_t||_2 - \sum_{i=1} \frac{||\mathbf{Y}^i \mathbf{A}_t - \mathbf{X}^i \mathbf{B}_t||_2^2}{2||\mathbf{Y}^i \mathbf{A}_t - \mathbf{X}^i \mathbf{B}_t||_2}$$

Therefore, it can be known that

$$\sum_{i=1}^{n} ||\mathbf{Y}^i \mathbf{A}_{(t+1)} - \mathbf{X}^i \mathbf{B}_{(t+1)}||_2 + \alpha \sum_{i=1}^{d} ||\mathbf{B}_{(t+1)}^i||_2 \leq$$

$$\sum_{i=1}^{n} ||\mathbf{Y}^i \mathbf{A}_t - \mathbf{X}^i \mathbf{B}_t||_2 + \alpha \sum_{i=1}^{d} ||\mathbf{B}_t^i||_2$$

According to above discussion, it can be demonstrated that the value of objective function decreases in each iteration. Meantime, the objective function is a convex function, which indicates that it will converge to the global optimum solution [28,41].

4   Experiments

In this section, we will analyze the performance of the proposed LFR_SS algorithm with the state-of-the-art algorithms on extensive multi-output datasets. Firstly, we briefly introduce the datasets used in our experiment. Then, we summarize the comparison algorithms and the experimental settings. Finally, we analyze the experimental results and get a conclusion.

4.1   Datasets and Comparison Algorithms

To evaluate the effectiveness of our proposed LFR_SS method, extensive experiments are conducted on real multi-output datasets. EDM [17] dataset predicts two output variables via using 16 feature variables representing mean values and deviations of the observed quantities of the considered machining parameters. ATP7d [22] dataset regards the details of flights as input variables and the minimum prices of next 7 days for 6 flight preferences are output variables. In dataset OES10 [22], the feature variables are employment types, and 16 output targets are a subset of categories above the 50 % threshold. SF1 [3] datasets predicts three potential types of Solar Flare from the ten feature variables describing active regions on the sun. WQ [12] dataset infers 16 chemicals from 14 biological parameters of river water quality. SCM20D [23] includes 16 regression targets, each target corresponding to the mean price of 20-days in the future. The datasets used in our experiments are briefly summarized in Table 2.

Table 2. Datasets used in our experiment. Where n, d, and c denote the number of samples, features and outputs, respectively.

Data sets	Samples (n)	Features (d)	Outputs (c)
EDM	154	16	2
ATP7d	296	411	6
SF1	323	10	3
WQ	1060	16	14
OES10	403	298	16
SCM20D	1503	61	16

We compared the performance of the LFR_SS with the comparison algorithms. Firstly, we performed the tasks of multi-output regression by using the algorithms which only has the process of sparse learning but without the low-rank constraint, *i.e.*, SMART algorithm, LSG21 algorithm and CSFS algorithm. Secondly, we considered the performance of the algorithms only has a low-rank regression but without sparse learning, *i.e.*, RRR algorithm. Finally, we compared the proposed method with the algorithms own both the low-rank constraint and sparse learning, *i.e.*, SLRR algorithm, but different from our proposed method as they do not perform the sample selection. The representative comparison methods are briefly summarized as below.

- SMART [24] algorithm includes both $\ell_{2,1}$-norm and ℓ_1-norm regularizer for feature selection, however without imposing the low-rank constraint on the rank of coefficient matrix.
- LSG21 [6] algorithm has the process of sparse learning via imposing the $\ell_{2,1}$-norm on coefficient matrix, but does not adopt the low-rank constraint.
- CSFS [10] algorithm conducts the feature selection via the $\ell_{2,1}$-norm regularizer, but neglects the low-rank structure of high-dimensional data.
- RRR [16] algorithm is a original linear regression modal, but different from traditional linear regression, because RRR method has a low-rank constraint on the rank of coefficient matrix.
- SLRR [5] algorithm not only uses a low-rank constraint to catch the low-rank structure of data, but also utilizes the $\ell_{2,1}$-norm on coefficient matrix for feature selection.

4.2   Experimental Settings

According to standard 10-fold cross validation. Firstly, we randomly parted each dataset into 10 groups. Then, we selected one group for testing and the others for training, repeat the all steps ten times to avoid the possible bias. Finally, we reported the experimental results of all methods. For the model selection, we set the tuning parameter $\alpha \in \{10^{-5}, ..., 10^5\}$, the rank of coefficient matrix $r \in \{1, ..., \min(d, c)\}$, and $(c, g) \in \{10^{-5}, ..., 10^5\}$ in the libSVM toolbox.

Two classical metrics are used to evaluate the performance of different methods [4], *i.e.*, average Correlation Coefficient (short for aCC) and average Root Mean Square Error (short for aRMSE). Specially, the aCC is often utilized to reflect the correlation between predicted outputs and the ground truth outputs, *i.e.*, the bigger aCC is, the predicted outputs are more closed to its corresponding ground truth outputs, and the model achieves more faithful results. The aRMSE can effectively reflect the stability of a algorithm, the smaller aRMSE is, the stability of the algorithm is better. Let $\mathbf{Y}_i$ and $\widehat{\mathbf{Y}}_i$ be the truth outputs and the predicted outputs, respectively. $\bar{\mathbf{Y}}_i$ and $\widetilde{\mathbf{Y}}_i$ are the means of truth outputs and the means of predicted outputs, respectively. And *ntest* means the number of testing samples. Their definition [4] as below

$$aCC = \frac{1}{d}\sum_{i=1}^{d} \frac{\sum_{j=1}^{ntest}\left(\mathbf{Y}_i^j - \bar{\mathbf{Y}}_i\right)\left(\widehat{\mathbf{Y}}_i^j - \widetilde{\mathbf{Y}}_i\right)}{\sqrt{\sum_{j=1}^{ntest}\left(\mathbf{Y}_i^j - \bar{\mathbf{Y}}_i\right)^2 \sum_{j=1}^{ntest}\left(\widehat{\mathbf{Y}}_i^j - \widetilde{\mathbf{Y}}_i\right)^2}} \tag{12}$$

$$aRMSE = \frac{1}{d} \sum_{i=1}^{d} \sqrt{\frac{\left(\mathbf{Y}_i^j - \widehat{\mathbf{Y}}_i^j\right)^2}{ntest}} \tag{13}$$

4.3   Regression Results

In our experiments, we set all the values of parameters in comparing algorithms as the experimental setting description in their papers. According to the framework of 10-fold cross validation, we will get a result on each run, then calculate average result in the whole ten runs. As a result, we summarize the performances of all comparing methods in Figs. 2 and 3, in terms of aCC and aRMSE respectively.

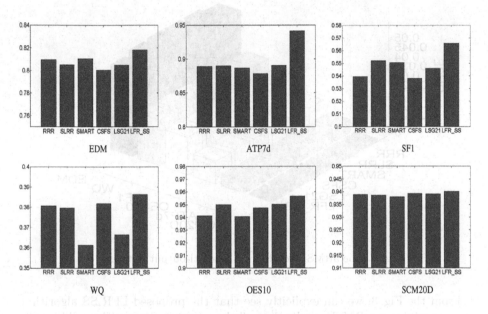

Fig. 2. The aCC results on six multi-output datasets

As can be seen from Fig. 2, the proposed LFR_SS method outperformed all the comparing methods in terms of the aCC. Note that, the RRR algorithm sometimes obtained smaller aCC results than other comparison algorithms in our experiment. This may be attributed to the RRR algorithm can make use of the low-rank constraint to find the inherent low-rank structure in data, but can not get the sparsity of coefficient matrix. It is hard to perform the task of multiple-output regression for high-dimensional data, because there are Always noise in data.

The SLRR algorithm could get better aCC results than the other comparison algorithms. This method has the low-rank constraint and the process of feature selection via leveraging a regularization term on coefficient matrix, *i.e.*, imposing the $\ell_{2,1}$-norm on coefficient matrix to result in row-sparse of the matrix.

While the SMART algorithm, CSFS algorithm and LSG21 algorithm always obtained better aCC results than the RRR algorithm, but sometimes had worse aCC results than the LRRR algorithm and the SLRR algorithm. The reason of which might because these three algorithms all owned the process of feature selection, but without low-rank constraint to utilize the correlation of multiple outputs for the regression model.

Moreover, we can know the aRMSE results of all the algorithms in our experiment from Fig. 3.

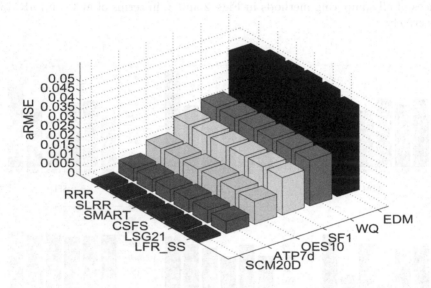

Fig. 3. The aRMSE results on six multi-output datasets

From the Fig. 3, we can explicitly see that the proposed LFR_SS algorithm achieved minimum aRMSE results than all the comparison algorithms. This validates that our proposed LFR_SS algorithm has the best stability when comparing to all the comparison algorithms.

According to the above analysis, we can make some conclusions that (1) the LFR_SS obtains the best results, which demonstrates that our proposed method is effective for conducting multi-output regression. The reasons of this result is because the LFR_SS algorithm performs the multi-output regression by fully utilizing the feature selection to eliminate the noise and the redundant features in data, imposing the low-rank constraint on the rank of coefficient matrix to exploit the correlation of multiple outputs, and combining the $\ell_{2,1}$-norm with loss function to select the important samples and remove the outliers to train the multi-output regression model. (2) the SLRR algorithm always can achieve a better result when comparing to the other comparison algorithms. This may illustrate that the algorithms, which own the low-rank constraint on the rank of coefficient matrix and the process of feature selection, have the advantage of fully using the inhered correlation of data. (3) The RRR method sometimes has a worse result than the SMART algorithm, the CSFS algorithm and the LSG21 algorithm. which may imply that either the algorithms only owning the low-rank constraint for seeking the correlation among multiple outputs, or the algorithms only conducting the feature selection without the low-rank constraint, just consider one aspect of the correlation structure inhered in data. The aCC results effectively validates that the proposed method can lead to the best faithful results over all the comparison methods for the task of multi-output regression.

5   Conclusion

To address the issue of multi-output regression for big scale of high-dimensional data. In this paper, we propose a new method, $i.e.$, LFR_SS algorithm by taking into account the correlation of outputs, feature-output, and samples. Firstly, this method utilizes a low-rank constraint on the rank of the coefficient matrix to obtain the correlation structure of outputs. Then, it combines the $\ell_{2,1}$-norm with the loss function to ensure remove the outliers to learn the multi-output regression model. Furthermore, it imposes an $\ell_{2,1}$-norm on the coefficient matrix to conduct feature selection and eliminate the redundant features. To ensure the multi-output variables share the same low-rank regression model, an orthogonal subspace learning is exploited to keep the low-rank structure of data by rotating the results of feature selection. Consequence, the extensive experimental results on datasets showed the LFR_SS algorithm is effective to perform the task of multi-output regression for high-dimensional data.

Acknowledgement. This work was supported in part by the China "1000-Plan" National Distinguished Professorship; the National Natural Science Foundation of China (Grants No: 61263035, 61573270, and 61672177); the China 973 Program (Grant No: 2013CB329404); the China Key Research Program (Grant No: 2016YFB1000905); the Guangxi Natural Science Foundation (Grant No: 2015GXNSFCB139011); the Innovation Project of Guangxi Graduate Education (Grants No: YCSZ2016046 and YCSZ2016045); the Guangxi Higher Institutions Program of Introducing 100 High-Level Overseas Talents; the Guangxi Collaborative Innovation Center of Multi-Source Information Integration and Intelligent Processing; and the Guangxi Bagui Scholar Teams for Innovation and Research Project.

References

1. Anderson, T.W.: Estimating linear restrictions on regression coefficients for multivariate normal distributions. Ann. Math. Stat. **22**, 327–351 (1951)
2. Argyriou, A., Evgeniou, T., Pontil, M.: Convex multi-task feature learning. Mach. Learn. **73**(3), 243–272 (2008)
3. Bache, K., Lichman, M.: Uci machine learning repository (2015)

4. Borchani, H., Varando, G., Bielza, C., Larrañaga, P.: A survey on multi-output regression. Wiley Interdisc. Rev. Data Mining Knowl. Discov. **5**(5), 216–233 (2015)
5. Cai, X., Ding, C., Nie, F., Huang, H.: On the equivalent of low-rank linear regressions and linear discriminant analysis based regressions. In: ACM SIGKDD International Conference on Knowledge Discovery and Data Mining, pp. 1124–1132 (2013)
6. Cai, X., Nie, F., Cai, W., Huang, H.: New graph structured sparsity model for multi-label image annotations, pp. 801–808 (2013)
7. Cands, E.J., Recht, B.: Exact matrix completion via convex optimization. Found. Comput. Math. **9**(6), 717–772 (2008)
8. Cao, J., Wu, Z., Wu, J.: Scaling up cosine interesting pattern discovery: a depth-first method. Inf. Sci. **266**(5), 31–46 (2014)
9. Cao, J., Wu, Z., Wu, J., Xiong, H.: Sail: Summation-based incremental learning for information-theoretic text clustering. IEEE Trans. Cybern. **43**(2), 570–584 (2013)
10. Chang, X., Nie, F., Yang, Y., Huang, H.: A convex formulation for semi-supervised multi-label feature selection. In: AAAI Conference on Artificial Intelligence, pp. 1171–1177 (2014)
11. Cheng, B., Liu, G., Wang, J., Huang, Z., Yan, S.: Multi-task low-rank affinity pursuit for image segmentation. In: International Conference on Computer Vision, pp. 2439–2446 (2011)
12. Džeroski, S., Demšar, D., Grbović, J.: Predicting chemical parameters of river water quality from bioindicator data. Appl. Intell. **13**(1), 7–17 (2000)
13. Gao, L., Song, J., Nie, F., Yan, Y.: Optimal graph learning with partial tags and multiple features for image and video annotation. In: CVPR (2015)
14. Gao, L., Song, J., Shao, J., Zhu, X., Shen, H.: Zero-shot image categorization by image correlation exploration. In: ICMR, pp. 487–490 (2015)
15. Gower, J.C., Dijksterhuis, G.B.: Procrustes problems. Oxford University Press (2004)
16. Izenman, A.J.: Reduced-rank regression for the multivariate linear model. J. Multivar. Anal. **5**(2), 248–264 (1975)
17. Karali, A., Bratko, I.: First order regression. Mach. Learn. **26**(26), 147–176 (1997)
18. Nie, F., Huang, H., Cai, X., Ding, C.H.Q.: Efficient and robust feature selection via joint l2,1-norms minimization. In: Conference on Neural Information Processing Systems 2010, pp. 1813–1821 (2010)
19. Qin, Y., Zhang, S., Zhu, X., Zhang, J., Zhang, C.: Semi-parametric optimization for missing data imputation. Appl. Intell. **27**(1), 79–88 (2007)
20. Rai, P., Kumar, A., Iii, H.D.: Simultaneously leveraging output and task structures for multiple-output regression. In: Advances in Neural Information Processing Systems, pp. 3185–3193 (2012)
21. Rothman, A.J., Ji, Z.: Sparse multivariate regression with covariance estimation. J. Comput. Graphical Stat. **19**(4), 947–962 (2010)
22. Spyromitros-Xioufis, E., Tsoumakas, G., Groves, W., Vlahavas, I.: Multi-label classification methods for multi-target regression. Computer Science (2014)
23. Spyromitros-Xioufis, E., Tsoumakas, G., Groves, W., Vlahavas, I.: Multi-target regression via input space expansion: treating targets as inputs. Mach. Learn., 1–44 (2016)
24. Wang, H., Nie, F., Huang, H., Risacher, S., Ding, C., Saykin, A.J., Shen, L.: Sparse multi-task regression and feature selection to identify brain imaging predictors for memory performance. In: IEEE International Conference on Computer Vision, pp. 557–562 (2010)

25. Wu, X., Zhang, C., Zhang, S.: Efficient mining of both positive and negative association rules. ACM Trans. Inf. Syst. (TOIS) **22**(3), 381–405 (2004)
26. Wu, X., Zhang, C., Zhang, S.: Database classification for multi-database mining. Inf. Syst. **30**(1), 71–88 (2005)
27. Wu, X., Zhang, S.: Synthesizing high-frequency rules from different data sources. IEEE Trans. Knowl. Data Eng. **15**(2), 353–367 (2003)
28. Zhang, C., Qin, Y., Zhu, X., Zhang, J., Zhang, S.: Clustering-based missing value imputation for data preprocessing. In: IEEE International Conference on Industrial Informatics, pp. 1081–1086 (2006)
29. Zhang, S., Cheng, D., Zong, M., Gao, L.: Self-representation nearest neighbor search for classification. Neurocomputing **195**, 137–142 (2016)
30. Zhang, S., Li, X., Zong, M., Cheng, D., Gao, L.: Learning k for knn classification. ACM Trans. Intell. Syst. Technol. (2016, Accepted)
31. Zhang, S., Qin, Z., Ling, C.X., Sheng, S.: "missing is useful": Missing values in cost-sensitive decision trees. IEEE Trans. Knowl. Data Eng. **17**(12), 1689–1693 (2005)
32. Zhang, S., Wu, X., Zhang, C.: Multi-database mining. IEEE Comput. Intell. Bull. **2**(1), 5–13 (2003)
33. Zhang, S., Zhang, C., Yang, Q.: Data preparation for data mining. Appl. Artif. Intell. **17**(5–6), 375–381 (2003)
34. Zhang, S., Zhang, J., Zhang, C.: Edua: an efficient algorithm for dynamic database mining. Inf. Sci. **177**(13), 2756–2767 (2007)
35. Zhu, X., Huang, Z., Cheng, H., Cui, J., Shen, H.T.: Sparse hashing for fast multimedia search. ACM Trans. Inf. Syst. (TOIS) **31**(2), 9 (2013)
36. Zhu, X., Huang, Z., Yang, Y., Shen, H.T., Xu, C., Luo, J.: Self-taught dimensionality reduction on the high-dimensional small-sized data. Pattern Recogn. **46**(1), 215–229 (2013)
37. Zhu, X., Li, X., Zhang, S.: Block-row sparse multiview multilabel learning for image classification. IEEE Trans. Cybern. **46**(2), 450–461 (2016)
38. Zhu, X., Li, X., Zhang, S., Ju, C., Wu, X.: Robust joint graph sparse coding for unsupervised spectral feature selection. IEEE Trans. Neural Netw. Learn. Syst., 1–13 (2016)
39. Zhu, X., Wu, X., Ding, W., Zhang, S.: Feature selection by joint graph sparse coding. In: Proceedings of the 2013 Siam International Conference on Data Mining, pp. 803–811. SIAM (2013)
40. Zhu, X., Zhang, J., Zhang, S.: Mixed-norm regression for visual classification. In: Motoda, H., Wu, Z., Cao, L., Zaiane, O., Yao, M., Wang, W. (eds.) ADMA 2013. LNCS (LNAI), vol. 8346, pp. 265–276. Springer, Heidelberg (2013). doi:10.1007/978-3-642-53914-5_23
41. Zhu, X., Zhang, S., Jin, Z., Zhang, Z., Xu, Z.: Missing value estimation for mixed-attribute data sets. IEEE Trans. Knowl. Data Eng. **23**(1), 110–121 (2011)
42. Zhu, X., Zhang, S., Zhang, J., Zhang, C.: Cost-sensitive imputing missing values with ordering. AAAI Press **2**, 1922–1923 (2007)
43. Zhu, Y., Lucey, S.: Convolutional sparse coding for trajectory reconstruction. IEEE Trans. Pattern Anal. Mach. Intell. **37**(3), 529–540 (2015)

Biologically Inspired Pattern Recognition for E-nose Sensors

Sanad Al-Maskari[1](✉), Wenping Guo[2,3], and Xiaoming Zhao[3]

[1] School of Information Technology and Electrical Engineering,
The University of Queensland, Brisbane, Australia
s.almaskari@uq.edu.au
[2] College of Computer and Information, Hohai University, Nanjing 210098, China
gwp@hhu.edu.cn
[3] Institute of Intelligent Information Processing,
Taizhou University, Taizhou 317000, China
tzxyzxm@163.com

Abstract. The high sensitivity, stability, selectivity and adaptivity of mailman olfactory system is a result of a large number of olfactory receptors feeding into an extensive layers of neural processing units. Olfactory receptor cells (ORC) contribute significantly in the sense of smells. Bloodhounds have four billion ORC making them ideal for tracking while human have about 30 million ORC. E-nose stability, sensitivity and selectivity have been a challenging task. We hypothesize that appropriate signal processing with an increase number of sensory receptors can significantly improve odour recognition in e-nose. Adding physical receptors to e-nose is costly and can increase system complexity. Therefore, we propose an Artificial Olfactory Receptor Cells Model (AORCM) inspired by neural circuits of the vertebrate olfactory system to improve e-nose performance. Secondly, we introduce and adaptation layer to cope with drift and unknown changes. The major layers in our model are the sensory transduction layer, sensory adaptation layer, artificial olfactory receptors layer (AORL) and artificial olfactory cortex layer (AOCL). Each layer in the proposed system is biologically inspired by the mammalian olfactory system. The experiments are executed using chemo-sensory arrays data generated over three-year period. The propose model resulted in a better performance and stability compared to other models. To our knowledge, e-nose stability, selectivity and sensitivity are still unsolved problem. Our paper provides a new approach in improving e-nose pattern recognition over long period of time.

Keywords: E-nose · Receptors · Drift · Neural · Noise · Classification · Adaptive · Sensors · Chemical

1 Introduction

The olfactory system provides a critical means of survival in mammals. It is used to find food, sense danger, identify mates and track preys. The biological

© Springer International Publishing AG 2016
J. Li et al. (Eds.): ADMA 2016, LNAI 10086, pp. 142–155, 2016.
DOI: 10.1007/978-3-319-49586-6_10

olfactory system relies on olfactory receptors (ORs) to detect odour in the environment, and each receptor type is encoded by a single OR gene. Our ability as human to smell is remarkable, still we are unable to provide exact description of what we smell. Our sense of vision has dominated all other sense including our sense of smell. We are incapable of describing a smell with details. The best we can do is to relate a smell to something similar. Although the human olfactory system didn't evolve to be similar to an elephant or bloodhound, technology enabled us to invent new ways to sense and monitor the environment around us. With the help of **Electronic nose (e-nose)**, chemical analystes and odour molecules in the environment can be monitored [1, 2]. E-nose systems have been used in a wide range of applications including: food industry, agriculture, air quality and environment monitoring, odour monitoring in a poultry shed, medicine, water and waste water quality control [1–4]. Despite their good reputation, their practical values are affected by their poor stability, making them very vulnerable to drift and noise. These vulnerabilities can be resolved using periodic recalibration which in turn increases their cost significantly, making them overly complex. Compensation for the need of periodic recalibration and ability to postpone the recalibration period using machine learning will need to solve the aforementioned dynamic issues. In general, the performance of e-noses is affected by following factors.

- **Sensor drift** is the gradual and unpredictable variation of the chemo-sensory signal responses when exposed to the same analyte under identical conditions [5–8]. Such variations could be caused by sensor poisoning, ageing and environmental changes such as humidity, temperature, pressure and system sampling non-specific adsorption [9].
- **Noise** can be defined as any unwanted effect that obscures the measurement of a detected signal. Noise can be caused by system circuit faults, sensor poisoning, sensor ageing, and environmental effects. Because sensor drift is not deterministic, it is difficult, if not impossible, to distinguish it from noise and *vice versa* [10, 11].
- **Gas Mixtures:** e-nose sensors installed in outdoor environment are exposed to various gases and their mixtures, making their detection a difficult task.
- **Natural Environment** poses a major challenge with fluctuation in humidity, temperature, wind direction and wind speed. These rapid changes prevent sensors from making accurate measurement.

In this paper our objectives are to provide an effective learning model to discriminate between different contaminants. In order to achieve such objective the learning model must be robust and accurate. An ideal learning model should be able to identify containments accurately and tolerate and expected environmental changes. We believe the best way to address such issues is to capture some properties of the mailman olfactory system. The data is generated using 4×4 sensory arrays, which represent four types of Olfactory Receptor Neurons (ORN) and each ORN consist of four receptors. In the first stage we create a transduction layer responsible for feature extraction, data pre-processing and denoising. Secondly, we created a sensory adaptation layer to enable e-nose to cope with unknown changes in the ambient environment mimicking mammalian olfactory system adaptation capabilities. The output of the adaptation layer is feed into the artificial olfactory receptor layer (AORL) to maximize sensory information and reduce the impact of irrelevant inputs. Finally, the AORNs project their axons into the artificial olfactory bulb to predict odour identity. The model is tested using random sequence of data with different gases and concentrations which was generated over 18 month. The study attempts to simulate natural odour application in real-world scenarios as much as possible. Our results demonstrate that our model can significantly enhance and improve e-nose odour identification and classification.

2   Related Work

The ultimate goal of electronic e-nose is to identify contaminants or gases with high accuracies and efficiencies. The chemo-sensors have the ability to identify specific gases or mixture of gases using metal oxide sensors. These metal oxide sensors can detect gases accurately in a closed environment. Unfortunately their selectivity and sensitivity will degreed over time and their prediction performance can degrade significantly in an outdoor environment. Univariate and multivariate methods have been proposed in the literature to enhance the prediction performance for e-noses by correcting drift. Some researchers used PCA to find drift direction in order to correct the drift. Some studies propose to use Component Correction Analysis (CCA) to identify drift direction [12]. In [13] a supervised Orthogonal Signal Correction (OSC) method is shown to be better than other correction methods. Drift correction methods make the assumption that sensor drift is linear; where in real environment, it is found to be nonlinear [12,14]. Furthermore, by removing some components from the original signals an important portion of the data can be lost which can impact the prediction performance.

An alternative approach to correction meted is to use unsupervised or supervised soft based methods. In the literature, several machine learning algorithms have been proposed including Self Organizing Maps (SOMs), multiple SOM, SVM, and neural networks [6–8,15–17].

Assuming that data exhibit similar distributions can be misleading alternatively assuming that source and target data have different distributions is also misleading. In reality, source and target data could drift closer or far a way from the source data in a non deterministic way. This is due to the fact that we know very little about the underlining processes responsible for generating the target data and we know only few parameters about the real process. Therefore, it is impossible to accurately model real world target distributions using the initial system state.

3 Methodology

3.1 Problem Definition

Consider a multi-class classification problem with a set of features x produced by a gas sensor. Let $S_1, \ldots, S_t$ be a batch of examples received from e-nose sensor at time interval t. S_t consists of labeled L and unlabeled U examples, $S = L \cup U$. We assume that L_t represent our source domain with labeled data such that $L_t = \{(x_{1t}, y_{1t}), \ldots, (x_{nt}, y_{nt})\}$ are of size tn where $x_{nt} \in R^N$ and $y_{nt} \in \{1, \ldots, M\}$ where M refers to the class label (gas type) of instance x_{nt} and $m \in M$. We make the assumption that our target domain contains unlabeled examples of size tn and given by $U_t = \{x_{1t}, \ldots, x_{nt}\}$. Therefore, each example can be represented by $< S, Y, N >$ where $Y \in \{1, \ldots, M\}$, indicating the class label of a labeled example $x_{nt} \in S$; $N \in \{0, 1\}$, indicating whether a given example S is labeled ($N = 1$) or unlabelled ($N = 0$). Each Instance in L is labeled, represented by $< S, Y \in M, N = 1 >$. Consequently, each instance in U is unlabeled and can be represented by $< S, Y =?, N = 0 >$. We assume that the source domain is labeled (L) and the target domain is unlabeled (U). In standard supervised learning problems both L and U are drawn from similar probability distributions where $P(x, y) = P(y|x).P(x)$. Therefore the objective of supervised learning is to train classifiers f_m using the training set $T(L)$ to estimate a target distribution $\hat{P}(x, y)$ such that unlabeled gas data U can be correctly classified with high accuracies. If the data distribution of the source domain L and target domain U are different such that $P(L) \neq P(U)$, then standard supervised learning methods will suffer significant performance degradation.

Because the data produced by e-nose sensors doesn't follow the standard supervised learning assumption we introduce a sensory adaptation layer. We consider that the source domain could have a different marginal distribution than the target domain. The source domain L can be represented by a marginal distribution $P^s(x, y) = P^s(y|x).P^s(x)$ and we will refer to it as P. The target domain U can be represented by a marginal distribution Q such that $P^t(x, y) = P^t(y|x).P^t(x)$. Applying domain adaptation methods can be useful in cases where $P \neq Q$ and P have some correlation with Q. Attempting to map the source domain and target domain where $P = Q$ can impact classifier performance. Similarly, attempting to map P to Q when they have no correlation is not useful for predicting U. Therefore, in our approach we propose to use a distance measure H to evaluate the similarities between P and Q before performing adaptation. In real applications assuming that P and Q either identical or dissimilar is not realistic. It is reasonable to assume that the target domain can have a dynamic behaviour resulting in fluctuating degrees of dissimilarities between P and Q. In this case we can assume that finding intrinsic relationship between P and Q is useful only if $H > \phi$. Creating a good learning model for our problem will depend on multiple factors including similarities, dissimilarities and correlations between P and Q.

Our objective is to train a learning model f_m using source domain training set $T(L)$ to predict unknown target domain data U. The classification decision process is made by applying all classifiers represented by f_c for $c \in \{1, \ldots, C\}$ to new unlabeled sample $x_{nt} \in U$ and predicting the gas type $\hat{y}_{nt}$ for which the corresponding classifier reports the highest confidence score:

$$\hat{y}_{nt} \in \underset{c \in \{1, \ldots, C\}}{\operatorname{argmax}} f_c(x_{nt}) \text{ where } \Phi_{nt} = \underset{c \in \{1, \ldots, C\}}{\max} \Phi_c(x_{nt}). \tag{1}$$

3.2 E-nose Olfactory System

The objective of e-nose system is to detect gases or odors in the atmosphere see Fig. 1. The ability to detect and identify different gases using e-nose still remains a challenging problem. Sensor selectivity for a specific gas can degrade overtime due to sensor aging and poisoning. Furthermore, as explained previously, sensors are subject to high uncertainties due to environmental changes. Moreover, chemo-sensors suffer from drift and noise. Therefore, the identification models should consider all the variabilities surrounding e-nose sensors. It is highly important to mention that not all gases detected by an enose can have a smell. For instance, sulfur dioxide So_2 will smell like rotten egg in low concentrations, but at high concentrations human are no capable of smelling this toxic gas. The major concern in our case is to be able to detect and identify different gases with high accuracies. To do that effectively, we mimic the mammalian olfactory system by introducing an Artificial Olfactory Receptor Cells (AORC). Secondly, transduction layer is used for data processing and feature extractions. Thirdly, an adaptation layer is introduce to cope with unknown and dynamic changes. Finally, the decision is made by an olfactory bulb layer after receiving the final outputs from the AORC layer.

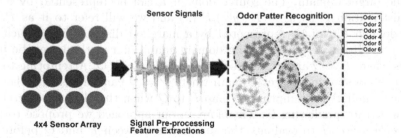

Fig. 1. Typical E-nose odor prediction process.

3.3 E-nose Adaptation with AORC

Learning in the Presence of Noise and Drift: *Problem 1:* How to create and adaptive model to cope with unknown changes such as noise and drift?

The inherent characteristics of most chemical sensor technologies and the way they are engineered to react to input signals raises a challenging problem for any learning model. Sensor drift, high levels of environmental noise, and highly variable ambient conditions have a profound impact on chemical gas sensor output signals. The poor stability, selectivity, and sensitivity of these sensors are major contributors to their performance issues [9,11,18]. Our proposed AORCM model Fig. 2 attempt to adapt to changes by introducing Adaptation Layer, Artificial Olfactory Receptor Cells and a Score Function which is used to update the AORCs.

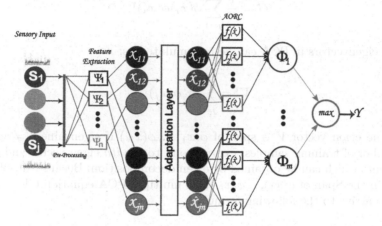

Fig. 2. Adaptive Olfactory Receptor Cells model(AORCM).

The adaptation layer is inspired by a class of methods recently proposed for unsupervised adaptations on the data manifold [19–21]. Because it is not feasible in real world scenarios to assume specific transformations between two domains, we employ unsupervised adaptive method. In such case no assumptions are made about the unknown target domain, instead we attempt to learn a transformation G from the unlabel data without the need of manually labelling target data.

Let O and F be an orthonormal matrices in $\mathbb{R}^{n \times d}$ with orthogonal column vectors where $O^{\mathsf{T}} O = I$ and $F^{\mathsf{T}} F = I$

The smallest principle angel θ between subspaces O and F where $\theta_i(O, F) = \theta_i \in [0, \pi/2]$ can be defined recursively [22] by:

$$cos(\theta_i) = \max_{u \in O} \max_{u \in F} u^{\mathsf{T}} v = u_i^{\mathsf{T}} v_i \qquad (2)$$

subject to:
$||u|| = ||v|| = 1, u_i^{\mathsf{T}} u_i = 0, v_i^{\mathsf{T}} v_i = 0, i = 1, ..., d-1$

By performing Singular Value decomposition on O and F

$$O^{\mathsf{T}} F = U_1 \Sigma V^{\mathsf{T}} \qquad (3)$$

the principle angles and vectors can be computed
$\theta_i = arccos\sigma_i, u_i = (OU_1)., i, v_i = (FV)., i,$
where $\Sigma = diag(\sigma_1, \sigma_2, .., \sigma_i)$ and θ_i is the i-th diagonal element of Σ.

Principle angles in kernel space. Kernels provide a way of mapping observed data to high dimensional feature space F ($\phi: X \rightarrow F$). If we assume that our source and target data can be mapped into a higher dimensional feature space $\phi(x1), ..., \phi(x_N)$ the covariance matrix can be written as:

$$\bar{C} = \frac{1}{N} \sum_{j=1}^{N} \phi(x_j)\phi(x_j)^{\mathsf{T}} \tag{4}$$

The eigenvectors in PCA can be reformulated as:

$$V^i = \sum_{j=1}^{N} \alpha_j^i \phi(x_j) \tag{5}$$

where the eigen vector $V \in$ span of $\phi(x_1), .., \phi(x_N)$ if eigenvalue $\lambda \neq 0$, N is the number of training samples, i correspond to the ith eigen vector and α_j^i are coefficients which can be obtained by eigen decomposition. Because all solutions of V is in the Span of $\phi(x_1), .., \phi(x_N)$, we multiply PCA equation $CV = \lambda V$ by $\phi(x_r)$ to arrive to the following formula:

$$(\phi(x_r).\bar{C}V) = \lambda(\phi(x_r).V) \text{ for all } r = 1, ..., N \tag{6}$$

Substituting $\bar{C}$ and (5) into (6) and defining N × N Grammatrix K by $K_ij := (\phi(x_i).\phi(x_j)) = K(x_i, x_j)$ we arrive to the following expression which can be solved by the eigenvalue problem:

$$N\lambda\alpha = K\alpha \tag{7}$$

where $\alpha = [\alpha_1, ..., \alpha_N]^{\mathsf{T}}$
Finally, our data can be projected into k-dimensional subspace using the following transformation matrix:

$$\bar{T}_k = K(x_i, x_j)V_k \tag{8}$$

The centered kernel matrix $\bar{K}$ is given by:

$$\bar{K} = K - \frac{1}{N}.I.K - K.I.\frac{1}{N} + \frac{1}{N}.I.K.I \tag{9}$$

where I is N by N unit matrix. For more details about KPCA refer to [23]. In this paper all input data are normalized using:

$$\frac{1}{N} \sum_{i=1}^{N} \phi(x_i) = 0 \tag{10}$$

We map the input data matrix M into higher dimensional feature space using a Gaussian Kernel function:

$$K(x_i, x_j) = exp\left(\frac{-||x_i - x_j||^2}{\sigma^2}\right) \tag{11}$$

After sorting the eigenvalues λ in descending order the importance rate IR can be calculated. The importance rate IR can be calculated by the first k eigenvalues λ:

$$IR = \left(\frac{\sum_{j=1}^{k} \lambda_j}{\sum_{j=1}^{n} \lambda_j}\right) * 100 \tag{12}$$

where n refers to total number of training eigenvectors. k should be calculated according to

$$k = argmin\{IR > \varphi\}. \tag{13}$$

The threshold φ used in this paper is 98 %. Therefore, optimal projection eigenvectors are selected based on k.
The principle angles between source and target subspaces P and Q can be computed using their principle vectors found by KPCA. The principle vectors of V and F can be computed according to (5)

$$V^p = \sum_{j=1}^{N1} \alpha_j^p \phi(P_j) \tag{14}$$

$$F^q = \sum_{l=1}^{N2} \beta_l^q \phi(Q_l) \tag{15}$$

where p and $q = [1...k]$, $N1$ and $N2$ refers to the total number of training samples and k refer to the number of principle vectors. Form (14) and (15) the basis matrices for the source and destination subspaces are $V = [V^1 V^2 ... V^k]$ and $F = [F^1 F^2 ... F^k]$. The projection Z of two kernel subspaces V and F therefore is

$$Z(p, q) = (V^p)^{\intercal} F^q) = \sum_{j=1}^{N1} \sum_{l=1}^{N2} \alpha_j^p \beta_l^q K(P_j, Q_l) = (\alpha^p)^{\intercal} K \beta^q \tag{16}$$

where $\alpha^p = [\alpha_1^p...\alpha_{N1}^p]^\top$ and $\beta^p = [\beta_1^q...\beta_{N2}^q]^\top$.

$$Overall : Z = V^\top F = \Delta^\top K \Xi \qquad (17)$$

where $\Delta = [\alpha^1...\alpha^k]^\top$, $\Xi = [\beta^1...\beta^k]$ and K is the N × N Grammatrix of input vectors.

The principle angle of Z can be calculated using cs decomposition From (3), the principle angles of Z subspaces $(\theta_i(V, F))$ can be calculated using SVD therefore $\theta = V^\top F$

Algorithm 1. Adaptive Olfactory Receptors Algorithm

1: **Input:**
   P_n: *Source data set* $P = (Xs_1, Xs_2, ..., Xs_d)$.
   Q_n: *Target data set* $Q = (Xt_1, Xt_2, ..., Xt_d)$.
   d: Subspace dimensions
2: **Output:** *new feature space G.*
3: *Normalize source and target data P and Q* (10).
4: *Generate all subspaces V and F for P and Q using KPCA according to equation* (14) *and* (15).
5: *Calculate similarities H between V and F using Hellinger distance.*
6: *if H > η then exit.*
7: *Select d subspaces corresponding to top k projection vectors according to* (13).
8: *Calculate the kernel projection matrix Z using equation* (16).
9: *perform CS decomposition of Z refer to equation* (3)
10: *Calculate the principle angles for Z refer to equation* (2)
11: *Generate new feature space G from Z using GFK kernel.*
12: **return** *G*

---

4   Experiments and Evaluation

In this paper we use a public data set provided by Vergara *et al.* [5]. The data set is generated over a three year period. The main focus of this public dataset is to emulate sensor drift in an enclosed environment using metal oxide sensors. A Total of 13,910 samples were collected over 36 month for six gases. In our experiments we consider the following setting:

Setting 1: classifier is trained on batch1 and tested on the remaining batches. To test the ability of classifiers to provide consistent results in noisy and drifting environment, we only use the first batch generated in the first two month to train the models.

4.1   E-nose Data

This paper attempt to classify 6 type of containments including Toluene(C_7H_8), Ammonia (NH_3), Acetaldehyde (C_2H_4O), Acetone (C_3H_6O), Ethylene (C_2H_4), and Ethanol (C_2H_6O) (Fig. 3).

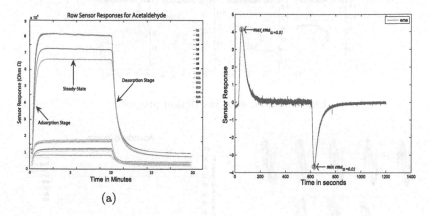

(a)

Fig. 3. (a) 16 MOX sensor responses for an e-nose when exposed to 100 ppm of Acetaldehyde. (b) Shows the exponential moving average analysis for 223 ppmv of Ethylene when $\alpha = 0.01$.

4.2   Synthetic Data Set

We generated synthetic data set to illustrate the concept of drift and how it can effect the classification performance of any classifier. The data generated is in a crescent shape with two classes C1 and C2. We applied multiple rotations to the original data set with different rotation degrees. A total of 9 data sets are generated refer to Fig. 4.

4.3   Analysis and Results

The main objective of the synthetic data set is to illustrate the relationship between classification accuracy and drift. From Fig. 5 we can conclude that as the rotation degree which represent drift increases the classification accuracies decreases. Furthermore, the hellinger distance for the two distributions increases as the rotation increases. For instance the H distance between batch 1 and batch 2 is 0.28 and between 1 & 9 is 0.82. Therefore, H distance can be used as an indicator to trigger the adaptation process.

In this experiment different classifiers are trained to predict various gases using E-nose data under *Setting 1*. We compared AORCM with KNN Ensemble, Bagging Tree (TB), Bagging tree with surrogate (TBS), Discriminant analysis, Fuzzy SVM and Kernel Fuzzy SVM [7,8]. Figure 6 shows various classifiers

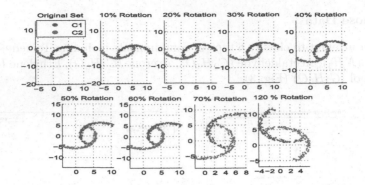

Fig. 4. E-nose odor prediction process.

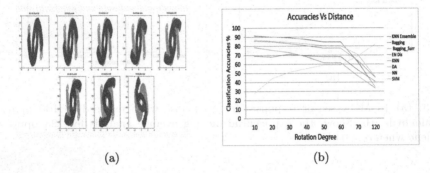

(a)   (b)

Fig. 5. (a) In the first and second data sets the classifier provide a very accurate fit but as the measured distance increases accuracies drops. (b) Shows classifications accuracies verses rotation degrees.

performance for the mentioned models. A total of 20 runs is executed and the average results is calculated. We use 10 fold cross validation to evaluate all the models then we use batch 2 to 9 as a test sets. The result demonstrate AORCM performance superiority as shown in Fig. 6. From this analysis the following statements are made:

- AORCM performed better than other classifiers in most cases with total average classification accuracy of 67.8 % compared to Discriminant analysis with average score of 64.64 %. Batch 10 is the last batch generated during the 3 years experiment and it has maximum drift. In the instance (1→10) our proposed method achieved best results 53.59 %. Clearly our proposed method is effective in dealing with noise and drift.

- Discriminant Analysis classifier performed better only in two cases first one is (1→4) and (1→8). The AORCM permeance is effected due to class imbalance problem. Batch 4 has few labels for Ethanol and Ethylene. Therefore, AORCM results could be improved by addressing the class imbalance issue. Overall, the performance of AORCM is better than other classifiers.

Classification Accuracies

	AORCM	ensKNN	TB	TBWS	DISCR	FSVM	MKFSVM
1->2	83.20	38.37	66.72	64.25	70.82	79.90	78.94
1->3	80.90	46.39	56.49	58.61	72.64	72.07	71.69
1->4	65.84	44.86	37.89	52.15	72.67	67.70	65.22
1->5	75.63	33.92	52.28	75.61	68.02	61.93	60.41
1->6	84.96	54.86	61.78	73.89	78.70	80.00	82.30
1->7	66.98	44.41	55.74	64.45	58.73	46.22	46.58
1->8	50.00	29.75	37.07	54.82	46.26	40.14	38.44
1->9	49.79	34.18	50.85	51.74	67.45	62.13	62.13
1->10	53.59	41.80	43.13	50.41	46.49	36.18	36.29
AVG	67.88	40.95	51.33	60.66	64.64	60.70	60.22

(a)

(b)

Fig. 6. AORCM classification accuracy compared to other models. The learners are only trained using the first batch and tested on the remaining batches.

Figure 7 shows the Artificial Olfactory Cortex Layer (AOCL) decision response image generated from different AORCs sensory response signals when exposed to six different gases including Acetone, Toluene and Ammonia. Dark blue areas indicate passive response from the AOC while the light yellow area indicate a very active response for the given AORC signal.

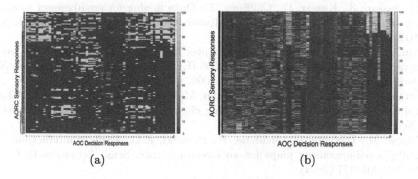

(a)   (b)

Fig. 7. (a) AOC Decision response when using few AORCs sensory signals.(b) AOC Decision response when using larger number of AORCs sensory response signals. (Color figure online)

5   Conclusions

Predicting odours and gases using e-nose is very challenging because the data is noisy, chaotic and contain drift. The prediction is even harder when e-nose is exposed to outdoor environment. This paper presented a new approach inspired by biological systems to improve e-nose pattern recognition. The proposed approach uses artificial olfactory receptors combined with adaptation layer to improve the prediction process. The experiments conducted show the superiority of our method compared to alternative state of the art techniques. Furthermore, our paper introduces adaptation trigger mechanism to maximize performance. Future work will focus on class imbalance problem for e-nose data. Secondly, we will examine the generalization abilities of our model when unseen instances are observed.

Acknowledgment. This work was partially supported by the following research grants: (1) No. LY14F020036 from the Natural Science Foundation of Zhejiang Province, China; (2) No. BK20141420 from the Natural Science Foundation of Jiangsu Province, China; (3) No. 61272261 from the Natural Science Foundation of China.

References

1. Nimsuk, N., Nakamoto, T.: Study on the odor classification in dynamical concentration robust against humidity and temperature changes. Sens. Actuators B: Chem. **134**(1), 252–257 (2008)
2. Al-Maskari, S., Saini, D., Omar, W.: Cyber infrastructure and data quality for environmental pollution control - in Oman. In: Proceedings of the 2010 DAMD International Conference on Data Analysis, Data Quality and Metada, vol. 71 (2010)
3. Al Maskari, S., Kumar, D., Chiffings, T.: Data mining for environment monitoring. In: International Conference on Software Technology and Engineering (ICSTE 2012). ASME Press (2012)
4. Pan, L., Yang, S.X.: A new electronic nose for downwind livestock farm odour measurement. In: Proceedings of the 2006 IEEE International Conference on Networking, Sensing and Control, ICNSC 2006, pp. 410–415 (2006)
5. Vergara, A., Vembu, S., Ayhan, T., Ryan, M.A., Homer, M.L., Huerta, R.: Chemical gas sensor drift compensation using classifier ensembles. Sens. Actuators B: Chem. **166–167**, 320–329 (2012)
6. Zuppa, M., Distante, C., Siciliano, P., Persaud, K.C.: Drift counteraction with multiple self-organising maps for an electronic nose. Sens. Actuators B: Chem. **98**(23), 305–317 (2004)
7. Al-Maskari, S., Li, X., Liu, Q.: An effective approach to handling noise and drift in electronic noses. In: Databases Theory and Applications - 25th Australasian Database Conference, ADC 2014, Brisbane, QLD, Australia, 14–16 July 2014, pp. 223–230 (2014)
8. Al-Maskari, S., Bélisle, E., Li, X., Digabel, S., Nawahda, A., Zhong, J.: Classification with quantification for air quality monitoring. In: Bailey, J., Khan, L., Washio, T., Dobbie, G., Huang, J.Z., Wang, R. (eds.) PAKDD 2016. LNCS (LNAI), vol. 9651, pp. 578–590. Springer, Heidelberg (2016). doi:10.1007/978-3-319-31753-3_46
9. Sharma, R., Chan, P., Tang, Z., Yan, G., Hsing, I., Sin, J.: Investigation of stability and reliability of tin oxide thin-film for integrated micro-machined gas sensor devices. Sens. Actuators B: Chem. **81**(1), 9–16 (2001)

10. Goodner, K.L., Dreher, J., Rouseff, R.L.: The dangers of creating false classifications due to noise in electronic nose and similar multivariate analyses. Sens. Actuators B: Chem. **80**(3), 261–266 (2001)

11. Tian, F., Yang, S.X., Dong, K.: Circuit and noise analysis of odorant gas sensors in an e-nose. Sensors **5**(1), 85–96 (2005)

12. Romain, A., Nicolas, J.: Long term stability of metal oxide-based gas sensors for e-nose environmental applications: an overview. Sens. Actuators B: Chem. **146**(2), 502–506 (2010). Selected Papers from the 13th International Symposium on Olfaction and Electronic Nose ISOEN 2009

13. Padilla, M., Perera, A., Montoliu, I., Chaudry, A., Persaud, K., Marco, S.: Drift compensation of gas sensor array data by orthogonal signal correction. Chemometr. Intell. Lab. Syst. **100**(1), 28–35 (2010)

14. Zhang, L., Tian, F., Nie, H., Dang, L., Li, G., Ye, Q., Kadri, C.: Classification of multiple indoor air contaminants by an electronic nose and a hybrid support vector machine. Sens. Actuators B: Chem. **174**, 114–125 (2012)

15. Marco, S., Pardo, A., Ortega, A., Samitier, J.: Gas identification with tin oxide sensor array and self organizing maps: adaptive correction of sensor drifts. In: Instrumentation and Measurement Technology Conference, IMTC 1997. Proceedings of Sensing, Processing, Networking, vol. 2, pp. 904–907. IEEE, May 1997

16. Nahar, V., Al-Maskari, S., Li, X., Pang, C.: Semi-supervised learning for cyber-bullying detection in social networks. In: Wang, H., Sharaf, M.A. (eds.) ADC 2014. LNCS, vol. 8506, pp. 160–171. Springer, Heidelberg (2014). doi:10.1007/978-3-319-08608-8_14

17. Bishop, C.M.: Neural Networks for Pattern Recognition. Oxford University Press Inc., New York (1995)

18. Liu, H., Tang, Z.: Metal oxide gas sensor drift compensation using a dynamic classifier ensemble based on fitting. Sensors **13**(7), 9160–9173 (2013)

19. Gallivan, K., Srivastava, A., Liu, X., Van Dooren, P., et al.: Efficient algorithms for inferences on grassmann manifolds. In: 2003 IEEE Workshop on Statistical Signal Processing, pp. 315–318. IEEE (2003)

20. Gopalan, R., Li, R., Chellappa, R.: Domain adaptation for object recognition: an unsupervised approach. In: 2011 IEEE International Conference on Computer Vision (ICCV), pp. 999–1006. IEEE (2011)

21. Gong, B., Shi, Y., Sha, F., Grauman, K.: Geodesic flow kernel for unsupervised domain adaptation. In: 2012 IEEE Conference on Computer Vision and Pattern Recognition (CVPR), pp. 2066–2073. IEEE (2012)

22. Björck, K., Golub, G.H.: Numerical methods for computing angles between linear subspaces. Math. Comput. **27**(123), 579–594 (1973)

23. Mika, S., Schölkopf, B., Smola, A.J., Müller, K.R., Scholz, M., Rätsch, G.: Kernel pca and de-noising in feature spaces. In: NIPS, vol. 4, p. 7. Citeseer (1998)

Addressing Class Imbalance and Cost Sensitivity in Software Defect Prediction by Combining Domain Costs and Balancing Costs

Michael J. Siers[✉] and Md Zahidul Islam

School of Computing and Mathematics, Charles Sturt University, Bathurst, Australia
{msiers,zislam}@csu.edu.au

Abstract. Effective methods for identification of software defects help minimize the business costs of software development. Classification methods can be used to perform software defect prediction. When cost-sensitive methods are used, the predictions are optimized for business cost. The data sets used as input for these methods typically suffer from the class imbalance problem. That is, there are many more defect-free code examples than defective code examples to learn from. This negatively impacts the classifier's ability to correctly predict defective code examples. Cost-sensitive classification can also be used to mitigate the affects of the class imbalance problem by setting the costs to reflect the level of imbalance in the training data set. Through an experimental process, we have developed a method for combining these two different types of costs. We demonstrate that by using our proposed approach, we can produce more cost effective predictions than several recent cost-sensitive methods used for software defect prediction. Furthermore, we examine the software defect prediction models built by our method and present the discovered insights.

Keywords: Cost sensitive · Software defect prediction · Class imbalance

1 Introduction

Software defect prediction (SDP) is the task of predicting which sections of code are likely to be defective. Within the literature, these sections of code are referred to as modules. For a module to be considered as defective, it must contain at least one bug. Each function/method is commonly considered as one module. SDP has been studied in many areas of software such as web applications [12], mobile applications [15] and SDP for a specific programmer [9].

SDP can be performed both cost-sensitively and non-cost-sensitively. When performed non-cost-sensitively, the SDP system aims to make as many correct predictions as possible. That is, it is optimizing for accuracy. However, incorrectly predicting a defective module as defect-free is very costly for a software development business. A cost-sensitive SDP system is aware of this cost and is less likely to predict a module as defect-free when it is actually defective.

© Springer International Publishing AG 2016
J. Li et al. (Eds.): ADMA 2016, LNAI 10086, pp. 156–171, 2016.
DOI: 10.1007/978-3-319-49586-6_11

Table 1. SDP costs

Cost	Business scenario	Typical value
TP	A defective module is correctly predicted as defective. The software development business must now assign resources to fixing the defective module	$1000
TN	A defect-free module is correctly predicted as containing no defects. No action is required	$0
FP	A defect-free module is incorrectly predicted as containing defects. After assigning resources to fix the module, the software development business learns that the module is actually defect-free. The typical value associated with this cost is the same as TP	$1000
FN	A defective module is incorrectly predicted as defect-free. This means that the defect will remain within the program. This can cause lots of damage to the program and potentially hinder the software development process at a later stage. The typical value associated with this cost is 5–8 times as much the cost of TP	$5000 to $8000

Typically, four costs need to be defined for a cost-sensitive SDP system. These are true positive (TP), true negative (TN), false positive (FP) and false negative (FN). An explanation of each of these costs is given in Table 1. The table includes typical values for these costs which are based on previous studies in SDP [10,16,18,19].

To build a software defect prediction system, a classification method such as C4.5 [13], EXPLORE [7], or SysFor [8] is used. These classification methods require a training dataset D_T as input. D_T is comprised of a set of records $R = \{R_0, R_1, \ldots, R_{|R-1|}\}$ and a set of attributes $A = \{A_0, A_1, \ldots, A_{|A-1|}\}$. Each record is comprised of a set of values $R_i = \{V_0, V_1, \ldots, V_{|A-1|}\}$ where each value corresponds to an attribute in A. Therefore, D_T can be thought of a table where each $R_i \in R$ is a row and each $A_j \in A$ is a column. One of the $A_j \in A$ is considered as the class attribute. This class attribute is the value of interest and is the value that the trained classifier will predict for a new, unseen record. In the case of SDP each $R_i \in R$ represents a module, each $A_j \in A$ represents a software measure, and the class attribute is whether or not the module is defective.

A D_T used to train a SDP system often suffers from the class imbalance problem. When a D_T contains many more records with one class value than records with a different class value, it is considered to be class imbalanced. More specifically, a SDP D_T typically contains many more examples of defect-free modules than defective modules. In this case, we refer to the defect class as the minority class, and the defect-free class as the majority class. This class imbalance problem causes a SDP system to have poor prediction performance on defective modules. Thus, a traditional classification method is more likely to make many FN predictions which is very costly to a business as described in Table 1.

One method for dealing with the class imbalance problem is to use a cost-sensitive classification method. This is done by setting the value of the FN cost to be I times the value of the FP cost [20]. I is the number of majority class records divided by the number of minority class records. Since there should be no penalty for a correct prediction, the values of TP and TN costs are set to 0. This way, the costs reflect the level of imbalance. A cost-sensitive classifier trained using these costs will produce non-cost-sensitive predictions, but will not be biased towards predicting records as the majority class. In this study, we refer to these cost values as balancing costs. On the other hand, we refer to the cost values described in Table 1 as the domain costs.

In Sect. 3, we first propose a cost-sensitive framework which uses both balancing costs and domain costs. We then experiment different possible methods for combining a balancing cost-matrix and a domain cost-matrix into a single cost-matrix for creating a cost-sensitive SDP system. Our framework consists of a recent cost-sensitive classification method CSForest [18,19], and a recent method for calculating cluster-specific balancing costs named Standoff [20]. The framework also uses a recent class-specific clustering method named RBClust [21] as input for Standoff. In Sect. 4 this framework is compared against recent cost-sensitive methods proposed for SDP on real world SDP data sets. Since our method produces logic rules, we present some of the insights discovered on the real world SDP data sets in Sect. 4.3. Section 2 details the existing work which is related to this study.

1.1   Main Contributions of This Study

- We experimentally compare different possible methods for combining domain costs and balancing costs in cost-sensitive software defect prediction.
- Based on our experiments, we propose a new cost-sensitive method for combining domain costs and balancing costs in software defect prediction.
- We propose a framework which calculates the balancing costs and uses our proposed cost-combining method. This framework uses recently published methods. We also modify one of these methods so that it may use multiple cost-matrices as input.
- Our experiments show that the proposed framework outperforms several existing methods when applied to the software defect prediction problem.
- We present insights into software defect prediction based on the rules discovered using our framework.

2   Related Work

2.1   Measuring Source Code

As mentioned in Sect. 1, the attributes within an SDP data set are software measures. For example, the Halstead complexity measures [5] can be used as the attributes. There are twelve Halstead complexity measures in total. The simplest four measures are the total and distinct number of operators, and the total and distinct number of operands. From these measures, the eight other measures can be calculated. Since Halstead only counts the operators and operands, it does not consider the logic of the program. Cyclomatic complexity [11] is a measure of how many paths of execution can be taken through a program. Both Halstead's and the cyclomatic complexity measures were used to generate the publicly available NASA MDP SDP datasets [1].

2.2 Sampling Techniques

There exist several different families of techniques for treating an imbalanced data set. One highly popular family is *sampling*. Undersampling techniques remove records from the majority class, and oversampling records increase the number of majority class records. A highly popular technique for oversampling technique is *SMOTE* [3]. For each minority record, SMOTE, or *Synthetic Minority Oversampling TEchnique* chooses a random nearest neighbor. Then, a new *synthetic* record is created whose attributes' values are randomly set between the minority record and the chosen nearest neighbor. For example, if the minority record $R_m = \{10, 4, 32, 12\}$ and the random nearest neighbour is $R_{nn} = \{4, 3, 12, 15\}$, then the new synthetic record might be $R_s = \{6, 3.5, 31, 13\}$. This process can be repeated multiple times for each record based on a user defined parameter usually written as %.

There have been many extensions of SMOTE within the literature [2,6,14]. Adasyn [6] extends SMOTE by calculating the number of synthetic records that should be generated from each existing minority record. The more majority records surrounding the minority record, the more synthetic records are generated from that minority record. In the original publication, Adasyn was found to outperform SMOTE. By calculating the number of new records to generate, Adasyn eliminates the need for the % parameter.

A recent undersampling approach is *Inverse Random Undersampling* or *IRUS* [22]. The core concept of IRUS is to invert the imbalance of the data set. This is achieved by randomly removing enough records from the majority class such that the minority class becomes the majority class. A technique known as *Bagging* is then used to create an ensemble of classifiers. The author's found that IRUS outperformed several existing methods including the previously mentioned SMOTE.

2.3 Classification Methods

Decision trees are a family of classification methods such as C4.5 [13] and CSTree [10,16]. Decision trees are a data structure of nodes and branches. The initial node, known as the root node represents all records within a training data set D_T. Depending on the algorithm being used, a splitting point is calculated which partitions the records represented in the root node into several other nodes. The process is repeated for every node until a stopping criterion is met. The nodes with no further branches coming from them are considered leaf nodes. Each branch represents the corresponding splitting condition. For example, a branch could be $A_3 > 30.5$. The node which this branch leads to will only have records for which attribute A_3's value is greater than 30.5.

The advantage of decision trees over other classification methods is that they are easy to understand by anyone. A decision tree outputs a set of logic rules which can be used to classify new records. Extracting a logic rule from a decision tree is performed by tracing a path from the root node to a leaf node and recording the branch's conditions. For example, a logic rule could look like:

IF $A_2 > 500$ AND $A_6 > 200$ AND $A_9 = $ "red" THEN POSITIVE.

The class value from the logic rule is the majority class value in the leaf node. To classify a new record, the record is tested against the conditions of each logic rule until a rule is found for which the record satisfies all conditions. The classification is then taken as the value written after "THEN". By reading the extracted logic rules, we can gain insights into the data. For example, a previous SDP study [19] extracted the following rule:

"If a module's Halstead length is greater than 44 and the number of blank lines is less than 6 then assuming the module is defect free is the safer choice since it is 2.3 times more costly to treat it as defective."

The core difference between most decision tree algorithms is how the splitting point is calculated. For example, C4.5 calculates the *gain ratio* [13] for each possible split, then chooses the split with the highest gain ratio. In this way, C4.5 optimizes for classification accuracy. Since CSTree aims to optimize for cost, it's splitting points are calculated using *expected misclassification cost (EMC)*. EMC is calculated using two simple equations. These are the the cost of predicting all records within a data set as positive, and all as negative (Written as C_P and C_N respectively). These values are calculated as shown in Eqs. 1 and 2 where C_{TP} is the cost of a true positive, N_{TP} is the number of true positives and similarly for FP, FN, and TN.

$$C_P = N_{TP} \times C_{TP} + N_{FP} \times C_{FP} \qquad (1)$$

$$C_N = N_{TN} \times C_{TN} + N_{FN} \times C_{FN}. \qquad (2)$$

Decision forest algorithms such as SysFor [8] and CSForest [18,19] build multiple decision trees and classify new records by combining the individual tree classifications. SysFor is a non-cost-sensitive technique and builds multiple C4.5 trees. This is done by first finding the set of best splitting points for the original dataset, and using each as the root node split for a decision tree. The normal C4.5 procedure continues after this split. Since SysFor finds the splits based on gain ratio, the first SysFor tree is identical to the tree produced by C4.5. CSForest is similar to SysFor however it uses EMC instead of gain ratio and CSTree instead of C4.5. Therefore it is a cost-sensitive algorithm.

CSForest has been shown to outperform CSTree for the SDP problem [18,19]. However, CSForest does not account for the class imbalance problem. The author's of CSForest also proposed an extension of CSForest called BCSForest [19] which incorporated an oversampling method into the CSForest algorithm. The experimental results demonstrated that BCSForest did not consistently outperform CSForest for the SDP problem. However since it did outperform CSForest in some cases, it demonstrated that there is room for improvement by incorporating a class imbalance treatment strategy.

2.4 Cost-Sensitive Classification for Class Imbalance Treatment

Recall from Sect. 1 the two types of costs that can be used as input for cost-sensitive classification: *domain costs* and *balancing costs*. Standoff [20] is a method for producing a classifier using balancing costs. This means that the resulting classifier is not cost-sensitive, but it is not as negatively affected by the class imbalance problem. Standoff does not use a single cost matrix for all records. Instead, it first finds the set of clusters within the minority class records and the set of clusters within the majority class records. These clusters are called class-specific-clusters [21]. A cost-matrix is then generated for each cluster. These cost-matrices are then used as input for a cost-sensitive classification algorithm. The authors of Standoff modified the cost-sensitive algorithm MetaCost [4] to use multiple cost-matrices for use in their study. Many cost-sensitive algorithms have been proposed in the literature since MetaCost. However, there has been no study on the effectiveness of Standoff when using a more recent cost-sensitive algorithm.

RBClust [21] is a fast method for finding class-specific-clusters. Since it is specifically designed for finding class-specific-clusters, it can be used in the Standoff algorithm. However, the effectiveness of doing so has not yet been studied within the literature.

3   Our Framework: BCF

We propose *Balanced Cost Framework (BCF)*, a classification framework for addressing class imbalance and cost-sensitivity. BCF is shown in Fig. 1. Within this Figure, blue rectangles represent processes, and orange ellipses represent inputs or outputs.

Standoff generates the balancing cost-matrices. However to do this, Standoff requires the class-specific-clusters to be found, and the training dataset to be supplied. In BCF, the class-specific-clusters are found from the training dataset using RBClust. Once the balancing cost-matrices are found through Standoff, BCF combines then with the user defined domain cost matrix. This cost-combining method is defined in the following section. Finally, these cost-matrices are used as input into our modified CSForest algorithm. The original CSForest algorithm was presented in the previous section.

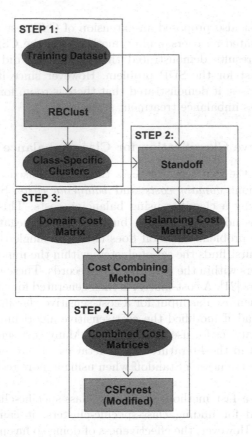

Fig. 1. The proposed framework: BCF (Color figure online)

3.1 Step 1: Generation of Class Specific Clusters (CSCs)

In Step 1, our framework uses the recent class specific clustering algorithm
RBClust [21]. This step is shown in the section of Fig. 1 labeled *Step 1*. RBClust
takes the training dataset as input and produces a set of class specific clusters.
In addition, RBClust requires two parameters to be set. The first parameter m is
a rule based classification algorithm. The second parameter θ controls the minimum size of a cluster. We use the same settings as used in the original RBClust
paper [21], m = C4.5 [13] and $\theta = 0.02$. Also, as advised, we turn off pruning in
C4.5 since overfitting is an advantage for RBClust.

3.2 Step 2: Calculation of Record Specific Balancing Costs

After finding the class specific clusters in Step 1, they are passed as input to the
Standoff [20] algorithm. The original training dataset is also passed to Standoff.
The Standoff algorithm then calculates the balancing costs for each record. We
slightly modify the Standoff algorithm. Originally, the Standoff algorithm used

the balancing costs in the cost-sensitive algorithm *MetaCost* [4]. However, in our implementation, we only use Standoff to find the balancing costs. In Step 3, these costs are combined with the domain costs. Finally, in Step 4, we use a modified CSForest [18,19] instead of MetaCost for training a cost-sensitive classifier using the combined costs.

The Standoff algorithm requires the setting of three parameters, D_T - the training dataset, L a cost-sensitive algorithm, and G a clustering method. Our framework accounts for all three of the Standoff parameters. Firstly, the training dataset is passed. Since we are only using Standoff for generating the balancing cost matrices in Step 2, we do not need to specify L. Lastly, since the class-specific-clusters are found in Step 1, we do not need to specify G. It should be noted that the original Standoff [20] required a clustering algorithm to run twice, once on the majority class records, and once on the minority class records. However, RBClust does not need to separate the data into the majority and minority subsets. Therefore, by using RBClust in our framework, we eliminate the need to run the clustering algorithm twice.

3.3   Step 3: Combination of Domain Costs and Balancing Costs

In this subsection, we empirically compare different methods for combining balancing cost matrices with a domain cost matrix. In Sect. 4, we use the most reliable combination method in our proposed framework.

We will now define some notations for the domain cost matrix and the balancing cost matrices. For the domain costs, we write the TP, TN, FP and FN costs as C_{TP}^{δ}, C_{TN}^{δ}, C_{FP}^{δ}, and C_{FN}^{δ} respectively. Similarly for balancing costs, we replace δ with β. Finally, the combined costs replace δ with γ.

Two Simple Combination Methods: We now experiment two simple techniques. These techniques involve either adding or multiplying the domain costs and balancing costs. The addition technique computes the combined costs as the addition of the corresponding domain and balancing costs. For example $C_{TP}^{\gamma} = C_{TP}^{\delta} + C_{TP}^{\beta}$. The difference between the addition technique and the multiplication technique is that the corresponding costs are multiplied instead of added.

These two simple techniques are compared against a CSForest classifier. Since CSForest is a recent cost-sensitive method proposed for SDP, it acts as a reasonable baseline for performance. The data set used for evaluation is MC2′, a publicly available software defect prediction data set [1]. We measure the effectiveness of the resulting classifier in total cost. The equation for total cost is given in Eq. 3. Note that total cost is calculated using the domain costs only. We use 10-fold stratified cross validation in order to reduce variance in this analysis.

$$TotalCost = (N_{TP} \times C_{TP}^{\delta}) + (N_{FP} \times C_{FP}^{\delta}) + (N_{TN} \times C_{TN}^{\delta}) + (N_{FN} \times C_{FN}^{\delta}) \quad (3)$$

We can observe from Table 2 that addition and multiplication methods are not lowering total cost from the CSForest classifier. Thus, the results for addition and multiplication are not promising.

Table 2. MC2′ total cost comparison (Lower the better)

CSForest	Addition	Multiplication
135	136	134

Using Balancing Costs in the Prediction Phase: Typically, cost-sensitive methods use domain-costs in both the classifier training process and the classifier prediction process. In our framework, we are using CSForest which first uses the costs in training the classifier. The costs are then used in CSVoting when classifying a new record.

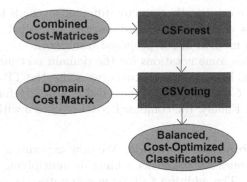

Fig. 2. Combined cost-matrices in CSForest, domain cost-matrix in CSVoting

Table 3. MC2′ total cost comparison (using only domain costs in CSVoting)

CSForest	Addition	Multiplication
135	124	132

In Table 2, the combined cost-matrices were used in both CSForest and CSVoting. However, it is possible that a better result could be achieved by only using the combined matrices once. Thus, we analyse the performance of using the combined cost-matrices when building the CSForest classifier, but only the domain costs in CSVoting. This way, the classifications are optimized for total cost in the CSVoting step. This process is illustrated in Fig. 2.

We now compare the above approach against using the combined cost-matrices for both CSForest and CSVoting. Note that in Table 2, we used the latter approach, but in Table 3 we use the former. Thus, a comparison can be made between the two tables. We also include the total cost of a CSForest classifier for comparison.

In comparison to Table 2, the multiplication technique performs slightly better, and the addition technique is performing much better. Therefore, we decide to use only domain costs in CSVoting. Before deciding whether to use the addition or multiplication technique, we first compare with a slightly more intelligent approach.

Only Changing the FP and FN Costs: We now analyse an alternative to the two simple techniques of addition and multiplication. When generating balancing costs, it is the ratio between the FP and FN costs that is important [20]. Therefore, we decide to change only the values of FP and FN in the domain costs to generate the combined cost-matrices. This allows us to maintain the domain costs of TP and TN classifications.

To represent both the balancing cost-matrices and the domain cost-matrix for the FP and FN costs, we introduce a weight a and b for the balancing and domain cost matrix respectively. Therefore, the combined cost matrix can be written as shown in Table 4. When setting balancing cost-matrices C_{FP}^{β} can be set arbitrarily since C_{FN}^{β} is calculated as a factor of C_{FP}^{β}. In this study, we set $C_{FP}^{\delta} = 1$. When using cost-sensitive classification to deal with class imbalance, C_{FP}^{β} is typically set to 1 [20]. Therefore, in order to simplify our calculations, we set C_{FP}^{γ} to C_{FP}^{δ}, that is, both of these values are one.

Table 4. Combining only FP and FN costs

		Actual	
		Positive	Negative
Predicted	Positive	C_{TP}^{δ}	C_{FP}^{δ}
	Negative	$a \times (C_{FN}^{\beta}) + b \times (C_{FN}^{\delta})$	C_{TN}^{δ}

We experiment with applying different sets of values for weights a and b where $a + b = 1.0$. They range from 0.0 to 1.0 and with pitch width 0.1. For example, when $a = 0.5$ and $b = 0.5$, this is the same as taking the average between C_{FN}^{β} and C_{FN}^{δ} for computing C_{FN}^{γ}. Figure 3 shows a clear trend for the best combination to be at $a = 0.5$ and $b = 0.5$.

When using this approach for combining the cost-matrices, we find that we can further reduce the total cost in the MC2′ data set to 122. Finally, in comparison with the CSForest classifier, our approach generates classifications which are approximately 10% less costly than CSForest. Therefore, we choose to use the combined cost-matrix in Table 4 for $a = 0.5$ and $b = 0.5$ and only the domain costs in CSVoting as shown in Fig. 2. This figure also shows a second degree trend curve. This curve helps to further illustrate the trend of performance for the different weight settings.

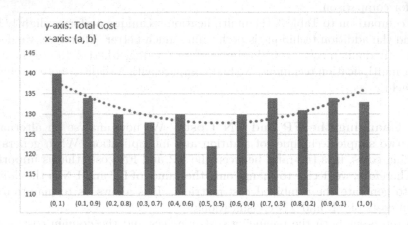

Fig. 3. Total cost comparison for different weight combinations

3.4 Step 4: Cost-Sensitive Classification Using Modified CSForest

CSForest [18,19] is a recent cost-sensitive classification method. It requires a single cost-matrix as input and thus does not handle multiple cost-matrices. This section details our modification of CSForest to use multiple cost-matrices rather than a single unchanging cost-matrix. This is done so that we can use the balancing costs generated by Standoff. Since CSForest trains a set of CSTree [10,16] decision trees, this modification is also applied to the CSTree algorithm.

A core step of CSForest is to calculate the cost of labeling an entire set of records as either negative or positive. By modifying this step to use multiple cost-matrices, this will in turn modify CSForest to use multiple cost-matrices. Each record has an associated cost-matrix. Thus, each record has a set of associated costs for TP, TN, FP, and FN classifications. Let $C_{TP_i}^{\beta}$ be the cost associated with a true positive classification for the ith record from the set of records R. Similarly, we can also write TN_i, FP_i and FN_i. Also, let the set of positive class records and the set of negative class records be written as P and N respectively. The modified equation for calculating the cost of labeling a set of records as positive can be written as follows.

$$C_P^{\beta} = \sum_{i=1}^{|R|} \begin{cases} C_{TP_i}^{\beta} & \text{if } R_i \in P \\ C_{FP_i}^{\beta} & \text{if } R_i \in N \end{cases} \tag{4}$$

Similarly, we can write the cost of labeling a set of records as negative as follows:

$$C_N = \sum_{i=1}^{|R|} \begin{cases} C_{FN_i}^{\beta} & \text{if } R_i \in P \\ C_{TN_i}^{\beta} & \text{if } R_i \in N \end{cases} \tag{5}$$

By using the equations defined above for C_P and C_N in Eqs. 4 and 5, we are modifying CSForest to use record specific cost-matrices. Since CSForest builds a forest of CSTree decision trees, we also use Eqs. 4 and 5 when building each CSTree.

4 Experiments

In this section, we compare the effectiveness of the proposed framework against existing methods.

4.1 Experimental Setup

Compared Methods: SDP could be performed with or without addressing cost-sensitivity and/or class imbalance. Therefore we chose to compare our proposed framework with methods which address one, both or neither of these issues. This can be illustrated by a Venn diagram as shown in Fig. 4. The Venn diagram has segmented the methods into three areas, those that address class imbalance, those that address cost-sensitivity, and methods for classification. Note that the domain costs and balancing costs are inputs to cost-sensitive algorithms, but they are not methods by themselves. However, they are shown here for clarity. Figure 4 also helps to illustrate an important concept of the proposed framework. That is, it utilizes both the domain costs and the balancing costs in the classification process.

We compare our framework against seven existing methods on four publicly available NASA SDP [1] data sets. Recently, the cleanliness of these data sets were criticized and a data cleaning approach was proposed [17]. We use the cleaned versions of the data sets. In order to reduce variance, we have used stratified 10 cross-fold validation. All compared methods use their default parameter settings. We use the following domain costs: $C_{TP}^{\delta} = 1$, $C_{TN}^{\delta} = 0$, $C_{FP}^{\delta} = 1$ and $C_{FN}^{\delta} = 5$. This is consistent with some published studies on SDP [18,19].

4.2 Results and Discussion

The results from the experimental setup described in the previous subsection are shown in Table 5. We choose to discuss these results by closely examining several matchups described as follows:

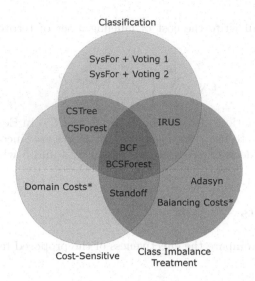

Fig. 4. Venn diagram: families of the compared methods

Table 5. Total cost comparison (Lower the better)

Data set	BCSForest	CSForest	CSTree	IRUS	Adasyn + C4.5	SysFor + Voting 1	SysFor + Voting 2	BCF
KC3'	166	174	170	157	155	156	160	**153**
MC2'	128	135	128	125	132	167	160	**123**
MW1'	128	136	173	179	147	122	130	**118**
CM1'	215	206	218	228	208	212	214	**203**
AVERAGE	159.25	162.75	172.25	172.25	160.5	164.25	166	**149.25**

Our Proposed Framework BCF vs all Other Methods: We find that in all four SDP data sets, BCF achieves the lowest total cost. On average, BCF provides an approximately 6.3 % improvement upon the next best method BCS-Forest [19]. This result suggests that BCF is a useful technique for cost-sensitive classification in SDP.

Cost-Sensitive Methods vs Non-cost-sensitive Methods: From Table 5 we can further calculate that the average performance of the cost-sensitive methods and non-cost-sensitive methods as 160.875 and 165.75 respectively. Although cost-sensitive methods provide a lower cost on average, this is not always the case. For example, in data set KC3', the non-cost-sensitive methods achieve approximately 5.2 times lower cost than the cost-sensitive methods.

Methods Which Address Class Imbalance vs Those that Don't: Similar to the previous comparison, we can further calculate the averages of those methods that address class imbalance and those that don't. These averages are

160.3125 and 166.3125 respectively. This difference in averages demonstrates that addressing class imbalance further reduces cost on average by approximately 3.61 %.

4.3 Extracted Knowledge

The study in which CSForest was proposed [19] also provided some of the knowledge that was extracted from the logic rules which CSForest generated. In Sect. 4.2 we showed that our proposed framework outperformed CSForest. Therefore, we have chosen to also present some of the logic rules which were generated by our proposed framework. The following knowledge has been extracted from the data sets MC2', MW1', and KC3'. We have simply chosen the three logic rules which we think are the most interesting and understandable as follows.

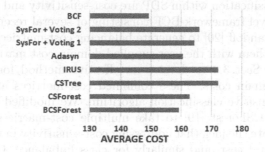

Fig. 5. Average total cost over all data sets (Lower the better)

Extracted Knowledge 1: If the number of conditions is greater than 15 and the number of blank lines is less than 11, then searching the module for bugs is 74 % less costly than assuming it contains none.

Extracted Knowledge 2: Interestingly, one rule found that out of the 17 modules in MC2' which had Halstead Length < 41, none contained bugs. This suggests that such a module is not worth the time to search for bugs in this module. Since the cost of assuming such a module is defect-free is 0, we cannot express it as a percentage of searching the module for defects. However, we can calculate the difference between these costs. If we multiply each cost used in our SDP domain cost-matrix by 50 we can get a realistic figure. This is supported by one study which multiplied a similar set of costs by $50 to achieve a "reasonable amount". Using Eq. 1 we can calculate that the cost of searching a module which has Halstead Length < 41 as $0 \times 0 + 1 \times 50 = \$50$. Therefore, for each such module searched for bugs, $50 is wasted.

Extracted Knowledge 3: If a module's Halstead Effort is less than 40456.31 and the number of unique operands is less than 21 then the probability that the module is defective is only approximately 7 % higher than defect-free. Halstead Effort can be translated into an estimate of seconds taken to program by dividing by 18 [5]. Therefore, we can extend this knowledge to the following: If a module takes less than 37 minutes to program and contains less than 21 unique operands then it should be searched for bugs since assuming it is defect-free is approximately 166 % more costly. It is interesting to note that even when there is a very small difference in probability between the module between defective, and defect-free (7 %), the cost difference is still significant (166 %). Note that trivial modules with very small Halstead Effort and number of unique operands still satisfy this rule such as getters and setters in object-oriented code.

5 Conclusion

Two issues for classification within SDP are cost-sensitivity and class imbalance. We proposed a novel framework BCF consisting of several recent methods. Our framework uses Standoff [20] to generate balancing cost-matrices for all records. It then combines them with the user-supplied domain cost matrix. We explored experimentally in Sect. 3.3 to develop an effective method for combining the balancing and domain costs. These combined cost-matrices are then used as input to a cost-sensitive classification algorithm. We modified the recent cost-sensitive method CSForest [19] to take multiple cost-matrices as input. Our experiments validate the idea that addressing cost-sensitivity in the classification can reduce the total cost, and similarly for class imbalance. Our experiments provide evidence that BCF can outperform several recent methods.

Since our framework builds decision forests, we are able to perform knowledge discovery on the studied NASA MDP data sets. In Sect. 4.3 we shared three insights which were found using our proposed framework. Our future work involves further experimental exploration to determine what factors affect the optimal method for combining balancing and domain costs.

References

1. Nasa-software defect datasets. http://openscience.us/repo/defect/mccabehalstedl. Accessed 24 July 2016
2. Bunkhumpornpat, C., Sinapiromsaran, K., Lursinsap, C.: Dbsmote: density-based synthetic minority over-sampling technique. Appl. Intell. **36**(3), 664–684 (2012)
3. Chawla, N.V., Bowyer, K.W., Hall, L.O., Kegelmeyer, W.P.: Smote: synthetic minority over-sampling technique. J. Artif. Intell. Res. **16**, 321–357 (2002)
4. Domingos, P.: Metacost: A general method for making classifiers cost-sensitive. In: Proceedings of the fifth ACM SIGKDD International Conference on Knowledge Discovery and Data Mining, pp. 155–164. ACM (1999)
5. Halstead, M.H.: Elements of software science. North-Holland (1977)
6. He, H., Bai, Y., Garcia, E.A., Li, S.: Adasyn: adaptive synthetic sampling approach for imbalanced learning. In: 2008 IEEE International Joint Conference on Neural Networks (IEEE World Congress on Computational Intelligence), pp. 1322–1328. IEEE (2008)

7. Islam, M.Z.: EXPLORE: a novel decision tree classification algorithm. In: MacKinnon, L.M. (ed.) BNCOD 2010. LNCS, vol. 6121, pp. 55–71. Springer, Heidelberg (2012). doi:10.1007/978-3-642-25704-9_7

8. Islam, Z., Giggins, H.: Knowledge discovery through sysfor: a systematically developed forest of multiple decision trees. In: Proceedings of the Ninth Australasian Data Mining Conference, vol. 121, pp. 195–204. Australian Computer Society, Inc. (2011)

9. Jiang, T., Tan, L., Kim, S.: Personalized defect prediction. In: 2013 IEEE/ACM 28th International Conference on Automated Software Engineering (ASE), pp. 279–289. IEEE (2013)

10. Ling, C.X., Sheng, V.S., Bruckhaus, T., Madhavji, N.H.: Maximum profit mining and its application in software development. In: Proceedings of the 12th ACM SIGKDD International Conference on Knowledge Discovery and Data Mining, pp. 929–934. ACM (2006)

11. McCabe, T.J.: A complexity measure. IEEE Trans. Softw. Eng. 4, 308–320 (1976)

12. Öztürk, M.M., Cavusoglu, U., Zengin, A.: A novel defect prediction method for web pages using k-means++. Expert Syst. Appl. 42(19), 6496–6506 (2015)

13. Quinlan, J.R.: C4. 5: programs for machine learning. Elsevier (2014)

14. Ramentol, E., Caballero, Y., Bello, R., Herrera, F.: Smote-rsb*: a hybrid preprocessing approach based on oversampling and undersampling for high imbalanced data-sets using smote and rough sets theory. Knowl. Inf. Syst. 33(2), 245–265 (2012)

15. Ricky, M.Y., Purnomo, F., Yulianto, B.: Mobile application software defect prediction. In: 2016 IEEE Symposium on Service-Oriented System Engineering (SOSE), pp. 307–313. IEEE (2016)

16. Sheng, V.S., Gu, B., Fang, W., Wu, J.: Cost-sensitive learning for defect escalation. Knowl. Based Syst. 66, 146–155 (2014)

17. Shepperd, M., Song, Q., Sun, Z., Mair, C.: Data quality: some comments on the nasa software defect datasets. IEEE Trans. Softw. Eng. 39(9), 1208–1215 (2013)

18. Siers, M.J., Islam, M.Z.: Cost sensitive decision forest and voting for software defect prediction. In: Pham, D.-N., Park, S.-B. (eds.) PRICAI 2014. LNCS (LNAI), vol. 8862, pp. 929–936. Springer, Heidelberg (2014). doi:10.1007/978-3-319-13560-1_80

19. Siers, M.J., Islam, M.Z.: Software defect prediction using a cost sensitive decision forest and voting, and a potential solution to the class imbalance problem. Inf. Syst. 51, 62–71 (2015)

20. Siers, M.J., Islam, M.Z.: Standoff-balancing: a novel class imbalance treatment method inspired by military strategy. In: Pfahringer, B., Renz, J. (eds.) AI 2015. LNCS (LNAI), vol. 9457, pp. 517–525. Springer, Heidelberg (2015). doi:10.1007/978-3-319-26350-2_46

21. Siers, M.J., Islam, M.Z.: Rbclust: High quality class-specific clustering using rule-based classification. In: European Symposium on Artificial Neural Networks, ESANN 2016 (2016, Accepted)

22. Tahir, M.A., Kittler, J., Yan, F.: Inverse random under sampling for class imbalance problem and its application to multi-label classification. Pattern Recogn. 45(10), 3738–3750 (2012)

Unsupervised Hypergraph Feature Selection with Low-Rank and Self-Representation Constraints

Wei He[1,2], Xiaofeng Zhu[1,2(✉)], Yonggang Li[1,2], Rongyao Hu[1,2], Yonghua Zhu[3], and Shichao Zhang[1,2(✉)]

[1] Guangxi Key Lab of Multi-source Information Mining & Security, Guangxi Normal University, Guilin 541004, Guangxi, China
[2] College of CS & IT, Guangxi Normal University, Guilin 541004, Guangxi, China
seanzhuxf@gmail.com
[3] Guangxi University, Nanning 530004, China

Abstract. Unsupervised feature selection is designed to select a subset of informative features from unlabeled data to avoid the issue of 'curse of dimensionality' and thus achieving efficient calculation and storage. In this paper, we integrate the feature-level self-representation property, a low-rank constraint, a hypergraph regularizer, and a sparsity inducing regularizer (*i.e.*, an $\ell_{2,1}$-norm regularizer) in a unified framework to conduct unsupervised feature selection. Specifically, we represent each feature by other features to rank the importance of features via the feature-level self-representation property. We then embed a low-rank constraint to consider the relations among features and a hypergarph regularizer to consider both the high-order relations and the local structure of the samples. We finally use an $\ell_{2,1}$-norm regularizer to result in low-sparsity to output informative features which satisfy the above constraints. The resulting feature selection model thus takes into account both the global structure of the samples (via the low-rank constraint) and the local structure of the data (via the hypergraph regularizer), rather than only considering each of them used in the previous studies. This enables the proposed model more robust than the previous models due to achieving the stable feature selection model. Experimental results on benchmark datasets showed that the proposed method effectively selected the most informative features by removing the adverse effect of redundant/nosiy features, compared to the state-of-the-art methods.

Keywords: Low-rank representation · Subspace learning · Feature self-representation · Hypergraph

1 Introduction

With the development of information technology, huge quantity of data has produced around our lives. They, however, also contain huge amounts of redundancy and noise. Moreover, these data are often high-dimensional and unlabeled

J. Li et al. (Eds.): ADMA 2016, LNAI 10086, pp. 172–187, 2016.
DOI: 10.1007/978-3-319-49586-6_12

[30,39], thus unsupervised dimensionality reduction methods have been devising to reduce the dimensions of high-dimensional unlabeled data by removing irrelevant features and finding a compact subspace [35].

Dimension reduction methods includes feature selection methods [36,39] and subspace learning methods [7]. Feature selection methods, such as t-test and sparse based regression, directly remove redundant or noisy features from original high-dimensional data by preserving crucial features and thus resulting in interpretable results, while subspace learning methods aim to transform original high-dimensional features into a low-dimensional space to remove the adverse impact of redundant/noisy features and thus resulting in stable results but prohibiting for interpretation. This obviously motivates us to integrate them in a unified framework to achieve both interpretable and stable results.

In this paper, we propose an unsupervised feature selection method by taking into account a low-rank constraint on the regression coefficient matrix, the high-order relationship of the samples, the local structure of the samples, and the feature-level self-representation property. Specifically, we reconstruct each feature by other features via the property of the feature-level self-representation to achieve minimal reconstruction error. We then add a low-rank constraint on the regression coefficient with the motivation of that noise and redundancy in high-dimensional data always increase the rank of the feature matrix and the low-rank constraint considering the relations among the response variables helps in finding the actual rank of the feature matrix [15,41]. We further employ a hypergraph regularizer to preserve the local structure of the samples and take into account the high-order relations among the samples. We finally use an $\ell_{2,1}$-norm regularizer to achieve group sparsity which penalizes all coefficients in the same row of the regression matrix together for joint selection or un-selection in predicting the response variables. In this way, the $\ell_{2,1}$-norm regularizer conducts feature selection by assigning the reconstruction coefficients of the features which satisfy the above constraints as non-zeros and assigning the reconstruction coefficients of the features which obey the above constraints as zeros.

Compared to the previous feature selection methods, the proposed method has the following contributions:

– We use the feature-level self-representation property to represent each feature by other features since there are no labeled data for unsupervised feature selection methods. Such a method considers the importance of features by their representation ability. Moreover, different from the traditional sample-level self-representation methods [21] representing each sample by other samples, this paper focuses on the feature level which is used to rank the importance of the features.
– The feature matrix often shows higher rank than its actual case due to containing noise and redundancy. Low-rank constraints are used to find the actual rank of the feature matrix and have been widely used for statistics [1,33,42] but seldom for feature selection. This paper pushes the low-rank constraint on the $\ell_{2,1}$-norm regularizer to achieve both subspace learning (*i.e.,* Linear Discriminative Analysis (LDA) [6]) and feature selection.

- Different from the graph Laplacian which preserves the local structure of the samples, *i.e.*, preserving the neighborhood relationship of two high-dimensional samples in the low-dimensional space, the proposed hypergraph regularizer is devised to preserve the high-order relation among the samples, *i.e.*, preserving the neighborhood relationship of several samples (any number of data points). Obviously, the graph Laplacian [8] is a special case of the hypergraph regularizer and the high-order relationship is thus more robust to discover the inherent complex relations among the samples than the graph Laplacian [8], which is very popular in machine learning and data mining [5].

The left parts of this paper are organized as follows: Sect. 2 introduces related work of feature selection methods and Sect. 3 gives the details of our proposed feature selection model. In Sects. 4 and 5, respectively, we analyze experimental results and conclude this paper.

2   Related Work

Feature selection methods are widely used for reducing the dimensions of data. They select a subset of features in accordance with criteria, such as distinguishing features with good characteristics and correlating to the predefined goal [29,37]. The traditional feature selection methods include filer methods [4,6,24], wrapper methods [17,25] and embedded methods [22,28]. Subspace learning methods aim to learn a particular smaller part of an objective space of high dimensional data having a specific desired property. The common subspace learning methods include: principal component analysis (PCA), neighborhood preserving embedding (NPE), linear discriminant analysis (LDA) and locality preserving projection (LPP).

Previous feature selection methods can be classified into three groups, *e.g.*, supervised feature selection, semi-supervised feature selection, and unsupervised feature selection. Supervised feature selection method is evaluated by known class labels and usually selects features according to the labels of training data. As supervised feature selection encloses labels to conduct discriminative analysis, it is usually able to select discriminative features. For example, Nie *et al.* [18] proposed a method imposes $\ell_{2,1}$-norm minimization on both regularization and loss function to selection the significant. Furthermore, Zhu *et al.* [40] utilized the correlation between indicator vectors, and to project features into a canonical space to consider the canonical relationships between features of different modalities. Cheng *et al.* [3] employed joint LDA and LPP regularization term to take both local and global structures of data distribution into account with subspace learning. And a $\ell_{2,1}$-norm regularization term is be used in their objective function to select relevant features.

Semi-supervised feature selection mainly utilizes a small number of labeled samples and a large number of unlabeled samples for training and classification. For example, Lv *et al.* [23] proposed a discriminative semi-supervised feature selection method based on the idea of manifold regularization. Wang *et al.* [27] proposed to first learn the class labels of unlabeled samples in a new feature subspace induced by the learned feature weights, and then to use the learned class labels to define the margins for feature weight learning.

In many data mining applications, all kinds of reasons, such as unknown labels and time-consuming to obtain labels, make it difficult obtain labels, thus unsupervised feature selection has shown to be effective in alleviating the irrelevant and redundant features [26,43]. Compared with supervised feature selection method and semi-supervised feature selection method, unsupervised feature selection lacks the label information. Previous unsupervised feature selection methods mainly utilize a number of evaluation indicators to remove the redundant features. For example, Liu et al. [16] combined Laplacian score with distance-based entropy measure to conduct unsupervised feature selection. Nie et al. [19] proposed to use a corresponding score calculated in a trace ratio to conduct feature selection. Recent years feature self-representation theory has been proposed to conduct unsupervised feature selection. Self-similarity is widely exist in nature [34]. A part of an object can be represented by other parts of itself. For the under processing data, a feature can be represented by a linear combination of other features of the data. Motivated by the this fact, Zhu et al. [38] proposed a method utilizes self-representation of features which were imposed $\ell_{2,1}$-norm on coefficients to enforce sparsity. It could select features by representing the data matrix over itself.

With the variety of multimodal data, the traditional graph structure can't fully represent the relations of data, while the hypergraph catch more and more attentions because of its unique characteristic. Yu et al. [32] illustrated two advantages of hypergraph, comparing with the graph. And then design an adaptive hypergraph learning method for transductive image classification to conquer two disadvantages of hypergraph, which is how to generate hyperedges and how to handle a large set of hyperedges. Huang et al. [11] apply the hypergraph in the unsupervised image categorization. They construct the hyperedges based on the respective features that selected by region of interest (ROI) of each image. Besides, They took use of all the hyperedges rather than local hyperedges to describe the inherent relation of pictures, which can enhance the pictures clustering performance by representing the shape and appearance in finer detail.

In recent years, low-rank theory has been proposed, and it has been wildly applied in subspace learning and segmentation. Liu et al. [14] proposed low-rank representation(LRR) segmentation method by seeking the lowest-rank representation from all the representation to recover a low-rank data matrix from acquired observations. The LRR optimization problem is convex and computational procedure can be finished in polynomial time. Peng et al. [20] employed local affinity and distant repulsion on low-rank representation to construct the graph, which can preserve the local topological structure. They derive an efficient approach from augmented Lagrange multiplier method to optimize the model.

3   Approach

3.1   Notations

In this paper, matrices are denoted as boldface uppercase letters and vectors are written as boldface lowercases letters. The i-th row and j-th column of a data matrix $\mathbf{X}$ are denoted as $\mathbf{x}^i$ and $\mathbf{x}_j$, respectively. The Frobenius norm of a matrix $\mathbf{X}$ is defined as $||\mathbf{X}||_F = \sqrt{\sum_i ||\mathbf{x}_i||_2^2}$, and the $\ell_{2,1}$-norm of $\mathbf{X}$ is denoted as $||\mathbf{X}||_{2,1} = \sum_i ||\mathbf{x}_i||_2 = \sum_i \sqrt{\sum_j x_{ij}^2}$. We further denote the transpose operator, the trace operator, and the inverse, of a matrix $\mathbf{X}$, as $\mathbf{X}^T$, $tr(\mathbf{X})$ and $\mathbf{X}^{-1}$, respectively.

3.2   Method

Denote $\mathbf{X} \in \mathbb{R}^{n \times d}$ and $\mathbf{Y} = [\mathbf{y}_1, \mathbf{y}_2, ..., \mathbf{y}_c] \in \mathbb{R}^{n \times c}$, respectively, as the feature matrix and the response matrix, traditional methods [36] usually formulate the linear relationship between the feature matrix and the response matrix as follows:

$$\min_{\hat{\mathbf{W}}} ||\mathbf{Y} - \mathbf{X}\hat{\mathbf{W}}||_F^2 + \lambda ||\hat{\mathbf{W}}||_F^2 \qquad (1)$$

where $\hat{\mathbf{W}} \in \mathbb{R}^{d \times c}$ denotes the regression coefficient matrix and λ is the tuning parameter. Equation (1) can be solved by the Ordinary Least Square (OLS) methods but has the following limitations: (1) the OLS method outputs dense results but we expect to yield sparse results for conducting feature selection since not all features are useful for learning tasks; (2) there are no labeled data in unsupervised learning thus we should replace $\mathbf{Y}$ by other variables; (3) noise and redundancy in $\mathbf{X}$ usually increase its rank [14]. To address these issues, in this paper, we first propose the following objective function:

$$\min_{\mathbf{W},\mathbf{b}} ||\mathbf{X} - \mathbf{X}\mathbf{W} - \mathbf{e}\mathbf{b}||_F^2 + \lambda ||\mathbf{W}||_{2,1} \qquad (2)$$

where $\mathbf{W} \in \mathbb{R}^{d \times d}$ denotes the regression coefficient matrix, $\mathbf{b} \in \mathbb{R}^{1 \times d}$ is the bias term, and $\mathbf{e}$ is a column vector whose elements are ones.

Equation (2) addresses the first two issues by (1) replacing the Frobenius norm regularizer (i.e., the second term in Eq. (1)) with an $\ell_{2,1}$-norm regularizer; and (2) replacing the response matrix $\mathbf{Y}$ with the feature matrix $\mathbf{X}$. In Eq. (2), the first change leads to a feature-level self-representation, i.e., $||\mathbf{X} - \mathbf{X}\mathbf{W}||_F^2 = \sum_{i=1}^n ||\mathbf{x}^i - \mathbf{X}\mathbf{w}_i||_2^2$, where each feature $\mathbf{x}^i$ is represented by other features $\mathbf{X} = [\mathbf{x}^1, ..., \mathbf{x}^n]$, i.e., $\mathbf{x}^i \approx \sum_{j=1}^n \mathbf{x}^j w_{ij}$ (where w_{ij} measures the distance between the i-th feature and the j-th feature). The $\ell_{2,1}$-norm regularizer on $\mathbf{W}$ penalizes $\mathbf{W}$ in a way that encourages joint row sparsity, i.e., each row of $\mathbf{W}$ is either all zeros or all non-zeros, to select or un-select the corresponding features in $\mathbf{X}$.

We further use a low-rank constraint to address the third issue by using a product of two r-rank matrices to express $\mathbf{W}$ as: $\mathbf{W} = \mathbf{A}\mathbf{B}$, where $\mathbf{A} \in \mathbb{R}^{d \times r}$, $\mathbf{B} \in \mathbb{R}^{r \times d}$ and $r < d$. This makes $\mathbf{W}$ have low-rank r and thus Eq. (2) becomes:

$$\min_{\mathbf{A},\mathbf{B},\mathbf{b}} ||\mathbf{X} - \mathbf{X}\mathbf{A}\mathbf{B} - \mathbf{e}\mathbf{b}||_F^2 + \lambda ||\mathbf{A}\mathbf{B}||_{2,1} \qquad (3)$$

Equation (3) implies that the rank of the predicted matrix $\mathbf{XAB} \in \mathbb{R}^{n \times r}$ (*i.e.*, $\mathbf{X} = \mathbf{XAB} - \mathbf{eb}$) is less than r. That is, each of d columns of $\mathbf{X}$ can be linearly represented by less than r latent features $\mathbf{X}$. This actually considers the correlations among the features to conduct subspace learning on $\mathbf{X}$ and then outputs its low-dimensional representation $\mathbf{XAB}$. Therefore, the low-rank constraint on the coefficient matrix conducts subspace learning on the feature matrix by considering the correlations among the features. Moreover, the constraint $\|\mathbf{AB}\|_{2,1}$ can be further proved to preserve the global structure of the data (*i.e.*, conducting Linear Discriminant Analysis (LDA) [3]) in the optimization part of this paper.

Besides, preserving the local structure of the samples has also been regarded as very important for subspace learning. In this paper, we consider to preserve both the global and the local topological structures of the samples in the feature selection model. Regarding that each sample of the dataset is not isolated to others, *i.e.*, there is a topological structure among the samples, we expect to preserve such structure to effectively identify the noise, outliers and redundancy in the data set by the following formulation:

$$\min_{\mathbf{A},\mathbf{B}} \sum_{i,j} (\mathbf{x}_i \mathbf{AB} - \mathbf{x}_j \mathbf{AB})^2 s_{ij} \tag{4}$$

where $\mathbf{S} = [s_{i,j}] \in \mathbb{R}^{n \times n}$ is the similarity matrix to measure the distance between the i-th sample and the j-th sample. In the seminal work Locality Preserving Projection (LPP) [8], $s_{i,j}$ is defined via heat kernel $H(x_i, x_j) = \exp[-\frac{\|x_i - x_j\|^2}{\sigma}]$, where $\sigma \in \mathbb{R}^+$, to measure the similarity of two samples. However, such similarity cannot measure other relationships, such as the pair-wise sample relationship, which has been successfully used to characterize the high-order relations of the data and thus discovering the inherent complex relations among the data [2,10]. In this paper, we construct a hypergraph to achieve such a goal to replace the previous graph.

Specifically, a hypergraph is defined as a triple $\mathbf{G_H} = (\mathbf{V}, \mathbf{E}, \mathbf{w}_H)$, where $\mathbf{V} = \{\mathbf{x}_1, \mathbf{x}_2, \ldots, \mathbf{x}_n\} \in \mathbf{R}^{d \times n}$, is the data point, $\mathbf{E}$ represents hyper-edges and $\mathbf{w}$ assign a real value to each edge. Different from normal graph edges (only connect two points), a hyper-edge can contain any number of data points, so that the high order relationships of samples can be achieve to improve the performance of the model. Figure 1 illustrates a hypergraph example, where the hyper-edges are $\mathbf{E} = \{e_1 = \{v_1, v_2, v_9, v_{10}\}\}, e_2 = \{v_5, v_8, v_{12}\}, e_3 = \{v_5, v_6, v_7, v_{11}\}, e_4 = \{v_2, v_3, v_4, v_6, v_{10}\}\}$. Given the constructed hypergraph, we change Eq. (4) into (5) as follows:

$$\begin{aligned}
&\frac{1}{2} \sum_{i,j} (\mathbf{x}_i \mathbf{AB} - \mathbf{x}_j \mathbf{AB})^2 s_{ij} \\
&= \sum_i (\mathbf{A}^T \mathbf{B}^T \mathbf{x}_i d_{ij} \mathbf{x}_i{}^T \mathbf{AB}) - \sum_{ij} (\mathbf{A}^T \mathbf{B}^T \mathbf{x}_i s_{ij} \mathbf{x}_i{}^T \mathbf{AB}) \\
&= tr(\mathbf{A}^T \mathbf{B}^T \mathbf{XDX}^T \mathbf{AB}) - tr(\mathbf{A}^T \mathbf{B}^T \mathbf{XSX}^T \mathbf{AB})
\end{aligned} \tag{5}$$

where $\mathbf{S} = \mathbf{I} - \mathbf{D}_v^{-\frac{1}{2}} \mathbf{SW}_H \mathbf{D}_e^{-1} \mathbf{S}^T \mathbf{D}_v^{-\frac{1}{2}} \in \mathbb{R}^{n \times n}$, $\mathbf{I} \in \mathbb{R}^{n \times n}$ is an identity matrix, $\mathbf{D}_v$ and $\mathbf{D}_e$, respectively, are diagonal matrices for representing the row and the

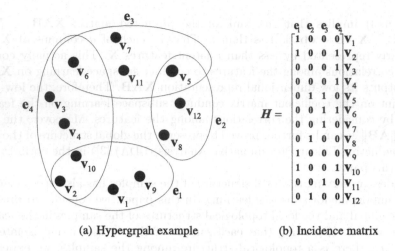

(a) Hypergrpah example   (b) Incidence matrix

Fig. 1. An illustration of a Hypergraph.

column sum of matrix $\mathbf{S}$, $\mathbf{W}_H$ is a diagonal matrix and its element represents the weight of hyperedges, and $\mathbf{D}$ is a diagonal matrix and its element is the sum of one row of $\mathbf{S}$. We embed Eq. (5) into (4) to obtain our final objective function as follows:

$$\min_{\mathbf{A},\mathbf{B},\mathbf{b}} ||\mathbf{X} - \mathbf{XAB} - \mathbf{eb}||_F^2 + \alpha tr(\mathbf{B}^T \mathbf{A}^T \mathbf{X}^T \mathbf{LXAB}) + \beta ||\mathbf{AB}||_{2,1} \qquad (6)$$

where $\mathbf{L} = \mathbf{D} - \mathbf{S}$ is the Laplacian matrix.

Equation (6) uses the feature-level self-representation loss function (*i.e.*, $||\mathbf{X} - \mathbf{XAB} - \mathbf{eb}||_F^2$) to rank the importance of the features by representing each features with the others, different from the traditional sample-level self-representation loss function representing each sample by all samples. Equation (6) also pushes a low-rank constraint on the regression coefficient (*i.e.*, $\mathbf{AB}$) to conduct subspace learning (*e.g.*, LDA) for preserving the global structure of the samples via taking into account the relations among features. In this way, the proposed feature selection model considers the relations of features via two different kinds of methods and thus enabling to result in a stable model. Besides, Eq. (6) further uses a hypergraph based regularizer to take into account the high-order relations among the samples to preserve the local structure of the samples. Therefore, our proposed method embeds subspace learning (including preserving both the global and the local structures of the samples) into the feature selection model to achieve interpretable and stable results.

3.3   Optimization

We employ the framework of Iteratively Reweighted Least Square (IRLS) [8] to solve Eq. (6), by iteratively optimizing one of the parameters (*i.e.*, $\mathbf{A}$, $\mathbf{B}$, and

b with other parameters fixed. Specifically, we iteratively conduct the following steps until one of the predefined stopping criteria is satisfied.

(i) Update b by fixing A and B. By setting the derivative of Eq. (6) with respect to **b** to zero, we have:

$$2\mathbf{e}^T\mathbf{X} + 2\mathbf{e}^T\mathbf{eb} - 2\mathbf{e}^T\mathbf{XAB} = 0 \tag{7}$$

After a simple mathematical transformation, we have:

$$\mathbf{b} = \frac{1}{n}\mathbf{e}^T\mathbf{X} - \frac{1}{n}\mathbf{e}^T\mathbf{XAB}. \tag{8}$$

By substituting Eq. (8) into (6), we have:

$$\min_{\mathbf{A},\mathbf{B}} \|\mathbf{X} - \mathbf{XAB} - \mathbf{e}(\frac{1}{n}\mathbf{e}^T\mathbf{X} - \frac{1}{n}\mathbf{e}^T\mathbf{XAB}\|_F^2 \\ + \alpha tr(\mathbf{B}^T\mathbf{A}^T\mathbf{X}^T\mathbf{LXAB}) + \beta\|\mathbf{AB}\|_{2,1} \tag{9}$$

where $\mathbf{H} = \mathbf{I} - \frac{1}{n}\mathbf{ee}^T \in \mathbb{R}^{n\times n}$ and $\mathbf{I} \in \mathbb{R}^{n\times n}$ is an identity matrix, so Eq. (6) can be rewritten as

$$\min_{\mathbf{A},\mathbf{B}} \|\mathbf{HX} - \mathbf{HXAB}\|_F^2 + \alpha tr(\mathbf{B}^T\mathbf{A}^T\mathbf{X}^T\mathbf{LXAB}) + \beta\|\mathbf{AB}\|_{2,1}. \tag{10}$$

(ii) Update B by fixing b and A. By setting the derivative of Eq. (10) with respect to **B** to zero, we have:

$$\mathbf{B} = (\mathbf{A}^T(\mathbf{X}^T\mathbf{HXA} + \alpha\mathbf{X}^T\mathbf{LX} + \beta\hat{\mathbf{D}})\mathbf{A})^{-1}\mathbf{A}^T\mathbf{X}^T\mathbf{HX} \tag{11}$$

where $\hat{\mathbf{D}} \in \mathbb{R}^{d\times d}$ is a diagonal matrix with the element $\hat{d}_{ii} = \frac{1}{2\times\|\mathbf{w}^i\|^2}, i = 1, ..., d$.

(iii) Update A by fixing b and B. By setting the derivative of Eq. (10) with respect to **A** to zero, we change Eq. (10) to:

$$\min_{\mathbf{A}} \quad tr(\mathbf{X}^T\mathbf{H} - \mathbf{X}^T\mathbf{HXAM}^{-1}\mathbf{A}^T\mathbf{X}^T\mathbf{H})(\mathbf{HX} - \mathbf{HXAM}^{-1}\mathbf{A}^T\mathbf{X}^T\mathbf{HX}) \\ + \alpha tr(\mathbf{X}^T\mathbf{HXAM}^{-1}\mathbf{A}^T\mathbf{X}^T\mathbf{HLHXAM}^{-1}\mathbf{A}^T\mathbf{X}^T\mathbf{HHX}) \\ + \beta tr(\mathbf{X}^T\mathbf{HHXAM}^{-1}\mathbf{A}^T\hat{\mathbf{D}}\mathbf{AM}^{-1}\mathbf{A}^T\mathbf{X}^T\mathbf{HHX}) \tag{12}$$

where $\mathbf{M} = \mathbf{A}^T(\mathbf{X}^T\mathbf{HX} + \alpha\mathbf{X}^T\mathbf{LX} + \beta\hat{\mathbf{D}})\mathbf{A}$ and thus $\mathbf{B} = \mathbf{M}^{-1}\mathbf{A}^T\mathbf{XHX}^T$. We further change Eq. (12) to the following:

$$\max_{\mathbf{A}} \quad tr(\mathbf{A}^T(\mathbf{X}^T\mathbf{HX} + \alpha\mathbf{X}^T\mathbf{LX} + \beta\hat{\mathbf{D}})\mathbf{A})^{-1}\mathbf{A}^T\mathbf{X}^T\mathbf{HXX}^T\mathbf{HXA}) \tag{13}$$

which is equivalent to:

$$\max_{\mathbf{A}} \frac{\mathbf{X}^T\mathbf{HXX}^T\mathbf{HX}}{\mathbf{X}^T\mathbf{HX} + \alpha\mathbf{X}^T\mathbf{LX} + \beta\hat{\mathbf{D}}} \tag{14}$$

By choosing the top ranked eigenvectors of Eq. (14), we yield the optimal **A** and thus yielding the optimal **B** via Eq. (11). It is noteworthy that Eq. (14) actually conducts LDA. We list the pseudo of our proposed optimization method to solve Eq. (6) in Algorithm 1.

Algorithm 1. The pseudo of solving the Eq. (6)

Input: X $\in \mathbb{R}^{n \times d}$ and parameters α and β.
Output: A $\in \mathbb{R}^{d \times r}$ and **B** $\in \mathbb{R}^{r \times c}$
1: Initialize: **A** $\in \mathbb{R}^{d \times r}$, **B** $\in \mathbb{R}^{r \times c}$.
2: **repeat**
3:    Compute $\hat{\mathbf{D}}$ according to $\hat{d}_{ii} = \frac{1}{2 \times \|\mathbf{w}^i\|^2}$;
4:    Update **b** via Eq. (8);
5:    Update **B** via Eq. (11);
6:    Update **A** by eigen-decomposition of Eq. (14);
7: **until** Eq. (6) *converges*;

---

4 Experiments

In this section, we regard our proposed feature selection model as Self-representation Hypergraph Low-rank dimensionality Reduction (SHLR for short) method to compare with the comparison methods in terms of classification performance. Specifically, we first used each dimensionality reduction method to map original high-dimensional data into low-dimensional space, and then used the resulting reduced data to conduct classification with Support Vector Machine (SVM) via the LIBSVM toolbox[1].

4.1 Experimental Settings

We evaluated the proposed method by comparing with the competing methods on three benchmark data sets. The selected datasets include ORL and Umist[2] and LungCancer[3], where both ORL and LungCaner were set for binary classification tasks and Umist was set for the multi-classification task. We summarized the details of the used data sets in Table 1.

We compared the proposed SHLR method with the following state-of-the-art methods:

- **NFS:** Non Feature Selection(NFS) uses original data to conduct classification. We use NFS as a benchmark method to demonstrate if dimensionality reduction methods are meaningful in real applications.
- **LE:** Laplacian Eigenmaps(LE) [13] build a similar graph (corresponding matrix) to reconstruct the local feature data manifold. The goal of LE was designed to preserve the global structures of data.
- **LPP:** Locality Preserving Projection(LPP) [31] preserves the neighborhood relation of each sample to conduct subspace learning. Hence, LPP is a local subspace learning method.

---

[1] Available at http://www.csie.ntu.edu.tw/~cjlin/libsvm/.
[2] Available at http://see.xidian.edu.cn/vipsl/database_Face.html.
[3] Available at http://archive.ics.uci.edu/ml/.

- **RFS**: Robust feature selection [18] uses a robust loss function to remove the impact of noise and an $\ell_{2,1}$-norm regularization to conduct feature selection.
- **JELSR**: Embedding Learning and Sparse Regression [9] adding the $\ell_{2,1}$-norm regularization to local liner regression and joint embedding learning and sparse regression to perform feature selection.

In the comparison methods, NFS does not conduct dimensionality reduction and can be regarded as the baseline of other dimensionality reduction methods, such as LE, LPP, RFS, JELSR and our SHLR. In dimensionality reduction methods, the methods (such as LPP, and LE) belong to subspace learning, while feature selection methods include RFS and JELSR. RFS does not consider subspace learning into the feature selection model. In contrast, our method devised a feature-level self-representation loss function to present each feature by other features.

In experimental setting part, following the settings in [3,12], we use a 10-fold cross-validation method in each experiment. Specifically, in each round, the original data were randomly divided into 10 parts where 9 of them for training and remaining one part for testing. We repeated the whole process 10 times to avoid the possible bias during data set partitioning for cross-validation. The final result was computed by averaging results from all experiments.

For the model selection, we set the parameters α and β both in a range of $\{0.01, 1, 10, 1000\}$, $c \in \{2^{-5}, \ldots, 2^5\}$, and $g \in \{2^{-5}, \ldots, 2^5\}$ in SVM by a 5-fold inner cross-validation. For fair comparison, we also conducted both 5-fold inner cross-validation and 10-fold outer cross-validation to conduct model selection for each comparison method.

In this paper, we used classification accuracy (ACC) as evaluation metric to measure the performance of binary classification and multi-class classification. Specifically, we defined classification accuracy (ACC) as follows:

$$ACC = \frac{N_{correct}}{N} \tag{15}$$

where N means the number of samples in a data set and $N_{correct}$ means correct classified samples. We further used other three additional measures (*e.g.*, sensitivity (SEN), specificity(SPE) and Area Under Curve (AUC)) to evaluate binary classification. In binary classification, the outcomes were denoted as either Positive (P) or Negative (N). We further parted the results into four groups, *i.e.*, True Positive (TP), False Positive (FP), True Negative (TN), and False Negative (FN). Therefor, we defined SEN as:

$$SEN = \frac{N_{TP}}{N_P} \tag{16}$$

where N_{TP} means the number of TP, and N_P means the number of actual positive labels.

SPC was defined as:

$$SPC = \frac{N_{TN}}{N_N} \tag{17}$$

Table 1. Summary of the benchmark datasets

Datasets	Instances	Features	Classes	Types of data
Umist	575	644	20	Multivariate
ORL	1943	3289	2	Multivariate
LungCancer	32	56	2	Multivariate

where N_{TN} means the number of TN, and N_N means the number of actual negative labels.

AUC was defined as:

$$AUC = \frac{S_0 - \frac{N_P(N_P+1)}{2}}{N_P N_N} \tag{18}$$

where $S_0 = \sum_{i=1}^{N_p} r_i$ and r_i is the i-th positive sample sort position.

4.2   Parameter Sensitivity

To evaluate the sensitivity of the parameter, *i.e.,* α and β in our objective function, we varied both the value of α and β in a range of $\{0.01, 1, 10, 1000\}$. We took ACC as the metric and showed the results in Fig. 2.

From Fig. 2, we knew that the proposed objective function is sensitive to the parameter's setting. That is, the proposed method achieved different classification accuracy with different values of α and β. This indicated that it is necessary to conduct proper dimensionality reduction in our experiments. Moreover, we should tune the parameter to improve the classification results.

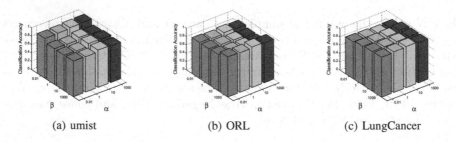

(a) umist   (b) ORL   (c) LungCancer

Fig. 2. Classification accuracy of the proposed method with different parameter setting on different data sets.

4.3 Experimental Results

We reported classification accuracy of all methods in Table 2. Figure 3 showed classification accuracy with standard deviations of each experiment for all methods on different data sets.

Table 2. Classification accuracy (average ACC of 10 times run ± STD) of all methods. The best results were emphasized through bold-face.

Data Sets	NFS	LE	LPP	RFS	JELSR	SHLR
Umist	0.904 ± 0.037	0.952 ± 0.029	0.965 ± 0.019	0.975 ± 0.016	0.971 ± 0.015	**0.986 ± 0.003**
ORL	0.883 ± 0.066	0.906 ± 0.069	0.928 ± 0.060	0.941 ± 0.033	0.947 ± 0.048	**0.973 ± 0.031**
LungCancer	0.718 ± 0.056	0.766 ± 0.055	0.779 ± 0.060	0.773 ± 0.054	0.793 ± 0.035	**0.810 ± 0.026**
Average	0.835 ± 0.274	0.754 ± 0.157	0.780 ± 0.046	0.896 ± 0.034	0.903 ± 0.033	**0.923 ± 0.020**

It is observed that the proposed method achieves the best performance, compared to all the comparison methods. Besides, we had the following observations:

- It is necessary to carry out feature selection on the high-dimensional data before performing classification. According to Table 2, we knew that all feature selection methods were better than NFS, which directly conducted classification on original high-dimensional data. For example, the worst feature selection method, *i.e.*, LE, increased on average classification accuracy by 3.7 %, than NFS, while our method improved by 12.2 % to NFS.
- Our propose method can achieve more robust or stable performance, compared to the comparison methods. From Table 2, our SHLR method achieved minimal standard deviation in each run of all experiments, compared to all the comparison methods. For example, our method outputted the standard deviation of 0.003 on data set Umist, while the best performance among all comparison methods (*i.e.*, RFS) was 0.015.

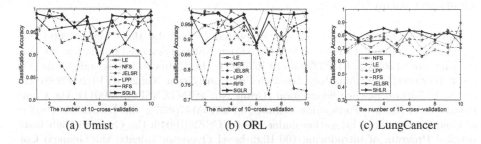

(a) Umist   (b) ORL   (c) LungCancer

Fig. 3. Classification accuracy of each run in all experiments for all methods.

In binary classification tasks, we also employed the evaluation metrics, such as SENsitivity(SEN), SPEcificity (SPE), and the Area Under an receiver operating

Table 3. Classification performances (%) of binary classification for all methods on different data sets. The best results were emphasized through bold-face. (SEN: SENsitivity, SPE: SPEcificity, and AUC: Area Under the receiver operating characteristic Curve).

Data Sets	Metrics	NFS	LE	LPP	RFS	JELSR	SHLR
ORL	SEN	89.2	90.9	93.5	93.8	93.8	**95.8**
	SPE	91.5	92.9	94.5	94.7	96.5	**96.6**
	AUC	91.9	93.8	96.6	96.2	97.4	**97.8**
LungCancer	SEN	88.7	89.6	92.2	92.4	92.6	**94.7**
	SPE	90.5	91.4	93.2	93.4	95.4	**95.7**
	AUC	90.8	92.4	95.2	95.2	96.2	**96.3**

characteristic Curve (AUC), to compare all the methods, and we reported the results in Table 3.

From Table 3, our proposed method still achieved the best performance than all the comparison methods. For example, the proposed method achieved the classification sensitivity of 95.8 %, specificity of 96.6 %, and AUC of 97.8 %, on data set ORL.

5 Conclusion

In this paper, we proposed a new unsupervised feature selection method by embedding the property of feature-level self-representation, a low-rank constraint on the regression coefficient, the hypergraph regularizer, and an ℓ_{21}-norm regularizer into a unified framework. Experimental results showed the advantages of the proposed method over the comparison methods on both binary classification and multi-class classification.

In our future work, we will extend the proposed method to conduct semi-supervised feature selection.

Acknowledgement. This work was supported in part by the China "1000-Plan" National Distinguished Professorship; the Nation Natural Science Foundation of China (Grants No: 61263035, 61573270 and 61672177), the China 973 Program (Grant No: 2013CB329404); the China Key Research Program (Grant No: 2016YFB1000905); the Guangxi Natural Science Foundation (Grant No: 2015GXNSFCB139011); the China Postdoctoral Science Foundation (Grant No: 2015M570837); the Innovation Project of Guangxi Graduate Education under grant YCSZ2016046; the Guangxi High Institutions' Program of Introducing 100 High-Level Overseas Talents; the Guangxi Collaborative Innovation Center of Multi-Source Information Integration and Intelligent Processing; and the Guangxi "Bagui" Teams for Innovation and Research, and the project "Application and Research of Big Data Fusion in Inter-City Traffic Integration of The Xijiang River - Pearl River Economic Belt(da shu jv rong he zai xijiang zhujiang jing ji dai cheng ji jiao tong yi ti hua zhong de ying yong yu yan jiu)".

References

1. Cai, X., Ding, C., Nie, F., Huang, H.: On the equivalent of low-rank linear regressions and linear discriminant analysis based regressions. In: SIGKDD, pp. 1124–1132 (2013)
2. Cao, J., Wu, Z., Wu, J.: Scaling up cosine interesting pattern discovery: a depth-first method. Inf. Sci. **266**(5), 31–46 (2014)
3. Cheng, D., Zhang, S., Liu, X., Sun, K., Zong, M.: Feature selection by combining subspace learning with sparse representation. Multimedia Syst., 1–7 (2015)
4. Gao, L., Song, J., Nie, F., Yan, Y., Sebe, N., Tao Shen, H.: Optimal graph learning with partial tags and multiple features for image and video annotation. In: Proceedings of the IEEE Conference on Computer Vision and Pattern Recognition, pp. 4371–4379 (2015)
5. Gao, L.L., Song, J., Shao, J., Zhu, X., Shen, H.T.: Zero-shot image categorization by image correlation exploration. In: ICMR, pp. 487–490 (2015)
6. Gheyas, I.A., Smith, L.S.: Feature subset selection in large dimensionality domains. Pattern Recogn. **43**(1), 5–13 (2010)
7. Gu, Q., Li, Z., Han, J.: Joint feature selection and subspace learning. IJCAI **22**, 1294–1299 (2011)
8. He, X., Cai, D., Niyogi, P.: Laplacian score for feature selection. In: NIPS, pp. 507–514 (2005)
9. Hou, C., Nie, F., Li, X., Yi, D., Wu, Y.: Joint embedding learning and sparse regression: a framework for unsupervised feature selection. IEEE Trans. Cybern. **44**(6), 793–804 (2013)
10. Hu, R., Zhu, X., Cheng, D., He, W., Yan, Y., Song, J., Zhang, S.: Graph self-representation method for unsupervised feature selection. Neurocomputing (2016)
11. Huang, Y., Liu, Q., Lv, F., Gong, Y., Metaxas, D.N.: Unsupervised image categorization by hypergraph partition. IEEE Trans. Pattern Anal. Mach. Intell. **33**(6), 1266–1273 (2011)
12. Jie, C., Wu, Z., Wu, J., Hui, X.: Sail: summation-based incremental learning for information-theoretic text clustering. ieee trans. syst. man cybern. part b cybern. **43**(2), 570–584 (2013). A Publication of the IEEE Systems Man & Cybernetics Society
13. Lewandowski, M., Makris, D., Velastin, S., Nebel, J.-C.: Structural Laplacian eigenmaps for modeling sets of multivariate sequences. IEEE Trans. Cybern. **44**(6), 936–949 (2014)
14. Liu, G., Lin, Z., Yan, S., Sun, J., Yu, Y., Ma, Y.: Robust recovery of subspace structures by low-rank representation. IEEE Trans. Softw. Eng. **35** (2013)
15. Liu, G., Lin, Z., Yu, Y.: Robust subspace segmentation by low-rank representation. In: CVPR, pp. 663–670 (2010)
16. Liu, R., Yang, N., Ding, X., Ma, L.: An unsupervised feature selection algorithm: Laplacian score combined with distance-based entropy measure. In: IITA, pp. 65–68 (2009)
17. Maugis, C., Celeux, G., Martin-Magniette, M.L.: Variable selection for clustering with gaussian mixture models. Biometrics **65**(3), 701–709 (2009)
18. Nie, F., Huang, H., Cai, X., Ding, C.H.: Efficient and robust feature selection via joint $\ell_{2,1}$-norms minimization. In: NIPS, pp. 1813–1821 (2010)
19. Nie, F., Xiang, S., Jia, Y., Zhang, C., Yan, S.: Trace ratio criterion for feature selection. In: AAAI, pp. 671–676 (2008)

20. Peng, Y., Long, X., Lu, B.L.: Graph based semi-supervised learning via structure preserving low-rank representation. Neural Process. Lett. **41**(3), 389–406 (2015)
21. Qin, Y., Zhang, S., Zhu, X., Zhang, J., Zhang, C.: Semi-parametric optimization for missing data imputation. Appl. Intell. **27**(1), 79–88 (2007)
22. Shi, X., Guo, Z., Lai, Z., Yang, Y., Bao, Z., Zhang, D.: A framework of joint graph embedding and sparse regression for dimensionality reduction. IEEE Trans. Image Process. **24**(4), 1341–1355 (2015). A Publication of the IEEE Signal Processing Society
23. Sunzhong, L.V., Jiang, H., Zhao, L., Wang, D., Fan, M.: Manifold based fisher method for semi-supervised feature selection. In: FSKD, pp. 664–668 (2013)
24. Tabakhi, S., Moradi, P., Akhlaghian, F.: An unsupervised feature selection algorithm based on ant colony optimization. Eng. Appl. Artif. Intell. **32**, 112–123 (2014)
25. Unler, A., Murat, A., Chinnam, R.B.: mr2PSO: a maximum relevance minimum redundancy feature selection method based on swarm intelligence for support vector machine classification. Inf. Sci. **181**(20), 4625–4641 (2011)
26. Wang, D., Nie, F., Huang, H.: Unsupervised feature selection via unified trace ratio formulation and K-means clustering (TRACK). In: Calders, T., Esposito, F., Hüllermeier, E., Meo, R. (eds.) ECML PKDD 2014. LNCS (LNAI), vol. 8726, pp. 306–321. Springer, Heidelberg (2014). doi:10.1007/978-3-662-44845-8_20
27. Wang, J.Y., Yao, J., Sun, Y.: Semi-supervised local-learning-based feature selection. In: IJCNN, pp. 1942–1948 (2014)
28. Wen, J., Lai, Z., Wong, W.K., Cui, J., Wan, M.: Optimal feature selection for robust classification via $\ell_{2,1}$-norms regularization. In: ICPR, pp. 517–521 (2014)
29. Wu, X., Zhang, C., Zhang, S.: Efficient mining of both positive and negative association rules. ACM Trans. Inf. Syst. **22**(3), 381–405 (2004)
30. Wu, X., Zhang, S.: Synthesizing high-frequency rules from different data sources. IEEE Trans. Knowl. Data Eng. **15**(2), 353–367 (2003)
31. Xu, Y., Song, F., Feng, G., Zhao, Y.: A novel local preserving projection scheme for use with face recognition. Expert Syst. Appl. **37**(9), 6718–6721 (2010)
32. Yu, J., Tao, D., Wang, M.: Adaptive hypergraph learning and its application in image classification. IEEE Trans. Image Process. **21**(7), 3262–3272 (2012)
33. Zhang, C., Qin, Y., Zhu, X., Zhang, J., Zhang, S.: Clustering-based missing value imputation for data preprocessing. In: IEEE International Conference on Industrial Informatics, pp. 1081–1086 (2006)
34. Zhang, S., Cheng, D., Zong, M., Gao, L.: Self-representation nearest neighbor search for classification. Neurocomputing **195**, 137–142 (2016)
35. Zhang, S., Li, X., Zong, M., Zhu, X., Cheng, D.: Learning k for KNN classification. ACM Transactions on Intelligent Systems and Technology (2016)
36. Zhang, S., Wu, X., Zhang, C.: Multi-database mining. **2**, 5–13 (2003)
37. Zhao, Z., Wang, L., Liu, H., Ye, J.: On similarity preserving feature selection. IEEE Trans. Knowl. Data Eng. **25**(3), 619–632 (2013)
38. Zhu, P., Zuo, W., Zhang, L., Hu, Q., Shiu, S.C.: Unsupervised feature selection by regularized self-representation. Pattern Recogn. **48**(2), 438–446 (2015)
39. Zhu, X., Huang, Z., Shen, H.T., Cheng, J., Xu, C.: Dimensionality reduction by mixed kernel canonical correlation analysis. Pattern Recogn. **45**(8), 3003–3016 (2012)
40. Zhu, X., Suk, H.-I., Shen, D.: Sparse discriminative feature selection for multi-class Alzheimer's disease classification. In: Wu, G., Zhang, D., Zhou, L. (eds.) MLMI 2014. LNCS, vol. 8679, pp. 157–164. Springer, Heidelberg (2014). doi:10.1007/978-3-319-10581-9_20

41. Zhu, X., Zhang, S., Jin, Z., Zhang, Z., Xu, Z.: Missing value estimation for mixed-attribute data sets. IEEE Trans. Knowl. Data Eng. **23**(1), 110–121 (2011)
42. Zhu, X., Zhang, S., Zhang, J., Zhang, C.: Cost-sensitive imputing missing values with ordering. In: AAAI Conference on Artificial Intelligence, 22–26 July 2007, Vancouver, British Columbia, Canada, pp. 1922–1923 (2007)
43. Zhu, Y., Lucey, S.: Convolutional sparse coding for trajectory reconstruction. IEEE Trans. Pattern Anal. Mach. Intell. **37**(3), 529–540 (2013)

Improving Cytogenetic Search with GPUs Using Different String Matching Schemes

Chantana Chantrapornchai[1]([⊠]) and Chidchanok Choksuchat[2]

[1] Department of Computer Engineering, Kasetsart University, Bangkok, Thailand
fengcnc@ku.ac.th
[2] Department of Computing, Silpakorn University, Nakorn Pathom, Thailand
cchoksuchat@hotmail.com

Abstract. Cytogenetic data involves analysis of chromosomes structure using karyotyping. Current cytogenetic data of patients in a hospital are very large. A physician needs to search and analyses these data for typical aberration. This paper presents the approach to speedup large cytogenetic data search with GPUs. It utilizes the parallel threads in GPUs which concurrently look for a typical string. The two search schemes are parallelized and their performances are compared. Different search scheme can be parallelized in the same manner. The experimental results show that the speedup up to 15 times can be achieved compared to the sequential version even for large number of strings searched. With the help of shared memory, parallel string search can be improved further by 8%. However, the shared memory has the limited size which cannot hold large number of strings. The percentage of data transfer time can be reduced if more strings are searched; i.e. more work load per thread. Using a more optimized string matching scheme leads to lower speedup due to the overhead of precomputing of state tables and occupies more GPU memory due to these tables. Thus, due to the nature of GPUs that have many concurrent threads, separate and smaller memory, the simple algorithm for a thread may be good enough. The optimization may be focused on the GPU-related issues such as memory coalesce, thread divergence etc. to improve the speedup further. We also present the application of the GPU string search for finding typical aberrations, and extracting the relevant data from patients' record for further analysis.

Keywords: String matching · Cytogenetic · Karyotypes · Text processing · Parallel processing · GPU · CUDA

1 Introduction

Patients' diagnosis records are typically large and separately stored in many departments using different formats such as database, text, images, spreadsheet etc. For example, in one hospital case study, the patient's profiled are stored in one software, and laboratory results are stored by another software.

© Springer International Publishing AG 2016
J. Li et al. (Eds.): ADMA 2016, LNAI 10086, pp. 188–202, 2016.
DOI: 10.1007/978-3-319-49586-6_13

The physician needs to look at the laboratory results, writes down the diagnosis and saves the patient visit records. The process to integrate the patients' laboratory diagnosis results to the patient profiles are very tedious since there are lots of patients' records and different data format must be considered.

In this research, we are interested to provide health information extraction, integration and visualisation framework for the patients data. The framework utilizes the high performance computing for information extraction, and analysis. We particularly focus on the information search where concurrency is employed for searching particular data. The data may be a string of typical aberration which indicates the certain symptom. For example, a physician may find the abnormality in patient chromosomes indicating two trans-locations between chromosome 9 and one chromosome 22, written as $t(9;22)$. Such aberration string may imply Chronic Myelogenous Leukemia (CML). From the experiments, using the GPUs can speedup the search by 15 %, when considering the shared memory, 8 % more improvement can be achieved. The optimal search in sequential may not perform well on the GPU platform.

Searching strings in large data are time consuming. Patient diagnosis data are usually stored in the format such as CSV or text, and database. Buying a computer server may be an expensive solution to store the database. For a cheaper solution, multi-threads are utilized over a many-core computer platform. To speedup the search, concurrent finding is needed. The concurrency can be implemented into two levels. The first one is at the task level and the database is split and stored separately in two or more computers connected them as a cluster. The second approach is to utilise thread levels. In this paper, we rely on the Graphic Processing Units (GPUs) which are affordable and available in many desktop PCs and run the parallel threads to find certain string.

To utilise the GPUs for searching, we consider the following questions in this work.

- GPUs have their owns memory ranges from 4GB-24GB depending on the model. The GPU memory is separated from the computer DRAM where it GPUs are located in, called *host*. To run threads on GPUs, the data must be copied from the host memory to the GPU memory. The transfer time can be large depending on the number of transfers and the size of data transferred. How do we optimize the data transfer to GPUs?
- Which algorithms are suitable for searching in GPUs? How do we parallelize the algorithms and divide the data for processing ?
- Does the fast algorithm on CPU also perform well with GPUs threads?

The paper is organized as follows: the next section gives the background on GPU architecture and related work. In Sect. 3, we present the parallel brute force string matching algorithm and its extension to other string matching algorithm, i.e., finite automaton. Experimental results are shown in Sects. 4 and 5 concludes the work.

2   Background

Graphics Processing Units (GPUs) are modern processors that contain many thousands of cores which can be used for general-purpose computations. To utilize GPUs, proper programming frameworks are needed. Compute Unified Device Architecture (CUDA) is one of them to program NVIDIA GPUs [1]. In the CUDA architecture, threads are organized in so-called grids and each grid is divided into blocks. Threads in a block are executed concurrently.

In CUDA, GPU cores are grouped into Streaming Multiprocessor (SM) units. One GPU usually contains 1–26 SMs. The memory hierarchy of GPU consists of several levels: *shared, global, texture, constant, register*, and *local*. Global memory (off-chip) can be accessed by all threads in all blocks while the shared memory (on-chip) can be accessed only by threads in the same block. Texture memory and constant memory are read-only memory and memory spaces are cached; however, the texture cache is optimised for 2D spatial locality. Global memory has the largest size, varying from 2 GB to 24 GB depending on current GPU models. While the access time of shared memory is much lower, the shared memory usually has a limited size of up to 112 KB. While most computations utilize the global memory since it is large, the shared memory is used for small and frequently accessed data. For that, the data must be explicitly copied from the global memory to shared memory before accessing it.

The GPU memory transfer latency can be a serious obstacle for improving the execution time: the data from the main memory must be copied to the GPU global memory, and/or from the global memory to the shared memory before the GPUs can start computing. Efficient algorithms are designed in a way to ensure that the data can retain inside the GPU memory as long as possible to reduce transfer time, thus reducing execution time.

2.1   Related Work

Many previous works were done to implement regular expression in parallel in various applications. Ficara et al. proposed parallel matching of regular expression which traverses a DFA in parallel on increase efficiency of synchronization between threads [2]. Mythkowicz, Musuvathi and Schulte implemented regular expression search in parallel which focuses on data parallelism [3]. The data and FA are divided and the gather and Enumerative algorithm are used similar to prefix sum algorithm. The benchmarks tested were Snort regular expression, Huffman decoding, Tokenization on 16-core Intel 2.67 GHz Xeon. Luchaup et al. proposed speculative pattern matching [4]. This work finds patterns against the network attack. The algorithm for checking a pattern in parallel and guessing the initial states was developed. The next states were also guessed. Since the network attack has different signatures, the number of states is more which requires the computation for DFAs in parallel. The data are divided into equal chunks and let them flow to be scanned in DFAs simultaneously. The authors tested 7 algorithms on a Pentium M at 1.5 GHz, Intel Core 2 2.4 GHz and Xeon E5520 2.27 GHz.

Also, in bioinformatics or bioengineering areas, there were attempts to implement string searching using parallel technology. Daz-Uriarte and Rueda presented the web application for analyzing aCGH data which was based on Message Passaging Interface (MPI) [5]. CNA (Copy Number Alteration) on Genomic DNA which relates diseases in human including cancers are compared in the array. Eight methods for checking CNA, finding gain and loss were implemented including CBS (Circular Binary Segmentation) [6], R package for GLAD (Gain and Loss Analysis of DNA) [7], CGH Segmentation [8], Hidden Markov Model. Fridlyand et al. presented data from (GO-Gene Ontology terms, PubMed citations, KEGG- Kyoto Encyclopaedia of Genes and Genomes and Reactome pathways) presented gene data and set of genes that have CNA [9].

Ohlebush and Kurtz presented the method for DNA string matching that is space efficient [10]. They compared rare maximal exact between multiple sequences. The suffix tree of a reference sequence was constructed and other sequences were compared against this suffix tree. The results from exact match for each pair were combined into multiple exact matches. The method was used to find a block of genes. Gang et al. presented the evaluation of parallel suffix tree implementations and suffix tree arrays [11] to speed up sequence matching. The experiments were done on Intel Core i7 3770 K quad-core and NVIDIA GeForce GTX680 GPU, where the suffix array takes 20–30 % space of suffix trees. The coalesced binary search and tile optimisation makes suffix array faster than using suffix tree for CUDA implementation.

Bayesian phylogenetic inference presented a tree of life which is a kind of species for known DNA sequencing in a package, MrBayes (http://mrbayes-gpu. sourceforge.net/). The parallel versions of Metropolis coupled Markov chain Monte Carlo (MC3) algorithm on various platforms including GPU were demonstrated. The extension, MrBayes MC3, utilizes quad-cure CPU along with GPU. The experiments show that CPU together with GPU enables 5.4 times speedup when testing against large data and 2 GPUs.

McKenna et al. demonstrated Genome Analysis Toolkit (GATK) which is a framework for analyzing DNA sequence using functional programming and MapReduce [12]. The framework was easy to use and can analyze various NGS which were used in 1000 Genomes Project and The Cancer Genome Atlas.

Rao et al. compared efficiency of string matching on multi-core, Brute force, Boyer Moore and KMP which tested against DNA data [13]. Prasad et al. (2010) proposed a method to find various bit patterns and tested against DNA pattern [14]. The algorithm, BNDM, transforms pattern string to bits and construct an NFA for it. The authors experimented on NCBI data set (24 MB) but did not tested the parallel version. Trans et al. again developed parallel bit pattern matching which extends to work with multiple fixed-length bit patterns [15]. The word is divided into n equal parts to be simultaneously searched. The bit pattern array was used as the table and the index file is used to store the beginning position to search. The author implemented using OpenCL and tested against one 480-core GPU and 8-multicore. They measured the number of words found per second.

Evangelia and Ross analyzed string matching with GPUs in database applications [16]. They studied KMP and BM algorithms. They considered to optimize the string matching algorithm based on GPU memory architecture: addressing divergence, addressing cache and memory pressure.

In [17], concurrent threads in NodeJS were used to search different records in the database. The user query was split into subqueries and the thread was forked to work with each subquery. Then, the results from each thread are merged as a final result.

In this work, we apply the string matching in [18], finite automata (FA), and parallelize it to find the cytogenetic data. We compare with the parallel brute force (BF) styles using concurrent threads in GPUs. The experiments show that such string matching algorithm can be speedup about 3 times while the brute force one can be speedup about 7 times.

3   Algorithms

We adopted Brute-Force String matching (BF) as a prototype. However, other kinds of string matching may be used. We will demonstrate next to integrate the FA style (Aho Corasick) into it. The modification to parallel version is as following: (1) Each thread i searches the string at a particular position i. (2) Rather than having one thread compare one string, and the thread skips the rest of the block and compares with the position in the next thread block for the whole *Data* size n. (3) To increase the workload, the thread compares several strings at the same time. This modified version is called, *Brute Force string matching on GPUs with Multiple keywords* (BFG-M).

3.1   Parallel Brute-Force String Matching

Algorithm 1 shows the logic of our parallel search for a set of *Keywords*. The whole *Data* file is read from a file at line 3, and is copied to the GPU memory at line 4. *Data* are kept in the GPU memory for searching multiple times. Keywords are transferred to the GPU memory in line 5. At line 7, the algorithm calls *GPUSearch* in 2. Then, the found results are copied back in line 8.

In Algorithm 1, keywords are transferred to the GPU memory in a pack as in Fig. 1. If we perform the match for one keyword at a time, *positionArray* is copied back from GPU memory for every keyword. Thus, the transfer overhead would be $M \times |Data|$, where $|Data|$ is the size of *Data* which may be very large ($> 2\,\text{GB}$) for M total number of keywords. We optimize the number of transfers since we found that transferring a large data set one time uses less time than transferring small amount of data multiple times. We also observed that there is a match, position i will match only one string starting at i. We pack all the keywords in *Keywords* and transfer the array once to the GPU memory. When the algorithm finishes the search, *positionArray* is transferred back once.

Due to the nature of query, many keyword strings are searched in one SPARQL-like query such as the listing:

Algorithm 1. Parallel search with BFG-M

Input: *dataArray, Keywords,*
Output: *positionArray*

1 Allocate device memory for *dataArray, Keywords, positionArray.*
2 **while** *not EOF* **do**
3   Read the data file chunk in *dataArray.*
4   Copy *dataArray* to the GPU memory.
5   Copy *Keywords,* to the GPU memory.
6   *positionArray* (initialized to false) and copy all *keywords* as *Keywords,*to the
      GPU memory.
7   Call *GPUSearch* BFG-M (Algorithm 2) with *dataArray, Keywords,* and
      positionArray.
8   Copy *positionArray* back to CPU and return the sum of total positions
      found.
9 **end**
10 Free all the memory.

```
SELECT   ?subject   ?predicate ?object
WHERE {   { FILTER ( regex (str (?object) "XXY")) . } UNION
    { FILTER ( regex (str (?object)  "Leukemia")) . } }
```

The above listing looks for a record with string "XXY" or "Leukemia". The list of keywords is packed in array, *Keywords,* and padded as a block of 256 characters in Fig. 1. and transferred to GPU memory in line 5 of Algorithm 1. One of the drawback for this approach is the length of each string must be fixed. To speedup the search time, the keywords must be copied to shared memory of GPUs. However, the shared memory size is limited. The number of simultaneous keywords searched will be limited too.

Keywords

Fig. 1. Packing keywords

Algorithm 2 searches multiple strings *Keywords* in the GPU memory by multithreads. At line 1, a *keyword* is extracted from *Keywords,* and compared in a loop body in line 3. Thread i compares the string starting at position i as in Fig. 2. Line 13 accumulates the number of matches. The thread marks at its own position i in *positionArray* to eliminate race condition. *positionArray* keeps the results; i.e., $positonArray[i]+=1$ means there is a *keyword* matched starting at *dataArray*[i].

To utilize the GPU threads more efficiently, the work item per thread must be enlarged. In GPUs, all threads in a block run at the same time. Each thread

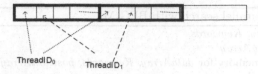

Fig. 2. Workload assignment.

has a unique thread ID with in a block. We allow the thread to search multiple positions skipped by a block size $i+ = blockDim * gridDim$, since the data is usually much larger than the number of threads in a block, For example, when the block size is 1024, and the data is 1 GB, for one block, a thread will start comparing at its own thread ID. In Algorithm 2, thread 0 will compare data at 0,1024,2048,... and thread 1 compares the data at 1, 1025, 2049, ... etc. as in Fig. 2.

Algorithm 2. GPUSearch BFG-M Keywords

Input: *dataArray, Keywords,*
Output: *positionArray*

```
1  for  each keyword in Keywords, do
2      j = 0
3      for i = threadIdx; i < n; i+ = blockDim * gridDim do
4          if dataArray[i] == keyword[0] then
5              j = 1
6              while i + j < n and j < m do
7                  if dataArray[i+j] ≠ keyword[j] then
8                      break
9                  end
10             end
11         end
12         if j == m then
13             positionArray[i] + = 1
14         end
15     end
16 end
```

3.2 Parallel FA String Matching

If we would like to integrate different matching style, the task parallelism may be done differently. For example, in the finite state machine style, one needs to construct the finite automation for each keyword. Then there are consideration for the construction: (1) whether the construction should be done in CPU or

GPUs, (2) whether the state table should be kept in the GPU global memory, or shared memory, (3) what is the size of the state table for a typical string?

For question (1), if the construction is done in CPU, the table must be transferred into GPU memory. The cost to transfer will be equal to what is addressed by question (3). For question (2), the shared memory size is limited, to keep state tables for all strings may not be possible. The size of state table for each string can be estimated by $256 \times M$, where M is the string length and 256 is the possible number of characters. For GPUs, we need to estimate the string length for preallocation of the GPU memory. If assume M is 256, total size of one table is 64 K which is too large for the typical shared memory (48 K). One possibility is to build all strings' tables on the CPU host and transfers all tables to the GPU global memory. Another possibility is to use shared memory and let the thread collaboratively build the table for one string and there will be no need to transfer the table. However, this needs the construction for each string and need synchronization for each thread before going to the next string. Our experiment used the latter approach. We assume the M is 120 so we can fit the table in shared memory. The table is built before each string search is called as in Algorithm 3 (FAG-M). For different kinds of string matching with pre-computed tables, like Boyer Moore, KMP can be done in similar way.

In Algorithm 3, each thread in a block perform lines 8–28 to construct a state table for each string $keyword$. M is the string length of $keyword$, NO_OF_CHARS is 256, and N is the total length of $dataArray$. The table is stored in $\_d\_TF$ which is in shared memory and accessed as a one-dimensional array.

Consider the work load. Each thread is responsible for consecutive data of $work\_size$ in line 6, as shown in Fig. 3, where $work\_size = 256$. The table is constructed for each string. After the table is constructed at line 32, each thread loops to the array and moves to the next state based on current $dataArray[i]$. If the thread reaches the state M, it is found. Based on this work division style, it is noted that we bound the overlap size to M, for each thread, as in $start + work\_size + M$ in line 32. Using this data division has a drawback. The accessed to memory in GPUs are not coalesced, which can reduce the total performance.

Fig. 3. Assign different workloads.

Algorithm 3. GPUSearch FAG-*M* Keywords

Input: *dataArray, Keywords,*
Output: *positionArray*

```
 1  for  each keyword in Keywords do
 2        p = threadIdx.x
 3        state = 0
 4        start = blockIdx.x * blockDim.x + threadIdx.x
 5        total_worker = blockDim.x * gridDim.x
 6        work_size = N/total_worker
 7        M = strlen(keyword)
 8        for (; p < (M + 1) * NO_OF_CHARS; p+ = blockDim.x) do
 9            row = (p/(NO_OF_CHARS)) /* get row index */
10            col = (p%(NO_OF_CHARS)) /* get column index */
11            if (state < M&&col == keyword[row]) then
12                _d_TF[p] = row + 1;
13            end
14            else
15                for (ns = row; ns > 0; ns − −) do
16                    if (keyword[ns-1] == col) then
17                        for (i = 0; i < ns − 1; i + +) do
18                            if (keyword[i]! = keyword[row − ns + 1 + i]) then
19                                break;
20                            end
21                        end
22                        if (i == ns − 1) then
23                            _d_TF[p] = ns
24                        end
25                    end
26                end
27            end
28        end
29        /* end constructing state table*/
30        _syncthreads()
31        start = start * work_size
32        for (i = start; i < start + work_size + M&&i < N; i + +) do
33            if i >= N then
34                break the loop
35            end
36            state = _d_TF[state * NO_OF_CHARS + dataArray[i]]
37            if state == M then
38                positionArray[i] = i − M + 1
39            end
40        end
41  end
```

3.3   Application to Triple Inference

Consider an N Triple file which contains a row of *<subject predicate object>*. Each term *subject, predicate*, or *object* can be literal or URI which can be linked to other terms. Such N Triple data can be inferred based on some standard predicate property such as *subClassOf*. For instance, if we find triple *<A subClassOf B>* and *<B subClassOf C>* then we can infer new triple *<A subClassOf C>*. This is sometimes called *transitive property*. Suppose we would like to infer for a given a set of subjects, we can concatenate each subject to the predicate *subClassOf* as in Fig. 4(a) and for each concatenated string the FA is constructed in Fig. 4(b). In Fig. 4(c), GPU threads find *Keywords*. The found positions are recorded in *positionArray*. Then, we only keep triples that are found. The object terms of these triples are kept in a hash table to remove redundancy and they are used as subjects in the next round.

4   Experiments

In the experiments, we demonstrate the following aspect to answer our problem description: (1) the transfer cost for the search algorithm and how it is affected (2) the speedup of both algorithms: BFG-M and FAG-M compared to sequential version.

The tested machine has the following specification: Intel(R) Core(TM) $i7 - 5820K$ CPU @ 3.30 GHz, 6 cores, and 16 GB RAM. An NVIDIA Tesla K40 card was used. The card contains 15 Multiprocessors, 192 CUDA cores per MP (totally 2,880 CUDA cores), 48 K shared memory, with maximum clock rate 745 MHz (0.75 GHz). Memory bus width is 384-bit. Total amount of global memory is 12,288 MB. We performed the test using a thread block size of 1024, a grid size of 480, one stream was used.

4.1   MESH 2016 Benchmark

The first experiment tested the performance of string matching by looking for keywords in quotes in MESH 2016 data set which are in N Triple format [19]. MESH 2016 N Triple file has size 1.6 G and contains 12 M triples. It is a vocabulary set, Medical Subject Headings (MeSH) thesaurus. It is used for online retrieval, indexer, or for cataloging.

The keyword strings were randomly generated as 10,100,1000, and 10000 strings. The data file was read by a chunk of 1 G into *dataArray*. Figure 5(a) is the speedup of BFG-M where we stored keywords in various memory types. For 1,000 and 10,000 strings, we did not test the case of shared memory since the number of keywords are too large to store. The execution is compared with the BF sequential version. When increasing more number of strings search, we gain more speedup. For 1000 and 10000 strings, the speedup of global memory case is 15.2 and 15.3 respectively and the speedup of unified memory case is 14.9 and 15.4 respectively. The maximum speedup gained is around 15 times.

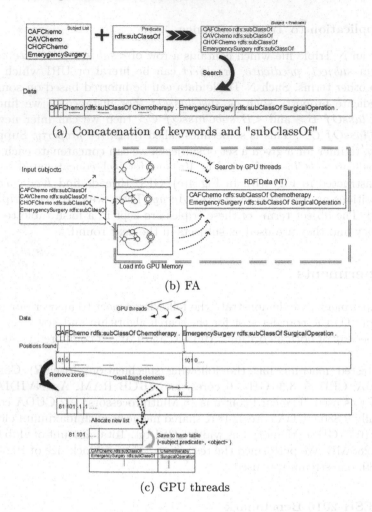

(a) Concatenation of keywords and "subClassOf"

(b) FA

(c) GPU threads

Fig. 4. Triple inferences.

Figure 5(b) is the transferred time which is the summation of *dataArray* (host to device), *Keywords* (host to device) and *positionArray* (device to host). It is the percentage of the data transfer time to total execution time. Unified memory case has the least data transfer time while shared memory case has the most data transfer time since it needs to transfer data to global memory and copy from global to shared memory. When more keyword strings are considered, the percentage of data time is reduced. That is increasing work item in GPU processing will be an advantage.

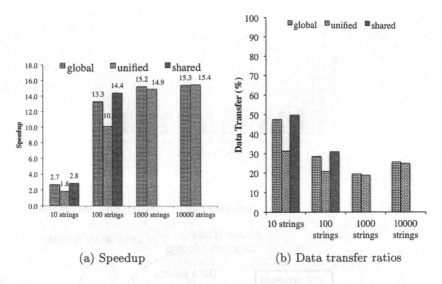

(a) Speedup   (b) Data transfer ratios

Fig. 5. Search random keywords for BFG-M

4.2 Bio2RDF Benchmark

Bio2RDF is a Linked data for the life sciences (http://download.bio2rdf.org/release/3/release.html) [20]. The data sets were in NQuad format and were transformed into N Triples. We combined some of the N Triples data set and generated the bigger data set of size 1.2 G, with 8 M triples. We selected random subjects for different sizes. Figure 6 is a speedup of each parallel search (BF and FA) compared to BF (sequential). "GPU-BF(1)" is the case for finding one string with BFG-M, "GPU-FA(1)" is the case for finding one string with FAG-M, and "GPU-BF(2)" is the case for finding two chained strings with BFG-M as in Sect. 3.3 (transitivity of $subClassOf$). It is found that BFG-M gives consistent speedup for the cases of 1 string and 2 strings. The speedup is around 7–8 times. BFG-M also gains more speedup that FAG-M.

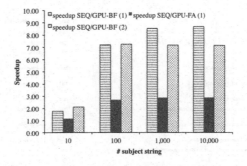

Fig. 6. Speedup sequential VS. BF and FA.

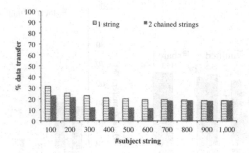

Fig. 7. Data transfer ratios of between 1 and 2 predicates.

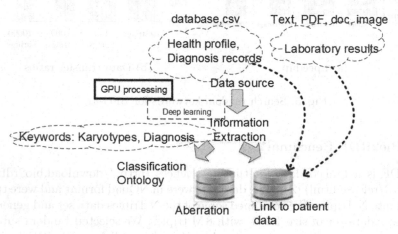

Fig. 8. Application framework

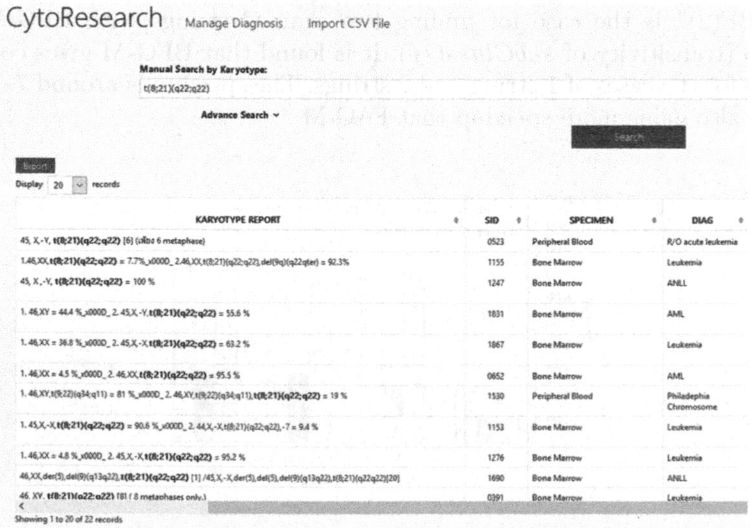

Fig. 9. Sample user interface.

In Fig. 7, we measured the percentage of data transfer time. The case of "1 string" refers to finding one string and the case of "2 chained string" refers to finding the two chained strings. It is seen that when performing more computation, the percentage of data transfer time is reduced.

5   Conclusion and Future Work

Particularly, for a classification of aberration types, one would like to find specific karyotypes. Such karyotypes are string symbols which are written in a patient record by a physician. In this research, we propose the concurrent search algorithm which looks for specific strings in large text data. The use of different string matching schemes is explored. We suggest the parallel implementation of these schemes. We demonstrate the application to *bio2rdf* (http://bio2rdf.org) data set to find specific disease or karyotype name, etc.

The experimental results show the consistent speedup for parallel version over the sequential version. However, when using a more optimized string matching algorithm, the speedup is less. This is due to the cost of preprocessing which may not be worth for GPU architecture and computing styles. For example, in FA style, it is needed to compute the automata before the search can begin. To compute the automata, we need memory storage for storing state machines. The size of the memory required to store in GPU memory is $256 \times M$, $M = strlen(keyword)$. The preprocess of constructing this FA has the complexity $O(\frac{M^2}{p})$, where p is total threads in a block.

We will integrate the GPU string processing into the application of classification and finding aberration as in Fig. 8. The finite state machine style can be easily expanded to handle regular expression search. The relevant information such as diagnosis, images, following records about the specific karyotypes are extracted and stored as linked data. Figure 9 shows a sample user interface for finding a karyotype string "t(8;21)(q22;q22)". t(8;21)(q22;q22) is the translocation which is one of the most common structural aberration in Acute Myeloid Leukemia (AML). AML is a cancer of the myeloid line of blood cells, the most common Acute Leukemia affecting adults, found in 5–12% of AML and in one-third of karyotypically abnormal M2 cases according to the French-American-British (FAB) classification [21].

Acknowledgment. This work was supported by the following institutes and research programs: NVIDIA Hardware grant, Kasetsart University Research and Development Funding (39.59) and Faculty of Engineering at Kasetsart University Research funding contract no. 57/12/MATE. We are also thankful to Assoc. Prof. Dr. Budsaba Rerkamnuaychoke, Human Genetic Laboratory, Department of Pathology, Faculty of Medicine Ramathibodi Hospital, Mahidol University, Bangkok, Thailand, for the blinded test diagnosis data.

References

1. NVIDIA: NVIDIA GPU programming guide (2015). https://developer.nvidia.com/nvidia-gpu-programming-guide. Accessed July 2015

2. Ficara, D., Antichi, G.: Scaling regular expression matching performance in parallel systems through sampling techniques. In: Proceedings of IEEE Global Telecommunications Conference (GLOBECOM 2011) (2011)
3. Mytkowicz, T., Musuvathi, M., Schulte, W.: Data-parallel finite-state machines. SIGPLAN Not. **49**(4), 529–542 (2014)
4. Luchaup, D., Smith, R., et al.: Speculative parallel pattern matching. Trans. Info. For. Sec **6**(2), 438–451 (2011)
5. Daz-Uriarte, R., Rueda, O.M.: ADaCGH: a parallelized web-based application and R package for the analysis of aCGH data. PloS one **2**(8) e737, 438–451 (2007)
6. Olshen, A., et al.: Circular binary segmentation for the analysis of array? Based DNA copy number data. Biostatistics **5**(4), 557–572 (2004)
7. Hup, P., et al.: Analysis of array CGH data: from signal ratio to gain and loss of DNA regions. Bioinformatics **20**(18), 3413–3422 (2004)
8. Picard, F., et al.: A statistical approach for array CGH data analysis. BMC Bioinformatics **6**(1), 27 (2005)
9. Fridlyand, J., et al.: Hidden Markov models approach to the analysis of array CGH data. J. Multivariate Anal. **90**(1), 132–153 (2004)
10. Ohlebusch, E., Kurtz, S.: Space efficient computation of rare maximal exact matches between multiple sequences. J. Comput. Biol. **15**(4), 357–377 (2008)
11. Liao, G., et al.: Ultra-fast multiple genome sequence matching using GPU. arXiv preprint arXiv:1303.3692 (2013)
12. McKenna, A., et al.: The genome analysis toolkit: a mapreduce framework for analyzing next-generation dna sequencing data. Genome Res. **20**(9), 1297–1303 (2010)
13. Rao, C., Raju, K.B., Raju, S.V.: Parallel string matching with multi core processors-a comparative study for gene sequences. Global J. Comput. Sci. Technol. Hardware Comput. **13**(1), 27–41 (2013)
14. Prasad, R., Agarwal, S., Yadav, I., Singh, B.: A fast bit-parallel multi-patterns string matching algorithm for biological sequences. In: Proceedings of the International Symposium on Biocomputing, ISB 2010, pp. 46:1–46:4. ACM (2010)
15. Tran, T.T., Giraud, M., Varré, J.-S.: Bit-Parallel multiple pattern matching. In: Wyrzykowski, R., Dongarra, J., Karczewski, K., Waśniewski, J. (eds.) PPAM 2011. LNCS, vol. 7204, pp. 292–301. Springer, Heidelberg (2012). doi:10.1007/978-3-642-31500-8_30
16. Sitaridi, E.A.: Ross, K.A.: GPU-accelerated string matching for database applications. The VLDB Journal, pp. 1–22 (2015)
17. Chantrapornchai, C., Navpatanitch, S., Choksuchat, C.: Parallel patient karyotype information system using multi-threads. Appl. Med. Informatics **37**(3), 39–48 (2015)
18. Charras, C., Lecroq, T.: Handbook of Exact String Matching Algorithms. King's College Publications, London (2004)
19. NLM: Introduction to mesh - 2016 (2016). https://www.nlm.nih.gov/mesh/introduction.html
20. Dumontier, M., Callahanand, A., Cruz-Toledo, J., Ansell, P., Emonet, V., Belleau, F., Arnaud, D.: Bio2RDF release 3: a larger connected network of linked data for the life sciences. In: Proceedings of the 2014 International Conference on Posters & #38; Demonstrations Track, ISWC-PD 2014 vol. 1272, pp. 401–404 (2014). CEUR-WS.org
21. Flandrin, G.: Classification of acute myeloid leukemias (2005). http://atlasgeneticsoncology.org/Anomalies/ClassifAMLID1238.html

CEIoT: A Framework for Interlinking Smart Things in the Internet of Things

Ali Shemshadi[1]($\boxtimes$), Quan Z. Sheng[1], Yongrui Qin[2], and Ali Alzubaidi[3]

[1] School of Computer Science, The University of Adelaide,
Adelaide, SA 5005, Australia
{ali.shemshadi,michael.sheng}@adelaide.edu.au
[2] School of Computing and Engineering, University of Huddersfield, Huddersfield, UK
yongrui.qin@hud.ac.uk
[3] College of Computer Science - Al Lith,
Umm Al-Qura University, Mecca, Saudi Arabia
aakzubaidi@uqu.edu.sa

Abstract. In the emerging Internet of Things (IoT) environment, things are interconnected but not interlinked. Interlinking relevant things offers great opportunities to discover implicit relationships and enable potential interactions among things. To achieve this goal, implicit correlations between things need to be discovered. However, little work has been done on this important direction and the lack of correlation discovery has inevitably limited the power of interlinking things in IoT. With the rapidly growing number of things that are connected to the Internet, there are increasing needs for correlations formation and discovery so as to support interlinking relevant things together effectively. In this paper, we propose a novel approach based on Multi-Agent Systems (MAS) architecture to extract correlations between smart things. Our MAS system is able to identify correlations on demand due to the autonomous behaviors of object agents. Specifically, we introduce a novel open-sourced framework, namely CEIoT, to extract correlations in the context of IoT. Based on the attributes of things our IoT dataset, we identify three types of correlations in our system and propose a new approach to extract and represent the correlations between things. We implement our architecture using Java Agent Development Framework (JADE) and conduct experimental studies on both synthetic and real-world datasets. The results demonstrate that our approach can extract the correlations at a much higher speed than the naive pairwise computation method.

Keywords: Internet of Things · Correlation · Multi-Agent System

1 Introduction

With advances in the enabling technologies, such as Radio Frequency Identification (RFID), sensors, power harvest and IPv6, nowadays people can easily connect their everyday objects to the Internet. Thus, the paradigm of Internet of Things is a compelling and shifting vision of the future Internet. In the

© Springer International Publishing AG 2016
J. Li et al. (Eds.): ADMA 2016, LNAI 10086, pp. 203–218, 2016.
DOI: 10.1007/978-3-319-49586-6_14

recent years, this paradigm has been tremendously growing. It is predicted that by 2020, billions of devices will be connected to the Internet [7]. Even at the present time, numerous cloud based platforms are providing services for connecting and managing smart things. Taking advantage of the mashup paradigm, IoT data is commonly visualized and presented through simple Web mashups. For instance, Fig. 1 shows an example of Web of Things mashup from the ThingSpeak platform[1]. As shown, none of the parts of the present mashup is referring to other mashups or correlated things on the Web. As a result, correlations remain implicit and IoT resources may remain isolated from each other if they are not interlinked. Interlinking relevant smart things will trigger improved user navigation as well as providing means for future IoT search engines. This is a very critical issue which resembles the role of hyperlinks in the Web.

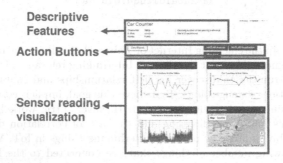

Fig. 1. Example of Web mashup from ThingSpeak platform

A hyperlink is a reference to Web resources that the reader can directly follow by clicking on it. Usually, hyperlinks are associated with a textual description about their target, which is called hypertext. They play a key role in interlinking Web resources and provide navigation between different Web pages for users. Web crawlers are navigated using those hyperlinks in Web-based documents [3]. To enable the interlinking between resources in the context of interconnected networks of things, one eminent issue is that the traditional approach of interlinking Web documents, cannot fully unravel the benefits of interlinking IoT. Table 1 summarizes the differences between interlinking IoT resources vs. hyperlinks in the traditional Web.

Due to the highly dynamic and heterogeneous nature of IoT, correlations between entities may quickly become outdated due to the frequent changes in the status of things. Thus, one eminent issue is how to effectively and efficiently establish and maintain the links to the correlated resources. As the above Table shows, IoT-links have different requirements from hyperlinks in the traditional Web. Automatic maintenance of IoT-links requires a solid understanding of the implicit correlations between IoT resources in the physical world.

---

[1] https://thingspeak.com/channels/38629.

Table 1. Requirements of traditional WWW hyperlinks vs. novel IoT-links

	Hyperlinks	IoT-Links
Establishment	Manual and static	Automatic and on-demand
Term	Long-lasting and fixed	Short term and highly dynamic
Connection Types	Simple (single type)	Various types
Weighted	No	Yes
Node Types	Web pages	Heterogeneous resources
Users	Human users and crawlers	Smart things, human users and crawlers

Although every pair of smart things around the world could potentially be correlated, in the context of IoT, correlations may not necessarily share the same type or the same weight. Thus, several types of correlations can be identified between interconnected things [2]. For instance, the type of the correlation that exists between two things that belong to the same person (owned by correlation) can be different from the correlation between two objects that are present in the same physical area (co-located correlation). Moreover, correlations of the same type, may not necessarily have the same weight. For example, the weight of co-located things may vary based on their distance from each other.

In this paper we propose the CEIoT (**C**orrelations **E**xtractor for **IoT**), a framework to facilitate automated correlation extraction for smart things in IoT. We use Multi-Agent System (MAS) architecture to design and implement our approach in order to be able to simulate the behaviors of smart things in real-world. We use MAS architecture mainly due to enable the solution to benefit from autonomous behaviors of the agents. Our framework regulates the extraction of different types of correlations. The correlation types that we cover in this paper are defined as follows [2]:

– Ownership object relationship (OOR): correlates objects with the same owner.
– Co-location object relationship (CLOR): correlates objects that are physically close to each other.
– Category based object relationship (CBOR): correlates objects which have the same describing tags.

We select the above correlation types mainly due to the nature and the structure of the IoT data that we have crawled using the ThingSeek crawler engine [11]. Each record in our dataset contains descriptive fields and sensor output fields. The sensor output in our dataset is includes the *data stream* field and the descriptive fields in our dataset include service description tags (which are defined by the owner and contain *tags, data stream unit, data stream symbol, type* and etc.), location (*latitude* and *longitude*) and owner (*user*). Due to diversity of the sensors in IoT and the lack of an standard ontology that correlates readings of various types of sensors, we cannot easily compare the values of sensor readings. For example, if sensor A reports the temperature as $12°C$ and sensor B reports it as $12°F$ sharing the same value 12 does not imply that their reading is similar. Thus, we opt out Sensor Reading Similarity [13] and rather focus on the descriptive fields.

In our CEIoT framework, we provide correlations with normalized weights to enable our model to represent more details from the real-world. Thus, our approach employs a weighted undirected graph to model the heterogeneous network of things. In this paper, we assume that each thing is registered only to one network and belongs to one user only. We only focus on the publicly available things. We design a distributed and scalable framework to support the correlation extraction and use Open Linked Data to present the extracted correlations. Our contributions are summarized as follows:

- We propose our CEIoT framework to extract correlations in IoT. We support correlations with different types. We use a distributed architecture to enable our CEIoT framework to estimate the weights. To the best of our knowledge, existing approaches are limited in terms of the diversity and scale.
- We define the process of correlation discovery for IoT. We propose two novel algorithms for extracting and one algorithm for integrating the extracted correlations. In the CEIoT framework, we localize CBORs and estimate the weights of correlations for CLORs to increase the efficiency of the correlation extraction process. We increase the efficiency of correlation extraction and integration using a distributed architecture.
- We conduct experiments to evaluate our approach. We crawl an IoT platform to obtain the real-world data. We use both synthetic and real-world datasets to demonstrate the efficiency and effectiveness of our framework.

Our paper is organized as follows.

2   The CEIoT Approach

In this section, we present the details of our CEIoT architecture for automated extraction and representation of the correlations between things in IoT.

2.1   Correlation Discovery Process

Today, there are many online IoT platforms such as Xively[2] and Paraimpu[3]. Despite of their large scale and complexity, no means has been deployed to analyze or present the correlations between things. This includes the correlations of things of both inter and intra data sources.

We define correlation discovery as the process of extraction and representation of correlations of any types that exist between the resources in the IoT. Figure 2(a) illustrates the process of correlation discovery for the IoT consisting of four different phases: *Collection, Extraction, Integration* and *Presentation*. In the first phase, things' data is collected via RESTful application interfaces

---

[2] https://xively.com.
[3] https://www.paraimpu.com.

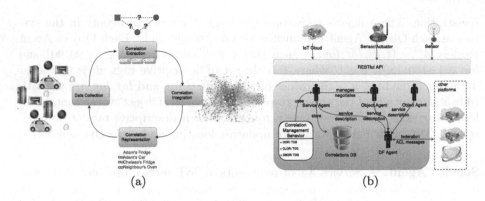

Fig. 2. (a) Different phases of the correlation discovery process in IoT; and (b) The CEIoT framework

and maintained on a server. In the next step, *Extraction*, the similarity of given object pairs is examined based on different measures and criteria to extract the correlations. During the *Integration* phase, all of the extracted correlations are integrated to form a *Things Correlations Graph* (*TCG*) [12], which resembles the graphs in the traditional social networks. Finally, in the last phase, the edges of the TCG are converted into IoT hyperlinks.

2.2 Framework Architecture and System Entities

Our CEIoT framework architecture is inspired by MAS framework. We design and implement a set of agent classes with built-in behaviors which facilitate the simulation of important entity types and their interactions in IoT correlation extraction problem. In the next step, agents are instantiated and deploy pre-designed communication protocols to interact and submit/receive messages. An overview of the CEIoT framework is shown in Fig. 2(b). Each platform in our framework operates independently from other platforms as IoT platforms operate in the real-world. The figure shows the main types of agents in each platform and how their instances interact with other parts of the system such as smart things, database and other platforms. Each platform maintains its own correlation database and Data Facilitator (DF) Service. As shown, agent classes include Service Agent, Object Agent and DF Agent. Due to the huge complexity of the nature of human user behaviors, we do not simulate them in our system and leave it for future works in this area. Existing agent classes and their roles are described in the following.

Object Agent. Object Agents are the main building block of the system; the smart things. These agents maintain the characteristics of the "things" that are connected to IoT and contain necessary behaviors to facilitate interconnections with other agents such as updating characteristics/readings and service registration. These agents constitute the largest number of agents in the system as each Object Agent is launched for one "thing" only. Each Object Agent models $A_i^o = (t, dt, lat, lon, u)$ such that $t \subset T$, $dt \in \mathbb{R}$, $lat \in \lfloor -90, 90 \rceil$ and $lon \in \lfloor -180, 180 \rceil$, $u \in U$ where T is the set of descriptive tags, dt is the latest datastream reading, u is the owner of the smart thing and lat, lon are the latitude and longitude of the object, respectively. Also, the Object Agent contains at least three default behaviors. One is to register their descriptive tags (t) into DF service. Two other behaviors are for updating location and other characteristics.

Service Agent. A Service Agent represents an IoT service provider (IoT platform) which facilitates the management of Object Agents. There is only one service agent per platform. It is responsible for coordinating and managing all agents present in its container as well as correlation discovery. These agents maintain all of the necessary information about their corresponding IoT platform. This includes host URL, port, Agent Communication Channel's address, platform ID and the DF Agent. Moreover, the Service Agent can launch, suspend or destroy Object Agents if required. The main responsibility of the Service Agent is to enquire other agents and update the correlation database frequently.

DF Agent. DF service facilitates the address book of each platform. Intra-platform agents can enquire the DF service to find agents with the specified services. The DF Agent stores tuples of services and agent URIs in the form of $\{(t, a)\}$ such that $t \in T$ and $a \in A$ where T is the set of descriptive tags and A is the set of all agents in the platform. Usually, DF service is provided only for the platform agents internally and is not designed to be shared across multiple platforms. Hence, inter-platform agent communication cannot be established in this situation. To avoid having a number of isolated MAS platforms, we devise a medium to share DF data across authorized platforms.

2.3 Correlation Extraction

Correlation extraction is the key step in the IoT correlation discovery process. In this step, our aim is to set up a efficient approach to extract the three types of correlations discussed earlier. As each type of correlation is discovered independently, firstly, we propose a separate approach for each correlation type. Then secondly, we investigate how we can integrate the process and the results.

CLOR. Given a pair of Object Agents (A_i^o, A_j^o) and a threshold $t \in (0, 1]$, CLOR can be defined as follows:

$$\texttt{clor}(A_i^o, A_j^o) = \begin{cases} \frac{\Delta(A_i^o.l, A_j^o.l)}{max\{\Delta(A^o.l, A^o.l)\}} & if \ \Delta(A_i^o.l, A_j^o.l) \geq t; \\ 0 & otherwise \end{cases}$$

where $\Delta : (latitude, longitude)^2 \rightarrow R$ returns the distance between two points (Manhattan, Euclidean, Haversine and etc.).

Algorithm 1. EXTRACT_CLOR

---

input : Granularity level l, max level l_m, current sub-tree rectangle T, *global* adjacency matrix M, set of object agents A

1   **if** $|A^o \in T| \geq 2$ **then**

2     $\forall (A_i^o, A_j^o) \in T : M(A_i^o, A_j^o) = M(A_i^o, A_j^o) + \frac{2^{l_m}}{\mathsf{D}(t)}$

3   **if** $l \leq l_m$ **then**

4       $T' = \mathsf{subtrees}(T)$

5   **foreach** $t \in T'$ **do if** $|A^o \in t| \geq 2$ **then**

6 | call EXTRACT_CLOR($l + 1, l_m, M$)

7

A naive approach for extracting CLORs would require mutual comparison between every pair of things. Therefore, the complexity for extracting the CLORs for N things using a naive approach is $\mathbf{O}(N^2)$, which is not suitable for a large number of things as in IoT. We introduce a weight estimation strategy for CLORs. We use R-Tree data structure and capping the distance granularity to extract CLORs. For distance granularity limit, we consider the area of the parent rectangle (T) with a diagonal length $\mathsf{D}(T)$ in which the distance of objects is bounded by $0 \leq d \leq \mathsf{D}(T)$. Thus, if we limit the granularity of the distance to $\mathsf{D}(T)$ (means that the distance can only be 0 or $\mathsf{D}(T)$), then the CLOR between all objects located in T form a complete graph $G_T = (V, E, w)$ where V is the set of objects, $E = V \times V$ is the set of edges defined between all objects in T and $w = \mathsf{D}(T)$. For a better precision of the weights w, we can divide the area T and strengthen the weights of edges in the same sub-areas. Thus, objects in the same sub-area $t \in T'$, will have a maximum distance of $\mathsf{D}(t)$ yielding a correlation which is $\frac{\mathsf{D}(T)}{\mathsf{D}(t)}$ stronger. For example, objects surrounded by a rectangle with $1\,km$ diagonal have a correlation twice stronger than objects surrounded by a larger rectangle with a $2\,km$ diagonal. Figure 3(a) depicts this idea.

Algorithm 1 describes our approach in further details. The Extract_CLOR is a recursive algorithm to estimate the strength of correlations between object agents. In the first level, the algorithm is launched with initial adjacency matrix $M = \{0\}_{m,m}$ where m is the number of object agents. In each recursion round (level l), the algorithm assigns correlation weights to all object agents included in the target area T. Thereafter, the algorithm would stop recursion for empty areas or if it reaches to the maximum level of granularity. The order of the algorithm would mainly depend on the distance granularity level rather than the number of object agents.

OOR. Given a pair of object agents (A_i^o, A_j^o), they have an OOR if and only if

$$\mathsf{oor}(A_i^o, A_j^o) = \begin{cases} 1 & if\ A_i^o.u = A_j^o.u; \\ 0 & otherwise \end{cases}$$

where $\mathsf{oor} : A^o \times A^o \to [0, 1]$ is the OOR score function. To obtain the correlation defined above, each Service Agent can reach the federated DFs and enquire the existing agents as well as their owners. The result set can be sorted based on the owners using a quick sort algorithm. Thus, OORs can be constructed for agents with the same owners which are in the same group. The complexity of such algorithm would be $\mathcal{O}(n)$ if a hashmap is used. The results are indexed by Service Agent to accelerate the retrieval.

CBOR. As defined earlier, each Object Agent A^o is assigned a set of textual tags $A^o.t = \{t_1, t_2, ..., t_k\}$ where each tag denotes a descriptive feature of the object such as its functionality or datastream unit. For instance, an Air Quality Egg which is designed to measures the indoor air quality and temperature, can be assigned textual tags such as "^oC", "air quality" and "indoor". The tags are assigned by the users of the IoT platform and thus, can vary significantly based on their count, keyword selection, dictation and used symbols. We assume that the tag set for each object can be used for the purpose of categorization. For a given pair of Object Agents (A_i^o, A_j^o), a text similarity function $\sigma : T^2 \to [0, 1]$ and a similarity threshold $\tau \in (0, 1]$ we define the CBOR as follows:

$$\text{cbor}(A_i^o, A_j^o) = \begin{cases} \prod \sigma(A_i^o.t, A_j^o.t) & \text{if } \sigma(A_i^o.t, A_j^o.t) > \tau; \\ 0 & \text{otherwise} \end{cases}$$

where $\text{cbor} : A^2 \to [0, 1]$ is the weight function of the CBOR correlation between (A_i^o, A_j^o). Algorithm 2 shows our approach to identify and extract CBORs amongst a set of given Object Agents. Using a naive approach for finding the similarity between all pairs of object agents is time consuming and complex. Due to limiting the recursions and through the use of dynamic programming, the runtime of this algorithm would be linear. Thereupon, the three scenarios of searching for similar objects using CBOR are:

- There is no matching result and a Null value returned.
- The number of objects in the list $\leq$ *Max_results*. Thus, all objects in the list are returned.
- The number of objects exceeds the max criterion. Provided that the list is descendingly sorted, a sub list with size of *Max_results* is cut from the first element and retrieved as an answer for the query indicating that there is high potentiality to connect, correlate, cooperate, and any action can be taken with these objects.

2.4   Correlation Integration

In the Integration phase, different TCGs are merged to form a universal graph. In the Aggregated Correlations Graph (ACG), each edge has a weight that indicates the strength of their correlation. For a given set of Object Agents A, we assume that the result of the correlation extraction process in all types of correlations maintain the same format and size. Thus, the result of CLOR, OOR and CBOR correlations are modelled as $M_{|A|\times|A|}^L$, $M_{|A|\times|A|}^O$ and $M_{|A|\times|A|}^B$, respectively. In the

---

Algorithm 2. EXTRACT_CBOR

   Input   : Requester ID, Type, Sensor_set, Actuators_set, Max_results
   Output: Results-list : List of relevant objects agents
1 Results = Search DF based on Type
2 **if** $Results \neq \emptyset$ **then**
3   **foreach** $Object \in result$ **do**
4   **if** $Sensors \neq \emptyset \parallel Actuators \neq \emptyset$ **then**
5             $S = \{Sensors\}$
6             $A = \{Actuators\}$
7             $S\_SIM(Sensor\_set, S) = \lceil Sensor\_set \cap S \rceil / \lceil Sensor\_set \cup S \rceil$
8             $A\_SIM(Actuators\_set, A) =$
                  $\lceil Actuators\_set \cap A \rceil / \lceil Actuators\_set \cup A \rceil$
9             $Similarity = S\_SIM + A\_SIM$
10   Assign Similarity to Object
11   Add Object to *Results-list*
12   Sort *Results-list* descendingly
13   **if** $\{Results - list\} \geq Max\_results$ **then**
14   Shrink the result by excluding elements from *Max_results* until the end
           of the list
15   **return** *Results-list*
16 **else**
17   **return** Null

---

simplest form of integration, TCGs of different types can be integrated through a weighted matrix integration as follows:

$$\overline{M} = \frac{1}{3} \sum_{M \in \{M^L, M^O, M^B\}} w_i.M \qquad (1)$$

where $w_i \in [0, 1]$ is the weight that is assigned to each correlation and $\sum w_i = 3$. However, this kind of correlation may result in the loss of correlation types and not very suitable for cases in which the type of each correlation is important.

2.5 Correlation Representation

Correlation representation is a part of the process in which all extracted correlations are presented in a standard format. We use RDF to represent correlations in IoT. Thus, we can maintain the connections between things while a large portion of them are quickly evolving. Through the use of specifically designed ontologies along with RDF correlations, our approach can be deployed to empower pattern queries for IoT search engines. In this regard, we use triple space computing (TSC) to facilitate the communication and store relationships triples. Figure 3(b) depicts an example of how things can be correlated using RDF triples, where each object is considered as a resource. Each statement is identified via a unique URI. A statement consists of three elements: *subject, predicate*, and *object*. A *subject* (a thing) is linked with an *object* (another thing).

The connection between two *objects* is called a *predicate*. The *predicate* explains the relationship between the *subject* and the *object* of the statement.

Two objects will have OOR if and only if they have the same owner. The result is retrieved from federated DFs as graph of RDF triples similar to the Listing 1.1. In this type of relationships, we point out the possibility of connecting objects under different regimes based on a criterion such as a common owner.

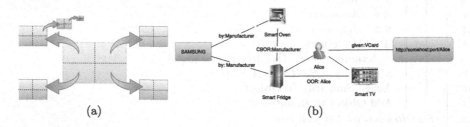

(a)   (b)

Fig. 3. (a) The surrounding rectangular area recursively breaks into four rectangles; and (b) An example of 2 types of relationships established among objects A, B, and C based on their common owners (OOR) and common tags for production batch (CBOR)

Listing 1.1. Example of OOR correlated object agents

```
<rdf:RDF
    xmlns:rdf="http://www.w3.org/1999/02/22-rdf-syntax-ns#"
    xmlns:RT="http://uqucs.com/RT"
    xmlns:vcard="http://www.w3.org/2001/vcard-rdf/3.0#"
    xmlns:dc="http://purl.org/dc/elements/1.1/">
  <rdf:Description rdf:about="http://uqucs.com/objectAgent10@Platform2">
    <RT:OOR>objectAgent43@Platform2</RT:OOR>
    <RT:OOR>objectAgent18@Platform2</RT:OOR>
    <RT:OOR>objectAgent5@Platform1</RT:OOR>
    <RT:OOR>objectAgent61@Platform2</RT:OOR>
    <RT:OOR>objectAgent42@Platform2</RT:OOR>
    <RT:OOR>objectAgent29@Platform1</RT:OOR>
    <RT:OOR>objectAgent48@Platform2</RT:OOR>
    <vcard:N>Tahani</vcard:N>
  </rdf:Description>
</rdf:RDF>
```

3   Experimental Results

In this section, we present the evaluation results for the proposed CEIoT framework. We conduct the experiments on a PC with a Core i7 2.20 GHz, 4 GB memory and Windows 7 64-bit. The details of the used datasets are as follows:

1. Synthetic dataset: we simulate a set of four IoT service providers where each service provider is supplied with one Service Agents and 1,000 Object Agents. Furthermore, each service provider is initialized on a separate platform, which can run on an independent machine or share the same machine with other service providers. We use this simulation to evaluate the framework on a distributed infrastructure and for multiple service providers. Figure 4(a) shows four RMA GUIs visualizing all four platforms.

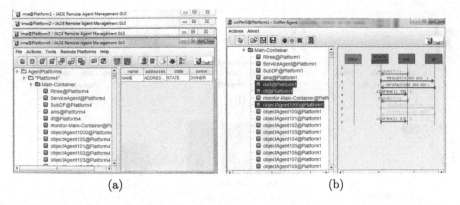

(a)   (b)

Fig. 4. (a) RMA view of launched platforms for each service agent; (b) Sniffed communications between two object agents

2. Real-world dataset: we crawled the datastream of public sensors on Xively[4] using the ThingSeek crawler [11], previously known as Pachube, which is a pioneering IoT platforms on the Internet with one of the largest collections of publicly available sensors. The dataset contains around 67,000 things and their most recent sensor readings. However, after filtering records with incomplete data, only 11,894 records remain in our dataset. A primary analysis of the tag sets reveals that the tags are scattered (Fig. 5(a)) but densely focused on some tags (Fig. 5(b)). Moreover, only less than 10 % of the tags have been assigned to more than 60 % of things (Fig. 5(c)). Thus, a single label based approach without considering location based correlations will not be helpful in application.

We use tools provided by *RMA GUI* such as *Sniffer* and *Dummy* for debugging and testing purposes. Dummy agent is used to communicate with agents in the platform and command them to execute specific behaviors. We prepare the agents with *Cyclic behaviors* that are responsible for receiving these commands and their execution. Otherwise, these behaviors are blocked until a command is received to prevent infinite loop. Blocking the behaviors in agents does not block the entire agent. Additional aim of implementing agents behaviors and communication is to use them as self-explained examples of how agents can be communicated and commanded by end users of the framework using third-party agent such as *Dummy* agent.

3.1 System Performance

Due to the scale and dynamics of the IoT, the overall performance of the system is important. Mainly, the following steps are paved by the system: (i) launching all independent platforms; (ii) launching all agents mentioned above; (iii) federating DF service; (iv) communication for exchanging stetting information; (v)

---

[4] https://xively.com/.

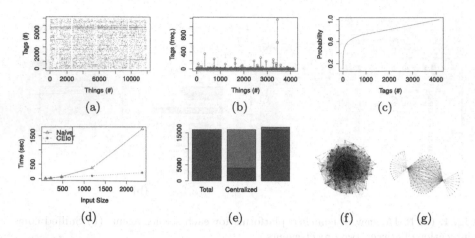

Fig. 5. (a) Things and tags relation; (b) Tags frequency; (c) Probability of reusing a tag; (d) Runtime; (e) Transacted messages for the synthetic dataset; and (e), (f) CLOR graph for things with different thresholds (Color figure online)

generating RDFs and message de/serialization; and (vi) R-tree insertion. Using the fixed parameters with an RMA launched for each platform, it takes the system roughly 25 s to perform all the mentioned key tasks. This is due to that each *Service Agent* waits for about 10 s to ensure all platforms are established, as well as additional 10 s for federation. The 20 s delay is used to prepare all platforms for the experiment.

We launch 4,000 object agents which are distributed over four different platforms. For the experiment, each object agent has a cycle behavior that receives a message with a *Request* per-formative act. It recognizes commands OOR, and CBOR. Otherwise, it responds with a *non-understood* message associated with commands which it can understand. If the commands are understood, it complies with them to search for other agents matching the type in the command.

The communication with Object Agent is performed as expected (Fig. 4(b)). The figure shows the UML sequence for sending relationship requests by the dummy agent (d0) to an object agent. Agent d0 is used in our experiment for requesting object agents to perform a relationship on demand which are OOR or CBOR. Lines 1–4 shows that sending the OOR request, the agent received the request, then it searches the DF in Line 2, and DF returns the results to the agent in Line 3. Finally, the agent confirms to the dummy agent the success of the relationship establishment. Additionally, there should appear the RDF description for the relationships of the object agent. The same process is done with CBOR in lines 5–8. In OOR relationships extraction, it takes 200 *ms* to retrieve the results and correlate them with the requester object as an RDF triple. However, in the CBOR it took approximately 1,300 *ms*. For CLOR correlation, we observe the proposed system's response under a varying number of input sizes to ensure the system's scalability. We compare our approach with the naive method. As Fig. 5(d) shows, the naive approach may take less time for results with small sizes. However, as the size of the input increases, for inputs with 500 or more things, our algorithm's runtime outperforms the naive approach. We find out that using the naive approach for an input size with the real-world dataset would be impractical, particularly when we consider the scope and the dynamics of the IoT.

3.2 Things Correlation Graph

For Algorithm 1, we apply different thresholds to simulate a search engine harvesting the graph. For example, it can be interesting in relationship matching a certain criteria. Thus, the threshold is used to reduce the search space and to limit it to connections with values equal or larger than the threshold. We visualize this by passing the symmetric weighted graph to the *GraphVis class*. The graph is the file holding all CLOR relationships among the Object Agents resulted from Algorithm 1. The higher the value of a threshold, the smaller is the search space. Figures 5(f) and (g) illustrates the samples of graphs produced based on different thresholds, respectively.

3.3 Message Volume

One of the key factors is to minimize the number of transacted messages between the machines. In our CEIoT framework, the messages that should be transacted between different platforms are summarized. Thus, we expect a dramatic reduction in the number of transacted messaged compare to a centralized scenario. Figure 5(e) shows the number of both internal and external messages which are transacted between agents during the experiment on the synthetic dataset. In this Figure, the As it is shown, although the red stack (top) shows the number of inter-platform messages and the blue (bottom) shows the number of intra-platform messages in our experiment. As shown, despite the fact that using the CEIoT approach in a distributed mode increases the number of transacted messages by a small amount, it may enhance the total throughput of the system by reducing the ratio of inter-platform messages which are quite expensive. Moreover, in the distributed mode the messages are being processed by a larger number of machines than the centralized approach.

4   Related Work

There are some studies which promote that the IoT can be implemented as the Internet of agents [16]. This is due to the ability of agents to mimic human activities such as willingness to achieve certain goals, and the social ability to interact with each other and with human as well. In a proposal by Fortino et al. [5], MAS is also used in developing smart environments for objects. The heterogeneity and disparity can intensify the problem of dealing with smart things in an effective way. Things data is never under centralized control. Hence, datasets are usually stored in distributed locations. Using central location for data storage is an obstacle for materializing an effective solution. The diversity of sensors infrastructures means that, data are structured and documented in different ways, which complicates combining their datasets in an easy way. Thus, one of the solutions called *Concinnity* [9] takes the advantage of Semantic Web technologies such as RDF, Ontology, and SPARQL.

There is an interesting paradigm called *Triple Space Computing* [4] that has the potential to facilitate storage and communication in the Internet of Things context. It is basically a dedicated Web for machines, which is combined of space-based computing and the Semantic Web. It uses RDF triples for data representation to exchange knowledge using shared space, just like HTML representation in human-driven Web [6]. TSC provides asynchronous communication such that consumers neither need to recognize each other through identifiers, nor need to concurrently consume data.

A provisional approach is the Social Internet of Things (SIoT) [1,2,8]. To socialize things in the IoT, unlike the solutions for socializing smart things that depend totally on their owners relationships, this approach seeks a solution to enable smart things to be interlinked by themselves. Objects have their own profiles and IDs, so they can discover other objects of interest and establish a friendship with each other according to their owners constraints. Additionally, this approach defines some types of relationships established among objects. For instance, *Parental Object Relationship* (*POR*), *Co-Location Object relationship* (*CLOR*), *Co-work Object Relationship* (*C-WOR*), and *Social Object Relationship* (*SOR*). However, no realistic and scalable solution given on how to identify and extract these relationships. Correlating things in IoT has various benefits such as better navigability experience, query result diversification and matters of an interest can be effectively discovered with the least effort possible [12]. As discussed before, Atzori et al. [2] discuss the characteristics of *Social Internet of Things* and define polices for relationships establishment among connected things. Unlike the approaches that socializes things according to what relationships their owners [8,10], this approach is more focused on things as key players in relationships establishment, which limits the roles of their owners to managing them and setting appropriate rules for their relationships. However, to the best of our knowledge, this vision has not been implemented yet. Additionally, currently no technical details have been given on how to automate the establishment of relationships between objects when they are aware of each other. Automated correlation extraction is limited to one type of correlation for a small number of objects [14,15].

5   Conclusion

One of the missing components in the IoT is something similar to hyperlinks in the World Wide Web. Unlike the traditional Web, establishing correlations in IoT can be challenging as it must be automated. In this paper, we have proposed the CEIoT framework for extraction and representation of correlations between things in the IoT. The framework is implemented in Java and is open-sourced.

We have used a distributed architecture and local correlation filtering to stabilize the performance of the library in different conditions. One of the limitations of our work is dismissing incomplete and uncertain data entries. We plan to extend the CEIoT framework to support incomplete and uncertain data to address a large portion of the records in our dataset.

References

1. Atzori, L., Iera, A., Morabito, G.: SIoT: giving a social structure to the internet of things. Commun. Lett. **15**(11), 1193–1195 (2011)
2. Atzori, L., Iera, A., Morabito, G., Nitti, M.: The social internet of things (SIOT)-when social networks meet the internet of things: concept, architecture and network characterization. Comput. Netw. **56**(16), 3594–3608 (2012)
3. Brin, S., Page, L.: The anatomy of a large-scale hypertextual web search engine. In: Proceeding of the 7th International World-Wide Web Conference (WWW) (1998). http://ilpubs.stanford.edu:8090/361/
4. Fensel, D., Krummenacher, R., Shafiq, O., Kuehn, E., Riemer, J., Ding, Y., Draxler, B.: Tsc-triple space computing. e & i. Elektrotechnik und Informationstechnik **124**(1–2), 31–38 (2007)
5. Fortino, G., Guerrieri, A., Russo, W.: Agent-oriented smart objects development. In: Proceeding of the 16th International Conference on Computer Supported Cooperative Work in Design (CSCWD 2012), pp. 907–912. IEEE (2012)
6. Gómez-Goiri, A., López-de Ipiña, D.: On the complementarity of triple spaces and the web of things. In: Proceeding of the 2nd International Workshop on Web of Things, p. 12. ACM (2011)
7. Gubbi, J., Buyya, R., Marusic, S., Palaniswami, M.: Internet of things (IoT): a vision, architectural elements, and future directions. Future Gener. Comput. Syst. **29**(7), 1645–1660 (2013)
8. Guinard, D., Fischer, M., Trifa, V.: Sharing using social networks in a composable web of things. In: Proceeding of the 8th IEEE International Conference on Pervasive Computing and Communications Workshops (PERCOM 2010), pp. 702–707. IEEE (2010)
9. Lee, C.H., Birch, D., Wu, C., Silva, D., Tsinalis, O., Li, Y., Yan, S., Ghanem, M., Guo, Y.: Building a generic platform for big sensor data application. In: Proceeding of 2013 IEEE International Conference on Big Data (IEEE BigData), pp. 94–102. IEEE (2013)
10. Pintus, A., Carboni, D., Piras, A.: The anatomy of a large scale social web for internet enabled objects. In: Proceeding of the 2nd International Workshop on Web of Things, p. 6. ACM (2011)
11. Shemshadi, A., Sheng, Q.Z., Qin, Y.: Thingseek: a crawler and search engine for the internet of things. In: Proceeding of the 39th International ACM SIGIR Conference on Research and Development in Information Retrieval (SIGIR), pp. 1149–1152. ACM (2016)
12. Shemshadi, A., Yao, L., Qin, Y., Sheng, Q.Z., Zhang, Y.: ECS: a framework for diversified and relevant search in the internet of things. In: Wang, J., Cellary, W., Wang, D., Wang, H., Chen, S.-C., Li, T., Zhang, Y. (eds.) WISE 2015. LNCS, vol. 9418, pp. 448–462. Springer, Heidelberg (2015). doi:10.1007/978-3-319-26190-4_30
13. Truong, C., Römer, K., Chen, K.: Sensor similarity search in the web of things. In: 2012 IEEE International Symposium on World of Wireless, Mobile and Multimedia Networks (WoWMoM), pp. 1–6. IEEE (2012)

14. Yao, L., Sheng, Q.Z.: Correlation discovery in web of things. In: Proceeding of the 22nd International Conference on World Wide Web Companion (WWW 2013), pp. 215–216. International World Wide Web Conferences Steering Committee (2013)
15. Yao, L., Sheng, Q.Z., Gao, B.J., Ngu, A.H., Li, X.: A model for discovering correlations of ubiquitous things. In: Proceeding of 13th International Conferernce on Data Mining (ICDM 2013), pp. 1253–1258. IEEE (2013)
16. Yu, H., Shen, Z., Leung, C.: From internet of things to internet of agents. In: Proc. of the 2013 IEEE International Conference on Green Computing and Communications (GreenCom), pp. 1054–1057. IEEE (2013)

Adopting Hybrid Descriptors to Recognise Leaf Images for Automatic Plant Specie Identification

Ali A. Al-kharaz[1,2]([✉]), Xiaohui Tao[1], Ji Zhang[1], and Raid Lafta[1]

[1] School of Agricultural, Computational and Environmental Sciences,
University of Southern Queensland, Toowoomba, Australia
{Ali.Al-kharaz,xtao,ji.zhang,RaidLuaibi.Lafta}@usq.edu.au
[2] Department of IT, Technical College of Management,
Baghdad Foundation of Technical Education, Baghdad, Iraq

Abstract. In recent years, leaf image recognition and classification has become one of the most important subjects in computer vision. Many approaches have been proposed to recognise and classify leaf images relying on features extraction and selection algorithms. In this paper, a concept of distinctive hybrid descriptor is proposed consisting of both global and local features. HSV Colour histogram (HSV-CH) is extracted from leaf images as the global features, whereas Local Binary Pattern after two level wavelet decomposition (WavLBP) is extracted to represent the local characteristics of leaf images. A hybrid method, namely "Hybrid Descriptor" (HD), is then proposed considering both the global and local features. The proposed method has been empirically evaluated using four data sets of leaf images with 256×256 pixels. Experimental results indicate that the performance of proposed method is promising – the HD outperformed typical leaf image recognising approaches as baseline models in experiments. The presented work makes clear, significant contribution to knowledge advancement in leaf recognition and image classification.

Keywords: Leaf image · Local feature · Global feature · Colour histogram · Texture · LBP · Wavelet

1 Introduction

Leaves of plant have become an interesting application of pattern recognition and classification. Many researches adopted leaf descriptors to identify plant species automatically. Colour is one of the most widely used low-level visual descriptor and is invariant to image size and orientation. With the selection of colour descriptor, the underlying colour space is also greatly helpful. Colour histogram is invariant to orientation and scale. Such unique characteristics make colour and colour histogram powerful in image recognition and classification. Based on such features, various approaches have been proposed to recognize different kinds of leaves in image. However, most of these approaches suffered from poor performance (e.g., low accuracy rate) and more or less failed in practice.

© Springer International Publishing AG 2016
J. Li et al. (Eds.): ADMA 2016, LNAI 10086, pp. 219–233, 2016.
DOI: 10.1007/978-3-319-49586-6_15

The RGB space was used by many researchers to extract the colour features in an image. Unfortunately, the RGB is not well suited in colour description for human interpretation and thus, daily practice [18]. In [11] the research recognises the leaf shape using Centroid Contour Distance (CCD) as shape descriptor. Centroid Contour Distance is a contour-based approach to represent image shapes only exploiting only boundary information. The approach calculates the distance between the midpoint and the points on the edge to the corresponding interval angle. Several leaf classification systems have incorporated texture features to improve the performance, such as that in [21] used entropy, homogeneity and contraction derived from digital wavelet transform (DWT). With the success, the approach has become one of the most important and powerful tools in image processing. Kadir et al. [17] used Polar Fourier Transform and three different kinds of geometric features to represent shapes. They also adopted statistical methods such as mean, standard deviation, and skewness to represent colour features and extracted texture features from GLCMs, and creatively added vein features into leaf identification in order to improve performance. Furthermore, Zulkifli et al. [32] compared the effectiveness of Zernike Moment Invariant (ZMI), Legendre Moment Invariant (LMI) and Tchebichef Moment Invariant (TMI) in feature extraction from leaf images. The features extracted by using the most effective moment invariant technique are then adopted to help classifying images using the General Regression Neural Network (GRNN). Combination of features derived from shape, vein, colour and texture of leaf images is also proposed with PCA to convert the features into orthogonal features in leaf identification system [16]. Another similar attempt was done by Liu et al. [20], who used combination of texture features and shape features for identification and used deep belief networks (DBNs) as the classifier. Texture features are derived from local binary patterns, Gabor filters and grey level co-occurrence matrix while shape feature vector is modelled using Hu Moment invariants and Fourier descriptors.

In this paper, we present a method for recognising the specie of plants from their leafs. In the work, local characteristic features like LBP is first extracted after decomposition by Haar wavelet. The features are combined with global features such as HSV colour histogram for yields promising performance with high level of accuracy and precision. The proposed method has achieved recognition performance and outperformed baseline models representing state-of-the-art leaf image recognition approaches. The work makes clear, significant contribution to knowledge advancement in leaf image classification and recognition of plant species.

The rest of this paper is organised as follows. The related work will be discussed in Sect. 2. The research problem will be formally defined in Sect. 3, followed by the proposed HD method delivered in Sect. 4. After that, the experimental evaluation and the results will be discussed in Sect. 5. Finally, Sect. 6 will conclude the paper and highlight the future work.

2   Related Work

Many works have been conducted on plant specie identification relying on leaf recognition. After using the histogram equalisation and ROI segmentation for enhancing, VijayaLakshmi and Mohan [27] used the Haralick Texture with Gabor and Shape Based Features such as area, centroid and orientation with colour features after convert RGB colour to HSV colour system. Finally they used the Fuzzy Relevance Vector Machine (FRVM) to characterise the type of leaves.

Alternatively, Du et al. [4] extracted Digital Morphology Features (DMF) from the contours of leaves. The DMF generally included Geometrical Features (GF) and invariable Moment Features (MF). Such features were then used with Move Median Centers (MMC) to train an effective classifier. Another similar work was completed by Wang et al. [28], who also used MMC and shape features (Geometric Features and Hu moment) to recognize leaf images. In [2], Gabor wavelet filters were exploited to extract texture filters in a foliar surface to improve the performance of plant classification.

Another great achievement was done by Pornpanomchai et al. [25]. They built a system to recognize Thai herb leaves in images (THLI). The system extracted 13 different kinds of features from the leaf images and then employed k-nearest neighbour (k-NN) in the recognition process. In [31], Wang and colleagues introduced Pulse-coupled neural network (PCNN), a new artificial neural network model for feature extraction. They firstly extracted leaf features by using PCNN, and then classified images by Support Vector Machines (SVMs). Leaves can also be classified using their structural properties. For instance, a leaf usually consists of triangular pieces that protrude around a polygon. Taking advantage of such structural properties, Im et al. [15] classified leaf images adopting statistical methods, along with variations of leaf contours.

The texture features of leaves have also been used in many works for leaf recognition. Ehsanirad et al. [6] extracted the Gray-Level Co-occurrence matrix (GLCM) and used Principal Component Analysis (PCA) algorithm to classify plants relying on leaf images. Furthermore, Gu et al. [8] attempted to recognise leaves using segmentation of a leaf's skeleton based on the combination of wavelet transform Gaussian interpolation. They also used k-nearest neighbor ($k = 1$) combined with a radial basis probabilistic neural network (RBPNN) as the classifier to recognize leaves on the basis of run-length features extracted from the skeleton.

Ahmed and colleagues [1] used comparison table between different methods for identification and classification of leaf images, with deep analysis for advantages and disadvantages in different methods, respectively. In [19], Liu and colleagues used wavelet decomposition and local characteristic of LBP to extract features for face recognition. For different practices, Du et al. [5] used wavelet domain local binary pattern features for writer identification. In a different work, Handa and Agarwal [10] compared different algorithms used in plant classification based on leaf recognition images and the accuracy for each one.

These related work, however, still have room for improvement because they considered either global or local features. A hypothesis of having a hybrid descriptor considering both the global and local features then motivated us in the work presented in the paper.

3 Research Problem

Let $\mathcal{IMG} = \{img_i \in \mathbb{IMG}, i = 1, \ldots, m\}$ be a set of images; $\mathcal{C} = \{c_1, \ldots, c_K\}$ be a set of classes, where $K = |\mathcal{C}|$. Assuming there is available a training set $\mathcal{IMG}_t = \{img_j \in \mathbb{IMG}, j = m + 1, \ldots, n\}$ with $y_j^k = \{0, 1\}, k = 1, \ldots, K$ provided for describing the likelihood of img_j belonging to class c_k, our research problem is how to learn an efficient binary prediction function $f(y^k|img)$ and use it to classify $img_i \in \mathcal{IMG}$.

4 Descriptors

An innovative method, namely *Hybrid Descriptors*, is proposed in this paper tackling the research problem defined in Sect. 3. The proposed method consists of three components: global feature extraction, local feature extraction, and hybrid descriptor generation for final classification of images $\mathcal{IMG}$. In this section, we will introduce the proposed method in detail. For the sake of easy discussion, we will refer to *Hybrid Descriptor* by just *HD*.

4.1 Global Feature Extraction

Global features are the set of features extracted from the whole image. Global features have been widely used with success in image classification. A typical global feature is colour histogram [9]. In this work, we used colour histogram as the global feature despite the visual difference between the gradients in plant leaves.

The RGB is a commonly used colour system, and is ideally suited for hardware implementation such as colour monitors. Unfortunately, the RGB is not well suited to specify colours because it is not practical for human interpretation. Contrary, the HSV (hue, saturation, value) model is an ideal tool for developing image processing algorithms based on colour descriptions. The HSV is deemed more natural and intuitive to humans [3,18]. For such a reason, images are normally converted from an RGB space to HSV colour space by using the following equations [23]:

$$H = COS^{-1} \frac{\frac{1}{2}[(R - G) + (R - B)]}{\sqrt{(R - G)^2 + (R - B)(G - B)}} \tag{1}$$

$$S = 1 - \frac{3}{R + G + B}(min(R, G, B)) \tag{2}$$

$$V = \frac{1}{3}(R + G + B) \tag{3}$$

The R, G, B represent *red, green* and *blue* components respectively with value between 0–255. In order to obtain the value of H from 0° to 360° and the value of S and V between 0 and 1, the following Equations are executed [23]:

$$H = (\frac{H}{255} \times 360)\%360 \tag{4}$$

$$V = V/255 \tag{5}$$

$$S = S/255 \tag{6}$$

where % is the modular operator that gives the reminder after dividing ($\frac{H}{255} \times$ 360) by 360.

Generally, for a given colour image, the number of actual colours only occupies a small proportion of the total number of colours in the entire colour space. Therefore, the hue component, which represents the colour information, is uniformly divided into eight coarse partitions. Similarly, the saturation and intensity components are divided into three partitions, respectively. Consequently, the global colour histogram can be calculated as follows [18]:

$$H = \begin{cases} 0, & \text{if } h \in [316, 360] \\ 1, & \text{if } h \in [1, 45] \\ 2, & \text{if } h \in [46, 90] \\ 3, & \text{if } h \in [91, 135] \\ 4, & \text{if } h \in [136, 180] \\ 5, & \text{if } h \in [181, 225] \\ 6, & \text{if } h \in [226, 270] \\ 7, & \text{if } h \in [271, 315] \end{cases} \tag{7}$$

$$S = \begin{cases} 0, & \text{if } s \in [0, 0.3) \\ 1, & \text{if } s \in [0.3, 0.7] \\ 2, & \text{if } s \in (0.7, 1] \end{cases} \tag{8}$$

$$V = \begin{cases} 0, & \text{if } v \in [0, 0.3) \\ 1, & \text{if } v \in [0.3, 0.7] \\ 2, & \text{if } v \in (0.7, 1] \end{cases} \tag{9}$$

The quantisation of the number of colours into several bins is done in order to decrease the number of colour used in feature matching. We propose the scheme to produce only 14 bins colour.

4.2   Local Feature Extraction

The local feature of objects are widely used in image matching and classification. A local feature descriptor takes into account the regions or objects to describe the image. In this work, the original leaf images are first decomposed using Haar wavelet before extracting local features. Wavelet transform is one of the best tools to determine where the low frequency and high frequency is. It involves in compression for decomposing the image into approximation and detail. The approximation sub-image shows the general trend of pixel values, and the three detail sub-images show the vertical, horizontal and diagonal details or changes in the image [7].

In wavelet transformation, low-pass filtering is conducted by averaging two adjacent pixel values, whereas the difference between two adjacent pixel values figured out for high pass filtering, as a result, it produces four sub-bands as the output of the first level Haar wavelet. The four sub-bands are LL_1, HL_1, LH_1 and HH_1 [14]. The process can be repeated to compute multiple scale decomposition, as in the two scales Wavelet shown in Fig. 1. The LL sub-band contains a rough description of the original image and is hence, called the approximation sub-band. The HH sub-band contains the high-frequency components along the diagonals. LH contains mostly the vertical detail information. HL represents the horizontal detail information. The sub-bands HL, LH and HH are called the detail sub-bands since they add the high-frequency detail to the approximation image [26]. Therefore, wavelet decomposes an image by reducing the resolutions of its sub-images and helps reduce the computational complexity in the proposed system and demonstrates that the image with 64×64 resolution is sufficient to recognize leaf image [22].

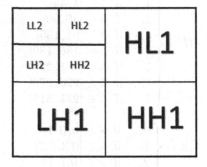

Fig. 1. Wavelet coefficient structure [22]

Local features are then extracted from the decomposed leaf images. Many local feature descriptors have been proposed in the past. One of the most influential descriptors is local binary patterns (LBP) [24], which will be adopted in our work. The LBP is a simple but efficient, powerful operator to describe the local image pattern (image texture). It has been used in many areas such as image retrieval, automatic face recognition and detection, and medical image analysis, etc. [12,22]. The LBP value is first obtained from the neighbourhood circular pixels using the central pixel. The value is then multiplied by binary weighting as final. The equations are as follows [30]:

$$LBP_{P,R}(x_c, y_c) = \sum_{i=0}^{p-1} s(g_P - g_c)2^p \qquad (10)$$

$$s(x) = \begin{cases} 1 \text{ if } x \geq 0 \\ 0 \text{ if } x < 0 \end{cases} \tag{11}$$

Where x_c and y_c are the coordinate of center pixel, P is circular sampling points or neighbourhood pixels of radius of R, g_p is grey scale value of P, g_c is centre pixels and s (sign) is threshold function. Examples of the circular neighbourhood are illustrated in Fig. 2 [13, 30].

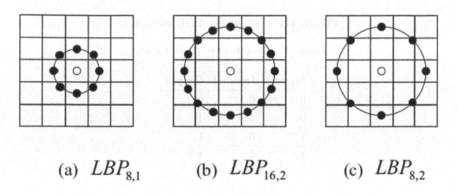

(a) $LBP_{8,1}$ (b) $LBP_{16,2}$ (c) $LBP_{8,2}$

Fig. 2. Circular neighbourhood for LBP [30]

4.3 Hybrid Descriptor

The framework of the proposed Hybrid Descriptor (HD) is illustrated in Fig. 3. The first step of the proposed method is to resize all images and make them to 256 × 256 pixels. The two different sets of feature vectors are then extracted from the resized images.

The colour histogram is used to extract global features in RGB colour space. It has a large number of bins because an RGB histogram model with 256 bins per channel has around 16.7 million degrees of freedom (256 × 256 × 256 bins). As a result, we used HSV colour histogram to reduce bins number for each image by used Eqs. 1–9 and extracted global features.

As aforementioned, we rely on LBP for local features as it is a powerful technique to describe leaf texture. However, it is time consuming to process all pixels in images because the window size is fixed [29]. To overcome such a problem, we exploit Haar wavelet, specifically, the sub-band LL_2 domain. The LL_2 domain is the approximation coefficients of wavelet decomposition, which contains most of energy and represents the low frequency information of a leaf image. Some of HSV colour histogram features are weak because lack of chromaticity, on the other hand LBP features represent local features only so features are combined with HSV colour histogram features by using (OR) gate digital truths in order to substitute the lack or local features by combination features for capturing a robust feature vector. Any rules resulting a minimum distance error

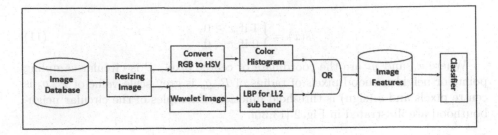

Fig. 3. The Framework of HD method

Table 1. The OR gate truth table

Input A	Input B	Output
0	0	0
0	1	1
1	0	1
1	1	1

then becomes the classifier to classify the leaf image to the corresponding plant specie.

Table 1 presents the (OR) gate truths table; Figure 4 depicts the minimum distance classifiers.

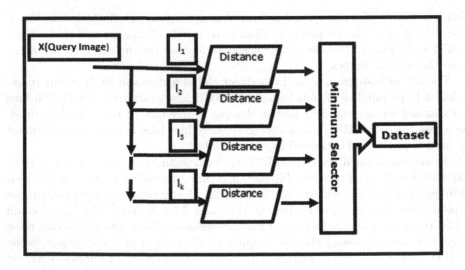

Fig. 4. The minimum distance classifier

5   Experimental Evaluation

5.1   Experimental Design

The experimental evaluation of proposed leaf image classification method is designed following the general framework of pattern classification. The CLEF 2011 image dataset has been used in the experiment. The data set consisted of four different plant species; *Cornus_mas, Magnolia_denudata, Ulmus_glabra*, and *Ulmas_ parvifolia*. The data set was divided into two subsets, the training set and testing set. The training set contained 60 leaf images, in which each plant specie had 15 leaf images with various sizes, directions and surfaces. The testing set included 40 leaf images, in which each specie shared ten. Classifiers were learned from the training set by using the features extracted from the leaf images. When a query leaf image was given from the testing set, hybrid descriptors were then extracted and used to compare with the classifiers to classify the leaf image to its belonging specie. Figure 5 shows some of image samples of the training set and Fig. 6 depicts the step-by-step dataflow in the experiment.

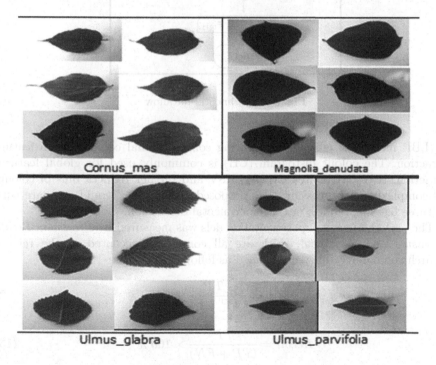

Fig. 5. Sample leaf images in experimental dataset

Three typical leaf classification and recognition methods were selected as the baseline models in experiments; LBP [24], WavLBP [7], and HSV-CH [23]. The LBP method is simple, efficient to describe texture in leaf images. The

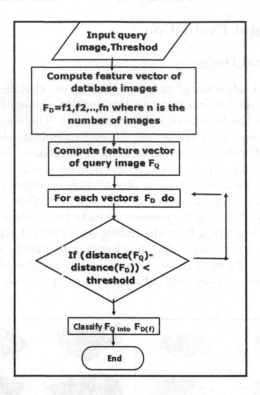

Fig. 6. Experimental dataflow

WavLBP method is efficient in reducing computational complexity in feature extraction. The colour histogram (CH) is commonly used for global features extraction. HSV-CH invites CH into HSV and further reduces bins of colours and complexity in processing. The proposed HD method would compare with the three baseline models in the experimental evaluation.

The performance of experimental models was measured by *Precision, Recall, F-measure* and *Accuracy*, which are all commonly accepted in the related research community. They are defined as follows:

$$Precision = \frac{TP}{(TP + FP)} \times 100\,\% \tag{12}$$

$$Recall = \frac{TP}{(TP + FN)} \times 100\,\% \tag{13}$$

$$Fmeasure = 2.\frac{P.R}{P + R} \times 100\,\% \tag{14}$$

$$Accuracy = \frac{(TP + FN)}{\mathcal{D}} \times 100\,\% \tag{15}$$

TN (True Negative) denotes the case of a negative sample being predicted negative (e.g., a *non-Cornus_mas* leaf image being classified into the complement class of *Cornus_mas* correctly); TP (True Positive) refers to the case a positive sample being predicted positive (e.g., a *Cornus_mas* leaf image being classified into the class of *Cornus_mas* correctly); FN (False Negative) refers to the case that a positive sample being predicted negative (e.g., a *Cornus_mas* leaf image being classified into the complement class of *Cornus_mas* incorrectly); and FP (False Positive) denotes the case that a negative sample being predicted positive (e.g., a *non-Cornus_mas* leaf image being classified into the class of *Cornus_mas* incorrectly).

5.2   Experimental Result Analysis

In this section, we discuss the results of the experimental evaluation conducted on our proposed system. We first evaluated the performance of our method when we implemented it with individual method such as LBP, WavLBP and HSV-CH. After evaluating the effectiveness of individual methods, we then investigated the performance of our proposed method by comparing with baseline models. The experimental results are presented in Figs. 7, 8, 9 and 10 and Table 2.

Table 2. Comparison of average results

Method	Precision	Recall	F-measure	Accuracy
LBP	79.56	21.39	32.89	87.5
WavLBP	85.22	21.22	32.23	87.5
HSV-CH	84.46	22.53	35.20	91.25
HD	**93.33**	**24.03**	**38.11**	**96.25**

Performance of the Proposed Leaf Recognition System. In this experiment, we evaluated our proposed system aiming at improving the performance of leaf image recognition systems. From the results shown in Table 2, one may see that there are noticeably increasing performance in terms of precision and recall for our proposed method when compared to baseline methods. The system with proposed method achieved 93.33 % and 24.03 % for precision and recall, respectively. The proposed method also yields a higher accuracy value compared to others. This is because the proposed method based on two feature descriptors: local features relying on LBP taken out from LL_2 sub band wavelet, and global features represented by colour histogram in natural colour to human HSV.

By further aggregating the obtained results from Table 2, Figs. 7, 8, 9 and 10 illustrate the detailed results in Precision, Recall, F-measure and accuracy for each method conducted on four different plant species. The results reveal that the proposed method is capable of recognising the leaf images of all four different species with high level of accuracy and precision.

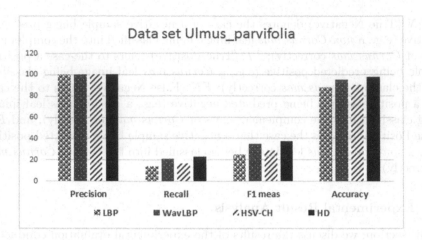

Fig. 7. Detailed experimental results in four methods for Ulmus_parvifolia dataset

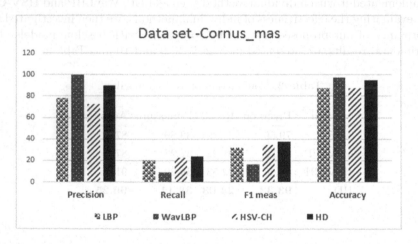

Fig. 8. Detailed experimental results in four methods for Cornus_mas dataset

Performance of the Leaf Recognition System with Different Methods.
We first explored the performance of the experimental system employing baseline
methods such as LBP, WavLBP and HSV-CH. Table 2 shows their averaged
experimental results. From the results, we can see that the LBP method yields
high degrees of precision and recall. However, the highest precision and recall
results were achieved when applying Wavelet or HSV colour histogram methods
before using LBP method. Such an observation reveals the difference of baseline
models in terms of their capacity of leaf images recognition and provides practical
justification for the development of our proposed method.

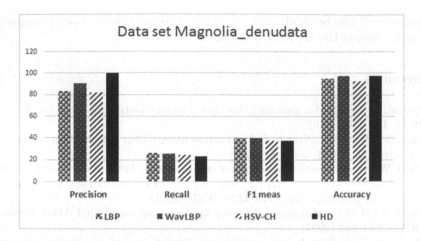

Fig. 9. Detailed experimental results in four methods for Magnolia_denudata dataset

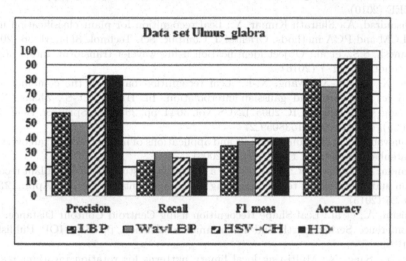

Fig. 10. Detailed experimental results in four methods for Ulmus_glabra dataset

6   Conclusions

Aiming at improving the performance of images recognition and classification, a hybrid descriptors method has been introduced in this paper to recognise and classify leaf images for plant species. The methods adopts a hybrid descriptor combining both global and local image features extracted from leaf images. Experimental results show that the proposed system yields promising performance with high level of accuracy and precision. The work has made contributions to knowledge advancement in leaf image classification and plant specie recognition. In the future, the performance of our proposed system will be further improved by using additional combination methods. Further experimental evaluations will also be conducted using large, extensive datasets and comparing with more state-of-the-art related methods.

References

1. Ahmed, N., et al.: An automatic leaf based plant identification system. Sci. Int. **28**(1), 427–430 (2016)
2. Casanova, D., et al.: Plant leaf identification using Gabor wavelets. Int. J. Imaging Syst. Technol. **19**(3), 236–243 (2009)
3. Chen, W., et al.: Identifying computer graphics using HSV colour model and statistical moments of characteristic functions. In: IEEE International Conference on Multimedia and Expo, pp. 1123–1126. IEEE (2007)
4. Du, J.-X., et al.: Leaf shape based plant species recognition. Appl. Math. Comput. **185**(2), 883–893 (2007)
5. Du, L., et al.: Wavelet domain local binary pattern features for writer identification. In: 20th International Conference on Pattern Recognition (ICPR), pp. 3691–3694. IEEE (2010)
6. Ehsanirad, A., Sharath Kumar, Y.: Leaf recognition for plant classification using GLCM and PCA methods. Oriental J. Comput. Sci. Technol. **3**(1), 31–36 (2010)
7. Ganesh, S.S., et al.: Object identification using wavelet transform. Indian J. Sci. Technol. **9**(5), 1–7 (2016)
8. Gu, X., Du, J.-X., Wang, X.-F.: Leaf recognition based on the combination of wavelet transform and gaussian interpolation. In: Huang, D.-S., Zhang, X.-P., Huang, G.-B. (eds.) ICIC 2005. LNCS, vol. 3644, pp. 253–262. Springer, Heidelberg (2005). doi:10.1007/11538059_27
9. Halawani, A., et al.: Fundamentals and applications of image retrieval: an overview. Datenbank-Spektrum **18**(14–23), 6 (2006)
10. Handa, A., Agarwal, R.: A review and a comparative study of various plant recognition and classification techniques using leaf images. Int. J. Comput. Appl. **123**(2), 20–25 (2015)
11. Hasim, A., et al.: Leaf Shape Recognition using Centroid Contour Distance. IOP Conference Series: Earth and Environmental Science, pp. 1-8. IOP Publishing (2016)
12. He, Y., Sang, N.: Multi-ring local binary patterns for rotation invariant texture classification. Neural Comput. Appl. **22**, 793–802 (2013)
13. Herdiyeni, Y., Santoni, M.M.: Combination of morphological, local binary pattern variance and colour moments features for Indonesian medicinal plants identification. In: 2012 International Conference on Advanced Computer Science and Information Systems (ICACSIS), pp. 255-259. IEEE (2012)
14. Ibrahim, A.A.: Iris recognition using Haar wavelet transform. J. Al-Nahrain Univ. Sci. **17**(1), 180–186 (2014)
15. Im, C., et al.: Recognizing plant species by normalized leaf shapes. In: Vision Interface, pp. 397–404 (1999)
16. Kadir, A., et al.: Performance improvement of leaf identification system using principal component analysis. Int. J. Adv. Sci. Technol. **44**, 113–124 (2012)
17. Kadir, A., et al.: Neural network application on foliage plant identification, 29(9), 15–22 (2013). ArXiv preprint arXiv:1311.5829
18. Kang, J., Zhang, W.: A framework for image retrieval with hybrid features. In: 24th Chinese Control and Decision Conference (CCDC), pp. 1326–1330. IEEE (2012)

19. Liu, X., et al.: Face features extraction based on multi-scale LBP. In: 2nd International Conference on Signal Processing Systems (ICSPS), pp. 438–441. IEEE (2010)

20. Liu, N., Kan, J.-M.: Plant leaf identification based on the multi-feature fusion and deep belief networks method. J. Beijing For. Univ. **3**, 14 (2016)

21. Man, Q.-K., Zheng, C.-H., Wang, X.-F., Lin, F.-Y.: Recognition of plant leaves using support vector machine. In: Huang, D.-S., Wunsch II, D.C., Wang, X.F., Lin, F.-Y. (eds.) Advanced Intelligent Computing Theories and Applications. With Aspects of Contemporary Intelligent Computing Techniques. Communications in Computer and Information Science, vol. 15, pp. 192–199. Springer, Heidelberg (2008)

22. Mohamed, A.A., et al.: Avatar face recognition using wavelet transform and hierarchical multi-scale LBP. In: 10th International Conference on Machine Learning and Applications and Workshops (ICMLA). IEEE (2011)

23. Nirapure, D., Reddy, U.: AST retrieval of images using filtered HSV colour level detection. Int. J. Emerg. Technol. Adv. Eng. **3**(8), 414–419 (2013)

24. Ojala, T., et al.: Multiresolution gray-scale and rotation invariant texture classification with local binary patterns. IEEE Trans. Pattern Anal. Mach. Intell. **24**(7), 971–987 (2002)

25. Pornpanomchai, C., et al.: Thai herb leaf image recognition system (THLIRS). Kasetsart J. (Nat. Sci.) **45**, 551–562 (2011)

26. Ruchika, M.S., Singh, A.R.: Compression of medical images using wavelet transforms. Int. J. Soft Comput. Eng. (IJSCE), pp. 2231–2307 (2012)

27. VijayaLakshmi, B., Mohan, V.: Kernel-based PSO and FRVM: an automatic plant leaf type detection using texture, shape, and colour features. Comput. Electron. Agric. **125**, 99–112 (2016)

28. Wang, X.-F., Du, J.-X., Zhang, G.-J.: Recognition of leaf images based on shape features using a hypersphere classifier. In: Huang, D.-S., Zhang, X.-P., Huang, G.-B. (eds.) ICIC 2005. LNCS, vol. 3644, pp. 87–96. Springer, Heidelberg (2005). doi:10.1007/11538059_10

29. Wang, Y., et al.: LBP texture analysis based on the local adaptive Niblack algorithm. In: IEEE Congress on Image and Signal Processing (CISP 2008), pp. 777–780 (2008)

30. Wang, Y.-D., et al.: Hand vein recognition based on multi-scale LBP and wavelet. In: IEEE International Conference on Wavelet Analysis and Pattern Recognition (ICWAPR), pp. 214–218 (2011)

31. Wang, Z., et al.: Leaf recognition based on PCNN. Neural Comput. Appl. **27**(4), 899–908 (2016)

32. Zulkifli, Z., et al.: Plant leaf identification using moment invariants & general regression neural network. In: IEEE 11th International Conference on Hybrid Intelligent Systems (HIS), pp. 430–435 (2011)

Efficient Mining of Pan-Correlation Patterns from Time Course Data

Qian Liu[1], Jinyan Li[1(✉)], Limsoon Wong[2], and Kotagiri Ramamohanarao[3]

[1] Advanced Analytics Institute, University of Technology Sydney,
81 Broadway, Ultimo, NSW 2007, Australia
`jinyan.li@uts.edu.au`
[2] School of Computing, National University of Singapore,
3 Computing Drive, Singapore 117417, Singapore
[3] Department of Computing and Information Systems,
The University of Melbourne, Parkville, VIC 3010, Australia

Abstract. There are different types of correlation patterns between the variables of a time course data set, such as positive correlations, negative correlations, time-lagged correlations, and those correlations containing small interrupted gaps. Usually, these correlations are maintained only on a subset of time points rather than on the whole span of the time points which are traditionally required for correlation definition. As these types of patterns underline different trends of data movement, mining all of them is an important step to gain a broad insight into the dependencies of the variables. In this work, we prove that these diverse types of correlation patterns can be all represented by a generalized form of positive correlation patterns. We also prove a correspondence between positive correlation patterns and sequential patterns. We then present an efficient single-scan algorithm for mining all of these types of correlations. This *"pan-correlation"* mining algorithm is evaluated on synthetic time course data sets, as well as on yeast cell cycle gene expression data sets. The results indicate that: (i) our mining algorithm has linear time increment in terms of increasing number of variables; (ii) negative correlation patterns are abundant in real-world data sets; and (iii) correlation patterns with time lags and gaps are also abundant. Existing methods have only discovered incomplete forms of many of these patterns, and have missed some important patterns completely.

1 Introduction

Time course data have been involved in many real-world applications, especially in the fields of finance, healthcare and biomedicine. This work investigates the mining algorithm of correlation patterns. A correlation pattern is defined as a series of highly correlated data movement trends between two sets of variables on some *subset* of time points (not necessarily on the whole span of the time points). The two basic types of correlations are the positive correlation or the negative correlation. A pattern of positive correlation is a set of variables showing the same direction in their data movements. On the other hand, the data changes

© Springer International Publishing AG 2016
J. Li et al. (Eds.): ADMA 2016, LNAI 10086, pp. 234–249, 2016.
DOI: 10.1007/978-3-319-49586-6_16

of one set of variables in a negative correlation pattern go jointly up or down whenever the value changes of the other set of variables move in the opposite direction. Variables in time course data also have time-dependent interactions. The influence of a variable on other variables sometimes may not be immediate. Instead, it is going to be effective only after some time delay, leading to time-lagged correlation (positive or negative). Thus, there are four types of correlation patterns: the basic positive and negative correlation patterns (i.e., synchronized correlations without time delay), and time-lagged positive and negative correlation patterns. Time course data in real-world applications may also contain a small amount of unknown noise and errors. These noise and errors can interrupt the time continuity of a correlation, leading to gaps in the correlation. Gaps complicate the complexity of correlation mining, because the gaps can happen at any time point, and the length of a gap is unknown.

This work introduces a new type of correlation pattern, named "pan-correlation patterns", to maximize the sequence of coherent data movements in one pattern. A pan-correlation pattern consists of a maximized sub-list V_0 of variables, where all the listed variables are associated with a segment of time points having the same length, such that V_0 can be divided into two not necessarily mutually-exclusive lists of variables V_1 and V_2, satisfying: (i) every pair of variables within V_1 are positively correlated, or time-lag positively correlated, or time-lag positively correlated with gaps; (ii) every pair of variables within V_2 are positively correlated, or time-lag positively correlated, or time-lag positively correlated with gaps; and (iii) every pair of variables between V_1 and V_2 are negatively correlated, or time-lag negatively correlated, or time-lag negatively correlated with gaps. V_1 or V_2 can be empty—in this case, a pan-correlation pattern is simplified as a positive pan-correlation pattern. By our definition, a pan-correlation pattern can cover all of the following characteristics: the basic positively/negatively correlated data movement trends, time-lag effects, and noise/error gaps. However, mining significant pan-correlation patterns is a problem of high complexity. Existing methods are not capable of conducting the mining of pan-correlation patterns. They may only be able to detect a special subtype of pan-correlation patterns, for example, positive correlation patterns by [3,6], or negative correlation patterns by [7,13], or both positive and negative correlations by [4,14], or time-lagged positive correlation patterns by [2,5].

Our work introduces an efficient algorithm for mining significant pan-correlation patterns. We proposed three critical ideas. First, we prove that all the different types of correlation patterns can be represented by a generalized form of positive correlation patterns—viz. pan-correlation patterns. Based on this theory, we can focus on the mining of all positive correlations. Second, the time course data set is transformed into a sequential data set containing sequences of "up", "down", and "no-change", which are the three movement trends of variables. With this data discretization idea [7,9,10], the pan-correlation mining problem can be converted into a sequential pattern mining problem. Central to how we enable the representation of the different types of negative correlation patterns through the generalized form of positive correlation patterns is that we make an opposite-mirror copy [8] of the original sequential data set and then add it to the original data. A cost of adding the mirror copy of the sequential data is that many redundant patterns are produced. Thus, our third new idea is to modify a sequential pattern mining algorithm to efficiently prune redundant patterns in the mining process. Our pan-correlation mining algorithm is tested on synthetic time course data sets and four microarray gene expression time course data sets. The synthetic data sets are used to demonstrate the efficiency of our algorithm. The experiments on the gene expression data show that negative correlation patterns are indeed abundant in real-world data sets, and that patterns with different time delays and gaps are common. It is worth noting that pan-correlation patterns are not a kind of pairwise correlation patterns, and it is of high time complexity, if not impossible, to use traditional algorithms, clustering or pairwise correlation, to mine pan-correlation patterns.

The rest of the paper is organized as follows. We define the six types of correlation patterns and their closure property in Sect. 2. We then describe our pan-correlation mining algorithm in Sect. 3. After that, we present results of our pan-correlation mining algorithm on synthetic and also real-life gene expression time course data sets in Sect. 4.

2 Problem Formulation

Let V be a set of N_V variables $v_1, v_2, \ldots, v_{N_V}$. Let T be a set of N_T consecutive time points $t_1, t_2, \ldots, t_{N_T}$. Here, t_j and t_{j+1} in N_T are two ordered consecutive time points with $t_j \prec t_{j+1}$, indicating that t_j precedes t_{j+1}. Let $m_{i,j}$ denote the value of variable v_i at time point t_j. A time course data set is then defined by the data matrix $M = [m_{i,j}]_{N_V \times N_T}$.

2.1 Correlation Patterns: Definitions

Definition 1. *A positive correlation pattern p is a pair comprising a subset V_0 of variables in V and a continuous segment T_p of time points in T such that, for every pair of consecutive time points from t_j to t_{j+1} in T_p, the values of all variables in V_0 decrease or increase simultaneously. A positive correlation p is written as $p = \langle V_0, T_p \rangle$.*

It is possible that the magnitude of the 'decrease' or 'increase' is very small. These slight decrease or increase movements are both considered as 'no-change'. Therefore, these time segments will be considered to have the same value movement in a positive correlation when the variables change their movements very slightly. More formally, for each v_i, we say the value of v_i increases (decreases) from consecutive time point t_j to t_{j+1} if it changes by at least δ_i, where δ_i is some specified threshold. Under this assumption, a pattern containing only 'no-change' provides less information for high correlation. Therefore, we require that a correlation pattern must contain at least one significant decrease/increase

movements. This convention is applied on all definitions, lemmas, and propositions in this work. Please note that δs might also result in information lost in
correlation pattern. Optimal δ's threshold should consider the tradeoff between
insignificant, noise change and information lost. There is no gold standard to
provide the best δs. Thus, the threshold of δs could be specified by users based
on domain knowledge.

Definition 2 (Cf. [7]). *A negative correlation pattern* n *is a triplet comprising
two non-overlapping subsets* V_1 *and* V_2 *of variables in* V *and a continuous segment* T_n *of time points in* T *such that, for every pair of consecutive time points
from* t_j *to* t_{j+1} *in* T_n, *the values of all variables in* V_1 *decrease while the values
of all variables in* V_2 *increase, and vice versa. A negative correlation* n *is written
as* $n = \langle (V_1, V_2), T_n \rangle$.

These two definitions describe a synchronized pace of value change without time
delay. In fact, some variables in the data matrix M may have influence on others,
but the effect may not take place immediately (i.e., after some time delay).

Definition 3. *A time-lagged negative correlation pattern* kn *is a pair of dis-
tinct lists* $\{(v_{x_1}, T_p^1), \ldots, (v_{x_h}, T_p^h)\}$ *and* $\{(v_{y_1}, T_q^1), \ldots, (v_{y_g}, T_q^g)\}$, *such that:
(i)* $V_1 = \{v_{x_1}, \ldots, v_{x_h}\}$ *and* $V_2 = \{v_{y_1}, \ldots, v_{y_g}\}$ *are two possibly overlapping
lists of not necessarily distinct variables of* V; *(ii)* $T_K^1 = \{T_p^1, \ldots, T_p^h\}$ *and
* $T_K^2 = \{T_q^1, \ldots, T_q^g\}$ *are two lists of* h *and* g *continuous time segments of the
same length in* T; *(iii) for every* $1 \leq r < |T_p^1|$ *and for every* $v_{x_i} \in V_1$, *the
value of* v_{x_i} *increases (decreases) from the* rth *time point in* T_p^i *to the* $(r+1)$th
time point in T_p^i *if and only if for all other* $v_{x_j} \in V_1$, *the value of* v_{x_j} *increases
(decreases) from the* rth *time point in* T_p^j *to the* $(r+1)$th *time point in* T_p^j; *(iv) for
every* $1 \leq r < |T_p^1|$ *and for every* $v_{y_i} \in V_2$, *the value of* v_{y_i} *increases (decreases)
from the* rth *time point in* T_q^i *to the* $(r+1)$th *time point in* T_q^i *if and only if for
all other* $v_{y_j} \in V_2$, *the value of* v_{y_j} *increases (decreases) from the* rth *time point
in* T_q^j *to the* $(r+1)$th *time point in* T_q^j; *and (v) for every* $1 \leq r < |T_p^1|$, *for every
* $v_{x_i} \in V_1$, *and for every* $v_{y_j} \in V_2$, *the value of* v_{x_i} *increases (decreases) from the
* rth *time point in* T_p^i *to the* $(r+1)$th *time point in* T_p^i *if and only if the value
of* v_{y_j} *decreases (increases) from the* rth *time point in* T_q^j *to the* $(r+1)$th *time
point in* T_q^j. *For convenience, a time-lagged negative correlation* kn *is written as
* $kn = \langle (V_1, V_2), (T_K^1, T_K^2) \rangle$.

When V_1 and T_K^1, or V_2 and T_K^2, are empty, kn is a time-lagged positive
correlation, denoted by kp.

A time segment can be extended into a discontinuous time segment to tolerate
some small amount of noise. For example, $T_p = [1, 2, 3, 4, 7, 8, 9, 10]$ is a discontinuous time segment containing a gap of length 2 between 4 and 7. The first 4
time points of T_p are continuous from 1 to 4, and the next 4 time points are con-
tinuous from 7 to 10. The pattern $p = \{(v, T_p = [1, 2, 3, 4, 7, 8, 9, 10]), (v', T_p' = [1, 2, 3, 4, 5, 6, 7])\}$ is defined as a positive correlation pattern with gaps if the
changes of the values of v for any two consecutive time points of $[1, 2, 3, 4]$ are

in the same direction as the changes of the values of v' for $[1, 2, 3, 4]$, and the changes of the values of v for any two consecutive time points of $[7, 8, 9, 10]$ are in the same direction as the changes of the values of v' for $[4, 5, 6, 7]$. The data movement trends between the time points 4 and 7 in v are not considered due to the gap.

Next, we introduce the definitions for (time-lagged) positive/negative correlation patterns that contain gaps. A pair of consecutive time points t_i and t_{i+1} is denoted as $tpp_{(i,i+1)}$. In this work, all time-point pairs are pairs of consecutive time points. Let $Tpp = \{tpp_{(i_j, i_j+1)} \mid j = 1, 2, \ldots, h\}$ be an ordered list of h time-point pairs, where $t_{i_j} \prec t_{i_{j+1}}$. Tpp is continuous if and only if for every $1 \le k \le h$, $i_k + 1 = i_{k+1}$. Otherwise, Tpp is discontinuous and contains gaps. A continuous Tpp corresponds to a continuous time segment. For example, $\{tpp_{(1,2)}, tpp_{(2,3)}, tpp_{(3,4)}\}$ corresponds to time segment $\{t_1, t_2, t_3, t_4\}$. A discontinuous Tpp may also corresponds to a continuous time segment. For example, $\{tpp_{(1,2)}, tpp_{(3,4)}\}$ correspond to time segment $\{t_1, t_2, t_3, t_4\}$. So, a time segment alone is not sufficient to define the data movements on the time-point pairs and the movement gaps.

Definition 4. *A negative pan-correlation pattern is a time-lagged negative correlation pattern with gaps. That is, it is a pair of distinct lists $\{(v_{x_1}, Tpp_p^1), \ldots, (v_{x_h}, Tpp_p^h)\}$ and $\{(v_{y_1}, Tpp_q^1), \ldots, (v_{y_g}, Tpp_q^g)\}$, such that: (i) $\mathcal{V}_1 = \{v_{x_1}, \ldots, v_{x_h}\}$ and $\mathcal{V}_2 = \{v_{y_1}, \ldots, v_{y_g}\}$ are two possibly overlapping lists of not necessarily distinct variables in V; (ii) $\mathcal{TPP}_K^1 = \{Tpp_p^1, \ldots, Tpp_p^h\}$ and $\mathcal{TPP}_K^2 = \{Tpp_q^1, \ldots, Tpp_q^g\}$ are two lists of time-point-pair lists all with the same length and possibly containing different gaps; (iii) for every $1 \le r < |Tpp_p^1|$ and for every $v_{x_i} \in \mathcal{V}_1$, the value of v_{x_i} increases (decreases) at the rth time-point pair in Tpp_p^i if and only if for all other $v_{x_j} \in \mathcal{V}_1$, the value of v_{x_j} increases (decreases) at the rth time-point pair in Tpp_p^j; (iv) for every $1 \le r < |Tpp_p^1|$ and for every $v_{y_i} \in \mathcal{V}_2$, the value of v_{y_i} increases (decreases) at the rth time-point pair in Tpp_q^i if and only if for all other $v_{y_j} \in \mathcal{V}_2$, the value of v_{y_j} increases (decreases) at the rth time-point pair in Tpp_q^j; (v) for every $1 \le r < |Tpp_p^1|$, for every $v_{x_i} \in \mathcal{V}_1$, and for every $v_{y_j} \in \mathcal{V}_2$, the value of v_{x_i} increases (decreases) at the rth time-point pair in Tpp_p^i if and only if the value of v_{y_j} decreases (increases) at the rth time-point pair in Tpp_q^j. For convenience, a negative pan-correlation pattern $\mathcal{C}$ is written as $\mathcal{C} = \langle (\mathcal{V}_1, \mathcal{V}_2), (\mathcal{TPP}_K^1, \mathcal{TPP}_K^2) \rangle$.*

When $\mathcal{V}_1$ and $\mathcal{TPP}_K^1$, or $\mathcal{V}_2$ and $\mathcal{TPP}_K^2$, are empty, $\mathcal{C}$ is a positive pan-correlation. Moreover, every continuous time segment $T*$ in the definitions from Definition 1 to 3 can be converted into a continuous Tpp. Thus all correlation patterns by these definitions can be rewritten by using time-point-pair list Tpp to replace time segment $T*$.

There are a huge number of positive and negative pan-correlation patterns in the data matrix M. However, we are only interested in those patterns that are closed. A pattern $\mathcal{C} = \langle (\mathcal{V}_1, \mathcal{V}_2), (\mathcal{TPP}_K^1, \mathcal{TPP}_K^2) \rangle$ is closed if (i) any time-point-pair list in $\mathcal{TPP}_K^1$ and $\mathcal{TPP}_K^2$ cannot be enlarged to include more time-point pairs without breaking the underlying correlation among the variables in $\mathcal{V}_1$ and

$\mathcal{V}_2$, and (ii) the list of variables $\mathcal{V}_1$ and $\mathcal{V}_2$ cannot be enlarged to include more variables as there can be no other variable that correlates positively or negatively to those in $\mathcal{V}_1$ and $\mathcal{V}_2$ over the same number of time-point pairs in $\mathcal{TPP}_K^1$ and $\mathcal{TPP}_K^2$. Every positive (negative) pan-correlation pattern can be derived from some closed positive (negative) pan-correlation patterns by deleting variables and/or deleting time-point pairs. Thus, the set of all closed positive (negative) pan-correlation patterns forms a lossless and non-redundant representation of positive (negative) pan-correlation patterns. These patterns are called closed patterns and more specifically $\mathbb{C}$-, $\mathbb{CP}$-, and $\mathbb{CN}$-closed patterns.

The following relationships between the various types of pan-correlation patterns can be easily proved.

Proposition 1. *Let $\mathcal{C} = \langle(\mathcal{V}_1, \mathcal{V}_2), (\mathcal{TPP}_K^1, \mathcal{TPP}_K^2)\rangle$, $\mathcal{C}_1 = \langle\mathcal{V}_1, \mathcal{TPP}_K^1\rangle$, and $\mathcal{C}_2 = \langle\mathcal{V}_2, \mathcal{TPP}_K^2\rangle$. Then*

- *$\mathcal{C}$ is in $\mathbb{CN}$ implies both $\mathcal{C}_1$ and $\mathcal{C}_2$ are in $\mathbb{CP}$.*
- *$\mathcal{C}$ is in $\mathbb{CN}$ if, and only if, $\mathcal{C}' = \langle(\mathcal{V}_2, \mathcal{V}_1), (\mathcal{TPP}_K^2, \mathcal{TPP}_K^1)\rangle$ is in $\mathbb{CN}$.*
- *$\mathcal{C}$ is closed in $\mathbb{C}$ if, and only if, it is closed in $\mathbb{CN}$.*
- *$\mathcal{C}_1' = \langle(\mathcal{V}_1, \{\}), (\mathcal{TPP}_K^1, \{\})\rangle$ is closed in $\mathbb{CN}$ implies $\mathcal{C}_1$ is closed in $\mathbb{CP}$.*
- *$\mathcal{C}$ is closed in $\mathbb{C}$ implies for $i \in \{1, 2\}$, for every (closed) pattern $\mathcal{C}' = \langle\mathcal{V}', \mathcal{TPP}'\rangle$ in $\mathbb{CP}$ where $\mathcal{C}_i \sqsubseteq_p \mathcal{C}'$, it is the case that $\mathcal{V}_i = \mathcal{V}'$ (Note that $\mathcal{C}_i = \mathcal{C}'$ does not hold).*

The second point of Proposition 1 implies some degree of redundancy, as the two patterns $\mathcal{C}$ and $\mathcal{C}'$ capture the same correlation information. We will deal with this redundancy later in Sect. 3.

2.2 Unified Representation of All Correlation Patterns

Let $V*$ be a set of variables $v_1*, v_2*, ..., v_{N_V}*$. Let $m_{i,j}* = -m_{i,j}$ denote the value of variable v_i* at time point t_j, and this value of v_i* at time point t_j is the negation of the value of v_i at time point t_j. A negated time course data set is then defined by the data matrix $M* = [m_{i,j}*]_{N_V \times N_T}$. It is also called a mirror-copy of M. Clearly, whenever the value of v_i increases (decreases) from time point t_j to time point t_{j+1}, the value of v_i* decreases (increases) from time point t_j to time point t_{j+1}. I.e., the value of v_i* moves in the opposite direction of v_i. Let M' be the matrix obtained by adding the negated data matrix $M*$ to the original data matrix M (details are given in Sect. 3.2). Then, the lemma below follows from this observation and can be easily proved.

Lemma 1. *$\mathcal{C} = \langle(\mathcal{V}_1 = \{v_{x_1}, \ldots, v_{x_h}\}, \mathcal{V}_2 = \{v_{y_1}, \ldots, v_{y_g}\}), (\mathcal{TPP}_K^1 = \{Tpp_p^1, \ldots, Tpp_p^h\}, \mathcal{TPP}_K^2 = \{Tpp_q^1, \ldots, Tpp_q^g\})\rangle$ is in $\mathbb{CN}$ in the data matrix M if, and only if, $\mathcal{C}* = \langle\mathcal{V} = \{v_{x_1}, \ldots, v_{x_h}, v_{y_1}*, \ldots, v_{y_g}*\}, \mathcal{TPP} = \{Tpp_p^1, \ldots, Tpp_p^h, Tpp_q^1, \ldots, Tpp_q^g\}\rangle$ is in $\mathbb{CP}$ in the data matrix M'.*

Based on the equivalence above, for $\mathcal{C}$ in $\mathbb{CN}$ with regard to M, we write $\mathcal{C}*$ for its counterpart in $\mathbb{CP}$ with regard to M'.

Every closed $\mathbb{CP}$ pattern in the data matrix M' is in a one-to-one correspondence with a closed $\mathbb{CN}$ pattern (also a closed $\mathbb{C}$ pattern) in the data matrix M.

Theorem 1. $\mathcal{C} = \langle (\mathcal{V}_1 = \{v_{x_1}, \ldots, v_{x_h}\}, \mathcal{V}_2 = \{v_{y_1}, \ldots, v_{y_g}\}), (\mathcal{TPP}^1_K = \{Tpp^1_p, \ldots, Tpp^h_p\}, \mathcal{TPP}^2_K = \{Tpp^1_q, \ldots, Tpp^g_q\}) \rangle$ *is closed in the data matrix M if, and only if, $\mathcal{C}*$ is closed in the data matrix M'. Thus, $\mathbb{C}$-closed patterns in M are in one-to-one correspondence with $\mathbb{CP}$-closed patterns in M'. (Proof is omitted due to page limitation.)*

3 Mining Algorithms

Our efficient mining of all significant pan-correlation patterns consists of the following four components.

3.1 Transform Time-Course Data Set M into Sequential Transaction Data Set S

Given a time-course data set $M = [m_{i,j}]_{N_V \times N_T}$, let $s_{i,j}$ be the value movement of the variable v_i between time point t_j and $t_{j+1} (= t_j+1)$. Specifically, $s_{i,j}$ is U (up) if $m_{i,j+1} \geq m_{i,j} + \delta_i$, and is D (down) if $m_{i,j+1} \leq m_{i,j} - \delta_i$, and is O otherwise. Let $R_i = \{s_{i,1}, s_{i,2}, \cdots, s_{i,N_T-1}\}$ be the sequence of all value movements of $v_i \in V$. Let $S = [s_{i,j}]_{N_V \times (N_T-1)}$ be a sequential transaction data set which is easily transformed from M. S has the same variables V as M does, but each variable in S has $N_T - 1$ sequential value movements. In the transformation, δ_i is used to define the scale of the variable v_i's value movement in M. δ_i for $v_i \in V$ is set as twenty percents of the absolute difference between the second maximum value of $m_{i,j}$ and the second minimum value of $m_{i,j}$, $1 \leq j \leq N_T$. The maximum value and the minimum value are discarded to avoid some outlier values of v_i in M.

We view S as a set of sequential transactions. And each row R_i in S corresponds to a sequential transaction and is viewed a sequence of value movements (U, D, and O). Given any variable $v_i \in V$ and any ordered set of time-point pairs $Tpp = \{tpp_{(i_j, i_j+1)} \mid j = 1, 2, \ldots, h\}$. Let $f(v_i, Tpp)$ be the list $\{s_{i,i_1}, \ldots, s_{i,i_h}\}$. Thus, $f(v_i, Tpp)$ gives the value movements of v_i during Tpp. We write $f'(v_i, Tpp)$ to denote the list obtained by flipping every U to D and every D to U in $f(v_i, Tpp)$. In S, a sequential pattern is a list of value movements (U, D, and O). A sequential pattern $sp = \{s_1, \ldots, s_h\}$ is said to occur in a sequential transaction R_i if there is a list of time-point pairs $Tpp = \{tpp_{(i_j, i_j+1)} \mid j = 1, 2, \ldots, h\}$, such that $f(v_i, Tpp) = sp$. That is, the value movements specified in the pattern sp occur in the transaction R_i in the same order as they appear in sp, possibly separated by other value movements. We write $supp(sp, S)$ to denote the support of the sequential pattern sp in S.

The space of all sequential patterns occurring in S is denoted by $\mathbb{SP}$. A closed sequential pattern in $\mathbb{SP}$ is defined below, which is similar to those in previous works [12].

Definition 5. *Let sp and sp' be two sequential patterns. We say $sp \leq sp'$ in $\mathbb{SP}$ if, and only if, sp is a subsequence of sp' or is identical to sp', and $supp(sp, S) = supp(sp', S)$. The closed patterns of $\mathbb{SP}$ are the maximal patterns in $\mathbb{SP}$ according to this partial order.*

It is obvious that $f(v_{x_i}, Tpp^i) = f(v_{x_j}, Tpp^j)$ for $1 \leq i, j \leq h$, for any pattern $C = \langle \mathcal{V} = \{v_{x_1}, ..., v_{x_h}\}, \mathcal{TPP} = \{Tpp^1, ..., Tpp^h\}\rangle$ in $\mathbb{CP}$ in M. The following easily-proved property connects the closed patterns in $\mathbb{SP}$ of S and those in $\mathbb{CP}$ of M.

Proposition 2. *For every $\mathbb{SP}$-closed pattern sp in S, there is a unique $\mathbb{CP}$-closed pattern $C = \langle \mathcal{V} = \{v_{x_1}, ..., v_{x_h}\}, \mathcal{TPP} = \{Tpp^1, ..., Tpp^h\}\rangle$ in M, such that $sp = f(v_{x_i}, Tpp^i)$ for $1 \leq i \leq h$. And for every $\mathbb{CP}$-closed pattern $C = \langle \mathcal{V} = \{v_{x_1}, ..., v_{x_h}\}, \mathcal{TPP} = \{Tpp^1, ..., Tpp^h\}\rangle$ in M, there is a $\mathbb{SP}$-closed pattern sp in S, such that $sp = f(v_{x_i}, Tpp^i)$ for $1 \leq i \leq h$. Thus, $\mathbb{SP}$-closed patterns in S are in one-to-one correspondence with $\mathbb{CP}$-closed patterns of M.*

3.2 Opposite Mirror Copy of S

In $S = [s_{i,j}]_{N_V \times (N_T - 1)}$, a positive correlation pattern is denoted by one sequence of value movements, while a negative correlation pattern is denoted by two sequences of value movements whose value movements are opposite to each other at every position, U vs. D, and D vs. U. To make available in S the unified formulation of positive and negative correlation patterns, an opposite mirror copy of each transaction in S is created and added into S. This data management technique was similarly used by [8] for mining biclusters.

Given the value movements of v_i in S, i.e., $R_i = \{s_{i,1}, s_{i,2}, ..., s_{i,N_T-1}\}$, let its opposite mirror copy be $R_{*i} = \{s_{*i,1}, s_{*i,2}, ..., s_{*i,N_T-1}\}$ where $s_{*i,j}$ is up if $s_{i,j}$ is down, $s_{*i,j}$ is down if $s_{i,j}$ is up, and otherwise $s_{*i,j} = s_{i,j}$. The opposite mirror copy of all transactions in S are added into S. The new transaction data set is denoted by $S' = [s'_{i,j}]_{2N_V \times (N_T-1)}$, where all R_is of v_is are indexed from 0 to $2N_V$-2 with step 2 in S', and all R_{*i}s are indexed from 1 to $2N_V$-1 with step 2. This index strategy is used later. S' is also the sequential transaction data set derived from M'.

Then, the crucial theorem below follows immediately from Theorem 1 and Proposition 2.

Theorem 2. *$\mathbb{SP}$-closed patterns in S' are in one-to-one correspondence with $\mathbb{C}$-closed patterns in M.*

3.3 Mine Frequent Closed Sequential Value Movements in S'

All $\mathbb{SP}$-closed patterns in S' can be detected using efficient algorithms of mining closed sequential patterns. After that, given a $\mathbb{SP}$-closed pattern in S', by Theorem 2, there is a corresponding $\mathbb{CP}$-closed pattern in M', i.e., a $\mathbb{C}$-closed pattern in M. Then, all pan-correlation patterns can be easily obtained from these frequent closed sequential value movements by restoring the time-point pair information and the transaction id information: given a $\mathbb{SP}$-closed pattern sp and its $supp(sp, S)$ with $\{v_{x_1}, ..., v_{x_h}, v_{y_1}*, ..., v_{y_g}*\}$, the variables from V of M' are grouped in one set while those from $V*$ are grouped in another set, indicating the negative correlation between the two sets; then, the time-point pair information associated with sp is detected by matching sp with each variable $v_{x_i} \in supp(sp, S)$ where there might be multiple matches in v_{x_i}, indicating multiple occurrence of sp in v_{x_i}.

3.4 Opposite Mirror Copy Causes Redundancy in Patterns

In M', every pan-correlation pattern has a mirror image that carries the same information. For example, a negative correlation pattern $\mathcal{C} = \langle(\mathcal{V}_1, \mathcal{V}_2), (\mathcal{TPP}_1, \mathcal{TPP}_2)\rangle$ in M can be represented by $\mathcal{C}* = \langle\mathcal{V}_1 \cup \mathcal{V}_2*, \mathcal{TPP}_1 \cup \mathcal{TPP}_2\rangle$ or $\mathcal{C}*' = \langle\mathcal{V}_1*\cup\mathcal{V}_2, \mathcal{TPP}_1\cup\mathcal{TPP}_2\rangle$ in M'. Here, $\mathcal{V}_1*$ is the negation of $\mathcal{V}_1$ and $\mathcal{V}_2*$ is the negation of $\mathcal{V}_2$. Correspondingly in S' from M', $sp = f(v_{x_i}, Tpp^i)$ for $v_{x_i} \in \mathcal{V}_1 \cup \mathcal{V}_2*$ and $sp' = f(v_{y_j}, Tpp^j)$ for $v_{y_j} \in \mathcal{V}_1 * \cup\mathcal{V}_2$, and $sp \neq sp'$. Thus, $\mathcal{C}$ is mined twice in terms of sp or sp' in S'. And once one of sp and sp' is known, there is no need to mine the other because the other can be produced according to the flip relationship between their components. Thus, sp and sp' are redundant. It is easily proved that a closed $\mathcal{C}$ correlation pattern in M is always detected twice in terms of sp and sp' in S'.

Fortunately, each pair of redundant patterns has some unique property below. Without loss of generality, let a pair of redundant patterns $\mathcal{C}* = \langle(v_{x_1}, Tpp_p^1), ..., (v_{x_h}, Tpp_p^h), (v_{y_1}*, Tpp_q^1), ..., (v_{y_g}*, Tpp_q^g)\rangle$ and $\mathcal{C}*' = \langle(v_{x_1}*, Tpp_p^1), ..., (v_{x_h}*, Tpp_p^h), (v_{y_1}, Tpp_q^1), ..., (v_{y_g}, Tpp_q^g)\rangle$ on M' both capture the same information as $\mathcal{C} = \langle(\mathcal{V}_1 = \{v_{x_1}, ..., v_{x_h}\}, \mathcal{V}_2 = \{v_{y_1}, ..., v_{y_g}\}), (\mathcal{TPP}_K^1 = \{Tpp_p^1, ..., Tpp_p^h\}, \mathcal{TPP}_K^2 = \{Tpp_q^1, ..., Tpp_q^g\})\rangle$ in M. Here $v*$ is the negation of v. Then, we rewrite $\mathcal{C}* = \{(v_{z_1}, Tpp^1), ..., (v_{z_{h+g}}, Tpp^{h+g})\}$ and $\mathcal{C}*' = \{(v'_{w_1}, Tpp^{1'}), ..., (v'_{w_{h+g}}, Tpp^{h+g'})\}$. In $\mathcal{C}*$ and $\mathcal{C}*'$, assume that all pairs of $(v_{z_*}, Tpp*)$ are ordered first according to the transaction indexes of v_{z_*} and then according to the time-point pairs in $Tpp*$. After that, it is easily proved that $v_{z_1} = v_{w_1}*$, or $z_1 = w_1$ and $Tpp^1 \leq Tpp^{1'}$, or vice versa.

Thus to avoid producing redundant $\mathbb{SP}$-closed patterns in S', we must modify the algorithm for mining sequential value movements. We apply two constraints below to prune the redundant patterns. (i) On a sub-dataset $S'_s \subseteq S'$ with the ascending order of the indexes of all transactions on S' (Please refer to Sect. 3.2 for the detail of the indexes), assume R_{x_j} is the first transaction on S'_s, i.e., the transaction with the minimum transaction index. If R_{x_j} is produced from a $v_i* \in V*$, all sequential patterns on S'_s are redundant and thus the search of new sequential patterns on S'_s should be pruned. (ii) Otherwise, given a frequent value movement e (i.e. a value movement U, D or O) on S'_s, let $R_{x_{min1}}$ be the transaction with the minimum id where e occurs, and pos_1 be the first occurrence position of e in $R_{x_{min1}}$; let $R_{x_{min2}}$ be the transaction with the second minimum id where e occurs, and pos_2 be the first occurrence position of e in $R_{x_{min2}}$. If $R_{x_{min1}}$ is produced from $v_i \in V$ and $R_{x_{min2}}$ is produced from $v_i* \in V*$ and

$pos_1 > pos_2$, the search in the branch of frequent sequential patterns adding e is redundant and should be pruned. The lemma below is easily proved.

Lemma 2. *Our pruning strategy can guarantee that the closed sequential patterns detected are complete and non-redundant in S'. (Proof is omitted due to page limitation.)*

3.5   Parameter Setting

Three parameters, min_V, min_{TPP} and max_O, are used to prune trivial correlation patterns. In a given pan-correlation pattern $\mathcal{C} = \langle \mathcal{V}, \mathcal{TPP} \rangle$, min_V is the minimum size of $\mathcal{V}$, min_{TPP} is the minimum size of $Tpp \in \mathcal{TPP}$, and max_O is the maximum number of O contained.

3.6   An Illustrative Example

Figure 1 illustrates how our algorithm works. A time-course data set M has six variables and eight time points. M is shown in Fig. 1(a) and visualized in Fig. 1(b). Figure 1(b) does not easily show a very nice pan-correlation between the six variables. But our algorithm can discover a good negative correlation pattern among the six variables.

By our algorithm, M is firstly discretized to obtain a sequential data S in the first part of Fig. 1(c). Then the opposite mirror copy of all sequences in S, as shown in the second part of Fig. 1(c), is constructed using the strategy in Sect. 3.2. All sequences in Fig. 1(c) comprise S' in Fig. 1(d). With $min_V = 5$ and $min_{TPP} = 5$, two pan-correlation patterns are available in S', as shown in Fig. 1(e). It can be clearly seen from Fig. 1(e) that these two pan-correlation patterns are the same in the original data M, which can be represented in Fig. 1(f). Our algorithm can prune the redundancy and only outputs this pattern (visualized in Fig. 1(g)). If the gaps are merged and the time points lagged are ignored (for visualization only), this correlation pattern is shown in Fig. 1(h).

4   Performance Evaluation and Application

Our algorithm was tested on both synthetic time-course data sets and real-world time-course data sets of biomedical gene expression. In the implementation, we modified the source code of BIDE+ [12] for detecting pan-correlation patterns by integrating our pruning strategies.

4.1   Efficiency and Scalability Results on Synthetic Data Sets

Two series of synthetic data sets are used. The first series of data sets have the same number of time points but have an increasing number of variables. The second series of data sets have the same number of variables but have an

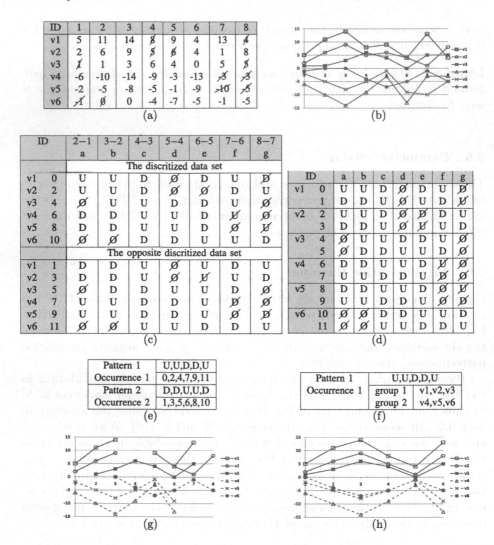

Fig. 1. An illustrative example of our algorithm. (a) An example of time-course data set M. (b) The plot of the example data set. (c) The discretized data set. (d) The combined data set using the opposite mirror copy strategy. (e) The negative pan-correlation patterns. (f) The pattern matching in the original data. (g) The plot of the pattern with gaps and lagged time points. (h) The plot of the pattern merging gaps and ignoring lagged time points (for visualization only). The strike-through numbers, U̶, Ø and D̶ indicate those values and value movements not in the detected patterns in (e). From (c) to (f), U indicates Up-changed, O no change, while D Down-changed.

increasing number of time points. The values in these data sets are randomly chosen from $\{-150, -148, -146, \ldots, 150\}$. The efficiency of BIDE+ without our pruning strategies is also evaluated on the mirror-copy datasets of the synthetic data. This performance is used for the comparison to show the contribution of our algorithm.

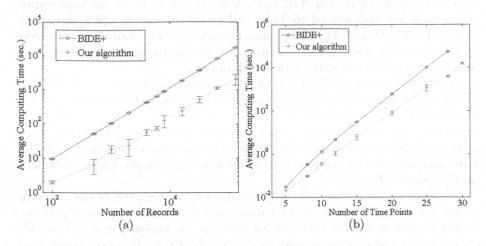

Fig. 2. The assessment on the synthetic data. Both min_{TPP} and min_V are set to 2, and max_O to the number of time points. (a) The computing time (sec.) when the number of variables increases. (b) The computing time (sec.) when the number of time points increases.

Our algorithm was applied to the first series of data sets to see its scalability when the variable size increases. We set the number of time points as $N_T = 20$, and increase N_V from 100 to 500, 1,000, 2,000, 4,000, 6,000, 8,000, 16,000, 32,000, 64,000, and to 128,000. The data at each N_V are randomly produced three times to avoid some randomization effect. The average computing time costs are shown in Fig. 2(a). It can be seen that the computing time cost by our algorithm increases very slowly. It has approximately linear increment of time complexity with increasing N_V. In particular when $N_V = 128,000$, the average computing time is about 30 min. Figure 2(a) also shows that BIDE+ without our pruning strategies is more than nine times slower than our algorithm when $N_V = 128,000$.

Both our algorithm and BIDE+ without our pruning strategies were also applied to the second series of synthetic data sets to examine its scalability when the size of time points increases. We keep the number of variables always as $N_V = 5000$ and randomly produce data sets with N_T varying from 5 to 8, 10, 12, 15, 20, 25, 28, and to 30. The data sets of each N_T are also randomly produced three times to avoid the randomization effect. The average computing time costs are shown in Fig. 2(b). The computing costs increase exponentially when the number of time points N_T increases. Again, Fig. 2(b) suggests that BIDE+ without our pruning strategies is more than 14 times slower than our algorithm when $N_T = 28$. In conclusion, our algorithm is much faster than sequential pattern mining algorithms to detect pan-correlation patterns.

4.2 Application in Time-Course Gene Expression Data

Our algorithm was also evaluated on four real-life microarray gene expression data sets: *alpha, cdc15, elu* [11], and *cdc28* [1]. All of them are time-course gene expression data related to Yeast cell cycle. *elu, cdc28, alpha* and *cdc15* involve 14, 17, 18 and 24 time points, respectively. The four data sets have 5,114 common available genes. our algorithm is able to detect significant pan-correlation patterns efficiently with less than 7 min.

At the min_{TPP} level of 70 % of N_T (i.e., spanning at least 10, 12, 13 and 17 time-point pairs in *elu, cdc28, alpha* and *cdc15* respectively), our algorithm detects 1,934 $\mathbb{C}$ pan-correlation patterns in *elu*, 5,942 in *cdc28*, 13,693 in *alpha* and 139,811 in *cdc15*. Because $\mathbb{C}$ pan-correlation patterns may overlap very much, we filter out overlapping patterns. This filtering results in 588, 2,392, 3,191 and 9,501 non-overlapping $\mathbb{C}$ correlation patterns in *elu, cdc28, alpha* and *cdc15*, respectively.

We examine the correlation coefficient, positive or negative, of the variables in our pan-correlation patterns to demonstrate that highly correlated patterns cannot be observed if the time lagging effect or the broken gap is not considered. Given a pan-correlation pattern $\mathcal{C} = \langle \mathcal{V}, \mathcal{TPP} \rangle$, its Pearson's correlation coefficient PCC is calculated by $PCC = \frac{\sum_{v_{x_i} \in \mathcal{V}, v_{x_j} \in \mathcal{V}, x_i \neq x_j} abs(p(v_{x_i}, v_{x_j}))}{(\|\mathcal{V}\| \times (\|\mathcal{V}\| - 1))}$, where $abs(*)$ returns the absolute value of $*$, $p(v_{x_i}, v_{x_j})$ is the Pearson's correlation coefficient between the value movements of two variables v_{x_i} and v_{x_j} on all time points in the original time-course data, and $\|\mathcal{V}\|$ is the number of unique variables in $\mathcal{V}$.

In comparison, we also calculate PCC only on $\mathcal{TPP}$, and call it $PCC^{\mathcal{TPP}}$. $PCC^{\mathcal{TPP}}$ is also calculated by the above equation except that $p(v_{x_i}, v_{x_j})$ is computed only on those time-point pairs involving in $\mathcal{TPP}$. When PCC or $PCC^{\mathcal{TPP}}$ is 1, it means that all the variables in $\mathcal{V}$ are correlated ideally with each other. When PCC or $PCC^{\mathcal{TPP}}$ is 0, there is completely no correlation

Table 1. PCC and $PCC^{\mathcal{TPP}}$ on four time-course gene expression data.

Dataset		min[a]	mean[a]	std[a]	max[a]
elu	PCC	0.191	0.294	0.026	0.450
	$PCC^{\mathcal{TPP}}$	0.719	0.832	0.022	0.923
cdc28	PCC	0.069	0.264	0.036	0.483
	$PCC^{\mathcal{TPP}}$	0.657	0.827	0.028	0.919
alpha	PCC	0.133	0.299	0.048	0.565
	$PCC^{\mathcal{TPP}}$	0.685	0.832	0.029	0.936
cdc15	PCC	0.122	0.347	0.083	0.799
	$PCC^{\mathcal{TPP}}$	0.620	0.826	0.034	0.933

[a]: The minimum, mean, standard deviation and maximum PCC or $PCC^{\mathcal{TPP}}$ of all pan-correlation patterns in each data set.

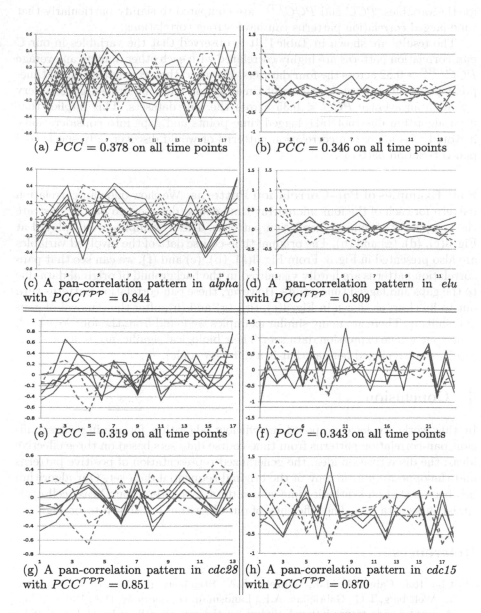

(a) $\hat{PCC} = 0.378$ on all time points

(b) $\hat{PCC} = 0.346$ on all time points

(c) A pan-correlation pattern in *alpha* with $PCC^{TPP} = 0.844$

(d) A pan-correlation pattern in *elu* with $PCC^{TPP} = 0.809$

(e) $\hat{PCC} = 0.319$ on all time points

(f) $\hat{PCC} = 0.343$ on all time points

(g) A pan-correlation pattern in *cdc28* with $PCC^{TPP} = 0.851$

(h) A pan-correlation pattern in *cdc15* with $PCC^{TPP} = 0.870$

Fig. 3. Four examples of pan-correlation patterns with two sets of variables: one set with solid blue line and the other with dashed red line. (a), (b), (e) and (f): The original time-course data of the involved variables in the four pan-correlation pattern examples on *alpha*, *elu*, *cdc28* and *cdc15* data set of Yeast cell cycle, respectively. (c), (d), (g) and (h): The corresponding pan-correlation pattern with smoothing after removing time-lagged points and gaps. Small errors may be in the pattern due to smoothing. (Color figure online)

for the variables. PCC and PCC^{TPP} are compared to signify particularly that time-lagged correlation patterns can have strong correlations.

The results are shown in Table 1. It is observed that the variables in our $\mathbb{C}$ pan-correlation patterns are highly correlated with each other, having an average $PCC^{TPP} > 0.82$ across the four datasets. However, their correlation on all time-point pairs without consideration of time lagging effect or broken gaps is very low with an average $PCC < 0.35$ across the four datasets. This implies that if an algorithm does not take lagged time points and gaps into considerations, it would miss many pan-correlation patterns or would discover only specialized pan-correlation patterns.

Four Examples of Pan-Correlation Patterns. We show one pan-correlation pattern for each of the four microarray time-course data sets to partly illustrate the complexity of mining correlation patterns. These examples are displayed at Fig. 3(c), (d), (g) and (h). The original time-course data of the involved variables are also presented in Fig. 3. From Fig. 3(a), (b), (e) and (f), we can see that pan-correlation patterns are hardly visualized in the background of original data due to the gaps and lagged time points. However, these pan-correlation patterns turn out to be clear, as shown in Fig. 3(c), (d), (g) and (h), after the removal of gaps and shifting. There are many similar examples we found from the four Yeast cell time-course gene expression data sets. Their biological significance is strong (the result is not reported here as that is a different topic).

5   Conclusion

In this work, we have proposed an efficient algorithm for mining all significant pan-correlation patterns from time-course data sets based on three effective ideas: the discretization idea, the generalized representation of positive patterns and the opposite-mirror copy of the original sequential data set. Our algorithm has been tested on synthetic time-course data sets and on four Yeast cell cycle time-course data sets. The efficiency of our algorithm has shown to be high.

References

1. Cho, R.J., Campbell, M.J., Winzeler, E.A., Steinmetz, L., Conway, A., Wodicka, L., Wolfsberg, T.G., Gabrielian, A.E., Landsman, D., Lockhart, D.J., Davis, R.W.: A genome-wide transcriptional analysis of the mitotic cell cycle. Mol. Cell **2**(1), 65–73 (1998)
2. Chuang, C.L., Jen, C.H., Chen, C.M., Shieh, G.S.: A pattern recognition approach to infer time-lagged genetic interactions. Bioinformatics **24**(9), 1183–1190 (2008)
3. Getz, G., Levine, E., Domany, E.: Coupled two-way clustering analysis of gene microarray data. Proc. Nat. Acad. Sci. **97**(22), 12079–12084 (2000)
4. Ji, L., Tan, K.L.: Mining gene expression data for positive and negative co-regulated gene clusters. Bioinformatics **20**(16), 2711–2718 (2004)

5. Ji, L., Tan, K.L.: Identifying time-lagged gene clusters using gene expression data. Bioinformatics **21**(4), 509–516 (2005)
6. Jiang, D., Pei, J., Ramanathan, M., Tang, C., Zhang, A.: Mining coherent gene clusters from gene-sample-time microarray data. In: Proceedings of the Tenth ACM SIGKDD International Conference on Knowledge Discovery and Data Mining (KDD 2004), pp. 430–439. ACM, New York (2004)
7. Li, J., Liu, Q., Zeng, T.: Negative correlations in collaboration: concepts and algorithms. In: KDD, pp. 463–472 (2010)
8. Madeira, S., Oliveira, A.: A polynomial time biclustering algorithm for finding approximate expression patterns in gene expression time series. Algorithms Mol. Biol. **4**(1), 8 (2009)
9. Madeira, S.C., Teixeira, M.C., Sa-Correia, I., Oliveira, A.L.: Identification of regulatory modules in time series gene expression data using a linear time biclustering algorithm. IEEE/ACM Trans. Comput. Biol. Bioinform. **7**(1), 153–165 (2010)
10. Roy, S., Bhattacharyya, D.K., Kalita, J.K.: CoBi: pattern based co-regulated biclustering of gene expression data. Pattern Recognit. Lett. **34**(14), 1669–1678 (2013)
11. Spellman, P.T., Sherlock, G., Zhang, M.Q., Iyer, V.R., Anders, K., Eisen, M.B., Brown, P.O., Botstein, D., Futcher, B.: Comprehensive identification of cell cycle-cregulated genes of the yeast saccharomyces cerevisiae by microarray hybridization. Mol. Biol. Cell **9**(12), 3273–3297 (1998)
12. Wang, J., Han, J.: BIDE: efficient mining of frequent closed sequences. In: 20th International Conference on Data Engineering, Proceedings, pp. 79–90 (2004)
13. Zeng, T., Li, J.: Maximization of negative correlations in time-course gene expression data for enhancing understanding of molecular pathways. Nucleic Acids Res. **38**(1), e1 (2010)
14. Zhao, Y., Yu, J., Wang, G., Chen, L., Wang, B., Yu, G.: Maximal coregulated gene clustering. IEEE Trans. Knowl. Data Eng. **20**(1), 83–98 (2008)

Recognizing Daily Living Activity
Using Embedded Sensors in Smartphones:
A Data-Driven Approach

Wenjie Ruan[1](✉), Leon Chea[1], Quan Z. Sheng[1], and Lina Yao[2]

[1] School of Computer Science, The University of Adelaide, Adelaide, Australia
{wenjie.ruan,leon.chea,michael.sheng}@adelaide.edu.au
[2] School of Computer Science and Engineering, UNSW, Kensington, Australia
lina.yao@unsw.edu.au

Abstract. Smartphones are widely available commercial devices and using them as a basis to creates the possibility of future widespread usage and potential applications. This paper utilizes the embedded sensors in a smartphone to recognise a number of common human actions and postures. We group the range of all possible human actions into five basic action classes, namely walking, standing, sitting, crouching and lying. We also consider the postures pertaining to three of the above actions, including standing postures (backward, straight, forward and bend), sitting postures (lean, upright, slouch and rest) and lying postures (back, side and stomach) . Training data was collected through a number of people performing a sequence of these actions and postures with a smartphone in their shirt pockets. We analysed and compared three classification algorithms, namely k Nearest Neighbour (kNN), Decision Tree Learning (DTL) and Linear Discriminant Analysis (LDA) in terms of classification accuracy and efficiency (training time as well as classification time). kNN performed the best overall compared to the other two and is believed to be the most appropriate classification algorithm to use for this task. The developed system is in the form of an Android app. Our system can real-time accesses the motion data from the three sensors and on-line classifies a particular action or posture using the kNN algorithm. It successfully recognizes the specified actions and postures with very high precision and recall values of generally above 96 %.

1 Introduction

Today, smartphones have become a natural part of our daily life; we rely on it more than ever. Its functionalities are diverse. Yet, its full potential is to be unleashed; one of these is the power of its in-built sensors. Figure 1 illustrates several of the many embedded sensors commonly found in modern smartphones. These sensors include the accelerometer, gyroscope and magnetometer, all of which are commonly found in modern COTS (Commercial off-the-shelf) smartphones. Besides smartphones, sensors (and in particular, motion sensors) play an important role in the design of many "smart" products. Though they have various applications such as security and games, there are also perhaps less obvious

© Springer International Publishing AG 2016
J. Li et al. (Eds.): ADMA 2016, LNAI 10086, pp. 250–265, 2016.
DOI: 10.1007/978-3-319-49586-6_17

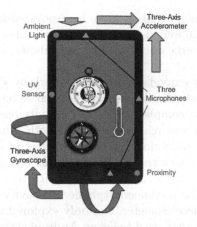

Fig. 1. Embedded sensors in a smartphone

ones such as human activity recognition by detecting changes in movement of different human body parts. Smartphones are a widely available commercial device and using it as a basis for human activity recognition creates the possibility of widespread usage and potential applications. This can also allow for large-scale data mining and significantly accelerate research in the fields of behavioural and social sciences.

Prior research has already been devoted to determining the effectiveness of sensors in the field of activity recognition. One of the first work in this area, Bao and Intille [4], investigated the performance of recognition algorithms using five accelerometers attached around the body, achieving a high overall accuracy rate of 84%. Further research has expanded on this field of research, adding in different methods of data collection (varying the number and types of sensors used as well as their position on the body), different models for data classification as well as different lists of activities to cover.

To date, these efforts cover a wide range of scenarios in which such a system may be used. For example, Kwapisz et al. [13] explore this in an outdoor environment with common actions that reflect different changes in body movement and posture, whilst Hung et al. [10] focus on recognising social actions at informal gatherings. Ermes et al. [6] take an interesting approach to this area by classifying a wide range of actions related to sports, including rowing, running, football and cycling. Some works are more focused on drawing conclusions regarding human behaviour. The work by Hung et al. [9] is one example of this. They use a worn accelerometer to track body movement with the aim of detecting conversing groups in a dense social setting, and from this, analyse social behaviour in these groups, such as dominance, leadership and cohesion.

However, much less work has focused on targeting this research to practical everyday situations. For example, Yang et al. [25] required the sensors to be distributed all around the body, and also require them to be strictly oriented in their proper positions in order for accuracy to not be compromised. It is clearly impractical for such a system to be incorporated for use in our everyday activities. If such a system were designed for widespread or commercial purposes, the more important criteria would include availability, accessibility, flexibility and ease of use.

Therefore, this paper extends beyond prior work by simplifying the activity recognition process in order for this field of research to be practical for everyday applications. We aim to accomplish the task of activity recognition whilst relying on widely available non-research based devices with minimal intrusion to our everyday activities, yet maintain an acceptable level of classification accuracy and efficiency. Hence, the smartphone stands out as being the most appropriate for this task. This goal has been investigated with much success by Kwapisz et al. [13], where they also used a smartphone in activity recognition. However, they only accessed the accelerometer, and only explored a few outdoor motion-based activities. In this paper, we develop an Android system that can accurately and efficiently recognize basic human actions and postures using a few embedded sensors in a smart-phone, including accelerometer, gyroscope and magnetometer. We intend to use more of the smartphone's sensors, and explore a wider coverage of possible human daily actions. Our main contributions are summarized as the following:

- We address a real-time human activity recognition (HAR) problem using a COTS smartphone. Our approach is light-weight, low-cost, and unobtrusive in the sense that only a smartphone is put in the pocket. Our proposed approach relaxes the requirement that people need to wear multiple devices (*e.g.,* sensors or transceivers) for daily activity recognition.
- We compare a series of classification methods including k Nearest Neighbors, Decision Tree and LDA in terms of recognizing accuracy and computation efficiency, which paves a way to deploy the machine learning technique for a practical, daily using and less computation-demanding human activity recognition.
- We conduct extensive experiments to validate our proposed approach. The experimental results demonstrate our system can achieves up to 100 % accuracy in real-world environments. In particular, we implement the system in an Android smartphone and release the APK (android application package) file and the Demo video, making an important step forward for a real-time, practical HAR system.

The rest of the paper is organized as follows. In Sect. 2, we illustrate our HAR system in terms of hardware and activity list. We describe our proposed approach in Sect. 3. In Sects. 4 and 5, we report the experimental results. We overview the related work in Sect. 6 and wrap up the paper in Sect. 7 with conclusion and some future research discussions.

2   System Overview

This section provides an outline of our proposed system. We first give a brief overview of the embedded smartphone sensors, then define an appropriate scenario in which the activities list for this system is devised.

2.1   Built-In Sensors

The sensors used in this paper include the accelerometer, the gyroscope and the magnetometer. These are commonly found in almost all modern smartphones, and provide a decent baseline for distinguishing between actions. An Android phone will be used as supplier of the required sensors because it is very popular, open-source, widely available and most importantly, easily accessible. Unlike many prior work, our system does not require any separate sensors, thus there is less freedom in deciding the location to attach the sensor. In an attempt to keep our work practical, we only consider realistic and common places to carry a smartphone. Some of these include pockets (shirt, front or rear pants), carry bag, cases attached around the waist or simply in the hand. Weighing up the commonness of each of these together with their perceivable effectiveness at being used in the process of action recognition, it is believed that shirt pockets is the best option for this system; it is a reasonably common place for a smartphone and this location is very effective at differentiating almost all postures since it can easily detect changes and movements in the upper body. Hence, *shirt pockets* will be the chosen placement of the smartphone in our work.

Fig. 2. List of actions and postures to be recognized

2.2 Defining Activity List

Prior efforts have investigated numerous activities in many different scenarios. In this paper, the aim is not to improve the existing models or methods, but rather to move this research area towards a more practical focus. Thus, whereas past works defined a list of related activities primarily suited for the research purpose, our system attempts to cover the range of possible actions one may be performing. We first consider all the possible states the human body may be in, and then define the five following basic actions that can cover all these states:

Walk: Body in motion
Stand: Stationary vertical position with 180° straight knees
Sit: Stationary vertical position with 90° bent knees
Crouch: Stationary vertical position with knees bent less than 90°
Lie: Stationary horizontal position

Although these actions may have clear precise definitions in a normal every-day context, here we loosen the definition slightly in order for each action to act as a class and encompass many more similar actions providing that the above definition is satisfied (for example, under the above definition, a squat would also be considered as a crouch). This eliminates the need to specify unnecessarily many actions, yet allow for coverage of the all the possible human actions.

Given a particular action class, it is possible to separate the encompassed states into what we define as posture. In this context, postures are simply states that are variations of the same action. In this paper, we consider the postures for standing, sitting and lying.

Standing postures considered include:

Backward: Standing position with backwards lean
Straight: Straight standing position
Forward: Standing position with forwards lean
Bend: Standing position with body bent forwards to about 90°

Sitting postures considered in this paper include:

Lean: Sitting position with backwards lean
Upright: Upright sitting position
Slouch: Sitting position with forward slouch
Rest: Sitting position with forward lean to rest on some surface

Lying postures considered in this paper include:

Back: Lying position with chest facing upwards
Side: Lying position with chest perpendicular to bed
Stomach: Lying position with chest facing downwards

Given any one posture, a person can also perform a range of activities. For example, a leaning posture whilst sitting may include activities such as watching, reading, and many others. However, preliminary testing indicated that distinguishing activities with the same posture was very hard to achieve given the setup of our system (with only one sensor at chest position). Thus, we will only consider recognizing the different actions and postures. Figure 2 illustrates the set of defined actions and postures that are to be trained for use with the system.

3   Methodology

3.1   Collection of Training Data

An intermediate version of the final system was developed to perform the training data collection. This intermediate system involves setting up the sensors outlined in Sect. 2.1, and then periodically accessing these sensors and writing the sensor data to a CSV file (Fig. 3).

To collect the required training data, a number of people were asked to perform a sequence of actions and postures specified in Sect. 2.2. They had the training system installed onto an Android smartphone and this was placed in their shirt pockets.

We treat each data point as an instantaneous reading from each of the three sensors in all three x-, y-, z-dimensions; this gives a 9-dimensional data point. For each of the specified actions and postures, we collected almost 1000 data points each or in some cases, a sufficient number for that particular action to be

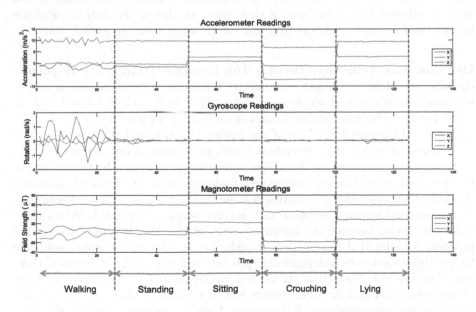

Fig. 3. Sensors readings for each of the basic actions

distinguished. The sensor readings of these actions do differ in one or more of the 9-dimensions based on our pilot experiments, supporting our assumption.

3.2 Classification Algorithms

This paper explores three well-known classification algorithms, namely k-Nearest Neighbor, Decision Tree Learning and Linear Discriminant Analysis.

k **Nearest Neighbour.** k Nearest Neighbour (kNN) is a non-parametric classification algorithm, and one of the simplest to implement. It assigns the output class as the majority vote of k of its neighbours. The neighbours can be computed through a variety of distance functions; the one used for this paper is euclidean distance. Given a set of training sensor data and a testing sensor data, the action label is estimated from the training samples whose observation sensor reading has the minimal distance when compared with the testing observation. Assuming we have a training dataset $\mathbf{T} = \{(\mathbf{s}_1, y_1), (\mathbf{s}_2, y_2), ..., (\mathbf{s}_N, y_N)\}$ with N samples, where $\mathbf{s}_i \in \mathbb{R}^D$ is the sensor readings, $y_i \in \mathbf{l} = \{l_1, ..., l_J\}$ is the corresponding action label. Then, given a distance measuring method and a testing sensor readings $\mathbf{o}$, we can search its k nearest neighbors, represented by $N_k(\mathbf{o})$. Finally, the testing data is classified by a majority vote of its neighbors, being assigned to a most-common location label y^* among its k nearest neighbors:

$$y^* = \arg\max_{l_j} \sum_{\mathbf{s}_i \in N_k(\mathbf{o})} \mathbb{I}(y_i = l_j) \tag{1}$$

where $i = 1, 2, ..., N; j = 1, 2, ..., J$; $\mathbb{I}$ is an indicator function that equals to 1 if $y_i = l_j$, otherwise 0. In the case of tied votes, we choose the nearest neighbor among the k nearest neighbors to break the tie when using an even k value.

Decision Tree Learning. Decision Tree Learning (DTL) is a very popular classification algorithm based on inductive inference. A decision tree or a classification tree is a tree in which each internal (non-leaf) node is labeled with an input feature. The arcs coming from a node labeled with a feature are labeled with each of the possible values of the feature. Each leaf of the tree is labeled with a class or a probability distribution over the classes. A decision tree is built using features of the training data. New instances are classified by traversing the tree from root node to a leaf (where each node represents one feature). However, this paper aims to accurately distinguish multiple activities, which substantially is a supervised learning problem with several outputs to predict. When there is no correlation between the outputs, a very simple way to solve this kind of problem is to build J independent models, i.e. one for each output, and then to use those models to independently predict each one of the J outputs. However, because it is likely that the output values related to the same input are themselves correlated, an often better way is to build a single model capable of predicting simultaneously all J outputs. First, it requires lower training time since only a single estimator is built. Second, the generalization accuracy of the resulting estimator may often be increased. With regard to decision trees, we adopt this strategy to support multi-output problems.

Linear Discriminant Analysis. Linear Discriminant Analysis (LDA) is another classification method based on features. The idea is to find linear combinations of features of the training data that produces an optimal separation of the classes. It is most commonly used as dimensionality reduction technique in the pre-processing step for classification applications. The goal is to project a dataset onto a lower-dimensional space with good class-separability in order avoid over-fitting (*i.e., curse of dimensionality*) and also reduce computational costs. This gives a lower dimensionality, yet retain the important information that is used to distinguish between the data.

4 Evaluation

4.1 Comparison of Different Methods

These three algorithms were analyzed in terms implementation, accuracy of classification and efficiency of the algorithm (*i.e.,* time of model training and testing). This was accomplished in MATLAB.

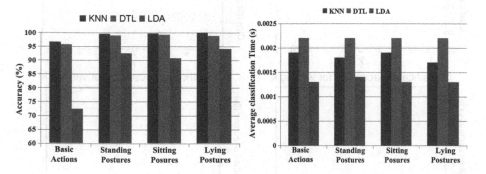

Fig. 4. Comparison of different classification algorithms

Fig. 5. Comparison of the average classification time

Figure 4 demonstrates overall performance comparison of each algorithm using the training data collected in each of the four detection modes. *k*NN appears to give the best overall performance compared to DTL and LDA. DTL has lower accuracy and higher training and classification time (in all four detection modes) compared to *k*NN, thus is concluded to be inferior to *k*NN in all aspects (Fig. 5). Comparing with *k*NN, LDA has slightly lower accuracy but a considerably lower classification time and hence, may be useful for speeding up the classification. However, classification time was not perceived to be causing any performance issues in this system. Therefore, *k*NN is the best option of the three (Fig. 6).

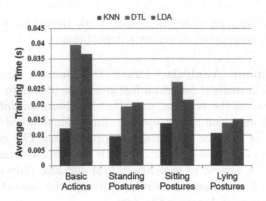

Fig. 6. Comparison of the average training time

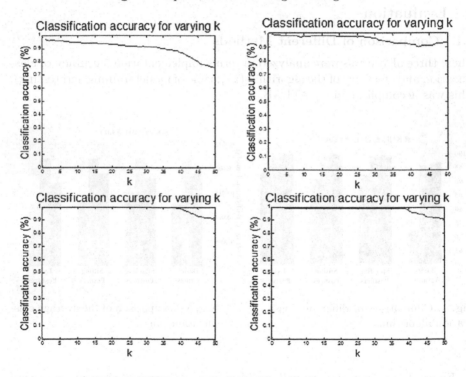

Fig. 7. Effect of varying k on classification accuracy

4.2 Optimal Selection of Parameters

From Sect. 3.2, it was decided to use kNN for the classification process in the final system. The main parameter, k (i.e., number of neighbours to use), can significantly affect the accuracy of the prediction, and hence, will require some analysis to determine the optimal value for our paper. This was achieved with MATLAB using 10-fold cross-validation. The parameter, k, was varied from

1–500, and the prediction accuracy was recorded for each of these values. Among four detection modes, there exhibits a clear trend that accuracy decreases as the value of k increases, with the maximum accuracy achieved at $k = 1$ (Fig. 7).

This is quite an important observation. The fact that all cases show the same trend and that the optimal value of k is always 1 suggests that this may be a recurring feature for our system (providing that the data continues to be generated in the same way, thus maintaining approximately constant size and noise). Although this is not sufficient evidence to draw a definitive conclusion, it can nevertheless still form a basis for further exploration. The implication of this observation is that expansions can be made to our system and/or the data sets without the need to re-calculate the optimal value for k.

4.3   Development of Real-Time HAR System

The final stage of the system involved integrating the material in the previous sections and completing the development of the system.

Figure 8 presents the system in its completed state[1]. Every second, the system retrieves the sensor readings and using kNN on the stored training data, it predicts the action performed, and automatically updates the word and renders the image as well as output the appropriate sound corresponding to that action. The implementation of the classification algorithm is sourced from a modified version [15] of the open-source library, WEKA.

Fig. 8. Developed system as Android app

A setting menu was added to enable the user to change the detection mode, mute sound, and/or switch to training mode. The first was accomplished by replacing the current training data set with the new training data set from memory. The last option simply brings the APP to the intermediate state developed in Sect. 3.1 and is used to obtain further training data.

---

[1] Our android APP is available for download at: https://drive.google.com/file/d/0Bwk_YqDcv7VsaEZySXoyN2ttM2c/view?usp=sharing.

260 W. Ruan et al.

5 In-situ Experiments

With a completed system, we performed numerous tests, similar to the process in which the training data was collected, except that we are now interested in the output action of the system. We evaluated the system's performance in each of the detection modes using the precision and recall metric. The results of this analysis are shown in Tables 1, 2, 3 and 4.

Table 1. Basic actions

		Predicted					Recall (%)
		Walk	Stand	Sit	Crouch	Lie	
Actual	Walk	307	36	5	1	0	87.97
	Stand	15	326	6	0	0	93.95
	Sit	3	2	342	0	0	98.56
	Crouch	0	0	0	346	0	100.00
	Lie	0	0	0	1	349	99.71
Precision (%)		94.46	89.56	96.88	99.43	100.00	

Table 2. Standing postures

		Predicted				Recall (%)
		Backward	Straight	Forward	Bend	
Actual	Backward	223	1	0	0	99.55
	Straight	2	221	0	0	99.10
	Forward	0	0	223	0	100.00
	Bend	0	0	0	224	100.00
Precision (%)		99.11	99.55	100.00	100.00	

Overall, the prediction and recall values for all the models are very high, indicating that the system is able to accurately recognize the specified actions and postures with minimal error. Of the four detection modes, the basic actions category appeared to be the least accurate, with some slight confusion between the similar actions of walking, standing and sitting. Nevertheless, the achieved accuracy is acceptable for most applications[2].

---

[2] Video demo of our system available at:http://cs.adelaide.edu.au/~wenjie/HRAphone .mp4.

Table 3. Sitting postures

		Predicted				Recall (%)
		Lean	Upright	Slouch	Rest	
Actual	Lean	399	0	0	0	100.00
	Upright	0	399	0	0	100.00
	Slouch	0	4	394	0	98.99
	Rest	0	0	0	395	100.00
Precision (%)		100.00	99.01	100.00	100.00	

Table 4. Lying postures

		Predicted			Recall (%)
		Back	Side	Stomach	
Actual	Back	220	0	0	100.00
	Side	0	222	0	100.00
	Stomach	0	1	223	99.55
Precision (%)		100.00	99.55	100.00	

6   Related Work

The goal of activity recognition is to detect human physical activities from the data collected through various sensors. There are generally three main research directions: *(i)* attaching multiple extra sensors and RFID tags on human body, *(ii)* deploying sensors or trans-receivers in the environment and people do not have to carry them, and *(iii)* utilizing a COTS smart-phone that almost everyone has without add extra cost and location constraints.

6.1   Wearable Sensors Based HAR

Wearable sensors such as accelerometers and gyroscopes are commonly used for recognizing activities [4,12]. For example, the authors in [11] design a network of three-axis accelerometers distributed over a user's body. Activities can then be inferred by learning information provided by accelerometers about the orientation and movement of the corresponding body parts. Bao and Intille [4], investigated the performance of recognition algorithms using five accelerometers attached around the body, achieving a high overall accuracy rate of 84%. Apart from sensors, RFID has been increasingly explored in HAR systems. Some research efforts propose to realize human activity recognition by combining RFID passive tags with traditional sensors (e.g., accelerometers) [5,17,23]. Other efforts dedicate to exploit the potential of using "pure" RFID techniques for activity recognition [16,26]. For example, Wang et al. [24] use RFID radio patterns to extract both spatial and temporal features, which are in turn used to characterize various activities. Asadzadeh et al. [3] propose to recognize gesture with passive tags by combining with multiple subtags to tackle uncertainty of the RFID readings. However, all of these efforts usually require people to carry the sensors or tags, even RFID readers (e.g., wearing a bracelet). In summary, wearable sensor based approaches have obvious disadvantages including discomfort of wires attached to the body as well as the irritability that comes from wearing sensors for a long duration.

6.2 Environmental Sensors Based HAR

As a result, some research efforts of exploring environmental sensor based HAR (also called *Device-free HAR*) have emerged recently [19,20]. Such approaches exploit radio transmitters installed in environment, and people are free from carrying any receiver or transmitter. Most device-free approaches concentrate on analyzing and learning distribution of received signal strength or radio links. For example, Ichnaea [21] realizes the device-free human motion tracking by exploring several installed wireless networks, in which it first uses statistical anomaly detection methods to achieve its detection capability and then employs an anomaly scores-based particle filter model and a human motion model to track a single entity in the monitored area. Zhang et al. [29] develop a tag-free human sensing approach using RFID tag array. More recently, the authors of [8] and [22] propose device-free activity recognition approaches using sensor arrays. RF-Care [27,28] proposes to recognize human falls and activities in a device-free manner based on a passive RFID array. WiVi [1,2] uses ISAR technique to track the RF beam, enabling a through-wall human posture recognition. Though promising, such HAR systems however require extra hardwares and also put a strict constraint to human mobilities (*i.e.*, being limited to the area that environmental sensor are deployed).

6.3 Smartphone Based HAR

Recently, smartphone-based HAR systems are also very popular due to its low-cost and being less intrusive [6,10,14,18]. These methods aim to utilize the accelerometers and gyroscopes embedded in smarphones to recognize human activities [13]. The HAR system present in this paper belongs to such technique category. Comparing to other two techniques, smartphone-based approach has two advantages: (*i*) it does not need more hardware hence without adding any financial burden; (*ii*) it substantially relaxes the requirement of human motion areas, unlike the environmental sensor based systems that assume the target user always locates in a specific area; and (*iii*) human daily activity contexts recognized by a smartphone can be much easier to be integrated into modern advanced IoT (*Internet of Things*) infrastructures considering the built-in Internet-connectivity, computation and storage capabilities in smartphones.

Until today, many smartphone-based attempts have been exploited. For example, Kwapisz et al. [13] introduce a system that uses phone-based accelerometers to perform activity recognition, which first collects labeled accelerometer data from twenty-nine users and then uses the resulting training data to induce a predictive model for recognition. While Hung et al. [10] propose to adopt the accelerometer to automatically recognized socially relevant actions, including in speaking, stepping, drinking and laughing. Henpraserttae et al. [7] proposed a method (using a transformation matrix to project sensor data) that allows data collected from different positions around the body to rectify into one universal coordinate system. Ermes et al. [6] take an interesting approach to this area by classifying a wide range of actions related to sports, including rowing, running, football and cycling. Some researchers also aim to mine useful social behaviors such as Hung et al. [9] use a worn accelerometer to track body movement for detecting conversing groups in a dense social setting, and from this, further analyzing social behavior in these groups, including dominance, leadership and cohesion.

However, much less work has focused on targeting this research to practical everyday situations. It is clearly impractical for such a system to be incorporated for use in our everyday activities. If such a system were designed for widespread or commercial purposes, the more important criteria would include availability, accessibility, flexibility and ease of use. Therefore, our system extends beyond prior work by simplifying the activity recognition process in order for this field of research to be practical for everyday applications. We aim to accomplish the task of activity recognition whilst relying on widely available non-research based devices with minimal intrusion to our everyday activities, yet maintain an acceptable level of classification accuracy and efficiency.

7   Conclusion

In this paper, we have demonstrated the possibility of using sensors embedded in smartphones to accurately perform the task of action recognition. We defined the actions and postures to cover the range of possible states the human body may be in, and have differentiated between these with very high precision and recall values using k Nearest Neighbour as the classification algorithm.

Our approach towards practicability has proven feasible in this paper. The fact that we can now obtain action and posture data without any human effort can allow for large-scale (and possibly more accurate) data collection. This can open up a number of possibilities and applications, including human activity monitoring (for health and/or research purposes) as well as enhanced features for leisurely apps such as games involving body part movements and automatic action identification systems to be incorporated into social networking apps.

Yet, there is still much room for improvement. One limitation of this system has a requirement that the phone must be placed in the shirt pocket. This is clearly impractical in widespread usage since different people may carry their phone in different ways, which may be a worthwhile exploration to our system.

Another area to improve upon is to devise a method to recognize activities associated with each posture. As mentioned earlier, this is very difficult (perhaps even impossible) given the current setup. This is because given the same posture, activities mainly differ in arm movements; this cannot be detected with simply a sensor on the chest. One possibility may be to access additional sensors in the smartphone such as sound or light, to provide more information in classifying certain activities. Overall, the ways to extend this system is vast, yet the potential applications are even greater.

References

1. Adib, F., Hsu, C.Y., Mao, H., Katabi, D., Durand, F.: Capturing the human figure through a wall. ACM Trans. Graph. (TOG) **34**(6), 219 (2015)
2. Adib, F., Katabi, D.: See through walls with wifi!. In: Proceedings of the ACM SIGCOMM 2013 Conference (SIGCOMM 2013), pp. 75–86 (2013)
3. Asadzadeh, P., Kulik, L., Tanin, E.: Gesture recognition using RFID technology. Pers. Ubiquit. Comput. **16**(3), 225–234 (2012)
4. Bao, L., Intille, S.S.: Activity recognition from user-annotated acceleration data. In: Ferscha, A., Mattern, F. (eds.) Pervasive 2004. LNCS, vol. 3001, pp. 1–17. Springer, Heidelberg (2004). doi:10.1007/978-3-540-24646-6_1
5. Buettner, M., Prasad, R., Philipose, M., Wetherall, D.: Recognizing daily activities with RFID-based sensors. In: Proceedings of 11th ACM International Conference on Ubiquitous Computing (UbiComp), pp. 51–60 (2009)
6. Ermes, M., Pärkkä, J., Mäntyjärvi, J., Korhonen, I.: Detection of daily activities and sports with wearable sensors in controlled and uncontrolled conditions. IEEE Trans. Inf. Technol. Biomed. **12**(1), 20–26 (2008)
7. Henpraserttae, A., Thiemjarus, S., Marukatat, S.: Accurate activity recognition using a mobile phone regardless of device orientation and location. In: 2011 International Conference on Body Sensor Networks, pp. 41–46. IEEE (2011)
8. Hong, J., Ohtsuki, T.: Ambient intelligence sensing using array sensor: device-free radio based approach. In: Proceedings of ACM Conference on Pervasive and Ubiquitous Computing Adjunct Publication (2013)
9. Hung, H., Englebienne, G., Cabrera Quiros, L.: Detecting conversing groups with a single worn accelerometer. In: Proceedings of the 16th International Conference on Multimodal Interaction, pp. 84–91. ACM (2014)
10. Hung, H., Englebienne, G., Kools, J.: Classifying social actions with a single accelerometer. In: Proceedings of the 2013 ACM International Joint Conference on Pervasive and Ubiquitous Computing, pp. 207–210. ACM (2013)
11. Kern, N., Schiele, B., Junker, H., Lukowicz, P., Tröster, G.: Wearable sensing to annotate meeting recordings. Pers. Ubiquit. Comput. **7**(5), 263–274 (2003)
12. Krishnan, N.C., Panchanathan, S.: Analysis of low resolution accelerometer data for continuous human activity recognition. In: Proceedings of IEEE International Conference on Acoustics, Speech and Signal Processing (ICASSP), pp. 3337–3340. IEEE (2008)
13. Kwapisz, J.R., Weiss, G.M., Moore, S.A.: Activity recognition using cell phone accelerometers. ACM SigKDD Explor. Newslett. **12**(2), 74–82 (2011)
14. Lane, N.D.. et al.: Bewell: a smartphone application to monitor, model and promote wellbeing. In: Proceedings of 5th International ICST Conference on Pervasive Computing Technologies for Healthcare, pp. 23–26 (2011)
15. Marsan, R.: Weka for android. GitHubRepository (2011). https://github.com/rjmarsan/weka-for-android/
16. Ruan, W.: Unobtrusive human localization and activity recognition for supporting independent living of the elderly. In: Proceedings of 2016 IEEE International Conference on Pervasive Computing and Communication Workshops (PerCom Workshops), pp. 1–3 (2016)

17. Ruan, W., Sheng, Q.Z., Yao, L., Gu, T., Ruta, M., Shangguan, L.: Device-free indoor localization and tracking through human-object interactions. In: 2016 IEEE 16th International Symposium on A World of Wireless, Mobile and Multimedia Networks (WoWMoM), June 2016
18. Ruan, W., Sheng, Q.Z., Yang, L., Gu, T., Xu, P., Shangguan, L.: Audiogest: enabling fine-grained hand gesture detection by decoding echo signals. In: The 2016 ACM International Joint Conference on Pervasive and Ubiquitous Computing (UbiComp 2016) (2016)
19. Ruan, W., Yao, L., Sheng, Q.Z., Falkner, N.J.G., Li, X.: Tagtrack: device-free localization and tracking using passive RFID tags. In: Proceedings of the 11th International Conference on Mobile and Ubiquitous Systems: Computing, Networking and Services (MobiQuitous 2014), pp. 80–89 (2014)
20. Ruan, W., Yao, L., Sheng, Q.Z., et al.: Tagfall: towards unobstructive fine-grained fall detection based on UHF passive RFID tags. In: The International Conference on Mobile and Ubiquitous Systems: Computing, Networking and Services (MobiQuitous 2015), pp. 140–149 (2015)
21. Saeed, A., Kosba, A.E., Youssef, M.: Ichnaea: a low-overhead robust WLAN device-free passive localization system. IEEE J. Sel. Topics Signal Process. **8**(1), 5–15 (2014)
22. Sigg, S., Scholz, M., Shi, S., Ji, Y., Beigl, M.: Rf-sensing of activities from non-cooperative subjects in device-free recognition systems using ambient and local signals. IEEE Trans. Mob. Comput. (TMC) **13**(4), 907–920 (2014)
23. Stikic, M., et al.: ADL recognition based on the combination of RFID and accelerometer sensing. In: Proceedings of International Conference Pervasive Computing Technologies for Healthcare (2008)
24. Wang, L., Gu, T., Xie, H., Tao, X., Lu, J., Huang, Y.: A wearable RFID system for real-time activity recognition using radio patterns. In: Proceedings of the 10th International Conference on Mobile and Ubiquitous Systems: Computing, Networking and Services (MobiQuitous) (2013)
25. Yang, A.Y., Iyengar, S., Kuryloski, P., Jafari, R.: Distributed segmentation and classification of human actions using a wearable motion sensor network. In: IEEE Computer Society Conference on Computer Vision and Pattern Recognition Workshops (CVPRW 2008), pp. 1–8. IEEE (2008)
26. Yao, L., Sheng, Q.Z., Ruan, W., Li, X., Wang, S., Yang, Z.: Unobtrusive posture recognition via online learning of multi-dimensional RFID received signal strength. In: Proceedings of IEEE 21st International Conference on Parallel and Distributed Systems (ICPADS 2015), pp. 116–123 (2015)
27. Yao, L., Ruan, W., Sheng, Q.Z., Falkner, N.J.G., Li, X.: Exploring tag-free RFID-based passive localization and tracking via learning-based probabilistic approaches. In: Proceedings of 23rd ACM International Conference on Information and Knowledge Management (CIKM) (2014)
28. Yao, L., Sheng, Q.Z., Li, X., Wang, S., Gu, T., Ruan, W., Zou, W.: Freedom: online activity recognition via dictionary-based sparse representation of RFID sensing data. In: 2015 IEEE International Conference on Data Mining (ICDM), pp. 1087–1092. IEEE (2015)
29. Zhang, D., Zhou, J., Guo, M., Cao, J., Li, T.: Tasa: tag-free activity sensing using RFID tag arrays. IEEE Trans. Parallel Distrib. Syst. (TPDS) **22**(4), 558–570 (2011)

Dynamic Reverse Furthest Neighbor Querying Algorithm of Moving Objects

Bohan Li[1,2,5(✉)], Chao Zhang[1,6(✉)], Weitong Chen[5], Yingbao Yang[3], Shaohong Feng[4], Qiqian Zhang[3], Weiwei Yuan[1,2], and Dongjing Li[1,7]

[1] College of Computer Science and Technology, Nanjing University of Aeronautics and Astronautics, Nanjing, China
{bhli,zhangchao0607}@nuaa.edu.cn
[2] Collaborative Innovation Center of Novel Software Technology and Industrialization, Nanjing, China
[3] College of Civil Aviation, Nanjing University of Aeronautics and Astronautics, Nanjing, China
[4] College of Electronic and Information Engineering, NUAA, Nanjing, China
[5] School of Information Technology and Electrical Engineering, University of Queensland, Brisbane, Australia
[6] Jiangsu Easymap Geographic Information Technology Corp., Ltd, Nanjing, China
[7] Department of Payment and Settlement, JD Finance, Nanjing, China

Abstract. With the development of wireless communications and positioning technologies, locations of moving objects are highly demanding services. The assumption of static data is majorly applied on previous researches on reverse furthest neighbor queries. However, the data are dynamic property in the real world. Even, the data-aware are uncertain due to the limitation of measuring equipment or the delay of data communication. To effectively find the influence of querying a large number of moving objects existing in boundary area vs querying results of global query area, we put forward dynamic reverse furthest neighbor query algorithms and probabilistic reverse furthest neighbor query algorithms. These algorithms can solve the query of weak influence set for moving objects. Furthermore, we investigate the uncertain moving objects model and define a probabilistic reverse furthest neighbor query, and then present a half-plane pruning for individual moving objects and spatial pruning method for uncertain moving objects. The experimental results show that the algorithm is effective, efficient and scalable in different distribution and volume of data sets.

Keywords: Moving objects · Reverse furthest neighbor · Uncertain moving object · Pruning · Long tail

This work is supported in part by National Natural Science Foundation of China (41301407), Natural Science Foundation of Jiangsu Province (BK20130819), CSC (201406835051), Innovation Funding (NJ20160028), Innovative and Entrepreneurial PHD(2015).

© Springer International Publishing AG 2016
J. Li et al. (Eds.): ADMA 2016, LNAI 10086, pp. 266–279, 2016.
DOI: 10.1007/978-3-319-49586-6_18

1 Introduction

With the increasing volume of static points and complexity of the moving objects in spatial data sets, it is challenging on effective and efficient querying for weak influence objects. Consider an example of opening a new franchise in Australia; there are regulations that to reduce the influence (marginal profit) [1] to the existing ones; as another example, the deployment and query processing in large sensor networks often require the design of location-aware algorithms for moving objects.

The motivation of this study the dynamic Reverse furthest neighbor (RFN) queries is largely inspired by an extensively studied query type, dynamic Reverse nearest neighbor (RNN) queries [2, 3]. Intuitively, the objective of dynamic RFN queries it to find weak influence moving objects in both frequent update and changing location environments. Which are beyond the existing RFN, aimed at data points in a static environment, in spatial queries.

Reverse furthest neighbor (RFN) query is originally one of the methods of spatial database data, which is a counterpart of reverse nearest neighbor query (RNN). To the best of our knowledge, currently, rare research has been done on RFN and its variants query, while paying more attention to NN and RNN problems [2–6, 9]. Furthermore, kNN and RkNN are widespread and practical, but they are not without problems [7, 8, 10]. Among those are lacking in processing dissimilarity data and ignoring the weak influence set. Those are both aspects which are not considered in NN series concerning. Regarding the RFN query, as pointed out by the researchers of facility search [18, 19]. Actually, RFN has a broad application prospect in facility location and solving long tail of the information [20].

The remainder of the paper is organized as follows: Sect. 2 surveys the related work on Euclidean space of query technology. Section 3 introduces the formal problem definitions. Section 4 studies algorithms for the queries proposed in Sect. 3. An extensive experimental evaluation follows in Sect. 4. Section 5 concludes this word and goes into a perspective of the future research.

2 Related Work

NN and RNN search has been thoroughly studied. The task is usually conducted in two different approaches: Probabilistic Nearest Neighbour Queries (PNN) and Probabilistic Reverse Nearest Neighbour Queries (PRNN).

Probabilistic Reverse Nearest Neighbor Queries. In [4], Lian et al. Firstly, they propose probabilistic reverse nearest neighbor (PRNN) query. The algorithm uses geometric pruning that significantly reduces the PRNN search space yet without introducing any false dismissals. In [5], Cheema et al. formalize probabilistic reverse nearest neighbor query that is to retrieve the objects from the uncertain data that have a higher probability than a given threshold to be the RNN of an uncertain query object. The algorithm imports antipodal corners and normalized half-space to filter false objects. In [8], Li et al. propose probabilistic reverse k-nearest neighbors query. The algorithm supports arbitrary values of k on the basis of two pruning strategies, namely spatial pruning and probabilistic pruning. In [9], Xu et al. present a general framework for answering interval reverse nearest neighbor queries (IRNN) on uncertain moving objects with Markov correlations. In the first phase, they apply space pruning and probability pruning techniques, which reduce the search space significantly. In the second phase, they propose an approach termed probability decomposition verification (PDV) algorithm to verify whether each unpruned object is an IRNN of the query object. In [10], Emrich et al. propose two types of RNN queries based on a well-established model for uncertain spatio-temporal data based on stochastic process, namely the Markov model. The experimental results show that the algorithm has a better query performance advantage.

Reverse Furthest Neighbor Queries. Query type of reverse furthest neighbor (RFN) was first slightly mentioned in [9], mainly to solve the related issues in the weak influence sets. In [12], Yao et al. proposed the R-tree based algorithms for both monochromatic and bichromatic versions of the RFN queries. They proposed the proposed the progressive furthest cell (PFC) algorithm and the convex hull furthest cell (CHFC) algorithm to handle RFN query. In [11], Jianquan Liu et al. select external pivots to construct metric indexes, and employ the triangle inequality to do efficient pruning by using the metric indexes. In [13], Jianquan Liu et al. consider the arbitrary RFN query that is without the constraint of its location. They figure out a non-trivial safe area to guarantee the efficiency of query processing. They design an efficient algorithm to answer the RFN query without extra cost of filtering or refinement, when q is located in such safe area. However, for the study of reverse furthest neighbor algorithm relatively still cannot meet the query requirement. Existing research has not put forward relatively efficient algorithm.

3 Preliminaries

3.1 Uncertain Moving Object Model

The popular model of uncertain data is the model of location update. Each moving object stores a location information in the database.

Definition 1 (Uncertainty Region [13]): An uncertainty region of an object O_i at time t, denoted by $U_i(t)$, is a closed region such that O_i can be found only inside this region.

Definition 2 (Uncertainty Probability Density Function [13]): The uncertainty probability density function of an object O_i, denoted by $PDF_i(x,t)$, is a probability density function of O_i's location x at time t.

Assume that the moving object O_i at time t can only be in the uncertainty region $U_i(t)$. The probability density function has the property that $\int_{U_i(t)} PDF_i(x,t) = 1$; otherwise = 0, for the outside of uncertainty region.

Definition 3 (P∃RFNQ): A probabilistic ∃ reverse furthest neighbor query retrieves all objects $o \in O$ having a sufficiently high probability to be the reverse furthest neighbor of q for at least one point of time $t \in T$, formally: $P\exists RFNQ(q, O, T, \tau) = \{o \in O : P\exists RFN(o, q, O, T) \geq \tau\}$.

Where $dist(x, y)$ is a distance of function defined on spatial points, typically the Euclidean distance. The query returns all objects from the database having a probability greater τ to have q as their probabilistic $\exists$ furthest neighbor.

Definition 4 (P∀RFN): A probabilistic RFN query retrieves all objects $o \in O$ having a sufficiently high probability to be the RFN of q for the entire set of timestamps T, formally

$$P\forall RFNQ(q, O, T, \tau) = \{o \in O : P\forall RFN(o, q, O, T) \geq \tau\}.$$

Where $P\forall RFN(o, q, O, T) = P\{t \in T, \forall o' \in O \backslash o : dist\,(o(t), q(t)) \geq dist(o(t),$
$o'(t))\}$.

3.2 Dynamic RFN Query Algorithm

In [14], TPL pruning technology is first presented to traverses the R-tree in best-first manner which retrieve potential candidates in an ascending order of their distance to the query point q because the RNNs are likely to be near q (Fig. 1).

Definition 5 (Dynamic RFN): Given a query object q, moving objects P and a certain time t. The reverse furthest neighbor of q to a data set P in a certain time t is defined as:

$$RFN(q, t) = \{p(t) \in P | \exists p'(t) \in P, dist(p(t), p'(t)) > dist(p(t), q(t))\}.$$

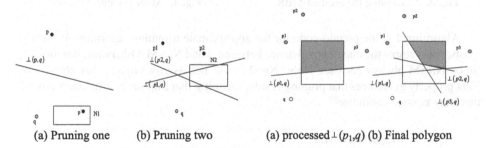

(a) Pruning one	(b) Pruning two	(a) processed $\perp (p_1,q)$ (b) Final polygon

Fig. 1. Illustration of half-plane pruning   **Fig. 2.** Computing the residual region

In general, if $p_1, p_2, ..., p_n$ are n moving objects, the any node whose MBR falls inside $\bigcup_{i=1 \sim n} PLq(p_i, q)$ cannot contain any RFN results. Let the residual region Nres be the area of node N outside $\bigcup_{i=1 \sim n} PLq(p_i, q)$ (i.e., the part of the MBR that many contain candidate RFNs of q). Then, N can be pruned if and only if Nres $= \Phi$. Consider Fig. 2 (a) that contains three moving objects $p1, p2, p3$. We compute the residual region Nres by trimming N with each bisector in turn. Initially, we set Nres $=$ N and use $\perp(p_1, q)$, after that Nres becomes the shaded trapezoid. Figure 2(b) shows the final Nres after processing all bisectors. Given p_1, p_2 and p_3, Nres is the only part of the node MBR N that may contain DRFNs of q.

The computation of Nres causes two cases. First, in the worst case, each bisector may introduce an additional vertex to Nres. Consequently, the trimming of the i-th $(1 \leq i \leq n)$ bisector takes $O(i)$ time because it may need to examine all edges in the previous Nres. Thus, the total processing cost is $O(n_2)$, i.e., quadratic to the number of half-planes. Second, this method cannot scale with the dimensionality because computing the intersection of a half-space and a hyper-polyhedron becomes increasingly complex. Motivated by this, we use a simpler alternative that requires only $O(n)$ time. The idea is to bound Nres by a residual MBR NresM. Initially NresM is set to N and then it is trimmed incrementally by each bisector. Figure 3(a) shows trimming with $\perp(p1, q)$, where instead of keeping the exact shape of Nres, just compute NresM(i.e., the shaded rectangle). Figures 3(b), (c) illustrate the residual MBRs after trimming with $\perp(p2, q), \perp(p3, q)$, respectively. Trimmed MBRs can be efficiently computed using the clipping algorithm.

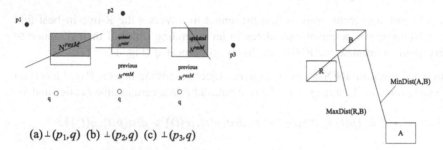

(a) $\perp(p_1, q)$ (b) $\perp(p_2, q)$ (c) $\perp(p_3, q)$

Fig. 3. Computing the residual MBR   **Fig. 4.** MBR pruning example

Algorithm 1 is the pseudo-code for the approximate trimming algorithm. If NresM exists, trim returns the maximum distance between q and NresM. Otherwise, it return ∞. Since NresM always encloses Nres, NresM $= \Phi$ necessarily implies that Nres $= \Phi$. This property guarantees that pruning is safe, meaning that trim never eliminates a node that may contain candidates.

Algorithm1. Trim
Input: Query object q, data moving objects P$(p_1, p_2,..., p_n)$,   N(t) is a rectangle being trimmed   **Output:** The maximum distance between q and NresM   1.NresM=N(t);   2.**for** i=1 to n   3. NresM=clipping(NresM,PL$p_i(p_i,q)$);   4.   if NresM= Φ 5.   then return ∞ ;   6. **end for**   7. **return** maxdist(NresM,q);

Existing moving objects in the process of fast query processing complex object space or uncertain objects usually use the minimum bounding rectangle MBR to trim the space. A spatial control method is proposed in the literature [15]. That is, A, B, R for three rectangular, if the distance from A to R to less than B to R, A dominate B. This section defines a space pruning method suitable for RFN. As shown in Fig. 4, the minimum distance from A to B is greater than the maximum distance from R to B, then we say that A prunes the B.

Definition 6 (Spatial Pruning): Let A, B, R be rectangles. If for all moving objects $b \in B$ it holds that every moving object $a \in A$ is further to b than any moving object $r \in R$, the rectangle A prunes B. The definition is as follows: $SP(A, R, B) = \forall a \in A, b \in B.r \in R : \text{dist}(a, b) > \text{dist}(r, b)$

$$Dom(A, R, B) = \sum_{i=1}^{d} \min_{b \in \{B_i^{min}, B_i^{max}\}} \left(MinDist(A_i, b_i)^2 - MaxDist(R_i, b_i)^2 \right) > 0$$

The above formula is evolved from the following equivalent formula 1–6, which is used in the following paper. The formula greatly improves the cutting efficiency of MBR and reduces the time complexity.

$$\forall a \in A, \forall b \in B, \forall r \in R : dist(a, b) > dist(r, b) \quad \Leftrightarrow$$
$$\forall b \in B : MinDist(A, b) > MaxDist(R, b)$$

The equivalence is represented by A, B, R rectangles. If for all moving objects $b \in B$ it holds that every moving object $a \in A$ is further to b than any moving object $r \in R$, then it holds that for all moving objects $b \in B$ has A to b minimum distance greater than R to b maximum distance.

Equivalence 1:

$$MinDist(A, B) = \sqrt{\sum_{i=1}^{d} \begin{cases} \left| A_i^{min} - B_i^{max} \right|^2, & \text{if } A_i^{min} > B_i^{max} \\ \left| B_i^{min} - A_i^{max} \right|^2, & \text{if } B_i^{min} > A_i^{max} \\ 0, & \text{else} \end{cases}} \quad (1)$$

$$MaxDist(A, B) = \sqrt{\sum_{i=1}^{d} \begin{cases} \left| A_i^{max} - B_i^{min} \right|^2, & \text{if } A_i^{mid} \geq B_i^{mid} \\ \left| B_i^{max} - A_i^{min} \right|^2, & \text{if } B_i^{mid} > A_i^{mid} \end{cases}}$$

Equivalence 2:

$$\forall b \in B : MinDist(A, b) > MaxDist(R, b) \quad \Leftrightarrow$$
$$\forall b \in B : \sqrt{\sum_{i=1}^{d} MinDist(A_i, b_i)^2} > \sqrt{\sum_{i=1}^{d} MaxDist(R_i, b_i)^2} \quad (2)$$

If all moving objects $b \in B$ has A to b minimum distance greater than R to b maximum distance, then it holds that for all dimensions has A to b minimum distance greater than R to b maximum distance.

Equivalence 3:

$$\forall b \in B : \sqrt{\sum_{i=1}^{d} MinDist(A_i, b_i)^2} > \sqrt{\sum_{i=1}^{d} MaxDist(R_i, b_i)^2} \qquad \Leftrightarrow$$

$$\forall b \in B : \sum_{i=1}^{d} (MinDist(A_i, b_i)^2 - MaxDist(R_i, b_i)^2) > 0 \qquad (3)$$

The equivalence is represented by A, B, R rectangles. If for all moving objects $b \in B$ on all dimensions has A to b minimum distance greater than R to b maximum distance, then it holds that all dimensions of A to b minimum distance squared minus R to b the maximum sum of squares and the resulting results are greater than zero.

Equivalence 4:

$$\forall b \in B : \sum_{i=1}^{d} (MinDist(A_i, b_i)^2 - MaxDist(R_i, b_i)^2) > 0 \qquad \Leftrightarrow$$

$$min_{b \in B}\left(\sum_{i=1}^{d} (MinDist(A_i, b_i)^2 - MaxDist(R_i, b_i)^2)\right) > 0 \qquad (4)$$

Proof. Instead of considering all possible $b \in B$, it is sufficient to consider only that moving object $b \in B$ which maximizes the inequality.

Lemma 1. Let $F : R^d \rightarrow R$ be a function that is summed by treating each dimension independently, i.e. There exists a function $f : R \rightarrow R$ such that $F(o) = \sum_{i=1}^{d} f(o_i)$

Also, let $B \subseteq R^d$ be a rectangle and $\sigma := argmin_{b \in B}(F(b))$ be the object in B that minimizes F. Then, the following holds:

$$min_{b \in B}\left(\sum_{i=1}^{d} f(b_i)\right) = \sum_{i=1}^{d} min_{b \in B}(f(b_i))$$

Proof. Let $F(b) = \sum_{i=1}^{d} f(b_i)$, then $min_{b \in B}\left(\sum_{i=1}^{d} f(b_i)\right) = min_{b \in B}(F(b))$. Let $\sigma := argmin_{b \in B}(F(b))$, then $min_{b \in B}(F(b)) = F(\sigma)$. Let $F(\sigma) = \sum_{i=1}^{d} f(\sigma_i)$, function to obtain the minimum value. The sum of the minimum values in each dimension, then $\sum_{i=1}^{d} f(\sigma_i) = \sum_{i=1}^{d} min_{b_i \in B_i}(f(b_i))$.

Equivalence 5:

$$min_{b \in B}\left(\sum_{i=1}^{d} (MinDist(A_i, b_i)^2 - MaxDist(R_i, b_i)^2)\right) > 0$$

$$\Leftrightarrow \sum_{i=1}^{d} min_{b_i \in B_i}(MinDist(A_i, b_i)^2 - MaxDist(R_i, b_i)^2) > 0 \qquad (5)$$

The formula is hereby replaced by Lemma 1. Let $F(b) = MinDist(A, b) - MaxDist(R, b)$.

Lemma 2. Let A and B be intervals. The function $f : R \to R$ defined as $f(x) = MinDist(A, x)^2 - MaxDist(R, x)^2$ has no local minimum.

Lemma 3. Let $f : R \to R$ be a function that has no local minimum and $I = [I_{min}, I_{max}] \subset R$ be an arbitrary finite interval. The value that minimizes f in the interval I must be either I_{min} or I_{max}, i.e. $argmin_{i \in I}(f(i)) \in \{I_{min}, I_{max}\}$

Proof. Let $P \in [I_{start}, I_{end}]$ be the value that minimizes f in I, i.e. $p = argmin_{i \in I}(f(a))$. Then, $\forall i \in I : f(i) \geq f(p), f(I_{min}) \geq f(p)$ and $f(I_{max}) \geq f(p)$. Note that $f(I_{min}) > f(p)$ and $f(I_{max}) > f(p)$ cannot both be true, because this would be a contradiction to the assumption that f has no local minimum. Thus it must either hold that $f(I_{start}) = f(p)$ or $f(I_{end}) = f(p)$, i.e. $I_{min} = argmin_{i \in I}(f(x))$.

Equivalence 6:

$$\sum\nolimits_{i=1}^{d} \min_{b_i \in B_i} (MinDist(A_i, b_i)^2 - MaxDist(R_i, b_i)^2) > 0$$

$$\Leftrightarrow \sum_{i=1}^{d} \min_{b_i \in \{B_i^{min}, B_i^{max}\}} (MinDist(A_i, b_i)^2 - MaxDist(R_i, b_i)^2) > 0 \tag{6}$$

According to Theorems 2 and 3, the equivalent formula 6 can get the function in any finite interval with no local minimum. That means if the function obtains the minimum value, the independent variable takes the interval of the endpoint.

Dynamic Reverse Furthest Neighbor Querying Algorithm of Moving Objects. We adopt a two-step framework that retrieves a set of candidates for DRFN (filtering step) and then removes the false misses (refinement step). Algorithm traverses the TPR-tree in a best-first manner, retrieving potential candidates in descending order of their distance to the query object q because the DRFNs are likely to be further q. Each pruned entry is inserted in a refinement set S_{rfn}. In the refinement step, the entries of S_{rfn} are utilized to eliminate false hits. As shown in Fig. 5, the query result contains only moving object p_1.

Initially, the algorithm visits the root of the TPR-tree and inserts its entries $N8$, $N7$ into a heap H sorted on their maxdist from q. Then it de-heaps $N8$, visits its child node and inserts into H the corresponding entries: $H = \{N4, N7, N6, N5\}$. The next node accessed is $N4$, where the first point $p7$ (i.e., the one furthest to q) has $dist(p7,q) > dist$ $(N7,q)$ ($N7$ is at the top of the heap) and is added to the candidate set S_{cnd}. The second point $p8$ in $N4$ lines in $PL_{p7}(p7,q)$ and is inserted into the candidate set S_{cnd}. The next de-heaped entry is $N7$. Double check whether $N7$ can be pruned. Since part of $N7$ lies in $PL_{p7}(p7,q)$, it has to be visited. Its child nodes $N2$ and $N3$ fall completely out of $PL_{p7}(p7,q)$. Therefore, they cannot contain any candidates and are added to S_{rfn}. $N1$ falls partially in $PL_{p7}(p7,q)$, i.e., trim will return a maxdist $(N1^{resM}, q)$ that is different from ∞. Thus, $N1$ is inserted into H. Then, the next de-heaped entry is $N1$, its child point $p1$ insert $S_{cnd} = \{p7, p8, p1\}$ and point $p3$ insert $S_{rfn} = \{N3, N2, p3\}$. The next heap entry $N6$ lies in $PL_q(p7,q)$ and is added to S_{rfn}. The last de-heaped entry is $N5$, it child point $p9$ is inserted into $S_{cnd} = \{p7,p8,p1,p9\}$ as it lies in $PL_{p1}(p1,q)$, and point is inserted into $S_{rfn} = \{N3,N2,p3,p11,p12,p10\}$ as it lies in $PL_q(p1,q)$. The filtering step terminates when $H = \Phi$. The contents of the heap at each phase of the filtering process are shown in Table 1.

Probabilistic Reverse Furthest Neighbor Queries on Uncertain Moving Object. In order to discover in the boundary region of the existence of a large number of uncertain moving objects of widely distributed global query objects influence, proposed probabilistic RNN query algorithm to solve uncertain moving objects of the influence of the weak set problem. The goal of this algorithm is to search the candidate object for each time interval t in the query time interval T.

Algorithm 2. TSF

Input: query object q, time interval T, TPU-tree
Output: Candidate sets and their corresponding influence
              sets after cutting

Variable description: $t \in T$, max-heap H order by maxi-
                          mum distance to q

1. Init $S_{cnd}{}^t = {}_\varphi$, $S_{ver}{}^{cnd,t} = {}_\varphi$, $S_{prn} = {}_\varphi$;

2. **for each** $t \in T$
3. Insert root entry of TPU-tree to H
4. **While** H is not empty
5.   (e(t),key)=de-heap H
6.   **if** $\exists$ e2 $\in H \cup S_{prn} \cup S_{cnd}{}^t$: $Dom(e2,q,\phi)$ then
7. $S_{prn} = S_{prn} \cup \{e\}$;
8.   **else if** e is directory entry
9.   For each child ch in e
10.   Insert ch in H
11.   **else if** e is leaf entry
12. $S_{cnd}{}^t = S_{cnd}{}^t \cup \{e\}$
13.   **end if**
14. **end while**
15. **for each** $cnd \in S_{cnd}{}^t$
16.   **if** $\exists$ le: $Dom(le,q,\phi)$ then
17.   **continue;**
18.   **end if**
19. $S_{ver}{}^{cnd,t} = \{le : \neg Dom(le,q,\phi)\} \wedge \neg Dom(q,le,\phi)$;
20. **end for**
21. $S_{cnd} = \bigcap_{t \in T} S_{cnd}{}^t$
22. **for each** $cnd \in S_{cnd}{}^t$
23. $S_{ver}{}^{cnd} = \bigcup_{t \in T} S_{ver}{}^{cnd,t}$
24. **end for**
25. **return** $(S_{cnd}, S_{ver}{}^{cnd})$

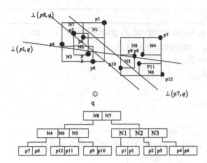

Fig. 5. Filter example

Table 1. During filtering the heap content

Action	Heap	S_{cnd}	S_{rfn}
visit root	$\{N8,N7\}$	Φ	Φ
visit N8	$\{N4,N7,N6,N5\}$	Φ	Φ
visit N4	$\{N7,N6,N5\}$	$\{p7,p8\}$	Φ
visit N7	$\{N1,N6,N5\}$	$\{p7,p8\}$	$\{N3,N2\}$
visit N1	$\{N6,N5\}$	$\{p7,p8,p1\}$	$\{N3,N2,p3\}$
visit N6	$\{N5\}$	$\{p7,p8,p1\}$	$\{N3,N2,p3,p11,p12\}$
visit N5	Φ	$\{p7,p8,p1,p9\}$	$\{N3,N2,p3,p11,p12,p10\}$

We initialize three empty sets. S_{cnd}^{t} contains all RFN candidates which are found during query processing. $S_{ver}^{cnd,t}$ contains influence objects which are needed for each candidate find the set of objects for the refinement step. S_{prn} contains moving objects or entries which have been verified not to contain candidates. When H is not empty, for each entry e, which is de-heaped from H, the algorithm checks whether e can be pruned. If there is another object in the heap H, S_{prn} or S_{cnd}^{t}, e2 prunes q(e2 is further to e than q), which implies that e cannot be RFN of q. So e can be putted S_{prn} sets. If e is a directory entry, the child node of e is inserted into the S_{cnd}^{t} sets. If e is a leaf node, e is inserted into the S_{cnd}^{t} sets. When the heap H is empty, check each candidate object in the S_{cnd}^{t}. If another object le prunes e in the S_{cnd}^{t}, e can be discarded. When performing all the time stamp, the result is required to be combined to get the final result. In the case of a P∀RFN we intersect the candidate sets for each moving object in time. In the case of a P∃RFN we have to unify the results in this step.

Algorithm3. TSR

Input: query object q, time interval T, probability τ, candidate sets S_{cnd}, influence sets S_{ver}^{cnd}

Output: result sets S_{cnd}

1. for each cnd $\in S_{cnd}$
2.   if cnd.p$<\tau$
3. $S_{cnd}=S_{cnd}$-{cnd};
4.   else
5.   for all $ifl \in S_{ver}^{cnd}$ do
6.   if $\exists t \in T$: dist(cnd(t),ifl(t))>dist(cnd(t)>q(t))
7.   then $S_{cnd}=S_{cnd}$-{cnd};
8.   break;
9.   end if
10.   end for
11. end for
12. return S_{cnd}

As shown in Algorithm 3, first the probability density function of each moving object is calculated by probability density function. The moving objects are filtered out by comparing the probability density function of each moving object with the given probability value τ. And then to meet the conditions of the moving object for comparison, it traverses the corresponding influence set of each candidate. The algorithm is the implementation of the $P\forall RFN$ algorithm, which can be applied to the implementation of $P\exists RFN$, which only needs to be targeted at a certain point in time, rather than the intersection of candidate objects in time T.

4 Experiments

4.1 Evaluation of DRFN

Our experimental environment: Intel(R) Core(TM) i7-3770 3.40 GHz CPU, 4 GB memory, Windows 7 operating system and Visual Studio. In the TPR-tree and heap files, the page size is 4 KB. In order to verify the feasibility and effectiveness of the proposed algorithm, varied tests are given. The data generated by the random data generator is used in the experiment. Through the procedures were randomly generated 128 KB, 256 KB, 512 KB, 1024 KB, and so on a number of different scale data sets. The moving object is randomly produced in the two-dimensional plan. The size of the plan is 500*500 unit coordinates.

As shown in Fig. 6, FTPL and BFS algorithm in different size of the moving object data set for a certain period of time the performance of continuous query. As can be seen from the graph, the FTPL algorithm outperforms the BFS algorithm in the moving object data set with different density. However, BFS algorithm is a direct comparison of the distance between the two moving objects, so the computation cost is high.

As shown in Fig. 7, comparison between FTPL algorithm and BFS algorithm at the cost of node access. In order to be more intuitive, we use the ratio of FTPL and BFS experimental results to explain the differences in cost. From the curve in results, we can see that the node access cost is always less than 1. This shows that the improved

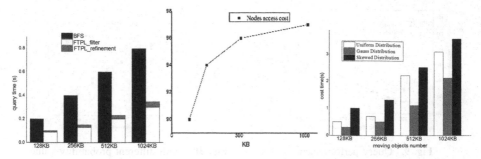

Fig. 6. Query cost in 2D space   **Fig. 7.** Node access times   **Fig. 8.** Query cost in varying data sets

algorithm proposed in this paper is lower than the BFS algorithm in node access. As seen from the Fig. 8, with the increase of the data set, the total cost of the query of the three kinds of distribution is a rising trend. the Gauss distribution of the query time is not large. This is because the distribution of the Gauss data set is concentrated distribution, for pruning MBR advantages. So the query cost is the lowest among the three.

4.2   Evaluation: P∀RFN and P∃RFN

The experimental data set is generated by the spatial data generator, and the motion of the simulated 100 k moving objects is generated in the spatial region of the 2000*2000. An uncertain region of the moving objects is a circle with a radius of 20, and the moving object speed is controlled at [20,30], and the probability density of the moving objects is distributed evenly.

(1) Comparison of different volume data sets

First, the number of different moving objects on the query efficiency, namely uncertain moving objects of probabilistic reverse furthest neighbor query cost. As shown in Fig. 9, with the increase of the density of moving objects, more moving objects become PRFN query results. If the number of moving objects becomes larger, the number of candidate objects and the number of affected objects will increase. The resulting probabilistic reverse furthest neighbor is to meet a certain probability results. In the refinement step, we first compare the probability value. For P∃RFN queries, all candidate objects to be trimmed for a single point of time. The P∀RFN only need to prune all the candidate objects of a time period T. So the cost of P∃RFN query is larger than the P∀RFN query. In addition, more candidate objects will increase the complexity of P∃RFN and P∀RFN queries.

(2) Comparison of different τ

When the query data set is the same, with the given probability to be different, it has a certain effect on the query efficiency. As shown in Fig. 10, With the increase of the probability value, the query cost of P∃RFN and P∀RFN is gradually reduced. With the increase of the probability value, the algorithm is the filter out

278   B. Li et al.

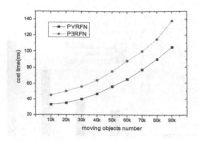 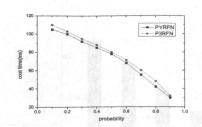

Fig. 9. Query performance   **Fig. 10.** with different probability value

of the candidate object is also gradually increased. Then the query cost is correspondingly reduced. As the number of moving objects in the refinement step is different, the cost of $P\exists RFN$ query is larger than that of $P\forall RFN$.

5 Conclusion and Future Work

According to the certain data sets and uncertain data sets, we give relevant definitions and corresponding pruning strategy. Our work solves the dynamic RFN query with uncertain frequent updates objects base on TPR-tree. TSF algorithm implement the PRFN query with the previous steps of filtrating and refinement. We conducted extensive Experiment on three varied distributed datasets and the experimental result show that our proposed algorithm improved the effectiveness and efficiency of queries.

Extend our work to process RFN queries on Crowdsourcing with more discussions and additional experiments. Moreover, we are planning to conduct the extended research with recommend systems which can be continuous interactions with hierarchical users.

References

1. Liu, J., Chen, H., Furuse, K., Kitagawa, H.: An efficient algorithm for arbitrary reverse furthest neighbor queries. In: Sheng, Q.Z., Wang, G., Jensen, C.S., Xu, G. (eds.) APWeb 2012. LNCS, vol. 7235, pp. 60–72. Springer, Heidelberg (2012). doi:10.1007/978-3-642-29253-8_6
2. Korn, F., Muthukrishnan, S.: Influence sets based on reverse nearest neighbor queries. ACM Sigmod Record **29**(2), 201–212 (2000)
3. Lian, X., Chen, L.: Probabilistic group nearest neighbor queries in uncertain databases. IEEE Trans. Knowl. Data Eng. **20**(6), 809–824 (2008)
4. Lian, X., Chen, L.: Efficient processing of probabilistic reverse nearest neighbor queries over uncertain data. VLDB J. **18**(18), 787–808 (2009)
5. Cheema, M.A., Lin, X., Wang, W., et al.: Probabilistic reverse nearest neighbor queries on uncertain data. IEEE Trans. Knowl. Data Eng. **22**(4), 550–564 (2010)

6. Zhu, J., Wang, X., Li, Y.: Predictive nearest neighbor queries over uncertain spatial-temporal data. In: Cai, Z., Wang, C., Cheng, S., Wang, H., Gao, H. (eds.) WASA 2014. LNCS, vol. 8491, pp. 424–435. Springer, Heidelberg (2014). doi:10.1007/978-3-319-07782-6_39
7. Li, K., Malik, J.: Fast k-Nearest neighbour search via dynamic continuous indexing. arXiv preprint arXiv:1512.00442 (2015)
8. Li, J., Wang, B., Wang, G.: Efficient probabilistic reverse k-nearest neighbors query processing on uncertain data. In: Meng, W., Feng, L., Bressan, S., Winiwarter, W., Song, W. (eds.) DASFAA 2013. LNCS, vol. 7825, pp. 456–471. Springer, Heidelberg (2013). doi:10.1007/978-3-642-37487-6_34
9. Xu, C., Gu, Y., Chen, L., et al.: Interval reverse nearest neighbor queries on uncertain data with markov correlations. In: Data Engineering ICDE 2013, pp. 170–181 (2013)
10. Emrich, T., Kriegel, H.-P., Mamoulis, N., Niedermayer, J., Renz, M., Züfle, A.: Reverse-nearest neighbor queries on uncertain moving object trajectories. In: Bhowmick, S.S., Dyreson, C.E., Jensen, C.S., Lee, M.L., Muliantara, A., Thalheim, B. (eds.) DASFAA 2014. LNCS, vol. 8422, pp. 92–107. Springer, Heidelberg (2014). doi:10.1007/978-3-319-05813-9_7
11. Trajcevski, G., Tamassia, R., Ding, H., et al.: Continuous probabilistic nearest-neighbor queries for uncertain trajectories. In: Proceedings of the 12th International Conference on Extending Database Technology: Advances in Database Technology, pp. 874–885. ACM (2009)
12. Yao, B., Li, F., Kumar, P.: Reverse furthest neighbors in spatial databases. In: ICDE 2009, pp. 664–675 (2009)
13. Liu, J., Chen, H., Furuse, K., Kitagawa, H.: An efficient algorithm for reverse furthest neighbors query with metric index. In: Bringas, P.G., Hameurlain, A., Quirchmayr, G. (eds.) DEXA 2010. LNCS, vol. 6262, pp. 437–451. Springer, Heidelberg (2010). doi:10.1007/978-3-642-15251-1_34
14. Cheng, R., Prabhakar, S., Kalashnikov, D.V.: Querying imprecise data in moving object environments. IEEE Trans. Knowl. Data Eng. 16(9), 1112–1127 (2004)
15. Tao, Y., Papadias, D., Lian, X.: Reverse kNN search in arbitrary dimensionality. In: Proceedings of VLDB Endowment, pp. 744–755 (2004)
16. Emrich, T., Kriegel, H.P., Ger, P., et al.: Boosting spatial pruning: on optimal pruning of MBRs. In: SIGMOD, pp. 39–50 (2010)
17. Said, A., et al.: User-centric evaluation of a k-furthest neighbor collaborative filtering recommender algorithm. In: Proceedings of the 2013 conference on computer supported cooperative work. ACM (2013)
18. Averbakh, I., Bereg, S.: Facility location problems with uncertainty on the plane. Discrete Optim. 2(1), 3–4 (2005)
19. Cabello, S., Díaz-Báñez, J.M., Langerman, S., Seara, C., Ventura, I.: Facility location problems in the plane based on reverse nearest neighbor queries. Eur. J. Oper. Res. 202(1), 99–106 (2010)
20. Yin, H., Cui, B., Li, J., Yao, J., Chen, C.: Challenging the long tail recommendation. Proc. VLDB Endowment 5(9), 896–907 (2012)

Research Papers

Relative Neighborhood Graphs Uncover the Dynamics of Social Media Engagement

Natalie Jane de Vries[1], Ahmed Shamsul Arefin[1], Luke Mathieson[1],
Benjamin Lucas[2,3], and Pablo Moscato[1(✉)]

[1] Faculty of Engineering and Built Environment, School of Electrical Engineering
and Computer Science, The University of Newcastle, Callaghan, NSW, Australia
{Natalie.deVries,Ahmed.Arefin,Luke.Mathieson,
Pablo.Moscato}@newcastle.edu.au
[2] Maastricht University, Limburg, The Netherlands
B.Lucas@maastrichtuniversity.nl
[3] Business Intelligence and Smart Services Institute (BISS) Heerlen,
Limburg, The Netherlands
http://www.newcastle.edu.au, http://www.maastrict.edu.nl,
http://www.biss-institute.nl/

Abstract. In this paper, we examine if the Relative Neighborhood
Graph (RNG) can reveal related dynamics of page-level social media
metrics. A statistical analysis is also provided to illustrate the application of the method in two other datasets (the Indo-European Language
dataset and the Shakespearean Era Text dataset). Using social media
metrics on the world's 'top check-in locations' Facebook pages dataset,
the statistical analysis reveals coherent dynamical patterns. In the largest
cluster, the categories 'Gym', 'Fitness Center', and 'Sports and Recreation' appear closely linked together in the RNG. Taken together, our
study validates our expectation that RNGs can provide a "parameter-
free" mathematical formalization of proximity. Our approach gives useful
insights on user behaviour in social media page-level metrics as well as
other applications.

Keywords: Social networking · Clustering · Proximity graph · Minimum spanning tree · Relative neighborhood graph

1 Introduction and Background

The popularity of social media by consumers has led to a greater commercial
need for understanding the different network structures that bind entities in
these novel online environments. Large datasets are generated in the online
world, which provide opportunities for generating a large base of web intelligence through data mining techniques. More specifically, social media analytics
is concerned with collecting, monitoring, analyzing, summarizing and visualizing social media data [30]. With social media networking data in particular,
clustering is a commonly used analytical method for finding groups of similar

© Springer International Publishing AG 2016
J. Li et al. (Eds.): ADMA 2016, LNAI 10086, pp. 283–297, 2016.
DOI: 10.1007/978-3-319-49586-6_19

objects [17]. As stated by Muhenbach and Lallich [26], analyzing social networks is a challenging data mining problem as the information that is hidden is implicit within the relationships in the network. However, although it is a difficult task, powerful benefits can be reaped by brands when leveraging social media analytics outcomes and creating real business value in doing so [15].

Previously, we have used a method called the MST-kNN clustering algorithm [19] and its Paraclique [8] variant - the MST-kNN with Paracliques [4] - to cluster social media brand pages based on their metrics data [24], consumer behavior data [12], online survey data [27] and computational stylistics and authorship attribution [5]. This methodology combines information from the Minimum Spanning Tree (MST) with that of a k-Nearest Neighbor graph (kNN). The result is a set of trees which span all the objects of interest. For the later method (kNN), a value for k needs to be set as a parameter.

In this contribution we explore a less "parameter-dependent" approach by introducing a proximity graph called the Relative Neighborhood Graph (RNG). Moreover, this study provides a preliminary application of a RNG in a social science setting, particularly in a social network analysis context of online behavior. Previously, RNGs have successfully been used in analyzing other Internet graphs as in [14] and could have applications in computational social sciences where the concept of strong ties is being revisited [22]. In this contribution we show the usefulness of RNGs in the analysis of social networks and uncovering hidden dynamic engagement relationships.

A RNG is defined as an undirected graph defined on a set of nodes, which have associated inter-nodes distances. An edge in the RNG connects two nodes p and q whenever there does not exist a third node r that is closer to both p and q than they are to each other [28] (note: the RNG was first introduced in terms of points on the Euclidean plane). Given this definition, it is clear that a RNG provides an additional insight into the similarity of the objects but modulated by the "local" neighborhood of an object. An important property to notice is that the RNG is a superset of a MST, this means that every edge of the MST can be found in the RNG. The characteristics and methodological details of RNGs are further outlined in the following section.

The result of the MST-kNN clustering algorithm is a forest (in graph theory terms, a set of trees). Such a forest can be explored in multiple ways; firstly, through simply inspecting the resulting sets of trees (as we have previously done in other studies) or, as in this work, by referring to the original similarity matrix and computing the RNGs for each set of vertices that belong to the same tree. This approach will deliver a more detailed structure for each of the partitions. One further motivation for this study is the previous finding of possible seasonal components in the large dataset containing social media metrics for Facebook location brand pages we will use in this study. In [24], we outline how this dataset was downloaded from the Facebook API and previously analyzed using clustering techniques. Due to the timeframe of the data collection matching the northern hemispheres spring, trends relating to this seasonal variation such as the high degree of gyms and fitness centers in one particular cluster versus all others have been observed. This phenomenon and the exact results from this previous study are further investigated here thus providing a good example to explore the use of RNGs in a social science networking study and elaborating on web intelligence findings specifically to social networking media. However, before analyzing the Facebook Pages dataset, we test the proposed method on two other datasets, namely; the Indo-European languages dataset and the Shakespearean-era Text Corpus dataset. Details are provided below.

2   Methodology

Datasets Used in This Study

84 Indo-European Languages Dataset. This dataset originated from a computational linguistic study of 84 Indo-European languages. This dataset is digitally provided as a 84 × 84 distance matrix by Dyen et al. [13]. Distances between languages were computed by calculating mean percent difference of cognancy in 200 Swadesh words [13]. Specifically, Dyen et al. used a list of basic vocabulary and estimated historical relationships (similarity) between two languages by computing the ratio of number of words (cognates) shared by them and the total number of words. The replacement rates of each word in the vocabulary are also considered. By considering the above ratio and replacement rates, they generated the so-called separation time between pairs of languages, which they provided as a distance-matrix of 84 languages. Extended explanations can be found in a previous publication by Mahata et al. [25].

256 Shakespearean Era Plays and Poems. This dataset is from a computational stylistic study of 256 plays and poems from the Shakespearean era, containing texts of authorship from the 16th and 17th centuries [5]. The machine-readable texts of the plays and poems are held in an archive by the Centre for Literary and Linguistic Computing at The University of Newcastle. A software tool called Intelligent Archive (IA) by Craig and Whipp [10] has been used to pre-process the corpus. The IA creates sub-corpora and generates counts of word-forms according to a parameterized user input, taking into account the variations in spelling commonly found in 16th and 17th century plays and poems, in addition to facilitating disambiguation of words by both context and frequency. The tool identified in total a set of approximately 66,907 unique words in the 256 texts, which is available in the form of a $66,907 \times 256$ matrix in [5]. Using this, we produced a complete weighted graph (i.e., a distance matrix of size 256 × 256) where all plays and poems are connected to each other. The weight of the connection between two works, i.e. the edge weights of the graph, corresponded to the pair-wise Jensen-Shannon divergence (JSD) [16] between the frequencies of words in two documents. This method and computation of the pair-wise JSD has been explained in [5].

Worlds Top Check-in locations Facebook Pages Dataset. Finally, a large Facebook social media metrics dataset is used to prove the utility of our method in a social networking setting as well as to show the scalability of the method presented in this paper. This dataset takes some of the most popular "check-in" locations Facebook pages and includes various page-level metrics as features. The total number of page 'Likes', 'Shares' and 'Talking About' counts are included in the study (for total of 15,625 pages). These three metrics were collected four times consecutively each week for a month. As stated, this dataset has previously been examined [24] with our clustering (partitioning) algorithm named MST-kNN in which more information on this dataset and how it was obtained is provided. In this study however, we aim to provide a deeper insight and uncover more hidden structures in the data through the introduction of an RNG. In doing so, we aim at unveiling possible seasonal trends in social media usage reflected by brands online metrics.

The MST-kNN Method. As a basis for the clustering method, a distance matrix first needs to be computed. This process is described in detail in [24] where it is explained how a correlation matrix can be employed to compute a distance matrix. In our prior work, we found that using a cosine-based correlation lead to the distance matrix that had the best results in the social media data. It is these clustering results that will be used in this study for comparison. First we will provide a short description of the MST-kNN process.

The MST-kNN clustering comes from intersecting edge sets of a Minimum Spanning Tree and a k-Nearest Neighbor graph (MST-kNN) [19]. This is a non-parametric graph partitioning algorithm that has been successfully applied in bioinformatics [7] and other areas [1–3,19–21]. The algorithm has as input a weighted, undirected, complete graph G, based on the distance matrix as stated and computes two proximity graphs: a minimum spanning tree GMST and a k-nearest neighbor graph GkNN, where the value of k is automatically determined by Eq. (1).

$$k = \min\{\lceil \ln(n) \rceil, \max\{k' \mid \text{G}k\text{NN is not connected}\}\} \tag{1}$$

Then, all edges in GMST are inspected. If for a given pair of nodes (x, y), connected by an edge in G, if neither x is one of the k nearest neighbors of y, nor y is one of the k nearest neighbors of x, the edge that is connecting this pair of nodes is eliminated from GMST. Deleting the first edge results in a new graph G which is now a spanning forest. This algorithm continues to apply these rules recursively to each tree in G until no further partition is possible. The outcome of this process is a set of trees which represent the resulting partition of the node set in clusters.

Relative Neighbourhood Graph. In this paper we explore the more "parameter-less" method of exploiting RNGs to uncover network and cluster structure. Here we provide the methodological details of using the relative neighborhood graph, as well as a technical result that the RNG of a graph is a superset of any MST of that graph. A RNG will be computed for each cluster in each of the graphs. However, for the Facebook metrics data, we only present and discuss the RNG of the largest cluster resulting from the MST-kNN clustering algorithm on the complete dataset. This cluster was termed Cluster 4 in our previous publication [24] and contains 2,104 nodes.

Theorem 1. *Given a graph G with edge weights drawn from a totally ordered set W, any minimum spanning tree T of G is a subgraph of the relative neighborhood graph R of G.*

Proof. Let $\omega : E(G) \to W$ be the weight function for G. An edge uv is included in R if and only if there is no vertex w adjacent to both u and v such that $\omega(wv) < \omega(uv)$ and $\omega(uw) < \omega(uv)$. Assume for contradiction that there exists a minimum spanning tree T of G such that at least one edge in $E(T)$ is not in the relative neighborhood graph R of G. Let uv be that edge. As uv is not in $E(R)$, there exists a vertex w such that uw and wv are edges of $G, \omega(wv) < \omega(uv)$ and $\omega(uw) < \omega(uv)$. As T is a tree we also have that either wv or uw is not in $E(T)$ (or neither are). Without loss of generality we may assume that wv is not in $E(T)$ and that after removal of uv from $E(T)$, w and v are in different disconnected subtrees of T. We can obtain a spanning tree with lower weight than T by removing uv from $E(T)$ and adding wv. This contradicts the minimality of T, and hence uv must be in the relative neighborhood graph R. More generally we may take the output of Kruskal's algorithm as the definition of a minimum spanning tree in the case where addition is not well defined for the elements of W. In this case the same argument holds as Kruskal's algorithm would not have pick uv to be in the minimum spanning tree.   □

Similar theorems have been previously proven for geometric graphs [28], however Theorem 1 covers general graphs with an arbitrary edge weighting scheme.

Statistical Validation. In order to statistically validate the results, we conduct a random permutation test. As stated, the clustering results taken from previous studies to be used in this study were from the cosine distance measure, as it produced the lowest (best) p-value in the statistical test of difference. The idea of a random permutation test is to randomly permute ("mix up") the graph node labels, which, in turn, correspond to the given categories, and to see how frequently we observe nodes of the same category linked by an edge or nodes of a different category linked by an edge. Each of the edges are scored; a self-score when a category is linked to itself and a diff-score when it is linked to a different category. The process is repeated a 1,000 times and this produces 1,000 sets of scores per graph with self-scores and diff-scores. Subsequently, the Wilcoxon signed-rank test and the Kruskal-Wallis test are used as statistical difference tests to obtain p-values for interpretation [18].

Implementation and Visualisation of Results. We implement the MST-kNN and its proposed RNG variant in R using the igraph package [11]. For computing the MST and MST-kNN we use the minimum.spanning.tree (Prim's algorithm) function. The outcomes were stored as GML (Graph Modelling Language) files and used as direct input to the randomization process. The randomization and statistical computation software was written in C++ using the

Standard Template Library. Finally, we use the graph visualization tool yEd [29] to display, manipulate and visualize our results.

3   Results

Language Data Results. Existing clusters of this dataset that have previously been published [4,6] and are taken as individual datasets from the computation of an RNG for each cluster. They are combined in one graph shown in Fig. 1. Table 1 shows the results of the random permutation test. As can be seen, for all three datasets, the method including an RNG produces the most significant outcomes. In the case of the language dataset, the Wilcoxon signed rank test produces a significant result at the 99 % confidence interval. This is due to the low number of edges connecting two languages of different language groups as shown in Fig. 1.

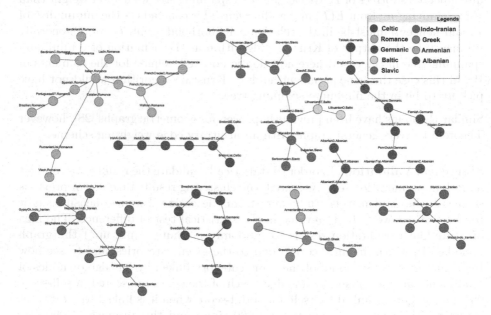

Fig. 1. The results of the MST-kNN+RNG of the Indo-European languages dataset. In this figure, all RNGs for each cluster are combined in one image. The legends show the different colors for categories. An almost perfect partitioning of the different language categories can be seen. (Color figure online)

Shakespearean Era Plays and Poems Results. As with the Language Dataset, existing clusters of this dataset that have previously been published [5] are taken as individual datasets from the computation of an RNG for each cluster. In Fig. 2 the RNGs of all the plays and poems clusters are combined in one image. The colors of the nodes correspond to their respective authors with

Table 1. Randomisation test

	Datasets	Wilcoxon test	Kruskal Wallis test
84 Indo-European Languages	MST-kNN	1.04E-07	2.09-E-07
	MST-kNN+Paracliques	1.02E−07	2.04E−07
	MST-kNN+RNG	**1.01E−07**	**2.01E−07**
Shakespeare (256 plays and poems)	MST-kNN	4.39E−12	8.77E−12
	MST-kNN+Paracliques	1.58E−11	3.17E−11
	MST-kNN+RNG	**1.17E−12**	**2.34E−12**
Facebook datas largest cluster (2104 pages)	MST-kNN	0.358256	0.716513
	MST-kNN+Paracliques	0.380797	0.761593
	MST-kNN+RNG	0.283297	0.566594

Shakespeares works represented in purple. White refers to Other authors and circles refer to poems whereas squares refer to plays. As can be seen, still a fairly good separation between authors occurs, however there are also links between authors. The extra links between authors show which authors are similar to each other in terms of their writing style and word usage.

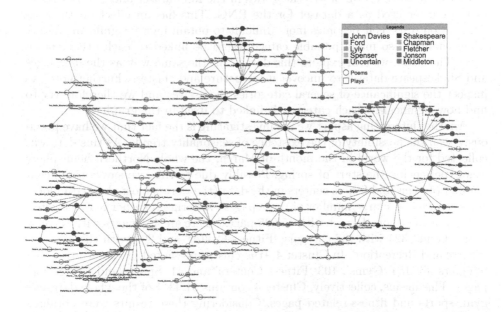

Fig. 2. The results of the MST-kNN+RNG of the Shakespearean Era Text Corpus dataset. In this figure, all RNGs for each cluster are combined in one image. Authors and their corresponding colors are shown in the legend. Round nodes in the graph refer to poems whereas square nodes refer to plays. (Color figure online)

Furthermore, the objective of studies about Shakespearean Era plays and poems is often to identify the unknown or anonymous authors of various poems and plays. This said, having uncovered more structure in this dataset, with a "parameter-less" approach, means that perhaps more of these works can be explored by looking at which authors they are paired to. In the figure, red nodes stand for those works of which authorship is uncertain, and as can be seen they are often connected to authors and not all to the same author. In previous publications some of these findings have been discussed [5] and these show the usefulness of clustering the Shakespearean Era text corpus.

Finally, as shown in Table 1, the RNG method again produced the lowest p-value with the most significant output at the 99 % confidence interval. Showing it found a better structure in the graph in terms of linking works (nodes) of one author to works of the same author whilst minimizing links going across works (nodes) of different authors. The successful use of the RNG in these two datasets now leads us to investigate the larger Facebook metrics dataset for the purpose of web intelligence generation and viable real-life applications in business analytics.

Facebook Social Media Metrics Results. Before analyzing the Facebook social media metric results, we must note, as stated in our previous publication [24], that the number of categories (labels) in the Facebook metrics dataset is very large with a total of 478 categories in the full dataset and 257 categories in the cluster used as a dataset for the RNG. This has an effect on the randomization tests as it becomes more difficult to obtain highly significant results when there are so many possible categories to be linked to each other via an edge. That said, we still explore this dataset in the same way as the languages and Shakespeare dataset to uncover its structure and clusters. Furthermore, we inspect the significance of several categories themselves and we therefore try to find how significantly each category is linked to itself in the graph.

Another interesting aspect we are investigating is the fact that we have previously seen the possibility of the existence of seasonality trends. As this data was collected for the world in the month of March, the whole Northern hemisphere was experiencing the start of spring. Possibly due to this, we saw a significant number of gyms and fitness centers in the data (the worlds most checked-in locations) as well as a large number of these pages clustering together in one cluster. In fact, in the whole dataset with 15,625 pages, there are 421 pages with the category 'Gym', 324 with the category 'Fitness Center' and 203 with the category 'Sports and Recreation'. In Cluster 4, (the cluster presented in this study) these numbers are 154 'Gyms', 103 'Fitness Centers' and 41 'Sports and Recreation' pages. This means, collectively, Cluster 4 contains 31.43 % of the whole dataset's gym, sports and fitness-related pages. Considering these results were obtained using a cosine-based distance matrix for input of the clustering algorithm, this means that objects were similar if they experienced similar trends. It is therefore interesting to observe so many gyms and fitness-related pages to come together in one cluster indicating that this cluster contains some grouping which may be related to a seasonal trend.

Table 2. Individual category randomisation test - best results

Category	Wilcoxon p-value	Kruskal-Wallis p-value
Zoo and Aquarium	0.0427	0.0851
Gym	0.0432	0.0861
Specialty Grocery Store	0.0433	0.0864
Malaysian Restaurant	0.0439	0.0875
Continental Restaurant	0.0445	0.0888

The Facebook Pages Metrics Dataset Cluster 4 RNG is shown is Fig. 3. Several of the largest categories are colored whilst other categories are left white for purpose of visualization. A small section of the graph in Fig. 3 is highlighted and shown in Fig. 4. This section shows a large number of gyms and fitness-related pages clustered together which means they are relative neighbors of one another.

In this highlighted section we can see the high level of proximity a lot of gym and fitness-related pages are to each other. However, even though visually some categories can be seen together, when looking at the statistical test results in Table 1, no significant results were found using any of the methods in terms of p-values from the Wilcoxon-Signed-Rank test or the Kruskal-Wallis test. For this reason we explored the Facebook dataset further and ran the random permutation test and significance testing for each individual category to find whether the p-values would improve and to see which category is the "best performing" in terms of being partitioned. Table 2 shows the results for the five "best performing" categories for Cluster 4 (the dataset used in this study).

As can be seen in Table 2, when interpreting the Wilcoxon-signed-rank test results, the "best performing" categories yield slightly significant results when investigated individually. This result may indicate that these categories of Facebook brand pages are most homogenous in the dataset. Furthermore, it is interesting to note that the second-most significant result was found for the category 'Gym' meaning this category is fairly homogenous in terms of its social media metrics on Facebook. This result indicated that our suspicion of having a high degree of gyms and fitness-center pages that were highly similar to each other, due to possible seasonal trends, can be satisfied. Furthermore, it is interesting to see which other categories produce significant results; especially considering that the statistical test for the whole graph did not produce significant results. Although the categories 'Zoo and Aquarium', 'Specialty Grocery Store', 'Malaysian Restaurant' and 'Continental Restaurant' are very small, this still indicates that these categories are more similar to each other than others.

In Table 3 we show the results for the largest five categories of the Facebook data cluster investigated in this study. As can be seen, the category 'Gym' appears here again as it is the second largest category. The other large categories presented in this table do not have significant partitioning results, possibly indicating higher levels of heterogeneity in these categories social media metrics. One reason could be their large size and the varied amount of customers interacting

Fig. 3. Result of computing the RNG of pages that belong to the largest cluster from the MST-kNN output. There are 2,104 nodes and 3,012 edges. Several categories are colored. Bright green refers to nodes with the category 'Gym', light green is 'Sports and Recreation', and 'Fitness Center' are colored in mint green. Restaurants are colored in orange (please note these may still have their own sub-categories such as 'American Restaurant' or 'Thai Restaurant') and 'Movie Theatre' and 'Theatres' are colored in darker and lighter red respectively. Finally, the category 'Airport' is colored in purple with 'Airport-Terminal' colored in slightly brighter purple. All other categories are left white. The red rectangle indicates the section where a 'zoomed shot' is taken presented in Fig. 4 (Color figure online)

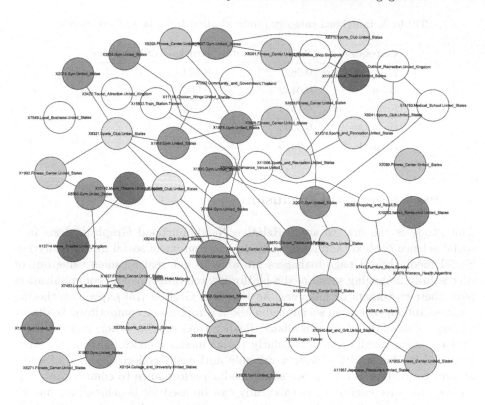

Fig. 4. A highlight taken from the RNG shown in Fig. 3 zoomed in for further inspection. A selection of gyms and fitness-related pages appear in this highlight. As in Fig. 3, bright green refers to 'Gyms', light green to 'Sports and Recreation', mint green to 'Fitness Center', red to '(Movie) Theatres' and 'Restaurants' are in orange. White nodes correspond to any 'Other' category. (Color figure online)

with these pages, therefore creating heterogeneous behaviors in the social media metrics. This result emphasizes the significance of the 'Gym' category further as it is the only large category with a statistically significant partitioning.

Moreover, we can also see from these results that individual brand pages can look at where they are located in a RNG, or how the category they are in is "scattered' across the whole RNG and use this knowledge in their online strategy building. For example, a specific type of restaurant may find that it is relative neighbors with more businesses from other categories. This means for this restaurant they should not look at fellow restaurants for comparison or competition, but rather, at other businesses who are its relative neighbors as it these pages it is similar to. Furthermore, for Facebook brand pages in the 'Gym and Fitness' sector these results indicate the opposite meaning that their category is somewhat homogenous and they should look at their direct competitors when investigating their social media metrics performance. These results are further elaborated on in the following section.

Table 3. Individual category randomisation test - largest categories

Category	Wilcoxon p-value	Kruskal-Wallis p-value
Movie Theatre	0.2627	0.5243
Gym	0.0432	0.0861
Fitness Center	0.1574	0.3139
Hospital	0.3595	0.7176
Airport	0.9475	0.1053

4   Discussion and Conclusion

This study is the first to apply Relative Neighbourhood Graph analysis in a social science context, and specifically, in the context of social media analytics. Social media brand page managers are able to get real business value out of successfully managing social media platforms, relationships and communication with their customers or followers. As Fan and Gordon [15] explain, gathering business intelligence from social media about suppliers or competitors, is almost as easy as monitoring its own affairs. Rather than seeking other, more difficult sources, social media and particularly social media metrics, such as the ones analyzed in this study, provide a valuable and easily-collected source of information to analyze and compare social media performance to competitors. The analysis we have conducted in this study can be used for benchmarking oneself to competitors, finding new competitors that a business did not know they had, or simply finding other businesses and their Facebook brand pages to benchmark and compare oneself to; including any page around the world.

The specific example of gyms and fitness-related pages in the Facebook pages metrics-based graph provides such an instance for comparison and benchmarking. Although this data was only collected for a one-month period, some initial insights are available and make the potential uses of this method clear. There are quite a few connections going between the different fitness-related pages. This means that a company that identifies itself as "just a gym" may not realise the comparison and benchmarking opportunity with sports and recreation centres, fitness centres or even other pages it has been linked to such as other local small businesses.

To practitioners and brand managers, the methodological process presented in this study provides a potential tool to include in their business monitoring, product life-cycle or marketing campaign analysis and their overall social media analytics processes. In fact, proximity graph-based techniques, such as the use of a RNG in this study, facilitate surveillance of a potentially otherwise unknown competitive landscape in the social media space. Social media brand pages can now inspect this competitive landscape through methods like these which existing "analytics" or monitoring tools do not allow. Furthermore, Fan and Gordon [15] point out the continued challenge of analyzing the growing amounts of data related to social media pages coming from new features such as location-based services. This study addresses this challenge by having taken the largest "check-in" location pages around the world and applying a highly scalable partitioning analysis method to this data. This exploratory study of a RNG in the social scientific context therefore provides us with future research suggestions in this field.

Future research recommendations include, for example, the use of alternative similarity functions for computing correlations between page-level metrics. We have based the present work on the cosine-based correlation, however further studies may include other robust measures of correlation [9]. These robust techniques can handle both large and structural outliers present in the data.

Another recommendation is to study the dynamical behavior of page-level metrics in conjunction with the introduction of a marketing campaign or during the product/service life cycle. Therefore, following more of a process that includes capture, understand and present findings in the social media analytics strategy [15]. In doing so, larger datasets may be used as they include longer periods of time expanding the length of the time-series metrics.

Larger datasets will require fast and highly-scalable methods to work efficiently. In the area of proximity graphs such as RNGs, parallelization is an option for dealing with this challenge as the computation required for each edge is independent of the results for other edges. This allows a simple parallel approach that can make use of highly parallel technologies such as GPU computing, which excel at simple, repetitive, high-volume computation [23]. This makes our general approach feasible even for very large scale datasets.

Finally, simple additions to studies like these can be made by incorporating data from a multitude of social media platforms and thus increasing the analytical span of the method and improving results for the brand managers. This would also increase the number of features and information available to investigate creating for a more comprehensive analysis of a business online competitive landscape.

In conclusion, this exploratory social media data mining study provides us with a viable path for continued research using proximity graphs to analyze online social media metrics data. The RNG approach presented in this study proves to be a successful tool in analyzing network graphs and with the indication of significant results for the individual Facebook pages categories, a useful tool for generating business intelligence for brands and companies social media page strategies.

Acknowledgments. The authors would like to thank Shannon Hochkins and Social-bakers (www.socialbakers.com) for technical and data collection support. PM is funded by ARC Future Fellowship FT120100060. The funder had no role in this study or the preparation of this manuscript.

References

1. Arefin, A.S., Inostroza-Ponta, M., Mathieson, L., Berretta, R., Moscato, P.: Clustering nodes in large-scale biological networks using external memory algorithms. In: Xiang, Y., Cuzzocrea, A., Hobbs, M., Zhou, W. (eds.) ICA3PP 2011. LNCS, vol. 7017, pp. 375–386. Springer, Heidelberg (2011). doi:10.1007/978-3-642-24669-2_36

2. Arefin, A.S., Riveros, C., Berretta, R., Moscato, P.: GPU-FS-kNN: a software tool for fast and scalable kNN computation using GPUs. PLOS ONE **7**(8), e44000 (2012)
3. Arefin, A.S., Riveros, C., Berretta, R., Moscato, P.: kNN-Borůvka-GPU: a fast and scalable MST construction from kNN graphs on GPU. In: Murgante, B., Gervasi, O., Misra, S., Nedjah, N., Rocha, A.M.A.C., Taniar, D., Apduhan, B.O. (eds.) ICCSA 2012. LNCS, vol. 7333, pp. 71–86. Springer, Heidelberg (2012). doi:10.1007/978-3-642-31125-3_6
4. Arefin, A.S., Riveros, C., Berretta, R., Moscato, P.: The MST-kNN with paracliques. In: Chalup, S.K., Blair, A.D., Randall, M. (eds.) ACALCI 2015. LNCS (LNAI), vol. 8955, pp. 373–386. Springer, Heidelberg (2015). doi:10.1007/978-3-319-14803-8_29
5. Arefin, A.S., Vimieiro, R., Riveros, C., Craig, H., Moscato, P.: An information theoretic clustering approach for unveiling authorship affinities in Shakespearean era plays and poems. PLOS ONE **9**(10), e111445 (2014)
6. Bryant, D., Filimon, F., Gray, R.D., Untangling our past: languages, trees, splits and networks. In: The Evolution of Cultural Diversity: A Phylogenetic Approach, pp. 67–83 (2005)
7. Capp, A., Inostroza-Ponta, M., Bill, D., Moscato, P., Lai, C., Christie, D., Lamb, D., Turner, S., Joseph, D., Matthews, J.: is there more than one proctitis syndrome? a revisitation using data from the trog 96.01 trial. Radiother. Oncol. **90**(3), 400–407 (2009)
8. Chesler, E.J., Langston, M.A.: Combinatorial genetic regulatory network analysis tools for high throughput transcriptomic data. In: Eskin, E., Ideker, T., Raphael, B., Workman, C. (eds.) RRG/RSB -2005. LNCS, vol. 4023, pp. 150–165. Springer, Heidelberg (2007). doi:10.1007/978-3-540-48540-7_13
9. Chilson, J., Ng, R.T., Wagner, A., Zamar, R.H.: Parallel computation of high-dimensional robust correlation and covariance matrices. Algorithmica **45**(3), 403–431 (2006)
10. Craig, H., Whipp, R.: Old spellings, new methods: automated procedures for indeterminate linguistic data. Literacy Linguist. Comput. **25**(1), 37–52 (2010)
11. Csardi, G., Nepusz, T.: The igraph software package for complex network research. Int. J. Complex Syst. **1695**(5), 1–9 (2006)
12. Jane, N., de Vries, A., Arefin, S., Moscato, P.: Gauging heterogeneity in online consumer behaviour data: a proximity graph approach. In: Socialcom and BDCloud, pp. 485–492. IEEE (2014)
13. Dyen, I., Kruskal, J.B., Black, P.: An Indoeuropean classification: a lexicostatistical experiment. Trans. Am. Philos. Soc. **82**(5), 1–132 (1992)
14. Escalante, O., Perez, T., Solano, J., Stojmenovic, I.: RNG-based searching and broadcasting over internet graphs and peerto-peer computing systems. In: The 3rd ACS/IEEE International Conference on Computer Systems and Applications. IEEE (2005)
15. Fan, W., Gordon, M.D.: The power of social media analytics. Commun. ACM **57**(6), 74–81 (2014)
16. Grosse, I., Bernaola-Galván, P., Carpena, P., Román-Roldán, R., Oliver, J., Stanley, H.E.: Analysis of symbolic sequences using the Jensen-Shannon divergence. Phys. Rev. E **65**(4), 041905 (2002)
17. Gundecha, P., Liu, H.: Mining social media: a brief introduction. Tutorials Oper. Res. **1**(4), 1–17 (2012)
18. Hollander, M., Wolfe, D.A., Chicken, E.: Nonparametric Statistical Methods, vol. 751. Wiley, New York (2013)

19. Inostroza-Ponta, M., Berretta, R., Mendes, A., Moscato, P.: An automatic graph layout procedure to visualize correlated data. In: Bramer, M. (ed.) Artificial Intelligence in Theory and Practice, IFIP 19th World Computer Congress, TC 12: IFIP AI 2006 Stream, IFIP, vol. 217, 21-24 August 2006, Santiago, Chile (2006)pages 179–188. Springer, 2006
20. Inostroza-Ponta, M., Berretta, R., Moscato, P.: QAPgrid: a two level QAP-based approach for large-scale data analysis and visualization. PLOS ONE 6(1), e14468 (2011)
21. Inostroza-Ponta, M., Mendes, A., Berretta, R., Moscato, P.: An integrated QAP-based approach to visualize patterns of gene expression similarity. In: Randall, M., Abbass, H.A., Wiles, J. (eds.) ACAL 2007. LNCS (LNAI), vol. 4828, pp. 156–167. Springer, Heidelberg (2007). doi:10.1007/978-3-540-76931-6_14
22. Jones, J.J., Settle, J.E., Bond, R.M., Fariss, C.J., Marlow, C., Fowler, J.H.: Inferring tie strength from online directed behavior. PLOS One 8(1), 1–6 (2013)
23. Liu, Y., Sui, Z., Kang, C., Gao, Y.: Uncovering patterns of inter-urban trip and spatial interaction from social media check-in data. PLOS ONE 9(1), e86026 (2014)
24. Lucas, B., Arefin, A.S., Jane, N., de Vries, R., Beretta, J.C., Moscato, P.: Engagement in motion: exploring short term dynamics in page-level social media metrics. In: SocialCom and BDCloud, pp. 334–341. IEEE (2014)
25. Mahata, P., Costa, W., Cotta, C., Moscato, P.: Hierarchical clustering, languages and cancer. In: Rothlauf, F., Branke, J., Cagnoni, S., Costa, E., Cotta, C., Drechsler, R., Lutton, E., Machado, P., Moore, J.H., Romero, J., Smith, G.D., Squillero, G., Takagi, H. (eds.) EvoWorkshops 2006. LNCS, vol. 3907, pp. 67–78. Springer, Heidelberg (2006). doi:10.1007/11732242_7
26. Muhlenbach, F., Lallich, S.: Discovering research communities by clustering bibliographical data. In: Huang, J.X. et al. (eds.) 2010 IEEE/WIC/ACM International Conference on Web Intelligence, Toronto, Canada, 31 August–03 September 2010, pp. 500–507. Computer Society (2010)
27. Naeni, L.M., Jane, N., de Vries, R., Reis, A.S., Arefin, R.B., Moscato, P., Identifying communities of trust, confidence in the charity, not-for-profit sector: a memetic algorithm approach. In: BDCLOUD, pp. 500–507. IEEE (2014)
28. Toussaint, G.T.: The relative neighbourhood graph of a finite planar set. Pattern Recogn. 12(4), 261–268 (1980)
29. Wiese, R., Eiglsperger, M., Kaufmann, M.: yFiles-visualization and automatic layout of graphs. In: Junger, M., Mutzel, P. (eds.) Graph Drawing Software, pp. 173–191. Springer, Heidelberg (2004)
30. Zeng, D., Chen, H., Lusch, R., Li, S.H.: Social media analytics, intelligence. IEEE Intell. Syst. 25(6), 13–16 (2010)

An Ensemble Approach for Better Truth Discovery

Xiu Susie Fang[1]([✉]), Quan Z. Sheng[1], and Xianzhi Wang[2]

[1] School of Computer Science, The University of Adelaide,
Adelaide, SA 5005, Australia
{xiu.fang,michael.sheng}@adelaide.edu.au
[2] School of Computer Science and Engineering, UNSW Australia,
Sydney, NSW 2052, Australia
xianzhi.wang@unsw.edu.au

Abstract. Truth discovery is a hot research topic in the Big Data era, with the goal of identifying true values from the conflicting data provided by multiple sources on the same data items. Previously, many methods have been proposed to tackle this issue. However, none of the existing methods is a clear winner that consistently outperforms the others due to the varied characteristics of different methods. In addition, in some cases, an improved method may not even beat its original version as a result of the bias introduced by limited ground truths or different features of the applied datasets. To realize an approach that achieves better and robust overall performance, we propose to fully leverage the advantages of existing methods by extracting truth from the prediction results of these existing truth discovery methods. In particular, we first distinguish between the *single-truth* and *multi-truth* discovery problems and formally define the ensemble truth discovery problem. Then, we analyze the feasibility of the ensemble approach, and derive two models, i.e., *serial model* and *parallel model*, to implement the approach, and to further tackle the above two types of truth discovery problems. Extensive experiments over three large real-world datasets and various synthetic datasets demonstrate the effectiveness of our approach.

Keywords: Truth discovery · Big data · Multi-truths · Ensemble approach

1 Introduction

In the *Big Data* era, various sources may provide description of the same data items (i.e., properties of certain objects). Due to the existence of possible errors, out-of-date data, and missing records, the data collected from different sources may conflict. This makes it of paramount importance to discover the truth from these data to facilitate reliable knowledge discovery and decision making. To this end, tremendous research efforts have been paid to the truth discovery problem from both artificial intelligence and database communities under the

© Springer International Publishing AG 2016
J. Li et al. (Eds.): ADMA 2016, LNAI 10086, pp. 298–311, 2016.
DOI: 10.1007/978-3-319-49586-6_20

topics of *information corroboration* [6], *information credibility* [13], *conflicting data integration* [5], *fact-checking* [7], *data fusion* [8], and *knowledge fusion* [4].

Despite the various truth discovery methods, such as those handling different data types (e.g., categorical and continuous data), and *source dependency* (e.g., copying relation among sources), those considering *source quality* (e.g., source accuracy/recall, specificity, sensitive, and freshness of data) and *object properties* (e.g., the difficulty of and relation between data objects), and those taking into account *value implications* (e.g., *complementary vote*[1]) and *truth properties* (e.g., *multiple truths* and "unknown" truths), no single method can fit or constantly outperform the others in all application scenarios [11] (our experiments on three real-world datasets and various synthetic datasets validate this conclusion). In addition, a recent investigation [10] shows that even an improved method does not always beat its original version.

Although an appropriate truth discovery method can be selected for each specific scenario [11,14], it is challenging to find a method that achieves generally good performance due to the technical limitations and biases of each specific method. As the ensemble approach has been proven to be effective for enhancing the robustness and overall performance of algorithms in many disciplines [2], in this paper, we study on the feasibility of ensembling existing methods for better truth discovery. Realizing such an ensemble truth discovery approach is a tricky task due to the complexity and diversity of existing truth discovery methods. In a nutshell, we make the following contributions in this paper:

- We distinguish between two types of truth discovery problems, i.e., the *single-truth* and *multi-truth* discovery problems, and formally define the ensemble truth discovery problem.
- We analyze the feasibility of the ensemble truth discovery approach, and propose two models, i.e., *serial* and *parallel model*, to implement the approach.
- We empirically evaluate our ensemble approach. Extensive experimental results show that our approach outperforms traditional methods on both real-world and synthetic datasets. In particular, the synthetic datasets with complete ground truths show the improved performance of the ensemble approaches without being biased by the sparsity of limited ground truths.

The rest of the paper is structured as follows. Section 2 reviews the related work. Section 3 defines the ensemble truth discovery problem. Section 4 analyzes the feasibility of the ensemble approach and presents two implementation models, namely the serial and parallel models. Finally, we report the experimental results in Sect. 5, and provide some concluding remarks in Sect. 6.

2   Related Work

Truth discovery has been actively studied by the data integration community in the last few years. Early methods for tackling this issue consist of taking

---

[1] If a source claims value(s) for a certain object, it implicitly votes against other candidate values of this object.

the mean, median for continuous data, and majority voting for categorical data. These methods commonly neglect sources' quality differences, treat every source equally, and are therefore inaccurate in cases where the majority of sources provide false values. Based on this consideration, various methods incorporate source quality by applying a general principle: a source is more trustworthy if it provides more truths; meanwhile, a value has a bigger possibility of being selected as truth if it is claimed by more high-quality sources. The existing truth discovery methods generally fall into three groups.

The *iterative* methods predict truths and estimate source reliability itera-tively until certain convergence conditions are met. Typical work in this cate-gory includes: *TruthFinder* [17], which applies a Bayesian analysis to conduct the iterative processes. *AccuSim* proposed by Dong et al. [5,10] incorporates the implication of value similarity. They further extend AccuSim by additionally con-sidering the copying relations among sources and introduce *AccuCopy. Average-Log, Investment,* and *PooledInvestment* are developed by Pasternack et al. [12] in order to prevent sources that make more claims from obtaining higher quality weights. *Cosine* and *2-Estimates* are proposed by Galland et al. [6] to adopt complementary vote, they further introduce an improved method 3-Estimates by incorporating "hardness of fact". *SSTF* [16] is a semi-supervised method, which refers to a small set of labeled truths as an additional input data. To relax the single-truth assumption, Wang et al. [15] introduce a Bayesian framework based method *MBM* for multi-truth discovery, in which they also incorporate a finer-grained copy detection technique. The second group is about *optimization based* methods. Both *CRH* [8] and *CATD* [9] model the truth discovery problem as an optimization problem. They differ in that the former is specially designed for handling heterogeneous data while the latter for the long-tail data. The third group is about *probabilistic graphical model based* methods. Methods in this cat-egory typically model truths as latent variables. For example, Zhao et al. design a *Gaussian Truth Model (GTM)* [20] for continuous data. *Latent Truth model (LTM)* [19] models source reliability using two metrics, i.e., specificity and sen-sitive, for multi-truth discovery. *Latent Credibility Assessment (LCA)* [13] addi-tionally considers more factors such as the probability of guessing to facilitate more accurate truth discovery.

A recent survey [10] tests the performance of several methods on two real-world datasets, which shows that no single method always outperforms the others, and nearly half of the mistakes in the best truth discovery results can be avoided if the trustworthiness of sources is known in apriori. More surveys and experimental studies in [14] and [11] show the potential of improving the usability and repeata-bility of existing truth discovery methods via an ensemble approach. To the best of our knowledge, [1] is the only work that applies an ensemble approach in truth dis-covery. It proposes two ensemble methods, i.e., *Uniform Weight Ensemble (UWE)* and *Adjusted Weight Ensemble (AWE)*, and proves that the ensemble approach can generally mitigate the biases introduced by sparse ground truth and outper-form the traditional methods. Our work is the first to formally define the ensemble truth discovery problem and to provide in-depth comparisons of different ensem-ble methods over both single-truth and multi-truth scenarios.

3   Problem Formulation

For the input of truth discovery, suppose M data sources (e.g., "*Wikipedia*"), $\mathbf{S} = \{S_1, S_2, \ldots, S_m\}$, provide values on N data items (e.g., "*the cast of Harry Potter*"), $\mathbf{D} = \{D_1, D_2, \ldots, D_n\}$. This input data can be visualized as an $M \times N$ data matrix (Fig. 1(a)). Each cell represents a *claim* that describes the value(s) claimed by a source on a data item (e.g., a claim "*July 9, 1956*" for the data item "*the birthday of Tom Hanks*" provided by source "*Wikipedia*"). The values in the cells of the same columns may conflict due to the different reliability of sources. The objective of the truth discovery problem is to predict the truth(s) for each data item (corresponding to a column), given the noisy data matrix, while estimating the reliability of each source (corresponding to a row). Since the numbers of true values may vary among data items in practice, e.g., "*the birthday of Tom Hanks*" contains only one date, but "*the cast of Harry Potter*" includes a team of actors, the truth discovery problem can be classified into two categories: (i) if we make the single-truth assumption by treating the values in each cell (claim) of the matrix as a joint single value, we have the *single-truth discovery problem*; and (ii) if we relax the assumption by treating each distinct value individually, meaning either each cell or the truths may involve several values, we have the *multi-truth discovery problem*. LTM [19] and MBM [15] are the only two methods that are applicable for multi-truth discovery, while all the rest belongs to single-truth discovery methods.

The input of the ensemble truth discovery problem can be formulated as adding a third dimension to the aforementioned data matrix, resulting in a cube (see Fig. 1(b)). The third dimension represents different truth discovery methods, which is denoted as $\mathbf{M} = \{M_1, M_2, \ldots, M_l\}$. Each cell of the cube contains values and their corresponding labels (true or false) provided by the corresponding method. For the single-truth discovery methods, they provide the same label to the value(s) in the same cell, while the multi-truth discovery methods label the value(s) individually. As the methods may have differed performance given a specific application scenario, their results may be conflicting and of varied quality. We formally define the ensemble truth discovery problem as follows:

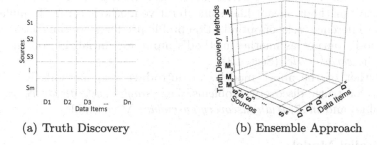

(a) Truth Discovery   (b) Ensemble Approach

Fig. 1. Input dimension comparison of the original and ensemble truth discovery

Ensemble Truth Discovery Problem. Given a 3-dimensional matrix (or cube), **L** truth discovery methods provide boolean labels on values claimed by **M** sources on **N** data items, the objective is to predict the truth of the **N** data items, while estimating the quality of different methods and sources.

4   Ensemble Approaches

4.1   Feasibility Analysis

Berti-Equille implements four approaches including *Simple Bayesian Ensemble* (SBE) [3], *Majority Voting* (MVE), *Uniform Weight* (UWE) and *Adjust Weight* (AWE) ensembles for combining twelve single-truth discovery methods. These approaches are straightforward, which simply unify the outputs of existing methods to the format of a triple {data item, true value(s), veracity score} and combine them directly. Although they are applicable for most of the existing methods, they neglect the useful intermediate results, such as source reliability obtained by the truth discovery methods, thus resulting in limited performance. Moreover, as one of the twelve combined methods, LTM is a special method which incorporates the enriched meaning of source reliability and can tackle multi-truth discovery problem. Naively combining LTM with other single-truth discovery methods and neglecting the two categories of truth discovery problems may further deteriorate the effectiveness of ensemble approaches. In this section, we analyze the feasibility of the ensemble approach and present the possible ways of ensembling the existing methods as follows.

Parallel Model. Although the output formats of existing truth discovery methods vary from one another, they can be transformed into the same format. Therefore, a possible way to ensemble the existing methods is to combine their outputs in a different manner, i.e., *parallel model* (to be detailed in Sect. 4.2).

Serial Model. As aforementioned, the existing methods realize truth discovery following the same general principle. Despite their different ways of implementations, they are generally mutually convertible in their ways of implementations. In particular, both the parameter inference in probabilistic graphical model based methods and the coordinate descent in optimization based methods require updating rules iteratively, which show their potential to be converted into iterative methods; meanwhile, some iterative methods can be formulated as parameter inference tasks or optimization problems. Thus, we can consider using one method's output as another method's input for initializing on the priors, forming the *serial model* (Sect. 4.3).

For either of the above models, we introduce two methods for the two categories of truth discovery problems, i.e., *single-truth discovery ensemble* (S-ensemble) and *multi-truth discovery ensemble* (M-ensemble).

4.2   Parallel Model

The parallel model unifies the format of and combines their outputs to ensemble existing methods. The ensemble truth discovery problem differs from the traditional truth discovery problem in that it takes 3-dimensional rather than 2-dimensional matrix data as inputs. To realize the parallel ensemble model, we first reduce the dimension of the ensemble problem by regarding each distinctive (*Source, Method*) pair as a virtual data source. Therefore, a value associated with a large number of (*Source, Method*) pairs indicates that it is either supported by many sources or predicted as truth by various truth discovery methods. As each method only provides Boolean values to the values provided by sources, we can further remove the values labeled as false to reduce the solution space. After such reduction, the ensemble problem is converted into a traditional truth discovery problem and can be handled using existing methods.

Parallel S-ensemble. This approach first runs all the existing methods and formulates their outputs into a 3-dimensional matrix. Then, it trims the matrix by applying the above-mentioned reduction operations. Finally, it applies one of the existing truth discovery methods on this trimmed matrix to deliver the final results. We call these parallel S-ensemble methods "*PS-Method*" (e.g., PS-Accu). Specially, though there is no copying relation among the original methods, there might be complex latent relations among the sources. In such cases, the source dependence-aware methods, e.g., AccuCopy, are applicable for implementing the ensemble. This is another difference between our work and UWE/AWE, as they simply ensemble the outputs of the methods, and consider the methods to be combined as virtual sources without considering data sources. Thus, they neglect the copying relations among sources.

Parallel M-ensemble. This approach first revises the existing methods under the single-truth assumption so that they can be applied to the multi-truth discovery scenario[2]. In particular, it treats the values in each cell of the matrix individually, and run the original methods to output source reliability. Then, it counts the number of values provided by each source on each data item, and calculates the truth probability of each number as follows:

$$P_{D_i}^*(n) = |s_{D_i}| \sqrt{\prod_{n_s=n, s \in S_{D_i}} A(s) \prod_{n_s \neq n, s \in S_{D_i}} (1 - A(s))} \qquad (1)$$

where $P_{D_i}^*(n)$ is the unnormalized probability[3] of truth number n of data item D_i, S_{D_i} is the set of sources which provides values on D_i, n_s is the number provided by source s, and $A(s)$ is the reliability of s. For each data item, it chooses the number with the biggest probability as the number of true values (denoted as N) and output the top-N values instead of choosing the value with the biggest confidence score as the outputs. It revises, if necessary, and runs all the truth discovery methods, formulates and trims their outputs as a 3-dimensional matrix. Finally, both the existing multi-truth discovery methods (LTM or MBM) and the revised single-truth discovery methods can be applied to this matrix to address the ensemble problem. We call these parallel M-ensemble methods "*PM-Method*" (e.g., PM-Accu).

---

[2] Hereafter we call the revised methods the modified single-truth discovery methods.

[3] Such values are then normalized to represent probabilities.

4.3   Serial Model

As an alternative, we can sequentially combine the existing methods, i.e., using one method's outputs as another method's a priori inputs to implement the ensemble approach leading to the serial ensemble model. Here, we simply omit the consideration of the impact of different orders of the single-truth discovery methods on the performance of the ensemble approach, but leave further research on this issue to our future work.

Most existing methods initialize source reliability by assigning uniform weights among the sources. There are some potential disadvantages of the uniform initialization: firstly, with uniform initialization of source reliability, the performance of methods may rely on the majority. This strategy works well for the case that the majority of sources are good. However, the real scenarios usually are not the case, as sources may copy from each other or provide out-of-date information. Moreover, when we apply truth discovery on challenging tasks, such as information extraction and knowledge graph construction, most of the sources are unreliable. For example [18] describes that in their task that "62 % of the true responses are produced only by one or two of the 18 systems (sources)"; secondly, for the scenario where tie cases (i.e., each source claims a unique value on a data item) exists, the results of the methods using uniform initialization are generally unrepeatable. This is because, for the tie cases, the methods would perform voting or averaging like operations and choose a random value as the truth at the beginning of the iteration, leading to randomized source reliability estimation. In contrast, "knowing the precise trustworthiness of sources can fix nearly half of the mistakes in the best fusion results" [10]. Both the above observations motivate us to ensemble existing methods based on a serial ensemble model, which utilizes the source reliability predicted by one method as the prior for initializing another method.

Serial S-ensemble. The sequence of combining the existing methods is a permutation problem. In this paper, we randomly choose the methods one by one, and use the source reliability predicted by a method to initialize its direct successor method. We call the serial S-ensemble methods "*SS-#*" (e.g., SS-3).

Serial M-ensemble. We adapt the single methods, when necessary, by using the same operations designed for parallel M-ensemble. Then, we run the revised methods in the same order as applied for serial S-ensemble. Similarly, we call the serial M-ensemble methods "*SM-#*" (e.g., SM-3).

5   Experiments

5.1   Experimental Setup

We compared our approaches with three groups of truth discovery methods.

Original Single-Truth Discovery Methods (STD). We chose five typical and competitive algorithms from this category for the comparison. Note that Sums was revised by incorporating complementary vote.

– *Voting.* For each item, it predicts the most frequently provided claim as the estimated truth(s) without iteration.
– *Sums, Avg-Log, TruthFinder, 2-Estimates.* All these methods iteratively evaluate source reliability and claims alternately from each other using different calculation methods.

Multi-Truth Discovery Methods (MTD). There are two existing multi-truth discovery methods:

– *LTM.* Based on a probabilistic graphical model, it recognizes a value as true if its veracity score exceeds 0.5.
– *MBM.* This method incorporates a new mutual exclusion definition for multi-truth discovery from the reformatted claims.

Modified Single-Truth Discovery Methods (MMTD). We adapted four representative single-truth discovery methods for the multi-truth scenario by applying the operations described in Sect. 4.2, resulting in four new methods, namely *Voting**, *Sums**, *Average-Log**, *TruthFinder**, and *2-Estimates**.

Based on the above representative methods, we derived methods following our ensemble approaches as follows:

– *Parallel S-Ensemble Group.* It contains five methods, i.e., *PS-Voting, PS-Sums, PS-AvgLog, PS-TruthFinder*, and *PS-Estimates*.
– *Parallel M-Ensemble Group.* It consists of seven methods, i.e., *PM-LTM, PM-MBM, PM-Voting**, *PM-Sums**, *PM-AvgLog**, *PM-TruthFinder**, and *PM-2Estimates**.
– *Serial S-Ensemble Group.* As Voting does not consider source reliability, we combined the other four single-truth discovery methods and implemented *SS-4*. We combined the four methods in the following order: Sums, Avg-Log, TruthFinder, and 2-Estimates[4], and compared *SS-1* through *SS-4* by gradually adding one method each time in Sect. 5.4.
– *Serial M-Ensemble Group.* We combined six methods in the following order: Sum*, Avg-Log*, TruthFinder*, 2-Estimates*, LTM, and MBM, to implement *SM-6*. We chose this order for the same reason as SS-4). We compared *SM-1* through *SM-6* in Sect. 5.4.

We implemented all the above methods in Java 7 and ran experiments on 3 PCs with Intel Core i7-5600 processor (3.20 GH × 8) and 16 GB RAM. The methods were evaluated in terms of three metrics, including *precision*, which is the average percentage of the true positives returned by the methods in the set of all predicted true values on all values of all data items, *recall*, which is the average percentage of the true positives returned by the methods in the set of ground truths on all values of all data items, and F_1 *score*, which is the harmonic mean of precision and recall, from which we can see the comprehensive performance of all the compared methods.

---

[4] We chose this order because it is the increasing order of precision of these four methods performed on three real-world datasets in [15].

5.2 Experiments on Real-World Datasets

In this section, we present the evaluation of our ensemble approaches with respect to the existing methods on three real-world datasets (namely *Book dataset* [17], *Biography dataset* [12], and *Movie dataset* [15], described in Table 1), where we have removed the duplicated and invalid records to clean the original datasets.

Table 2 shows the evaluation results. For each single method group (i.e., single-truth discovery method group and multi-truth discovery method group, including the modified single-truth methods), no methods consistently outperformed the others on all the real-world datasets, which is consistent with the previous survey studies [11]. Among those single methods, Voting almost always achieved the best precision. As the data items in all the three real-world datasets

Table 1. Characteristics of three real-world datasets

Book dataset	Biography dataset	Movie dataset
# sources (Websites): 649	# sources (users): 55,259	sources (Websites): 16
# claims: 13,659	# claims: 227,584	# claims: 33,194
attribute: author names	attribute: children	attribute: director names
# objects (books): 664	# objects (person): 2,579	# objects (movies): 6,402
ground truths count (GT):	ground truths count (GT):	ground truths count (GT):
86 books (12.95 %)	2,578 person (99.9 %)	200 movies(3.12 %)
Avg. Coverage per source: 0.0317	Avg. Coverage per source:0.0016	Avg. Coverage per source: 0.0625
Avg. # distinct values per data item	Avg. # distinct values per data item	Avg. # distinct values per data item
(conf): 3.2	(conf): 2.45	(conf): 1.2
Avg. # claims per source: 21.05	Avg. # claims per source: 4.12	Avg. # claims per source: 2074.62

Table 2. Method comparison on real-world datasets and synthetic datasets (The best performance values in each method group are in bold. We consider multi-truth discovery methods and modified single-truth methods as one group. The best performance values among our ensemble approaches are highlighted in the gray background).

Group	Method	Book			Biography			Movie			Syn.(R)	Syn.(80P)
		Prec.	Recall	F₁	Prec.	Recall	F₁	Prec.	Recall	F₁	Corr. Rate	Corr. Rate
STD	Voting	**0.837**	0.328	0.471	0.876	0.855	0.865	**0.91**	0.292	0.442	**0.321**	0.581
	Sums	**0.837**	0.54	0.656	0.859	0.881	0.87	0.847	0.591	0.696	0.319	0.623
	AvgL.	0.826	0.605	0.698	0.904	0.886	0.895	0.847	0.643	0.731	0.317	0.58
	TruthF.	**0.837**	0.605	0.702	0.905	0.886	0.895	0.847	**0.71**	**0.772**	**0.32**	0.62
	Est.	**0.837**	**0.621**	**0.713**	**0.908**	**0.888**	**0.898**	0.863	0.692	0.768	0.319	**0.626**
MTD	LTM	0.826	0.651	0.728	0.91	0.88	0.895	0.812	0.813	0.812	0.225	0.223
	MBM	0.826	**0.744**	**0.783**	**0.915**	**0.89**	**0.902**	**0.852**	**0.833**	**0.842**	**0.32**	**0.533**
MMTD	Voting*	0.756	0.638	0.692	0.873	0.851	0.862	0.864	0.523	0.652	0.318	0.586
	Sums*	0.826	0.644	0.724	0.905	0.887	0.896	0.81	0.534	0.644	0.319	0.623
	AvgL*	0.663	0.709	0.685	0.88	**0.89**	0.885	0.812	0.65	0.722	0.317	0.58
	TruthF.*	0.698	0.709	0.703	0.876	0.88	0.878	0.853	0.723	0.783	**0.32**	0.623
	Est.*	**0.826**	**0.734**	**0.777**	0.89	0.88	0.885	**0.865**	0.722	0.787	0.319	**0.626**
PS-ens.	PS-Voting	**0.837**	0.63	0.719	**0.905**	**0.886**	**0.895**	0.915	0.75	0.824	**0.323**	**0.632**
	PS-Sums	**0.837**	0.64	**0.725**	**0.905**	**0.886**	**0.895**	0.92	0.78	0.844	0.322	0.631
	PS-AvgL.	**0.837**	0.638	0.724	**0.905**	**0.886**	**0.895**	0.92	0.78	0.844	0.322	**0.632**
	PS-TruthF.	**0.837**	0.64	**0.725**	**0.905**	**0.886**	**0.895**	**0.927**	0.792	0.854	0.322	0.631
	PS-Est.	**0.837**	0.64	**0.725**	**0.905**	**0.886**	**0.895**	0.925	**0.816**	**0.867**	0.322	0.631
PM-ens.	PM-Voting*	**0.86**	0.754	0.804	0.91	0.9	0.905	0.899	0.821	0.858	0.321	0.627
	PM-Sums*	0.827	0.751	0.787	0.91	0.89	0.9	0.883	0.833	0.857	0.32	0.627
	PM-AvgLog*	0.829	0.763	0.795	0.915	0.897	0.906	0.886	0.833	0.859	**0.325**	0.623
	PM-TruthF.*	0.834	0.791	0.812	0.91	0.9	0.905	0.886	0.854	0.87	0.322	0.626
	PM-Est.*	0.842	0.766	0.802	0.92	0.89	0.905	0.904	0.846	0.874	0.32	0.626
	PM-LTM	0.837	0.808	0.822	0.93	0.91	0.92	0.91	**0.86**	0.884	0.322	0.623
	PM-MBM	**0.86**	**0.812**	**0.836**	**0.93**	**0.92**	**0.925**	**0.922**	0.85	**0.885**	0.32	**0.628**
SS-ens.	SS-4	0.837	0.721	0.775	0.91	0.9	0.905	0.87	0.753	0.807	**0.325**	0.628
SM-ens.	SM-6	0.836	0.764	0.798	0.93	0.92	0.925	0.913	0.866	0.889	0.321	0.563

involve multiple true values, LTM and MBM generally achieved better performance than the original single-truth discovery methods, esp. in recall and F_1 score. The modified single-truth discovery methods also achieved relatively higher precision and recall than their original methods. The original single-truth discovery methods showed higher precision but achieved lower recall than multi-truth discovery methods. This indicates that the original single-truth discovery methods tend to underestimate the number of true values.

Both our parallel ensemble methods, i.e., PM and PS, returned better results than the element methods. The serial ensemble methods, i.e., SS-4 and SM-6, also showed relatively better performance. In particular, both PM and SM-6 (resp., PS and SS-4) outperformed the original multiple (resp., single) truth discovery methods they combined in terms of precision, recall and F_1 score on all the three real-world datasets. In our experiments, five single-truth discovery methods are combined for PS and seven multi-truth discovery methods are combined for PM. The obtained 3-dimensional matrices are not significantly different from each other, which resulted in the outcome that all PM and PS methods show similar performance. Due to the existence of multiple true values in the datasets, PM and SM-6 methods performed better than PS and SS-4 methods. However, neither the SM-6 nor the PM methods could consistently dominate the other, and the results are different among different datasets. Similar situations occurred when we compared SS-4 with PS. Further performance studies of SS and SM will be presented in Sect. 5.4.

5.3 Experiments on Synthetic Datasets

Due to the limited ground truths of real-world datasets, the performance evaluation may be biased by the available ground truth. In this section, we present the comparison of our approaches with the element methods on synthetic datasets with a wide spectrum of distribution settings and complete ground truths. We first generated synthetic datasets by applying the dataset generator proposed by Waguih et al. [14]. This generator contains six parameters that can be configured to simulate a wide spectrum of truth discovery scenarios. Three parameters, namely the number of sources (M), the number of data items (N), and the number of distinct values per data item (V), determine the scale of the generated dataset, while the other three parameters, source coverage (cov), ground truth distribution per source (GT), and distinct value distribution per data item ($conf$), determine the characteristics of the generated dataset.

We fixed the scale parameters by setting $M = 50$, $N = 1,000$, and $V = 20$, configured both cov and $conf$ to follow exponential distributions. In particular, we chose two distributions (i.e., the random[5] and 80-pessimistic[6] distributions) for GT. We chose these distributions as they are closest to the real world scenarios. Specifically, for the exponential distribution of $conf$, the majority of data

---

[5] Random ground truth distribution per source means the number of true positive claims per source is random.

[6] 80-pessimistic ground truth distribution per source means 80 % of the sources provide 20 % true positive claims, while 20 % of the sources provide 80 % true positive claims.

items have few distinct values while few data items have many conflicts. For the case of exponential source coverage, most sources claim values for few data items whereas few sources cover the majority of data items. When we face with the challenging task of information extraction and knowledge base construction, the majority of sources are always error-prone, and truths are maintained by the minority. Therefore, random and 80-pessimistic *GT* distributions are more representative. Based on the above configurations, we obtained two types of synthetic datasets, namely *Synthetic(R)* and *Synthetic(80P)*, each containing 10 datasets. The metrics of each method were measured as the average of 10 executions over the 10 datasets included by the same dataset type.

Table 2 shows the performance comparison of different methods on the synthetic datasets. As each data item in the synthetic datasets has only one single true value, every method predicted values for all the data items. In this case, we specially measure the methods in terms of *correct rate* by computing the percentage of matched values between each method's output and ground truths. Specifically, the experimental results show almost the same pattern with those on the real-world datasets, which confirms that the ensemble approaches indeed lead to more accurate truth discovery. As sources in Synthetic(R) claim random numbers of true positive values, all methods returned low-quality results for this dataset with correct rate kept around 0.32. Our ensemble methods only showed slightly better performance. The multi-truth discovery methods, especially LTM, failed to return good results on both datasets, where each data item has only one single true value. This is also the reason why SM-6 and PM methods performed worse than SS-4 and PS.

5.4   Impact of Method Numbers on Serial Ensemble Model

To analyze the impact of the number of methods (which are used to derive the ensemble approaches) on the two serial ensemble models (i.e., SS and SM), we conducted experiments on all the above datasets. In particular, we studied the performance of the serial ensemble methods by gradually adding one method each time. We combined the existing methods in the same order as described in Sect. 4.3, where SS-1 is the same as Sums, the source reliability output by Sums was used as the input of AverageLog to realize SS-2. Following a similar way, we further added TruthFinder and 2-Estimates to implement SS-3 and SS-4. Similarly, we gradually combined Sums*, AverageLog*, TruthFinder*, 2-Estimates*, LTM, and MBM to form SM-1 through SM-6. Through the above procedures, we finally obtained four SS methods (from SS-1 to SS-4) and six SM methods (from SM-1 to SM-6).

Figure 2 shows the performance of SS, SM, and the applied existing methods. In particular, the precision, recall and F_1 score of SS and SM fluctuated on all the real-world datasets, and the correct rate of them fluctuated on all the synthetic datasets, while we gradually combined more methods. Each serial ensemble method outperformed the last combined method except the special case of SS-1 (exactly Sums) and SM-1 (exactly Sums*), where the two methods are the same. This indicates that naively and serially combining more methods

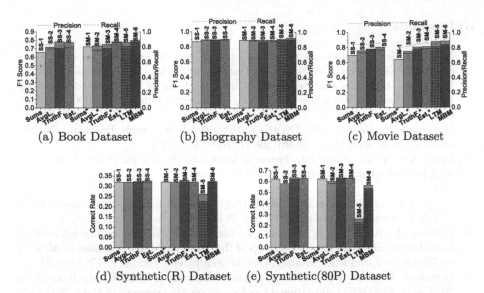

(a) Book Dataset   (b) Biography Dataset   (c) Movie Dataset

(d) Synthetic(R) Dataset   (e) Synthetic(80P) Dataset

Fig. 2. Impact of combining different numbers of single methods on SS and SM. The offsets on the precision and recall lines are the corresponding precision and recall of the corresponding SS and SM methods, while the upper bounds of the stack columns are the corresponding F_1 score of the corresponding SS and SM methods.

does not necessarily improve the effectiveness of the serial ensemble methods in a proportional manner. However, the accuracy of a single-truth discovery method could be improved by using the source reliability predicted by other methods as inputs. This indicates parallel ensemble model is generally better than serial ensemble model in obtaining the best ensemble performance.

6   Conclusion

In this paper, we focus on the problem of ensembling the existing truth discovery methods for more robust and consistent truth discovery. Several surveys have shown that a "one-fits-all" truth discovery method is not achievable due to the limitations of the existing methods. Therefore, combing various competing methods could be an effective alternative for conducting high-quality truth discovery. Given very few research efforts have been conducted on this issue, we analyze the feasibility of such an ensemble approach. We propose two novel models, namely *serial model* and *parallel model*, for combining the truth discovery methods. We further present several implementations based on the above models for both single-truth and multi-truth discovery problems. Extensive experiments over three real-world datasets and various synthetic datasets demonstrate the effectiveness of our ensemble approaches.

References

1. Berti-Equille, L.: Data veracity estimation with ensembling truth discovery methods. In: IEEE Big Data Workshop on Data Quality Issues in Big Data (2015)
2. Dietterich, T.G.: Ensemble methods in machine learning. In: Proceedings of the First International Workshop on Multiple Classifier Systems (MCS 2000), Cagliari, Italy (2000)
3. Domingos, P., Pazzani, M.: On the optimality of the simple Bayesian classifier under zero-one loss. Mach. Learn. **29**(2), 103–130 (1997)
4. Dong, X.L., et al.: From data fusion to knowledge fusion. In: Proceedings of the 40th International Conference on Very Large Data Bases (VLDB 2014), Hangzhou, China (2014)
5. Dong, X.L., et al.: Integrating conflicting data: the role of source dependence. VLDB Endowment (PVLDB) **2**(1), 550–561 (2009)
6. Galland, A., Abiteboul, S., Marian, A., Senellart, P.: Corroborating information from disagreeing views. In: Proceedings of the Third ACM International Conference on Web Search and Data Mining (WSDM 2010), New York, NY, USA (2010)
7. Goasdoué, F., et al.: Fact checking and analyzing the web. In: Proceedings of the 2013 ACM SIGMOD International Conference on Management of Data (SIGMOD 2013), New York, NY, USA (2013)
8. Li, Q., et al.: Resolving conflicts in heterogeneous data by truth discovery and source reliability estimation. In: Proceedings of the 2014 ACM SIGMOD International Conference on Management of Data (SIGMOD 2014), Snowbird, Utah, USA (2014)
9. Li, Q., et al.: A confidence-aware approach for truth discovery on long-tail data. VLDB Endowment (PVLDB) **8**(4), 425–436 (2015)
10. Li, X., et al.: Truth finding on the deep web: is the problem solved? VLDB Endowment (PVLDB) **6**(2), 97–108 (2013)
11. Li, Y., Gao, J., Meng, C., Li, Q., Su, L., Zhao, B., Fan, W., Han, J.: A survey on truth discovery. ACM SIGKDD Explor. Newsl. (2016)
12. Pasternack, J., Roth, D.: Knowing what to believe (when you already know something). In: Proceedings of the 23rd International Conference on Computational Linguistics (COLING 2010), Stroudsburg, PA, USA (2010)
13. Pasternack, J., Roth, D.: Latent credibility analysis. In: Proceedings of the 22nd International World Wide Web Conference (WWW 2013), Rio de Janeiro, Brazil (2013)
14. Waguih, D.A., Berti-Equille, L.: Truth discovery algorithms: an experimental evaluation. CoRR abs/1409.6428 (2014)
15. Wang, X., et al.: An integrated Bayesian approach for effective multi-truth discovery. In: Proceedings of the 24th ACM International Conference on Information and Knowledge Management (CIKM 2015), Melbourne, Australia (2015)
16. Yin, X., Tan, W.: Semi-supervised truth discovery. In: Proceedings of the 20th International World Wide Web Conference (WWW 2011), Hyderabad, India (2011)
17. Yin, X., et al.: Truth discovery with multiple conflicting information providers on the web. In: Proceedings of the 13th ACM SIGKDD International Conference on Knowledge Discovery and Data Mining (KDD 2007), San Jose, California, USA (2007)

18. Yu, D., et al.: The wisdom of minority: unsupervised slot filling validation based on multi-dimensional truth-finding. In: Proceedings of the International Conference on Computational Linguistics (COLING 2014), Dublin, Ireland (2014)
19. Zhao, B., et al.: A Bayesian approach to discovering truth from conflicting sources for data integration. VLDB Endowment (PVLDB) 5(6), 550–561 (2012)
20. Zhao, B., Han, J.: A probabilistic model for estimating real-valued truth from conflicting sources. In: Proceedings of 10th International Workshop on Quality in Databases (QDB 2012), Instanbul, Turkey (2012)

Single Classifier Selection for Ensemble Learning

Guangtao Wang, Xiaomei Yang$^{(\boxtimes)}$, and Xiaoyan Zhu

Deparment of Computer Science and Technology,
Xi'an Jiaotong University, Xi'an 710049, China
yangxiaomei@stu.xjtu.edu.cn

Abstract. Ensemble classification is one of representative learning techniques in the field of machine learning, which combines a set of single classifiers together aiming at achieving better classification performance. Not every arbitrary set of single classifiers can obtain a good ensemble classifier. The efficient and necessary condition to construct an accurate ensemble classifier is that the single classifiers should be accurate and diverse. In this paper, we first formally give the definitions of accurate and diverse classifiers and put forward metrics to quantify the accuracy and diversity of the single classifiers; afterwards, we propose a novel parameter-free method to pick up a set of accurate and diverse single classifiers for ensemble. The experimental results on real world data sets show the effectiveness of the proposed method which could improve the performance of the representative ensemble classifier Bagging.

Keywords: Classifiication · Ensemble learning · Accurate and diverse classifiers

1 Introduction

Classification has been one of the research hotspots for years in the field of machine learning, and a number of classification algorithms have been proposed. The ensemble classifier, which integrates a set of single classifiers in a specific way (e.g., weighted or unweighted voting) to make a prediction for a new instance, has been demonstrated to be often much more accurate than the single classifiers [5,7,8]. This is attributed to that the ensemble classifier can partly overcome the following three issues encountering by the single classifier: the statistical, computational and representational problems [5,7]. There have been different kinds of ensemble classification methods, such as Bagging [2], Boost [10] and RandomForest [13], etc. And the experimental results showed the advantages of these ensemble classifiers in terms of classification accuracy. However, this does not mean that the combination of any set of single classifiers will achieve better classification performance. For example, let A_E be an ensemble classifier constructed over three single classifiers A_1, A_2 and A_3, if both of A_1 and A_2 have poor generalization ability and always make same predictions for any new instance, by voting the predicted results of these three single classifiers as the prediction of A_E, the performance of A_E depends on either A_1 or A_2 and so

© Springer International Publishing AG 2016
J. Li et al. (Eds.): ADMA 2016, LNAI 10086, pp. 312–328, 2016.
DOI: 10.1007/978-3-319-49586-6_21

will be poor. Researchers have done a lot of work on how to guarantee the performance of the ensemble classifier [5,7,12]. Based on these research works, there is a corollary to guide the construction of an accurate ensemble classifier.

Corollary 1. *The efficient and necessary condition to construct an accurate ensemble classifier is that the single (or base) classifiers are accurate and diverse/independent of each other.*[1]

Corollary 1 tells us the key point to construct an accurate ensemble classifier is to find a set of accurate and diverse single classifiers. [6] compared three well-known decision trees based ensemble classifiers with respect to Boost, Bagging and Randforest in the viewpoints of this corollary, and stated that the diversity among the single classifiers plays a critical role in constructing a better ensemble classifier. And the accuracy requirement of the single classifier is moderate, i.e., the prediction of the single classifier being correct is better than random. This phenomenon has also been recognized in [4,17] about classifier combination. And the diverse ensemble classifier has a better potential to improve the accuracy than non-diverse ensemble classifier [3,15,19]. Therefore, there will be a notable question: How to define or evaluate the independence/diversity among the single classifiers for ensemble?

In the field of ensemble learning, it is usually difficult to evaluate the diversity between two classifiers directly. Researchers usually resort to measuring the diversity based on the prediction/classification results of the classifiers on a given test data. The computation of the diversity depends on the expression of the prediction results of the classifiers [15].

1. A numeric vector which records the predicted posterior probabilities of all class labels. e.g., for a classification problem with k class labels, the prediction results are a vector with k probability values [20,22].
2. Class label which directly indicates the predicted result [5,6,14].
3. Correct/incorrect decision which records whether the predicted label is correct or not [11,13,16].

For the numeric vector based expression, one of the assumptions is that a classifier outputs independent estimates of the posterior probabilities. However, this is usually not the case since all these posterior probabilities sum up to a constant 1. Moreover, not all the classifiers can directly output the posterior probabilities of the class labels. Thus, the researchers usually define the diversity function ψ in terms of either class labels or correct/incorrect decisions [5,6,11,13–16]. There have been several metrics proposed based on either the class labels or correct/incorrect decisions to assess the diversity between different models [4,6,15,18]. These metrics can guide us to identify the diverse single classifiers. Yet, Kuncheva et al. [15] have stated that, in order to guarantee the improvement over the performance of single classifiers, there exists a minimum threshold value for each of these diversity metrics to pick up the diverse single classifiers for ensemble.

---

[1] In the field of ensemble learning, the independent single classifier is generally called diverse one, and the independence is also named diversity.

Unfortunately, the existing researches usually just supply the function to evaluate the diversity between different classifiers, but do not give any effective approach to preassign the threshold. Furthermore, the threshold will vary with different classification problems being solved [4,6,15,18]. In this paper, we present a statistical method which can not only quantize the diversity between different classifiers but also can set the threshold adaptively. Different from the existing diversity evaluation functions acting on the prediction results in terms of either the class labels or the correct/incorrect decisions, the proposed diversity evaluation method concerns both of them, which will make full use of prediction results. Afterwards, with the help of the proposed diversity evaluation function, we further propose a method for picking up accurate and diverse single classifiers for ensemble. The experimental results on real world classification data sets demonstrate the effectiveness of the proposed method.

The rest of the paper is organized as follows. Section 2 gives the formal definitions of accurate and diverse classifiers, and proposes a new metric to evaluate the diversity between different classifiers. Section 3 introduces the single classifier selection method. Section 4 conducts the experimental study of the proposed method and Sect. 5 concludes our work.

2   Definitions of Accurate and Diverse Classifiers for Ensemble

2.1   Accurate Classifier for Ensemble

let A_E be an ensemble classifier constructed with n single classifiers $\{A_1, A_2, \cdots, A_n\}$, and $pr_{A_i} (1 \leq i \leq n)$ be the probability of A_i to make an error prediction on a new coming instance. Then, an accurate classifier for ensemble can be defined as follow.

Definition 1. Accurate classifier *A single classifier $A_i (1 \leq i \leq n)$ is accurate if and only if $pr_{A_i} < 1/2$.*

Definition 1 tells us that a single classifier is accurate if and only if its probability to make an error prediction is less than $1/2$. The rationality of this definition can be demonstrated as follows.

- First, we construct another ensemble classifier $\hat{A}_E$ over n other single classifiers $\{\hat{A}_1, \hat{A}_2, \cdots, \hat{A}_n\}$. Where each single classifier $\hat{A}_i$ $(1 \leq i \leq n)$ has the identical probability to make an error prediction. And this probability is defined as the maximum value of $\{pr_{A_i} : 1 \leq i \leq n\}$. i.e., $pr_{\hat{A}_i} = \max\{pr_{A_1}, pr_{A_2}, \cdots, pr_{A_n}\} (1 \leq i \leq n)$.
- Then, by voting the predictions of single classifiers, we can get that $pr_{A_E} \leq pr_{\hat{A}_E}$ since A_E is constructed over a set of single classifiers with lower possibility to make an error prediction. Suppose that the single classifiers of the ensemble classifier (A_E or $\hat{A}_E$) are independent of each other, we can get a discrete random variable X following binomial distribution $(n, pr_{\hat{A}_i})$, Where

X indicates the number of single classifiers in $\{\hat{A}_1, \hat{A}_2, \cdots, \hat{A}_n\}$ which make an error prediction. According to binomial distribution, we can get:

$$pr_{A_E} \leq pr_{\hat{A}_E} = Pr(X \geq \lceil n/2 \rceil) = \sum_{i=\lceil n/2 \rceil}^{n} \binom{n}{i} (pr_{\hat{A}_i})^i (1 - pr_{\hat{A}_i})^{n-i}; \quad (1)$$

Where $\lceil n/2 \rceil$ denotes the smallest integer being greater than (or equal to) $n/2$.

- Finally, according to Chernoff's inequality [1], for binomial distribution (n, p),

$$pr_{A_E} \leq Pr(X \geq k) \leq exp(-n \cdot D(\frac{k}{n} \parallel p)), \ if \ and \ only \ if \ p < \frac{k}{n} < 1; \quad (2)$$

Where $D(a \parallel p)$ is the relative entropy between two Bernoulli distributions with parameters a and p, and define as $D(a \parallel p) = a \cdot log\frac{a}{p} + (1 - a) \cdot log\frac{1-a}{1-p}$. Corresponding to the ensemble classifier $\hat{A}_E$, $p = pr_{\hat{A}_i}, k = \lceil n/2 \rceil$ and so $a = \frac{k}{n} \geq 1/2$. Therefore, if and only if $pr_{\hat{A}_i} < 1/2$, by voting the predictions of $\{\hat{A}_1, \hat{A}_2, \cdots, \hat{A}_n\}$ as the prediction of $\hat{A}_E$, Eq. 2 will be always true. And pr_{A_E} will be bounded by $exp(-n \cdot D(\frac{k}{n} \parallel P))$. Moreover, in the case of $p = pr_{\hat{A}_i} < 1/2$, with the increasing of n, the value of $exp(-n \cdot D(\frac{k}{n} \parallel P))$ will approach 0 since $D(\frac{k}{n}) > 0$. That is, the more single classifiers used, the smaller the probability of A_E to make an error prediction. This will be a very good property for ensemble classifier. In a word, all these conclusions will be true under $pr_{\hat{A}_i} = max\{pr_{A_1}, pr_{A_2}, \cdots, pr_{A_n}\} < 1/2$ So in order to get an accurate ensemble classifier A_E over $\{A_1, A_2, \cdots, A_n\}$, pr_{A_i} should be less than $1/2$.

In the analysis above, it is noteworthy that, without the independence/diversity among the base models $\{A_1, A_2, \cdots, A_n\}$, the random variable X in Eq. 1 will not follow binomial distribution and further the Eq. 2 might be false. This will result in that pr_{A_E} might be no-converging or converging too slowly. This implies that the diversity of the single classifiers plays a critical role in constructing an accurate ensemble classifier. Therefore, there will be a notable question: How to define or evaluate the independence/ diversity between the single classifiers?.

2.2 Diverse Classifier for Ensemble

As we know, in order to evaluate the diversity of different classifiers, researchers usually resort to the prediction results of the classifiers since it is quite difficult to evaluate the diversity of different classifiers directly. Suppose that A_i and A_j $(1 \leq i \neq j \leq n)$ are two different single classifier models, and R_i and R_j are the prediction/classification results of A_i and A_j on a given test data set D, we can give the definition of diverse classifier for ensemble as follow.

Definition 2. Diverse classifier *Two single classifiers A_i and A_j are diverse with each other if and only if $\psi(R_i, R_j) < \delta$, where ψ is a function which computes the diversity between A_i and A_j based on R_i and R_j, and δ is a given minimum threshold.*

According to Definition 2, two critical dimensions should be considered: one is to find a function ψ to evaluate the diversity between two classifiers, and the other is to set a proper threshold δ to pick up the diverse classifiers. However, the existing diversity evaluation functions usually just give the metric to quality the diversity between two classifiers, but do not supply the threshold to further identify whether two classifiers are diverse with each other. Moreover, the proper threshold setting will vary with different classification problems. And there is still no effective way to preassign the threshold.

In order to handle this problem, in this section, we propose a statistical method which can not only quantize the diversity of different classifiers, but also adaptively set the proper threshold δ. Moreover, different from the existing diversity evaluation functions acting on the prediction results in terms of either the class labels or the correct/incorrect decisions, the proposed diversity evaluation method concerns both of them, which will make full use of the prediction results.

Suppose there are two different classifiers A_1 and A_2, and a data set D with K class labels $\{C_1, C_2, \cdots, C_K\}$ $(K \geq 2)$ then we can construct a $K \times K$ contingency table (see Table 1) based on the class labels and correct/incorrect decisions predicted by A_1 and A_2 on D.

Table 1. The $K \times K$ contingency table

Classified label	C_1	C_2	$\cdots$	C_K	Total
C_1	$N_{1,1}$	$N_{1,2}$	$\cdots$	$N_{1,K}$	$N_{1,*}$
C_2	$N_{2,1}$	$N_{2,2}$	$\cdots$	$N_{2,K}$	$N_{2,*}$
$\vdots$	$\vdots$	$\vdots$	$\vdots$	$\vdots$	$\vdots$
C_K	$N_{K,1}$	$N_{K,2}$	$\cdots$	$N_{K,K}$	$N_{K,*}$
Total	$N_{*,1}$	$N_{*,2}$	$\cdots$	$N_{*,K}$	N

In Table 1, $N_{i,j}$ $(1 \leq i, j \leq K)$ is the number of test instances incorrectly classified as C_i and C_j by A_1 and A_2, respectively. $N_{i,*} = \sum_{j=1}^{K} N_{i,j}(1 \leq i \leq K)$, $N_{*,j} = \sum_{i=1}^{K} N_{i,j}(1 \leq j \leq K)$ and $N = \sum_{i=1}^{K} N_{i,*} = \sum_{j=1}^{k} N_{*,j} = \sum_{i=1}^{K} \sum_{j=1}^{k} N_{i,j}$. N is the total number of instances of D which are incorrectly predicted by either A_1 or A_2.

With this $K \times K$ contingency table, we can get the diverse measure κ by Eq. 3. The greater the value of $|\kappa|$, the lower the diversity between A_1 and A_2.

$$\kappa = \frac{\Theta_1 - \Theta_2}{1 - \Theta_2}; \tag{3}$$

Where $\Theta_1 = \sum_{i=1}^{K} \frac{N_{i,i}}{N}$ and $\Theta_2 = \sum_{i=1}^{K} \frac{N_{i,*}}{N} \times \frac{N_{*,i}}{N}$.

According to the contingency table analysis, the joint frequency distribution of C_i and C_j predicted by A_1 and A_2 is $\frac{N_{i,j}}{N}$, and $\frac{N_{i,*}}{N}$ and $\frac{N_{*,j}}{N}$ correspond to the marginal frequency distributions of C_i and C_j. The rationality of κ being able to evaluate how strong the independence between A_1 and A_2 is demonstrated as follows.

1. Suppose that A_1 and A_2 are independent of each other, the expected joint distribution of C_i and C_j would be $\frac{N_{i,*}}{N} \times \frac{N_{*,i}}{N}$. In Eq. 3, the numerator $\Theta_1 - \Theta_2$ can also be represented as $\sum_{i=1}^{K} (\frac{N_{i,j}}{N} - \frac{N_{i,*}}{N} \times \frac{N_{*,j}}{N})$. Therefore, in the case that A_1 is independent of A_2, and $\Theta_1 - \Theta_2$ will be quite close to 0 in practice.

2. If A_1 is positively related to A_2, an instance being predicted as $C_i (1 \leq i \leq K)$ by A_1 means that it is more likely that A_2 classifies the instance into C_i as well. This will increase the value of $\frac{N_{i,i}}{N}$, i.e., the elements on the main diagonal of the contingency table. Otherwise, if A_1 is negatively related to or independent with A_2, the instance will be predicted as different classes by these two classifiers. This will reduce the value of $\frac{N_{i,i}}{N}$. So we can get that the value of $N_{i,i}$ can reflect the dependence of two different learners. That is why we define the metric κ by the elements on the main diagonal of the contingency table.

3. The denominator $1 - \Theta_2$ of κ plays a role to limit the value of κ into the range $[-1, 1]$. If the classification results of A_1 and A_2 on the test data are always identical, $\sum_{i=1}^{K} N_{i,i} = N$, so $\Theta_1 = 1$ and κ achieves its maximum value 1. If A_1 and A_2 are independent of each other, κ will be 0 or quite near 0. And $\kappa < 0/ > 0$ means A_1 is negatively/positively related to A_2. The greater the value of $|\kappa|$, the stronger the dependence between A_1 and A_2.

In practical application, the metric κ is estimated via the prediction results of the classifiers on only a sample data set rather than the whole population. This might also be the reason why there needs a threshold δ in Definition 2. Therefore, we need to further understand the statistical significance of κ, including the statistical significance of $\kappa \neq 0$ and its confidence interval. And the confidence interval will be set as the threshold δ in Definition 2.

A statistic $Z = \kappa/\delta$ following standard normal distribution $N(0, 1)$ has been proposed to test whether the two classifiers are independent of each other, where δ is the standard error of κ. The detail of its calculation can be found in [9]. However, the statistic Z is so conservative that suffers from high Type I error. That is, the stastic Z has high probability on failing to detect the true independent classifiers [21]. Furthermore, the δ constructed based on such a conservative statistic Z will also be too relaxed. This could also be the reason why the test statistic Z is rarely reported.

For this purpose, we need to find a statistic with stronger test power than Z for significant test of κ. As we know, the distribution of the class labels predicted by a classifier on a K-class classification problem would follow either binomial $(K = 2)$ or multi-nominal $(K > 2)$ distribution. Both of binomial and multi-nominal distributions are derived from exponential family of distributions. Meanwhile, inspired by the idea that the independence between two variables, which follow the well-known exponential distribution (i.e., normal distribution), is usually statistically tested by a t-stastic, we attempt to employ a t-stastic in Eq. 4 to test the significance of $\kappa \neq 0$, and further determine whether two classifiers A_1 and A_2 are independent of each other or not.

$$t = \frac{\kappa}{\sqrt{\frac{1 - \kappa^2}{N - 2}}};\tag{4}$$

The t statistic follows the Student's t-distribution with freedom of degree $N - 2$ under the null hypothesis that A_1 and A_2 are independent of each other. If the t- statistic test accepts the null hypothesis under given significance level α, we can conclude that the measure κ has no significant difference with 0, i.e., A_1 and A_2 are independent of each other.

Next, we will demonstrate that the statistic t is a better choice than Z for significance test of $\kappa \neq 0$ by the simulation experiments.

1. As we know, the multi-class ($K > 2$) problem can be equivalently transformed into multiple binary-class ($K = 2$) problems. Thus, in our experiments, without loss of generality, we only focus on simulating the predictions of a binary-class classifier which follow binomial distribution $B(N, p)$, where N denote the size the test data D and p is the probability of a given class label predicted by the classifier. With this in mind, we can setup the simulation procedure by following steps.

Step 1 Simulating predictions of the independent classifiers A_1 and A_2 by two in dependent binomial distributions $B(N, p_1)$ and $B(N, p_2)$, where $0 < p_1, p_2 < 1$.

Step 2 Performing two-tail statistical tests on these predictions to detect whether A_1 and A_2 are independent of each other by t and Z statistics under given significant level $\alpha = 0.05$. Then recording whether these tests make wrong conclusions that A_1 and A_2 are not independent.

Step 3 Repeating Step (1) and (2) M times (M should be large enough) and calculating the error rate to make wrong conclusions by tests with t and Z statistics, respectively. Here, the error rate is calculated by M_x/M where M_x denotes the times of wrong conclusions making by x statistic and $x = t$ or Z.

2. Figure 1 shows comparison results of t and Z statistics in terms of error rate. (i) In Fig. 1(a), we set the number of simulations $M = 10000$, the size of test data $n = 1000$ and p_1 and p_2 are randomly picked up from $(0, 1)$ independently. From this figure, we can get that the error rate inducing by t statistic is much smaller than (generally half of) that inducing by Z statistic. (ii) In Fig. 1(b), we set the number of simulations $M = 10000$, the sizes of test data vary from $n = 20$ to 2000 with step $= 10$. From this figure, we can observe that, the Type I error induced by t statistic is always smaller than that induced by Z statistic under different sizes of test data, and also smaller than the given significance level α. This denotes the robustness of the statistic t on the size of test data.

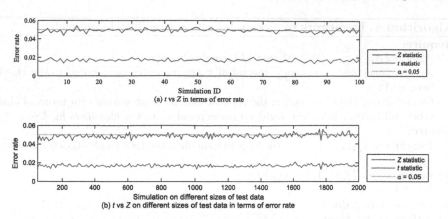

Fig. 1. Comparison of t and Z statistics in terms of error rate (Color figure online)

In summary, by the simulation experiments, we can get that t statistic in Eq. 4 works well on detecting the independence between different classifiers.

According to Eq. 4, its inverse can be calculated as follow

$$\kappa = \frac{t}{\sqrt{N - 2 + t^2}}; \tag{5}$$

where t statistic follows student distribution with degree of freedom $N - 2$.

Let t_c denote the critical value of student distribution with degree of freedom $N - 2$ under a given significance level α (e.g., $\alpha = 0.05$), then we can get the confidence interval of $\kappa = 0$ as $[-\frac{t_c}{\sqrt{N-2+t_c^2}}, \frac{t_c}{\sqrt{N-2+t_c^2}}]$. If κ calculated between two single classifiers falls into this interval, we can conclude that these two classifiers are statistically independent of each other under the given significance level α. Based on this confidence interval, we can set the minimum threshold $\delta = \frac{t_c}{\sqrt{N-2+t_{c2}}}$. And the diverse classifiers can be detected by comparing κ with δ directly. That is, two classifiers A_1 and A_2 are diverse with each other if and only if $|\kappa| < \delta$.

3   Picking Up Single Classifiers for Ensemble

According to Corollary 1, the necessary and sufficient condition to construct an accurate ensemble classifier is that the single classifiers are accurate and diverse. Thus, the core of ensemble classifier construction is a process to find out the accurate and diverse models from a given set of single classifiers $\{A_1, A_2, \cdots, A_k\}$. Algorithm 1 gives the single classifier picking up method over these k single classifiers.

Algorithm 1 consists of two filters: (i) accuracy based filter and (ii) diversity based filter. The former (lines 2–6) is used to find out the accurate single

320   G. Wang et al.

Algorithm 1. Framework to pick up the single classifiers for ensemble

Require:

$ASet = \{A_1, A_2, \cdots, A_k\}$: the set of k single classifiers;

$Accs = \{acc_1, acc_2, \cdots, acc_k\}$: the set of k accuracies corresponding to the classifiers in $ASet$;

$Outs = \{out_1, out_2, \cdots, out_k\}$: the set of k classification outputs (in terms of class label and correct/incorrect decision) corresponding to the classifiers in $ASet$;

Ensure:

$FlagList = \{f_1, f_2, \cdots, f_k\}$: the flags to indicate whether a single classifier is picked up;

1: Initialization: $\{f_i \leftarrow 1, 1 \leq i \leq k\}$;

{Part 1: accuracy based filter}

2: **for** $i \leftarrow 1$ to k **do**

3:     **if** $acc_i < 0.5$ **then**

4:         $f_i \leftarrow 0$

5:     **end if**

6: **end for**

{Part 2: diversity based filter}

7: $IX = $ sort($Accs$, "descend")$\{IX$ is the indices of the single classifiers sorted by $Accs$ in a descending order}

8: **for** $i \leftarrow 1$ to $k-1$ **do**

9:     $Idi \leftarrow IX[i]$;

10:     **if** $f_{Idi} == 1$ **then**

11:         **for** $j \leftarrow (i+1)$ to k **do**

12:             $Idj \leftarrow IX[j]$;

13:             **if** $f_{Idj} == 1$ **then**

14:                 $\kappa = $ DiverComp(Out_{Idi}, Out_{Idj});

                    {According to Eq. 3} of κ

15:                 **if** $\kappa \geq \delta$ **then**

16:                     $f_{Idj} \leftarrow 0\{\delta$ is set to $\frac{t_c}{\sqrt{N-2+t_c^2}}$ according to Eq. 5}

17:                 **end if**

18:             **end if**

19:         **end for**

20:     **end if**

21: **end for**

---

classifiers. And the classifiers whose classification accuracy being smaller than 0.5 are filtered out (i.e., set the corresponding flags in $FlagList$ as 0) according to Definition 1 in Sect. 2.1. The later (lines 7–21) aims at identifying the diverse classifiers. In this filter, if $|\kappa|$ between two classifiers is greater than predefined threshold δ according to Definition 2 in Sect. 2.2, the classifier with lower classification accuracy will be filtered out. Where the threshold δ is set according to Eq. 5. Afterward, the flag list $FlagList$ records whether the classifiers in $\{A_i, 1 \leq i \leq k\}$ are identified as the single classifiers for ensemble, where $f_i = 1/0$ means the classifier A_i is picked up/filtered out. Finally, an ensemble classifier can be constructed based on the picked up single classifiers in specific way, such as voting or weighted voting.

4 Experimental Study

In this section, we experimentally evaluate the proposed method to pick up single classifiers for the well-known ensemble method over the UCI data sets and then compare the classification accuracy with that of two traditional ensemble method (i.e., RandomForest and Boosting). Next, we introduce the experimental study from the following aspects: benchmark data set, experimental setup, and experiment results and analysis.

4.1 Benchmark Data Set

To study the performance of the proposed method in picking up single classifiers for ensemble, we employed 12 benchmark data sets which are available from UC Irvine Machine Learning Repository. The statistical information of these data sets is summarized in Table 2[2]. It is noted that the number of instances of these data sets is not large enough even quite small. This is due to that, in the situation of constructing classification models on small data sets, it will be easy to suffer from statistical problem for single classifier, and the ensemble classifier would be a better choice.

Table 2. Summary of the 12 benchmark data sets.

Dataset ID	Dataset Name	F	I	T
1	CAR	6	1728	4
2	COLIC-ORIG	23	368	2
3	HABERMAN	4	306	2
4	HAYES-ROTH-TEST	5	28	4
5	HAYES-ROTH-TRAIN	5	132	4
6	HEART-C	14	303	5
7	HEART-H	14	294	5
8	IRIS	4	150	3
9	LIVER-DISORDERS	7	345	2
10	POSTOPERATIVE-PATIENT	9	90	3
11	TAE	6	151	3
12	VOTE	17	435	2

---

[2] In the table, "F", "I" and "T" denote the numbers of features, instances and target concept values, respectively.

4.2   Experimental Setup

1. The representative and well-known ensemble method Bagging is employed as the baseline since that it has been stated in [6] that Bagging method has better performance in the situations with classification noise. And in our experiment, two commonly-used classification algorithms Decision Tree and Naive Bayes are used to constructed the single classifiers for Bagging. Thus, there will be two different kinds of Bagging based ensemble classifiers named Bag+DT and Bag+NB, respectively. The number of single classifiers used for original Bagging method is set as 10, 30, 50 and 70, respectively.
2. The performance of these ensemble classifiers before and after applying Algorithm 1 is compared in terms of classification accuracy of the ensemble classifier and number of single classifiers picked up for ensemble.

   For each data set D in Table 2, 5×10-fold crossvalidation is performed to estimate the performance of the ensemble classifiers. At first, the data set D is divided into 10 disjoint folds $\{D_i : 1 \leq i \leq 10\}$ where each D_i with a class distribution approximating that of D. Then, for each D_i, the rest 9 folds $\{D_1, \cdots, D_{i-1}, D_{i+1}, D_{10}\}$ are randomly splitted into two equal disjoint subsets D_{train} and D_{valid}. The single classifiers are trained using D_{train}, and filtered out based on the classification results on D_{valid}. Finally, the ensemble classifier is evaluated on the held out one fold D_i. This procedure is performed 5 times on D whose instances are randomly rearranged to make a stable performance estimation.

4.3   Experimental Results and Analysis

Comparison in Terms of Classification Performance. Table 3 compares the original ensemble classifier and the ensemble classifier constructed with Algorithm 1 in terms of classification accuracy (i.e., acc_{before} and acc_{after}) and average number of single classifiers (i.e., column N_{pick}) picked up. Meanwhile, in order to further demonstrate whether the proposed Algorithm 1 can effectively identify the accurate and diverse single classifiers for ensemble, we employ the non-parametric statistical method Wilcoxon Rank Sum Test to compare the classification accuracies of the ensemble classifiers before and after using Algorithm 1. The column p-$value$ in Table 3 gives the p-values of statistical test results for each data set under significance level $\alpha = 0.05$. The alternative hypothesis is that the ensemble classifier constructed with Algorithm 1 is better than the original ensemble classifier in terms of classification accuracy. The symbol "+" denotes the proposed method improves the classification accuracy of ensemble classifier significantly.

From Table 3, we can observe that:

1. After identifying the accurate and diverse single classifiers for ensemble by Algorithm 1, the average classification accuracy over the 12 data sets is improved either for Bag+DT or Bag+NB.

Table 3. Comparison between the original ensemble classifier and the ensemble classifier filtered by Algorithm 1

	Bagging + Decision tree														
Data ID	Number of single classifiers = 10				Number of single classifiers = 30				Number of single classifiers = 50				Number of single classifiers = 70			
	N_{pick}	acc_{before}	acc_{after}	$p\text{-}value$	N_{pick}	acc_{before}	acc_{after}	$p\text{-}value$	N_{pick}	acc_{before}	acc_{after}	$p\text{-}value$	N_{pick}	acc_{before}	acc_{after}	$p\text{-}value$
1	8.50	88.65	88.76	0.86	14.24	89.36	89.72	0.49	15.96	89.51	89.99	0.36	16.12	89.50	90.09	0.29
2	7.52	84.39	84.66	0.92	11.56	84.11	85.09	0.47	13.68	84.33	84.93	0.71	14.46	84.28	84.55	0.91
3	7.78	72.22	72.49	0.82	12.94	72.42	72.94	0.65	15.04	72.43	72.74	0.76	15.86	72.10	72.95	0.48
4	6.10	56.67	62.00	0.32	11.18	61.33	62.33	0.89	14.18	60.00	62.00	0.86	16.08	60.00	60.67	0.97
5	8.52	78.04	78.04	0.90	16.04	77.75	77.12	0.55	20.72	76.52	79.14	0.19	23.16	75.95	79.63	0.08
6	5.80	79.88	79.28	0.90	7.66	80.87	79.55	0.37	8.44	80.67	79.80	0.45	8.90	80.60	79.40	0.36
7	8.52	79.42	79.83	0.85	14.24	80.43	80.90	0.72	16.86	80.63	80.69	0.84	18.34	80.97	81.18	0.81
8	8.06	93.33	93.47	0.97	16.36	93.07	93.87	0.58	21.36	93.20	94.13	0.52	25.24	93.20	94.40	0.40
9	4.80	65.23	65.40	0.96	4.96	66.77	65.72	0.69	4.78	67.75	67.31	0.93	5.16	68.45	67.55	0.53
10	7.90	65.56	66.89	0.65	16.6	67.56	66.67	1.00	21.38	67.34	65.11	0.39	25.32	67.11	65.11	0.36
11	2.40	49.82	51.27	0.61	6.98	50.63	52.08	0.53	11.06	51.03	52.20	0.47	14.44	51.43	52.58	0.68
12	8.86	96.04	96.09	0.86	19.4	96.00	96.23	0.80	24.62	96.09	96.23	0.97	28.58	96.04	96.09	0.97
Average	7.11	75.77	76.52		12.68	76.69	77.02		15.67	76.62	77.02		17.64	76.64	77.02	
	Bagging + Naive Bayes														
Data ID	Number of single classifiers = 10				Number of single classifiers = 30				Number of single classifiers = 50				Number of single classifiers = 70			
	N_{pick}	acc_{before}	acc_{after}	$p\text{-}value$	N_{pick}	acc_{before}	acc_{after}	$p\text{-}value$	N_{pick}	acc_{before}	acc_{after}	$p\text{-}value$	N_{pick}	acc_{before}	acc_{after}	$p\text{-}value$
1	6.88	83.46	83.59	0.90	10.32	83.40	84.45	0.04+	11.02	83.59	84.85	0.02+	11.3	83.58	85.05	0.01+
2	4.66	79.46	79.84	0.75	5.86	79.29	80.27	0.53	6.26	79.46	80.88	0.20	6.72	79.29	80.66	0.31
3	5.98	75.15	74.82	0.75	8.06	75.21	74.62	0.62	9.22	75.28	74.30	0.36	10.40	75.15	74.37	0.57
4	5.76	53.00	52.33	0.96	16.04	56.00	55.33	1.00	23.28	57.33	59.67	0.61	30.46	56.00	55.33	0.99
5	7.94	65.15	64.68	0.85	17.04	63.85	64.88	0.53	21.22	64.15	66.23	0.34	23.54	64.45	67.33	0.21
6	6.02	83.62	83.76	0.95	9.1	83.56	83.55	0.99	10.08	83.69	83.82	0.92	10.82	83.56	83.88	0.97
7	4.84	83.08	83.35	0.90	5.52	83.42	84.04	0.67	6.1	83.28	83.84	0.76	6.06	83.21	84.11	0.62
8	9.10	94.53	94.40	0.88	20.58	94.93	94.80	0.93	28.36	95.07	94.93	0.92	35.12	94.93	95.07	0.88
9	5.54	54.79	58.95	0.03+	8.58	55.83	62.28	0.00+	10.26	56.52	61.53	0.01+	11.04	56.35	62.45	0.00
10	9.00	66.23	66.67	0.82	23.04	65.78	65.78	0.89	32.88	66.00	66.67	0.71	39.74	65.78	67.78	0.34
11	2.08	49.94	48.50	0.71	3.7	48.74	50.75	0.41	4.68	49.02	52.11	0.14	5.40	49.28	52.49	0.14
12	3.32	90.16	90.25	0.86	3.66	90.06	90.39	0.65	3.9	90.20	90.29	0.88	3.96	90.20	90.34	0.84
Average	5.93	73.21	73.43		10.96	73.34	74.26		13.94	73.63	74.93		16.21	73.48	74.90	

2. By statistically comparing the classification accuracies before and after filtering the single classifiers by Algorithm 1 on each data set, (i) for Bag+DT, the classification accuracy acc_{after} is statistically equal to acc_{before}. Meanwhile, the average number of single classifiers is significantly decreased. Such as, for ensemble classifiers constructed over 70 single classifiers, the average number of picked up single classifiers is only 17.64, and the classification accuracy does not reduce; (ii) for Bag+NB, the classification accuracy after filtering acc_{after} is statistically equal to or even superior to that before filtering acc_{before}. Meanwhile, the number of picked up single classifiers is significant less than the number of single classifiers in original ensemble classifier.

The above experimental results demonstrate that the proposed Algorithm 1 can effective pick up the accurate and diverse single classifiers for ensemble. i.e., the proposed algorithm can not only significantly reduce the number of single classifiers used for ensemble, but also guarantee or even improve the performance of ensemble classifiers (i.e., Bag+DT and Bag+NB).

Comparison with Traditional Ensemble Classifiers. Table 4[3] compares two traditional ensemble classifiers with the ensemble classifier constructed with Algorithm 1 in terms of classification accuracy. For consistency, when comparing with Bagging+DT(NB), the single classifiers for Boosting is also Decision Tree (Naive Bayes). What's more, we also use the same statistical method Wilcoxon Rank Sum Test to compare the classification accuracy and the symbol has the same meaning with that in Table 3.

[3] In the table, "acc_{RF}" "$acc_{BD}(acc_{BN})$" and "acc_{after}" denote the classification accuracy of RandomForest, Boosting+DT(Boosting+NB) and Bagging filtered by Algorithm 1.

Table 4. Comparison with traditional ensemble classifiers

Bagging + Decision tree

Data ID	Number of single classifiers 10					Number of single classifiers 30					Number of single classifiers 50					Number of single classifiers 70				
	acc_{RF}	acc_{BD}	acc_{after}	p_1-value	p_2-value	acc_{RF}	acc_{BD}	acc_{after}	p_1-value	p_2-value	acc_{RF}	acc_{BD}	acc_{after}	p_1-value	p_2-value	acc_{RF}	acc_{BD}	acc_{after}	p_1-value	p_2-value
1	93.16	95.93	88.76	0.00-	0.00-	94.13	96.57	89.72	0.00-	0.00-	94.27	96.67	89.99	0.00-	0.00-	94.55	96.92	90.09	0.00-	0.00-
2	69.74	66.03	84.66	0.00+	0.00+	70.39	66.03	85.09	0.00+	0.00+	70.17	66.03	84.93	0.00+	0.00+	70.66	66.03	84.55	0.00+	0.00+
3	74.98	74.65	72.49	0.25	0.28	75.43	74.65	72.94	0.25	0.32	75.43	74.65	72.74	0.23	0.30	75.69	74.65	72.95	0.23	0.32
4	44.00	50.00	62.00	0.09	0.19	44.67	50.00	62.33	0.10	0.18	44.00	50.00	62.00	0.09	0.19	44.00	50.00	60.67	0.11	0.22
5	61.23	62.09	78.04	0.00+	0.00+	61.09	61.80	79.12	0.00+	0.00+	60.80	61.80	79.14	0.00+	0.00+	60.80	61.80	79.63	0.00+	0.00+
6	81.46	79.41	79.28	0.26	0.49	82.31	80.08	79.55	0.20	0.44	82.50	80.08	79.80	0.20	0.47	82.63	80.08	79.40	0.16	0.42
7	80.22	82.19	79.83	0.46	0.24	79.19	81.65	80.90	0.31	0.41	79.00	81.65	80.69	0.31	0.39	79.00	81.65	81.18	0.26	0.44
8	95.47	96.00	93.47	0.21	0.14	95.73	95.07	93.87	0.22	0.32	95.20	95.07	94.13	0.34	0.36	95.87	95.07	94.40	0.27	0.40
9	62.97	63.20	65.40	0.26	0.28	63.20	63.20	65.72	0.25	0.25	63.20	63.20	67.31	0.14	0.14	63.20	63.20	67.55	0.12	0.12
10	59.11	58.67	66.89	0.15	0.13	59.78	59.11	66.67	0.18	0.15	60.45	59.11	65.11	0.27	0.21	60.67	59.11	65.11	0.28	0.21
11	40.69	41.09	51.27	0.04+	0.04+	40.69	41.09	52.08	0.03+	0.03+	40.83	41.09	52.20	0.03+	0.03+	41.09	41.09	52.58	0.02+	0.02+
12	96.00	95.25	96.09	0.47	0.27	96.55	95.12	96.23	0.40	0.21	96.41	95.07	96.23	0.45	0.20	96.50	95.07	96.09	0.38	0.23
Average	71.58	72.04	76.52			71.93	72.03	77.02			71.85	72.03	77.02			72.05	72.06	77.02		

Bagging + Naive Bayes

Data ID	Number of single classifiers 10					Number of single classifiers 30					Number of single classifiers 50					Number of single classifiers 70				
	acc_{RF}	acc_{BN}	acc_{after}	p_3-value	p_4-value	acc_{RF}	acc_{BN}	acc_{after}	p_3-value	p_4-value	acc_{RF}	acc_{BN}	acc_{after}	p_3-value	p_4-value	acc_{RF}	acc_{BN}	acc_{after}	p_3-value	p_4-value
1.00	93.16	90.35	83.59	0.00-	0.00-	94.13	90.35	84.45	0.00-	0.00-	94.27	90.35	84.85	0.00-	0.00-	94.55	90.35	85.05	0.00-	0.00-
2.00	69.74	70.87	79.84	0.00+	0.00+	70.39	71.16	80.27	0.00+	0.00+	70.17	69.63	80.88	0.00+	0.00+	70.66	70.01	80.66	0.00+	0.00+
3.00	74.98	73.80	74.82	0.48	0.39	75.43	73.80	74.62	0.41	0.41	75.43	73.80	74.30	0.38	0.45	75.69	73.80	74.37	0.36	0.44
4.00	44.00	40.00	52.33	0.28	0.19	44.67	40.00	55.33	0.22	0.14	44.00	40.00	59.67	0.13	0.08	44.00	40.00	55.33	0.21	0.14
5.00	61.23	61.09	64.68	0.29	0.28	61.09	61.09	64.88	0.27	0.27	60.80	61.09	66.23	0.19	0.20	60.80	61.09	67.33	0.14	0.15
6.00	81.46	84.14	83.76	0.23	0.45	82.31	84.14	83.55	0.35	0.43	82.50	84.14	83.82	0.34	0.46	82.63	84.14	83.88	0.35	0.47
7.00	80.22	85.06	83.35	0.17	0.29	79.19	84.99	84.04	0.06	0.38	79.00	84.99	83.84	0.07	0.36	79.00	84.99	84.11	0.05	0.39
8.00	95.47	95.73	94.40	0.34	0.29	95.73	95.73	94.80	0.35	0.35	95.20	95.73	94.93	0.46	0.37	95.87	95.73	95.07	0.37	0.39
9.00	62.97	63.20	58.95	0.15	0.14	63.20	63.20	62.28	0.41	0.41	63.20	63.20	61.53	0.33	0.33	63.20	63.20	62.45	0.42	0.42
10.00	59.11	67.11	66.67	0.15	0.48	59.78	67.11	65.78	0.21	0.43	60.45	67.11	66.67	0.20	0.48	60.67	67.11	67.78	0.17	0.46
11.00	40.69	41.09	48.50	0.10	0.11	40.69	41.09	50.75	0.04+	0.06	40.83	41.09	52.11	0.03+	0.03+	41.09	41.09	52.49	0.03+	0.03+
12.00	96.00	94.89	90.25	0.00-	0.00-	96.55	95.12	90.39	0.00-	0.00-	96.41	95.12	90.29	0.00-	0.00-	96.50	95.12	90.34	0.00-	0.00-
Average	71.58	72.28	73.43			71.93	72.32	74.26			71.85	72.19	74.93			72.05	72.22	74.90		

From Table 4, we can observe that: (1) for Bagging+DT, the classification accuracy acc_{after} is statistically equal or superior to that of RandomForest and Boosting+DT, especially over Datasets 2, 5 and 11; (2) for Bagging+NB, the classification accuracy acc_{after} is statistically equal to that of RandomForest and Boosting+DT. It fails over Datasets 1 and 12, but has better performance over Datasets 2 and 11.

In a word, comparing with traditional ensemble classifiers, the proposed Algorithm 1 can significantly reduce the number of single classifiers but not at the cost of classification accuracy.

Sensitive Analysis of Number of Single Classifiers for Ensemble. For ensemble learning, the number of single classifiers used for ensemble plays an important role. Different numbers of single classifiers will lead to different performance of the ensemble classifier. In this section, we give the analysis of the number of single classifiers on the performance of the ensemble classifier constructed with Algorithm 1. In this experiment, the number of single classifiers is set between 10 and 70 with step = 20.

Figure 2 gives the sensitive analysis of the number of single classifiers on the performance of the ensemble classifiers Bag+DT (see Fig. 2(a)) and Bag+NB (see Fig. 2(b)) before and after using Algorithm 1. From this figure we can observe that:

1. For Bag+DT, with the number of single classifiers increasing, for all data sets except for the 9th and 12th ones, the classification accuracy of the ensemble classifier using Algorithm 1 is greater than (on 8 data sets) or close to (on 2 data sets) that of the original ensemble classifiers. For the 9th and 12th data sets, although the accuracy of the ensemble classifier after filtering is a bit lower, the accuracy difference is not statistically significant.

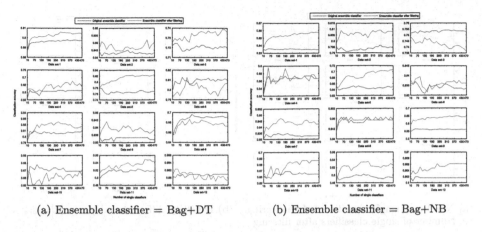

(a) Ensemble classifier = Bag+DT   (b) Ensemble classifier = Bag+NB

Fig. 2. Sensitive analysis of number of single classifiers on classification accuracy of Bagging (Color figure online)

2. For Bag+NB, with the number of single classifiers increasing, for all data sets except for the 3rd data set, the classification accuracy of the ensemble classifier after using Algorithm 1 is greater than that of the original ensemble classifier. However, for the 3rd data set, the accuracy difference is still not statistically significant.

Figure 3 shows how (i) the number of single classifiers picked up by Algorithm 1 (See Fig. 3(a)) and (ii) the diversity differences of the single classifiers before and after using Algorithm 1 (See Fig. 3(b)) vary with different number of single classifiers used in original Bagging classifier. From Fig. 3 we can observe that:

1. In Fig. 3(a), no matter for Bag+DT (red line) and Bag+NB (blue line), the number of single classifiers picked up by Algorithm 1 is significant smaller than the number of single classifiers used in original Bagging classifier. This means that Algorithm 1 can effectively reduce the number of single classifiers used to construct the ensemble classifier.
   With the number of single classifiers increasing, the number of picked up single classifiers first increases and then becomes stable or increases quite slowly. This indicates that, constructing more single classifiers does not mean that there will appear more accurate and diverse single classifiers. According to Corollary 1, the performance of an ensemble classifier depends on whether the single classifiers are accurate and diverse. Therefore, the phenomenon, that the number of picked up classifiers becoming stable with the number of single classifiers increasing, can be used to illustrate why the classification accuracy in Fig. 2 of the original ensemble classifier does not increase with the number of single classifiers increasing.
2. Figure 3(b) shows the difference of diversity of single classifiers before (i.e., $|\kappa|_{before}$) and after ($|\kappa|_{before}$) using Algorithm 1. Where the blue and red

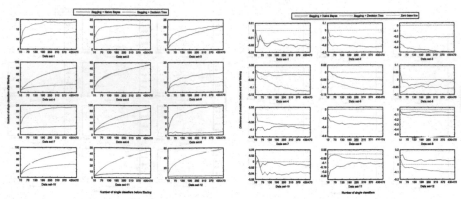

(a) Number of single classifiers before filtering   (b) Diversity difference $|\kappa|_{after} - |\kappa|_{before}$
vs. Number of single classifiers after filtering

Fig. 3. Sensitive analysis of number of single classifiers on (a) number of classifiers after filtering (b) the diversity differences before and after filtering (Color figure online)

lines stand for the diversity difference $|\kappa|_{after} - |\kappa|_{before}$ under Bag+NB and Bag+DT, respectively, and the green line stands for the zeros baseline.

According to Sect. 2, the diversity of single classifiers plays a critical role in improving the classification performance of an ensemble classifier. In this paper, we proposed a new metric $|\kappa|$ to evaluate the diversity between single classifiers. The smaller the value of $|\kappa|$ over a set of single classifiers, the more possible to construct a superior ensemble classifier based on these single classifiers. Figure 3(b) shows that for all the Bagging ensemble classifiers over all the data sets except for Bag+NB on the 12th data set, the value of $|\kappa|_{after} - |\kappa|_{before}$ is always smaller than 0 when the number of single classifiers is greater than 30. Meanwhile, this is consist with that in Fig. 2 the ensemble classifiers constructed with Algorithm 1 usually have better performance.

In a word, these phenomenons above mean that the proposed Algorithm 1 can effectively identify a good subset of single classifiers for ensemble while improving or reserving the classification performance of the ensemble classifier.

5   Conclusions

In this paper, we have proposed a new method to pick up a set of accurate and diverse single classifiers for ensemble learning. We have applied the proposed method on the representative ensemble learning method Bagging, and compared the classification performance of Bagging before and after using our method on the UCI data sets. The experimental results show that our proposed method can effectively identify a small subset of single classifiers for ensemble while improving the classification performance of Bagging on most data sets. For future work, we plan to apply our method on other kinds of ensemble learning methods, such as Boosting, RandomForest etc. to further assess its effectiveness.

Acknowledgements. This work is supported by the National Natural Science Foundation of China (Grant Nos. 61502378, 61402355), the Postdoctoral Science Foundation of China (Grant No. 2014M562417), the Program of State Key Software Engineering Laboratory, Wuhan University, China (Grant No. 2015Program17), the Fundamental Research Funds for the Central Universities (Grant No. xjj2014050) and the Shaanxi Province Postdoctoral Sustentation Fund, China.

References

1. Arratia, R., Gordon, L.: Tutorial on large deviations for the binomial distribution. Bull. Math. Biol. **51**(1), 125–131 (1989)
2. Breiman, L.: Bagging predictors. Mach. Learn. **24**(2), 123–140 (1996)
3. Brown, G., Wyatt, J., Harris, R., Yao, X.: Diversity creation methods: a survey and categorisation. Inf. Fusion **6**(1), 5–20 (2005)
4. Cunningham, P., Carney, J.: Diversity versus quality in classification ensembles based on feature selection. In: López de Mántaras, R., Plaza, E. (eds.) ECML 2000. LNCS (LNAI), vol. 1810, pp. 109–116. Springer, Heidelberg (2000). doi:10. 1007/3-540-45164-1_12
5. Dietterich, T.G.: Ensemble methods in machine learning. In: Kittler, J., Roli, F. (eds.) MCS 2000. LNCS, vol. 1857, pp. 1–15. Springer, Heidelberg (2000). doi:10. 1007/3-540-45014-9_1
6. Dietterich, T.G.: An experimental comparison of three methods for constructing ensembles of decision trees: bagging, boosting, and randomization. Mach. Learn. **40**(2), 139–157 (2000)
7. Dietterich, T.G.: Ensemble learning. In: The Handbook of Brain Theory and Neural Networks, vol. 2, pp. 110–125 (2002)
8. Džeroski, S., Ženko, B.: Is combining classifiers with stacking better than selecting the best one? Mach. Learn. **54**(3), 255–273 (2004)
9. Fleiss, J.L., Cohen, J., Everitt, B.: Large sample standard errors of kappa and weighted kappa. Psychol. Bull. **72**(5), 323 (1969)
10. Freund, Y., Schapire, R.E., et al.: Experiments with a new boosting algorithm. In: ICML, vol. 96, pp. 148–156 (1996)
11. Giacinto, G., Roli, F.: Design of effective neural network ensembles for image classification purposes. Image Vis. Comput. **19**(9), 699–707 (2001)
12. Hansen, L.K., Salamon, P.: Neural network ensembles. IEEE Trans. Pattern Anal. Mach. Intell. **10**, 993–1001 (1990)
13. Ho, T.K.: The random subspace method for constructing decision forests. IEEE Trans. Pattern Anal. Mach. Intell. **20**(8), 832–844 (1998)
14. Kohavi, R., Wolpert, D.H., et al.: Bias plus variance decomposition for zero-one loss functions. In: ICML, vol. 96, pp. 275–283 (1996)
15. Kuncheva, L.I., Whitaker, C.J.: Measures of diversity in classifier ensembles and their relationship with the ensemble accuracy. Mach. Learn. **51**(2), 181–207 (2003)
16. Kuncheva, L.I., Whitaker, C.J., Shipp, C.A., Duin, R.P.: Limits on the majority vote accuracy in classifier fusion. Pattern Anal. Appl. **6**(1), 22–31 (2003)

328   G. Wang et al.

17. Lam, L.: Classifier combinations: implementations and theoretical issues. In: Kittler, J., Roli, F. (eds.) MCS 2000. LNCS, vol. 1857, pp. 77–86. Springer, Heidelberg (2000). doi:10.1007/3-540-45014-9_7
18. Lee, J.W., Giraud-Carrier, C.: Automatic selection of classification learning algorithms for data mining practitioners. Intell. Data Anal. **17**(4), 665–678 (2013)
19. Peterson, A.H., Martinez, T.R.: Estimating the potential for combining learning models. In: Proceedings of the ICML Workshop on Meta-learning, pp. 68–75 (2005)
20. Sharkey, A.J.: Linear and order statistics combiners for pattern classification. In: Sharkey, A.J. (ed.) Combining Artificial Neural Nets, pp. 127–161. Springer, London (1999)
21. Sim, J., Wright, C.C.: The kappa statistic in reliability studies: use, interpretation, and sample size requirements. Phys. Ther. **85**(3), 257–268 (2005)
22. Tumer, K., Ghosh, J.: Error correlation and error reduction in ensemble classifiers. Connection Sci. **8**(3–4), 385–404 (1996)

Community Detection in Dynamic Attributed Graphs

Gonzalo A. Bello[1], Steve Harenberg[1],
Abhishek Agrawal[1], and Nagiza F. Samatova[1,2](✉)

[1] North Carolina State University, Raleigh, NC, USA
samatova@csc.ncsu.edu
[2] Oak Ridge National Laboratory, Oak Ridge, TN, USA

Abstract. Community detection is one of the most widely studied tasks in network analysis because community structures are ubiquitous across real-world networks. These real-world networks are often both attributed and dynamic in nature. In this paper, we propose a community detection algorithm for dynamic attributed graphs that, unlike existing community detection methods, incorporates both temporal and attribute information along with the structural properties of the graph. Our proposed algorithm handles graphs with heterogeneous attribute types, as well as changes to both the structure and the attribute information, which is essential for its applicability to real-world networks. We evaluated our proposed algorithm on a variety of synthetically generated benchmark dynamic attributed graphs, as well as on large-scale real-world networks. The results obtained show that our proposed algorithm is able to identify graph partitions of high modularity and high attribute similarity more efficiently than state-of-the-art methods for community detection.

Keywords: Community detection · Attributed graph · Dynamic graph

1 Introduction

Community detection is one of the most important and widely studied tasks in the field of network analysis [9]. Community structures, representing groups of interacting objects, are ubiquitous across real-world networks and, as such, are of interest in many domains. For example, communities may represent groups of friends in social networks, groups of functionally associated proteins in biological networks, or spatiotemporal climate patterns in climate networks [2].

Traditionally, communities are defined as sets of vertices more densely connected among each other than to the rest of the graph. Therefore, most community detection methods identify communities by taking into account only the structural properties of the graph. However, complex systems represented by real-world networks inherently have multiple sources of information.

On one hand, objects in real-world networks often have individual properties, or *attributes*, that characterize them. By describing these objects as simple

330 G.A. Bello et al.

unattributed vertices we are potentially losing valuable information that may allow us to identify more meaningful communities [4]. For this reason, an active line of research in community detection focuses on identifying communities in *attributed graphs* by incorporating both the structural properties of the graph and the attribute information of the vertices [5,6,8,14,15].

Moreover, real-world networks are constantly evolving, in many cases rapidly and dramatically. Hence, another active line of research in community detection is concerned with identifying and tracking communities in *dynamic graphs* [1,7]. However, to the best of our knowledge, none of the methods proposed for community detection in dynamic graphs also incorporate attribute information, nor do they address the additional challenges inherent to the evolution of attributed graphs, such as the presence of changes in the attributes of the vertices. For this reason, in this paper, we propose an algorithm for detecting communities in dynamic attributed graphs that incorporates both temporal and attribute information along with the structural properties of the graph.

The main contributions of this paper are as follows. First, we propose an efficient algorithm for community detection in dynamic attributed graphs. Our proposed algorithm handles graphs with heterogeneous attribute types, as well as changes to both the structure and the attribute information, which is essential for its applicability to real-world networks. Second, we introduce a methodology to generate benchmark dynamic attributed graphs for testing community detection algorithms. And third, we evaluate our proposed algorithm on large real-world networks to test its accuracy and scalability.

2 Related Work

Several methods have been proposed to identify communities in attributed graphs [5,6,8,14,15]. One approach is to augment the structure of the graph by incorporating attribute information, and then apply a traditional community detection method in the augmented graph. For example, the Inc-Cluster algorithm [15] adds new vertices representing attribute values, while the CODICIL algorithm [14] adds new edges between vertices with similar attribute information. Another approach is to extend traditional community detection methods to take into account both the structure of the graph and the attribute information. For example, the I-Louvain algorithm [5] extends the Louvain method by introducing an inertia-based modularity function that measures the similarity between vertices based on the Euclidean distance between their attribute vectors.

However, these existing methods have important limitations. First, none of these methods explicitly identify communities in dynamic graphs, and thus would need to be run from scratch every time the graph changes. Second, some of these methods require computing pairwise similarities between all vertices in the graph [5,14], which leads to high time and space complexities that may not be suitable for large-scale graphs. And third, most of these methods are designed to handle only certain attribute types, whether binary, categorical

[14,15] or numeric [5], which hinders their applicability to real-world networks with heterogeneous attribute types.

3   Community Detection in Dynamic Attributed Graphs

In this section, we formally define the problem of community detection in dynamic attributed graphs (Sect. 3.1), describe our proposed community detection algorithm (Sect. 3.2), and introduce a methodology to generate benchmark dynamic attributed graphs (Sect. 3.3).

3.1   Problem Statement

Let $G = (V, E, X)$ be an *attributed graph*, where V is the set of vertices, E is the set of edges, and $X = \{x_1, x_2, ..., x_D\}$ is the set of attributes associated with the vertices in V. We define a *dynamic attributed graph* as an ordered set $\mathcal{G} = \{G_t\}_{t=1}^{T}$, where T is the number of time steps and G_t is the attributed graph at time step t. Note that we assume that corresponding vertices are labeled consistently across different time steps.

The problem of *community detection in dynamic attributed graphs* is defined as follows. Given a dynamic attributed graph $\mathcal{G}$, partition each attributed graph $G_t \in \mathcal{G}$ into a set of disjoint communities $C_t = \{c_{1t}, c_{2t}, ..., c_{K_t t}\}$, where $\bigcup_{i=1}^{K_t} c_{it} = V_t$ and $c_{it} \cap c_{jt} = \emptyset$ for all $c_{it}, c_{jt} \in C_t$ with $i \neq j$, such that vertices in the same community are more densely connected between each other and have more similar attribute values than those in different communities.

3.2   Algorithm for Community Detection in Dynamic Attributed Graphs

Measuring the Quality of the Partition of the Graph. A "good" partition of an attributed graph is one that maximizes both the structural quality of the partition and the similarity between the vertices in each community.

To measure the structural quality of the partition of the graph, we use the well-known *modularity* function, which is defined as the difference between the number of edges within the communities and the expected number of such edges in a random graph with the same degree distribution [12].

Definition 1 (Modularity). *Let* $G = (V, E)$ *be an unweighted, undirected graph partitioned into a set of disjoint communities* $C = \{c_1, c_2, ..., c_K\}$*. The* **modularity** *of the partition of* G *[11] is given by*

$$Q(C) = \frac{1}{2m} \sum_{v,w \in V} \left[A_{vw} - \frac{k_v k_w}{2m} \right] \delta(c_v, c_w) \tag{1}$$

where A is the adjacency matrix of G (i.e., $A_{vw} = 1$ if vertices v and w are adjacent and $A_{vw} = 0$ otherwise), m is the number of edges in G, k_v is the degree of vertex v, c_v is the community of vertex v, and $\delta(c_v, c_w)$ is the Kronecker delta function (i.e., $\delta(c_v, c_w) = 1$ if $c_v = c_w$ and $\delta(c_v, c_w) = 0$ otherwise).

To measure the similarity between the vertices in each community, we define the *attribute similarity* function as follows.

Definition 2 (Attribute Similarity). *Let $G = (V, E, X)$ be an attributed graph partitioned into a set of disjoint communities $C = \{c_1, c_2, ..., c_K\}$. The **attribute similarity** of the partition of G is given by*

$$S(C) = \frac{1}{K \cdot D} \left[\sum_{i-1}^{K} \sum_{j=1}^{D} \sum_{v, w \in c_i} \frac{sim(x_{jv}, x_{jw})}{|c_i|^2} \right] \tag{2}$$

where $K = |C|$ is the number of communities, $D = |X|$ is the number of attributes in G, x_{jv} is the value of the j-th attribute for vertex v, and $sim(x_{jv}, x_{jw})$ is a function of the similarity between x_{jv} and x_{jw} (simple matching coefficient for binary attributes, Jaccard index for categorical attributes, and the inverse of one plus the Euclidean distance for numeric attributes).

It is worth noting that, unlike other similarity measures for community detection in attributed graphs [5], the proposed attribute similarity function allows us to combine attributes of different types.

Detecting Communities in Attributed Graphs. The goal of our proposed algorithm for community detection in attributed graphs is to maximize both the modularity and the attribute similarity of the partition of the graph. Maximizing modularity is a computationally hard problem, but several heuristic algorithms have been proposed to efficiently identify graph partitions with high modularity [9]. Here, we adapt the Louvain method [3], a greedy local search algorithm, to detect communities in attributed graphs by optimizing a multi-objective function given by

$$\mathcal{F}(C) = \alpha \cdot \mathcal{Q}(C) + (1 - \alpha) \cdot \mathcal{S}(C) \tag{3}$$

where $\alpha \in (0, 1]$ is a *weighting parameter* that balances the trade-off between modularity and attribute similarity.

The adapted method for detecting communities in attributed graphs (see Algorithm 1) proceeds as follows. In the first phase of the algorithm, the vertices are initially assigned to singleton communities. Then, each vertex is iteratively and sequentially reassigned to the community that yields the highest positive gain in the multi-objective function $\mathcal{F}$ until no further improvement can be achieved. The gain in $\mathcal{F}$ resulting from assigning a vertex z to a community c_i is given by

$$\Delta\mathcal{F}(z, c_i) = \alpha \cdot \Delta\mathcal{Q}(z, c_i) + (1 - \alpha) \cdot \Delta\mathcal{S}(z, c_i) \tag{4}$$

where $\Delta\mathcal{Q}(z, c_i)$ and $\Delta\mathcal{S}(z, c_i)$ are the gains in modularity and attribute similarity, respectively. The gain in modularity is given by

$$\Delta\mathcal{Q}(z, c_i) = \frac{k_{z,c_i}}{2m} - \frac{k_z \cdot k_{c_i}}{2m^2} \tag{5}$$

where k_{z,c_i} is the number of edges between vertex z and vertices in community c_i, and k_{c_i} is the number of edges incident to vertices in community c_i. To efficiently compute the gain in attribute similarity resulting from assigning vertex z to community c_i, we approximate K as K', where K and K' are the number of communities before and after assigning z to c_i, respectively. Then, the gain in attribute similarity is given by

$$\Delta \mathcal{S}(z, c_i) = \frac{1}{K' \cdot D} \sum_{j=1}^{D} \left[\sum_{v,w \in c_i \cup \{z\}} \frac{sim(x_{jv}, x_{jw})}{(|c_i| + 1)^2} - \sum_{v,w \in c_i} \frac{sim(x_{jv}, x_{jw})}{|c_i|^2} \right] \quad (6)$$

In the second phase of the algorithm, a new graph is constructed by aggregating the vertices in each community into a single vertex. The number of edges between two vertices in this new graph is given by the sum of the edges between vertices in the two corresponding communities. Likewise, the attribute values of a vertex in this new graph are given by the modes (in the case of binary or categorical attributes) or the means (in the case of numeric attributes) of the attribute values of the vertices in the corresponding community. The first phase of the algorithm is then reapplied on this new graph.

These two phases are repeated iteratively until no further changes to the community structure can be made. Then, the partition that yields the highest value of the multi-objective function for the original graph is returned. Note, however, that there is no guarantee of the optimality of the partition. Furthermore, the output of the algorithm depends on the order in which the vertices are iterated over, although results indicate that this does not generally have a significant impact on the value obtained for the objective function [3].

Updating Communities in Dynamic Graphs. A common approach for finding communities in dynamic graphs is to iteratively apply a community detection algorithm at each time step t [1,7]. However, it is not necessary to run the algorithm from scratch at each time step to obtain a "good" partition of the graph. Initializing the partition with information from the previous time step before applying the community detection algorithm may be more efficient and has been shown to yield more stable communities across time steps [1].

Naively, we can initialize the partition of the graph at time step t with the final partition at the previous time step $t - 1$, but this may limit the ability of the algorithm to find "good" communities, particularly if the graph changes significantly between time steps. Other approaches include randomly removing vertices from the partition [1], or removing all vertices that changed between time steps $t - 1$ and t [7]. However, none of these approaches take into account the differences in the degree of change of the vertices. Vertices with a higher degree of change are more likely to have changed communities between time steps, and thus, should be removed from the partition with a higher probability.

For this reason, we define a scoring function to measure the degree of change of a vertex between two consecutive time steps, $t - 1$ and t, in terms of both its structure and its attribute information, where the trade-off between structural and attribute change is given by the weighting parameter α (see Definition 3).

Definition 3 (Vertex Change Score). *Let* $G_t = (V_t, E_t, X_t)$ *and* $G_{t-1} = (V_{t-1}, E_{t-1}, X_{t-1})$ *be attributed graphs at time steps* t *and* $t-1$, *respectively. The* **change score of a vertex** z, *such that* $z \in V' = V_t \cap V_{t-1}$, *is given by*

$$\mathcal{H}(z) = \alpha \left[\frac{\sum_{v \in V'} (A_{tvz} \,\&\, A_{t-1vz})}{\sum_{v \in V'} (A_{tvz} \mid A_{t-1vz})} \right] + (1-\alpha) \left[1 - \sum_{j=1}^{D} \frac{sim(x_{t_{jz}}, x_{t-1_{jz}})}{D} \right] \quad (7)$$

where A_t *is the adjacency matrix of* G_t, $x_{t_{jz}}$ *is the value of the* j-*th attribute for vertex* z *in* G_t, *and* $\&and \mid$ *are the AND and OR operators, respectively.*

We use this scoring function to identify a partial partition of the graph based on the degree of change of each vertex and a user-defined *update parameter* $\beta \in [0, 1]$. First, we set the current communities at time step t to be those at the previous time step $t-1$; that is, $C_t = C_{t-1}$. If $\beta = 0$, all vertices remain in their current communities. If $\beta = 1$, all vertices are removed from their current communities and assigned to singleton communities. If $\beta \in (0, 1)$, the top $\beta \times |V'|$ vertices with the highest change score are removed from their current communities and assigned to singleton communities, while the other $(1 - \beta) \times |V'|$ vertices remain in their current communities. Note that this only applies to vertices present in both V_{t-1} and V_t. Vertices present in V_t but not in V_{t-1} are assigned to singleton communities without pre-processing. Finally, we

---

Algorithm 1. Greedy local search algorithm for community detection in attributed graphs

---

    Input   : attributed graph at time t, $G_t = (V_t, E_t, X_t)$; vector of initial
                 community memberships at time t, C; weighting parameter, α
    Output: vector of community memberships at time t, C

1 $G = (V, E, X) \leftarrow G_t$
2 **do**
      /* Phase one: */
3   **do**
4   *num_moves* $\leftarrow 0$
5   **for each** $v \in V$ **do**
6   **for each** community $c \in C$ connected to v **do**
7                 $\Delta\mathcal{F}(v, c) \leftarrow \alpha \cdot \Delta\mathcal{Q}(v, c) + (1-\alpha) \cdot \Delta\mathcal{S}(v, c)$
8   **if** $\exists c \in C$ such that $\Delta\mathcal{F}(v, c) > 0$ **then**
9                     $C[v] \leftarrow arg\,max_{c \in C}\,\Delta\mathcal{F}(v, c)$
10   *num_moves* $\leftarrow$ *num_moves* $+1$
11   **while** *num_moves* > 0
      /* Phase two: */
12   Compute the value of the multi-objective function, $\mathcal{F}(C)$, for G_t
13   Build a new graph $G = (V, E, X)$ by aggregating each $c \in C$ into one vertex
14   Reinitialize $C[v] \leftarrow v$ **for each** $v \in V$
15 **while** at least one vertex is moved during phase one
16 **return** C with the highest value of the multi-objective function, $\mathcal{F}(C)$, for G_t

---

apply the community detection algorithm described in the previous section (see Algorithm 1) with the current communities as the initial partition of the graph.

3.3 Benchmark Dynamic Attributed Graphs for Testing Community Detection Algorithms

To evaluate our proposed community detection algorithm, we introduce a methodology to generate benchmark dynamic attributed graphs. This methodology extends current benchmarks for testing community detection algorithms in static graphs [8,10]. Our benchmark graph generation methodology is as follows.

For $t = 0$:

1. Generate an unattributed graph G_0 with n vertices partitioned into a set of disjoint communities C_0, such that the degree of the vertices of G_0 and the sizes of the communities in C_0 follow power law distributions, as described in [10]. The partitioning of G_0 is controlled by a user-defined *mixing parameter* $\mu \in [0, 1]$, which indicates the fraction of edges of each vertex with vertices outside its community.
2. For each vertex v, generate a set of D attribute values $\{x_{1v}, x_{2v}, ..., x_{Dv}\}$, such that each attribute value is sampled from a uniform distribution and all the vertices in the same community are assigned the same attribute value. For this paper, we limit our benchmark graph generation methodology to binary attributes. However, it could be further extended to include categorical and numeric attributes.
3. For each attribute i, draw a random sample of vertices S of size $\lambda \times n$, where $\lambda \in [0, 1]$ is a user-defined *noise parameter*. Introduce noise to attribute i by changing the value of x_{iv} for every vertex $v \in S$.

For $t = 1$ to T, where T is the number of time steps:

1. Set $G_t = G_{t-1}$ and $C_t = C_{t-1}$.
2. Draw a random sample of vertices S of size $\delta \times n$, where $\delta \in [0, 1]$ is a user-defined *change parameter*.
3. Move every vertex $v \in S$ to a different community.
4. Modify the edges of every vertex $v \in S$ such that $((1 - \mu) \times 100)\%$ of its neighbors belong to its new community. This ensures that the structure of G_t remains closely defined by the mixing parameter μ.
5. Modify the attribute values of every vertex $v \in S$. For each attribute i, v is assigned attribute value x_{iv} such that

$$\begin{cases} x_{iv} = x_c \text{ with probability } 1 - \lambda \\ x_{iv} \neq x_c \text{ with probability } \lambda \end{cases} \tag{8}$$

where x_c is the mode of the values for attribute i among the vertices in the new community of v. This ensures that the noise in G_t remains closely defined by the noise parameter λ.

4   Experimental Evaluation

In this section, we present the results obtained using our proposed algorithm on a variety of synthetically generated benchmark graphs (Sect. 4.1), as well as on large-scale real-world networks (Sect. 4.2).

4.1   Benchmark Graphs

We evaluated the performance of our proposed algorithm on dynamic attributed graphs generated using our benchmark graph generation methodology. We generated a diverse set of graphs by varying the number of vertices ($n = \{1000, 10000\}$), the mixing parameter ($\mu = \{0.1, 0.9\}$), the noise parameter ($\lambda = \{0.1, 0.9\}$), and the change parameter ($\delta = \{0.1, 0.9\}$). For simplicity, we considered the number of attributes ($D = 10$) and the number of time steps ($T = 10$) to be fixed. Moreover, for each combination of the parameters, we generated ten dynamic attributed graphs to account for non-determinism in the generation process.

Analysis of Weighting Parameter α. To analyze the impact of α on the performance of our proposed algorithm, we considered the change parameter δ to be fixed and varied the mixing parameter μ and the noise parameter λ. The modularity and attribute similarity of the graph partitions obtained are shown in Figs. 1 and 2, respectively.

When the mixing parameter is low ($\mu = 0.1$), the best partition of the graph is clearly defined by its structural properties. For this reason, incorporating attribute information does not have an impact on the performance of the algorithm, which remains constant regardless of the value of α. It is worth noting that, in these cases, incorporating attribute information does not degrade the quality of the communities with respect to the original modularity optimization algorithm ($\alpha = 1$), even if the noise parameter is high ($\lambda = 0.9$).

On the other hand, when the mixing parameter is high ($\mu = 0.9$), the best partition of the graph is not clearly defined by its structural properties. Thus, the partition obtained by the original modularity optimization algorithm ($\alpha = 1$) is of low modularity and low attribute similarity. In these cases, incorporating attribute information improves the quality of the communities in terms of attribute similarity at the expense of a minimal decrease in modularity. As expected, this improvement is inversely proportional to the value of α.

These results are in agreement with previous studies that indicate that the performance of structure-only community detection methods degrades with respect to that of attribute-only community detection methods when the graph has an ambiguous structure (i.e., mixing parameter of 0.6 or greater) [8].

Analysis of Update Parameter β. To analyze the impact of β on the performance of our proposed algorithm, we considered the mixing parameter μ and the noise parameter λ to be fixed and varied the change parameter δ. The running time of our proposed algorithm and the modularity and attribute similarity of the graph partitions obtained are shown in Figs. 3 and 4, respectively.

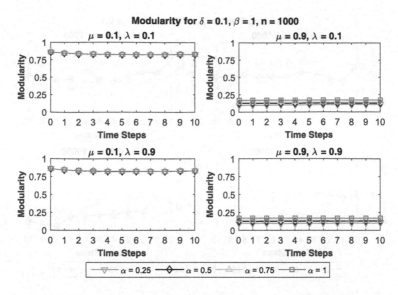

Fig. 1. Average *modularity* of the partitions obtained at each time step of the benchmark graphs using our proposed algorithm with update parameter $\beta = 1$ and multiple values of the weighting parameter α. Benchmark graphs were generated using the following parameters: number of vertices $n = 1000$, mixing parameter $\mu = \{0.1, 0.9\}$, noise parameter $\lambda = \{0.1, 0.9\}$, and change parameter $\delta = 0.1$.

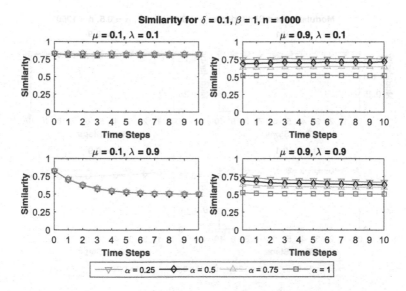

Fig. 2. Average *attribute similarity* of the partitions obtained at each time step of the benchmark graphs using our proposed algorithm with update parameter $\beta = 1$ and multiple values of the weighting parameter α. Benchmark graphs were generated using the following parameters: number of vertices $n = 1000$, mixing parameter $\mu = \{0.1, 0.9\}$, noise parameter $\lambda = \{0.1, 0.9\}$, and change parameter $\delta = 0.1$.

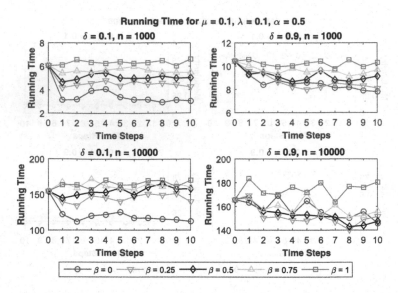

Fig. 3. Average *running time* (in seconds) at each time step of the benchmark graphs of our proposed algorithms with weighting parameter $\alpha = 0.5$ and multiple values of the update parameter β. Benchmark graphs were generated using the following parameters: number of vertices $n = \{1000, 10000\}$, mixing parameter $\mu = 0.1$, noise parameter $\lambda = 0.1$, and change parameter $\delta = \{0.1, 0.9\}$.

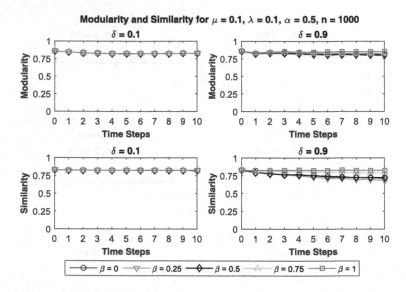

Fig. 4. Average *modularity* and *attribute similarity* of the partitions obtained at each time step of the benchmark graphs using our proposed algorithms with weighting parameter $\alpha = 0.5$ and multiple values of the update parameter β. Benchmark graphs were generated using the following parameters: number of vertices $n = 1000$, mixing parameter $\mu = 0.1$, noise parameter $\lambda = 0.1$, and change parameter $\delta = \{0.1, 0.9\}$.

We observe that initializing the partition of the graph at each time step before applying the community detection algorithm ($\beta < 1$) generally results in lower running times at a minimal performance cost in terms of modularity and attribute similarity. This is particularly true when the change parameter is low ($\delta = 0.1$), as the communities at time step t are expected to be very similar to those at time step $t-1$, and less so as the change parameter increases ($\delta = 0.9$).

Parameter Selection. The analysis of the weighting parameter α suggests that its optimal value is inversely proportional to the mixing parameter μ. Similarly, the analysis of the update parameter β suggests that its optimal value is directly proportional to the change parameter δ. However, the value of both μ and δ is given by the true partition of the graph, which is not known at the time of the parameter selection. For this reason, to select the optimal value of the weighting parameter α and the update parameter β, we would first need to estimate the value of the mixing parameter and the change parameter from the observable properties of the graph.

The mixing parameter has been shown to have a strong negative correlation with the local clustering coefficient [13]. Therefore, it can be estimated as a function of one minus the local clustering coefficient of the graph.

To estimate the change parameter, we first observe that the distribution of the vertex change score (see Definition 3) at a time step t is generally bimodal. More precisely, it is a two-component mixture distribution, where the component with the higher mean corresponds to the vertices changing community at time step t. Therefore, the change parameter can be estimated as the mixing proportion of the component with the higher mean obtained after fitting a two-component Gaussian mixture model to the vertex change scores using the Expectation Maximization (EM) algorithm.

Alternatively, the analysis of the parameters also suggests that simply setting the value of both α and β to 0.5 generally yields a good trade-off between efficiency and quality of the communities.

4.2   Real-World Networks

We also evaluated the performance of our proposed algorithm on large-scale dynamic attributed networks constructed from three real-world data sets.

The **DBLP**[1] data set provides publication records from 1991 to 2000. In the corresponding network, an edge between two vertices is present if the authors represented by those two vertices collaborated in a publication. Vertices have 19 categorical attributes representing each author's areas of publication (e.g., artificial intelligence, bioinformatics, security).

The **Yelp**[2] data set provides user reviews of a select set of businesses from 2004 to 2012. In the corresponding network, an edge between two vertices is

---

[1] dblp.uni-trier.de/xml.

[2] www.yelp.com/dataset_challenge.

Table 1. Number of vertices and number of edges per time step of real-world networks

Graph	Time Step							
DBLP	$t = 0$	$t = 1$	$t = 2$	$t = 3$	$t = 4$	$t = 5$	$t = 6$	$t = 7$
Num Vertices	13,782	26,359	38,797	51,176	64,366	78,089	93,142	110,065
Num Edges	33,528	71,636	114,950	159,154	213,806	273,188	341,330	426,548
Yelp	$t = 0$	$t = 1$	$t = 2$	$t = 3$	$t = 4$	$t = 5$	$t = 6$	$t = 7$
Num Vertices	7	380	2,488	8,683	20,132	36,895	61,338	97,039
Num Edges	14	8,286	90,626	461,468	1,469,412	3,146,770	5,772,004	10,372,332
TripAdvisor	$t = 0$	$t = 1$	$t = 2$	$t = 3$	$t = 4$	$t = 5$	$t = 6$	$t = 7$
Num Vertices	15	714	3,564	10,354	39,723	112,190	203,562	297,301
Num Edges	26	5,116	54,418	292,390	1,485,726	7,866,020	18,708,064	28,288,858

present if the users represented by those two vertices reviewed the same business. Vertices have 38 categorical attributes representing the type of businesses reviewed by each user (e.g., restaurants, shops, services), as well as a numeric attribute corresponding to the average rating assigned by each user.

The **TripAdvisor**[3] data set provides hotel reviews from 2002 to 2012. In the corresponding network, an edge between two vertices is present if the users represented by those two vertices reviewed the same hotel. Vertices have a numeric attribute corresponding to the average rating assigned by each user.

Experimental Setup. Dynamic attributed networks were constructed for each real-world data set using its first 8 years of data, such that time step $i - 1$ of each network corresponds to the data available up to the i-th year. The networks were preprocessed to remove all singleton vertices, as these would not be placed in any community. Further details of the networks are shown in Table 1.

We identified communities in these real-world networks using our proposed algorithm, as well as CODICIL [14] and I-Louvain [5], two state-of-the-art algorithms for community detection in attributed graphs. CODICIL and I-Louvain have been shown to outperform structure-only and attribute-only community detection methods, as well as other methods for community detection in attributed graphs, such as Inc-Cluster [15]. Since CODICIL and I-Louvain were not designed for dynamic graphs, we ran them from scratch at each time step. Moreover, as neither of these algorithms can handle both categorical and numeric attributes, we used CODICIL for the networks with categorical attributes, and I-Louvain for the networks with numeric attributes.

All experiments on real-world networks were performed on an Intel machine running RHEL Server 6.7 consisting of two hex-core E5645 processors and 64 GB DDR2 RAM. Our proposed algorithm was implemented in C++ and compiled with GCC 4.8.2 using the optimization flag -O3. For simplicity, we set the value of the weighting parameter α and the update parameter β to 0.5 for all experiments, as suggested in Sect. 4.1. For the state-of-the-art algorithms, we used Python implementations provided by the authors with its default parameters.

---

[3] times.cs.uiuc.edu/~wang296/Data.

Note that the CODICIL implementation returns an augmented graph that combines structural and attribute information, after which a traditional community detection method must be applied. To this end, we used the Louvain method. However, the Louvain method's time and space requirements were not included in CODICIL's time and space measurements.

The results obtained by each algorithm were compared in terms of the quality of the communities identified and the efficiency of the implementation. Due to the existence of multiple definitions of community, evaluating their quality requires the consideration of multiple metrics. To evaluate their structural properties, we computed the modularity and average density of the graph partition, and to evaluate the homogeneity of their attribute information, we computed the attribute similarity, as well as the average entropy [6] or average standard deviation, for the networks with categorical or numeric attributes, respectively. The results obtained are shown in Table 2.

Discussion of Results. With respect to the quality of the communities, our proposed algorithm obtained graph partitions with higher modularity than both state-of-the-art algorithms (an average improvement of 6 % and 8 % compared to CODICIL and I-Louvain, respectively). For the networks with categorical attributes, our proposed algorithm identified communities with higher average density and attribute similarity than CODICIL (an average improvement of 6 % and 5 %, respectively), but also higher average entropy. On the other hand, for the networks with numeric attributes, I-Louvain identified communities with higher density, higher attribute similarity, and lower average standard deviation than our proposed algorithm (an average improvement of 4 %, 1 %, and 2 %, respectively). It is worth noting that many communities identified by the algorithms were 2-cliques. Local metrics, such as average density, are likely to be skewed towards these smaller communities, and global metrics, such as modularity, might provide a better estimate of the quality of the graph partition as a whole.

With respect to the efficiency of the algorithms, the CODICIL implementation required, on average, 2000 times more time and 6 times more peak memory than the proposed algorithm, while the I-Louvain implementation required, on average, 900 times more time and 400 times more peak memory. Moreover, the CODICIL implementation was not able to run on 5 time steps in the allotted time of 5 h, while the I-Louvain implementation was not able to run on 7 time steps with the allotted memory of 64 GB. Even though these measurements are implementation-dependent, they empirically confirm the observations regarding the high time and space complexity of these algorithms made in Sect. 2.

Overall, the results obtained show that our proposed algorithm is able to efficiently identify communities of high quality in large-scale real-world networks. While the state-of-the-art algorithms obtained some graph partitions of higher average density and attribute similarity, their time and memory requirements were considerably higher. Therefore, we conclude that our proposed algorithm achieves a desired balance between efficiency and accuracy of the results.

Table 2. Results obtained (number of communities, modularity, average density, attribute similarity, average entropy or average standard deviation, peak memory in MB, and running time in seconds) using our proposed algorithm and a state-of-the-art community detection algorithm (CODICIL for networks with categorical attributes or I-Louvain for networks with numeric attributes) on large-scale real-world networks. Asterisks (∗) and daggers (†) indicate that the implementation of the algorithm did not run in the allotted time of 5 h or with the allotted memory of 64 GB, respectively. Best results are shown in bold and underlined.

Graph	Proposed Algorithm							State-of-the-art: CODICIL/I-Louvain						
	Num Com	Mod	Den	Att Sim	Ent/ Std	Peak Mem	Run Time	Num Com	Mod	Den	Att Sim	Ent/ Std	Peak Mem	Run Time
DBLP + Categorical Attributes														
$t=0$	4,006	1.00	0.91	0.95	0.22	**6.08**	**0.16**	4,023	1.00	0.91	0.95	**0.21**	22.37	518.38
$t=1$	6,019	0.99	0.88	0.94	0.42	**15.41**	**0.45**	6,322	0.99	0.88	0.94	**0.39**	40.27	2,011.42
$t=2$	7,075	0.98	0.86	0.93	0.68	**28.69**	**0.96**	8,026	0.98	**0.88**	0.93	**0.55**	60.82	4,209.35
$t=3$	7,662	0.97	0.86	0.92	0.91	**44.43**	**1.56**	9,322	0.97	**0.87**	**0.93**	**0.72**	78.45	7,401.62
$t=4$	8,132	0.95	0.86	0.92	1.12	**63.34**	**3.23**	10,449	0.95	**0.88**	**0.93**	**0.87**	100.88	10,397.90
$t=5$	8,655	0.94	0.87	0.92	1.27	**85.94**	**7.41**	11,566	**0.95**	**0.88**	**0.93**	**0.99**	124.93	17,271.63
$t=6$	9,152	0.93	0.87	0.92	1.41	113.00	**12.69**	∗	∗	∗	∗	∗	∗	∗
$t=7$	9,921	0.92	0.88	0.92	1.55	143.69	**25.29**	∗	∗	∗	∗	∗	∗	∗
Yelp + Categorical Attributes														
$t=0$	2	0.49	0.83	0.60	3.59	**1.52**	**0.00**	2	0.49	0.83	0.60	3.59	5.04	0.14
$t=1$	17	**0.40**	**0.66**	**0.65**	6.44	**1.75**	**0.01**	13	0.24	0.50	0.51	**5.81**	8.15	1.32
$t=2$	126	**0.58**	**0.79**	**0.74**	**5.44**	**3.10**	**0.38**	51	0.48	0.61	0.63	5.56	29.36	55.57
$t=3$	65	**0.66**	**0.82**	**0.83**	6.25	**7.80**	**2.48**	85	0.62	0.70	0.73	**5.90**	121.69	971.89
$t=4$	129	**0.69**	**0.91**	**0.91**	6.16	**18.99**	**59.84**	166	0.65	0.79	0.83	**5.77**	374.77	8,033.96
$t=5$	254	0.72	0.94	0.94	5.96	38.04	**179.60**	∗	∗	∗	∗	∗	∗	∗
$t=6$	464	0.67	0.96	0.95	5.89	67.53	**699.93**	∗	∗	∗	∗	∗	∗	∗
$t=7$	839	0.66	0.98	0.72	5.99	116.14	**299.81**	∗	∗	∗	∗	∗	∗	∗
Yelp + Numeric Attributes														
$t=0$	2	0.49	0.83	0.71	0.77	**1.52**	**0.00**	2	0.49	0.83	0.71	0.77	35.60	0.41
$t=1$	15	**0.40**	**0.65**	0.72	0.74	**1.76**	**0.01**	9	0.39	0.63	0.72	**0.73**	60.76	1.60
$t=2$	41	**0.61**	0.73	0.73	0.73	**3.04**	**0.27**	32	0.52	**0.86**	**0.76**	**0.72**	641.06	46.32
$t=3$	63	**0.68**	0.80	0.74	0.88	**7.80**	**0.25**	61	0.50	**0.93**	**0.77**	**0.82**	7,869.45	398.87
$t=4$	124	**0.68**	0.91	0.79	0.85	**18.98**	22.61	128	0.49	**0.96**	**0.80**	**0.81**	21,104.88	2,730.09
$t=5$	225	0.70	0.95	0.81	0.82	**38.07**	122.61	†	†	†	†	†	†	†
$t=6$	399	0.71	0.95	0.83	0.70	**67.80**	512.50	†	†	†	†	†	†	†
$t=7$	735	0.70	0.97	0.76	0.99	**115.97**	901.80	†	†	†	†	†	†	†
TripAdvisor + Numeric Attributes														
$t=0$	6	0.71	1.00	0.73	1.05	**1.64**	**0.00**	6	0.71	1.00	0.73	1.05	35.58	0.50
$t=1$	144	0.94	0.95	0.75	0.86	**1.97**	**0.01**	144	0.94	0.95	0.75	0.86	80.74	3.98
$t=2$	425	0.95	0.92	0.74	0.87	**3.31**	**0.05**	439	0.95	**0.93**	0.74	**0.86**	911.01	109.32
$t=3$	618	**0.95**	0.90	0.73	0.88	**8.28**	**0.25**	696	0.92	**0.94**	**0.74**	**0.87**	9,451.34	778.46
$t=4$	1,636	0.93	0.93	0.73	0.93	**28.10**	1.83	†	†	†	†	†	†	†
$t=5$	2,179	0.88	0.94	0.73	0.98	**94.06**	20.97	†	†	†	†	†	†	†
$t=6$	2,850	0.86	0.95	0.73	1.01	**203.46**	138.74	†	†	†	†	†	†	†
$t=7$	3,475	0.84	0.95	0.73	0.99	**334.98**	385.27	†	†	†	†	†	†	†

5   Conclusions

In this paper, we proposed an efficient algorithm for community detection in dynamic attributed graphs. While several methods exist to detect communities in attributed graphs [5, 6, 8, 14, 15] and in dynamic graphs [1, 7], this is, to the best of our knowledge, the first methodology to incorporate both temporal and attribute information along with the structural properties of the graph.

We evaluated our proposed algorithm on a variety of synthetically generated benchmark dynamic attributed graphs, as well as on large-scale real-world networks. The results obtained show that our proposed algorithm is able to identify graph partitions of high modularity and high attribute similarity more efficiently than state-of-the-art methods for community detection [5, 14] in terms of both space and time. Moreover, unlike these state-of-the-art methods, our proposed algorithm is able to handle multiple attribute types, which is essential given the heterogeneous nature of many real-world networks.

Acknowledgments. This material is based upon work supported in part by the Laboratory for Analytic Sciences, the U.S. Department of Energy, Office of Science, Advanced Scientific Computing Research, and NSF grant 1029711.

References

1. Aynaud, T., Guillaume, J.: Static community detection algorithms for evolving networks. In: Proceedings of the 8th International Symposium on Modeling and Optimization in Mobile, Ad Hoc and Wireless Networks, 2010, pp. 513–519 (2010)
2. Bello, G.A., Angus, M., Pedemane, N., Harlalka, J.K., Semazzi, F.H.M., Kumar, V., Samatova, N.F.: Response-guided community detection: application to climate index discovery. In: Appice, A., Rodrigues, P.P., Santos Costa, V., Gama, J., Jorge, A., Soares, C. (eds.) ECML PKDD 2015. LNCS (LNAI), vol. 9285, pp. 736–751. Springer, Heidelberg (2015). doi:10.1007/978-3-319-23525-7_45
3. Blondel, V., Guillaume, J., Lambiotte, R., Lefebvre, E.: Fast unfolding of communities in large networks. J. Stat. Mech. Theor. Exp. **2008**(10), P10008 (2008)
4. Bothorel, C., Cruz, J., Magnani, M., Micenkova, B.: Clustering attributed graphs: models, measures and methods. Netw. Sci. **3**(03), 408–444 (2015)
5. Combe, D., Largeron, C., Géry, M., Egyed-Zsigmond, E.: I-Louvain: an attributed graph clustering method. In: Fromont, E., Bie, T., Leeuwen, M. (eds.) IDA 2015. LNCS, vol. 9385, pp. 181–192. Springer, Heidelberg (2015). doi:10.1007/978-3-319-24465-5_16
6. Dang, T., Viennet, E.: Community detection based on structural and attribute similarities. In: International Conference on Digital Society, pp. 7–12 (2012)
7. Dinh, T., Xuan, Y., Thai, M.: Towards social-aware routing in dynamic communication networks. In: 2009 IEEE 28th IPCCC, pp. 161–168. IEEE (2009)
8. Elhadi, H., Agam, G.: Structure and attributes community detection: comparative analysis of composite, ensemble and selection methods. In: Proceedings of the 7th Workshop on Social Network Mining and Analysis, pp. 10:1–10:7 (2013)
9. Fortunato, S.: Community detection in graphs. Phys. Rep. **486**(3), 75–174 (2010)
10. Lancichinetti, A., Fortunato, S., Radicchi, F.: Benchmark graphs for testing community detection algorithms. Phys. Rev. E **78**(4), 046110 (2008)
11. Newman, M.: Analysis of weighted networks. Phys. Rev. E **70**(5), 056131 (2004)
12. Newman, M., Girvan, M.: Finding and evaluating community structure in networks. Phys. Rev. E **69**(2), 026113 (2004)
13. Peel, L.: Estimating network parameters for selecting community detection algorithms. In: 2010 13th Conference on Information Fusion, pp. 1–8. IEEE (2010)

14. Ruan, Y., Fuhry, D., Parthasarathy, S.: Efficient community detection in large networks using content and links. In: Proceedings of the 22nd International Conference on World Wide Web, pp. 1089–1098. ACM (2013)
15. Zhou, Y., Cheng, H., Yu, J.: Clustering large attributed graphs: an efficient incremental approach. In: 2010 IEEE 10th ICDM, pp. 689–698. IEEE (2010)

Secure Computation of Skyline Query in *MapReduce*

Asif Zaman[✉], Md. Anisuzzaman Siddique, Annisa, and Yasuhiko Morimoto

Graduate School of Engineering, Hiroshima University,
Kagamiyama 1-7-1, Higashi-Hiroshima 739-8521, Japan
{d140094,d144809}@hiroshima-u.ac.jp, anis_cst@yahoo.com,
morimoto@mis.hiroshima-u.ac.jp

Abstract. To select representative objects from a large scale database is an important step to understand the database. A skyline query, which retrieves a set of non-dominated objects, is one of popular methods for selecting representative objects. In this paper, we have considered a distributed algorithm for computing a skyline query in order to handle "big data". In conventional distributed algorithms for computing a skyline query, the values of each object of a local database have to be disclosed to another. Recently, we have to be aware of privacy in a database, in which such disclosures of privacy information in conventional distributed algorithms are not allowed. In this work, we propose a novel approach to compute the skyline in a multi-parties computing environment without disclosing individual values of objects to another party. Our method is designed to work in *MapReduce* framework − in *Hadoop* framework. Our experimental results confirm the effectiveness and scalability of the proposed secure skyline computation.

Keywords: Secure skyline · MapReduce · Hadoop · Information security · Data mining

1 Introduction

To select representative objects from a large-scale database is an important step to understand the database. A skyline query is one of popular methods for selecting representative objects. A skyline query retrieves a set of representative objects, each of which is not dominated by another object. For example, if we consider the issue of a financial investment: an investor tends to purchase the stock that can minimize the commission costs and predicted risks. As a result, the target can be formalized as finding the skyline stock with minimum cost and minimum risks. Figure 1 shows seven stock records with their costs (d_1) and risks (d_2). If we attempt to find a suitable suggestion list for our clients using skyline query, the result will be $\{B, D, G\}$. From Fig. 1(b), it is clearly understandable that no other object can dominate those three objects, hence, they are in the result of a skyline query. Skyline query attracts consistent attention in database research, due to its applications in decision making and analytic.

© Springer International Publishing AG 2016
J. Li et al. (Eds.): ADMA 2016, LNAI 10086, pp. 345–360, 2016.
DOI: 10.1007/978-3-319-49586-6_23

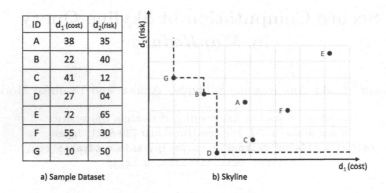

ID	d₁ (cost)	d₂ (risk)
A	38	35
B	22	40
C	41	12
D	27	04
E	72	65
F	55	30
G	10	50

a) Sample Dataset   b) Skyline

Fig. 1. A *skyline* problem

From a different angle, we have to be aware of privacy of individual object in a database. So far, a lot of algorithms for computing skyline query have been proposed, some of which are designed in distributed computing environment to be able to handle "big data" [1,12,16]. However, none of them considered privacy issues in a database. In this work, we have proposed a novel approach so that data can be processed in distributed manner, meanwhile privacy of individual object has been preserved.

Assume that many organizations have done some surveys about commission cost and risk prediction. Assume that each of the organizations has collected a same kind of privacy information of their customers. Since each organization does not want to disclose the database, each can not compute skyline query of the union of all organizations' databases but only compute skyline query of its own database. It is no doubt that the skyline of the union is more valuable than that of each.

Suppose, for example, two individual organization have done some market research and they have collected a database about commission cost and risk prediction. These information are sensitive and both the parties are not wanting to disclose the values of individual object. But, these two parties are willing to get the outcome of skyline query of the cost and the risk from the union of the two parties' database. In conventional skyline computation methods, it is not possible to get the result of skyline query result without disclosing the values to others. Proposed method can solve this issue, not only by preserving data privacy but also by doing the job in distributed *MapReduce* framework.

The remaining part of this paper is organized as follows. Section 2 reviews the related work. Section 3 discusses the notions and basic properties for *skyline* objectset and define our secure *skyline* computation problem. In Sect. 4, we specify details of our algorithms with proper examples and analysis. We experimentally confirm our algorithms in Sect. 5 under a variety of settings. Finally, Sect. 6 concludes this work and describes our future intention.

2   Related Work

Our work is motivated by previous studies of skyline query processing, multiparty secure computation, as well as *MapReduce* based query processing.

2.1   Skyline Query

Borzsonyi et al. have proposed the skyline operator over large databases and proposed three algorithms, which are *Block-Nested-Loops (BNL)*, *Divide-and-Conquer (D&C)*, and B-tree-based schemes [8]. BNL compares each object of the database with every other objects, and reports it as a result only if any other object does not dominate it. *D&C* divides the dataset into several partitions such that each partition can fit into memory. Local skyline objects for each individual partition are then computed by a main-memory skyline algorithm. The final skyline is obtained by merging the local skyline objects for each partition. Kossmann et al. improved the *D&C* algorithm and proposed nearest neighbor (NN) algorithm for efficiently pruning dominated objects by partitioning the data space iteratively based on the nearest objects in the space [13]. Chomicki et al. proposed *Sort-Filter-Skyline (SFS)* as a variant of BNL [9], which can improve BNL by presorting. The most efficient method so far is *Branch-and-Bound Skyline (BBS)*, proposed by Papadias et al., which is a progressive algorithm based on the *best-first nearest neighbor (BF-NN)* algorithm [15].

Recently, parallel computing paradigm becomes very popular for skyline computation. W.T. Balke et al. have introduced skyline queries in distributed environments in [6]. Vertically partitioned web information are supported by their work. Thereafter, within the literature, abundant studies achievements had been received to address distributed skyline queries. Both of Wang et al. and Chen et al. researched skyline queries in structured P2P networks, named BATON networks, where peers are responsible for a partial region of data space [23]. Rocha-Junior et al. [17] proposed a grid-based approach for distributed skyline processing (AGiDS), which assumes that each peer maintains a grid-based data summary structure for describing its data distribution. Arefin et al. [5] worked on agent based privacy skyline-set for distributed database but the problem solved in this paper is different from the conventional skyline query, which we are considering in this paper.

The *MapReduce* framework, which have been developed by Google, has become popular to process queries over big data due to its scalability and fault tolerance. In [26], Zhang et al. first proposed a preliminary way for skyline queries in *MapReduce* framework. Three skyline algorithms, called MR-BNL, MR-SFS and MR-Bitmap, were proposed by the author using the *MapReduce* framework. In [16], Park et al. introduced an efficient parallel algorithm, SKY-MR, for processing the skyline queries. In the SKY-MR algorithm, a sky-quad tree was introduced with a sample of the entire dataset and was utilized in the data partition and local pruning.

2.2 Multi-party Secure Computation

Multi-party computation problem, which was introduced by Yao [25] and extended by Goldreich, Micali, Wigderson [10] and many others, has attracted attention in privacy-aware computing environment. Secure function evaluation, as was introduced by Yao, allows a set $P = \{p_1, \cdots, p_n\}$ of n players/parties to compute an arbitrary agreed function of their private data, even if an adversary may corrupt and control some players/parties in various ways – the privacy of data will be preserved. Security in Multi-party Computation means that the parties' data remain secret (except for what is revealed by the intended results of the computation) and that the results of computation are guaranteed to be correct [10]. In general, Multi-party Computation protocols tend to be less efficient than specific purpose protocols.

In privacy-preserving data mining problems, there are another multi-party secure computation problems that have been discussed in the literature. Lindell and Agrawal proposed two different privacy preserving data mining problems. In the problem defined in Lindell's paper [14], two parties, both having non public databases, want to jointly conduct a data mining operation on the union of their two databases. The problem is how to compute the operation without disclosing their database to other parties, or any third party. On the other hand, in Agrawal's paper [3], the problem is as follows: one party say "Alice" is allowed to conduct data mining operation on a private database owned by another party say "Bob", the problem is how could Bob prevent Alice from accessing precise information in individual data records, while Alice is still able to conduct the data mining operations. Lindell and Pinkas used secure multi-party computation protocols to solve their problem, while Agrawal used the data perturbation method.

2.3 MapReduce Implementations of Skyline Query

Google's *MapReduce* [7,11,22] and its open source variant *Hadoop* [4] are powerful tools for building scalable parallel applications. Recently, *MapReduce* has attracted a lot of attention in handling "big data". There exist some recent works on large-scale skyline computation using *MapReduce* [12,20]. In [19,21], we have proposed a *MapReduce* based algorithm to process k-dominant skyline query. The k-dominant skyline query can reduce the number of retrieved objects by relaxing the dominance definition, in which an object more likely to be dominated.

However, all of the above methods cannot preserve privacy of individual object during the distributed computation. This paper propose a secure distributed computation on *MapReduce* to compute a skyline query. To the best of our knowledge, there is no such *MapReduce* based algorithm for secure skyline query so far.

3 Preliminaries

3.1 Dominance and Skyline

Given a dataset DS with d-dimensions $\{d_1, d_2, \cdots, d_d\}$ and n objects $\{O_1, O_2, \cdots, O_n\}$. We use $O_i.d_j$ to denote the j-th dimension value of object O_i. Without loss of generality, we assume that smaller value in each attribute is better.

Dominance: An object $O_i \in DS$ is said to dominate another object $O_j \in DS$, denoted as $O_i \prec O_j$, if $O_i.d_r \leq O_j.d_r$ $(1 \leq r \leq d)$ for all d dimensions and $O_i.d_t < O_j.d_t$ $(1 \leq t \leq d)$ for at least one attribute. We call such O_i as *dominant object* and such O_j as *dominated object* between O_i and O_j. For example, in Fig. 1(b) object F is dominated by object C.

Skyline: An object $O_i \in DS$ is said to be a *skyline object* of DS, if and only if there is no such object $O_j \in DS$ $(j \neq i)$ that dominates O_i. The skyline of DS, denoted by $Sky(DS)$, is the set of skyline objects in DS. For dataset shown in Fig. 1(a), object D dominates $\{A, C, F, E\}$ and objects $\{B, G\}$ are not dominated by any other objects in DS. Thus, skyline query will retrieve $Sky(DS) = \{B, D, G\}$ (see Fig. 1(b)).

3.2 Hadoop MapReduce

Hadoop is an open source implementation of the *MapReduce* framework, maintained by Apache Software Foundation [4]. This framework is designed to allow users to define a *MapReduce* job only by defining the *map* and *reduce* functions. In this framework, data are represented as $\langle key, value \rangle$ pairs and computations are distributed across a shared nothing cluster of autonomous machines. Jobs to be performed using the *MapReduce* framework mainly refer to two user-defined functions, called *Map* and *Reduce:*

$$Map(k_1, v_1) \rightarrow list(k_2, v_2)$$
$$Reduce(k_2, list(v_2)) \rightarrow list(v_3)$$

The *Map* function (sometimes called Mapper) processes on each $\langle key, value \rangle$ pair of input data, and produces intermediate $\langle key, value \rangle$ pairs. The intermediate $\langle key, value \rangle$ pairs are then sorted and grouped associated with the same intermediate *key*. The *Reduce* function (sometimes called Reducer) takes a *key* and a list of *values* for that *key*, applies the processing algorithm, and generates the final result. In other words, if we want to compute an algorithm in this scalable framework, we have to design the algorithm by *Map* and *Reduce* functions.

3.3 Multi-party Secure Skyline Problem

Let us consider a situation where several organizations have done some surveys about commission cost and risk prediction. We assume that each of the organizations has collected a same kind of privacy information of their customers.

Each organization wants to find the result of skyline query of the union of these organizations' databases. But, none of them is allowed to disclose values of their database to other organizations.

We call participant organizations of the skyline computation as parties. To accomplish the computation, we use a trusted third party as *Coordinator*. We assume that *Coordinator* is not vulnerable to intruder and will not disclose sensitive information.

To simplify the problem, we assume the number of participant parties is two and we denote the two parties as $DataNode_1$ and $DataNode_2$, respectively.

In order to ensure the privacy of communication between the *Coordinator* and *DataNodes*, the *Coordinator* uses *Public-key* cryptography, that is the *Coordinator* is equipped with a public key (key_1) and a private key (key_2). *DataNodes* are informed about the key_1 of *Coordinator* and before sending any data packet to *Coordinator*, it must be encrypted by key_1. Only *Coordinator* can decrypt the data using key_2. Similarly, both *DataNodes* use the *Symmetric-key cryptography* for their own data's privacy. Let the cryptography keys be key_3 and key_4 for $DataNode_1$ and $DataNode_2$, respectively.

4 Multi-party Secure Skyline Algorithm

The proposed algorithm consists of six steps: (1) Preparing the $\langle key, value\rangle$ pair, (2) Ordering with *MapReduce*, (3) Disguise the original order, (4) Return of disguised order values, (5) Merging and sorting, and (6) Skyline computation.

4.1 Preparing the $\langle key, value\rangle$ Pair

In order to avoid confusion, we will denote *encryption-key* for the cryptographic key to distinguish with the *key* of $\langle key, value\rangle$ pairs, which are used in MapReduce framework. In Fig. 2, each *DataNode* has a collection of two dimensional sensitive data. Both *DataNodes* perform *Symmetric-key cryptography*

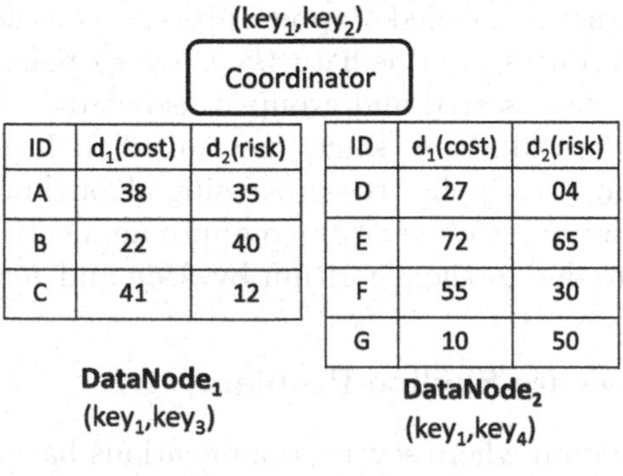

ID	d_1(cost)	d_2(risk)
A	38	35
B	22	40
C	41	12

ID	d_1(cost)	d_2(risk)
D	27	04
E	72	65
F	55	30
G	10	50

Fig. 2. Multi-party *skyline* computation

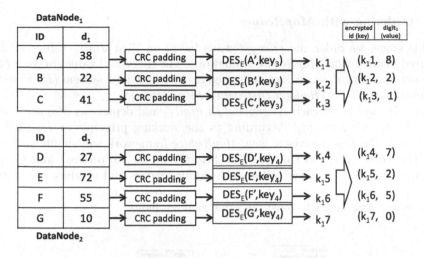

Fig. 3. Generation of $\langle key, value \rangle$ pair

(e.g. DES - Data Encryption Standard) with private *encryption-key* before sharing any data with *Coordinator*.

Figure 3 illustrates the encryption and how to generate $\langle key, value \rangle$ pair's of the first digit, $digit_1$, (the least significant digit). In the example, only the dimension d_1 is described though we perform the same encryption on all the sensitive dimensions. Each party encrypts their data as follows: (1) Each *DataNode* adds redundant bits for each *ID* by using CRC (*Cyclic Redundancy Check*) scheme [24]. We call this process as "CRC Padding". Assume that, for the example case given in Fig. 3, first record of $DataNode_1$ whose *ID* is "A", after padding modulo string becomes "A'". (2) *IDs* with padded CRC bits, are encrypted by the corresponding node's encryption key. In example, we generate "key" $(k_1 1)$ for the first record of $DataNode_1$ (whose original *ID* was "A", after padding it became "A'") by encrypting with key_3. Note that owner of the *encrypted_id* can decrypt but any other parties cannot. Though CRC scheme is usually used as a transmission error checking tool, we also used the CRC scheme to check whether a processed $\langle key, value \rangle$ is *DataNode*'s (is inherent in *DataNode*). (3) We generate $\langle key, value \rangle$ pairs of the first digit (the least significant digit). For example, $\langle key, value \rangle$ pair of the first record of $DataNode_1$ is $\langle k_1 1, 8 \rangle$ (as the $digit_1$ of dimension d_1 for object "A" is 8).

In our method, a key (an *encrypted_id*) of a $\langle key, value \rangle$ pair must be unique. However, there is a possibility that different records happen to have a same key (*encrypted_id*) though the possibility is extremely small. If such a collision is found by *Coordinator*, *Coordinator* forces all *DataNodes* to reproduce keys (*encrypted_ids*) with different "CRC Padding". Hence the possibility of conflicting in "CRC Padding" can be eliminated.

4.2 Ordering with *MapReduce*

In this stage, we order the *encrypted_ids* based on their $digit_1$ values in distributed manner with *MapReduce* framework. The encrypted values of data IDs and corresponding $digit_1$ values are stored in distributed file system (*DFS*) more specifically in *HDFS* (*Hadoop Distributed File System*).

The *Mapper* reads each $\langle encrypted\_id, digit_1 \rangle$ and depicts as *Mapper*'s output: $\langle digit_1, encrypted\_id \rangle$. According to the working principle of *MapReduce* framework, the $digit_1$ serves as *key*. *MapReduce* framework will shuffle and sort the $\langle key, value \rangle$ pairs so that the *key* ($digit_1$) values are ordered and tagged together. The *Reducer* layer collects the shuffled values and produce the sorted order of *encrypted_ids*. Note that several *encrypted_ids* may have same $digit_1$ value as *key* and they will have the same ranking index.

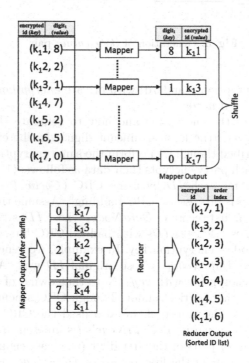

Fig. 4. Ordering with *MapReduce*

In Fig. 4, we can see that a *Mapper* takes the $\langle k_11, 8 \rangle$ pair as input and produces $\langle 8, k_11 \rangle$ as output. Similarly, a *Mapper* takes $\langle k_13, 1 \rangle$ and produces $\langle 1, k_13 \rangle$, and so on. After the shuffling, pairs are grouped by *key* and are fed into *Reducer* layer. *Reducer* produces the sorted order of *encrypted_id*.

4.3 Disguise the Original Order

In order to conceal the actual order ($\{1, 2, 3, ...\}$) from intruders, we had better to shift the density of the order distribution. The detailed idea of the order value transformation is discussed in [2]. Brief explanation of the transformation is as follows: First of all, we have to select a target distribution other than uniform distribution. Target distribution is a user specified data distribution function such as Gaussain or Zipf or similar distribution. After choosing the target distribution, we have to generate $|X|$ unique values from the target distribution where X is the collection of ordered sequence indices. Then, we sort the generated random values into a table T. The sorted i^{th} index value of ordered list (*encrypted_id*) is then given the order value of $T[i]$.

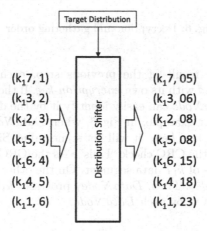

Fig. 5. Disguise the original order

In Fig. 5, our *Distribution Shifter* module receives the target distribution and generate 6 unique (because, $|X| = 6$) values and after ordering $T = \{05,\ 06,\ 08,\ 15,\ 18,\ 23\}$. From previous step, we know that $k_1 7$'s rank is 1 in our sorted list. We will replace this rank with the first value of T, i.e., the new order index of $k_1 7$ is 05. Similarly, $k_1 3$'s order index will be 06 and so on. Hence, we have been able to shift the order index while preserving the order sequence − which we call as disguised order.

4.4 Return of Disguised Order Values

Disguising the rank by the previous procedure makes the order of *encrypted_ids* difficult to find while it preserves the original order. After transferring distribution of the rank values, the results are sent back to both *DataNodes*. Note that other parties cannot infer the sensitive information since the original *IDs* are encrypted.

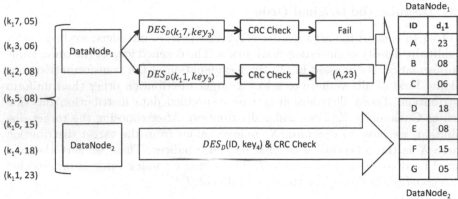

$d_1 1$: dimension d_1(cost)'s digit$_1$ disguised rank.

Fig. 6. Decryption and gathering order

After receiving the result of the previous step, each *DataNode* tries to decrypt the *encrypted_id* with its own *encryption-key*. If the *DataNode* owns the received data, the *DataNode* can easily identify it by the decrypted *ID* and the CRC code checking. For example, in Fig. 6, when *DataNode*$_1$ tries to decrepit $k_1 7$ with its encryption key, key_3, it will get some value. Since the $k_1 7$ does not belong to *DataNode*$_1$, the CRC check "Fails". As the CRC "Fails", *DataNode*$_1$ discards further process of $k_1 7$ data segment. On the other hand, $k_1 1$ passes the CRC check of *DataNode*$_1$. Then, *DataNode*$_1$ processes $k_1 1$ for the next round. Similar process is followed by each *DataNode*.

4.5   Merging and Sorting

After receiving the disguised order of $digit_1$, each *DataNode*'s next job is to merge those values with $digit_2$ values. These new values along with encrypted *IDs* are sent to *MapReduce* framework for sorting to get the order of $digit_2$ and $digit_1$. The encryption of *IDs* can be done, after CRC padding, by using the *encryption-key* as mentioned in Sect. 4.1. The *encryption-key* may be different to that used in $digit_1$. In Fig. 7 show the process of the $digit_2$ values. In *DataNode*$_1$, "A" has been encrypted to a new *encrypted_id* "$k_2 1$". Since $digit_1$'s disguised order of "A" is 23 and $digit_2$'s value is 3, *DataNode*$_1$ constructs the concatenated value 3.23, which makes the order of $digit_1$ preserved. Similarly, for "B", the concatenated value becomes 2.08. After calculating all the concatenated values of dimension d_1, we feed these data to *MapReduce* framework to get the order of *ID*'s. Similarly, we calculate the order of other dimension's values. These orders will be used to calculate a skyline query.

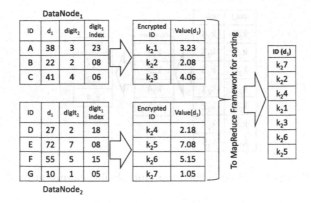

Fig. 7. Merge and sort

4.6 Skyline Computation

From the output of *MapReduce* in the previous step, the *Coordinator* get the order of data on each dimension. *Coordinator* uses these order to calculate candidates of skyline query. The detailed process of this skyline calculation, which does not use individual values but the orders, has originally published in [18]. In the work [18], the authors partition dataset vertically and sort each partition. Then, Coordinator collects (ID, Rank) pairs. Figure 8 show the example. The left table of the figure has the *encrypted-ids* ordered by each dimension rank value. $ID(d_1)$ is the sorted ID of d_1, which shows that object k_27 is the 1st and object k_22 is the 2nd and so forth. Similarly, object k_24 is the 1st on d_2 and object k_23 is the 2nd and so forth. *Coordinator* maintains a counter for each object. In the beginning, the counter for each object is set to zero. *Coordinator* reads the *encrypted-ids* from the 1st row and increments the counter values based on the row value. In the running example, the counter of k_27 and k_24 are incremented by the 1st row. Next, the counter of k_24 and k_23 are incremented by the 2nd row. Next, the counter of k_24 and k_25 are incremented by the 3rd row. If one of the counter becomes the number of dimensions, *Coordinator* will stop the increment procedure. In the running example, *Coordinator* stops the increment procedure when k_24's counter becomes 2 in the 3rd row. Then, *Coordinator* collects IDs whose counter value is not 0 as a candidate skyline. In the example, the candidate will be $\{k_27, k_22, k_24, k_23\}$. Note that other objects must be dominated by k_24 whose counter values is 2.

After the candidates are collected, we examine dominance for all pairs of the candidate objects. We use a sort filtering skyline (SFS) method [9] based on the order of the counter values. As the number of candidates are much smaller than the number of objects any skyline processing algorithm can check the dominance without spending a lot of time. In the example, we can find that k_23 is dominated by k_24 and *Coordinator* finalizes the skyline result as $\{k_27, k_22, k_24\}$ which are *encrypted-ids* of $\{G, B, D\}$, respectively.

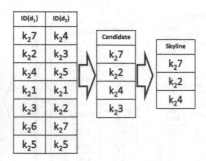

Fig. 8. Candidate generation and skyline computation

In Fig. 8, the left table, which has order for each dimension, contains *encrypted-ids*. In the figure, we explain the candidates generation procedure assuming $k_2 1$ in d_1 and $k_2 1$ in d_2 is the same. But, we have to note that the *encrypted-ids* for d_1 and those of d_2 are different, which prevents compromising relative merits of two anonymous records. In the candidate generation process, each *DataNode* discloses the identity of *encrypted-ids* to *Coordinator*. Only *Coordinator* knows the information of the left table. Each *DataNode* does not find the information of records in the table that are owned by others.

5   Experiments

This section reports our experimental results to examine the effectiveness and efficiency of proposed method. We have configured a cluster of 4 commodity PCs in a high speed Gigabit Ethernet networks, each of which has an Intel Core 2 Duo E8500 3.16 GHz CPU, 8 GB memory. We have compiled the source codes under *Java* V8. We have used *Hadoop* version 2.5.2 and 64 bit Cent-OS 7. We have set the replication parameter of *Hadoop* cluster to 2. Since none of the existing algorithm has considered the computation issue of secure skyline, we could not compare to others. Instead, we could check the scalability of ours and report a comparison of the time of skyline computation with the security enhancement and the time of skyline computation without the security enhancement. We have used synthetic dataset. The aim of our experiment design was to check computational overhead secure approach over traditional non-secure approach.

A. Effect of encryption process over data distribution: Figure 9(a) shows the effect of encryption process over data distribution. We have varied data size from 50 k to 250 k per node. Our experimental results for a single iteration of the process explained in Sect. 4.1 shows that the runtimes are identical for Anti-correlated, correlated and Independent dataset. Hence no effect of data distribution upon this step.

B. Ordering with *MapReduce*: Ordering with *MapReduce* is one of the vital time consuming part of the proposed work, described in Sect. 4.2. Figure 9(b)

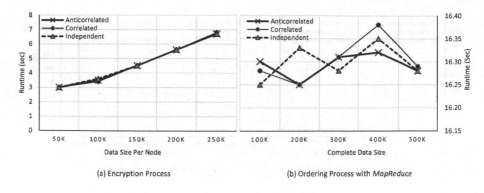

(a) Encryption Process

(b) Ordering Process with *MapReduce*

Fig. 9. Effect of encryption & ordering process

shows us the runtime comparison of different data distribution for *MapReduce* ordering. These runtimes are recorded for a single iteration. We varied data size from 100 k to 500 k. It shows that the ordering time varies very little (in terms of few milliseconds) for different distribution.

C. Disguising the Original Order: Figure 10(a) shows the execution time comparison for Disguising the original order among different data distribution. The process is described in Sect. 4.3. Our experimental results shows that this process is not affected by data distribution at all. As shown in figure, we have varied our dataset size from 100 k to 500 k and found the execution times are identical for Anti-Correlated, Correlated and Independent data distribution.

D. Candidate Selection and Skyline Computation: This process is described in Sect. 4.6. As shown in Fig. 10(b), this process is affected by data distribution. We varied our dataset size form 100 k to 500 k and found that this process is more efficient for Correlated dataset and less efficient for Anti-Correlated dataset. However, the performance for Independent dataset lies in between the performance for Anti-Correlated and Correlated dataset.

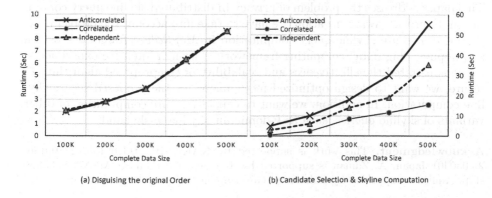

(a) Disguising the original Order

(b) Candidate Selection & Skyline Computation

Fig. 10. Disguising original order, candidate selection & skyline computation

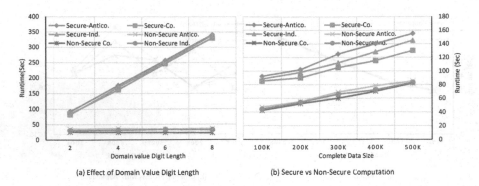

(a) Effect of Domain Value Digit Length   (b) Secure vs Non-Secure Computation

Fig. 11. Effect of domain value digit length & secure *vs* non-secure computation

E. Effect of domain value length: One of the interesting finding for our proposed algorithm is given in Fig. 11(a). In this experiment we have fixed the data size to 200k and used two dimensional data but varied the domain value digit length and record the execution time for complete calculation. We can see that the domain value digit length has considerable effect on our proposed method. However, in conventional non-secure skyline algorithms like SFS are not affected by such digit length.

F. Secure vs Non-secure Computation: Figure 11(b) gives us an idea about computational overhead due to impose of secure computation scheme. It is obvious that the secure approach requires some extra execution time. However, we can notice that the computational overhead seems to be an identical value though the data size increased linearly. Hence, it may not be considered as overhead when processing "big data". In this experiment we have fixed the domain value digit length to 3 and data dimension to 4.

6   Conclusion

This paper addresses the problem of privacy in distributed skyline query computation. In privacy aware situation, we have to take into account the problem. We proposed a secure skyline query computation in *MapReduce* framework, which is a popular "big data" computing framework. Through intensive experiments, we demonstrated the effectiveness and scalability of the proposed algorithm. In future, we want to design optimized mechanisms for the proposed secure skyline computation. In addition, we want to consider secure computation of other variants of skyline queries, such as k-dominant skyline, k-skyband.

Acknowledgment. This work is supported by KAKENHI (16K00155, 23500180, 25.03040) Japan. A. Zaman is supported by Japanese Government MEXT Scholarship. Annisa is supported by Indonesian Government DG-RSTHE scholarship.

References

1. Afrati, F.N., Koutris, P., Suciu, D., Ullman, J.D.: Parallel skyline queries. In: ICDT, pp. 274–284 (2012)
2. Agrawal, R., Kiernan, J., Srikant, R., Xu, Y.: Order preserving encryption for numeric data. In: ACM SIGMOD International Conference on Management of Data, pp. 563–574 (2004)
3. Agrawal, R., Srikant, R.: Privacy-preserving data mining. In: ACM SIGMOD International Conference on Management of Data, pp. 439–450. ACM (2000)
4. Apache: Apache hadoop (2010). http://hadoop.apache.org
5. Arefin, M.S., Morimoto, Y.: Privacy aware parallel computation of skyline sets queries from distributed databases. In: 2013 International Conference on Computing, Networking and Communications (ICNC), pp. 186–192 (2011)
6. Balke, W.-T., Güntzer, U., Zheng, J.X.: Efficient distributed skylining for web information systems. In: Bertino, E., Christodoulakis, S., Plexousakis, D., Christophides, V., Koubarakis, M., Böhm, K., Ferrari, E. (eds.) EDBT 2004. LNCS, vol. 2992, pp. 256–273. Springer, Heidelberg (2004). doi:10.1007/978-3-540-24741-8_16
7. Blanas, S., Patel, J.M., Ercegovac, V., Rao, J., Shekita, E.J., Tian, Y.: A comparison of join algorithms for log processing in MapReduce. In: SIGMOD, pp. 975–986 (2010)
8. Borzsonyi, S., Kossmann, D., Stocker, K.: The skyline operator. In: Proceedings of ICDE, pp. 421–430 (2001)
9. Chomicki, J., Godfrey, P., Gryz, J., Liang, D.: Skyline with presorting. In: Proceedings of ICDE, pp. 717–719 (2003)
10. Goldreich, O., Micali, S., Wigderson, A.: How to play any mental game. In: Proceedings of the Nineteenth Annual ACM Symposium on Theory of Computing, pp. 218–229. STOC 1987. ACM (1987)
11. Jiang, D., Tung, A.K.H., Chen, G.: Map-Join-Reduce: toward scalable and efficient data analysis on large clusters. In: IEEE TKDE, pp. 1299–1311 (2011)
12. Mullesgaard, K., Pedersen, H.L., Zhou, Y.: Efficient skyline computation in MapReduce. In: EDBT, pp. 37–48 (2014)
13. Kossmann, D., Ramsak, F., Rost, S.: Shooting stars in the sky: an online algorithm for skyline queries. In: Proceedings of VLDB, pp. 275–286 (2002)
14. Lindell, Y., Pinkas, B.: Privacy preserving data mining. In: Bellare, M. (ed.) CRYPTO 2000. LNCS, vol. 1880, pp. 36–54. Springer, Heidelberg (2000). doi:10.1007/3-540-44598-6_3
15. Papadias, D., Tao, Y., Fu, G., Seeger, B.: Progressive skyline computation in database systems. ACM Trans. Database Syst. **30**, 41–82 (2005)
16. Park, Y., Min, J.K., Shim, K.: Parallel computation of skyline and reverse skyline queries using MapReduce. Proc. VLDB Endow. **6**(14), 2002–2013 (2013)
17. Rocha-Junior, J.B., Vlachou, A., Doulkeridis, C., Nørvåg, K.: AGiDS: a grid-based strategy for distributed skyline query processing. In: Hameurlain, A., Tjoa, A.M. (eds.) Globe 2009. LNCS, vol. 5697, pp. 12–23. Springer, Heidelberg (2009). doi:10.1007/978-3-642-03715-3_2
18. Siddique, M.A., Tian, H., Morimoto, Y.: Distributed skyline computation of vertically splitted databases by using MapReduce. In: Han, W.-S., Lee, M.L., Muliantara, A., Sanjaya, N.A., Thalheim, B., Zhou, S. (eds.) DASFAA 2014. LNCS, vol. 8505, pp. 33–45. Springer, Heidelberg (2014). doi:10.1007/978-3-662-43984-5_3

19. Siddique, M.A., Tian, H., Morimoto, Y.: k-dominant skyline query computation in MapReduce environment. IEICE Trans. Inf. Syst. **98**, 1745–1361 (2015)
20. Tao, Y., Lin, W., Xiao, X.: Minimal MapReduce algorithm. In: Proceedings of SIGMOD, pp. 529–540 (2013)
21. Tian, H., Siddique, M.A., Morimoto, Y.: An efficient processing of k-dominant skyline query in MapReduce. In: Proceedings of ACM International Workshop on Bringing the Value of Big Data to Users (Data4U), pp. 29–35 (2014)
22. Vernica, R., Carey, M.J., Li, C.: Efficient parallel set-similarity joins using MapReduce. In: Proceedings of SIGMOD, pp. 495–506 (2010)
23. Wang, S., Ooi, B.C., Tung, A.K.H., Xu, L.: Efficient skyline query processing on peer-to-peer networks. In: 2007 IEEE 23rd International Conference on Data Engineering, pp. 1126–1135, April 2007
24. Williams, R.: A painless guide to CRC error detection algorithms (1996). ftp.rocksoft.com/papers/crc_v3.txt
25. Yao, A.C.: Protocols for secure computations. In: Proceedings of the 23rd Annual IEEE Symposium on Foundations of Computer Science, pp. 160–164 (1982)
26. Zhang, B., Zhou, S., Guan, J.: Adapting skyline computation to the MapReduce framework: algorithms and experiments. In: Xu, J., Yu, G., Zhou, S., Unland, R. (eds.) DASFAA 2011. LNCS, vol. 6637, pp. 403–414. Springer, Heidelberg (2011). doi:10.1007/978-3-642-20244-5_39

Recommending Features of Mobile Applications for Developer

Hong Yu, Yahong Lian, Shuotao Yang, Linlin Tian, and Xiaowei Zhao[✉]

School of Software, Dalian University of Technology, Dalian, China
vivian_dlut@163.com

Abstract. Features recommendation is an important technique for getting the requirements to develop and update mobile Apps and it has been one of the frontier study in requirements engineering. However, the mobile Apps' descriptions are always free-format and noisy, the classical features recommendation methods cannot be effectively applied to mobile Apps' features recommendation. In addition, most mobile Apps' source codes that contain API calling information can be obtained by software tools, which can accurately indicate the functional features. Therefore, this paper proposes a hybrid feature recommendation method of mobile Apps, which is based on both explicit description and implicit code information. A self-adaptive similarity measure and KNN is used to find relevant Apps, and functional features are extracted from the Apps and recommended for developers. Experimental results on four categories Apps show that the proposed features recommendation method with hybrid information is more effective than the classical method.

Keywords: mobile Apps · Features recommendation · Requirements engineering · Explicit information · Implicit information

1 Introduction

With the rapid development of information technology, the smart mobile application(App) is more and more prevalent and the number of mobile applications gets explosive growth. As data shows (http://www.appannie.com), the number of mobile applications in Google Play store has increased about sixty percent since July 2013 to June 2014, reached up to 1.5 million. And the number of download has increased about fifty percent since the fist quarter of 2013 to the first quarter of 2014, reached up to 150 million. Continuously increasing download reflects the large demand of Apps, and massive number of Apps triggers the fierce competition of Apps' market. For App developers, it is very important and worthy to automatically learn advantage of same type of Apps and explore optimal Apps' functional features.

Feature recommendation is one of the important methods in software requirement engineering [3, 8, 15]. In recent years, feature recommendation of traditional software for developers has got some developments [21]. However, feature recommendation of mobile Apps is on the way of studying. Feature recommendation of

© Springer International Publishing AG 2016
J. Li et al. (Eds.): ADMA 2016, LNAI 10086, pp. 361–373, 2016.
DOI: 10.1007/978-3-319-49586-6_24

traditional software is mainly based on the mining on the software description. Compared with traditional software, Apps' descriptions are free from structure, imprecise, and have some noises. On the other hand, we can get the API information of Apps what is impossible in traditional software by direct or indirect way. As in the App developing process, similar functions are often implemented by similar APIs, it is intuitive that we can recommend features of Apps by mining both the description and APIs.

Based on the above analysis, this paper presents a feature recommendation of Mobile Applications for developers, namely, developers input the description snippets or some APIs information of an App which is to be developed or upgraded, recommendation system will return the most relevant and valuable functional features to developers. The source data of the recommendation system can be downloaded from App stores, including descriptions, bytecodes and so on. We firstly use text process technique and Incremental Diffusive algorithm which proposed in [10] to learn the Apps' functional features, construct Apps-Features matrix, and then we employ decompilation technique to recognize implicit APIs information, construct Apps-API matrix. Finally, we propose to achieve the most relevant Apps to the query, and recommend the main functional features of the relevant Apps by a self-adaptable clustering method. The experiment results show that the recommendation method that utilize Apps' explicit information(description) and implicit information(APIs) outperforms the classical feature recommendation methods of traditional software.

2   Related Work

Recommending functional features of software products for developers is an important work in the requirements engineering. It can contribute to confirm software requirements, update software's function, support software product lines and software product reusing [3,8,15,17]. Different from recommending software for users [2,18,23], the functional feature recommendation method can be divided into two types: manual methods and automated methods. Manual methods mainly refer to the domain analysis techniques, such as feature oriented domain analysis [13] and feature oriented reuse method [12]. This type of methods depends on domain expert's comparative analysis on the existing specification or the description on websites, naturally, the ability of domain expert and the archived documents make a great impact on analysis results. Automated methods apply data mining and information retrieval techniques to recognize and recommend functional features. The classical automated methods include online topic recommendation forums [21] and on-demand functional feature recommendation [10]. Similar with manual methods, automated methods are also based on the existed documents, which lead to the limitation of recommended functional feature.

Recent years, collaborative topic models are very prevalent among recommendation system [11], All Collaborative Filtering methods share a capability to utilize the past ratings of users in order to predict or recommend new content that an individual user will like. However, in our paper, we want to recommend functional feature to the developers, not to users, on the other hand, it is very hard to collect developer rating on App rather than user rating. In the frame of matrix-factorization approaches [1], it relies on the idea that there is a small number of latent factors (features) that underly the preferences (interactions) between users and items. As these features are defined so as to fit at best the data, no obvious interpretation can be made of them and as a result, therefore, recommendations have no obvious explanation.

Different from standard functional description of traditional software, mobile Apps are free of publication and don't have standard product description layout. Therefore, traditional functional features recommendation methods cannot effectively work for App functional features recommendation. On the other hand, besides the descriptions, App stores provide more information for Apps such as reviews and downloads. And several techniques has been developed to analyze Apps' description, reviews and downloads for identifying functional features [5,16,22,24,25].

According to statistics in November 2014 of Strategy Analytics which is a famous research institution, IOS applications and Android applications accounted for 96 % of the mobile Apps market. These mobile Apps are available commonly through App stores which are maintained by Apple, Google, or Research in Motion. Although the App stores only distribute the byte code of the mobile apps, the Software Bertillonage technique [6] can be used to track code across mobile Apps. Code is an objective indication of Apps' features and it is hard to get for classical software. Ruiz et al. [19] shows that almost 23 % of the code classes inherit from a base class in the Android API, and 27 % of the classes inherit from a domain specific base class. McMillan et al. [14] uses the API calls to find similar Apps. It is intuitive that the same kind of Apps would have same or similar code information. However, the code information has never been integrated with the description information for features recommendation.

In this paper, we define Apps' descriptions as explicit information, and code information of Apps like APIs as implicit Information. As generalization of classical software features recommendation proposed by Negar Hariri, we design a mobile Apps feature recommendation method with explicit Information, named AFR - Explicit (Apps Features Recommendation with Explicit Information). The ultimate work of this paper is to put forward the mobile Apps functional feature recommendation method which integrates both the explicit and implicit information, named AFR - Hybrid (Apps Features Recommendation with Hybrid Information). Both AFR-Explicit and AFR-Hybrid would recommend functional features for developers. A case study on numbers of Apps with four categories shows that the hybrid information greatly enhances the accuracy in searching for similar Apps and also improves the precision and recall of functional feature recommendation.

3   Framework of Apps Features Recommendation with Hybrid Information

Based on the explicit and implicit App information, this paper presents the following functional feature recommendation framework in Fig. 1. It consists of three steps. The first step is to crawl raw data from the App store. Second step is to extract both the explicit and implicit information of Apps with clustering techniques. The third step is to search similar Apps corresponding to query Apps and recommend the most relevant functional features to developers.

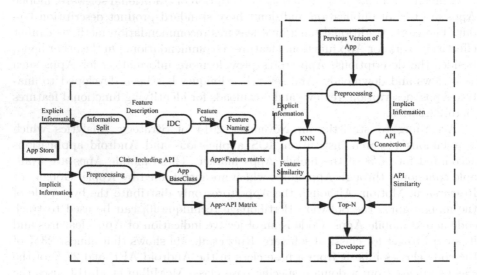

Fig. 1. The framework of the proposed App feature recommendation system

Different from the recommendation method of traditional software product, the functional feature recommendation method proposed in this paper integrates both explicit and implicit information of App, conquers the recommended deviation due to the incompleteness and non-objective of description information, and reinforces the accuracy and completeness of recommendation. The rest of this paper explains the second and third phases which are about Apps information analysis and Apps similarity calculating approach in more detail.

4   Mobile App Information Analysis

In order to recognize similar Apps' functional feature, we need to analyse both explicit and implicit information of mobile Apps. Since product description format of mobile Apps is not unified, we apply text process technique to extract Apps' product functional feature; API information of Apps are gained by decompiling technique. In this section, we take Android Apps in Google Play as example to illustrate how to process explicit and implicit information.

4.1   Explicit Information Analysis

We first construct functional feature repository by analysing explicit information. The functional feature recognition and extraction process from explicit information is as follows:

Preprocess. For product description of mobile Apps, removes the inference term, conducts statement decomposition, gets multiple functional feature description (the decomposed sentences or phrases), extracts keywords, represents functional feature as vector forms.

Mobile App description often includes one or several paragraphs with some keywords and inference terms. Keywords are the description words which can represent Apps' characteristic and its function, mainly noun, verb and adjective. Whereas, inference terms are those preposition or conjunction in the sentence which can be omitted, such as 'this', 'provides' and so on. Extracted keywords from App description then will compose functional feature frequency pattern sets. We apply tf-idf [20] method to calculate the weight of functional feature, and transform mobile Apps feature descriptions into word frequency vectors.

Recognize Functional Feature. After representing functional feature description as vectors, then we apply incremental diffusive clustering(IDC) algorithm to get feature class.

In the process of feature extraction, There may be many similar description represents the same functional feature or the same class of functional feature. These description may be compose of analogous or similar keywords but express the same meaning. In order to more clearly express Apps feature information, we should cluster the similar description, and in this paper we employ the Incremental Diffusive clustering [10] algorithm to obtain Apps' function features.

The Incremental Diffusive algorithm is composed of following steps:

- Adopt Spherical K-Means Algorithm [7] to cluster the functional feature description and form the candidate sets of functional feature classes.
- Select one of the "best" functional feature class (which has highest intra-class similarity)
- Recognize corresponding domain terms(refers to the keywords that its' functional feature centroid weight larger than certain threshold) of this functional feature set, and delete these terms from remainder feature sets.
- If the number of discovered functional feature classes is enough or there is not surplus functional feature description then stop; otherwise repeat 1–3 steps on the rest of functional feature description words.

There are 2 main stages in Spherical K-means. The first stage is similar to traditional K-means that randomly choose k centroids, every instance assign to the nearest cluster and update cluster centroids until stable and then stop iteration. The second stage is iterative optimization, we attempt to move a randomly selected feature request to the cluster for which it maximizes the overall cohesion gain. The second stage continues until no further optimization is possible. This post-processing stage has been shown to effectively improve the quality of clustering results [9].

When we carry on Incremental Diffusive clustering algorithm [10], we determine the ideal cluster number k by a technique designed by F. Can et al. [4] which consider the degrees to which each feature differentiates itself from other features. The modified formula is computed as:

$$k = \sum_{i=1}^{n} \frac{1}{|v_i|} \sum_{j=1}^{m} \frac{f_{i,j}^2}{N_j} \tag{1}$$

where $f_{i,j}$ is the frequency of term t_j appearing in functional feature f_i, N_j is the frequency of term t_j appearing in all functional features, v_i is the vector form of f_i. This method has already been shown to be efficient across real datasets [9].

Denomination of Functional Feature. It is to select keywords from functional feature class which can represent class, and give them a unified name.

Because the cluster center in the above process often does not corresponds to specific functional feature, we cannot use the cluster center to denominate functional feature. For one feature description in a functional feature class, we calculate similarity between the feature description and the cluster center, and sum the elements of the feature's term vector which is larger than threshold(set as 0.1 in this paper). We normalize the above values and add up. The functional feature description with the highest value is chosen as the name of the functional feature class.

Construct App-Feature Matrix. The column of the matrix represents Apps and the row of the matrix stands functional features and each element (i, j) is 0 or 1 which describes whether the ith App contains the jth functional feature.

4.2   Implicit Information (API) Processing

With the Software Bertillonage technique [6], we can get API information by crawling the bytecodes in App stores. This paper extracts Android Apps' API by following procedures.

- Obtain DEX format compile class (execution file of Dalvik) by carrying on Android virtual machine.
- Transform file into JAR compressed file using decompiling tool dex2jar[1]. Most of the App download from App store can be correctly transformed into JAR file, but some proportion cannot.

---

[1] http://code.google.com/p/dex2jarverifiedonJune2013.

- Apply JClassInfo[2] tool to extract APIs' call information from .class file. In our experiment, we reject those API classes that are not included in Android package.
- Generate Apps and APIs relation matrix (0–1 matrix).

5 Functional Feature Recommendation of Mobile App

To recommend functional features for developers according to the input query information (explicit information such as App's name or part of description), the similarity measure of two Apps is important. In this paper, we define two types of similarity related with explicit and implicit information respectively.

The functional feature similarity $AppFeaSim$ (a, b) is the cosine similarity of two Apps' explicit information, it is defined as (2):

$$AppFeaSim(a, b) = \frac{|F_a \cap F_b|}{\sqrt{|F_a| \cdot |F_b|}} \qquad (2)$$

where, F_a, F_b represents non-empty functional features set of App a and App b respectively.

The API similarity $AppAPISim$ (a, b) is the similarity of two Apps' common used API information, defined as (3).

$$AppAPISim(a, b) = \frac{|(A_a \cap A_n)|}{\sqrt{|A_a| \cdot |A_n|}} \qquad (3)$$

where, A_a, A_b represents non-empty APIs set of App a and App b respectively. Combining AppFeaSim(a,b) and AppAPISim(a,b), this paper presents the following self-adaptive similarity of two Apps:

$$AppSim(a, b) = \alpha * AppFeaSim(a, b)$$
$$+\beta * AppAPISim(a, b)$$
$$\alpha + \beta = 1 \qquad (4)$$

α and β are the self-adaptive coefficients and defined as:

$$\alpha = \frac{n_{F_a}/N_F}{n_{F_a}/N_F + n_{A_a}/N_A}, \beta = \frac{n_{A_a}/N_A}{n_{F_a}/N_F + n_{A_a}/N_A} \qquad (5)$$

where, N_F, N_A are the total number of functional features extracted from explicit information and the total number of APIs from all Apps, n_{F_a}, n_{A_a} are the number of functional features extracted from App a and the number of APIs from App a. Parameter α and β balances the two types of similarity.

With the above self-adaptive similarity measure, we can get the top-k most similar Apps of the developer's query. The functional feature sets of these k

---

[2] http://jclassinfo.sourceforge.net/.

Apps form the candidate functional feature set. We define the recommendation probability of functional feature f on the query of App a as follows:

$$pred(a, f) = \frac{\sum_{b \in KNN(a)} Simialrity(a,b) \cdot m_{b,f}}{\sum_{b \in KNN(a)} Simialrity(a,b)} \quad (6)$$

where, $KNN(a)$ is the k nearest neighbor of App a, $Similarity(a,b)$ is the similarity of App a and App b, and the definition of $m_{b,f}$ is as follows:

$$m_{b,f} = \begin{cases} 1 \text{ if App b has feature f} \\ 0 \text{ otherwise} \end{cases}$$

The parameter k in KNN is set with experience. If the amount of Apps is little, then k is a fixed value ($k = 10$ in this paper); otherwise we can set k as a special percentage of the Apps(5 % in this paper).

If the similarity $Similarity(a,b)$ takes the value of functional feature similarity $AppFeaSim(a,b)$, the feature recommendation method is considered as the direct extension of classical feature recommendation method, denoted as AFR-Explicit; if the similarity $Similarity(a,b)$ takes the value of API similarity $AppAPISim(a,b)$, the recommendation method only uses the implicit information of Apps and denote as AFR-Implicit. If the similarity $Similarity(a,b)$ takes the value of self-adaptive similarity $AppSim(a,b)$ proposed in the paper (Eq. 4), the recommendation method integrates both explicit and implicit information, denoted as AFR-Hybrid.

6   Experiment and Evaluation

To verify the efficiency of our proposed mobile Apps functional feature recommendation method AFR-Hybrid which integrates both explicit information and implicit information, we collect four common types of Apps in Google Play as experiment data shown in Table 1. The number of functional features is determined manually by domain experts.

6.1   Experiment Set Up

As the method used in traditional software recommendation system, we adopt the cross validation leave-one-out method and make some changes to evaluate

Table 1. Test data set

Data Type	The number of App	The number of cluster feature	The number of API class
Finance	164	40	736
Music	98	50	757
Shopping	56	30	953
Communication	123	40	1053

Table 2. An example of AFR-Hybrid

App's name	CalcPack Financial Calculators
Input functional feature	effect of inflation personal finance calculators results update as you enter the figures annualized returns
The expected functional features	life insurance calculator asset allocation growth compound interest
Recommendation functional features set (N = 10)	(allows you to do two very simple commercial real estate calculations) expenditure /income categories inventory entry/multiple item transactions/invoice entry reaches the take profit price life insurance calculator web services/soa (your mobile carrier s message and data rates may apply) enjoy freedom and peace of mind for the spain individual self-employment (consider the development of his accounts on the 10-year period)

the performance of our proposed recommendation method. The details are as follows:

- Randomly remove 1/2 or 2/3 functional features from the query App functional features set.
- Input the rest of functional features of the query App into the recommendation system.
- Evaluate the relevance of the recommended functional features.

Table 2 gives a concrete example on feature recommendation. For App "CalcPack Financial Calculators", we remove 3 of its' functional features, and input the rest four functional features into the recommendation system. We can get the top 10 relevant functional features recommended by AFR-Hybrid method. The accuracy can be evaluate by domain experts. In order to objectively evaluate the recommendation method, this paper applies 5-fold cross validation, that is to say, we divide one type of Apps into 5 equal parts S_1, S_2, S_3, S_4, S_5; and then use each part $S_i, i = 1, 2, ..5$ as test data while others $\bigcup_{j \neq i} S_j$ as train data.

6.2   Experiment Evaluation

We adopt Recall and Precision to evaluate experiment results. Recall is the fraction of relevant functional feature that can be located by recommendation method and Precision evaluates the quality of the recommendation. For each query App a in the test data set, we randomly remove its several functional features and then input residual part to the recommendation system.

For the query App a, the recall is computed as:

$$Recall(a) = \frac{\sharp number\ of\ R_N(a) \cap f_p(a)}{\sharp number\ of\ f_p(a)} \tag{7}$$

where $f_p(a)$ represents the randomly removed feature sets, and $R_N(a)$ notes the recommended features set. The recall on test dataset S_i is:

$$rec(S_i) = \sum_{a \in S_i} Recall(a)/|S_i| \tag{8}$$

The average recall is:

$$R = \sum_{i=1,\dots k} rec(S_i)/k \tag{9}$$

When recall is 1.0, it represents that the removed features are all recommended back; the higher the recall, the more removed features recommended back.

For the query App a, the precision is defined as:

$$Precision(a) = \frac{\sharp number\ of\ R_N^{'}(a)}{\sharp number\ of\ R_N(a)} \tag{10}$$

where we get relevant functional feature set by manual labelling, $R_N^{'}(a) \in R_N(a)$. In the experiment we invited 20 volunteers with domain knowledge and all volunteers have at least 1 year experience in mobile Apps developing. If more than two-thirds of the volunteers agree with the feature can meet the demand of query Apps, then it is thought to be a relevant features. The precision on test set S_i is defined as:

$$pre(S_i) = \sum_{a \in S_i} precision(a)/|S_i| \tag{11}$$

The average precision is:

$$P = \sum_{i=1,\dots k} pre(S_i)/k \tag{12}$$

6.3   Experiment Results and Analysis

The experiment results on the four types of Apps set with AFR-Explicit, AFR-Implicit and AFR-Hybrid recommendation methods on different number of recommendation features are shown in Table 3, 4 and 5.

According to the recall and precision, in the most cases, the AFR-Hybrid recommendation method which combines both explicit and implicit information performs better than the AFR-Explicit recommendation method that only

Table 3. Recall and precision on N = 5

Data type	Recall			Precision		
	AFR-Explicit	AFR-Implicit	AFR-Hybrid	AFR-Explicit	AFR-Implicit	AFR-Hybrid
Finance	0.3257	0.2817	*0.3415*	0.5872	0.3542	*0.6331*
Music	0.3789	0.4217	*0.5297*	0.4500	0.4648	*0.4962*
Shopping	0.4366	*0.5203*	0.4745	0.4999	*0.5709*	0.5613
Communication	0.3331	0.3401	*0.3528*	0.5287	0.4635	*0.5743*

Table 4. Recall and precision on N = 10

Data type	Recall			Precision		
	AFR-Explicit	AFR-Implicit	AFR-Hybrid	AFR-Explicit	AFR-Implicit	AFR-Hybrid
Finance	0.4123	0.3647	*0.4530*	0.5727	0.3755	*0.6035*
Music	0.4902	0.4881	*0.5717*	0.4133	0.4181	*0.4752*
Shopping	0.5506	0.5702	*0.5779*	0.4953	0.5152	*0.5608*
Communication	0.4713	0.4660	*0.5097*	0.5038	0.4373	*0.5509*

Table 5. Recall and precision on N = 15

Data type	Recall			Precision		
	AFR-Explicit	AFR-Implicit	AFR-Hybrid	AFR-Explicit	AFR-Implicit	AFR-Hybrid
Finance	0.5619	0.5018	*0.5775*	0.5128	0.4067	*0.5538*
Music	0.5940	0.5957	*0.6627*	0.3733	0.3781	*0.4419*
Shopping	0.6903	0.7259	*0.7267*	0.4733	0.4667	*0.5067*
Communication	0.6284	*0.6813*	0.6770	0.4889	0.4533	*0.5222*

uses explicit information and the AFR-Implicit method which only makes use of implicit information. In theory, with the increase of N, recall increases gradually, whereas precision decreases. The AFR-Explicit method basically conforms to this law. But AFR-Implicit method is not totally conformed to this law. It is because that the AFR-Implicit method only uses API information to measure the similarity between Apps. While APIs' similarity is not exactly equivalent to APPs' similarity, and sometimes those Apps with similar APIs may completely different. The proposed AFR-Hybrid method which adopts self-adaptive similarity measure and organically combines the Apps description (Explicit information) and API information (Implicit information) improves algorithm's stability in the base of recall and precision.

Compared with AFR-Explicit method, AFR-Hybrid method makes up the incompleteness of description information and avoids the impact of fishing Apps (Apps' description is identical, but function is completely different). Compared with AFR-Implicit method, AFR-Hybrid method can effectively refrain from the situation that APIs is similar but Apps is dissimilar, and meanwhile stabilize the method.

7   Conclusion

With the growing of mobile application market, App developers need to develop and upgrade Apps perfectly. Thus, how to determine mobile Apps' functional feature becomes critical. Based on a large amount of existing similar Apps and mining their functional feature to recommend is an effective way to address this problem. Different from traditional software feature recommendation, this paper puts forward a novel recommendation method which integrates both explicit and implicit information of Apps. As mobile application's API information can more accurately and objectively reflect Apps' function, the proposed method can overcome the noises in the product description, and also effectively improve the recall and precision of recommendation. On four types of experimental dataset including finance, music, shopping and communication, our proposed recommendation method AFR-Hybrid achieves competitive performance compared with traditional method. Besides, the functional tightness of classified Apps and APIs with similar function will have some impacts on our proposed recommendation method. In future, we will further mining the relationship between API and functional feature to improve the quality of App classification and functional feature recommendation.

References

1. Aleksandrova, M., Brun, A., Anne, B.: Search for user-related features in matrix factorization-based recommender systems. ECML-PKDD - Doctoral session (2014)
2. Bae, D., Han, K., Park, J., Yi, M.Y.: AppTrends: a graph-based mobile app. recommendation system using usage history. In: International Conference on Big Data and Smart Computing, pp. 210–216 (2015)
3. Borba, P., Teixeira, L., Gheyi, R.: A theory of software product line refinement. Theor. Comput. Sci. **455**(7), 2–30 (2012)
4. Can, F., Ozkarahan, E.A.: Concepts and effectiveness of the cover-coefficient-based clustering methodology for text databases. ACM Trans. Database Syst. Tods Homepage **15**(4), 483–517 (1990)
5. Davidsson, C., Moritz, S.: Utilizing implicit feedback and context to recommend mobile applications from first use. In: The Workshop on Context-Awareness in Retrieval and Recommendation, pp. 19–22 (2011)
6. Davies, J., German, D.M., Godfrey, M.W., Hindle, A.: Software bertillonage. Empirical Softw. Eng. **18**(6), 1195–1237 (2013)
7. Dhillon, I.S., Modha, D.S.: Concept decompositions for large sparse text data using clustering. In: Machine Learning, pp. 143–175 (2001)
8. Dhungana, D., Grnbacher, P., Rabiser, R.: The DOPLER meta-tool for decision-oriented variability modeling: a multiple case study. Autom. Softw. Eng. **18**(1), 77–114 (2011)
9. Duan, C., Cleland-Huang, J., Mobasher, B.: A consensus based approach to constrained clustering of software requirements. In: ACM Conference on Information and Knowledge Management, CIKM 2008, Napa Valley, California, USA, October, pp. 1073–1082 (2008)

10. Dumitru, H., Gibiec, M., Hariri, N., Cleland-Huang, J.: On-demand feature recommendations derived from mining public product descriptions. In: International Conference on Software Engineering, pp. 181–190 (2011)
11. Ekstrand, M.D., Riedl, J.T., Konstan, J.A.: Collaborative filtering recommender systems. Found. Trends Hum. Comput. Interact. **4**(2), 81–173 (2011)
12. Han, Y., Go, G., Kang, S., Lee, H.: A feature-oriented mobile software development framework to resolve the device fragmentation phenomenon for application developers in the mobile software ecosystem. In: Zhang, Y., Peng, L., Youn, C.-H. (eds.) Cloud Computing, vol. 167, pp. 189–199. Springer, Heidelberg (2016). doi:10.1007/978-3-319-38904-2_20
13. Kang, K.C.: Feature-oriented domain analysis. Alphascript Publishing (2010)
14. Mcmillan, C., Grechanik, M., Poshyvanyk, D.: Detecting similar software applications. In: International Conference on Software Engineering, pp. 364–374 (2012)
15. Olszak, A., Lazarova-Molnar, S., Jørgensen, B.N.: Evolution of Feature-Oriented Software: how to stay on course and avoid the Cliffs of modularity drift. In: Holzinger, A., Cardoso, J., Cordeiro, J., Libourel, T., Maciaszek, L.A., Sinderen, M. (eds.) ICSOFT 2014. CCIS, vol. 555, pp. 183–201. Springer, Heidelberg (2015). doi:10.1007/978-3-319-25579-8_11
16. Qiao, X., Chun, Y., Xiaofeng, L., Junliang, C.: A trust calculating algorithm based on social networking service users' context. Chin. J. Comput. **34**(12), 2404–2413 (2011)
17. Rahman, M.M., Roy, C.K.: Textrank based search term identification for software change tasks. In: IEEE International Conference on Software Analysis, Evolution and Reengineering, pp. 540–544 (2015)
18. Carmen Rodríguez-Hernández, M., Ilarri, S.: Towards a context-aware mobile recommendation architecture. In: Awan, I., Younas, M., Franch, X., Quer, C. (eds.) MobiWIS 2014. LNCS, vol. 8640, pp. 56–70. Springer, Heidelberg (2014). doi:10.1007/978-3-319-10359-4_5
19. Ruiz, I.J.M., Nagappan, M., Adams, B., Hassan, A.E.: Understanding reuse in the android market. In: IEEE 20th International Conference on Program Comprehension (ICPC), pp. 113–122 (2012)
20. Salton, G., Mcgill, M.J.: Introduction to modern information retrieval, mcgraw-hill, n.y. Theses (1983)
21. Shi, W., Sun, X., Li, B., Duan, Y., Liu, X.: Using feature-interface graph for automatic interface recommendation: A case study. In: International Conference on Advanced Cloud and Big Data, pp. 296–303 (2015)
22. Shi, Y.C., Meng, X.W., Zhang, Y.J., Wang, L.C.: Adaptive learning approach of contextual mobile user preferences. J. Softw. **23**(10), 2533–2549 (2012)
23. Sun, Y., Chong, W.K., Man, K.L., Rho, S., Xie, D.: Exploring critical success factors of mobile recommendation systems: the end user perspective. In: Yang, G.-C., Ao, S.-L., Huang, X., Castillo, O. (eds.) Transactions on Engineering Technologies, pp. 45–57. Springer, Singapore (2016). doi:10.1007/978-981-10-0551-0_4
24. Venu Gopalachari, M., Sammulal, P.: Personalized collaborative filtering recommender system using domain knowledge. In: International Conference on Computer and Communications Technologies, pp. 1–6 (2014)
25. Xu, F.L., Meng, X.W., Wang, L.C.: A collaborative filtering recommendation algorithm based on context similarity for mobile users. J. Electr. Inf. Technol. **33**(11), 2785–2789 (2011)

Adaptive Multi-objective Swarm Crossover Optimization for Imbalanced Data Classification

Jinyan Li[1(⊠)], Simon Fong[1(⊠)], Meng Yuan[1], and Raymond K. Wong[2]

[1] Department of Computer Information Science, University of Macau,
Macau SAR, China
{yb47432, ccfong, mb55525}@umac.mo
[2] School of Computer Science and Engineering,
University of New South Wales, Kensington, Australia
wong@cse.unsw.edu.au

Abstract. Training a classifier with imbalanced dataset where there are more data from the majority class than the minority class is a known problem in data mining research community. The resultant classifier would become under-fitted in recognizing test instances of minority class and over-fitted with overwhelming mediocre samples from the majority class. Many existing techniques have been tried, ranging from artificially boosting the amount of the minority class training samples such as SMOTE, downsizing the volume of the majority class samples, to modifying the classification induction algorithm in favour of the minority class. However, finding the optimal ratio between the samples from the two majority/minority class for building a classifier that has the best accuracy is tricky, due to the non-linear relationships between the attributes and the class labels. Merely rebalancing the sample sizes of the two classes to exact portions will often not produce the best result. Brute-force attempt to search for the perfect combination of majority/minority class samples for the best classification result is NP-hard. In this paper, a unified preprocessing approach is proposed, using stochastic swarm heuristics to cooperatively optimize the mixtures from the two classes by progressively rebuilding the training dataset is proposed. Our novel approach is shown to outperform the existing popular methods.

Keywords: Class rebalancing · Swarm optimization · Classification

1 Introduction

Imbalance dataset is referred to the phenomenon where there are far more samples in one class than in the other. Some data mining applications that would have to deal with imbalance datasets are those typically would have to be trained with large amount of common samples but with limited rare samples. They include big data analytics and text mining [1], forecasting natural disasters [2], fraud detection in transactions [3], target identification from satellite radar images [4], classifying biological anomalies [5] as well as computer-assisted medical diagnosis and treatment [6], just to name a few.

© Springer International Publishing AG 2016
J. Li et al. (Eds.): ADMA 2016, LNAI 10086, pp. 374–390, 2016.
DOI: 10.1007/978-3-319-49586-6_25

Imbalanced data classification has long been an important and challenging problem in data mining and machine learning [7]. Conventional supervised learning algorithms by greedy search are usually designed to embrace the imbalanced dataset without regards to the class balance ratio by default. Most original classification model induction algorithms were designed without the consideration of imbalance issue initially. Those trained models suffer from overfitting from the sheer volume of majority training data; the recognition power for identifying rare test samples is limited due to the lack of sufficient training (known as underfitting) given few minority samples are available. Additional performance metrics which are used to assess or judge whether a classification model is incompetent owing to imbalance training go beyond just accuracy. Some useful metrics which are based on the counts of true-positive, false-positive etc. include G-mean [8], F1 measure [9], Kappa statistics [10], AUC/ROC [11], Matthews correlation coefficient [12] and Balance error rate [13] that have been used in the literature.

Pre-processing styled rebalancing schemes have been proposed in the past, mainly in the aspects of artificially inflating the minority class data, resampling down the volume of the majority class data, or a combination of the two. It was already shown [10] that merely matching the quantities of the majority and minority data to equal, does not yield the highest possible classification performance.

In this paper, an adaptive rebalancing model as a preprocessing tool is proposed, by considering the drawbacks of the current methods for solving imbalanced classification problem. What data mining users desire as the features of an ideal rebalancing tool, in observation of the above-mentioned limitations are: high performance that is not only in accuracy but in other reliability measures as well, free of parameter calibration, joint rebalancing actions by increasing and decreasing the minority and majority samples respectively, and be able to complete the dual actions till reaching the best possible performance within a reasonable time.

Given a potentially very large number of instances in the original dataset, finding the best ratio between two majority and minority classes of data is a challenging combinational optimization problem. Without resorting to brute-force, swarm optimization is applied on each aspect of rebalancing – one on searching for the appropriate amount of majority instances, and the other one on estimating the best combo of control parameters (the intensity and how far that the neighbors of the minority samples are to be fabricated) with respect to enlarging the minority population size. Our proposed rebalancing method couples these two optimizations as an unified iterative approach which will progressively enhance the mixtures of the optimized data from the two swarm optimizations by crossing over their optimized results generation after generation until a good quality dataset is produced. This unified rebalancing approach is called Adaptive Multi-Objective Swarm Crossover Optimization (AMSCO), within which the optimization of majority instances is called Swarm Instance Selection (SIS), the optimization of minority instances is called Optimized Synthetic Minority Oversampling Technique (OSMOTE). The overall design of AMSCO is shown in Fig. 1.

In this new approach, Particle Swarm Optimization (PSO) algorithm is chosen as the core optimizer whose searching particles represent the solution candidates. The original dataset, after it was loaded for the first time, will become the current dataset, and be checked with respect to its quality by inferring a candidate classifier from it. Until the

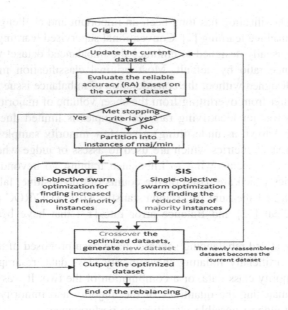

Fig. 1. Flow chart of AMSCO

performance of the candidate classifier meets the expected requirement, the current dataset will be subject to two parallel swarm optimizations for optimally increasing the minority samples and decreasing the majority samples. The two swarms operate independently because their candidate solutions are different in nature. Their outputs, however, are crossed over by selectively merging instances from the most competent optimized datasets into one that is framed by the size of the original dataset. The selected dataset in turn becomes the current dataset when the optimization cycle iterates. The dataset is checked by the criteria and passed to two swarm operations again, if it is still not good enough. Decision tree is used as the classification algorithm here which works like a wrapper approach in testing the goodness of the current dataset.

The advantages of AMSCO are: (1) progressively refining the dataset by rebalancing the instances through swarm optimizations to its best; (2) the imbalanced dataset becomes balanced in such a way that both accuracy and reliability are maximized; (3) the size of the resultant balanced dataset is controlled so the original dataset size is approximately preserved; (4) as a by-product, the base-learner used in the wrapper approach is trained upon finding the well balanced dataset. The classifier trained by the final dataset (which supposed to offer the best performance) can be instantly put into use; (5) compared to the tightly coupled swarm optimization techniques and other conventional rebalancing methods by linear search, AMSCO has superior in fast computational speed, accuracy and reliability.

The remainder of this paper is structured as follows. Section 2 reviews popular approaches that have been successfully employed in imbalanced dataset classification to certain extent. In Sect. 3, we elaborate the design of AMSCO. The benchmark datasets, experiment and its results are described in Sect. 4. Section 5 concludes this paper.

2 Related Works

Class rebalancing techniques can be summed up as two major categories. The first group concerns about data level, which re-sampling the two classes samples [14]. The second level pertains to alter the classification algorithms for imbalanced data classification. Researches proposed various resample techniques in data level. Random over-sampling [15] and under-sampling [16] are the simplest method. The former augments minority scale through copying its samples, and the latter randomly delete majority class samples to realize a balance. Typical under-sampling and over-sampling techniques may subside the disadvantage of information vacancy and over learning [17] respectively. One-side selection [18] techniques categorized majority class samples and eventually find out the safety samples to fulfill the under-sampling. The most popular sampling method is SMOTE (Synthetic Minority Oversampling Technique) [19] that usually achieve effective performance. The principle is letting the algorithm fabricate extra minority data into the dataset through observing and analyzing the characteristics of minority class sample' spatial structure. Assuming the oversampling rate is N (Eq. (1) synthesizes N times new minority class samples) and each minority class sample $x_i \in S_{minority}$. The other parameter k is used by the algorithm to examine k neighbors of x_i in minority class samples, then to randomly select x_t from the k neighbors using Eq. (1) to generate the synthetic data $x_{new,N}$:

$$x_{new,N} = x_i + rand[0,1] \times (x_t - x_i) \tag{1}$$

In Eq. (1) *rand* [0, 1] generates a random number between 0 and 1. N and k influence SMOTE to generate a suitable number of characteristic minority class samples. The second level of approaches solves the class imbalance problem during the training stage. It contains ensemble based techniques [20] and cost-sensitive learning approaches [21]. The basic idea of ensemble learning is that, a strong classifier will be voted and integrated by a series of weak classifiers after several rounds of iteration. The commonly used ensemble learning methods are bagging [22], boosting [23], and random forest [24]. SMOTEBoost [25] approach combines Adaboosting and SMOTE, and it offers high performances. Cost-sensitive learning assigns a different weight to each part of confusion matrix by the cost matrix. In general, the cost of misclassified minority class is the largest, thus in order to obtain a results with minimum cost, the classifier will be bias to minority class. The distinguished cost-sensitive learning algorithms are associated with boosting [26, 27] or Support Vector Machines (SVM) [28] to tackle class imbalance problem.

3 Design of AMSCO

The proposed AMSCO uses mainly two swarm optimization processes that run independently, for fixing the exceeding majority data instances and shortage of minority data instances respectively. The overall idea of AMSCO is to smash up the original dataset and to resemble it back again using only the qualified instances selected by the two swarm optimization processes. This is done by first dividing the original dataset into two groups – one group contains instances which are purely of majority class and the other group purely of minority class. Through the swarm search processes, the instances are randomly encoded into search particles which represent some candidate solutions, and the particles move iteratively from random positions towards some global optimum. The instances being selected from the swarm processes are being heuristically enhanced, leaving only some fittest ones at the end.

At the end of each iteration, the qualified instances which are represented by the fittest particles so far are gathered in buffers, in preparation of packing them into a new dataset by crossover operation. The new dataset after packed, will be taken as the current dataset, and subject to the optimization in the new iteration. By this way, the current dataset evolves in improving its quality, offering increasingly higher fitness iteration after iteration. Eventually the most refined dataset is outputted as the best dataset at the end of the whole preprocessing operation, by which the imbalance problem is solved and the final dataset is ready to induce a classifier that will have the best possible performance.

The intermediate dataset which is supposed to be the best thus far at each round of optimization, is assembled from four possible datasets. The compositions of the datasets are shown in Fig. 2.

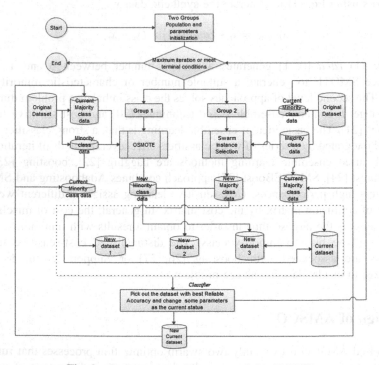

Fig. 2. Components and data around the AMSCO

The four possible datasets, which are candidates to be crossed over and packed into a new generation of current dataset are the three combinations of optimized majority dataset, optimized minority datasets, and the current dataset. From Fig. 2, they are mixed and enumerated as follow: (1) New dataset 1 = New Majority class data (after optimized by SIS on the majority instances) + Current Minority class data; (2) New dataset 2 = New Majority class data (after optimized by SIS on the majority instances) + New Minority class data (after optimized by OSMOTE on minority instances); (3) New dataset 3 = New Minority class data (after optimized by OSMOTE on minority instances) + Current Minority class data; and, (4) New dataset 4 = Current Majority class data + Current Minority class data.

The rationale behind crossing over different majority and minority portions of the optimized data is hoping to generate the new dataset that may have the perfect balance/ratio by fusing the best from the two optimized portions. Since the perfect ratio between the majority and minority instances is not known in advance, and it is short of a deterministic way (other rather brute-force) to compute the ratio, we resort to iterative heuristics in summing up the best instances from both majority and minority classes found so far for the improved version of dataset. It is worth noting a few remarks that: (1) An ideal balanced dataset is required to have both portions of majority and minority although the exact ratio is elusive; so in the four combos of new datasets, there must be certain instances that come from majority and minority classes; (2) The swarm optimization is purely probabilistic, there may be chances that both OSMOTE and SIS yield no better quality instances (none or even negative improvement), so New dataset 4 is needed as a backup; (3) The terminal conditions are the result doesn't change in the past defined iterations (convergence);The length of new minority class samples cannot too bigger than the new majority class samples (2 times are used in the experiment); or it achieves the maximum iteration (4) The sizes of the optimized datasets do vary from time to time; there are chances that they are shorter than the original dataset. When this happens the full dataset of the highest fitness will fill in, followed by all of the second best, and so forth until the new generation of current dataset matches the size of the original dataset. This is to prevent the assembled dataset overly shrinking or enlarging along the way. (5) OSMOTE and SIS are separate processes since the ways how the instances are selected for reduction and replication are different, though they share the same objectives and similar fitness evaluation functions.

OSMOTE and SIS which are the core of the proposed rebalancing model are based on Swarm intelligence optimization algorithm [29] which iteratively evolves a solution from randomly picked values to some global optimal result. Specifically, Particle Swarm Optimization (PSO) algorithm [30] is used in OSMOTE and SIS whose search agents imitate the flying patterns of birds. Assuming there is a population $X = (X_1, X_2, ..., X_n)$ which is grouped by n particles in D dimensional search space, the ith particle in this space is expressed as a vector X_i with D dimension, $X_i = (x_{i1}, x_{i2}, ..., x_{iD})^T$, and the position of the ith particle in the search space represents a potential solution. As the objective function, the program can calculate the corresponding fitness of position X_i of each particle, where the speed of the ith particle is $V_i = (V_{i1}, V_{i2}, ..., V_{iD})^T$, the extremum value of each agent is $P_i = (P_{i1}, P_{i2}, ..., P_{iD})^T$ and the extremum of the population is $P_g = (P_{g1}, P_{g2}, ..., P_{gD})^T$. In the process of iteration, the extremum value of each agent and the population will update their position and speed [31]. Equations (2) and (3) show the mathematical process as follows:

$$V_{id}^{k+1} = \omega * V_{id}^k + c_1 r_1 \left(P_{id}^k - X_{id}^k\right) + c_2 r_2 \left(P_{id}^k - X_{id}^k\right) \tag{2}$$

$$X_{id}^{k+1} = X_{id}^k + V_{id}^{k+1} \tag{3}$$

In the Eq. (2), ω is inertia weight; $d = 1, 2, ..., D$; $i = 1, 2, ..., n$; k is the current iteration time; c_1 and c_2 are non-negative constants as the velocity factor, r_1 and r_2 are random values between 0 to 1 and V_{id} is the particle speed.

Cooperation evolution draws thought from population coordination theory in the ecology. It simulates the mutual influence and mutual restriction between various populations in nature to strengthen performances of each population and global [32]. Generally, in multiple swarm collaboration algorithm, particles randomly be divided into M sub-groups, S_i, $1 \leq i \leq M$, collaboration through the information interaction between populations. There are three kinds of evaluation rules:

$$\text{ER1} : V_i = \omega * V_i + c_1 r_1 \left(P_i^s - X_i\right) + c_2 r_2 (P_i^{sg} - X_i) \tag{4}$$

$$\text{ER2} : V_i = \omega * V_i + c_1 r_1 \left(P_i^s - X_i\right) + c_2 r_2 (P_i^{sg} - X_i) + c_3 r_3 (P_i^g - X_i) \tag{5}$$

$$\text{ER3} : V_i = \omega * V_i + c_1 r_1 \left(P_i^s - X_i\right) + c_2 r_2 (P_i^g - X_i) \tag{6}$$

In Eqs. (4) to (6) P_i^s, P_i^{sg} and P_i^g are respectively stands for the best value of particle X_i, the best fitness of particle X_i's sub-group and the best fitness of all groups. Thus, multiple swarms collaboration method founded on reducing a problem into several or more sub-problems. The sub-groups demarcate and search the searching space in parallel and though communicating and sharing the best information between each sub-groups to find out the final global best solution for a short time. In the case of AMSCO, two sub-groups are constructed, for cooperatively optimizing two classes of data in OSMOTE and SIS, the best information are crossed over, until a final global best solution is obtained as a well-balanced dataset. More information regarding OSMOTE and SIS will be explained in the following sub-sections.

The fitness is defined as a product of accuracy and Kappa or Kappa statistics [33]. Kappa is chosen to estimate the credibility of a classification model. When a classifier suffers from imbalanced dataset, it has a sign of high accuracy but a low value (zero or even negative) of Kappa. Kappa is an efficient indicator to be fairly reflect the consistency of test data and the dependability of the classification model, so as to investigate whether the performances fall into a secure area. There are six degrees of interpretation for Kappa outcome ranging between −1 and 1 in mathematics [33]. Subzero part denotes that this model is worthless, and each of the other five levels is segmented by a 0.2 interval. These areas respectively stand for the strength of agreement in poor, slight, fair, moderate, substantial and almost prefect. The model typically has some credibility when its Kappa value exceeds over 0.4 [10, 34], and the credibility will increase with the improvement of Kappa statistics. For the fitness which is used to represent the goodness of the classification model, a composite performance criterion called Reliable Accuracy (RA) is defined, where $RA \; \alpha \, (Accuracy \times Kappa)$, of an induced model. The fitness function therefore is the performance evaluation of a decision tree, generated from a given dataset. Other classification algorithms could be optionally used in lieu of decision tree, though it is used because of its popularity. The fitness is RA that depends on the accuracy and Kappa which are defined as follow:

$$RA = \frac{P_o^2 - P_o P_c}{1 - P_c} \tag{7}$$

$$Accuracy = \frac{TP + TN}{P + N} \tag{8}$$

$$Kappa = \frac{P_o - P_c}{1 - P_c} \tag{9}$$

$$P_o = Accuracy = \frac{TP + TN}{c^+ + c^-} \tag{10}$$

$$P_c = \frac{(TP + FP) \times (TP + FN) + (FN + TN)(FP + TN)}{(c^+ + c^-)^2} \tag{10}$$

where TP, TN, FP, FN, C^+ and C^- are the counts of true-positive, true-negative, false-positive, false-negative, instances of positive/majority class and instances of negative/minority class respectively. P_o is the measure of the percentage of agreement, and P_c is the chance of agreement.

3.1 Optimized SMOTE for Over-Sampling Minority Instances

OSMOTE is extended from Synthetic Minority Oversampling Technique (SMOTE) [19], which is one of the most popular methods to over-sample the minority instances for rebalancing an imbalanced dataset. Its basic idea is to fabricate extra minority data into the dataset by inserting synthetic samples along the line segments connecting any or all of the k minority class nearest neighbors in the data space. In SMOTE, two parameters are required and they need to be manually set. One is the over-sampling rate of N which tells the algorithm to synthesize N times new minority class samples, e.g. $N = 2$ means the minority class data are going to be doubled. The other parameter k is used by the algorithm to inspect k nearest neighbours of each minority data to generate the synthetic data.

This method can effectively create synthetic examples increasing the population of minority instances, rather than by over-sampling with replacement. Depending upon the extent of over-sampling required by N, certain neighbors from the k nearest neighbors are chosen randomly. One obvious drawback is the lack of guideline recommending what the values of $N \in [1 \ldots \infty]$ and $k \in [1 \ldots (C^+ \text{ and } C^-)]$ should be used, for generating a rebalanced dataset that gives the highest classification performance. In many cases, these two parameters are arbitrarily chosen; though the resultant classification performance is improved, it is not maximized.

In our OSMOTE approach, the two parameters of the system, N which indicates the target amount of dataset after over-sampled, and k the range of reference neighbors for duplicating the minority data vectors, are to be optimized using PSO. Each particle of PSO swarm, p, is coded as a candidate solution with a pair of values within the ranges of N and k. So $p = \langle v, \kappa \rangle$, where $v \in [1\ldots\infty]$ and $\kappa \in [1\ldots(C^+ \text{ and } C^-)]$. The fitness function is the resultant decision tree built on a dataset after over-sampled by SMOTE with $N = v$ and $k = \kappa$. The RA performance is treated as fitness in this case, for evaluating the goodness of the chosen p such that $fitness = RA = fitness\_evaluation(DT(p))$.

OSMOTE is aimed at improving both accuracy and Kappa where these two values fluctuate dynamically during the search process. As such, it is a dynamic multi-objective algorithm. Without considering complex situations like Pareto, this dual objective algorithm actually just tries to maintain a certain high accuracy level while maximizing the Kappa which is deemed more importantly as credibility to its highest. Figure 3 shows a snapshot of the fluctuation patterns of accuracy and Kappa as the optimization progresses. In general, accuracy and Kappa follow similar trends but of different intensities in swinging ups and downs; and the overall trends are on the rise. One can see that in this example, accuracy and Kappa have both reached a very high value of approximately 1, at the 310th cycle of iteration. Since the two objectives are not opposing each other, a special type of optimization called the non-inferior set tactics [35] is adopted here and customized for this specific rebalancing task.

It first collects all the possible solutions of this multi-objective problem. These solutions in non-inferior set are satisfied with several update criteria. Three update criteria are used here regulating the OSMOTE for the particles evolution, optimization and convergence.

(1) Both accuracy and Kappa of the new particle must be better than the existing one;
(2) Either one of the accuracy or Kappa of the new particle must be better than the existing one, as well as the defined tolerance is larger than the absolute value of difference of the other measures such as F1, ROC and BER;
(3) The current threshold value of Kappa of the new particle must be greater than the older particle's Kappa value.

If a new particle meets any of the above three criteria, it will replace the current one into the next generation. Otherwise, this particle will follow its trajectory and move to a neighbor position. This algorithm progressively lifts up Kappa through updating by the

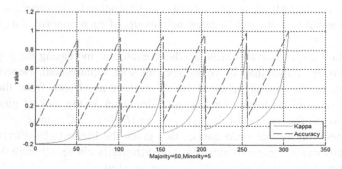

Fig. 3. Snapshot of fluctuating values of accuracy and Kappa during OSMOTE of a imbalanced dataset

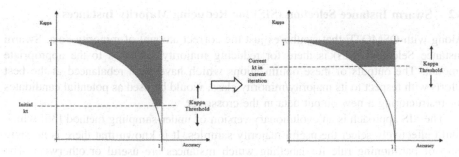

Fig. 4. Illustration of non-inferior region in the search space

constraint conditions. At the start, the initial threshold of Kappa, k_t, is set at 0.4 that is the second bottom confidence level of Kappa. In the last phase of each iteration, the average Kappa value in current non-inferior set will compare with the latest threshold value, the threshold will increase further if the average value increases, and vice versa. By doing so, the non-inferior region will progressively be reduced as the Kappa threshold is lifting up. Figure 4 illustrates this concept; the shaded region is the search space within which the particles of PSO move and scout for the highest values of accuracy and Kappa. The search time is significantly reduced when the non-inferior region (search space) is closing up by the variable, threshold Kappa. The ultimate objective of SMOTE is to achieve finding the maximum value of the reliable accuracy (*RA*) which is dependent both on Kappa and accuracy. It is noted that the constraints in preparing the search over the current dataset are based on these conditions: $100 \leq N \leq r \times size(majority\, data,\ C^{+})$, and $2 \leq k \leq size(minority\, data,\ C^{-})$ where r is the ratio of majority and minority classes in the original data.

The pseudo code of OSMOTE is listed in Algorithm 1.

Algorithm 1: Optimized SMOTE (OSMOTE)

Specify a Swarm intelligence algorithm *A* and a Classifier *C*
// Initialize the population of the *A* algorithm and the other related parameters
Define the floor value of Kappa: k_t
Define the scope of *K* and *N*
  1: Load dataset
  2: Initialize the particle's position and speed // through the SMOTE and *C* to get the Current Kappa and
      Current Accuracy as the initial population fitness
  3: **for** *i*= 1: Maximum number of iteration
  4:     Radom select the global best (*g_b*) from the non-inferior solution
  5:     Update the particles' speed and position //as the update condition, the new local best replace the older
  6:     **if** particle satisfied with the conditions
  7:       change the positions
  8:     **else** randomly select the position
  9:     **end if**
10:    Combine the new and old non-inferior solution set
11:    Filter the final results set by the update condition
12:    **if** there is no Kappa value bigger than k_t // update Kappa threshold
13:      reduce k_t
14:    **elseif** the mean of Kappa in the solution set is bigger than
15:      k_t+step length
16:      increase k_t
17:    **endif**
18: **end for**
19: **Output:** Get the final global best non-inferior solution

3.2   Swarm Instance Selection (SIS) for Reducing Majority Instances

Along with OSMOTE that syntheses just the correct amount of minority data, Swarm Instance Selection (SIS) is there for reducing majority instances to the appropriate amount. The outputs of these optimizations which have been rebalanced at the best efforts with respect to its majority/minority class, would be used as potential candidates for restructuring a new output data in the crossover.

The SIS approach is an evolutionary version of under-sampling method [36] which could effectively select the useful majority samples. It is known that there is no rigid ratio or partitioning rule for labelling which instances are useful or otherwise. The relations between different combinations of majority and minority class data with the predicted class are non-linear. Therefore, a wrapper approach, using the classifier (DT in our case) to tell which combinations of majority instances and minority instances can find us the best combination which offers the highest classification performance from the base learner.

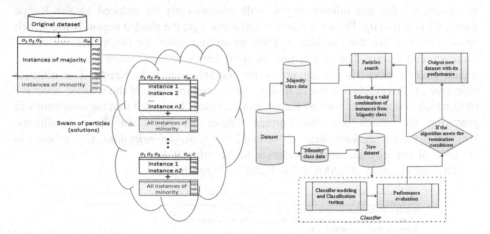

Fig. 5. Majority instances are selected into swarm particles in SIS

Fig. 6. Flow chart of Swarm Instance Selection

Given the sheer volume of instance, brute-force is not feasible. That is the reason why the majority data selection process is devised to be optimized by stochastic swarm search. In SIS each PSO particles encodes a candidate solution as two parts, one of which is a collection of instances randomly selected from the majority instances of the current dataset (or original dataset if it is the first cycle). Figure 5 shows the random selection concept. The other part of the particle is the whole of the minority group of instances from the current (or original) dataset. Like chromosomes in genetic algorithm, the particles will change in their collections towards a global best solution represented by the maximum performance from the base learner.

The size of the PSO particle is $Size(p) = Random\_selection(Data_{maj}) + Data_{min}$. The size of the swarm particle is also constrained by the minimum and maximum limits as follow: $Min \geq r \times Size(Data_{min})$, and $Max < r \times Size(Data_{maj})$, where r is the ratio of the minority data over majority data of the current dataset. Figure 6 shows the working procedure of Swarm Instance Selection approach.

In SIS, each randomly selected group from the majority class data will combine with all of the minority class data to create a new candidate dataset, which will be used to build a decision tree for performance testing. As SIS runs, the particles will eventually choose the best combination of majority class samples through the comparison of each particle's fitness over some time. Compared with other under-sampling methods, SIS incurs certain overhead in computational speed. However, SIS often can achieve a better solution (globally best) at the end, without using the naïve brute-force.

Both OSMOTE and SIS have the advantage of staying focused in tackling the imbalance problem with respect to one of their classes, through cooperation hand-in-hand via the crossover operation at the end of each iteration. Their respective class data are rebalanced while considering the best sampled data that were generated thus far. Conventional over-sampling or under-sampling techniques however focus only on their individual class size, neglecting about how the other class size and instances within might have been evolved or improved. The conventional over-sampling and under-sampling are conducted independently without regards of their counter-part. This is the prime advantage and unique difference between AMSCO and the traditional rebalancing algorithms. The pseudo codes of SIS is listed as follow:

Algorithm 2: Swarm Instance selection (SIS)

Specify a Swarm intelligence algorithm A and a Classifier C

```
1: Load dataset
2: Initialize the population of the A algorithm and the other
      related parameters
3: Using dataset to initialize the particle's position and speed
      //through classifier C get the Current Reliable Accuracy as the initial population fitness
4: for i= 1: Maximum number of iterations
5:       select the global best
6:       if the current best less than the new
7:       replace
8:       else
9:             randomly select the best
10:   end if
11:   if the terminal condition is met
12:         break
13:   end if
14: end for
15: Output: get the best combination of the instances to build a new dataset with its performances
```

---

4 Experiment

Eight rebalancing methods are used in the experimentation to evaluate the performance of the proposed methods in this paper versus the traditional ones. The first to compare is basic classification algorithm, decision tree, without any pre-processing of rebalancing, three methods are commonly used in algorithm level to the bias of classifier, and the other four are sampling methods which include the traditional over-sampling methods with the three swarm rebalancing algorithms.

- Decision Tree (DT): one of the most popular classifiers. It often shows good performance in imbalanced dataset classification. Many papers in the workshop of ICML 2013 investigated C4.5 with imbalanced dataset and it effectively increases the performance of sampling techniques from imbalanced dataset [37].
- Bagging: Bagging method + DT.
- AdaBoost.M1 (AdaBM1): AdaBM1 + DT.AdaBoost.M1 stands for Discrete Ada-Boost [38], which is classical boosting method.
- Cost-sensitive (CTS): CTS + DT. The values of cost matrix respectively matching to the elements of confusion matrix, TP and TN's cost are zero. FN denotes the misclassified minority class samples; its cost is 10. Misclassified majority class samples are FP which cost half of FN.
- Synthetic Minority Over-Sampling Technique (SMOTE): SMOTE + DT. The two parameters are manually selected taking the default values. The average value of its ten times operation is used as the final performance.
- Swarm Instance Selection algorithm (SIS): SIS + DT.
- Optimized SMOTE Algorithm (OSMOTE): OSMOTE + DT.
- SIS-OSMOTE: SIS-OSMOTE + DT. The two optimization processes are placed sequentially. OSMOTE will load the new dataset from SIS as the Kappa of SIS is greater than 0.3 or it reaches the maximum iteration. In this case, SIS and OSMOTE work independently without crossing over their optimized datasets.
- AMSCO: Adaptive Multi-Objective Swarm Crossover Optimization. Its logics are in Fig. 1.

For fair comparison, the maximum iteration of bagging, AdaBM1, SIS, OSMOTE, SIS-OSMOTE and AMSCO is standardized at 100. The amount of base classifiers of Bagging, the population of SIS, OSMOTE, SIS-OSMOTE and AMSCO is 20. Stratified 10-cross-validation is used as the verification and testing method. There are 30 imbalanced datasets being used for benchmarking as they are selected from 100 binary class imbalanced dataset from KEEL [39]. These datasets have low Kappa statistics. Table 1 lists the characteristics of these datasets, Maj and Min respectively denotes majority class and minority class, Imb.r is the imbalance ratio (majority/minority). The imbalance ratio ranges from 1.87 to 129.44. The simulation software is programmed by Matlab 2014b. The simulation computing platform is CPU: E5-1650 V2 @ 3.50 GHz, RAM: 32 GB.

For easy comparison, the average performance values across the 30 benchmark datasets are charted as histograms featuring vis-a-vis all the rebalancing algorithms, the swarm and the traditional. The results unanimously point to an observation that AMSCO has an edge over the performances outperforming the rest of them. Figure 7 shows the overall performance comparison chart.

Observing from the results, it is apparent that, there are rooms for improvement for decision tree that classifies the original datasets without any rebalancing, both the accuracy and Kappa are the lowest among all. In the four typical methods, SMOTE performs relatively well. In the ensemble learning, AdaBM1 is better than Bagging. Cost-sensitive learning is more effective than Bagging. Our individual class optimization methods using Swarms are much better than the traditional methods, demonstrating the usefulness of swarm optimization, except for SIS is worse than

Table 1. Characteristic of datasets used in experiment

Dataset	#Samples	Maj	Min	Imb.r	Dataset	#Samples	Maj	Min	Imb.r
abalone-17_vs_7-8-9-10	2338	2280	58	39.31	poker-8-9_vs_6	1485	1460	25	58.4
abalone-19_vs_10-11-12-13	1622	1590	32	49.69	poker-8_vs_6	1477	1460	17	85.88
abalone-20_vs_8-9-10	1916	1890	26	72.69	poker-9_vs_7	244	236	8	29.5
abalone-21_vs_8	581	567	14	40.5	vehicle1	846	629	217	2.9
abalone19	4174	4142	32	129.44	vehicle3	846	634	212	2.99
abalone9-18	731	689	42	16.4	winequality-red-3_vs_5	691	681	10	68.1
cleveland-0_vs_4	177	164	13	12.62	winequality-red-8_vs_x6-7	855	837	18	46.5
flare-F	1066	1023	43	23.79	winequality-red-8_vs_6	656	638	18	35.44
glass-0-1-4-6_vs_2	205	188	17	11.06	winequality-white-3-9_vs_5	1482	1457	25	58.28
glass-0-1-5_vs_2	172	155	17	9.12	winequality-white-9_vs_4	168	163	5	32.6
glass-0-1-6_vs_2	192	175	17	10.29	yeast-0-3-5-9_vs_7-8	506	456	50	9.12
glass2	214	197	17	11.59	yeast-0-5-6-7-9_vs_4	528	477	51	9.35
haberman	306	225	81	2.78	yeast-1-2-8-9_vs_7	947	917	30	30.57
pima	768	500	268	1.87	yeast-1-4-5-8_vs_7	693	663	30	22.1
poker-8-9_vs_5	2075	2050	25	82	yeast-1_vs_7	459	429	30	14.3

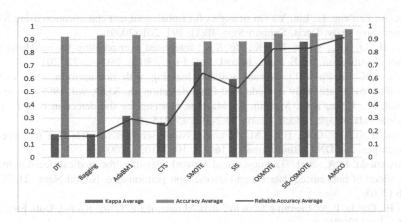

Fig. 7. Overall comparison of the rebalancing methods in terms of Kappa, accuracy and Reliable Accuracy

SMOTE. However, OSMOTE which focuses on minority data is more effective than SIS. Compared to the conventional methods, AMSCO's Kappa values stay above 0.9 in 26 out of 30 cases, achieve perfect performance at 1 in 3 out of 30 cases. AMSCO is relatively more stable too comparing to other methods. The average values of accuracy and Kappa are the highest, and the standard deviations of the two performance measures are the lowest for AMSCO.

5 Conclusions

This paper proposed AMSCO algorithm to tackle class imbalanced dataset in sampling two classes in parallel. AMSCO aims at rebalancing the dataset, improving classification model's credibility and preserving the high accuracy, within reasonable time.

It implements two swarm optimization algorithm to progressively find out the best performance for a specific classifier. One optimization process called OSMOTE is extended from SMOTE. It focuses on inflating the minority data to an appropriate amount. The other is called SIS which selects only the useful instances for filtering the majority data. Experimental results show that, AMSCO can significantly outperform a number of rebalancing methods in different categories. In our future works, AMSCO will be extended to solving imbalanced problem on multi-class classification.

Acknowledgement. The authors are thankful for the financial support from the Research Grant Temporal Data Stream Mining by Using Incrementally Optimized Very Fast Decision Forest (iOVFDF), Grant no. MYRG2015-00128-FST, offered by the University of Macau, FST, and RDAO.

References

1. Sun, A., Ee-Peng, L., Liu, Y.: On strategies for imbalanced text classification using SVM: a comparative study. Decis. Support Syst. **48**(1), 191–201 (2009)
2. Cao, H., Li, X.L., Woon, D.Y.K., Ng, S.K.: Integrated oversampling for imbalanced time series classification. IEEE Trans. Knowl. Data Eng. **25**(12), 2809–2822 (2013)
3. Chan, P.K., Stolfo, S.J.: Toward scalable learning with non-uniform class and cost distributions: a case study in credit card fraud detection. In: KDD, vol. 1998 (1998)
4. Kubat, M., Holte, R.C., Matwin, S.: Machine learning for the detection of oil spills in satellite radar images. Mach. Learn. **30**(2-3), 195–215 (1998)
5. Choe, W., Ersoy, O.K., Bina, M.: Neural network schemes for detecting rare events in human genomic DNA. Bioinformatics **16**(12), 1062–1072 (2000)
6. Mazurowski, M.A., et al.: Training neural network classifiers for medical decision making: the effects of imbalanced datasets on classification performance. Neural Netw. **21**(2), 427–436 (2008)
7. He, H., Garcia, E.A.: Learning from imbalanced data. IEEE Trans. Knowl. Data Eng. **21**(9), 1263–1284 (2009)
8. Tang, Y., Zhang, Y.Q., Chawla, N.V., Krasser, S.: SVMs modeling for highly imbalanced classification. IEEE Trans. Syst. Man Cybern. Part B Cybern. **39**(1), 281–288 (2009)
9. Guo, H., Viktor, H.L.: Learning from imbalanced data sets with boosting and data generation: the DataBoost-IM approach. ACM SIGKDD Explor. Newslett. **6**(1), 30–39 (2004)
10. Li, J., Fong, S., Mohammed, S., et al.: Improving the classification performance of biological imbalanced datasets by swarm optimization algorithms. J. Supercomput. **72**, 3708 (2016). doi:10.1007/s11227-015-1541-6
11. Chawla, N.V.: C4. 5 and imbalanced data sets: investigating the effect of sampling method, probabilistic estimate, and decision tree structure. In: Proceedings of the ICML, vol. 3 (2003)
12. Stone, E.A.: Predictor performance with stratified data and imbalanced classes. Nat. Methods **11**(8), 782–783 (2014)
13. Chen, Y.-W., Lin, C.-J.: Combining SVMs with various feature selection strategies. In: Guyon, I., Nikravesh, M., Gunn, S., Zadeh, L.A. (eds.) Feature Extraction: Foundations and Applications. Studies in Fuzziness and Soft Computing, pp. 315–324. Springer, Heidelberg (2006)

14. Wallace, B.C., et al.: Class imbalance, redux. In: 2011 IEEE 11th International Conference on Data Mining (ICDM). IEEE (2011)
15. Liu, A., Ghosh, J., Martin, C.E.: Generative oversampling for mining imbalanced datasets. In: DMIN (2007)
16. Batuwita, R., Palade, V: Efficient resampling methods for training support vector machines with imbalanced datasets. In: The 2010 International Joint Conference on Neural Networks (IJCNN). IEEE (2010)
17. Drummond, C., Holte, R.C.: C4. 5 class imbalance, and cost sensitivity: why under-sampling beats over-sampling. In: Workshop on learning from imbalanced datasets II, vol. 11 (2003)
18. Kubat, M., Matwin, S: Addressing the curse of imbalanced training sets: one-sided selection. In: ICML, vol. 97 (1997)
19. Chawla, N.V., Bowyer, K.W.: SMOTE: synthetic minority over-sampling technique. J. Artif. Intell. Res. **16**, 341–378 (2002)
20. Galar, M., et al.: A review on ensembles for the class imbalance problem: bagging-, boosting-, and hybrid-based approaches. IEEE Trans. Syst. Man Cybern. Part C Appl. Rev. **42**(4), 463–484 (2012)
21. Thai-Nghe, N., Gantner, Z., Schmidt-Thieme, L: Cost-sensitive learning methods for imbalanced data. In: The 2010 International Joint Conference on Neural Networks (IJCNN). IEEE (2010)
22. Zhu, X.: Lazy bagging for classifying imbalanced data. In: IEEE ICDM 2007, pp. 763–768 (2007)
23. Sun, Y., Kamel, M.S., Wang, Y.: Boosting for learning multiple classes with imbalanced class distribution. In: IEEE ICDM 2006, pp. 592–602 (2006)
24. del Río, S., et al.: On the use of MapReduce for imbalanced big data using random forest. Inf. Sci. **285**, 112–137 (2014)
25. Chawla, N.V., Lazarevic, A., Hall, L.O., Bowyer, K.W.: SMOTEBoost: improving prediction of the minority class in boosting. In: Lavrač, N., Gamberger, D., Todorovski, L., Blockeel, H. (eds.) PKDD 2003. LNCS (LNAI), vol. 2838, pp. 107–119. Springer, Heidelberg (2003). doi:10.1007/978-3-540-39804-2_12
26. Fan, W., et al.: AdaCost: misclassification cost-sensitive boosting. In: ICML, vol. 99 (1999)
27. Sun, Y., et al.: Cost-sensitive boosting for classification of imbalanced data. Pattern Recognit. **40**(12), 3358–3378 (2007)
28. Zadrozny, B., Langford, J., Abe, N.: Cost-sensitive learning by cost-proportionate example weighting. In: IEEE ICDM 2003, pp. 435–442 (2003)
29. Kennedy, J., et al.: Swarm Intelligence. Morgan Kaufmann, San Francisco (2001)
30. Poli, R., Kennedy, J., Blackwell, T.: Particle swarm optimization. Swarm Intell. **1**(1), 33–57 (2007)
31. Li, J., et al.: Adaptive swarm balancing algorithms for rare-event prediction in imbalanced healthcare data. Comput. Med. Imaging Graph (2016). http://dx.doi.org/10.1016/j.compmedimag.2016.05.001
32. Van den Bergh, F., Engelbrecht, A.P.: A cooperative approach to particle swarm optimization. IEEE Trans. Evol. Comput. **8**(3), 225–239 (2004)
33. Landis, J.R., Koch, G.G.: The measurement of observer agreement for categorical data. Biometrics **33**, 159–174 (1977)
34. Viera, A.J., Garrett, J.M.: Understanding interobserver agreement: the kappa statistic. Fam. Med. **37**(5), 360–363 (2005)
35. Fonseca, C.M., Fleming, P.J.: Multiobjective optimization and multiple constraint handling with evolutionary algorithms. I: a unified formulation. IEEE Trans. Syst. Man Cybern. Part A Syst. Hum. **28**(1), 26–37 (1998)

36. García, S., Herrera, F.: Evolutionary undersampling for classification with imbalanced datasets: proposals and taxonomy. Evol. Comput. **17**(3), 275–306 (2009)
37. Cieslak, D.A., Chawla, N.V.: Learning decision trees for unbalanced data. In: Daelemans, W., Goethals, B., Morik, K. (eds.) ECML PKDD 2008. LNCS (LNAI), vol. 5211, pp. 241–256. Springer, Heidelberg (2008). doi:10.1007/978-3-540-87479-9_34
38. Freund, Y., Schapire, R.E.: Experiments with a new boosting algorithm. In: ICML, vol. 96 (1996)
39. Alcalá, J., et al.: Keel data-mining software tool: data set repository, integration of algorithms and experimental analysis framework. J. Multiple Valued Logic Soft Comput. **17** (255-287), 11 (2010)

Causality-Guided Feature Selection

Mandar S. Chaudhary[1], Doel L. Gonzalez II[1], Gonzalo A. Bello[1],
Michael P. Angus[1], Dhara Desai[1], Steve Harenberg[1], P. Murali Doraiswamy[2],
Fredrick H.M. Semazzi[1], Vipin Kumar[3], and Nagiza F. Samatova[1,4(✉)], for the
Alzheimer's Disease Neuroimaging Initiative

[1] North Carolina State University, Raleigh, NC, USA
samatova@csc.ncsu.edu
[2] Duke University, Durham, NC, USA
[3] University of Minnesota, Twin Cities, MN, USA
[4] Oak Ridge National Laboratory, Oak Ridge, TN, USA

Abstract. Identifying meaningful features that drive a phenomenon
(response) of interest in complex systems of interconnected factors is
a challenging problem. Causal discovery methods have been previously
applied to estimate bounds on causal strengths of factors on a response
or to identify meaningful interactions between factors in complex systems, but these approaches have been used only for inferential purposes.
In contrast, we posit that interactions between factors with a potential
causal association on a given response could be viable candidates not
only for hypothesis generation but also for predictive modeling. In this
work, we propose a causality-guided feature selection methodology that
identifies factors having a potential cause-effect relationship in complex
systems, and selects features by clustering them based on their causal
strength with respect to the response. To this end, we estimate statistically significant causal effects on the response of factors taking part in
potential causal relationships, while addressing associated technical challenges, such as multicollinearity in the data. We validate the proposed
methodology for predicting response in five real-world datasets from the
domain of climate science and biology. The selected features show predictive skill and consistent performance across different domains.

1 Introduction

Complex systems, such as the climate and biological systems, are characterized
by an intricate interconnected network of interacting factors. These interactions
often represent causal relationships among the factors (predictors), as well as
between the factors and a phenomenon of interest (response). For example, in the

Data used in preparation of this article were obtained from the Alzheimers disease
Neuroimaging Initiative (ADNI) database (adni.loni.usc.edu). As such, the investigators within the ADNI contributed to the design and implementation of ADNI
and/or provided data but did not participate in analysis or writing of this report.
A complete listing of ADNI investigators can be found at http://adni.loni.usc.edu/
wp-content/uploads/how_to_apply/ADNI_Acknowledgement_List.pdf.

© Springer International Publishing AG 2016
J. Li et al. (Eds.): ADMA 2016, LNAI 10086, pp. 391–405, 2016.
DOI: 10.1007/978-3-319-49586-6_26

climate science domain, factors may represent large scale ocean and atmospheric patterns, summarized as time series called *climate indices*, whose interactions are known to influence extreme weather phenomena, such as droughts and floods [2, 3]. Similarly, in biology, factors may represent genes, the expression levels of which have been found to have an effect on a phenotype of interest, such as disease status [12]. Identifying meaningful factors in these complex systems that can be used to predict the response is a challenging task.

Traditionally, causality-driven methods have been applied to investigate causal relationships between variables. In the climate science domain, they have been used to construct causal graphs (Definition 1) that capture interactions among climate indices [8,9]. The relationships found in these causal graphs are further studied to confirm existing hypotheses and, if possible, generate new ones. On the other hand, in biology, cause-effect relationships are established by performing randomized gene knock-out experiments. Estimating bounds on the potential causal effects of genes on a phenotype has been helpful for prioritizing such experiments [12,13]. However, these approaches require further domain expertise to validate the results and do not focus on identifying predictive factors for any specific response.

Recent methods construct the local causal structure for a response to select predictive features [1,16]. These features are identified by utilizing the idea of constraint-based and score-based learning methods. However, these methods do not incorporate the causal effect of a predictor on the response for selecting predictive features and do not identify meaningful cause-effect relationships between variables in the system.

Consequently, we introduce the problem of *causality-guided feature selection* for identifying predictors with significant causal effect on the response. To do this, we construct causal graphs using a constraint-based learning algorithm, such as PC-stable [7], and leverage causal relationships in this graph to estimate the causal effect of a predictor on the response. We evaluate the stability of each predictor by performing a random permutation test to obtain a set of predictors having statistically significant causal effect on the response. In the end, we cluster the predictors and select features from each cluster with the most significant causal effect to form the new feature space.

Finally, we validate our proposed methodology on two motivational use cases in the domains of climate science and biology. Specifically, we apply this methodology to select features for predicting seasonal rainfall in the regions of African Sahel and East Africa, predicting riboflavin production rate in bacterium *B. Subtilis*, and predicting the cognitive score in male and female patients respectively. In climate science, the African Sahel region has been studied extensively following a series of severe droughts in the 1970s and 1980s [3]. Repeated droughts throughout the 2000s have led to a humanitarian crisis in the region, with approximately 10.3 million food-insecure people in 2013[1]. East Africa is a similarly vulnerable region, including within it Lake Victoria, a mostly precipitation-fed

---

[1] http://www.fao.org/fileadmin/user_upload/emergencies/docs/
SITUATION%20UPDATE%20Sahel%201%2007%202013.pdf.

resource for millions of people. In biology, identifying genes having significant causal effect on a phenotype of interest such as the riboflavin (vitamin B_2) production rate is a challenging task [4,13]. Another important task is to identify biomarkers that can be used to detect the phase of mild cognitive impairment (MCI) which is preceded by Alzheimers disease (AD) among individuals. Currently, no biomarkers have been validated for predicting the risk of AD, and hence there is a greater need to discover key biomarkers [5].

2   Problem Statement

Let $X = \{X_1, X_2, .., X_p, Y\}$ be a set of variables consisting of p predictors, $\{X_1, X_2, .., X_p\}$, and a response, Y. For example, for our use case in the climate science domain, X may be a set of p climate indices and Y may be seasonal rainfall at a target region (e.g., the African Sahel or East Africa). Informally, we define *causality-guided feature selection* as the task of selecting features based on the *potential causal relationships* among the variables in X, with the goal of improving the prediction of Y. To do this, we first introduce the concepts of *causal graph* and *causal effect*.

Definition 1 (Causal Graph). *Given a set of variables X, a causal graph $G = (V, E)$ is defined as a graph where $V = X$ is the set of nodes and E is the set of edges, such that each directed edge, $X_i \to X_j$, represents a potential causal relationship where X_i is a potential cause of X_j, and each undirected edge, $X_i - X_j$, or bidirected edge, $X_i \leftrightarrow X_j$, represents an ambiguous relationship between X_i and X_j.*

Definition 2 (Causal Effect). *Given a predictor X_i and a response Y in a causal graph G, the causal effect of X_i on Y is defined as the change in Y for a unit change in X_i. Then, the estimated causal effect of X_i on Y is given by the regression coefficient of X_i, θ_i, when Y is regressed on X_i and its parents S_i; that is,*

$$Y = \theta_i X_i + \theta_{S_i}^{\top} S_i + \epsilon_i \tag{1}$$

where ϵ_i is the residual of Y.

Finally, we formally define the problem of *causality-guided feature selection*: Given a set of variables X consisting of p predictors and a response Y, a causal graph G, and a set of causal effects of the predictors in X on Y, cluster the predictors in X based on their causal effect, and select predictors (i.e., features) from each cluster with the most statistically significant causal effect on the response.

3   Method

In this section, we describe our causality-guided feature selection methodology, as outlined in Algorithm 1. First, we use a constraint-based learning algorithm to construct a causal graph and select the potential causal relationships in the

---

Algorithm 1. Causality-Guided Feature Selection

Require: A set of variables $X = \{X_1, X_2, .., X_p, X_{p+1}\}$ consisting of p predictors,
    $\{X_1, X_2, ..., X_p\}$, and a response, $Y = X_{p+1}$
 1: Let $f_{new} = \emptyset$ be the new feature space
 2: Let G be a CPDAG constructed using the PC-stable algorithm (see Section 3.1)
 3: Let C be the set of potential causal relations in G
 4: Let $\Theta = \emptyset$ be a set of statistically significant causal effects
 5: Let Φ be a set of p-values of the statistically significant causal effects
 6: **for each** $c = X_i \rightarrow X_j$ **in** C **do**
 7:     Let $\Theta_i = \emptyset$
 8:     Let $\mathcal{G}$ be the set of Markov equivalent graphs generated for X_i
 9:     **for each** $g \in \mathcal{G}$ **do**
10:         Let $\theta_{X_{i,1}}$ be the causal effect of X_i on Y computed with PCR using X_i and
            its set of parents S_i in g (see Section 3.2)
11:         $[\theta_{X_{i,1}}, p\text{-}value_{\theta_{i,1}}]$=ASSESS_STABILITY($\theta_{X_{i,1}}$, X_i, S_i, Y)
12:         $\Theta_i = \Theta_i \cup \theta_{X_{i,1}}$
13:     **end for**
14:     **if** $\Theta_i \neq \emptyset$ **then**
15:         $\theta_i = \arg\min_{\theta \in \Theta_i} |\theta|$
16:         $\Theta = \Theta \cup \theta_i$
17:         $\Phi = \Phi \cup p\text{-}value_{\theta_i}$
18:     **end if**
19:     Repeat steps 7-18 for X_j and Θ_j if $X_j \neq Y$
20: **end for**
21: f_{new} =FEATURE_SELECTION(Θ, Φ, X)
22: **return** f_{new}

Algorithm 2. ASSESS_STABILITY

Require: A causal effect, $\theta_{X_{i,1}}$, a predictor, X_i, its parents S_i, and a response Y
 1: Let N=100
 2: Let Θ_{rand} be a set of randomized causal effects computed from N random permu-
    tations of the response (see Section 3.2)
 3: $p\text{-}value = \dfrac{\left|\{\theta'_m \in \Theta_{rand} \text{ s.t. } |\theta'_m| \geq |\theta_{X_{i,1}}|\}\right| + 1}{N+1}$
 4: **if** p-value < 0.05 **then**
 5:     **return** $[\theta_{X_{i,1}}, p\text{-}value]$
 6: **else**
 7:     **return** $\emptyset$
 8: **end if**

---

graph (Sect. 3.1). Second, for each predictor that takes part in a potential causal relationship, we compute its causal effect on the response using a methodology that addresses multicollinearity and then estimate the significance of this causal effect by performing a random permutation test (Sect. 3.2). Finally, we cluster the predictors based on their causal effect and select features from each cluster with the most statistically significant causal effect on the response (Sect. 3.3).

Algorithm 3. FEATURE_SELECTION

Require: A set of statistically significant causal effects, Θ, its corresponding set of p-values Φ, and set of predictors in X.

1: Perform K-Means clustering on Θ by identifying the optimal number of clusters, k, using the Elbow Method

2: **for each** cluster c_i **do**

3:     Let $\phi \subset \Phi$ be the p-values of statistically significant causal effects in c_i

4:     Select all predictors, $p_i \subset X$, in c_i whose causal effect has p-value equal to $min(\phi)$

5:     $f_{new} = f_{new} \cup p_i$

6: **end for**

7: **return** f_{new}

3.1 Constructing Causal Graphs and Selecting Potential Causal Relationships

We construct a graph of ambiguous and potential causal relationships using a constraint-based structure learning algorithm. Specifically, we use the PC-stable algorithm because of its ability to construct graphs with order-independent adjacency structure and to mitigate the effect of false positives edges [7]. In the next two paragraphs we present a summary of the constraint-based structure learning algorithm.

In the first step, the PC-stable algorithm constructs a completed undirected graph G over a set of variables X and initializes the size of the conditioning set, m, to zero. Next, for each variable $X_i \in X$, it stores the nodes adjacent to X_i in its adjacency set $a(X_i)$. For every pair of adjacent variables X_i and X_j in G, it checks whether the two variables are independent conditioned on S (i.e., $X_i \perp X_j | S$), such that $S \subseteq a(X_i)$ or $S \subseteq a(X_j)$ and $|S| = m$. If the variables are conditionally independent, the edge between them is removed from G and the conditioning variable(s), S, is stored in their separating set, $sepset(X_i, X_j)$ and $sepset(X_j, X_i)$. Similarly, the remaining pairs of adjacent variables are checked for conditional independence. This completes the first iteration of the conditional independence test. The adjacency set is updated for every variable, and the value of the conditioning set, m, is incremented by 1 in the next iteration. At the end of this step, the algorithm yields a *skeleton* of the causal graph, which contains undirected edges between variables that were not found to be conditionally independent. In our experiments, we use the Fisher's Z test to determine if two variables are conditionally independent at a significance threshold $\alpha = 0.05$.

In the next step, for every unshielded triple $(X_i - X_j - X_k)$ such that X_i and X_k are not adjacent, the algorithm orients $X_i - X_j - X_k$ into a v-structure, $X_i \rightarrow X_j \leftarrow X_k$, if and only if $X_j \notin sepset(X_i, X_k)$. Then the algorithm tries to orient as many remaining edges as possible using the following set of rules:

- Rule 1: Given $X_i \rightarrow X_j$ and $X_j - X_k$, orient $X_j - X_k$ to $X_j \rightarrow X_k$ such that X_i and X_k are not adjacent.

- Rule 2: Given $X_i - X_k$ and a chain $X_i \to X_j \to X_k$, orient $X_i - X_k$ into $X_i \to X_k$.
- Rule 3: Given two chains, $X_i - X_j \to X_k$ and $X_i - X_l \to X_k$, orient $X_i - X_k$ into $X_i \to X_k$.

The output of the PC-stable algorithm is a completed partially directed acyclic graph (CPDAG), which represents an approximation of the Markov equivalence class of the data. An estimated CPDAG can contain directed, undirected, and bidirected edges, as shown in Fig. 1a. A directed edge $X_1 \to Y$ represents a cause-effect relationship where X_1 is a potential cause of Y, an undirected edge $X_1 - X_2$ implies some association between the variables, and a bidirected edge $X_2 \leftrightarrow X_3$ represents a sampling error or a hidden common cause of X_2 and X_3 that is not present in the data.

Assumptions. In order to interpret a CPDAG causally, we consider the following three assumptions:

- The underlying causal structure is sparse and acyclic.
- The graph built by the PC-stable algorithm is a well approximated representation of the underlying probability distribution in the data (i.e., causal faithfulness).
- There is absence of confounding variables in the data (i.e., causal sufficiency); i.e., given any two variables X and Y in the data having a common cause Z, then Z is also present in the data.

In real-world scenarios, it is difficult to assume that a system is causally sufficient, since there might be factors that are difficult to observe and measure. As a result, we cannot prove the existence of the potential causal relationships in the CPDAG. Nonetheless, they can provide an approximation of the underlying interactions between the variables. There are causal inference algorithms such as the Fast Causal Inference (FCI) algorithm and the Really Fast Causal Inference (RFCI) that detect the presence of hidden causes [18]. For future work, we plan to explore and determine the applicability of such causal inference algorithms to our proposed methodology.

Once the CPDAG is constructed using the PC-stable algorithm, we select all the directed edges in the graph for further processing because they represent potential causal relationships between pairs of variables. Moreover, these directed edges are present across all the Markov equivalent DAGs that can be generated from the estimated CPDAG. Figure 1a shows a CPDAG and a set of Markov equivalent DAGs (Figs. 1c to e) generated from the CPDAG by orienting all the undirected and bidirected edges into directed edges such that no additional v-structure or cycle is created.

3.2   Estimating Causal Effects and Assessing Its Statistical Significance

We estimate the causal effect on the response of every predictor that takes part in a potential causal relationship. For example, consider a potential causal

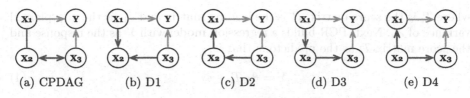

(a) CPDAG   (b) D1   (c) D2   (d) D3   (e) D4

Fig. 1. (a) A completed partially directed acyclic graph (CPDAG) and (b)–(e) the set of possible DAGs D1–D4. D1 is an invalid DAG as it contains an additional v-structure, $X_1 \rightarrow X_2 \leftarrow X_3$, that is not present in the estimated CPDAG. D2–D4 are valid DAGs and they belong to the same Markov equivalence class.

relationship $X_i \rightarrow X_j$, where an external intervention on predictor X_i leads to a unit change in X_i. To estimate the causal effect of X_i on Y, the Intervention calculus when DAG is Absent (IDA) method has been proposed, which explores the local neighborhood of X_i by changing every undirected and bidirected edge incident on X_i to a directed edge [13]. Thus, if there is a total of m undirected and bidirected edges incident on X_i, then up to 2^m Markov equivalent graphs can be generated. For each graph, the IDA method identifies the parents of X_i (i.e., predictors having a directed edge towards X_i). The response Y is then regressed on X_i and its parents, $S_i \subseteq \{X_1, X_2, ..., X_p, Y\} \setminus X_i$; that is,

$$Y = \theta_i X_i + \theta_{S_i}^\top S_i + \epsilon_i \tag{2}$$

where the regression coefficient of X_i, θ_i, is the estimated causal effect of X_i on Y for the corresponding Markov equivalent graph and ϵ_i is the residual of Y.

Thus, the IDA method yields a set of causal effects for X_i, $\Theta_i = \{\theta_i^1, \theta_i^2, ..., \theta_i^k\}$, across $k \leq 2^m$ Markov equivalent DAGs. The causal effect θ_i is 0 if $Y \in S_i$, since no change in X_i can have an effect on its parents; otherwise, θ_i is estimated as shown above.

For several real-world applications, such as our use cases in the domains of climate science and biology, the data exhibits a high degree of multicollinearity; that is, near-linear dependencies among the predictors. We determine the presence of multicollinearity by computing the variance inflation factor (VIF); for example, for our climate data set, a high VIF (i.e., greater than 10) indicates a high degree of multicollinearity in the data [15]. The presence of multicollinearity leads to a poor estimation of the coefficients in the linear regression models in Eq. 2 [14]. To address this issue, we perform Principal Component Regression (PCR), instead of linear regression, to estimate the causal effects.

Principal Component Regression. Given a Markov equivalent graph, let $\mathcal{X}_i$ be a matrix whose columns consist of predictor X_i and its parents (i.e., the predictors in S_i). Then, Principal Component Regression (PCR) computes n principal components of $\mathcal{X}_i$, where n is less than or equal to the number of columns of $\mathcal{X}_i$, using singular value decomposition (SVD) as follows:

$$\mathcal{X}_i = T_i P_i^\top + \epsilon_{\mathcal{X}_i} \tag{3}$$

where T_i is the score matrix, P_i is the loading matrix and $\epsilon_{\mathcal{X}_i}$ is the unexplained variance of $\mathcal{X}_i$. Next, PCR builds a regression model with Y as the response and the score matrix T_i as the predictors; that is,

$$Y = \theta_{T_i} T_i + \epsilon_{T_i} \tag{4}$$

We solve for the regression coefficients θ_{T_i} of the score matrix T_i using the least squares method as follows:

$$\theta_{T_i} = (T_i^\top T_i)^{-1} T_i^\top Y \tag{5}$$

The regression coefficients of the score matrix T_i, θ_{T_i}, are then used to compute the regression coefficients of the original predictor matrix $\mathcal{X}_i$ as follows:

$$\theta_{\mathcal{X}_i} = P_i \theta_{T_i} \tag{6}$$

where $\theta_{\mathcal{X}_i}$ contains the regression coefficients of X_i and of the predictors in S_i across n principal components. We then select, as the causal effect of X_i on Y, the regression coefficient of X_i that corresponds to the first principal component, $\theta_{X_{i,1}}$, since it captures the maximum variance of $\mathcal{X}_i$.

Statistical Significance Test. To assess the significance of a causal effect $\theta_{X_{i,1}}$, we perform the statistical test described in Algorithm 2. The response Y is randomly permuted $N = 100$ times and the corresponding set of randomized causal effects $\Theta_{rand} = \{\theta_1', \theta_2', ..., \theta_N'\}$ is computed using Eq. 6. A p-value is calculated to measure the probability that the magnitude of a randomized causal effect is greater than or equal to the magnitude of $\theta_{X_{i,1}}$. The causal effect $\theta_{X_{i,1}}$ is considered to be statistically significant if its p-value is less than 0.05.

This test is performed for each causal effect of X_i on Y computed across all the Markov equivalent graphs. The result is a set of statistically significant causal effects of X_i on Y, $\Theta_i = \{\theta_{X_{i,1}}^1, \theta_{X_{i,1}}^2, ..., \theta_{X_{i,1}}^l\}$, where $l \leq k$. Similarly, we compute the set of statistically significant causal effects of X_j on Y, Θ_j, for predictor X_j taking part in the potential causal relationship $X_i \rightarrow X_j$. From a set of statistically significant causal effects, Θ_i, we select θ_i, such that

$$\theta_i = \arg \min_{\theta \in \Theta_i} |\theta| \tag{7}$$

The causal effect θ_i represents the minimum change in the response Y for a unit change in X_i. We compute the causal effects θ_i for every predictor X_i that takes part in a potential causal relationship in the CPDAG. This results in a set of predictors having a statistically significant causal effect on the response.

3.3   Feature Selection via Clustering

Finally, as a feature selection technique, we group the predictors by clustering them based on their causal effects, as described in Algorithm 3. Specifically, we use the K-Means clustering method and identify the optimum number of clusters k using the Elbow Method. From each cluster, we select the predictors with the most statistically significant causal effect on the response i.e., the ones with the lowest p-value from the statistical test (see Sect. 3.2). By doing so, we minimize the number of redundant predictors having similar causal effect on the response.

4 Empirical Evaluation

In this section, we describe the application of our proposed causality-guided feature selection methodology. Specifically, our goal is to select features that improve the prediction of (1) seasonal rainfall at the African Sahel and East Africa regions, (2) riboflavin production rate, and (3) cognitive score of male and female patients.

4.1 Data Description

We used two real-world data sets from the climate science domain and three real-world data sets from the biology domain to demonstrate the performance of our methodology.

Climate Science: For the African Sahel region (ASR), we used seasonal rainfall from July to September as the response and monthly values from January to June of 34 climate indices as the predictors. Similarly, for the region of East Africa (EAR), we used seasonal rainfall from October to December as the response and monthly values from January to September of 33 climate indices as the predictors. Monthly rainfall data for both regions was obtained from the Gridded Precipitation Climatology Centre (GPCC) V6 data set [17]. Monthly data for 30 climate indices was obtained from the Earth System Research Laboratory (ESRL)[2], and 5 additional climate indices were constructed to incorporate local atmospheric conditions at each region. For our experiments, we used 57 years of data from 1951 to 2007.

Biology: For our first data set, we used publicly available data[3] containing 71 observations of 101 variables measuring the logarithm of the expression level of 100 genes and the riboflavin production rate (RPR) in the bacterium $B.$ $subtilis$ [4]. Our second and third data sets were collected from the Alzheimer's Disease Neuroimaging Initiative (ADNI). The data sets contained microarray gene expression data collected from 266 male patients and 219 female patients. The cognitive score of the subjects in each gender was used as the response variable, i.e., CS_M for male patients and CS_F for female patients.

---

[2] http://www.esrl.noaa.gov/psd/data/climateindices/list/.
[3] http://www.annualreviews.org/doi/suppl/10.1146/annurev-statistics-022513-115545/suppl_file/riboflavinv100.csv.

4.2   Data Preprocessing

The following pre-processing steps were performed after dividing the data into training and test sets using leave-one-out cross-validation (LOOCV),

1. **De-trending:** Given the temporal nature of the climate data sets, the predictors and the response variable were detrended to remove seasonal trends. This step was performed only on climate data sets.
2. **Normalization:** The predictors and the response variable were standardized using their z-scores. Note that the z-scores were computed using the average and the standard deviation from the training sets.

Due to the indefinite amount of time taken by the PC-stable algorithm for constructing causal graphs from the microarray gene expression data, we used the training set to select the top-100 genes correlated with the response as predictors.

4.3   Performance Comparison

We evaluate the performance of our causality-guided feature selection methodology by training classification models using C5.0 decision trees and regression models using the linear regression method. The response variables for ASR, EAR and RPR data sets were discretized into three categories: high, normal or low (i.e., values in the higher 66.7^{th} percentile, between the lower 33.3^{rd} and the higher 66.7^{th} percentile, and in the lower 33.3^{rd} percentile, respectively). For the two ADNI data sets, CS_M and CS_F, the cognitive score collected from the Mini-Mental State Exam (MMSE) was used as the continuous response variable. Additionally, there were two groups of patients within each gender based on the diagnostic information made available by ADNI. The first group comprised of controls (i.e., patients who did not suffer from Alzheimer's), and the second group comprised of patients suffering from late Mild Cognitive Impairment (LMCI). This diagnostic information for males and females was used as the discretized response variable in both the data sets. The decision trees were trained on these five data sets using the corresponding discretized response variables. The regression models were built using the z-scores of the response variables.

The performance of our proposed methodology was compared against eight feature selection methods, including univariate methods, two regression methods, and two local causal discovery-based methods. For the univariate methods, we selected the top-K predictors where the value of K was equal to the average number of features constructed by our proposed methodology across all the folds.

The classification accuracies of the feature selection methods on five real-world data sets are shown in Fig. 2a–e. We observe that our proposed methodology is the best performing method for ASR and EAR, and its accuracy is within 6 % of the best performing method for RPR, CS_M and CS_F. While Lasso outperforms all the methods on RPR and CS_M, it is the worst performing method for CS_F and our methodology has a minimum improvement of 8 % over Lasso for ASR and EAR. Similarly, MMHC and HITON have higher accuracy values

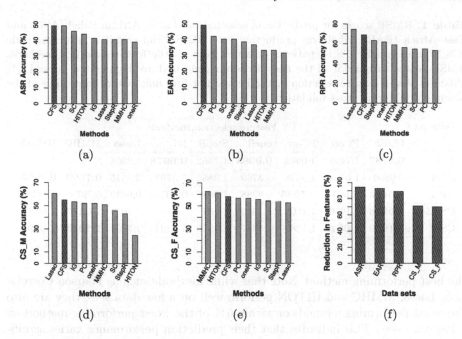

Fig. 2. Mean classification accuracy over leave-one-out cross validation (Accuracy) for the African Sahel (a) East Africa (b), *B. subtilis* (c), diagnostic information for Male (d) and Female (e) obtained from Causality-Guided Feature Selection (CFS), Max-Min Hill Climbing (MMHC), HITON Markov Blanket (HITON) Spearman Correlation (SCorr), Pearson Correlation (PCorr), oneR (oneR), Lasso Regression (Lasso), Stepwise Regression (StepR), and Information Gain (IG). (f) shows the percentage reduction in the features using CFS across five data sets.

while predicting CS_F, but their performance is as good as random guessing for EAR, and HITON is the worst performing method for CS_M.

We now compare the root mean squared errors (RMSE) for each method to evaluate the performance of the selected features on the regression models. Table 1 presents our findings that four out of five times the RMSE score of our proposed methodology is within 6 % of the best performing method. The remaining methods, however, don't show the same level of robustness in terms of the results obtained. For example, Pearson Correlation was among the top performing methods in terms of classification accuracy for ASR and EAR, but its performance is relatively poor when evaluating regression models. Furthermore, although Lasso has higher accuracy values for RPR and CS_M (see Fig. 2c–d), it is the worst performing method for EAR and CS_F. While MMHC and HITON have high accuracy values for CS_F, they had the worst RMSE scores for ASR as they could not find any feature from some of the training sets during cross-validation.

A summary of the results is shown in Table 2. We observe that the classification accuracy and RMSE score of our methodology is consistently within 6 % of

Table 1. RMSE scores for prediction of seasonal rainfall at African Sahel (ASR) and East Africa (EAR), riboflavin production rate (RPR) and cognitive score for male (CS_M) and female (CS_F) patients obtained from all the feature selection methods. RMSE scores within 6 % of the best performing method are highlighted in bold. (*) indicates that a feature selection method could not find any feature from some of the training sets during cross-validation

Data set	Feature selection methods								
	CDFS	PCorr	SCorr	oneR	StepR	IG	Lasso	MMHC	HITON
ASR	**0.9787**	1.1499	1.0942	**0.9985**	1.7362	**0.9978**	1.2304	*	*
EAR	0.9860	1.0720	1.0448	1.3769	1.3885	1.3751	3.5515	**0.7599**	**0.7655**
RPR	**0.5483**	**0.6112**	**0.5335**	0.8288	1.0956	0.6077	0.6432	0.7978	0.7114
CS_M	**1.0576**	1.1976	1.1805	1.1939	1.134	1.2086	1.2039	**1.0477**	**1.0148**
CS_F	**1.1079**	**1.0646**	1.1285	1.1337	1.1859	1.1321	1.2397	**1.1033**	**1.0996**

the best performing method. Note that while methods such as Pearson Correlation, Lasso, MMHC and HITON perform well on a few data sets, they are also the worst performing methods or within 6 % of the worst performing method on other data sets. This indicates that their prediction performance varies significantly with data sets from different domains, whereas our methodology shows predictive skill across all of the data sets in terms of classification accuracy and RMSE score. Moreover, we also observe a percentage reduction in the selected features by at least 70 % as compared to the original feature space across all the data sets (see Fig. 2f).

4.4   Time Complexity

The majority of the time taken by the proposed methodology is due to constructing a causal graph, computing causal effects and their significance. The asymptotic time complexity of the PC-stable algorithm is polynomial time on sparse graphs, i.e., $O(p^q)$ where p is the number of variables in the graph and q is the maximum number of vertices adjacent to any vertex in the graph such that $q = O(n^{1-b})$, where n is the sample size and $0 < b \leq 1$. To estimate a causal effect, the time complexity of linear regression is dominated by matrix multiplication; i.e., $O(p^2 n)$ if $n > p$ else it is $O(p^3)$. In the worst case, we may end up computing causal effect of every predictor on the response variable. As a result, when $n > p$ the time taken to compute the causal effect of each predictor on the response variable and its significance across all the Markov equivalent graphs is $O(2p \cdot 2^q \cdot (p^2 n + 100 p^2 n)) \approx O(2^{q+1} p \cdot c p^2 n)$ where 2^q is the number of Markov equivalent graphs and c is a constant. Thus, the total time complexity is $O(p^q + 2^{q+1} p \cdot c p^2 n)$. Similarly, for $n < p$ it is $O(p^q + 2^{q+1} p \cdot c p^3)$.

Table 2. A summary of the performance of all the feature selection methods in terms of classification accuracy and RMSE scores across all five data sets. A solid upper triangle indicates the best performing method, whereas a solid lower triangle indicates the worst performing method on a given data set. An upper triangle with pattern indicates that a method's performance was within 6 % of the accuracy and RMSE score of the best performing method, and a lower triangle with pattern indicates a method's performance was within 6 % of the accuracy and RMSE score of the worst performing method.

Performance Metrics	Data set	Feature Selection Methods								
		CFS	PCorr	SCorr	oneR	IG	StepR	Lasso	MMHC	HITON
Accuracy	ASR									
	EAR									
	RPR									
	CS_M									
	CS_F									
RMSE	ASR									
	EAR									
	RPR									
	CS_M									
	CS_F									

5   Related Work

The underlying behavior of a complex system can be attributed to the intricate network of potential cause-effect relationships between factors. Identifying meaningful predictors that can further the understanding of such complex systems is a challenging task. In the climate science domain, constraint-based structure learning methods for causal discovery have been applied to generate graphs of information flow (i.e., causal graphs) describing the interactions between four climate indices [8,9]. Similarly, in the domain of bioinformatics, causal inference algorithms have been applied to estimate the brain network structure from fMRI data and to explain the variations observed in high-throughput gene expression data [6,11]. Note that these methods focus on studying potential causal relationship using domain knowledge, which may not be feasible for high-dimensional systems.

Feature selection methods based on local causal structure learning identify variables within the Markov Blanket of a target variable [1,10,16]. HITON and Max-Min Hill Climbing (MMHC) are prominent examples of such feature selection methods, which employ the divide and conquer approach to find the parents, children and the spouses of the target variable [1]. However, these methods are based on constraint-based learning and do not incorporate the causal information between the variables to select the features. In this work, we have proposed a novel method that integrates the causal strength of the predictors on the target variable to select meaningful predictors. The results show that our method produces better predictive performance over these methods.

6 Conclusion

Causality-guided methods have been used in multiple domains to facilitate the understanding of complex systems. Traditionally, the application of these methods has been limited to descriptive and inferential purposes. In this work, we propose a causality-guided feature selection methodology that identifies predictors taking part in potential causal relationships and selects features based on their causal strength with respect to the response via clustering. We achieve this through a number of technical contributions, such as estimating the statistical significance of causal effects on the response, while addressing multicollinearity in the data. Our proposed methodology was found to perform consistently in terms of classification accuracy and RMSE score across real-world data sets from the domains of climate science and biology, suggesting that the newly selected features have predictive skill for the response.

Acknowledgments. This material is based upon work supported in part by the Laboratory for Analytic Sciences (LAS), the Department of Energy National Nuclear Security Administration under Award Number(s) DE-NA0002576 and NSF grant 1029711. Data collection and sharing for this project was funded by the Alzheimer's Disease Neuroimaging Initiative (ADNI) (National Institutes of Health Grant U01 AG024904) and DOD ADNI (Department of Defense award number W81XWH-12-2-0012). ADNI is funded by the National Institute on Aging, the National Institute of Biomedical Imaging and Bioengineering, and through generous contributions from the following: AbbVie, Alzheimers Association; Alzheimers Drug Discovery Foundation; Araclon Biotech; BioClinica, Inc.; Biogen; Bristol-Myers Squibb Company; CereSpir, Inc.; Cogstate; Eisai Inc.; Elan Pharmaceuticals, Inc.; Eli Lilly and Company; EuroImmun; F. Hoffmann-La Roche Ltd. and its affiliated company Genentech, Inc.; Fujirebio; GE Healthcare; IXICO Ltd.; Janssen Alzheimer Immunotherapy Research & Development, LLC.; Johnson & Johnson Pharmaceutical Research & Development LLC.; Lumosity; Lundbeck; Merck & Co., Inc.; Meso Scale Diagnostics, LLC.; NeuroRx Research; Neurotrack Technologies; Novartis Pharmaceuticals Corporation; Pfizer Inc.; Piramal Imaging; Servier; Takeda Pharmaceutical Company; and Transition Therapeutics. The Canadian Institutes of Health Research is providing funds to support ADNI clinical sites in Canada. Private sector contributions are facilitated by the Foundation for the National Institutes of Health (www.fnih.org). The grantee organization is the Northern California Institute for Research and Education, and the study is coordinated by the Alzheimers Therapeutic Research Institute at the University of Southern California. ADNI data are disseminated by the Laboratory for Neuro Imaging at the University of Southern California. PMD has received research grants and/or advisory fees from several government agencies, advocacy groups and pharmaceutical/imaging companies, and received a grant from ADNI to support data collection for this study. He also owns stock in several companies whose products are not discussed here.

References

1. Aliferis, C.F., Statnikov, A., Tsamardinos, I., Mani, S., Koutsoukos, X.D.: Local causal and markov blanket induction for causal discovery and feature selection for classification Part I: Algorithms and empirical evaluation. J. Mach. Learn. Res. **11**, 171–234 (2010)
2. Andrews, E., Antweiler, R.C., Neiman, P.J., Ralph, F.M.: Influence of enso on flood frequency along the california coast. J. Climate **17**(2), 337–348 (2004)
3. Bader, J., Latif, M.: The 1983 drought in the west sahel: a case study. Climate Dynam. **36**(3), 463–472 (2011)
4. Bühlmann, P., Kalisch, M., Meier, L.: High-dimensional statistics with a view toward applications in biology. Annu. Rev. Stat. Appl. **1**, 255–278 (2014)
5. Chen, Z., Padmanabhan, K., Rocha, A.M., Shpanskaya, Y., Mihelcic, J.R., Scott, K., Samatova, N.F.: Spice: discovery of phenotype-determining component interplays. BMC Syst. Biol. **6**(1), 1–19 (2012)
6. Chindelevitch, L., Ziemek, D., Enayetallah, A., Randhawa, R., Sidders, B., Brockel, C., Huang, E.S.: Causal reasoning on biological networks: interpreting transcriptional changes. Bioinformatics **28**(8), 1114–1121 (2012)
7. Colombo, D., Maathuis, M.H.: Order-independent constraint-based causal structure learning. J. Mach. Learn. Res. **15**(1), 3741–3782 (2014)
8. Ebert-Uphoff, I., Deng, Y.: Causal discovery for climate research using graphical models. J. Climate **25**(17), 5648–5665 (2012)
9. Ebert-Uphoff, I., Deng, Y.: Causal discovery from spatio-temporal data with applications to climate science. In: Proceedings of the 2014 13th International Conference on Machine Learning and Applications, pp. 606–613. IEEE (2014)
10. Guyon, I., Aliferis, C., Elisseeff, A.: Causal feature selection. Computational methods of feature selection, pp. 63–86 (2007)
11. Iyer, S.P., Shafran, I., Grayson, D., Gates, K., Nigg, J.T., Fair, D.A.: Inferring functional connectivity in mri using bayesian network structure learning with a modified pc algorithm. Neuroimage **75**, 165–175 (2013)
12. Maathuis, M.H., Colombo, D., Kalisch, M., Bühlmann, P.: Predicting causal effects in large-scale systems from observational data. Nat. Methods **7**(4), 247–248 (2010)
13. Maathuis, M.H., Kalisch, M., Bühlmann, P., et al.: Estimating high-dimensional intervention effects from observational data. Ann. Stat. **37**(6A), 3133–3164 (2009)
14. Montgomery, D.C., Peck, E.A., Vining, G.G.: Introduction to linear regression analysis. John Wiley & Sons (2015)
15. Neter, J., Kutner, M.H., Nachtsheim, C.J., Wasserman, W.: Applied linear statistical models, vol. 4. Irwin Chicago (1996)
16. Peña, J.M., Nilsson, R., Björkegren, J., Tegnér, J.: Towards scalable and data efficient learning of markov boundaries. Int. J. Approx. Reason. **45**(2), 211–232 (2007)
17. Schneider, U., Becker, A., Finger, P., Meyer-Christoffer, A., Rudolf, B., Ziese, M.: Gpcc full data reanalysis version 6.0 at 0.5: monthly land-surface precipitation from rain-gauges built on gts-based and historic data. FD_M_V6_050 (2011)
18. Spirtes, P., Glymour, C.N., Scheines, R.: Causation, Prediction, and Search, vol. 81. MIT press, Cambridge (2000)

Temporal Interaction Biased Community Detection in Social Networks

Noha Alduaiji[1(✉)], Jianxin Li[1], Amitava Datta[1], Xiaolu Lu[2], and Wei Liu[1]

[1] The University of Western Australia, Perth, WA 6009, Australia
Noha.Alduaiji@research.uwa.edu.au,
{Jianxin.Li,Amitava.Datta,Wei.Liu}@uwa.edu.au
[2] RMIT University, Melbourne, VIC 3000, Australia
Xiaolu.Lu@rmit.edu.au

Abstract. Community detection in social media is a fundamental problem in social data analytics in order to understand user relationships and improve social recommendations. Although the problem has been extensively investigated, most of the research examined communities based on static structure in social networks. Our findings within large social networks such as Twitter, show that only a few users have interactions or communications within any fixed time interval. It is not difficult to see that it makes more potential sense to find such active communities that are biased to temporal interactions of social users, rather than relying solely on static structure. Communities detected with this new perspective will provide time-variant social relationships or recommendations in social networks, which can greatly improve the applicability of social data analytics.

In this paper, we address the proposed problem of temporal interaction biased community detection using a three-step process. Firstly, we develop an activity biased weight model which gives higher weight to active edges or inactive edges in close proximity to active edges. Secondly, we redesign the activity biased community model by extending the classical density based community detection metric. Thirdly, we develop two different expansion-driven algorithms to find the activity biased densest community efficiently. Finally, we verify the effectiveness of the extended community metric and the efficiency of the algorithms using three real datasets.

Keywords: Social networks · Temporal community · Social interactions

1 Introduction

Detecting communities in real-world graphs such as large social networks, web graphs, and biological networks is a fundamental problem of considerable practical interest that has received a great deal of research and market attention [6]. These networks are commonly modelled as a graph containing vertices and edges. Community structure discovery can be regarded as the exploration of subgraphs

© Springer International Publishing AG 2016
J. Li et al. (Eds.): ADMA 2016, LNAI 10086, pp. 406–419, 2016.
DOI: 10.1007/978-3-319-49586-6_27

where vertices are densely connected among themselves and loosely connected to other vertices [2].

With the assumption that the whole graph is available, the subgraphs or communities can be detected by using techniques such as hierarchical clustering [14], spectral clustering [10], partitioned clustering [5], correlation-based clustering [15], and ranking-based algorithms [8]. Although the above techniques are very popular, at times it is better to explore local communities with particular requirements, rather than general communities based on the whole graph. For instance, for a given vertex in a graph, Cui et al. [3] targeted the best community the vertex (or the social user) belongs to. A similar problem was also studied by Wu et al. [17] whose aim was to find a community with a set of specified vertices, instead of only one. They proposed mechanism to reduce the local and global free rider effects in the process of identifying local communities for the discovery of robust local communities.

In this paper, we investigate the problem of local community detection from a new perspective. Based on our observation and study in Twitter, we found only a few social users had interactions within a specific period of time. For instance, in our tracked Twitter dataset we found that less than 30 % of users had actual interactions with their followers or friends as tracked by retweet actions or mentions. In this work, if two users have an edge linking each other with interactions within a certain period of time, their edge is denoted as *active*, otherwise it is denoted as *inactive*. On further inspection of the active edges, we observed that the frequency of interactions increased in diversity and the inactive edges became active after a period of time, if they were close to many other active edges. These observations motivated us to study the problem of *temporal interaction biased community detection*, denoted here as *TIB-community detection*. A TIB-community is defined as a subgraph with a maximum activity biased density score. In summary, a TIB-community consists mostly of active edges, a few inactive edges that are well connected to many active edges, and the associated endpoints (i.e., vertices).

Consequently, our proposed study of *TIB-community detection* metric is different to existing community detection metrics. With respect to community detection in weighted networks, Lu et al. [9] detected communities based on the idea that the two nodes should be in the same community if the edge weight between two nodes is high enough, or adapted it by using conductance. In addition, the user-to-user interaction was also considered for community detection in mobile networks by simply taking into account direct interaction. In [1], their k-clique community is found by applying the classical k-clique concept and checking if any member has direct interactions with $k - 1$ other members. However, unlike us, both of them failed to consider the influence of relations with frequent interactions with their neighbours where we assume the weight of an edge represents the frequency of the interaction of its endpoint nodes. Furthermore, their community detection does not consider the effect of biased density. Compared with the existing work of local community detection [3,17], our work offers greater applicability in tracking changes of large social networks over time, based on real interactions of social users. Our method allows us to find the most active communities in different time intervals even if the structure of the social network is unchanged, based on static links. These new aspects can improve social recommendations, link predictions and advertisement placement in social networks.

We make the following contributions in this paper:

- We investigate the local community detection from a new perspective, i.e., temporal interaction biased community detection. It is significant to track the change of social networks with time-variant interactions.
- We formalize the proposed problem of *TIB-community* and verify that TIB-community discovery is NP-complete. To address it, we develop two expansion-based community detection algorithms.
- We evaluate the efficiency of proposed algorithms and verify the appropriateness of the activity biased community detection metric by conducting experiments on three real datasets, Twitter, Facebook and Amazon.

The rest of this paper is organized as follows. We first review the existing community detection metrics in Sect. 2. In Sect. 3, we introduce the preliminaries and formally define the problem of temporal interaction biased community. We also verify the complexity of the proposed problem and explain it with examples. To efficiently detect the TIB-communities, we develop two solutions - one is breadth first search based TIB-community detection and another is clique based TIB-community detection, which are presented in Sect. 4. Finally, we show the experimental results in Sect. 5 and conclude our work in Sect. 6.

2   Related Work

Local community detection aims to find communities in a given graph that contains query nodes. Many local community detection methods have been proposed in recent years due to its wide applicability in many real-world large networks when global topologies are not, or impractical to obtain. Three common components exist in local community detection methods: first, the real world problem is modelled as either a binary or weighted network; second, a *goodness metric* is defined accordingly to characterise the desired features of a community, and third, an algorithm is proposed to search for the community to maximise the goodness metric. According to Cui et al. [3], such methods are referred to as *Community Search with Maximality constraint* (CSM), in contrast to Community Search with Threshold constraint (CST).

Most local community detection algorithms for weighted networks modify the goodness metric in binary cases to incorporate edge weights. For example, given k-nodes in a sub-graph $C = (V, E, W)$, a density metric that measures the internal connectedness of a sub-graph, can be defined as the ratio of the total edge weights to the number of vertices in the sub-graph:

$$D = \frac{\sum_{u,v \in V} w(u, v)}{|V|}, \tag{1}$$

where $w(u, v)$ is the weight between node u and v and $|V|$ is number of vertices $\in V$.

Our work builds on these existing methods in an attempt to deal with a specific snapshot of dynamic communities. Dynamic community detection is recognised as crucial, in time variant analysis of social network evolution [1]. We model the dynamic communication between social users during a specified time frame as a social graph. Instead of only carrying out structure analysis on the weighted social graph using static weights, we propagate the communication strength of active edges to their neighbourhood edges iteratively to ensure current weakly connected nodes still receive a chance to be included in the community if the neighbourhood edges are highly influenced by the active edges. Then we develop a greedy algorithm to detect the local community that maximises the density metric for the weighted graph in Eq. 1. A k-clique based algorithm is also implemented as a solution for comparison.

3   Problem Definition

The discovery of active and well connected communities can result in more effective information diffusion and targeted marketing. In most approaches for community detection, weakly connected edges are often disregarded, which leads to uncovering sparse communities only.

Active Edges and Biased Active Edges. In the context of detecting temporal interaction biased communities, we argue that, weakly connected edges are also important. They may only have a smaller influence in the current time interval, but may gain importance, especially when they are in a time-evolving and dynamic local community instead of a static one. In other words, they may become active after a certain time if their neighbourhood nodes are highly interactive. Therefore, we redefine the "active edges" and propose a new model to determine the potential weight of an edge.

Given a social graph $G(V, E, W)$, where V and E are the set of vertices and edges in the graph, respectively. $\forall e \in E$, it is associated with an initial weight $w(e) \in W$, which is determined within a time interval, and it indicates the frequency of interactions between two nodes. In previous studies, the associated weight of an edge indicates the extent to which the edge is active [16]. Consider the Twitter network as an example, an edge weighted 40 means it carries 40 interactions, and it is highly likely to be regarded as an active edge. But an inactive edge may be important, especially when a majority of its neighbours are highly active edges. In our work, we define an *active edge* based on the edge itself and its neighbours, denoted as *Biased Active Edge*. For example, consider an edge weighted 3, which is usually regarded as inactive in previous research literature. However, if it has an active neighbour, it may become a biased active edge since it will be reweighted by propagating the weights of its active neighbours. In order to effectively model the influence of neighbourhood activeness, the active probability of edges can be simulated through the following proposed *activity biased weight model*.

Activity Biased Weight Model and Density Criterion. Consider $\forall e \in E$, the model determines the weight of an edge based on a two-step process. The first step is to normalize the weight $w(e)$ of an edge e based the following criteria:

$$P(e) = \begin{cases} 1, & w(e) \geq 50 \\ w(e)/50, & 20 \leq w(e) < 50 \\ 0, & \text{otherwise} \end{cases} \tag{2}$$

where $0 \leq P(e) \leq 1$, and the non-zero values indicate different degrees of activity. When $P(e) = 1$, it indicates 100% active edge. 50 and 20 in the above equation are activities parameters that depend on a social network dataset. For example, based on our observation in our Twitter network dataset, active users are those who mention and retweet each other 50 times and more in certain time interval, whereas inactive users' interactions are less than 20 time. We consider this straightforward way to weight edges, which uses the observed interactions. Other more sophisticated ways such as Gaussian distribution are also applicable. The second step is to propagate the normalised weight of an active edge e_i to its neighbour e based on:

$$O(e) = 1 - \lambda^h \cdot P(e_i). \tag{3}$$

where λ ($0 < \lambda < 1$) is a decaying factor, h is the number of hops between the current edge and its neighbours, and $P(e_i)$ is the weight of an active neighbour of e. The computation of $O(e)$ is repeated for the next neighbour edge within h hops from e. Then the result of $O(e)$ is stored in a hash-table that will be used to calculate the final weight $f(e)$ for edge e:

$$f(e) = 1 - \prod_{e \in E} O(e). \tag{4}$$

$f(e)$ multiples the values in the hash-table to learn the new weight for the edge e. This learning model will take the neighbours of an edge into consideration, which helps to reweight an edge based on the activity of its neighbours as well.

We consider a concrete example in Fig. 1. The original weighted graph is given in Fig. 1(a), after the first step, which is applying $P(e)$, the weights are renormalized from 0 to 1 in Fig. 1(b). When Eq. 3 is carried out $\forall e \in E \wedge P(e) \neq 1$, the results are stored in the hash-table as shown in Fig. 1(c). We omitted some of the details, to avoid clutter. Finally, we used the hash-table to compute the final weight of each edge as shown in Fig. 1(d), which is computed using $f(e)$ Eq. 4.

It is important and crucial to evaluate the quality of the communities found based on our proposed scheme. In order to do so, we propose an *activity biased density* metric, which extends the work done by Wu et al. [17]. The density of an activity biased community is computed as follows:

$$\rho(C) = \frac{\sum_{e \in C} f(e)}{|C|}. \tag{5}$$

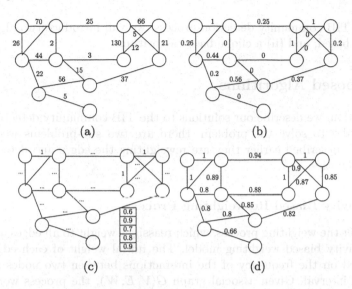

Fig. 1. An example weighted graph (a), and the same graph after applying $P(e)$(b), $O(e)$(c), and $f(x)$(d), respectively, with $\lambda = 0.5$.

That is the sum of the biased weight within the community divided by the size of the community, $|C|$.

In this paper we aim to find a local community with consideration of its potentially dynamic nature. Formally, the Temporal Interaction Biased (TIB) community detection problem is:

Problem 1 (TIB Community Detection). Find a community $C(V', E', W') \subseteq G$, which is a set of connected k cliques, such that: (i) $\forall e \in E'$, e is active or close to active edges, and (ii) $\rho(C)$ is maximized.

There are two key points in solving the TIB community detection problem: (i) Assign a proper weight considering not only current time interval, but influential probability also, and (ii) Find the set of cliques that can maximize our proposed density metric. We address the first point by adopting the activity biased weight model, which considers temporal interactions by taking into account the weight of neighbouring edges to an inactive edge. In order to solve the second point, we adopt the clique percolation method (CPMw) [4] to find community based on cliques C where $\rho(C)$ is maximized.

Like many graph search algorithms, the local community detection problem is NP-Complete. Many heuristic based greedy algorithms have been proposed and implemented. One of the most popular is the *clique percolation method* CPM [12], which builds up communities by joining overlapping k-cliques, where a clique is a fully connected sub-graph. For example, a k-clique where $k = 3$ can be visualised as a triangle. The TIB community includes finding cliques in the graph. Also, k cliques have to be assigned the maximum weight by our proposed model and maximum density by $\rho(C)$. The next section will give more details about the TIB community detection process, we will introduce two algorithms: (i) a node based, and (ii) a clique based algorithm.

4   Proposed Algorithms

In this section, we describe our solutions to the TIB community detection problem. In order to solve the problem, there are two sub-problems we need to address. As described earlier they are reweighting the edges and detecting the TIB communities.

4.1   Activity Biased Re-weighting Process

We consider the weighting process, which reassigns weight to all edges according to the activity biased weighting model. The initial weight of each edge, $w(e)$, is set based on the frequency of the interactions between two nodes within a fixed time interval. Given a social graph $G(V, E, W)$, the process works as in Algorithm 1.

Algorithm 1. Activity Biased Weight Model

Input: $G(V, E, W)$
Output: $\overline{W}$
    $\overline{W} \leftarrow \{\}, W \leftarrow \{\}$
2: **for** $\forall e \in E$ **do**
        $P(e) \leftarrow w(e)$ // Normalise the weight
4: **end for**
    for $P(e) < 1$ **do**
6:      $h \leftarrow 1, N \leftarrow \{\} HashTable = \{\}$
        while $h < 4$ **do**   // If it is less than four hops.
8:         $N \leftarrow N \cup FindNeighbor(e)$ // Find e neighbours and store them in set N
           for $\forall e_i \in N$ **do**
10:           $HashTable \leftarrow O(e)$ // Compute $O(e)$ based on Eq. 3
           end for
12:      $h \leftarrow h + 1$
        end while
14: **end for**
    for $\forall O(e) \in HashTable$ **do**
16:      $\overline{W} \leftarrow computef(e)$ // Carry out the final weight using Eq. 4
    end for
18: **return** $\overline{W}$

We first normalize the initial weight using $P(e)$, as shown in Lines 2 to 4. For every edge e whose $P(e) < 1$ we initialise: (1) h, the number of hops from the edge to its neighbours, (2) N which is a set containing e-neighbours and (3) hash-table for e (line 5–6). Then we find the neighbours of e using a breadth-first-search (BFS) and store them in N (line 8). We then calculate $O(e)$ and store the result in the hash-table (line 10). This process repeats until the maximum hop constraint $h < 4$ is reached. Next we take the set of weights $O(e)$ in the hash-table as input (lines 15), and output the $f(e)$ values for that edge e (lines 16). The process repeats until all edges are processed to output the set $\overline{W}$ with the biased weight (lines 18).

4.2   TIB Community Detection

The second subquestion is indeed a local community detection process based on the new weighting model. Thus, the result of Algorithm 1, $G(V, E, \overline{W})$, is the input for our TIB community detection. In order to solve this, we implement and compare two algorithms for detecting TIB communities.

Algorithm 2. BFS based TIB community detection

Input: $G(V, E, \overline{W})$
Output: A set of TIB communities $\mathcal{C} = \{C_0, C_1, \dots\}$
    $visited \leftarrow \{\ \}$, $prev\_r \leftarrow -\infty$ // Initialize visited node list and previous density
2: **for** $P(e) = 1$ **and** $e \notin visited$ **do**
      $visited \leftarrow visited \cup e$
4:      $N \leftarrow N \cup FindNeighbor(e)$ // Create BFS tree and find neighbors for e
      if $\rho(N) < prev\_r$ **then break**;
6:      **else**
          $prev\_r \leftarrow \rho(N)$
8:      **end if**
      $\mathcal{C} \leftarrow \mathcal{C} \cup N$ $N \leftarrow \{\ \}$, $prev\_r \leftarrow -\infty$ // Output set N as a community then reset set N and previous density
10: **end for**
    return $\mathcal{C}$

BFS based TIB Community Detection. In Algorithm 2, we start the detection process by creating the BFS tree for each unvisited active edge (Lines 2–4). Next we add the neighbour edge e to candidate $setN$ and compute the biased density using Eq. 5, for $set\ N$ (lines 5–8). If the density value increased with the last added edge, then we keep iterating and go to the next neighbour edge. Otherwise, we remove the edge and output $set\ N$ as a community C, as shown in line 9. The process repeats until all edges are exhausted, and a set of TIB communities can be identified.

Clique Based TIB Community Detection. The clique based TIB Community Detection Algorithm 3 describes the second TIB-Community detection scheme, which is based on cliques using a percolation method for a weighted graph (CPMw) [4].

    This approach has a high accuracy in finding highly dense communities. It takes $G(V, E, \overline{W})$ as an input, and identifies the overlapped community structures using connected cliques. The algorithm first obtains a set Λ of all maximal cliques that cannot be further extended beyond size k, as shown in Fig. 2(a). We consider all the adjacent cliques that share $k-1$ nodes in common. The original

CPMw considers cliques only with an intensity score higher than a threshold θ, and it is calculated as:

$$I(G) = (\prod_{e \in E} w_e)^{|E|}. \tag{6}$$

This function allows k-cliques to contain links weaker than threshold. Therefore, the resulting communities contain k-cliques with an intensity higher than I. However, in this experiment, we replace the intensity function $I(G)$ with our own biased density measure $\rho(C)$ from Eq. 5 to threshold limit the cliques. Our biased density measure can find TIB communities that are not necessarily a set of connected cliques. Next we compute the density $\rho(C)$ for the percolated cliques, as the union of maximally reachable k-cliques, where $\forall \lambda \in \Lambda'$, (lines 5–10). We also compute $\rho(\lambda)$, $\forall \lambda \in \Lambda'$, in order to compare the density to choose either the union of the clique or cliques themselves (shown in Fig. 2(b)), to be the final identified TIB communities (lines 12–17).

Algorithm 3. Clique based TIB community detection

Input: $G(V, E, \overline{W})$, k
Output: A set of TIB communities $\mathcal{C} = \{C_0, C_1, \dots\}$
$\quad \mathcal{C} \leftarrow \{\}, \Lambda' \leftarrow \{\}, r_c \leftarrow 0, R \leftarrow \{\}$ // Λ' is the set of percolate cliques.
2: $\Lambda \leftarrow kMaxClique(G)$ // Find all maximal size k cliques.
$\quad$ **for** $\forall \lambda \in \Lambda$ percolate **do** $\Lambda' \leftarrow \lambda \cup \Lambda'$
4: **end for**
$\quad$ **if** $|\Lambda'| \geq 2$ **then**
6: $\quad\quad r_c \leftarrow \rho(\Lambda'), r_\lambda \leftarrow 0$ // Compute density of Λ' as a threshold r_c.
$\quad\quad$ **for** $\lambda \in \Lambda'$ **do**
8: $\quad\quad\quad R \leftarrow R \cup \{\rho(\lambda), \lambda\}$ // Compute and store the density-clique pairs to R.
$\quad\quad\quad r_\lambda \leftarrow \max\{\rho(\lambda), r_\lambda\}$ // r_λ holds the current max density in the clique set.
10: $\quad\quad$ **end for**
$\quad$ **end if**
12: **if** $r_c > r_\lambda$ **then** $\mathcal{C} \leftarrow \Lambda'$ // Find the TIB communities based on density scores.
$\quad$ **else**
14: $\quad\quad$ **for** $\forall \{\rho(\lambda), \lambda\} \in R$ **and** $r_\lambda = \rho(\lambda)$ **do**
$\quad\quad\quad \mathcal{C} \leftarrow \mathcal{C} \cup \lambda$
16: $\quad\quad$ **end for**
$\quad$ **end if**
18: **return** $\mathcal{C}$

Algorithm Explanations. With reference to the example in Fig. 1(d), the clique-based algorithm first finds all the cliques in the graph and computes their density score, as shown in Fig. 2(d). Next, the score of percolated cliques are computed. The percolated cliques form one community only if their density score is higher than the density score for each clique, otherwise each clique is a community itself. Figure 2(b) shows both cases, where the two left percolated cliques have a lesser density score when united, while the right part is a contradicted case. The BFS based TIB community detection begins with edges have

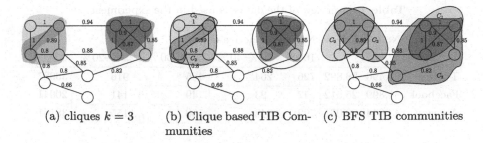

(a) cliques $k = 3$ (b) Clique based TIB Communities   (c) BFS TIB communities

Fig. 2. TIB community detection methods.

$P(e) = 1$, and add branches to a community only if it can maximize the density score. Otherwise, remove the edge from the community and output the current community. Figure 2(c) shows the TIB communities based on the BFS.

5 Experiments

Experimental Settings. This experiment has been carried out on Windows 7, 64-bit operating system IntelCore i7, CPU 3.4 GHs with 16 GB RAM. To test the proposed algorithms, we assess their performance on real networks such as, Twitter, Facebook and Amazon then perform experiments on CPMw to compare the results.

The experiment begins with the process of normalising the weights of edges so their maximum values are 1. The decay factor was set as 0.5 for all experiments, and the k clique value was set to 3. Then the normalized weight was propagated to their neighbours using the TIB-Weighted model. The model takes our dataset as the input graph G. To measure the model's performance, we assessed the time cost, memory cost, the probability of activities, and the density score of each detected community. We also compared our TIB-cliques algorithm with CPMw algorithm [13]. CPMw is community detection method with a high accuracy in detecting dense communities based on connected cliques, and has been tested on real network data.

Datasets Descriptions. Twitter dataset was used in this study. It was obtained from the Twitter Application Programming Interface (API), from Twitter Inc., which is free and available for researchers and Twitter application developers [7]. Seed users were randomly chosen to begin the data collection process. The nodes of the graph represent users of Twitter and the edges represent followership and interaction links. For example, edge (v, u) with weight $w(30)$ in the Twitter dataset represents an interaction involving user v following user u and who communicated using either a mention or a retweet or both, a minimum of 30 times. We collected two snapshots of Twitter datasets both of which took place between June 2015 and December 2015, about 20 weeks apart. Facebook [18] and Amazon datasets [11] are both publicly available. The weight in the Facebook dataset

Table 1. Overview of the datasets used in the experiment

Dataset	Nodes	Edges	Number of active edges				
			100 %	50 %–40 %	39 %–30 %	29 %–20 %	0 %
Twitter	25,490	38,892	726	264	486	916	30987
Facebook	1,899	25,512	42	23	49	141	20041
Amazon	43,644	50,061	792	348	560	1155	38161

describes the frequency of message exchanges between users, while weights in the Amazon dataset describes the frequency of product purchases by certain users. Note that weights in the Amazon dataset were randomly assigned to edges as the original dataset did not include weight information. Therefore, the random weight assignment is used to prove our approach. Dataset statistic information is shown in Table 1.

Experimental Results. In order to assess the TIB model, we conducted our experiments over the three real datasets and compared the outputs of the TIB model and the CPMw model. The effectiveness was measured by the detected communities, the number of active edges in those communities, as well as the dynamic effect. We also evaluated their performance in terms of the run time of the algorithms and the associated memory cost.

Effectiveness of TIB Models: Table 2 shows numbers of detected communities by TIB-cliques algorithm and CPMw. The results indicate that the number of communities detected by TIB-cliques algorithm is less than communities detected by CPMw. All of the communities detected by TIB-cliques algorithm, were a subset of CPMw communities. Moreover, TIB communities are all dense and contain mostly active edges. This result indicates that the biased density score and the biased weight model are both able to detect the most active and dense communities.

Table 2. Number of communities detected using the two different methods.

Method	Twitter	Facebook	Amazon
TIB	21	3	46
CPMw	29	18	358

Figures 3(a)–(c) shows the percentage of activities in the detected communities by TIB and CPMw in Twitter, Facebook, and Amazon, respectively. It can be seen that in all three datasets, TIB significantly outperforms CPMw in detecting active communities. Therefore, we can safely conclude that CPMw produces irrelevant communities due to the inclusion of more inactive edges.

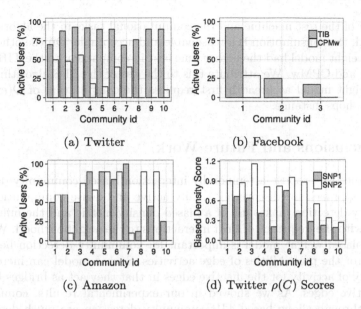

Fig. 3. Figure 3(a)–(c) show active users (%) per community on each dataset. Figure 3(d) shows the biased density score of top-10 Twitter communities.

Furthermore, we tested the effectiveness of TIB community detection with different time intervals. Figure 3(d) shows the biased density scores for Twitter communities at snapshot 1 and snapshot 2. The weight and density scores in snapshot 2 were significantly higher than that of snapshot 1 for the 10-top communities. This is because most members in such communities maintained their interactions over time. Consequently, it can be seen that our TIB community detection method and TIB density scores can help better track long living communities in real networks.

Efficiency of TIB Algorithms: We compared the overall running time and the memory cost of different methods, shown in Table 3. The cost of time and memory for TIB-cliques algorithm was almost three order of magnitudes (slower) than that of CPMw. However, the accuracy of detecting active and dense communities with TIB method is higher than that of CPMw. When the tested data

Table 3. The space and time cost on three datasets of different methods. "BiasedWM" is the biased weighting method, and the other two are same as previous.

	Time (Min)			Space (MiB)	
	BiasedWM	TIB	CPMw	TIB	CPMw
Twitter	2880	88.72	**0.007**	0.31	**0.21**
Facebook	15	74	**0.005**	0.13	**0.09**
Amazon	50	0.162	**0.004**	0.33	**0.32**

size, CPU of the test machine and the computational tasks in the algorithm are considered, the consumption is a reasonable and acceptable. On the other hand, the TIB weight model had the higher time cost in comparison to the TIB-cliques algorithm and CPMw. We attribute this to the use of the BFS algorithm in the biased weight model as breadth first requires the computation of $O(e)$ for all edges in h hops distance.

6   Conclusions and Future Work

In this paper we propose a temporally interaction biased community detection as a new approach for discovering locally active and dense communities. TIB-community detection makes use of a biased density metric and the influence of relations with the frequency of their interactions to the neighbourhood. We argue that the biased weight model is important for community detection because it helps mirror the probabilities of edge activities. Such a model can increase the probability of activity for the inactive edges in that they act as bridges between highly active edges. As we showed in our experimental results, communities detected through clique based TIB-community detection are much denser and more active. Therefore, we contend that our proposed TIB-community detection approach is very useful in many applications, e.g., dynamic interaction tracking, link predictions and advertisement placement in social networks.

In the future, we aim to explore how to optimizing the algorithms in order to accelerate the computational performance on parallel or cloud platforms.

Acknowledgments. This work was supported by the ARC Discovery Project under Grant No. DP160102114.

References

1. Chan, S.-Y., Hui, P., Xu, K.: Community detection of time-varying mobile social networks. In: Zhou, J. (ed.) Complex 2009. LNICSSITE, vol. 4, pp. 1154–1159. Springer, Heidelberg (2009). doi:10.1007/978-3-642-02466-5_115
2. Clauset, A., Newman, M.E.J., Moore, C.: Finding community structure in very large networks. Phys. Rev. E **70**(6), 66–111 (2004)
3. Cui, W., Xiao, Y., Wang, H., Wang, W.: Local search of communities in large graphs. In: Proceedings of SIGMOD, pp. 991–1002 (2014)
4. Farkas, I., Ábel, D., Palla, G., Vicsek, T.: Weighted network modules. New J. Phys. **9**(6), 180 (2007)
5. Fortunato, S.: Community detection in graphs. Phys. Rep. **486**(3), 75–174 (2010)
6. Girvan, M., Newman, M.E.: Community structure in social and biological networks. Proc. Nat. Acad Sci. **99**(12), 7821–7826 (2002)
7. Lim, K.H., Datta, A.: Following the follower: detecting communities with common interests on twitter. In: Proceedings of Conference on Hypertext and Social Media, pp. 317–318 (2012)
8. Liu, N.N., Yang, Q.: Eigenrank: a ranking-oriented approach to collaborative filtering. In: Proceedings of SIGIR, pp. 83–90 (2008)

9. Lu, Z., Wen, Y., Cao, G.: Community detection in weighted networks: algorithms and applications. In: Proceedings PerCom, pp. 179–184 (2013)
10. Luxburg, U.V.: A tutorial on spectral clustering. Stat. Comput. **17**(4), 395–416 (2007)
11. Opsahl, T., Panzarasa, P.: Clustering in weighted networks. Soc. Netw. **31**(2), 155–163 (2009)
12. Palla, G., Derényi, I., Farkas, I., Vicsek, T.: Uncovering the overlapping community structure of complex networks in nature and society. Nature **435**(7043), 814–818 (2005)
13. Price, T., Peña III, F.I., Cho, Y.-R.: Survey: enhancing protein complex prediction in ppi networks with go similarity weighting. Interdisc. Sci. Comput. Life Sci. **5**(3), 196–210 (2013)
14. Rattigan, M.J., Maier, M., Jensen, D.: Graph clustering with network structure indices. In: Proceedings of Conference on Machine Learning, pp. 783–790 (2007)
15. Su, Z., Yang, Q., Zhang, H., Xu, X., Hu, Y.: Correlation-based document clustering using web logs. In: Proceedings of HICSS, pp. 1–7 (2001)
16. Tsolmon, B., Lee, K.: Extracting social events based on timeline and user reliability analysis on twitter. In: Proceedings of CICLing, pp. 213–223 (2014)
17. Wu, Y., Jin, R., Li, J., Zhang, X.: Robust local community detection: on free rider effect and its elimination. Proc. VLDB Endow. **8**(7), 798–809 (2015)
18. Yang, J., Leskovec, J.: Defining and evaluating network communities based on ground-truth. Knowl. Info. Sys. **42**(1), 181–213 (2015)

Extracting Key Challenges in Achieving Sobriety Through Shared Subspace Learning

Haripriya Harikumar[1](✉), Thin Nguyen[1], Santu Rana[1], Sunil Gupta[1],
Ramachandra Kaimal[2], and Svetha Venkatesh[1]

[1] Centre for Pattern Recognition and Data Analytics, Deakin University,
Geelong, Australia
{hharikum,thin.nguyen,santu.rana,sunil.gupta,
svetha.venkatesh}@deakin.edu.au
[2] Computer Science and Engineering Department, Amrita University,
Kollam, India
mrkaimal@am.amrita.edu

Abstract. Alcohol abuse is quite common among all people without any age restrictions. The uncontrolled use of alcohol affects both the individual and society. Alcohol addiction leads to a huge increase in crime, suicide, health related problems and financial crisis. Research has shown that certain behavioral changes can be effective towards staying abstained. The analysis of behavioral changes of quitters and those who are at the beginning phase of quitting can be useful for reducing the issues related to alcohol addiction. Most of the conventional approaches are based on surveys and, therefore, expensive in both time and cost. Social media has lend itself as a source of large, diverse and unbiased data for analyzing social behaviors. Reddit is a social media platform where a large number of people communicate with each other. It has many different sub-groups called subreddits categorized based on the subject. We collected more than 40,000 self reported user's data from a subreddit called '/r/stopdrinking'. We divide the data into two groups, short-term with abstinent days less than 30 and long-term abstainers with abstinent days greater than 365 based on badge days at the time of post submission. Common and discriminative topics are extracted from the data using JS-NMF, a shared subspace non-negative matrix factorization method. The validity of the extracted topics are demonstrated through predictive performance.

Keywords: Social media · Reddit · Shared subspace · Topics · Alcohol addiction

1 Introduction

Excessive drinking is associated with serious consequences to the health and often leading to early death. According to Centers for Disease Control and Prevention[1] excessive alcohol use led to approximately 88,000 deaths and 2.5 million

---

[1] http://www.cdc.gov/.

© Springer International Publishing AG 2016
J. Li et al. (Eds.): ADMA 2016, LNAI 10086, pp. 420–433, 2016.
DOI: 10.1007/978-3-319-49586-6_28

years of potential life lost each year in the United States from 2006 – 2010, short-ening the lives of those who died by an average of 30 years. Excessive drinking was responsible for 1 in 10 deaths among working-age adults aged 20–64 years. The economic costs of excessive alcohol consumption in 2010 were estimated at $249 billion[2]. The high social cost manifests itself through increase in crimes, lives being lost and families falling apart. It usually takes strong motivation and long time to become successful in quitting. High relapse rate is common in early attempts to quit as strong withdrawal symptoms [15] often forces one to turn back to alcohol for a quick relief. It is therefore important to extract the challenges that people who has just started their journey faces and identify the variety of motivating factors and pathways that successful people have taken on their journey to long-term abstinence.

Various surveys and questionnaires based studies were conducted in alcohol and other substance abuse related research [6]. A sample of college students was selected for identifying the risky behavior [4] associated with alcohol addiction and its effects on their mental health [14]. Even though all these studies revealed some significant factors associated with alcohol consumption, the limited avail-ability of data leads to inconsistent conclusions. The emergence of social media provides huge amount of data for various such studies. Thus social media pro-vides a platform for creating new relationships that may influence the behaviors. Twitter data for the prediction of obesity and diabetes statistics [1] and Reddit data for identifying suicidal ideation [3] and the importance of social feedback [2] in member retention in a weight loss community are some studies which use social media data.

In this work we use data collected from Reddit. It is a social media platform where registered users can share their thoughts and experiences with other people by submitting content such as text posts or web links. They can announce a behavior they want to change, keep track of it and other users can provide social support. User posts are organized by areas of interest called 'subreddits'. The specific subreddit that we are interested in is related to users with a shared interest in quitting alcohol addiction is '/r/stopdrinking'. In this forum they discuss issues and experiences on their journey to sobriety. In addition to the posts users can also keep a badge of honor which counts the number of days that the user has self-reported to be sober. The badge resets when a relapse is reported. We crawled all the posts from this subreddit from the year 2011 to April 2016. It resulted in a total of 42,337 posts. We use the badge information to divide these posts into two groups. The first group is from users (referred to as 'Class 0') with the drinking abstinence period of at most 30 days while the second group (referred to as 'Class 1') is from users with the abstinence period of at least 365 days. The different time horizons are chosen to understand both the issues faced within the first month after deciding to quit and the characteristics associated with really long-term abstainers. This division results in a total of 10,551 posts where 6,910 posts belongs to 'Class 0' and the remaining 3,641 posts to 'Class 1'. Some example of posts from this group along with their badge

---

[2] http://www.cdc.gov/alcohol/fact-sheets/alcohol-use.htm.

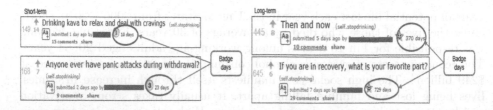

Fig. 1. The abstinence badges obtained for short-term and long-term abstainers in '/r/stopdrinking' subreddit.

labels are shown in Fig. 1. A recent study has analyzed Reddit data [12] for alcohol drinkers but is limited to only long-term abstinence. We note that it is important to understand not only the factors behind success but also the challenges faced during the early period of the quitting attempt.

We divide the data into two groups and our objective is to find the issues faced by short-term quitters, topics discussed mostly by the long-term quitters while also considering their common threads of discourse. We use a technique called Joint-Subspace Non-negative Matrix Factorization (JS-NMF) [7], which can learn shared and discriminative topics by jointly modeling any two data sources. JS-NMF has been applied to improve the performance of tag-based image and video retrieval in online social image/video sharing systems by leveraging an auxiliary data source. The capability of extracting shared and discriminative structures is utilized in our work for identifying the common and discriminative topics in the posts of long and short-term abstainers. Using JS-NMF we discover 3 groups of topics. One is the shared topics by two groups, and the other two are the discriminant topics of each group. The data for shared subspace is obtained by converting the raw posts into a vector of unigrams. We used a common vocabulary list for representing the posts vector belonging to each group. The discriminant topics reveal that short-term groups are suffering from withdrawal symptoms, cravings, dehydration and from change in food habits. On the contrary, the long-term group has positive perspective towards life and they motivate others for staying abstained. From the topics we can understand the success pathways to abstinence and key motivations on why they are staying abstained. It came as no surprise that the family and children are the main reasons for majority of them. However, the pathways they are following for staying abstained are more interesting - some of them mentioned the book 'Alcohol Anonymous' as being helpful, some are actively involved in stopdrinking meeting, joining chat, watching movies and following spiritual path. Predictive results are also shown to establish the validity of both the joint and individual topics.

2   Methods

2.1   Datasets

We collected data from an online community called Reddit. The community has 'redditors' who regularly post about their views, comments, self experiences and various interests. The Reddit has subgroup called 'subreddit' categorized based on the subject areas. We collected data from subreddit '/r/stopdrinking'. The '/r/stopdrinking' subreddit inspires alcoholics to control their habit of drinking. We used PRAW, Python Reddit API Wrapper for collecting posts and its associated meta data such as the badge information, score, post title, post content, username from Reddit. Each redditor can request for a badge, which can be used to count the number of abstinent days. We used the badge information, shown in Fig. 1, for grouping the posts into two categories. The posts which have a badge value less than 30 days at the time of posting belongs to *short-term* group and greater than 365 days at the time of posting belongs to *long-term* group. We removed the posts with badge days greater than 30 days and less than 365 days from further analysis as these posts contain information about the transition phase rather than behavioral difference. In our experience, 30 days versus 365 days time frame seems to be sufficient to identify the behavioral differences between the two groups. Since our goal is to identify the patterns which characterizes the beginners and the sobers posts which belongs to these two time frame contain more information about their behavior difference than others.

We collected 42,337 posts from years 2011–2016 and created two groups (short-term and long-term) from them. The short-term group contains 6,910 posts and the long-term group contains 3,641 posts respectively.

Table 1. The "average similarity" of top-10 discriminative topics of each group with respect to the topics in the other group.

Short-term	Values	Long-term	Values
S1	0.025	L37	0.023
S47	0.027	L17	0.025
S45	0.029	L43	0.025
S44	0.030	L40	0.030
S18	0.032	L26	0.031
S24	0.033	L16	0.033
S17	0.034	L38	0.033
S41	0.035	L4	0.033
S4	0.036	L25	0.033
S5	0.036	L18	0.034

2.2   Feature Extraction

Data pre-processing: Data cleaning and pre-processing are important steps before feature extraction. A python library called Beautiful Soup, is used for removing HTML and XML tags from the raw posts. There is a set of words included in python nltk library, called stop word list and it can be updated by users based on their requirements. So for removing most common words we updated the stop word list by adding some extra words like 'drink', 'sober', 'would' etc. We used the stopword list for removing most frequent and less discriminative words from the raw posts.

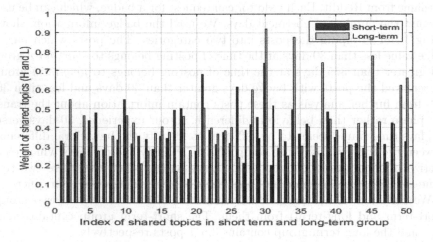

Fig. 2. Weight or relevance of shared topics in short-term and long-term group.

Unigram extraction: Unigram is a widely used and simple representation of documents [5] even though we cannot identify semantics and co-occurrence of words. We use Scikit-learn [11] to extract unigrams from the pre-processed posts. We extract unigrams that are at least 3 characters long and occur a minimum of 20 times in whole corpus. This process led to a vocabulary with a total of 2,241 unigrams for short-term group and 1,698 unigrams for long-term group. The combined vocabulary has 2,149 unigrams. We converted each post into a vector of unigrams, where an element represents the frequency of the corresponding unigram. After the conversion process we removed the post vectors that have non-zero counts for less than or equal to 5 unigrams. Finally we got a term-by-document matrix of size 2149×6256 for short-term group and of size 2149×2960 for long-term group.

2.3   Topic Extraction

The behavior and the thoughts of the users submitting posts may differ depending up on the days of sobriety. So our goal is to find and analyze the common

Algorithm 1. Shared and discriminative topic extraction using JS-NMF.

1. **Input**: Dataset X_{pre} with dimension $M_1 \times N_1$, Y_{pre} with dimension $M_2 \times N_2$, V_X with dimension $1 \times M_1$ and V_Y with dimension $1 \times M_2$.
2. Construct a common vocabulary V by taking a union of V_X and V_Y.
3. Set M as the length of vocabulary V.
4. Create X with dimension $M \times N_1$ using X_{pre} and V. Similarly, create Y with dimension $M \times N_2$ using Y_{pre} and V.
5. Compute $W, U, V, H, L = $ JS-NMF(X, Y, K, R_1, R_2) following Algorithm 1 of [7].
6. Compute the common topic weights as $F = H(1:K,:)$ or $L(1:K,:)$.
7. Sort the vector $\sum_n F_{:,n}$ in descending order and store the indices as a set S_{XY}.
8. Compute z_X as $z_X = U^T V \mathbf{1}$ where $\mathbf{1}$ is a vector of all ones of appropriate length.
9. Compute z_Y as $z_Y = V^T U \mathbf{1}$ where $\mathbf{1}$ is a vector of all ones of appropriate length.
10. Sort z_X in ascending order and store the indices as a set S_X.
11. Sort z_Y in ascending order and store the indices as a set S_Y.
12. **Output**: S_X, S_Y and S_{XY}.

topics and the discriminant topics among the two groups. To this end, we created two dictionaries, one for short-term and the other for the long-term group. We used Joint Shared Non-negative Matrix Factorization (JS-NMF) [7] for extracting the most discriminating topics and common topics from each group. JS-NMF is a non-negative matrix factorization based method that can jointly model two data sources and capture their common and discriminative topics. However, to model the data sources jointly, it needs a common vocabulary for the two sources. In our application, we created a common vocabulary by merging the vocabularies of short-term group and long-term group through their union. Let us denote the size of the common vocabulary by M. Let X be a term-by-document matrix with N_1 documents and M terms (or unigrams) and Y be a dictionary N_2 documents and M terms, JS-NMF decomposes the matrices jointly as,

$$X \approx [W \mid U] H \qquad (1)$$

$$Y \approx [W \mid V] L \qquad (2)$$

where W is a $M \times K$ matrix contains the shared or common topics, U and V represent the discriminative topics and are of sizes $M \times (R_1 - K)$ and $M \times (R_2 - K)$ respectively. The matrices H and L are weight matrices of topics and are of sizes $R_1 \times N_1$ and $R_2 \times N_2$ respectively.

Although JS-NMF can capture discriminative topics through matrices U and V, depending on the number of basis vectors specified, some of the topics in U may be quite similar to the topics in V [8]. We detect the similarity between the topics captured by U and V through a matrix $U^T V$. To find the most discriminative topics we sum each row of the matrix $U^T V$ to get a vector z_X such that $z_X = U^T V \mathbf{1}$ where $\mathbf{1}$ is a vector of all ones. Each element of z_X can be thought of as "average similarity" of the corresponding topic in U with all the topics in V. Sorting the elements of z_X in ascending order sorts the topics of U in the decreasing order of discrimination with the topics of V. A similar ordering over the discriminative topics in V (with respect to U) can be found by utilizing the matrix $V^T U$ instead of $U^T V$. Using this process, we use top 10 discriminative topics of U and top 10 discriminative topics from V.

Like the discriminative topics, we also refine the set of common topics. For this, we consider the sub-matrices $H(1 : K, :)$ and $L(1 : K, :)$, which represent the weights of common topics for the two groups. We select top 10 rows with the highest row sums from each group and choose the topics that lie in the intersection of the top 10 sets of both short-term and long-term group.

Algorithm 1 provides a step-by-step procedure for extracting common and discriminative topics from two groups using JS-NMF. Table 1 shows the "average similarity" of top 10 topics from each group with respect to the topics in the other group. Table 2 shows the weight for the top 10 common topics in each group. We got three common topics in the intersection set (in bold).

2.4   Classification

Least Absolute Shrinkage and Selection Operator (Lasso) [13] is a predictive model that can be used to learn a parsimonious linear classifier. This model generated using Lasso can simultaneously select features from a large set of features and build a classifier. We use Lasso on the feature matrices H and L learnt using JS-NMF on Reddit posts from the two groups. We stratified our dataset into training and testing data in a 6:4 ratio.

A 10 fold cross validation available in scikit-learn is used in our work to fit 60 % training data. We used a combined feature space $[W|U|V]$ for Lasso. We found that the co-efficients of most of the features corresponding to W part are zero. This indicates that most of the predictive power is limited in the

Table 2. The "weight" of the top 10 common topics in each group where the first two columns are for short-term and the last two columns are for long-term group. The word cloud of intersection set (C30,C21,C34) is shown in Table 4.

Short-term	Values	Long-term	Values
C30	**0.740**	**C30**	**0.919**
C21	**0.681**	**C34**	**0.876**
C26	0.613	C45	0.776
C10	0.546	C38	0.743
C18	0.500	C50	0.661
C34	**0.494**	C49	0.624
C17	0.492	C28	0.598
C39	0.483	C2	0.546
C6	0.478	C31	0.515
C29	0.454	**C21**	**0.503**

discriminatory features $[U|V]$ while W part does not contribute much to discrimination between the two groups. This shows the effectiveness of discriminative topics learnt using JS-NMF.

For evaluating the performance of classification, we used two measures: Area Under ROC Curve (AUC) and F_1-measure. The higher the value of these measures, the better is the classification performance.

2.5   Classification Results

We used the remaining 40 % data as test data for evaluating the performance of the topics extracted by JS-NMF. The AUC and F_1-score obtained after running Lasso on the three feature spaces of dataset are shown in Table 3. The AUC obtained for the projected spaces $[W|U|V]$ and $[U|V]$, are high compared to the W. The classifier performs well in discriminant feature space. The common feature space has less predictive power compared to the discriminant ones. F_1-score also confirms that predictive power of the discriminant topics are high.

Table 3. The AUC and F_1-score obtained after applying Lasso on projected feature spaces extracted using JS-NMF.

Projected Space	AUC	F_1-socre
UV	0.747	0.399
WUV	0.749	0.403
W	0.605	0.15

3   Discussion

The JS-NMF analysis of the posts helps to answer some queries such as reasons behind abstinence and drinking, physical and mental effects of consuming alcohol, how can they remain sober and why they are sober for long period. The identified common as well as discriminant patterns of both groups are the answers we found by analyzing the posts. The word cloud of the topics shared by each group is shown in Table 4. The weight computed for the shared topics in short-term and long-term group is shown in Fig. 2. We used the weight to find out the most discriminative topics associated with each group. The word cloud of top 10 most discriminative topics from each group is shown in Tables 5 and 6.

3.1   Common Topics

Both the short-term and the long-term group have some common topics. This can be seen in common topics extracted using JS-NMF. The identified topics are categorized into two - reasons for drinking and withdrawal symptoms.

Table 4. Word cloud of common topics in both short-term and long-term abstainers group.

Topic	Word cloud
C30	reason thanksgiving young decisions words reasons quit nights anymore ...
C34	community sidebar text members thread questions idea page localmods folks using ...
C21	alcohol abuse body losedance stronger enjoy simply mental effects problem ...

Table 5. Word cloud of top-10 discriminative topics in short-term abstainers group.

Topic	Word cloud	Topic	Word cloud
S1	water soda tonic coffee noticed body	S24	doctor issues detox liver health withdrawal blood
S47	chocolate bar eating lose fact food children calories weight	S17	plan course anymore staying keeping
S45	miss christmas strange missing hungover shots beers	S41	brain body method practice trick listen voice
S44	dinner food instead fine ordered ready restaurant wanting birthday	S4	line began face quickly half
S18	past usually idea hurt beers worst completely fiance	S5	hour minutes spent birthday pulled

- *Reasons for drinking and stop drinking* - Both group have some reasons for drinking and stop drinking alcohol which is revealed in Topic C30. They drink because of the social situations like parties, social ineptitude, due to sleeping problem, the type of friend circles they have. This can be seen in posts submitted by long-term group ``One of the reasons I drank was a social ineptitude...'' and short-term group such as ``Sleep was another reason I drank, I often get racing thoughts and find it hard to fall asleep at night...''. The short-term group decided to quit drinking because of health problems which can be seen in posts ``I decided that the next half of life should be spent sober if for no other reason than to avoid the health problems I am guarunteed to create for myself...''. But comparing with short-term group, this long-term abstainers had some strong reasons to quit drinking like they have been in jail which is visible in some posts like ``However due to a number of reasons (2 duis, jail time, many blackouts, both parents are addicts, etc.) I can proudly say I am almost three years sober!...''.

- *Physical and mental effects* - Topic C21 is a common topic for short-term and long-term group. The physical and mental changes related to alcohol drinking are explained in some posts. Long-term group are experiencing the positive changes even though they had some heavy withdrawal issues in the beginning. ``Despite these, freedom from alcohol feels good, physically, mentally, and spiritually. I am more in touch with my body, and I'm able to improve it through exercise. ...''. But most of the beginners are dealing with withdrawal symptoms and some of them are enjoying the sober days as well. ``After two weeks my body felt like my own again...'' or ``I definitely had issues with panic attacks, shortness of breath, night sweats, and random'itching' attacks...''.

3.2 Discriminative Topics

The discriminative topics for short-term are grouped into three categories - dehydration and cravings, change in food habits and withdrawal symptoms. Similarly, from the long-term abstainer discriminant topics, we found some cues about the motivations behind their abstinence and how they are staying abstained.

Group 1 - Short-term abstainers

- *Dehydration and cravings* - Topic S1 and S45 discusses the issues faced by short-term abstainers in the beginning phase of abstinence. They can't completely recover from their cravings. Most of the people belonging to this group are trying to manage their urge by drinking coffee or soda or water during parties or special occasions. Some people have dehydration problem and they tend to drink a lot of water. These posts support the topics extracted ``I noticed when I quit I became very dehydrated....'' or ``I drink a lot of coffee now too. It helps with the cravings....'' or ``Both events went really well! I got through by drinking water, soda

Table 6. Word cloud of top-10 discriminative topics in long-term abstainers group.

Topic	Word cloud	Topic	Word cloud
L37	club **book** purpose american discuss religious suggestion	L16	**freedom** dreamed secret lose fear forward happiness tough thoughts given
L17	idea greater **faith** power reason human questions chapter	L38	spoke **prison** anger volunteer parole
L43	join **chat** webchat provided synced type future match eastern	L4	self **fear** resentment personal willing usually security tried
L40	matters greater father **spiritual** power religious experience principles	L25	father wants spiritual **husband** women children chapter
L26	action **steps** worked drugs fucked peace matter defects character	L18	unless family **wants** story doctor approach experience

water, coffee, and REALLY focusing on the moment....''. Topic S41 and
some posts reveals that the brain will trick and most of the time controls people. So in order to achieve sobriety, they should self-control the thoughts in their
brain.

- *Change in food habits* - Topic S47 and S44 are about the change in their food
  habits. Some people have cravings for food, chocolate and sugar. ''Alright,
  I'm over two weeks sober and am doing great battling the alcohol
  cravings. But, I've got one slight problem now... I'm having HORRIBLE
  sugar and chocolate cravings....''. Change in food habits can either have
  positive or negative influence. In a negative perspective, some people eat a
  lot and have over-eating problem which is evident from posts ''The last
  few days since getting sober have been horrible. I'm depressed and
  overeating....''. But some are quite happy because they are now eating
  healthy foods and doing exercises which can be seen in some post ''I eat
  healthy and exercise quite frequently now.....''.
- *Withdrawal symptoms* - People in the beginning phase of quitting are dealing with some serious physical and mental withdrawal symptoms like panic
  problem, blood pressure, sleeping and stomach problems which can be seen in
  Topic S24. They normally consult some doctors for these issues as mentioned in
  some posts ''I went to the doctor and he prescribed me some anti-anxiety
  medication to help with the withdrawal symptoms.....''. Doctors advise
  them to do some Liver function tests and blood test which is supported by post
  ''I saw a doctor today to discuss my high cholesterol and my high GGT

score and my bad `liver function results`.....'' and often prescribe medicines too for their relief.

Group 2 - Long-term abstainers

- *How to stay abstained* - Topic L37 captures the discussions about 'Alcohol Anonymous' book informally called 'The Big Book'. `''I was so fortunate in my arrival in AA. On day one my sponsor showed me that the book ''Alcoholics Anonymous'' is the story of how''`'over 100 men and women have `recoverED`...''`Thank God for that man`...''`. This book covers the story of some women and men recovered from alcohol addiction. There is also an 'Alcohol Anonymous' group [9,10] which motivates people to stopdrinking alcohol by some regular inspiring meetings and group activities. The 12-step program (Topic L26) followed by this group is good for most of the people for staying abstained. It is a set of guiding principles outlining a course of action for recovery from addiction or other behavioral problems. The stopdrinking meetings, chatting and watching movies in Topic L43 are also some good and effective ways for staying abstained. This is strongly supported by some posts such as `''Go to the chat and talk to people they will help you`....''`. Following the spiritual path (Topic L40) helped some long-term abstainers for their journey to long-term recovery. Topic L17 claims that mind is more important and have faith for achieving their final goal of stop drinking supported by the post `''The next thing I value most, what has made learning possible, is being on a spiritual path again`....''`.
- *Why staying abstained* - Family is one of the main reasons for staying abstained. Those who have children are more responsible about their life and they stopped drinking because their family wants. Sometimes they stop their drinking habits by a doctor's advice. Some are thinking about their parents and to make them happy they had chosen the stop drinking path, evident from posts like `''I'm grateful that my children have never seen me drink or drunk. I'm very grateful to be comfortable in my own skin. I'm also much healthier than I was 10 years ago`....''`. The Topic L4 reveals that one of the main reasons for drinking is fear and suggest others to avoid this for achieving sobriety and it is supported by the post `''The hardest thing to do has been to let go of fear, resentment, and regret. If I work to overcome the influence of these emotional responses in my life and try to build a foundation of trust and hope for the future, sobriety seems to come naturally`...''`. The sobers in the long-term group are now enjoying the freedom and happiness of their alcohol free life which is shown in Topic L16.

4 Conclusion

Our study focused on analyzing behavioral differences and common problems among short-term and long-term abstainers group. We collected more than 40,000 posts from stopdrinking subreddit. The abstinence badge, a reward given to the registered users in this community is used here to categorize whole posts into two groups. JS-NMF, a shared subspace non-negative matrix factorization technique for capturing shared and discriminant subspace for identifying the common and shared topics associated with each group. We have found that the withdrawal symptom is a common problem in the beginning phase of abstinence and most of them reached out for clinical care. Staying abstained is stressful and some topics in the short-term group reveals it. But long-term group motivates the others by their own experiences and giving some ways for continuing abstinence. We used Lasso for evaluating the predictive power of extracted topics and found that the discriminant feature space is performing well for classification. The patterns identified are strongly supported by some psychological studies as well. These result promises the large possibility of using machine learning techniques for large scale online data analysis.

Acknowledgment. This work is partially supported by the Telstra-Deakin Centre of Excellence in Big Data and Machine Learning.

References

1. Abbar, S., Mejova, Y., Weber, I.: You tweet what you eat: Studying food consumption through Twitter. In: Proceedings of the 33rd Annual ACM Conference on Human Factors in Computing Systems, pp. 3197–3206 (2015)
2. Cunha, T.O., Weber, I., Haddadi, H., Pappa, G.L.: The effect of social feedback in a reddit weight loss community. In: Proceedings of the 6th International Conference on Digital Health Conference, pp. 99–103 (2016)
3. De Choudhury, M., Kiciman, E., Dredze, M., Coppersmith, G., Kumar, M.: Discovering shifts to suicidal ideation from mental health content in social media. In: Proceedings of the 2016 CHI Conference on Human Factors in Computing Systems, pp. 2098–2110 (2016)
4. Ekpenyong, N.S., Aakpege, N.Y.: Alcohol consumption pattern and risky behaviour: a study of university of port harcourt. IOSR J. Humanit. Soc. Sci. (IOSR-JHSS) **19**(3), 1 (2014)
5. Feldman, R., Sanger, J.: The Text Mining Handbook: Advanced Approaches in Analyzing Unstructured Data. Cambridge University Press, New York (2007)
6. Gilpin, E.A., Pierce, J.P., Farkas, A.J.: Duration of smoking abstinence and success in quitting. J. Nat. Cancer Inst. **89**(8), 572 (1997)
7. Gupta, S.K., Phung, D., Adams, B., Tran, T., Venkatesh, S.: Nonnegative shared subspace learning and its application to social media retrieval. In: Proceedings of the 16th ACM SIGKDD International Conference on Knowledge Discovery and Data Mining, pp. 1169–1178 (2010)
8. Gupta, S.K., Phung, D., Adams, B., Venkatesh, S.: Regularized nonnegative shared subspace learning. Data Min. Knowl. Discov. **26**(1), 57–97 (2013)
9. Kelly, J.F., Hoeppner, B., Stout, R.L., Pagano, M.: Determining the relative importance of the mechanisms of behavior change within Alcoholics Anonymous: A multiple mediator analysis. Addiction **107**(2), 289–299 (2012)
10. Magura, S., McKean, J., Kosten, S., Tonigan, J.S.: A novel application of propensity score matching to estimate Alcoholics Anonymous effect on drinking outcomes. Drug Alcohol Depend. **129**(1), 54–59 (2013)

11. Pedregosa, F., Varoquaux, G., Gramfort, A., Michel, V., Thirion, B., Grisel, O., Blondel, M., Prettenhofer, P., Weiss, R., Dubourg, V.: Scikit-learn: machine learning in python. J. Mach. Learn. Res. **12**, 2825–2830 (2011)
12. Tamersoy, A., De Choudhury, M., Chau, D.H.: characterizing smoking and drinking abstinence from social media. In: Proceedings of the 26th ACM Conference on Hypertext & Social Media, pp. 139–148 (2015)
13. Tibshirani, R.: Regression shrinkage and selection via the lasso. J. Royal Stat. Soc. Ser. B (Methodological) **58**, 267–288 (1996)
14. Weitzman, E.R.: Poor mental health, depression, and associations with alcohol consumption, harm, and abuse in a national sample of young adults in college. J. Nerv. Mental Dis. **192**(4), 269–277 (2004)
15. Witteman, J., Post, H., Tarvainen, M., de Bruijn, A., Perna, E.D.S.F., Ramaekers, J.G., Wiers, R.W.: Cue reactivity and its relation to craving and relapse in alcohol dependence: A combined laboratory and field study. Psychopharmacology **232**(20), 3685–3696 (2015)

Unified Weighted Label Propagation Algorithm Using Connection Factor

Xin Wang[1,2(✉)], Songlei Jian[1,2], Kai Lu[1,2], and Xiaoping Wang[1,2]

[1] College of Computer, National University of Defense Technology,
Changsha, People's Republic of China
dywangxin@foxmail.com, {jiansonglei,kailu,xiaopingwang}@nudt.edu.cn
[2] Science and Technology on Parallel and Distributed Processing Laboratory,
National University of Defense Technology, Changsha, People's Republic of China

Abstract. With the social networks getting increasingly larger, fast community detection algorithms like the label propagation algorithm, are attracting more attention. But the label propagation algorithm deals vertices with no proper weight, which leads to the loss in the performance. We propose the connection factor of the vertex to measure its influence on the local connectivity. The connection factor can reveal the topological structure feature, and we propose a unified weight to modify the original label propagation algorithm. Experiments show that our Unified Weighted LPA has an average performance promotion from 5 % to 10 %, in the best case more than 30 %.

Keywords: Community detection · Label propagation algorithm · Connection factor

1 Introduction

The social network is getting increasingly larger with the development of the Internet, and the community detection has attracted more attention [7]. The goal of community detection is to cluster densely connected vertices into communities. Various kinds of approaches have been proposed, based on diverse criteria, such as the SimRank [3] and the Ncut [10]. One popular and influential algorithm is the label propagation algorithm [9].

However, the label propagation algorithm(LPA) has some weakness. One of the problems is that the LPA views vertices merely equal, which impacts the accuracy and quality. Such weakness limits the further application of the LPA. Many approaches have been proposed, but what they discuss is not detailed enough when it comes to the neighbours and the local topological feature.

In this paper, we will propose new criteria of the vertex, the connection factor, which is based on the influence upon the connectivity among its neighbours. The connection factor studies the local connectivity topological feature and can reflect whether the vertex is more likely to be near the centre of the community or not. In the combination of the connection factor and the degree, we make the

© Springer International Publishing AG 2016
J. Li et al. (Eds.): ADMA 2016, LNAI 10086, pp. 434–444, 2016.
DOI: 10.1007/978-3-319-49586-6_29

Unified Weighted Label Propagation Algorithm(UWLPA), and we can see the promotion in performance in some data experiments.

The rest of the paper is organised as follows: In the following section, we introduce the related works briefly. Section 3 contains the definition and calculation of the connection factor. Section 4 discusses the factor with corresponding topological structure, and gives the unified weight as the combination of the connection factor and degree, which is used in the Unified Weighted LPA. The data processing experiments are shown in Sect. 5. The whole paper is concluded in Sect. 6.

2   Related Works

The LPA is one popular and influential algorithm in the community detection [9]. The basic idea of LPA is rational. Every vertex is supposed to hold a label of the community it belongs to. The algorithm keeps updating the label in random order every turn, while the new label is the most popular one in its neighbours (if many candidates, choose one randomly). After the algorithm finishes, the vertices with the same label belong to the same community. The LPA has a near-linear time complexity and an outstanding scalability for large-scale networks.

However, the LPA also has many shortcomings. One problem is that, during the label updating, the LPA views all vertices purely equal and assigns the label to be the one held by the most neighbours, in ignorance of some priority. Then the result can depend more on the label updating order, which is randomly chosen. Consequently in practical application, the accuracy, quality and stability of the LPA are greatly impacted.

There are many modifications of the LPA [15]. Some try to improve the stability [12], some produce modified detailed algorithm procedures to promote the performance [13]. One more intuitive way is to make the vertex propagate label by some weight or priority to make the new label to be the one with the largest sum of the weight of neighbours, and such algorithms can propose a quite complex priority for the label propagation in many aspects, including the vertice, the edge and the community size [5].

Such complicated priority takes many aspects into consideration, but lacks a detailed discussion on one particular aspect, especially the neighbour vertices. As a matter of fact, most algorithms merely make the neighbour vertex weight to be the degree, but the degree is only part of the topological feature. The high-degree vertex may not only share edges within the cluster but also have connections with other clusters. It can lead to a limited promotion if the degree is considered only and we need some other weight based on the local topological structure.

In this paper, we propose the connection factor to measure the influence of the vertex upon the connectivity among its neighbours and to make the unified weight, which performs better in clustering accuracy and quality.

3   The Connection Factor

This section discusses the definition and calculation of the connection factor.

3.1   Definition of the Connection Factor

Definition 1 (The Undirected Graph). *Let the $G = (V, E)$ donates an undirected graph, where V is the set of vertices and E is the set of edges. We use $e = \{u, v\} \in E$ to indicate an edge between vertices u and v.*

Definition 2 (The Neighbour Set). *Given an undirected graph $G = (V, E)$, the neighbour set $\Gamma(u)$ of $u \in V$ is the set of its adjacent vertices.*

$$\Gamma(u) = \{ \, v \in V \mid \{u, v\} \in E \, \} \tag{1}$$

Definition 3 (The Local Subgraph). *Given an undirected graph $G = (V, E)$ and the object vertex $u \in V$, its local subgraph, $LG(u)$, is the subgraph whose vertex set is $\Gamma(u) \cup \{u\}$.*

$$LG(u) = (\, \Gamma(u) \cup \{u\}, \; LE(u) \,) \tag{2}$$
$$LE(u) = \{ \, e \mid e = \{i, j\} \in E, \; i, j \in \Gamma(u) \cup \{u\} \, \} \tag{3}$$

Definition 4 (The Neighbour Subgraph). *Given an undirected graph $G = (V, E)$ and the object vertex $u \in V$, its neighbour subgraph, $NG(u)$, is the subgraph whose vertex set is $\Gamma(u)$.*

$$NG(u) = (\, \Gamma(u), \; NE(u) \,) \tag{4}$$
$$NE(u) = \{ \, e \mid e = \{i, j\} \in E, \; i, j \in \Gamma(u) \} \tag{5}$$

The neighbour subgraph is the left part after removing the object vertex from local subgraph, which may become several connected components. The connection factor focuses on the difference of the local connectivity before and after the object vertex is removed, and we need to measure such connectivity.

Definition 5 (Connected Pair). *Given a graph, if two vertices are connected through a path in the graph, then such vertex pair makes the connected pair in that graph.*

The function δ is used to judge whether a pair of vertices $i, j \in V$ can form a connected pair in a given graph $G = (V, E)$.

$$\delta(i, j, G) = \begin{cases} 1, & \text{there is one path from } i \text{ to } j \text{ in } G \\ 0, & \text{otherwise} \end{cases} \tag{6}$$

After the removal, the connected component can be parted and the number of connected pairs diminishes. The division of the number of remaining connected pairs by the original number of connected pairs can represent the difference in connectivity, and the connection factor is defined so. In the following definition, the degree of u, $deg(u)$, is also used.

Definition 6 (The Connection Factor). *Given an undirected graph $G = (V, E)$ and the object vertex $u \in V$, the connection factor of u is the number of connected pairs in the neighbour subgraph divided by the number of connected pairs in the local subgraph, where the object vertex is not counted. The factor is defined 1 when the denominator is 0.*

$$
factor(u) = \begin{cases} \dfrac{\displaystyle\sum_{i,j\in\Gamma(u)} \delta(i,j,NG(u))}{\displaystyle\sum_{i,j\in\Gamma(u)} \delta(i,j,LG(u))}, & deg(u) \geq 2 \\[2em] 1, & otherwise \end{cases} \tag{7}
$$

Suppose we want to get the connection factor of a in (Fig. 1), the number of connected pairs is 15, and 6 pairs are left after the removal. As a result, the connected factor is 0.4 . Note that the object vertex is not counted when it comes to the number of connected pairs.

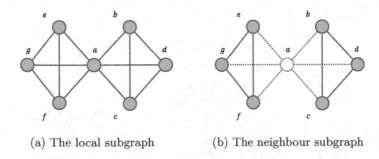

(a) The local subgraph   (b) The neighbour subgraph

Fig. 1. Example of the connection factor calculation

3.2   Calculation of the Connection Factor

In this subsection, we are going to discuss a simplified calculation of the connection factor.

If the degree is zero or one, the denominator is zero. There is no connected pair even before the removal. Such vertices are of little influence and we assign 1 to their connection factors.

If the degree is more than one, we can view the local subgraph as one connected component C_0, and the neighbour subgraph may split into many connected components $C_1, C_2, \ldots$ and C_k. The denominator can be simplified:

$$
C^2_{num(C_0)-1} = C^2_{deg(u)} = \frac{deg(u) * (deg(u) - 1)}{2}
$$

where $num()$ is the size of input set. And the numerator can be simplified as $\sum_{i=1}^{k} C^2_{num(C_i)}$.

To enumerate the connected components $C_1, C_2, \ldots$ and C_k, we can use the Breadth First Search whose starting vertex is randomly chosen. After finding one connected component, remove it from the current subgraph and continue searching until the subgraph is empty.

Then the calculation is simplified as Eq. (8), where the $C_1, C_2, \ldots$ and C_k are the enumerated connected components of the neighbour subgraph.

$$ factor(u) = \begin{cases} \dfrac{2 \sum\limits_{i=1}^{k} C^2_{num(C_i)}}{deg(u) * (deg(u) - 1)}, & deg(u) \geq 2 \\ \\ 1, & otherwise \end{cases} \tag{8} $$

4   The Connection Factor and the Unified Weight

This section discusses the connection factor with the topological feature and then gives the Unified Weighted LPA.

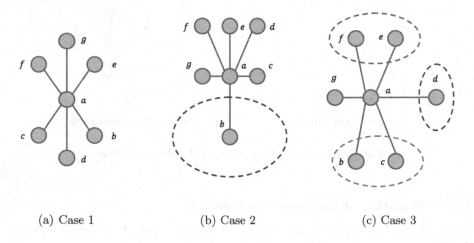

(a) Case 1   (b) Case 2   (c) Case 3

Fig. 2. Typical cases in which the factor is 0 and the degree is 6. The coloured circle indicates that there may be some community. (Color figure online)

4.1   Discussion of the Topological Feature

The low connection factor indicates that the object vertex is of great local connectivity influence, according to the definition. It is intuitive to give more weight(higher priority) to vertices with more influence on local connectivity and vice versa. As a matter of fact, however, such opinion is misleading. We will show that with typical cases for the maximum influence(when the connection factor is 0), with the degree fixed to be six.

Firstly we omit the case where neighbours have no extra edge (Fig. (2a)) since the neighbours have to accept the label of a and the weight is of no contribution. In a simple case (Fig. (2b)), one neighbour, b, can be more densely connected to *the outside world* for a, with the coloured circle indicating some community. Those nodes except b have a tendency to form one independent community as well as a tendency to be part of the outside community. In (Fig. 2c), when most neighbours have dense connections to the outside, a can hardly propagate its label to neighbours densely connected to the outside. And a is under the influence from its neighbours and the communities behind. Then we see a is likely to be near the border of communities in the last two cases.

It is destructive to give high weight to vertex a. In (Fig. 2b) there can be no more independent native community and in (Fig. 2c), several different communities may form one huge community, since a is near the border of communities and has a higher priority to force neighbours to accept and spread its label. On the contrary, a lower priority for vertex a will not result in such problem.

We have discussed that a low-factor vertex deserves a lower priority in label propagation. On the other hand, higher connection factor leads to a higher priority. The high factor shows that the neighbours are too densely connected to be easily influenced by the removal of object vertex, which indicates the vertex is near the centre of a community. A higher priority for such vertex is helpful to form a community quickly.

4.2   The Unified Weighted LPA

Although the connection factor can reveal the LPA priority, it is impacted by the degree. The correlation of high connection factor and community centre is strong only for high-degree vertices. The low-degree can achieve a high connection factor easily, but they are structurally unimportant and such high weight contributes nothing but noise.

Consequently, we give high priority only when the degree and connection factor are both high. The low degree can lead to the low priority. Then the weight can be unified of the connection factor and degree. We introduce the unified weight of vertex u in Eq. (9), where $deg(u)$ is the degree, and $factor(u)$ represents the connection factor. A constant parameter c is also introduced for flexibility.

$$w(u) = deg(u) * \exp(c * factor(u)) \tag{9}$$

Then we get the Unified Weighted Label Propagation Algorithm (UWLPA) in Algorithm 1.

5   Experiment

This section conducts experiments to evaluate the proposed algorithm.

Algorithm 1. The Unified Weighted Label Propagation Algorithm

1: **Input:** $G = (V, E)$, *maxturn*;
2: **for** each $v \in V$ **do**
3:     Initialize the label $C(v) = v$;
4:     Calculate the connection factor using Equation (8);
5:     Calculate the unified weight using Equation (9);
6: **end for**
7: $t = 1$;
8: **while** $t \leq maxturn$ **do**
9:     $flag = 0$;
10:    Arrange a random vertex updating order and keep it in X;
11:    **for** each $x \in X$ **do**
12:        Find the neighbour set $\Gamma(x)$ using Equation (1);
13:        Find all the labels held in $\Gamma(x)$ and keep them in L;
14:        **for** each $l \in L$ **do**
15:            $weight(l) = $ sum of weight of all the vertices holding l in $\Gamma(x)$;
16:        **end for**
17:        $c = $ the label with the highest weight in L, randomly choose one if many;
18:        **if** $C(x) \neq c$ **then**
19:            $C(x) = c$;
20:            $flag = 1$;
21:        **end if**
22:    **end for**
23:    **if** $flag \neq 1$ **then**
24:        **break**
25:    **end if**
26:    $t = t + 1$;
27: **end while**
28: **Output:** the label array C;

5.1 Experiment Setup

We are going to compare results of the UWLPA and LPA, and study the influence of parameters. Codes are implemented in Matlab according to Algorithm 1, where the weight is always 1 for the LPA.

Dataset. Table 1 shows six networks we use, the Zarachy network [14], the dataset of Polbooks [8], the Dolphins network [6], the network of Polblogs [1], the Arxiv High Energy Physics - Theory collaboration network (the Ca-HepTh) and the General Relativity and Quantum Cosmology collaboration network (the Ca-GrQc) [4]. Directed graph is treated undirected.

Evaluation Matrices. For the networks with the ground-truth, we use the NMI (*Normalized Mutual Information*) [11] and the RI (*Rand Index*) [2], both of which measure the similarity with the ground-truth and reveal the clustering accuracy. For the other two networks, we evaluate the quality by *modularity* [8], one popular measure of the structure of networks. Networks with high modularity have dense inner-connections, in other words, the clustering is of high quality.

Table 1. Basic information of datasets

Dataset	#Vertices	#Edges	#Classes	Type
Zarachy	34	78	2	Undirected
Polbooks	105	441	3	Undirected
Dolphins	62	159	2	Undirected
Polblogs	1224	19090	2	Directed
Ca-HepTh	9877	51971	Unknown	Undirected
Ca-GrQc	5242	28980	Unknown	Undirected

5.2 Accuracy Analysis

We fix $c = 1$ in Eq. (9) and evaluate the accuracy of the algorithm. We use these datasets with ground-truth and repeat 50 times for average performance.

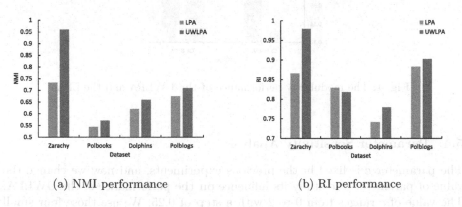

(a) NMI performance   (b) RI performance

Fig. 3. The accuracy performance of the UWLPA and the LPA

The Fig. (3a) presents the NMI performance. We can see the UWLPA performs better all the time. It achieves best dealing with the Zarachy for an average NMI of 0.96 and a promotion over 30 %. For the other three datasets, the promotion is about 5 % and 6 %, which is also satisfying. The results are similar for the RI, as shown in Fig. (3b). The UWLPA still performs best with an average RI of 0.98 and an average promotion of 13 % in the Zarachy. The promotion keeps 5 % in the Dolphins while it diminishes a little in the Polblogs. As for the Polbooks, the UWLPA performs 1.3 % worse, but if we take NMI into account, such loss in RI is acceptable.

Considering both NMI and RI, we can see the UWLPA has a fair promotion in the accuracy performance compared to the LPA.

5.3 Quality Analysis

We also fix $c = 1$ in Eq. (9) and evaluate the quality performance. We use the last two networks and compare the modularity. We run each algorithm with corresponding data for 10 times for average performance.

In (Fig. 4), UWLPA performs quite well both in the Ca-HepTh and in the Ca-GrQc. It has an average promotion of 5.56 % in the Ca-GrQc and an average promotion of 9.97 % in the Ca-HepTh, with the maximum promotion 11.49 %. And then we can conclude that the unified weight can lead to a good promotion in quality performance.

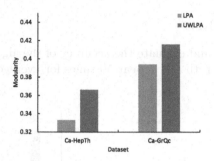

Fig. 4. The modularity performance of the UWLPA and the LPA

5.4 Parameter Sensitivity Analysis

The parameter c is fixed in the previous experiments, and now we change the value of parameter c to study its influence on the performance of the UWLPA. The value of c ranges from 0 to 2 with a step of 0.25. We use those four small networks and also repeat 50 times.

The average results of various datasets are shown in (Fig. 5). The line of NMI and the line of RI are mostly consistent. The Zarachy network shows an ideal curve where the performance first increases to the top and diminishes later. The result of Polblogs is similar, except that the performance suddenly drops when $c = 0.75$, breaking the top into two. When it comes to the Polbooks, the performance never diminishes after reaching the top. As for the Dolphins, however, the performance keeps going down as c grows, although the result is always better than the original LPA.

Then we can conclude that the influence of the parameter is diverse according to the various datasets and topological structures. And the unified weight can contribute to the promotion of algorithm performance most of the time, while the best performance requires a precise selection of the parameter.

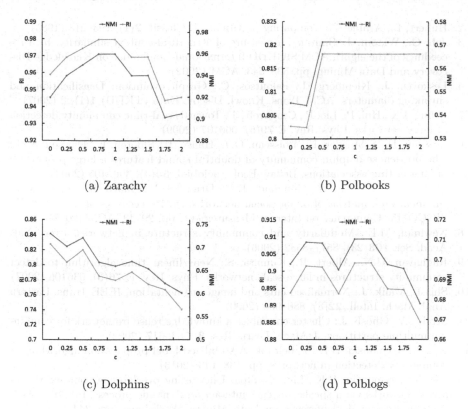

(a) Zarachy

(b) Polbooks

(c) Dolphins

(d) Polblogs

Fig. 5. Effect of the parameter c on UWLPA

6   Conclusions

We propose the connection factor with its simplified calculation, and then discuss the correlation of the connection factor and the label propagation weight. We provide the unified weight for the UWLPA. The experimental results show that the UWLPA has a promotion of 5 %, 10 % or even the higher. We plan to study the correlation between the promotion and the topological feature and try to combine the UWLPA with other LPA modification in the future work.

Acknowledgement. This research is supported by National High-tech R&D Program of China (863 Program) under Grants 2015AA01A301, by program for New Century Excellent Talents in University' by National Science Foundation (NSF) China 61272142, 61402492, 61402486, 61379146, 61272483, by the laboratory pre-research fund (9140C810106150C81001).

References

1. Adamic, L.A., Glance, N.: The political blogosphere and the 2004 us election: divided they blog. In: Proceedings of the 3rd International Workshop on Link Discovery, pp. 36–43. ACM (2005)

2. Hubert, L., Arabie, P.: Comparing partitions. J. Classif. **2**(1), 193–218 (1985)
3. Jeh, G., Widom, J.: Simrank: a measure of structural-context similarity. In: Proceedings of the eighth ACM SIGKDD International Conference on Knowledge Discovery and Data Mining, pp. 538–543. ACM (2002)
4. Leskovec, J., Kleinberg, J., Faloutsos, C.: Graph evolution: Densification and shrinking diameters. ACM Trans. Knowl. Discov. Data (TKDD) **1**(1), 2 (2007)
5. Leung, I.X., Hui, P., Lio, P., Crowcroft, J.: Towards real-time community detection in large networks. Phys. Rev. E **79**(6), 066107 (2009)
6. Lusseau, D., Schneider, K., Boisseau, O.J., Haase, P., Slooten, E., Dawson, S.M.: The bottlenose dolphin community of doubtful sound features a large proportion of long-lasting associations. Behav. Ecol. Sociobiol. **54**(4), 396–405 (2003)
7. Mislove, A., Marcon, M., Gummadi, K.P., Druschel, P., Bhattacharjee, B.: Measurement and analysis of online social networks. In: Proceedings of the 7th ACM SIGCOMM Conference on Internet Measurement, pp. 29–42. ACM (2007)
8. Newman, M.E.: Modularity and community structure in networks. Proc. Nat. Acad. Sci. **103**(23), 8577–8582 (2006)
9. Raghavan, U.N., Albert, R., Kumara, S.: Near linear time algorithm to detect community structures in large-scale networks. Phys. Rev. E **76**(3), 036106 (2007)
10. Shi, J., Malik, J.: Normalized cuts and image segmentation. IEEE Trans. Pattern Anal. Mach. Intell. **22**(8), 888–905 (2000)
11. Strehl, A., Ghosh, J.: Cluster ensembles-a knowledge reuse framework for combining multiple partitions. J. Mach. Learn. Res. **3**, 583–617 (2002)
12. Xie, J., Szymanski, B.K.: Labelrank: A stabilized label propagation algorithm for community detection in networks, pp. 138–143 (2013)
13. Xie, J., Szymanski, B.K., Liu, X.: Slpa: Uncovering overlapping communities in social networks via a speaker-listener interaction dynamic process. In: 2011 IEEE 11th International Conference on Data Mining Workshops, pp. 344–349. IEEE (2011)
14. Zachary, W.W.: An information flow model for conflict and fission in small groups. J. Anthropol. Res. **33**, 452–473 (1977)
15. Zhang, X., Fei, S., Song, C., Tian, X., Ao, Y.: Label propagation algorithm based on local cycles for community detection. Int. J. Mod. Phys. B **29**(5), 112–142 (2015)

MetricRec: Metric Learning for Cold-Start Recommendations

Furong Peng[1], Xuan Lu[2], Jianfeng Lu[1(✉)], Richard Yi-Da Xu[3], Cheng Luo[4], Chao Ma[1], and Jingyu Yang[1]

[1] School of Computer Science and Engineering, Nanjing University of Science and Technology, Nanjing 210094, Jiangsu, People's Republic of China
{pengfr,lujf,yangjy}@njust.edu.cn
[2] Department of Eletronic Information Engineering,
Shanxi University, Taiyuan 030013, Shanxi, People's Republic of China
xuanlu@sxu.edu.cn
[3] School of Computing and Communications,
University of Technology, Sydney, NSW 2008, Australia
yida.xu@uts.edu.au
[4] School of Electronics and Information Engineering,
Tongji University, Shanghai 201804, People's Republic of China
321luocheng@tongji.edu.cn

Abstract. Making recommendations for new users is a challenging task of cold-start recommendations due to the absence of historical ratings. When the attributes of users are available, such as age, occupation and gender, then new users' preference can be inferred. Inspired by the user based collaborative filtering in warm-start scenario, we propose using the similarity on attributes to conduct recommendations for new users. Two basic similarity metrics, cosine and Jaccard, are evaluated for cold-start. We also propose a novel recommendation model, MetricRec, that learns an interest-derived metric such that the users with similar interests are close to each other in the attribute space. As the MetricRec's feasible area is conic, we propose an efficient Interior-point Stochastic Gradient Descent (ISGD) method to optimize it. During the optimizing process, the metric is always guaranteed in the feasible area. Owing to the stochastic strategy, ISGD possesses scalability. Finally, the proposed models are assessed on two movie datasets, Movielens-100K and Movielens-1M. Experimental results demonstrate that MetricRec can effectively learn the interest-derived metric that is superior to cosine and Jaccard, and solve the cold-start problem effectively.

Keywords: Recommender system · Cold-start · Metric learning

1 Introduction

Recommendation systems are an important part of modern online services and stores such as Amazon, Delicious, and Youku. Many successful recommenders

© Springer International Publishing AG 2016
J. Li et al. (Eds.): ADMA 2016, LNAI 10086, pp. 445–458, 2016.
DOI: 10.1007/978-3-319-49586-6_30

are paradigms of collaborative filtering (CF) [3]. CF exploits users' historical behaviors to search neighbors and estimate the probability of liking an item based on their neighbors. When a recommender system is in operation, new users are added constantly. However, CF is incompetent to new users due to the absence of historical ratings. This poses a challenge that is often called the 'cold-start' recommendation problem.

Various auxiliary information was tried in order to deal with cold-start problems. Interview is one of the methods to obtain interests of new users [8]. CF can be qualified after obtaining the initial ratings (results of interviews). However, the interview processing is time consuming for both users and providers. In contrast, the passive initial information such as social networks [17], cross domain knowledge [22] and demographic information [1,14] is more convenient in practice. Since friends in social networks usually have similar interests, the profile of a new user can be created from his/her friends. The behaviors of new users in other domains are also useful for building profiles. However, new users in our recommender system may also be new to social networks and other domains. In this case, we focus on solving cold-start recommendations with demographic information, which can be obtained during user's registrations. For ease of description, we use attributes in place of demographic information.

Based on user attributes, many methods were proposed for cold-start recommendations. Park et al. [14] used a pairwise regression model to find the correlation between user attributes and ratings. Agarwal and Chen [1] supposed that user attributes are the prior of latent profiles, and proposed a linear regression model to predict new user profiles. Gantner et al. [6] also considered linear regression to link user attributes and profiles, and applied a Bayesian pair-wise regression as the loss function. Peng et al. [15] utilized an n-dimensional Markov random field as the prior of latent profiles for cold-start recommendations. Kula [10] extended the matrix factorization model [13] and proposed to learn the embeding features of attributes to inference the profile of new users and items. Differently from these methods, we propose learning an interest-derived metric of user attributes and utilizing the nearest neighbors to deal with cold-start recommendations, as illustrated in Fig. 1.

Inspired by user based collaborative filtering (UBCF) [18], we suppose that user's similarity can be estimated in the attribute space for cold-start recommendations. Two basic similarity metrics, cosine and Jaccard, are evaluated as measurements to find similar users. Analogous to UBCF, the neighbors of a new user are used to predict his/her interest. However, since the neighbors in the attribute space may have different interests, we propose a new model, MetricRec, that learns an interest-derived metric. Based on the metric, neighbors in the attribute space can have similar interest, and subsequently reasonable ratings of new users can be estimated from the corresponding neighbors. However, learning such a metric is inefficient because the feasible area is conic. To optimize MetricRec efficiently, we propose an Interior-point Stochastic Gradient Descent (ISGD) method that can adaptively choose a proper learning rate to keep parameters in the feasible area. The effectiveness of ISGD will be demonstrated in experiments.

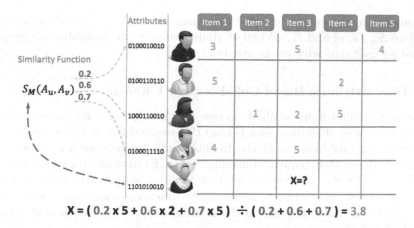

Fig. 1. The motivation of MetricRec. All users are correlated by their attributes. An interest-derived similarity function $S_M(\mathbf{A}_u, \mathbf{A}_v)$ makes the users with similar interests close to each other in the attribute space. The rating of the new user (in the last line of the rating matrix) on item 3 can be inferred by her neighbors (1st, 3rd and 4th users).

2   MetricRec Model

In this section, we first define necessary notations, then introduce the UBCF and UABCF, finally propose MetricRec model and ISGD. Suppose there are n users and m items, each user u has attributes $\mathbf{A}_u \in \mathbb{R}^d$, the user-item rating matrix is $\mathbf{R} \in \mathbb{R}^{n \times m}$. d is the number of attributes. The observed elements in $\mathbf{R}$ are indicated by $\mathbb{O}$. $\mathbf{R}_u$ indicates the ratings generated by user u. $\bar{R}_u$ is the average rating of user u. The estimated rating of user u and item i is denoted as $\hat{R}_{u,i}$.

2.1   User Based Collaborative Filtering

Suppose that users who acted similarly in history, will have similar behaviors in the feature. This is the basic assumption behind user based collaborative filtering (UBCF) method. Under the assumption, if we can successfully find similar users of a considered user, then we can infer the user's rating regarding to an item. Similar users are also referenced as neighbors. The metric of measuring user's similarity is crucial to the success of UBCF [11]. A commonly used similarity is cosine

$$S_{cos}(u,v) = \frac{< \mathbf{R}_u, \mathbf{R}_v >}{||\mathbf{R}_u|| \cdot ||\mathbf{R}_u||}, \tag{1}$$

where $< \cdot, \cdot >$ is inner product of two vectors, and $|| \cdot ||$ is the 2-norm. It can be easily verified that users with more similar historical ratings will get a higher cosine score. Through $S_{cos}(u, v)$, we can find a user's k-nearest neighbor set $\mathbb{N}_{ui}$ regarding to any item i. Then the rating of user u on item i can be estimated as

$$\hat{\mathbf{R}}_{ui} = \bar{R}_u + \frac{\sum_{v \in \mathbb{N}_{ui}} (\mathbf{R}_{vi} - \bar{R}_v) S_{cos}(u,v)}{\sum_{v \in \mathbb{N}_{ui}} S_{cos}(u,v)}. \tag{2}$$

Obviously, given a new user u to the recommendation system, it is impossible to calculate $S_{cos}(u, v)$ and $\bar{R}_u$. When attributes of users are available, we propose estimating user similarity in the attribute space.

2.2 User Attribute Based Collaborative Filtering

When user attributes are available, a new user's neighbors can be estimated in term of attributes. The basic idea behind this method is that users with similar attributes will rate items similarly. In this study, we propose a method (User Attributes Based Collaborative Filtering, UABCF) to deal with user cold-start problems. UABCF considers that users with more similar attributes will have more shared preference. Two basic similarity functions, cosine and Jaccard, can be considered as measurements in the attribute space. The cosine and Jaccard are calculated as

$$S_{Acos}(\mathbf{A}_u, \mathbf{A}_v) = \frac{<\mathbf{A}_u, \mathbf{A}_v>}{||\mathbf{A}_u|| \cdot ||\mathbf{A}_u||} \tag{3}$$

and

$$S_{AJacc}(\mathbf{A}_u, \mathbf{A}_v) = \frac{|\mathbf{A}_u \wedge \mathbf{A}_v|}{|\mathbf{A}_u \vee \mathbf{A}_v|}, \tag{4}$$

where $\wedge$ and $\vee$ are bit-wise *and, or* operators respectively, $|\cdot|$ is 1-norm. Moreover, when all attributes are category features, the hashing technique can be applied to accelerate the nearest neighbor search, and the Hamming similarity [12] can be considered as corresponding measurement.

With the two similarity functions on attributes, the rating of a new user u on an item i can be estimated as

$$\hat{\mathbf{R}}_{ui}^{Acos} = \frac{\sum_{v \in \mathbb{N}_{ui}} \mathbf{R}_{vi} S_{Acos}(\mathbf{A}_u, \mathbf{A}_v)}{\sum_{v \in \mathbb{N}_{ui}} S_{Acos}(\mathbf{A}_u, \mathbf{A}_v)} \tag{5}$$

and

$$\hat{\mathbf{R}}_{ui}^{AJacc} = \frac{\sum_{v \in \mathbb{N}_{ui}} \mathbf{R}_{vi} S_{AJacc}(\mathbf{A}_u, \mathbf{A}_v)}{\sum_{v \in \mathbb{N}_{ui}} S_{AJacc}(\mathbf{A}_u, \mathbf{A}_v)}. \tag{6}$$

The two similarity functions can be replaced by any other such as Pearson correlation coefficient and mean squared difference [11]. Fussing multiple similarity functions can also be applied to improve performances. In the following, we consider a more general method to learn a similarity function with metric learning methods [19].

2.3 Metric Learning

The user's attributes may have different impact on preference. For example, two users with different ages may have similar preferences on items, but may have different preferences with distinct occupations. Thus, treating all attributes equally may limit the performance of UABCF. If we could learn a similarity function according to user's preference, then performance of UABCF will be improved.

In this work, we try to identify an interest-derived measurement from observed ratings such that users are as close in the attribute space as their interests. A popular method for this task is metric learning [21] that utilizes Mahalanobis distance in place of Euclidean distance to learn a distance function. The Mahalanobis distance is defined as

$$d_{\mathbf{M}}(x, x') = \sqrt{(x - x')\mathbf{M}(x - x')^T}. \tag{7}$$

$\mathbf{M}$ is a symmetric positive definitive (SPD) matrix. When $\mathbf{M}$ is an identify matrix, the Mahalanobis distance is recovered to a Euclidean distance. As in UABCF, we need a similarity measurement not a distance measurement. Therefore, the exponent transform is applied [19] (which is also called kernel regression.)

$$S_{\mathbf{M}}(u, v) = \exp(-d_{\mathbf{M}}(\mathbf{A}_u, \mathbf{A}_v)^2). \tag{8}$$

$S_{\mathbf{M}}(u, v)$ ranges from 0 to 1, which is a similarity measurement. Following UABCF, we can estimate the ratings for a new user as

$$\hat{\mathbf{R}}_{ui}(\mathbf{M}) = \frac{\sum_{v \in \mathbb{N}_{ui}} \mathbf{R}_{vi} S_{\mathbf{M}}(u, v)}{\sum_{v \in \mathbb{N}_{ui}} S_{\mathbf{M}}(u, v)}. \tag{9}$$

$\mathbf{M}$ is a parameter to optimize such that the estimated ratings are as close as to the real observed values. To achieve the optimization, $\mathbf{M}$ will increase the similarity of users with shared ratings and decrease the similarity of those without shared ratings. As attributes similarity is aligned with ratings, $\mathbf{M}$ is referenced as interest-derived metric. With empirical risk minimization, we obtain the following loss function

$$\mathcal{L} = \sum_{u,i \in \mathcal{O}} \frac{1}{2} (R_{ui} - \hat{R}_{ui}(\mathbf{M}))^2$$

$$\text{subject to} \quad \mathbf{M} \succ 0, \mathbf{M} = \mathbf{M}^T. \tag{10}$$

Minimizing $\mathcal{L}$ with respect to $\mathbf{M}$ is challenging and inefficient [2], because $\mathbf{M}$ is required to be SPD. One trick of avoiding the SPD constraint is to redefine $\mathbf{M} = \mathbf{LL}^T$, ($\mathbf{L} \in \mathbb{R}^{d \times r}$). By this definition, $\mathbf{M}$ must be a SPD matrix, and solving $\mathcal{L}$ with respect to $\mathbf{L}$ is an unconstrained optimization problem. Consequently, Newton's method can be applied. However, it has been proved that solving $\mathbf{L}$ in place of $\mathbf{M}$ is easily trapped in local minimums [2]. In the following, we will optimize $\mathcal{L}$ with respect to $\mathbf{M}$ directly, and propose an efficient and scalable solution in the next section.

2.4   Interior-Point Stochastic Gradient Descent

Stochastic Gradient Descent (SGD) has been applied in many large scale optimizations such as deep learning [5] and matrix factorization [7]. The method iteratively updates parameters in gradient descent direction, and randomly chooses one sample to compute the gradient each time. As the gradient of one sample is easily affected by noise, several samples (mini-batch) can be chosen for calculating gradient. In our experiments, we set the mini-batch size to 5. Besides, SGD cannot comply with the constraint of SPD. Inspired by the interior-point method [4], we propose an Interior-point Stochastic Gradient Descent (ISGD) to adjust the learning rate such that $\mathbf{M}$ is always updated in the feasible area.

We first derive the gradient of $\mathcal{L}$, and introduce our learning method afterwards. Given several ratings $\mathbb{O}'$, the gradient of $\mathcal{L}$ with respect $\mathbf{M}$ is

$$\frac{\partial \mathcal{L}}{\partial \mathbf{M}} = \sum_{u,i \in \mathbb{O}'} (\hat{R}_{ui}(\mathbf{M}) - R_{ui}) \frac{\partial \hat{R}_{ui}(\mathbf{M})}{\partial \mathbf{M}} \tag{11a}$$

$$\frac{\partial \hat{R}_{ui}(\mathbf{M})}{\partial \mathbf{M}} = \frac{\sum_{v \in \mathbb{O}_i} (R_{vi} - \hat{R}_{ui}(\mathbf{M})) S'_{\mathbf{M}}(u,v)}{\sum_{v \in \mathbb{O}_i} S_{\mathbf{M}}(u,v)} \tag{11b}$$

$$S'_{\mathbf{M}}(u,v) = -S_{\mathbf{M}}(u,v)(A_u - A_v)^T (A_u - A_v), \tag{11c}$$

where $\mathbb{O}_i$ is a set of users who have rated item i. Obtaining the gradient, $\mathbf{M}$ can be updated by

$$\mathbf{M}^{(t)} = \mathbf{M}^{(t-1)} - \eta \frac{\partial \mathcal{L}}{\partial \mathbf{M}}, \tag{12}$$

where $\eta > 0$ is the learning rate. Usually, the initial value $\mathbf{M}^{(0)}$ is assigned with identity matrix $\mathbf{I}$. It can be easily verified that the incrementation $\frac{\partial \mathcal{L}}{\partial \mathbf{M}}$ is symmetric. Correspondingly, $\mathbf{M}^{(t)}$ is also symmetric. But $\mathbf{M}^{(t)} \succ 0$ can not be guaranteed. We apply the following strategy to make $\mathbf{M}^{(t)}$ be positive definitive.

The constraint $\mathbf{M}^{(t)} \succ 0$ can be satisfied by adjusting η. The initial value $\mathbf{M}^{(0)} = \mathbf{I}$ is a positive definitive matrix. Without loss of generality, suppose $\mathbf{M}^{(t-1)} \succ 0$. If $\frac{\partial \mathcal{L}}{\partial \mathbf{M}}$ is negative definitive, then $-\eta \frac{\partial \mathcal{L}}{\partial \mathbf{M}}$ is positive definitive. It is very easy to verify that the summation of two positive definitive matrices is also positive definitive. Consequently, $\mathbf{M}^{(t)}$ is positive definitive. If $\frac{\partial \mathcal{L}}{\partial \mathbf{M}}$ is not negative definitive, we decrease η until $\mathbf{M}^{(t-1)} - \eta \frac{\partial \mathcal{L}}{\partial \mathbf{M}} \succ 0$. In the worse case, η will be 0 and $\mathbf{M}$ will not be updated at t-th iteration. The next updating step will be continued.

Summarizing the above strategies, we obtain the ISGD algorithm as described in Algorithm 1.

3   Experiments

3.1   Experiment Setup

Two movie datasets, Movielens-100K and Movielens-1M[1], were used for evaluations. Both datasets contain user attributes including age, occupation and gender. The occupation and gender are category features, and the age was also discretized into 7 categories. Ratings range from 1 to 5. About 5 % ratings are observed in the rating matrix. The details are summarized in Table 1.

---

[1] https://grouplens.org/datasets/movielens/.

Algorithm 1. ISGD

Require:
 The user-item rating matrix $\mathbf{R}$;
 Attributes of all users $\mathbf{A}$;
 Number of iterations to be executed T;
Ensure:
 The interest-derived metric $\mathbf{M}$
1: initialize $\mathbf{M}^{(0)} = \mathbf{I}$, $\eta = 0.001$
2: **for** each $t \in [1, \ldots, T]$ **do**
3: draw $\mathcal{O}'$ from $\mathcal{O}$
4: compute $\frac{\partial \mathcal{L}}{\partial \mathbf{M}^{(t-1)}}$ with Eq. 11
5: $\eta' = \eta$
6: decrease η' until $\mathbf{M}^{(t-1)} - \eta' \frac{\partial \mathcal{L}}{\partial \mathbf{M}^{(t-1)}} \succ 0$
7: update $\mathbf{M}^{(t)} = \mathbf{M}^{(t-1)} - \eta' \frac{\partial \mathcal{L}}{\partial \mathbf{M}^{(t-1)}}$
8: **end for**
9: **return** $\mathbf{M}^{(T)}$;

Table 1. Statistical information of datasets

Datasets	#Users	#Items	#Ratings	Rating rage	Density
Movielens-100K	1,681	943	100,000	1–5	0.063
Movielens-1M	3,883	6,040	1,000,000	1–5	0.0511

Users were divided into 10 folds. Each time, one fold was regarded as new users and the rest as existing ones. The cold-start recommendation was simulated by recommending items to new users. Existing users were used to train models. The average result of 10 folds was reported.

Four metrics were used to quantify the accuracy of algorithms: nDCG [14], Precision, Recall and F1 [18].

$$nDCG_i = \frac{DCG_i^p}{IDCG_i^p}, \quad DCG_i^p = \sum_{l=1}^{p} \frac{2^{R_{il}} - 1}{log_2(1 + l)} \tag{13}$$

where p is the number of evaluated items; R_{il} is the rating of l-th item; $IDCG_p$ is the ideal value of DCG_p.

$$Precison = \frac{|\{relevant\ items\} \cap \{recommended\ items\}|}{|recommended\ items|}$$

$$Recall = \frac{|\{relevant\ items\} \cap \{recommended\ items\}|}{|relevant\ items|} \tag{14}$$

$$F1 = 2\frac{|\{relevant\ items\} \cap \{recommended\ items\}|}{|recommended\ items| + |relevant\ items|}.$$

When calculating Precision, Recall and F1, the ratings greater than or equal to 4 were considered as relevant ones. All four metrics range from 0 to 1. A higher score indicates a better performance. The four metrics were calculated on top 5,10,15 and 20 recommended items in our experiments.

3.2 Methods for Comparison

Some well-known baselines and state-of-the-art methods were used for comparisons. Their short introductions and settings are listed as follows.

- **RandomRec** recommends items by choosing randomly.
- **MostPopular** recommends items by the average rating. The higher score indicates more popular. The most popular items are recommended.
- **RLFM** (Regression-based Latent Factor Model [1]) is a matrix factorization based model that fits attributes with latent profiles. To save computing time, we solved it by alternating least square.
- **TKR** [9] utilizes user's keywords and item's tags for cold-start recommendations. In our experiments, this method was adapted by considering user attributes and item attributes as keywords and tags respectively.

3.3 Experimental Results

All the methods in our experiments were implemented in Python and are optimized by grid-search for their best performances. The regularization coefficient was searched in $\{10^n | n = -2, -1, 0, 1, 2, 3, 4\}$ and 0. The dimension of latent profile was searched from 5 to 30 with a step of 5. In the parameter searching process, the sum score of nDCG and F1 was regarded as the final indicator to measure performances. The dimension of RLFM's latent profile was set to 10 for both Movielens-100K and Movielens-1M, and its regularization coefficient was set to 10 for Movielens-100K and 100 for Movielens-1M. The regularization coefficient of TKR was set to 10 for Movielens-100K and 100 for Movielens-100M.

Two similarity measurements, cosine and Jaccard, were implemented to evaluate UABCF. They are referenced as *UABCF(cosine)* and *UABCF(Jaccard)* respectively. In our experiments, the number of neighbors was set to 200 for both Movielens-100K and Movielens-1M.

Tables 2 and 3 report the experimental results on Movielens-100K and Movielens-1M respectively. The best score in each group is highlighted in bold-face and the second best in italic. RandomRec gives the results of randomly choosing. Compared with RandomRec, we can access the improvement of designed methods from random choosing. From these tables, we can see that MostPopular generally outperformed RandomRec and was a good baseline. MostPopular can be considered as a simple rule-based method that recommends the most popular items for all new users. RLFM and TKR obtained higher performance than MostPopuar on both datasets, because the attributes can relfect user's preference. On Movielens-100K, TKR's performance was better than or equal to RLFM, since item's attributes were used for creating the user's and item's attribute correlation matrix of TKR. The extra information of item's attributes boosted TKR's performance. On Movielens-1M, more ratings

Table 2. Movielens-100K

Algorithms	nDCG				Precision				Recall				F1			
	@5	@10	@15	@20	@5	@10	@15	@20	@5	@10	@15	@20	@5	@10	@15	@20
RandomRec	0.4798	0.5029	0.5338	0.5670	0.5767	0.5749	0.5767	0.5780	0.0956	0.1905	0.2892	0.3884	0.1512	0.2535	0.3344	0.3989
MostPopular	0.6995	0.7069	0.7233	0.7430	0.8090	0.7828	0.7602	0.7400	0.1362	0.2580	0.3665	0.4635	0.2133	0.3413	0.4268	0.4872
RLFM	*0.7055*	**0.7118**	*0.7261*	*0.7460*	0.8114	*0.7879*	*0.7626*	*0.7403*	*0.1366*	*0.2608*	**0.3693**	**0.4658**	0.2143	*0.3449*	**0.4294**	*0.4889*
TKR	0.7038	0.7093	0.7259	0.7444	0.8124	0.7845	**0.7627**	0.7399	*0.1374*	0.2596	*0.3673*	0.4641	*0.2150*	0.3428	*0.4279*	0.4878
UABCF(cosine)	0.7040	0.7081	0.7239	0.7442	*0.8141*	0.7834	0.7597	0.7396	0.1372	0.2602	0.3665	0.4629	*0.2150*	0.3431	0.4268	0.4870
UABCF(Jaccard)	0.7050	0.7080	0.7237	0.7442	0.8135	0.7837	0.7590	0.7395	0.1370	0.2604	0.3662	0.4633	0.2147	0.3436	0.4266	0.4872
MetricRec	**0.7088**	*0.7117*	**0.7268**	**0.7468**	**0.8184**	**0.7889**	0.7615	**0.7421**	**0.1385**	**0.2615**	0.3666	*0.4655*	**0.2167**	**0.3454**	0.4272	**0.4893**

Boldface indicates the best score.
Italics indicates the second best score.

Table 3. Movielens-1M

Algorithms	nDCG				Precision				Recall				F1			
	@5	@10	@15	@20	@5	@10	@15	@20	@5	@10	@15	@20	@5	@10	@15	@20
RandomRec	0.6995	0.7069	0.7233	0.7430	0.8090	0.7828	0.7602	0.7400	0.1362	0.2580	0.3665	0.4635	0.2133	0.3413	0.4268	0.4872
MostPopular	0.7571	0.7498	0.7540	0.7631	0.8673	0.8434	0.8263	0.8085	0.1049	0.2008	0.2902	0.3708	0.1733	0.2884	0.3728	0.4350
RLFM	*0.7679*	*0.7589*	*0.7622*	*0.7703*	**0.8761**	**0.8523**	**0.8340**	**0.8150**	**0.1062**	**0.2032**	**0.2930**	**0.3736**	**0.1755**	**0.2918**	**0.3765**	**0.4385**
TKR	0.7562	0.7506	0.7547	0.7635	0.8665	0.8447	0.8273	0.8089	0.1049	0.2012	0.2903	0.3708	0.1733	0.2889	0.3730	0.4351
UABCF(*cosine*)	0.7672	0.7593	0.7618	0.7699	0.8736	0.8510	0.8321	0.8131	0.1058	*0.2028*	*0.2923*	0.3725	*0.1749*	0.2912	0.3756	0.4373
UABCF(*Jaccard*)	0.7671	0.7599	0.7620	0.7703	0.8735	*0.8517*	0.8320	0.8137	*0.1059*	0.2030	0.2924	0.3728	*0.1749*	*0.2915*	*0.3757*	0.4376
MetricRec	**0.7691**	**0.7611**	**0.7634**	**0.7714**	*0.8737*	0.8515	*0.8326*	*0.8149*	*0.1059*	*0.2028*	*0.2923*	*0.3733*	*0.1749*	0.2912	*0.3757*	*0.4383*

Boldface indicates the best score.

Italics indicates the second best score.

were available. RLFM achieved better performance, which is a attribute-aware matrix factorization model that first inference new user's latent profile from attributes, and conduct recommendations with item's latent profile afterwards.

The proposed UABCF is a kind of lazy learning methods that do not demand a training process. In the predicting phase, the nearest neighbors are considered as predictors. On both datasets, UABCF obtained promising results that were better than Mostpopular and close to RLFM and TKR. As the cosine and Jaccard were empirically given, they may not well capture the similarity of user's interest in the attribute space. With metric learning, an interest-derived metric was optimized for aligning user's interest similarity and user attributes similarity. From the two tables, we can see that the optimized metric (MetricRec) outperformed the cosine and Jaccard measurements. The improvement demonstrates that the optimized interest derived metric is superior to empirical metrics. Compared to RLFM and TKR, MetricRec obtained better performance on Movielens-100K and competitive performance on Movielens-1M.

In summary, the proposed UABCF obtained promising results on both datasets. By optimizing the interest derived metric, MetricRec achieved the best or second best result in our experiments.

3.4   Sensitivity to the Number of Neighbors

In most lazy neighbor-based methods, k-nearest neighbor classier as an example, the number of neighbors is an important factor regarding to performances. In Figs. 2 and 3, we plotted the sum of nDCG and F1 sores to demonstrate the impact of k.

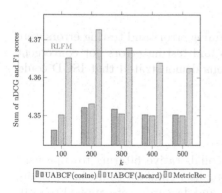
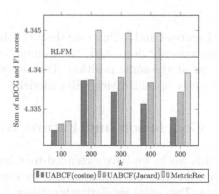

Fig. 2. Sum of nDCG and F1 versus neighbor size (k) on Movielens-100K

Fig. 3. Sum of nDCG and F1 versus neighbor size (k) on Movielens-1M

Three methods, UABCF(cosine), UABCF(Jaccard) and MetricRec, were reported in the results. We can see that $k = 200$ was the best value for three models. Performances increased from $k = 100$ to $k = 200$, and began to decrease after

$k > 200$. Results of $k < 200$ indicated the lack of support neighbors. Results of $k > 200$ indicated the neighbor size was so large that noise users were included in the neighborhood. The results also demonstrated that the interest-derived metric was superior to cosine and Jaccard.

3.5   Convergence of ISGD

In this section, we empirically study the convergence of ISGD. Taking the Movielens-100K as an example, we report the convergence of training errors and testing errors. Training errors versus iterations were plotted in Fig. 4, and testing errors in Fig. 5.

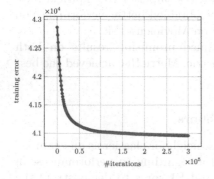

Fig. 4. Training error versus iterations on Movielens-100K

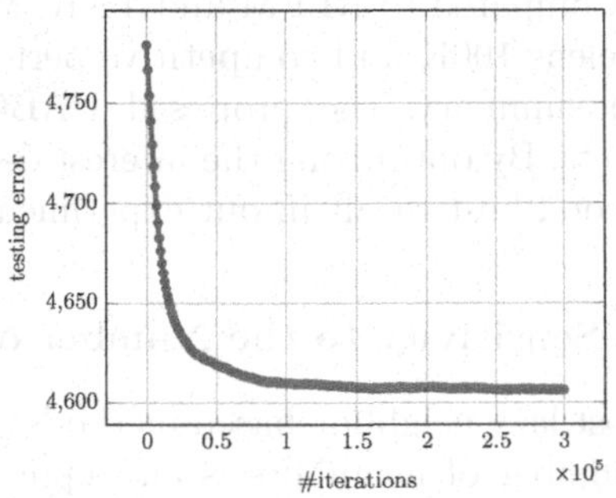

Fig. 5. Testing error versus iterations on Movielens-100K

Figures 4 and 5 illustrate that: (1) the training errors and testing errors were monotonously decreased, (2) training errors and testing errors got convergent almost at the same position. Two observations demonstrated that ISGD could efficiently optimize MetricRec model.

4   Conclusion and Future Work

In this work, we have extended user based collaborative filtering to cope with user cold-start recommendations by measuring user similarities in the attribute space. Two basic similarity functions, cosine and Jaccard, were applied to evaluate UABCF. As the two similarities may not capture user's interest in the attribute space, we have developed a metric learning model MetricRec. It learns an interest-derived metric such that users with similar interest are close to each other in the attribute space. Moreover, an efficient Interior-point stochastic gradient descent method was proposed to optimize MetricRec. Experimental results on two movie datasets demonstrated that UABCF can obtain promising results with cosine and Jaccard, and MetricRec can effectively deal with cold-start recommendations. It was also empirically demonstrated that ISGD can efficiently optimize MetricRec.

As the neighborhood-based collaborative filtering in MetricRec limited its performance, we plan to develop a model-based method (matrix factorization as an example) with metric learning incorporated in the future work. Besides, we will also consider the following directions. Evaluate the pairwise loss function [14], BPR loss function [16] and k-th order statistic loss [20] in cold-start scenarios. When item attributes are available, Metric leaning can be applied to item cold-start recommendations. Furthermore, MetricRec can also be extended to deal with recommending new items to new users.

References

1. Agarwal, D., Chen, B.C.: Regression-based latent factor models. In: Proceedings of the 15th ACM SIGKDD International Conference on Knowledge Discovery and Data Mining, pp. 19–28. ACM (2009)
2. Bellet, A., Habrard, A., Sebban, M.: A survey on metric learning for feature vectors and structured data. arXiv preprint arXiv:1306.6709 (2013)
3. Bobadilla, J., Ortega, F., Hernando, A., Gutiérrez, A.: Recommender systems survey. Knowl.-Based Syst. **46**, 109–132 (2013)
4. Boyd, S., Vandenberghe, L.: Convex Optimization. Cambridge University Press, New York (2004)
5. Dean, J., Corrado, G., Monga, R., Chen, K., Devin, M., Mao, M., Aurelio Ranzato, M., Senior, A., Tucker, P., Yang, K., Le, Q.V., Ng, A.Y.: Large scale distributed deep networks. In: Pereira, F., Burges, C.J.C., Bottou, L., Weinberger, K.Q. (eds.) Advances in Neural Information Processing Systems 25, pp. 1223–1231. Curran Associates, Inc. (2012)
6. Gantner, Z., Drumond, L., Freudenthaler, C., Rendle, S., Schmidt-Thieme, L.: Learning attribute-to-feature mappings for cold-start recommendations. In: 2010 IEEE 10th International Conference on Data Mining (2010)
7. Gemulla, R., Nijkamp, E., Haas, P.J., Sismanis, Y.: Large-scale matrix factorization with distributed stochastic gradient descent. In: Proceedings of the 17th ACM SIGKDD International Conference on Knowledge Discovery and Data Mining, KDD 2011, pp. 69–77. ACM, New York (2011)
8. Houlsby, N., Hernandez-lobato, J.M., Ghahramani, Z.: Cold-start active learning with robust ordinal matrix factorization. In: ICML (2014)
9. Ji, K., Shen, H.: Addressing cold-start: scalable recommendation with tags and keywords. Knowl.-Based Syst. **83**, 42–50 (2015)
10. Kula, M.: Metadata embeddings for user and item cold-start recommendations. arXiv preprint arXiv:1507.08439 (2015)
11. Liu, H., Hu, Z., Mian, A., Tian, H., Zhu, X.: A new user similarity model to improve the accuracy of collaborative filtering. Knowl.-Based Syst. **56**, 156–166 (2014)
12. Ma, C., Liu, C., Peng, F., Liu, J.: Multi-feature hashing tracking. Pattern Recogn. Lett. **69**, 62–71 (2016)
13. Mnih, A., Salakhutdinov, R.R.: Probabilistic matrix factorization. In: Platt, J.C., Koller, D., Singer, Y., Roweis, S.T. (eds.) Advances in Neural Information Processing Systems 20, pp. 1257–1264. Curran Associates, Inc. (2008)

14. Park, S.T., Chu, W.: Pairwise preference regression for cold-start recommendation. In: RecSys (2009)
15. Peng, F., Lu, J., Wang, Y., Yi-Da Xu, R., Ma, C., Yang, J.: N-dimensional markov random field prior for cold-start recommendation. J. Neurocomputing **191**, 187–199 (2016)
16. Rendle, S., Freudenthaler, C., Gantner, Z., Schmidt-Thieme, L.: Bpr: Bayesian personalized ranking from implicit feedback. In: Proceedings of the Twenty-Fifth Conference on Uncertainty in Artificial Intelligence, pp. 452–461. AUAI Press (2009)
17. Sedhain, S., Sanner, S., Braziunas, D., Xie, L., Christensen, J.: Social Collaborative Filtering for Cold-start Recommendations. In: RecSys (2014)
18. Su, X., Khoshgoftaar, T.M.: A Survey of Collaborative Filtering Techniques. Adv. Artif. Intell. **2009**, 4:2 (2009)
19. Weinberger, K.Q., Tesauro, G.: Metric learning for kernel regression. In: International Conference on Artificial Intelligence and Statistics, pp. 612–619 (2007)
20. Weston, J., Yee, H., Weiss, R.J.: Learning to rank recommendations with the k-order statistic loss. In: Proceedings of the 7th ACM Conference on Recommender Systems, pp. 245–248. ACM (2013)
21. Xing, E.P., Ng, A.Y., Jordan, M.I., Russell, S.: Distance metric learning with application to clustering with side-information. In: Advances in Neural Information Processing Systems **15**, 505–512 (2003)
22. Yan, M., Sang, J., Xu, C.: Unified YouTube video recommendation via cross-network collaboration. In: ICMR (2015)

Time Series Forecasting on Engineering Systems Using Recurrent Neural Networks

Dongxu Shao, Tianyou Zhang$^{(\boxtimes)}$, Kamal Mannar, and Yue Han

Accenture Analytics Innovation Center, Singapore, Singapore
{dongxu.shao,tianyou.zhang,kammal.mannar,laura.han}@accenture.com

Abstract. Modern large scale processing and manufacturing systems cover a wide array and large number of assets that need to work together to ensure that plant is generating output reliably and at the desired yield rate, such as the viscosity of quench oil in styrene cracking system. However, due to the complexity of the overall process, it is important to consider the entire plant as a network to identify deterioration patterns and forecast condition.

Instead to figure out the prediction from engineering perspective, we propose to leverage deep learning approach to predict the next state based on the historical information. Particularly, recurrent neural network (RNN) is selected in this paper as a basis for temporal forecasting. Considering the fact that there are multiple sub-systems running in parallel whose independence cannot be captured by a normal RNN, we design a LSTM (Long Short Term Memory) network for each subsystem and feed the outputs of LSTMs into a linear neural network layer for predicting viscosity one-hour ahead.

Keywords: Recurrent neural networks · LSTM · Time series forecasting

1 Introduction

Nowadays the fast development of information technologies allow industrial processing plants to deploy more efficient, inexpensive and multi-functional sensors to monitor the condition in real-time manner and forecast the system temporal status. However, forecasting the temporal properties such as product yield rate or power load from a large number of variables is challenging [1]. More interestingly, some equipment have several sub-systems running in parallel, so there is great need from industrial perspective to create an indicator for the condition of each sub-system [2].

To confront the challenge, our approach is to apply feature selection method, such as Vector Autoregression (VAR) model [3], to analyze the correlations across multiple time series and remove the redundant features. With the selected features, machine learning approaches could be then adopted for forecasting system status. The newly emerging recurrent neural networks [4,5] have found extensive significant applications in time series forecasting with wide application domain

© Springer International Publishing AG 2016
J. Li et al. (Eds.): ADMA 2016, LNAI 10086, pp. 459–471, 2016.
DOI: 10.1007/978-3-319-49586-6_31

including dynamic system identification and control, long-term predictions of chemical processes, and nuclear power plant condition monitoring [6–8]. Among different types of RNN models, long short term memory (LSTM) network [5] is promising presently due to its capability to avoid the long-term dependency problem, and will be adopted in our solution.

In addition to employing the recurrent neural networks to perform time series forecasting, and in order to invent an indicator for each sub-system in the processing equipment, we propose to upgrade the topology of the RNN to a hierarchical structure with two hierarchies. The first hierarchy consists of several independent RNN models, one for each sub-system, while the second hierarchy is another model to consolidate the first hierarchy outcomes for the forecasting. So the outputs of the RNN models in the first hierarchy can be considered as the condition indicators of the sub-systems, based on which the user can take actions accordingly.

Contribution: In this paper, we consider the problem of time series forecasting based on a large number of variables. More importantly, we take into account of real-time applications where some of the variables are obtained from several independent groups and a condition indicator is required for each group. To this end, a hierarchical model is developed on top of recurrent neural networks and validated by using real-world data.

2   Related Work

2.1   Time Series Forecasting

Time series forecasting is to learn from the previously observed time series to predict the future values in one or multiple steps ahead. It is one of the key fields of data analytics and plays the important role in many fields of science and engineering, such as economics, finance, meteorology, telecommunication, diagnostics and prognostics [1, 9]. Traditionally, time series forecasting was dominated by linear statistical models, including exponential smoothing, autoregressive integrated moving average (ARIMA) models [10], state space and structural models (e.g. Kalman filter) [11].

Other than the variants of AR models, artificial neural network (ANN) models have grown to be another focus in the field due to its nonlinearity and great adaptability. Many successes of ANN based time series forecasting have been documented in the literature, such as load forecasting in power system [12], prognostic modeling of industrial equipment [13]. Darbellay et al. [14] discovered that ANN was capable of modeling non-linear process with unknown functional relationship especially when the results were difficult to fit. Qi [15] concluded that ANN outperformed other models when the input data was kept as current as possible.

2.2   Recurrent Neural Networks

Among different type of ANN models, recurrent neural network (RNN) has attracted great interest among researchers since the late 1990s [4, 5]. Compared to the typical feed-forward ANN, RNN is distinguished by its feedback loops between neurons. Through those feedback loops, RNN uses its outputs of the previous step as part of inputs of the current step to process the sequence of inputs. As sequential information is preserved in the inner states, RNN is particularly good at modelling the sequential data, such as speech recognition [16], time series forecasting [6–8] and etc.

Based on the topology of feedback paths, there are different RNN models including long short term memory (LSTM) network [5]. LSTM is the trending RNN model presently because it resolves the vanishing gradient problem of the traditional RNN and is capable to work with the time series of long delay and varying frequency.

Lipton et al. [17] used LSTM to model the multivariate time series of clinical measurements and found that LSTM outperformed other methods including multilayer perceptron model. Pawowski et al. [18] applied LSTM to successfully detect the dangerous levels of methane concentration in a coal mine with the multivariate time series of sensor readings. Ma et al. [19] developed the LSTM model to predict the travel speed data based on the time series of traffic microwave detectors and showed that LSTM achieved the best prediction performance in terms of both accuracy and stability in comparison with other prevailing parametric and nonparametric algorithms. Xiong et al. [20] benchmarked LSTM model with linear Ridge/Lasso and autoregressive GARCH models on forecasting $S\&P$ 500 volatility and summarized that LSTM outperformed by at least 31 %.

3 Problem Background

3.1 Problem Statement

Consider a processing system with a number of sensors inside to monitor the real-time condition. Suppose the update frequency of all the sensors is identical and one of the sensor readings is the variable we want to predict. Let $D = \{(x_1, t_1), \ldots, (x_n, t_n)\}$ be the data set of sensor readings except the target variable, and let $y = \{(y_1, t_1), \ldots, (y_n, t_n)\}$ be the readings of the target variable. Since the time interval of the readings is identical, we omit the timestamps without causing any confusion. So the data set D can be rephrased as $D = \langle x_1, \ldots, x_n \rangle$, and the target variable is $y = \langle y_1, \ldots, y_n \rangle$.

Suppose there are M many sensors deployed in the equipment. So there are $M - 1$ many columns in D. The sensors could collect different types of the data, such as temperature, pressure, and flux. Some of these readings could be highly correlated. For example, the increment of the temperature in a water tank may probably result in the increase of the humidity reading. Such inter correlation will have negative impact on the predictive modeling [21] and should be removed as much as possible in the beginning of the solution. Moreover, the removal of redundant variables also reduces the computation complexity.

Another challenge of this task comes from the temporal aspect. First of all, each observation represents the status monitored by every sensor at the same

time, so there are time lags across these readings. In other words, the readings at
the same time are not corresponding to the same piece of material. For example,
suppose a water tank is connected to a valve and the flux at the valve is increased.
But the acceleration of the pressure increasing at the tank is not changed immediately. The time lag between these two readings is the time needed for the water
goes from the valve to the tank. Hence, x_i cannot be used to predict y_i directly.

Due to the fact that many processing systems have sub-systems running in
parallel, it is very interesting, important and innovative to design an indicator
for each sub-system to represent its condition. The impact of sub-systems to
the target variable may not be necessarily the same, so the indicators should be
invented individually. Moreover, the indicators should be used for the forecasting
task.

3.2 Data Description

The original data is collected from a styrene cracking equipment with four fur-
naces as illustrated in Fig. 1, in which the viscosity of quench oil plays an important role for system stabilization. High viscosity will yield a more stable vapour
blanket, and hence reducing the cooling rate, which has a significant negative
impact on the productivity [22].

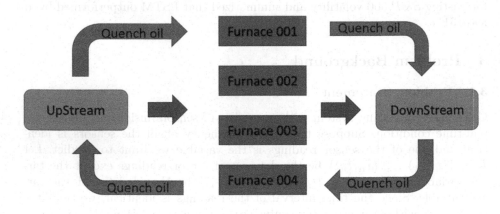

Fig. 1. Styrene cracking workflow

The duration of the collected data covers from 2014-01-01 00:00:00 to 2015-
10-31 23:00:00, and the update frequency of sensors is once per hour. So there are
totally 14172 observations. In their equipment, there are 283 sensors including
the sensor to monitor the viscosity. So the dimension of D is 14172×282. Here
is the number of sensors in each furnace:

In order to anonymize the data due its sensitivity to our clients, we scale
the viscosity into the range of $[0,1000]$. The temporal plot of the viscosity is
shown in Fig. 2. The viscosity of quench oil moves around a relatively low value

Table 1. Table of number of sensors

Furnace number	001	002	003	004
Number of sensors	63	76	73	70

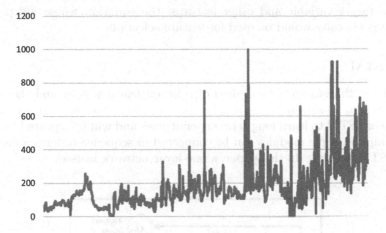

Fig. 2. Scaled viscosity of quench oil

but sometimes jumps sharply. There are some time points with low viscosity followed immediately by those with very high viscosity, rising challenges in forecasting. Moreover, it was found in cross correlation analysis, there are long-term dependencies (20–40 h lag) between furnace sensor values and viscosity. This phenomenon drove us to focus on LSTM model instead of the traditional RNN.

4 Preliminaries

4.1 VAR

Vector auto-regression (VAR) is a generalization of univariate auto-regression model (AR) to multi-dimension in the field of time series analysis. Commonly, it is applied to study the linear interdependencies among multiple time series. With VAR, relationship among different time series can be captured.

A p-th order VAR model, VAR(p) is defined as,

$$y_t = c + A_1 y_{t-1} + A_2 y_{t-2} + ... + A_p y_{t-p} + w_t \tag{1}$$

where y_{t-k} refers to k-th lag of y, c is a $k \times 1$ vector of constants, A_i is a coefficient matrix and error terms w_t.

To illustration the above concept, an example of VAR(1), the vector autoregressive model with 1 lag for two different time series denoted by $x_{t,1}, x_{t,2}$ is as follows:

$$x_{t,1} = c_1 + \theta_{1,1} x_{t-1,1} + \theta_{1,2} x_{t-1,2} + w_{t,1} \tag{2}$$

$$x_{t,2} = c_2 + \theta_{2,1} x_{t-1,1} + \theta_{2,2} x_{t-1,2} + w_{t,2} \tag{3}$$

In general, for time series data of m dimension, a VAR(p) model returns a linear system with m equations where the first p lags of each variable would be used as regression predictors for each variable. To understand the relationship between target variable and other features, the equation whose independent variable is viscosity would be used for feature selection.

4.2   LSTM

Consider at timestamp t, the values of selected features is x_t and the output indicator is h_t as shown in Fig. 3 [23]. The module in the rectangle of Fig. 3 is the core of LSTM to learn long-term dependencies and will be repeated for every timestamp. Multiple modules can be connected in sequence to form a multiple-layer LSTM network. In this paper, a one-layer network is used.

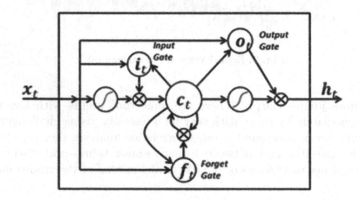

Fig. 3. LSTM recurrent neural network

When a new data x_t comes in, the first step is to determine what information will be thrown away from previous memory state c_{t-1}. This is done by the forget gate:

$$f_t = \sigma(W_f \cdot [h_{t-1}, x_t] + b_f), \tag{4}$$

where W_f is a matrix, b_f is a vector and σ is an aggregating function.

The second step is to determine what information will be extracted from the new data x_t for memory state update. This process consists of an input gate layer and a candidate layer for memory update. The computation of input gate layer is:

$$i_t = \sigma(W_i \cdot [h_{t-1}, x_t] + b_i), \tag{5}$$

and the candidate memory cell value is obtained by:

$$\hat{c}_t = \tanh(W_c \cdot [h_{t-1}, x_t] + b_c), \tag{6}$$

where W_i, W_c are matrices and b_i, b_c are vectors.

Given the variables having been computed so far, the new memory cell state follows from the combination of old state with some information dropped by the forget gate and the incremental update selected by the input layer and candidate layer:

$$c_t = f_t \cdot c_{t-1} + i_t \cdot \hat{c}_t. \tag{7}$$

The final step is to deliver the output based on the updated memory cell state. The output gate layer is designed to determine which part of the memory cell state will be used for output:

$$o_t = \sigma(W_o \cdot [h_{t-1}, x_t] + b_o), \tag{8}$$

where W_o is a matrix and b_o is a vector.

Then the final output is obtained by an aggregation of o_t and c_t:

$$h_t = o_t \cdot \tau(c_t), \tag{9}$$

where τ is an aggregating function.

4.3 SGD

Consider a typical machine learning task, where the training data set is $\{x_1, \ldots, x_n\}$, the target variable is $\{y_1, \ldots, y_n\}$, and the error measurement is Q. Suppose the algorithm to predict y from x is F using a vector of parameters Θ. Then the value of Θ should minimize the total error of the prediction. In other words, the aim is to find:

$$\arg\min_{\Theta} \sum_{j=1}^{n} Q(y_j, F(x_j, \Theta)). \tag{10}$$

Let Q_j denote $Q(y_j, F(x_j, \Theta))$. Then (10) can be rephrased as:

$$\arg\min_{\Theta} \sum_{j=1}^{n} Q_j. \tag{11}$$

The classical gradient descent method to solve this optimization problem is to update Θ by iterations until some stopping criteria is met:

$$\Theta_{new} = \Theta_{old} - \eta \sum_{j=1}^{n} \nabla Q_j(\Theta_{old}), \tag{12}$$

where η is a pre-defined step size.

While evaluating the sum-gradient may be very ineffective in many cases because of the large scale of data set and the complexity of the computation [24], the stochastic gradient descent method updates of Θ in each iteration is approximated by sweeping the training examples one by one:

$$\Theta_{j+1} = \Theta_j - \eta \nabla Q_j(\Theta_j), \tag{13}$$

where $\Theta_1 = \Theta_{old}$ and $\Theta_{n+1} = \Theta_{new}$.

5 Approach

Given the data and the problem stated as in the previous section, we propose to predict the target variable by leveraging recurrent neural networks and linear regression, following a preprocessing step to select features from the original data set.

5.1 Overall Workflow

The basic idea of our solution is to compute an indicator for each sub-system, based on which the prediction of the target variable is obtained. Figure 4 shows the general workflow.

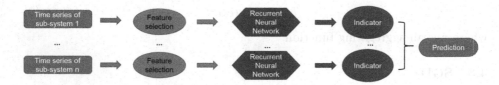

Fig. 4. Approach overview diagram

Within each sub-system, we will first compute an importance ranking list of the temporal sensor readings using VAR, and then select the top features in the list and feed them into the next step. The number of selected features may vary from case to case and we will do experiments on the different number of features to see the impact.

The recurrent neural network model will take a number of time series as input and employ a hidden state as the memory of the network to capture information about what happened in the past time period. The historical memory will be used to compute the output of this network (i.e., the indicator) together with the sensor readings of current timestamp. After this computation, the memory state will be updated for the next time period with the latest incoming data automatically.

Given the indicator for each sub-system, a regression model is needed to predict the target variable. In our solution, we choose the linear regression model.

As the last step, the numerous parameters in our solution to be tuned will be optimized.

5.2 Feature Selection

Feature selection is needed for selecting important features among multiple sensory data. In our solution, we propose to use VAR model as a base for linear dependencies removal.

Algorithm 1. VAR-based Feature Selection

1: Input: set of n_i readings in one sub-system D_i, target variable y, lag number p, number of selected features S.
2: Normalize the data between $[0, 1]$ column-wisely.
3: Use VAR(p) model to regress y based on normalized D_i.
4: Let $\langle c_1^j, \ldots, c_{n_i}^j \rangle$ be the coefficients for each feature with lag j obtained from VAR(p) model.
5: Let $c_l = \max\{|c_l^j| : j = 1, \ldots, p\}$.
6: Select a subset I_i of the feature indices with top S coefficients.
7: Output: $\{x \upharpoonright_{I_i} : x \in D_i\}$.

The first step of feature selection is to normalize each feature into the range $[0, 1]$ to remove the effect of different scale. Then the normalized data is fed into the VAR(p) model, where we set $p = 24$ (i.e., 24 h) to capture enough historical information.

The VAR model will give a coefficient for each feature with each lag c_l^j. We define the coefficient for a feature to be the maximum of the coefficients across all lags where users are free to customize their own function to aggregate the coefficients:

$$c_l = \max\{|c_l^j| : j = 1, \ldots, p\}. \tag{14}$$

Then we select features with top S coefficients. The entire process of feature selection is stated in Algorithm 1.

5.3 Predictive Modeling

Suppose x_t^i is the sensor readings from the i−th furnace at time t, and we adopt the LSTM model as in Fig. 3 to compute the indicator h_t^i. Suppose there are K nodes in the memory cell state. And we choose the aggregating functions to be:

$$\sigma(x) = \tau(x) = \max(0, x). \tag{15}$$

Let y_t be the target variable at timestamp t. The goal at this stage is to predict y_t based on all the information happening before.

As stated above, we use linear regression here to predict y_t. In time series analysis on target variable, we found that there is the dominant partial autocorrelation (PAC) at $t - 1$ (5.18 times of the second highest PAC at lag 18). Hence we include y_{t-1} as one of the predicting variables, and then the prediction of y_t is $\hat{y}_t$:

$$\hat{y}_t = m_0 \cdot y_{t-1} + \sum_{m=1}^{N} m_i \cdot h_t^i, \tag{16}$$

where $\langle m_0, \ldots m_N \rangle$ are the parameters to be optimized.

Let Θ be the collection of all the parameters used including $\langle m_0, \ldots m_N \rangle$, matrices W's and vectors b's for each sub-system. Let F be the function to compute $\hat{y}_t$ from observed data:

$$\hat{y}_t = F(y_{t-1}, D_t, \Theta), \tag{17}$$

where D_t denotes the input time series D up to timestamp t.

We adopt the mean absolute error (MAE) as the error measurement of the prediction:

$$MAE = \frac{\sum_{j=1}^{n} |y_{t_j} - \hat{y}_{t_j}|}{n}. \tag{18}$$

Let $Q_j(\Theta)$ be the $j - th$ absolute error $|y_{t_j} - F(y_{t-1}, D_t, \Theta)|$. Then the goal of the parameter optimization is to find:

$$\arg\min_{\Theta} \sum_{j=1}^{n} Q_j(\Theta). \tag{19}$$

Now the SGD method is employed to obtain optimal Θ as shown in Algorithm 2, which will also deliver $\hat{y}_{t_n}$ as the prediction of target variable at timestamp t, $h_{t_n}^i$ as the indicator for each sub-system, as well as the weight vector $\langle m_{t_n}^i, \ldots, m_{t_n}^M \rangle$.

Algorithm 2. RNN-based Forecasting

1: Input: output of feature selection for each sub-system B_i, target variable y, number of nodes in the hidden layer l, step size η, stopping criteria for SGD P, prediction function F and error function Q.
2: Initiate Θ and $X = \emptyset$.
3: **for** $j = 2$ to n **do**
4:    Let $X_{j-1} = \{y_{t_{j-1}}\} \cup (\bigcup_{i=1}^{n} B_i \lceil t_j)$.
5: **while** P is not met: **do**
6:    **for** $j = 1$ to $n - 1$ **do**
7:       Update Θ as $\Theta - \eta \nabla Q_j(\Theta)$.
8: Compute $o_{t_n}^i$ for each i and $\hat{y}_{t_n}$ using Θ.
9: Output: $\hat{y}_{t_n}$, $\langle m_{t_n}^1, \ldots, m_{t_n}^M \rangle$ and $\langle h_{t_n}^1, \ldots, h_{t_n}^M \rangle$.

---

6   Result Presentation

6.1   Experimental Evaluation

As described in Sect. 3, the whole data set consists of 14172 records with 282 features exclusive of the viscosity and it is split to 80:20 for training and testing correspondingly. In our proposed solution, the number of features to select and the number of nodes in the memory state are yet to determine. So experiments will be conducted with different values of these numbers.

The total number of features in each sub-system is around 70 as summarized in Table 1. Let A be the number of selected features in each sub-system. Then we will do experiments for $A = 10, 20, 30, 40, 50$. Meanwhile, we vary the number of hidden nodes $l = 100, 200, 300, 400, 500$. For parameters in SGD, we set the step size η to be 0.001, and the stopping criteria to be 500 iterations.

The experiments are conducted in Amazon cloud instance with 15 GB memory, and an 8-core 2.60 GHz processor. For each pair of number of features and number of nodes, we run our algorithm 100 times and take the average of the error and running time. All the measures described below refer to the testing dataset.

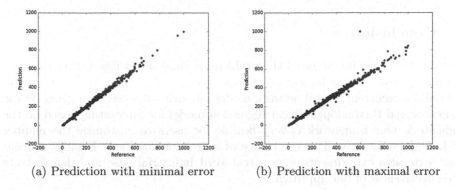

(a) Prediction with minimal error   (b) Prediction with maximal error

Fig. 5. Examples of prediction

Figure 5 shows how the predictions agree with the truth. The best prediction among the experimental results illustrated in Fig. 5(a) presents high agreement between the prediction and ground truth. In Fig. 5(b) which shows the result with maximal error, it can be observed that the points are gathered along a straight line but not $y = x$. In other words, this running of algorithm failed to find the optimal slope. Additionally, there are a number of points distributed far from the fitted straight line.

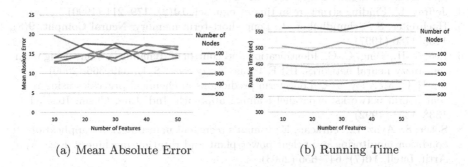

(a) Mean Absolute Error   (b) Running Time

Fig. 6. Performance comparison

The performance of the experiments is illustrated in Fig. 6. Since the viscosity has been scaled into [0, 1000], Fig. 6(a) shows that the mean absolute errors are quite small for all the parameter combinations, indicating that the model has great capability to forecast the viscosity. An additional interesting observation is that the difference of errors is ignorable for different combinations. The reason could be that the top 10 features is able to capture most information useful for viscosity prediction. Figure 6(b) tells that the running time is almost invariant to the number of features, while it increases significantly with the number of nodes.

7   Conclusion

In this paper, we investigated the problem of time series forecasting for engineering systems where indicators of sub-systems are needed. In our proposal, we build a recurrent neural network model for each sub-system to generate the indicator and then adopt a linear regression model for forecasting based on the indicators. Our framework is very flexible for users to customize the number of LSTM layers as well as the choice of regression function. Finally, we conduct extensive experiments with a real-word industrial case and demonstrate the effectiveness of our approach.

References

1. Sorjamaa, A., Hao, J., Reyhani, N., Ji, Y., Lendasse, A.: Methodology for long-term prediction of time series. Neurocomputing **70**(1618), 2861–2869 (2007). Neural Network Applications in Electrical Engineering Selected papers from the 3rd International Work-Conference on Artificial Neural Networks (IWANN 2005)
2. Nian, X., Wang, Z., Feng, Q.: A hybrid algorithm based on differential evolution and group search optimization and its application on ethylene cracking furnace. Chin. J. Chem. Eng. **21**(5), 537–543 (2013)
3. Hatemi-J, A.: Multivariate tests for autocorrelation in the stable and unstable VAR models. Econ. Model. **21**(4), 661–683 (2004)
4. Jeffrey, L.: Finding structure in time. Cogn. sci. **14**(2), 179–211 (1990)
5. Hochreiter, S., Schmidhuber, J.: Long short-term memory. Neural Comput. **9**(8), 1735–1780 (1997)
6. Lee, C.-H., Teng, C.-C.: Identification and control of dynamic systems using recurrent fuzzy neural networks. IEEE Trans. Fuzzy Syst. **8**(4), 349–366 (2000)
7. Su, H.T., McAvoy, T.J.: Long-term predictions of chemical processes using recurrent neural networks: a parallel training approach. Ind. Eng. Chem. Res. **31**(5), 1338–1352 (1992)
8. Şeker, S., Ayaz, E., Türkcan, E.: Elman's recurrent neural network applications to condition monitoring in nuclear power plant and rotating machinery. Eng. Appl. Artif. Intell. **16**(7), 647–656 (2003)
9. Bontempi, G., Ben Taieb, S., Borgne, Y.-A.: Machine learning strategies for time series forecasting. In: Aufaure, M.-A., Zimányi, E. (eds.) eBISS 2012. LNBIP, vol. 138, pp. 62–77. Springer, Heidelberg (2013). doi:10.1007/978-3-642-36318-4_3
10. Box, G.E.P., Pierce, D.A.: Distribution of residual autocorrelations in autoregressive-integrated moving average time series models. J. Am. Stat. Assoc. **65**(332), 1509–1526 (1970)

11. De Gooijer, J.G., Hyndman, R.J.: 25 years of time series forecasting. Int. J. Forecast. **22**(3), 443–473 (2006)
12. Hippert, H.S., Pedreira, C.E., Souza, R.C.: Neural networks for short-term load forecasting: a review and evaluation. IEEE Trans. Power Syst. **16**(1), 44–55 (2001)
13. Mi, L., Tan, W., Chen, R.: Multi-steps degradation process prediction for bearing based on improved back propagation neural network. In: Proceedings of the Institution of Mechanical Engineers, Part C: Journal of Mechanical Engineering Science, pp. 1544–1553 (2011)
14. Darbellay, G.A., Slama, M.: Forecasting the short-term demand for electricity: Do neural networks stand a better chance? Int. J. Forecast. **16**(1), 71–83 (2000)
15. Qi, M.: Predicting us recessions with leading indicators via neural network models. Int. J. Forecast. **17**(3), 383–401 (2001)
16. Sak, H., Senior, A.W., Beaufays, F.: Long short-term memory recurrent neural network architectures for large scale acoustic modeling. In: INTERSPEECH, pp. 338–342 (2014)
17. Lipton, Z.C., Kale, D.C., Elkan, C., Wetzell, R.: Learning to diagnose with lstm recurrent neural networks. arXiv preprint arXiv:1511.03677 (2015)
18. Pawłowski, K., Kurach, K.: Detecting methane outbreaks from time series data with deep neural networks. In: Yao, Y., Hu, Q., Yu, H., Grzymala-Busse, J.W. (eds.) RSFDGrC 2015. LNCS (LNAI), vol. 9437, pp. 475–484. Springer, Heidelberg (2015). doi:10.1007/978-3-319-25783-9_42
19. Ma, X., Tao, Z., Wang, Y., Haiyang, Y., Wang, Y.: Long short-term memory neural network for traffic speed prediction using remote microwave sensor data. Transp. Res. Part C: Emerg. Technol. **54**, 187–197 (2015)
20. Xiong, R., Nicholas, E.P., Shen, Y.: Deep learning stock volatilities with google domestic trends. arXiv preprint arXiv:1512.04916 (2015)
21. Liu, H., Dougherty, E., Dy, J.G., Torkkola, K., Tuv, E., Peng, H., Ding, C., Long, F., Berens, M., Parsons, L., et al.: Evolving feature selection. IEEE Intell. Syst. **20**(6), 64–76 (2005)
22. Pedišić, L., Matijević, B., Perić, B.: Influence of quenching oils composition on the cooling rate. In: 1st International Conference on Heat Treatment and Surface Engineering of Tools and Dies (2005)
23. Greff, K., Srivastava, R.K., Koutník, J., Steunebrink, B.R., Schmidhuber, J.: Lstm: a search space odyssey. arXiv preprint arXiv:1503.04069 (2015)
24. Bousquet, O., Bottou, L.: The tradeoffs of large scale learning. In: Advances in Neural Information Processing Systems, pp. 161–168 (2008)

EDAHT: An Expertise Degree Analysis Model for Mass Comments in the E-Commerce System

Jiang Zhong[1(⊠)], You Xiong[2], Weili Guo[2], and Jingyi Xie[2]

[1] Key Laboratory of Dependable Service Computing in Cyber Physical Society, Ministry of Education, Chongqing University, Chongqing 400030, China
zhongjiang@cqu.edu.cn
[2] College of Computer Science, Chongqing University, Chongqing 400030, China

Abstract. In order to help consumers to retrieve the most valuable information from amount of comments quickly, we present a method of evaluating the expertise degree of comments. Firstly, we propose an algorithm to construct automatically an attribute-word hierarchy tree from the massive comments data. Secondly, we develop an expertise degree analysis based on attribute-word hierarchy tree (EDAHT) to estimate the expertise degree of comments. The experiments results on 8,000 manual scoring data show that EDAHT model is high consistent with the manual scoring data, and this novel model is effective.

Keywords: Comment text · Expertise degree · Attribute hierarchy tree · Feature extraction

1 Introduction

There are large numbers of comments in almost every e-commerce system, such as Amazon, JD, Tmall and so on. According to the 33rd Statistical Report on China Internet Development report of CNNIC, in 2013, 41.1 % of the consumers take users' comments into account when they are shopping online. However, the comments qualities have great difference because of users' experiments, knowledge and attitudes. It is a challenge task to find the useful comments from mass comments.

Ortigosa et al. [1] propose a hybrid approach combining lexical-based and machine-learning techniques for sentiment analysis in Facebook and get a high accuracy (83.27 %). Angulakshmi et al. [2] present a domain sentiment word extraction approach based on the propagation of both known sentiment lexicon and extract product features for domain-specific opinion word discovery. Momeni et al. [3] compare the characteristics of useful comments on different platforms (YouTube and Flickr), use crowd-sourcing techniques to collect judgments of usefulness of comments and designate a comment as useful when they achieve majority agreement between coders to infer useful comments. Namihira et al. [4] propose a new method of providing an expertise score to calculate the information credibility depending on user's knowledge (expertise). Chen et al. [5] propose a method of making tweet recommendations based

© Springer International Publishing AG 2016
J. Li et al. (Eds.): ADMA 2016, LNAI 10086, pp. 472–480, 2016.
DOI: 10.1007/978-3-319-49586-6_32

on collaborative ranking to capture personal interests, and the method considers three major elements on Twitter: tweet topic level factors, user social relation factors and explicit features such as authority of the publisher and quality of the tweet.

However, to the best of our knowledge, there is no research on estimating the professional degree of comments. In this paper, we focus on the methods to evaluate the expertise degree of comments, which would help people select the useful comments according to the specialty aspect.

We explain the features used for measuring the expertise degree of comments in Sect. 2. We present experiments data sets and experiments setup in Sect. 3, conclude our study and outline the future directions in Sect. 4.

2 Analysis Model for Expertise Degree

In this section, we firstly explain the idea of expertise degree analysis model (EDA model) in e-commerce domain. Then, we present the analysis details of the EDA model.

2.1 Expertise Features of Comments

According to the common sense, we employ three features of comments to evaluate the expertise degree: comprehensiveness, specificity and cohesion [6, 7].

Comprehensiveness, which means that the comment evaluates the commodity or service from multiple aspects. The more commodity's attributes contained in comment, the more comprehensive the comment is. However, it is emphasized that the comment's length and the comprehensiveness are not necessarily positive correlations. For example, there are two comments on a phone as following.

> **Comment-1:** Good. Very beautiful, very beautiful, very beautiful, very beautiful, very beautiful.
> **Comment-2:** High resolution, perfect functions, long standby time.

Although the text of comment-1 is longer than comment-2, it used attributes repeatedly and only including appearance attribute. Whereas, the comment-2 evaluates the phone from three aspects, namely resolution, functions and standby time. Obviously, comment-2 is more comprehensive than comment-1.

Specificity, which reflects how many detail information about commodity a comment contains. The more details contained in the comment, the more specific the comment is. For example, there are two comments on a phone as following.

> **Comment-3:** Good service.
> **Comment-4:** Quick delivery, good service attitude and beautiful pack.

The above two comments evaluate the service quality of transactions. But it is obvious that comment-4 provides more detail information (i.e., delivery, service attitude and pack) than comment-3. Therefore, comment-4 is more specific than comment-3.

Cohesion, which reflects the associations among the attributes contained in a comment. If a comment contains more closely associated terms, it tends to be more cohesive. According to the Ransdell's definition [8], the association of a comment refers to its capacity to represent a plurality of mutually independent terms with respect to a specific domain. For example, there are three short comment fragments: C1 ("Color"), C2 ("Size and Color"), and C3 ("Color and Appearance"), C1 is more cohesive than C2 because C1 only consists of a single term. However, C2 is more cohesive than C3 because "Color" and "Size" are both sub-types of "Appearance" and are less mutually independent than the terms "Color" and "Appearance" contained in C3.

2.2   Construction Algorithm for Concept Hierarchy Tree

Attribute-word Hierarchy Tree (AHT) is similar as product categories tree, which attach a set of attribute-words for each node in product category tree. We can obtain Attribute-words from comment text by syntax rule. Each node in the AHT is a two-tuples $<c_{ij}, T\_S_{ij}>$, where c_{ij} represents the *jth* product category in level i and T_S_{ij} is set of attribute-words attached to category c_{ij}.

Figure 1 shows an example of AHT, the root node includes widely used attribute-words in comments text: *Brand, Quality* and *Price*. The node Office Stationery includes attribute-words: *Pen, Eraser*.

The key problem is how to determine which attribute-words should attach to every category node. In this paper, we employ the improved *Gini* coefficient method [9] to compute the relationship between the category and attribute-words.

Let $C_0, C_1, \cdots, C_m$ are product-categories sets of every level of product category tree, where C_0 is the root-level categories and C_m is the *mth*-level categories.

All comments text of a product are merged into one document, which is tagged by m classification labels same as product category names, and D is the documents set for all products. Let $T = \{t_1, t_2, \cdots, t_n\}$ is attribute-words set extracted from D [10–12].

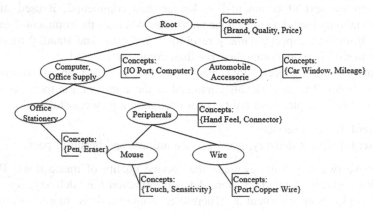

Fig. 1. A fragment of the AHT

We suppose there are n categories in category level C_i. Using vector $W_k = [w_1, w_2, w_3, \cdots, w_n]$ represents correlations between attribute-word t_k and every category, and w_j is computed as Eq. (1).

$$w_j = Gini(t_k, c_{ij}) = P(t_k|c_{ij})P(c_{ij}|t_k) \tag{1}$$

where $P(t_k|c_{ij})$ is the probability of that attribute-word t_k be contained in the documents of category c_{ij}, and $P(c_{ij}|t_k)$ is conditional probability that document will belong to category c_{ij} when it contained attribute-word t_k [9].

Sort the vector W_k according to the value from big to small and the result is $W_k' = [w_1', w_2', w_3', \cdots, w_n']$. Let δ be threshold value, w_1' represents the Gini coefficient between the attribute-word t_k and category c_{ij}. If $w_1' - w_2' > \delta$, we could attach t_k to category c_{ij}.

Algorithm: Gini Based Attribute Hierarchy Tree Construction

Input: comments data D, product categories $C_0, C_1, \cdots, C_m$, and attribute-words set T

Output: attribute-words set for each product category T_S

```
1  for i=m to 1
2    l=|C_i|
3    for k=1 to |T|
4      for j=1 to l
5        W_k[j]= Gini(t_j,c_1j)
6      end for
7      W'_k= [w'_1, w'_2,w'_3, ..., w'_l]= sort(W_k)
8      if (w'_1-w'_2>Δ) and (w'_1=w_z) then
9        T=T-{t_z}
10       T_S_ij = T_S_ij ∪ t_z
11     end if
12   end for
13 end for
14 T_S_01 =T
15 return {T_S}
```

The time complexity and space complexity of this algorithm are analyzed as follows. Suppose the maximum number of product categories at each level in the attribute-word hierarchy tree is l_max, and the number of attribute-word extracted is t, product number is d, the levels of the product category tree is m, the time complexity of function Gini is $O(l\_max \times d)$, then time complexity of constructing algorithm for

AHT is $O(m \times l\_max^2 \times t \times d)$. Normally product category number l_max and level product category level m are limitation constants, so the time complexity of algorithm is $O(t \times d)$.

2.3   Computing the Expertise Features Based on AHT

After we get the attribute hierarchy tree, we could use it to computing the expertise features described in Sect. 2.1.

In this paper, we employ the function $CComp(r_i)$ to measure comprehensiveness feature of the comment text r_i.

$$CComp(r_i) = \#attribute - words \ in \ r_i \qquad (2)$$

The function $CSpec(r_i)$ is used to measure the comment specificity feature. To make the specificity function sensitive to the average depths of the matching concepts, an exponential function is designed:

$$CSpec(r_i) = e^{-\frac{\sum_{i=1}^{n} depth(t_i)}{n}} \qquad (3)$$

In Eq. 3, n refers to the total number of attribute-word in comment r_i. The function $depth(t_i)$ is the level of product category which is the t_i attached in AHT.

We use the Leacock-Chodorow similarity function [13] to measure the cohesion feature of comments. The cohesion function of a comment is $CCohe(r_i)$ in Eq. (4).

$$CCohe(r_i) = \begin{cases} \frac{\sum_{i=1}^{n-1} \sum_{j=i+1}^{n} Sim(t_i, t_j)}{n(n-1)}, & n > 1 \\ 0, & otherwise \end{cases} \qquad (4)$$

Function $Sim(t_i, t_j)$ in Eq. 4 is used to calculate the Leacock-Chodorow semantic similarity between two attribute-words in AHT, which is described as Eq. (5).

$$Sim(t_i, t_j) = \begin{cases} -\ln \frac{len(t_i, t_j)}{2Depth_{max}}, & len(t_i, t_j) < Depth_{max} \\ 1, & len(t_i, t_j) = 0 \\ 0, & len(t_i, t_j) = Depth_{max} \end{cases} \qquad (5)$$

where $len(t_i, t_j)$ is the shortest path between these two attribute-words, and $Depth_{max}$ is the maximum level of the AHT.

If we get the above expertise features of comments, we could use machine learning methods, such as kNN, SVM classification methods to establish expertise degree analysis models.

3 Experiment Results and Analysis

3.1 Datasets and Evaluating Methods

In this paper, we collect real transaction data of a B2C platform in year 2012, where 18,495,086 comments, including 8 first-level product categories, 31 second-level product categories and 115 third-level product categories have been collected. These categories all come from commodities categories of the B2C platform. In order to evaluate the accuracy of the EDA model, we design a manual scoring system to collect the scoring data about the expertise degree of comments. We provide two ways to collect manual scoring.

1. **Five-Point way:** Scoring people rate expertise degree of each comment on five-point scale. In this way, 1 means the comment is the lowest-expertise level, 2 means the comment is the low-expertise level, 3 means the comment is the ordinary-expertise level, 4 means the comment is the high-expertise level and 5 means the comment is in the highest-expertise level.
2. **Comparison way:** Scoring people compare the expertise degree of a pair of comments to the same product. In this way, −1 means the expertise degree of first comment is higher than the second, 0 means the expertise degree of these two comments equal and 1 means the expertise degree of the first comment is lower than the second.

From the manual scoring system, we collected 5,393 scoring data for the single comment through Five-Point way, and collected 3,293 comparison results of comment-pairs of same product. A total of more than 8,000 scoring data is a relatively big data set of manual scoring.

Similar with classification problems, we use Precision, Recall, and F-Measure indexes to evaluate the performances of EDAHT.

$$Precision = \sum_{i=1}^{n} \frac{\# \ Correct \ Scoring \ Comments \ of \ P_i}{\# \ Comments \ of \ P_i} \times \omega_i \times 100\% \quad (6)$$

$$Recall = \frac{\# \ Correct \ Scoring \ Comments}{\# \ total \ Comments} \times 100\% \quad (7)$$

$$F - Measure = \frac{Precision \times Recall \times 2}{Precision + Recall} \quad (8)$$

$$\omega_i = \frac{\# \ Comments \ of \ P_i}{\# \ Comments \ of \ P_i} \quad (9)$$

3.2 Experiments Setup

Because of the manual scoring for the expertise degrees of comments are subjective, we design five experiments as following:

Exp1, exact comparison. Exactly compare the results between original manual scoring and classification models' prediction values.

Exp2, two-levels experiment, while scoring two is the boundary. That is, comments scoring as one to two are "low expertise level" and comments scoring from three to five are "high expertise level".

Exp3, two-levels experiment, while it remove the comments scoring three. The comments scoring as one to two are "low expertise level" and comments scoring from three to five are "high expertise level"

Exp4, two-levels experiment, while scoring three is the boundary.

Exp5, comparison experiment. Exactly compare the results between original manual comparison and classification models' prediction results.

In this paper, we use kNN, Naïve Bayes, Random Forrest and LibSVM methods to establish the expertise prediction models for comments data.

3.3   Experiment Results

Experiment results are shown in Fig. 2. It can be seen that among all the Naïve Bayes and LibSVM classifiers have the better results on all experiment data sets, and the kNN classifier has the worst results.

Although all classification models are not good in experiment **Exp1**, the two_levels experiments results show that all kinds of prediction models based on the expertise

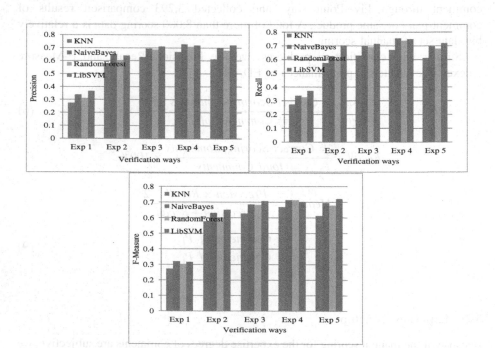

Fig. 2. Precision, Recall, and F-Measure in different verification ways

features could get nearly precision rate **0.7**. It is means that the expertise features using attributes hierarchy tree are efficiently. On the other hand, the features used in EDAHT are meaningful and interpretable to human beings.

Results are also shown that all classification models get better performances in **Exp4** than in **Exp2**, which means that comments with scoring three should be classified as "low expertise level" better than it should be classified as "high expertise level".

4 Conclusion

With the prosperity and development of e-commerce market, more and more people pay attention to the value of comments. In this paper, we present an expertise degree analysis model based on the attribute-word hierarchy tree for e-commerce system. We present comprehensiveness, specificity and cohesion expertise features, what are intuitive and easy to be interpreted. The experiments results show that although we only use three decision features, all classification prediction models could get satisfaction performances to some extents.

Acknowledgement. This work was supported in part by the National High-tech R&D Program of China (NO. 2015AA015308, Fundamental Research Funds for the Central Universities (NO. 106112014CDJZR188801).

References

1. Ortigosa, A., Martín, J.M., Carro, R.M.: Sentiment analysis in Facebook and its application to e-learning. J. Comput. Hum. Behav. **31**, 527–541 (2014)
2. Angulakshmi, G., Chezian, R.M.: Three level feature extraction for sentiment classification. An ISO 3297: 2007 Certified Organization (2014)
3. Momeni, E., Sageder, G.: An empirical analysis of characteristics of useful comments in social media. In: Proceedings of the 5th Annual ACM Web Science Conference, pp. 258–261. ACM (2013)
4. Namihira, Y., Segawa, N., Ikegami, Y., et al.: High precision credibility analysis of information on Twitter. In: 2013 International Conference on Signal-Image Technology & Internet-Based Systems (SITIS), pp. 909–915. IEEE (2013)
5. Chen, K., Chen, T., Zheng, G., et al.: Collaborative personalized tweet recommendation. In: Proceedings of the 35th International ACM SIGIR Conference on Research and Development in Information Retrieval, pp. 661–670. ACM (2012)
6. Yan, X., Lau, R.Y., Song, D., Li, X., Ma, J.: Toward a semantic granularity model for domain-specific information retrieval. ACM Trans. Inf. Syst. (TOIS) **29**(3), 15 (2011)
7. Wang, S., Chang, X., Li, X., Sheng, Q.Z., Chen, W.: Multi-task support vector machines for feature selection with shared knowledge discovery. Sig. Process. **120**, 746–753 (2016)
8. Ransdell, J.M.: Charles Peirce: The Idea of Representation. Columbia University, New York (1967)
9. Sun, A., Lim, E.P.: Hierarchical text classification and evaluation. In: Proceedings of IEEE International Conference on Data Mining 2001, ICDM 2001, pp. 521–528. IEEE (2001)

480   J. Zhong et al.

10. Hu, M., Liu, B.: Mining and summarizing customer reviews. In: Proceedings of the 2004 ACM SIGKDD International Conference on Knowledge Discovery and Data Mining, pp. 168–177. ACM Press, New York (2004)
11. Zhao, P., Li, X., Wang, K.: Feature extraction from micro-blogs for comparison of products and services. In: Lin, X., Manolopoulos, Y., Srivastava, D., Huang, G. (eds.) WISE 2013. LNCS, vol. 8180, pp. 82–91. Springer, Heidelberg (2013). doi:10.1007/978-3-642-41230-1_7
12. Chang, X., Shen, H., Wang, S., Liu, J., Li, X.: Semi-supervised feature analysis for multimedia annotation by mining label correlation. In: Tseng, V.S., Ho, T.B., Zhou, Z.-H., Chen, A.L.P., Kao, H.-Y. (eds.) PAKDD 2014. LNCS, vol. 8444, pp. 74–85. Springer, Heidelberg (2014). doi:10.1007/978-3-319-06605-9_7
13. Leacock, C., Chodorow, M.: Combining local context and WordNet similarity for word sense identification. J. WordNet Electron. Lexical Database **49**(2), 265–283 (1998)

A Scalable Document-Based Architecture for Text Analysis

Ciprian-Octavian Truică[1]([✉]), Jérôme Darmont[2], and Julien Velcin[2]

[1] Computer Science and Engineering Department, Faculty of Automatic Control and Computers, University Politehnica of Bucharest, Bucharest, Romania
`ciprian.truica@cs.pub.ro`
[2] Univ Lyon, Lumière, ERIC EA 3083, Lyon, France
`{jerome.darmont,julien.velcin}@univ-lyon2.fr`

Abstract. Analyzing textual data is a very challenging task because of the huge volume of data generated daily. Fundamental issues in text analysis include the lack of structure in document datasets, the need for various preprocessing steps and performance and scaling issues. Existing text analysis architectures partly solve these issues, providing restrictive data schemas, addressing only one aspect of text preprocessing and focusing on one single task when dealing with performance optimization. Thus, we propose in this paper a new generic text analysis architecture, where document structure is flexible, many preprocessing techniques are integrated and textual datasets are indexed for efficient access. We implement our conceptual architecture using both a relational and a document-oriented database. Our experiments demonstrate the feasibility of our approach and the superiority of the document-oriented logical and physical implementation.

Keywords: Text analytics · Indexing methods · Document-oriented databases

1 Introduction

A vast amount of textual data is generated daily and it is really challenging to develop efficient models and systems to enhance processing performance while doing accurate text analysis. The most fundamental challenges when working with large volumes of heterogeneous text datasets include the lack of structure of textual corpora, the various required preprocessing steps, the need for efficient access and the ability to scale up.

Structural issues may be addressed by resorting to textual data warehousing and On-Line Analytical Processing (OLAP). However, such approaches only partially solve the problem because they use a structured schema that falls short when applied to large, heterogeneous volumes of data. Moreover, using a predefined schema makes them extremely dataset-specific.

Moreover, when dealing with textual data, we distinguish different preprocessing levels: quite basic operations (e.g., cleaning HTML tags, tokenization,

© Springer International Publishing AG 2016
J. Li et al. (Eds.): ADMA 2016, LNAI 10086, pp. 481–494, 2016.
DOI: 10.1007/978-3-319-49586-6_33

language identification); intermediate operations (e.g., stemming, lemmatization, indexing); and advanced operations (e.g., part of speech tagging, named entity recognition, topic modeling). Each complexity layer in this process requires the previous layer and all operations must remain tractable in terms of memory and CPU time. To the best of our knowledge, no text analysis tool implements all layers, nor any processing workflow.

Finally, when working on performance and scaling issues, state-of-the-art research focuses on one aspect of text analysis, e.g., aggregation, top-k keyword extraction and text indexing. However, text processing techniques used in a single application may be many and, as we mention above, interdependent.

Hence, we present in this paper a scalable text analysis architecture that addresses all these issues. More precisely, we deal with the lack of structure by adopting a novel generic, document-oriented data model that allows storing heterogeneous textural corpora with no predefined structure. We also integrate in our framework all the preprocessing methods that are useful for information retrieval, data mining, text analysis and knowledge discovery. We also propose a new compact data structure to minimize index storage space and the response time of create, read, update and delete (CRUD) operations. Such indexes benefit to text preprocessing, querying and further analysis, and adequately contribute to global scaling.

The remainder of this paper is organized as follows. In Sect. 2, we discuss related works. In Sect. 3, we present the architecture and implementation of our approach. In Sect. 4, we experimentally validate our proposal. In Sect. 5, we finally conclude this paper and hint at future research.

2   Related Works

2.1   Text Cubes and OLAP

Extensive work on information retrieval (IR) and text analysis have been done using OLAP. Most proposals use Text Cubes for OLAPing multidimensional text databases [20]. Lin et al. focus on optimizing query processing and reducing storage costs of Text Cubes [12]. They experimentally show that average query time and storage cost are related to a cube's number of dimensions. Zhang et al. use Text Cubes for topic modeling [20] and experimentally show that their approach is much faster than computing each topic cube from scratch. Finally, Ding et al. address the problem of keyword search and top-k document ranking using Text Cubes [7]. Their algorithms perform well in terms of query response and memory cost when the number of search terms is small.

Ben Kraiem et al. propose a generic multidimensional model for OLAP on tweets [2]. Their experiments show some promising results for knowledge discovery when applying OLAP on a small corpus, but query performance decreases when data volume increases. Bringay et al. propose a data warehouse model to analyze large volumes of tweets [4]. They introduce different operators to identify trends using the top-k most significant words over a period of time for a specific geographical location, as well as the impact of hierarchies on such operators. Unfortunately, no time performance and storage cost analysis is provided.

In conclusion, research done so far on text analysis and OLAP focuses on small, structured datasets and scaling up is not guaranteed.

2.2   Text Preprocessing and Analysis

Managing morphological variation of search terms in IR has been quite extensively studied [10,11]. The main successful methods are stemming [14] and lemmatization, which are used to optimize search, minimize the space allocated to inverted indexes (Sect. 2.3) and, in the case of lemmatization, to add linguistic information. Lemmatization is useful for different types of advanced text analysis, e.g., named entity recognition, automatic domain specific multi-term extraction and part of speech (PoS) tagging. Moreover, lemmatization is easier of use than stemming, saves storage and improves retrieval performance [11].

Topic modeling is a statistical model for discovering hidden themes that occur in a collections of documents. In recent years, it has been extensively studied, showing the usefulness of analyzing latent topics and discovering topic patterns [3,16]. Popular approaches for topic modeling are latent semantic indexing (LSI) [6], latent Dirichlet allocation (LDA) [3], the non-parametric extension hierarchical Dirichlet process (HDP) [16] and non-negative matrix factorization (NMF) [1].

2.3   Document Indexing

Inverted indexes are data structures used in search engines, whose main purpose is to optimize query response speed. Basic inverted indexes store terms, a list of documents where each term appears and a weight. Weight measures the number of occurrences of the term in a document, e.g., raw term frequency/word co-occurrence (TF), normalized Term Frequency (TF_n), etc. In the various methods for managing inverted indexes, great emphasis is put on storage space reduction. For instance, a pruning algorithm based on term frequency-inverse document frequency (TF*IDF) can be used to minimize index size [19]. Yet, updating an inverted index is also a problem, because it is dependent on documents. The index must indeed be updated each time documents are added or deleted.

3   Proposed Approach and Implementation

3.1   Approach Overview

The approach we propose (Fig. 1) is subdivided into four steps: (1) clean and preprocess documents using natural language processing (NLP) and store the information in a database; (2) construct indexes; (3) analyze data, e.g., with topic modeling, etc.; (4) query and search data, extract top-k most relevant documents, create visualizations and analyses. We construct the inverted index,

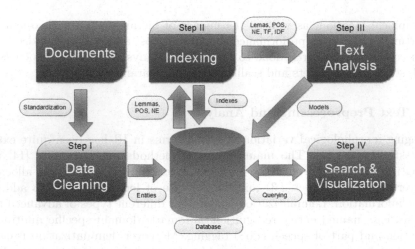

Fig. 1. System architecture

vocabulary, PoS and named entities (NE) indexes during the index construction step. Indexes may be used afterward by data mining, text analysis, search and visualization. The search engine sorts documents based on a ranking function (e.g., TF*IDF, Okapi or BM25) to extract the top-k documents.

To implement our document-oriented approach, we quite naturally rely on a document-oriented database management system (DODBMS). DODBMSs are a class of NoSQL systems that aim to store, manage and process data using a semi-structured model. DODBMSs encapsulate data in collections of documents [8]. A document can contain other nested documents, which turns out to be very flexible [17].

One feature of DODBMSs is that they are often optimized for create and read operations, while offering reduced functionality for update and delete queries. DODBMSs are designed to work with large amounts of data and the main focus is on the efficiency of data storage, access and analysis [13]. Another key feature of DODBMSs is the distribution of data across multiple sites. In particular, DODBMSs can horizontally scale CRUD operations throughput [5]. Moreover, decentralized data stores provide good mechanisms for fail-over, removing the single point of failure, due to their scalability and flexibility [9].

We selected MongoDB as our DODBMS, since it beats the best mean time performances for CRUD operations both in single and distributed environments [17]. Moreover, we also implemented our approach with PostgreSQL, to provide a point of comparison with a well-established, efficient relational database management system (RDBMS) (Sect. 4).

3.2   Data Models

We design a generic model to store heterogeneous text data using a data warehouse snowflake schema (Fig. 2). The central component of the model is the *docu-*

ments entity, where we store basic information and metadata about a document, e.g., timestamp, title, raw, clean and lemmatized text, etc. The *document_tags* entity is used to store metadata represented by tags, which can be existing tags, hashtags or at tags. The *vocabulary* entity links documents to information extracted or inferred from the text, which helps enhancing metadata with different weights and tags, e.g., PoS, TF, TF_n, lemmas, etc. The *named_entities* entity stores all the information about entities automatically extracted from the original corpus.

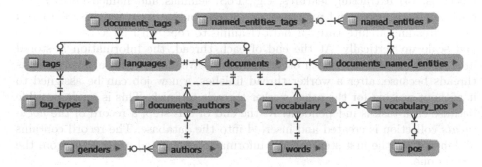

Fig. 2. Conceptual model

The DODBMS schemaless design takes all the information presented in the relational schema and stores it for each document in a record of the collection. Using this design, all one-to-many and many-to-many relationships become either vectors (e.g., *hashtags, at tags*) or nested documents (e.g., *words, named_entities*). Where the information is not present, these vectors and nested documents may be missing thanks to the flexibility of schemaless database design. A problem that arises is duplication, as multiple records can bear the same metadata, since all the information for a document is stored in one single record. The *vocabulary* entity is constructed as a separate collection. This entity is constructed dynamically, taking user input constraints into account, e.g. date, tags, search words, named-entities.

Interaction with the database is achieved through CRUD operations, aggregation functions and views. We use read operations for information extraction and data visualization. Aggregation functions are used for constructing indexes, searching and preprocessing data for text analysis. We make use of MapReduce for this purpose when using the document-oriented database architecture. Dynamically materialized cubes are constructed using views with aggregation functions, fine-graining query results using different measures, e.g., timestamps, locations, lemmas, tags, named entities.

3.3   Text Preprocessing

The data cleaning module serves three functions: (1) corpus standardization, (2) text preprocessing using NLP to enrich data, and (3) entity creation and information insertion into the database.

The entire corpus is standardized by determining all the fields of a document, including metadata and the labels of *documents*. Then, during the preprocessing step, the following techniques are applied: (1) text cleaning by removing HTML/XML tags and scripts; (2) language identification; (3) expanding contractions; (4) extracting features, e.g., PoS, lemmas and named entities; (5) removing stop words and punctuation; (6) computing term weights. We use a multithreading architecture for data cleaning to cope with large data volumes and scale up vertically. At the end of each thread, the information is stored in a dictionary, together with other metadata. We choose to use asynchronous threads because, after a worker thread finishes, a new job can be assigned to it without waiting for the other worker threads to finish. This is made possible because each task is independent. At the end of this step, a record of the *documents* collection is created and inserted into the database. The record contains all labels from the first step and the information extracted using NLP from the second one.

In the DODBMS implementation, a record stores all the information because its attributes are created dynamically. In contrast, the RDBMS architecture can only store predefined fields due to its rigid schema. Thus, undefined fields are omitted.

The RDBMS approach merges the data cleaning step with the index construction step, because many-to-many relationships between entities, translated as bridge tables, are indexes as well. We could not use a multithreading approach here because information could be lost. Multiple threads could indeed check at the same time whether the information is present and receive a negative response. A constraint violation error could appear and the transaction terminate by a *rollback*. If constraints are missing, then duplicate information could appear and this would impact text analysis.

3.4   Index Management

We propose several indexes for document aggregation, search, extraction of the top-k most signification terms and text analysis, e.g., topic modeling, document clustering. These new indexing structures minimize storage costs and maximize the time performance of CRUD operations.

Index construction in the DODBMS architecture is done using the MapReduce framework. Four indexes are created: (1) an *inverted index* that stores, for each term, a list of corresponding documents; (2) a *vocabulary*, a novel inverted index with additional information for each term in the corpus, e.g., list of documents where the term is found, the TF and TF_n of the term for a document and IDF; (3) a PoS index that stores the part of speech of each term; (4) a named-entity index used for storing named entities. There are no integrity constraints between these collections to improve query response time. Moreover, the structure proposed for the *vocabulary* facilitates query response time, aggregation and search (Fig. 3). MapReduce is used to construct all indexes. It is also central in aggregation queries needed by the search algorithms. To improve index construction and query response times, we horizontally scale the database, and by doing so add more MapReduce worker.

In the RDBMS architecture, indexes are the bridge tables translating many-to-many relationships between entities. The *vocabulary* is the bridge table between the *documents* table and the *words* table. The PoS index is the bridge table between the *vocabulary* table and the *pos* table. In this case, the index also contains the TF and IDF of each term.

```
{    "_id" : 791297891579,
     "word" : "hillary",
     "idf" : 4.39,
     "createdAt" : "2016-03-03 13:09:13.802",
     "docIDs" : [
          { "docID" : 628840524375, "count" : 4, "tf" : 0.64 },
          { "docID" : 628840524457, "count" : 6, "tf" : 0.70 },
          .....
          { "docID" : 628840524430, "count" : 3, "tf" : 0.71 } ] }
```

Fig. 3. Vocabulary index structure

The number of entries in the indexes constructed for the DODBMS is equal to the number of terms in the entire corpus. In the RDBMS, the inverted index has more entries, i.e., $\sum_{d \in D} | t : t \in d |$, where D is the corpus and $| t : t \in d |$ is the number of distinct terms that appear in document d.

Updating indexes in the DODBMS is based on document insertion date. The update method we use constructs an intermediary index for new documents, and then it updates the primary index by appending the new documents' ID and TF to existing labels. Then, the IDF of each term is updated for the whole index. When documents are deleted, we apply a bulk delete operation. In this case, a list of deleted document IDs is stored, which helps update the index structure by removing the deleted documents and then updating the IDF of each term.

Updating indexes in the RDBMS implementation is easier thanks to the database's structure. When documents are added, indexes are automatically updated based on the insertion date of the last added documents. When documents are removed, the corresponding index entries are also removed. For both operations, the IDF of each term must be recalculated.

4 Experimental Validation

In this section, we test each step of our approach and we compare the results achieved by the two instances we developed, i.e., the DODBMS version implemented with MongoDB and the RDBMS version implemented with PostgreSQL. Tests are done using a news corpus consisting of 110,000 articles[1], a corpus of 5,000,000 tweets[2] and a scientific corpus of 20,000 abstracts from [15]. The size of these corpora, some would argue, is rather small with respect to Big Data. Yet, it is sufficient to illustrate our architecture's good time performance. Text analysis is indeed not usually done on large corpora. Moreover, other corpora used in the literature are smaller, e.g., 3,000 documents [20], 2,013 records [12], 65,333 tweets [2], 1,801,810 tweets [4].

Our architecture can be deployed in a cloud environment if all the requirements are met, i.e., if Python packages, PostgreSQL, and MongoDB are available. Tests are done on machines that reside in an OpenStack private cloud platform. We purposely selected this hardware architecture and dataset sizes to show that our architecture can achieve good performance even on end-user workstations, as it is sometimes not desirable to send data online due to privacy issues. Moreover, end-users presumably cannot afford very powerful, parallel computers.

4.1 News Articles Corpus Experiments

The first set of experiments are done using two computers with the same hardware configuration: 4 GB RAM and 1 CPU with two 2.2 GHz cores. We choose this hardware architecture to show that our method gives good results on simple computers. Using the initial news articles corpus, seven corpora are created consisting of 100 to 110,000 documents. They are referred to as Corpus $i, i \in \{1, 2, ..., 7\}$. For comparison reasons, experiments are done using a single-thread approach.

Figure 4a presents the average time (in seconds) for populating the databases. Duplicate documents are removed in this step. This is done by checking whether an article already exists in the database based on its title. If the document does not exist, then a new record is added. Otherwise, tags are verified so that metadata are not omitted, as the same article could have more tags for different instances found in the corpus. The second set of tests evaluates the efficiency of text cleaning and index construction (Fig. 4b).

Figure 4c shows the total storage space (in MB) for all corpora. To respect database normalization in PostgreSQL, bridge tables materializing many-to-many relations have to be added. In MongoDB, such relationships translate into vectors or nested documents inside collections. For example, the *documents* collection contains the authors table as an array of nested documents and the tags table as an array. This brings the issue of duplicates, as we may have the same tags for different document that would be stored in each element of the

---

[1] http://www.corpora.heliohost.org.
[2] Collected with Twitter's tools at https://dev.twitter.com.

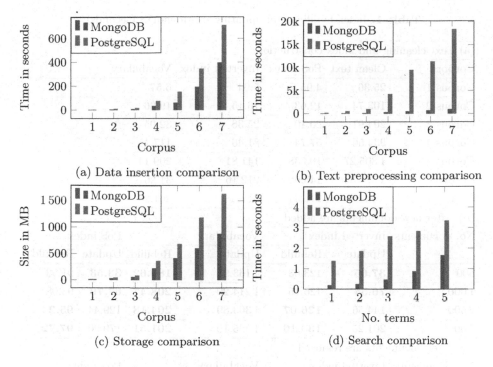

(a) Data insertion comparison

(b) Text preprocessing comparison

(c) Storage comparison

(d) Search comparison

Fig. 4. Performance comparison

collection. However, it is a small cost to pay as, using this structure, joins are removed, whereas join is the costliest operation in RDBMSs.

Experimental results show that MongoDB efficiently stores the data, minimizing storage space by 30 % with respect to PostgreSQL. Moreover, based on the number of records in each collection, from a computational point of view, a select operation performed on a smaller entity shows faster response times than one performed on an entity with a lot of records. For example, it is faster to query the *vocabulary* collection than to interrogate the *vocabulary* table, because the table contains more records than the collection.

Figure 4d presents the mean time for extracting the top-k documents. Tests are performed on Corpus 7 with $k = 20$. After each search, the database cache and buffers are cleared so that the comparison is accurate. MongoDB is from 86 % faster than PostgreSQL for one term-search to over 50 % faster for five terms.

Table 1a presents mean text cleaning and index construction times, as index construction is done separately in MongoDB. MapReduce functions were developed to further improve performance. Our results show that text cleaning and index creation is improved by 94 % with MongoDB (Fig. 4b). Moreover, index update is an important feature in a system where new documents are added or deleted. We use new corpora of 500 to 5,000 articles from Corpus 5 to test this feature in MongoDB. For comparison purposes, for each operation, we tested the

Table 1. Index building and updating in MongoDB (seconds)

(a) Text cleaning and index construction

Corpora	Clean text	PoS index	Inverted index	Vocabulary
Corpus 1	25.36	4.95	5.94	6.57
Corpus 2	103.74	12.63	16.15	19.26
Corpus 3	204.91	20.10	25.36	32.95
Corpus 4	934.56	57.74	81.46	117.15
Corpus 5	1 805.27	103.38	141.81	209.14
Corpus 6	2 524.33	148.69	212.16	312.61
Corpus 7	3 564.07	216.34	311.88	461.88

(b) After new documents are added

No. documents	Inverted Index		Vocabulary		PoS index	
	Update	Rebuild	Update	Rebuild	Update	Rebuild
500	**37.65**	123.63	1 163.95	**189.62**	**33.53**	91.52
1 000	**76.54**	126.16	1 214.20	**208.47**	**67.79**	93.08
2 500	144.56	**126.07**	1 303.89	**204.18**	129.44	**95.24**
5 000	201.26	**130.16**	1 395.10	**201.51**	179.90	**97.72**

(c) After documents are removed

No. documents	Inverted index		Vocabulary		PoS index	
	Update	Rebuild	Update	Rebuild	Update	Rebuild
500	**1.08**	122.36	1 129.35	**198.35**	91.33	91.91
1000	**1.38**	123.42	1 135.08	**199.20**	92.00	92.88
2500	**1.51**	126.64	1 160.26	**195.09**	93.15	94.52
5000	**1.60**	129.76	1 202.71	**203.01**	96.03	97.40

performance of updating and rebuilding the entire index. Updating the inverted index and the PoS index (Table 1b) works fast if the number of added documents is small, but time performance shifts for bigger corpora. Then, it is better to rebuild the entire index. If documents are deleted, it is faster to rebuild the inverted index (Table 1c). Little improvement is seen between updating and rebuilding the PoS index (Table 1c) when documents are deleted. Concerning vocabulary, it is faster to rebuild the entire index than to update it, because the IDF must be recomputed for each element in the collection (Tables 1b and c).

4.2 Twitter Corpus Experiments

This set of experiments is carried out using one machine with the following hardware configuration: 12 GB RAM and 3 CPU with 4 2.6 GHz cores. We choose this hardware configuration to prove that our architecture does not require spe-

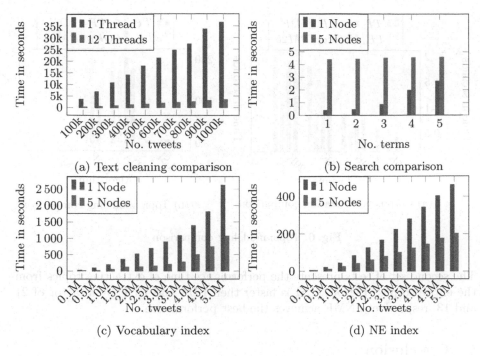

(a) Text cleaning comparison

(b) Search comparison

(c) Vocabulary index

(d) NE index

Fig. 5. Index construction comparison

cialized hardware to have good time performance. We work on 5,000,000 tweets in these experiments.

Figure 5a presents the results obtained when using a multithreading architecture. The improvement obtained from switching from a single thread to a 12-thread implementation is 90 %, lowering preprocessing time by a factor of 10. We can observe that the number of nodes used by MongoDB directly impacts performance and enhances response time, especially for large numbers of tweets. The construction time of the vocabulary index improves significantly, by over 59 % (Fig. 5c). The same happens with the named entities index, with an improvement over 40 % (Fig. 5d). Keyword search performance remains constant when we scale the database horizontally (Fig. 5b).

4.3 Scientific Articles Corpus Experiments

This set of experiments uses the scientific corpus and is carried out using the same hardware configuration as in Sect. 4.2. These experiments are designed to test the time performance for constructing the vectorization matrices and extracting topics. Figure 6a displays construction time for four different vectorization matrices, namely TF, TF_n, TF*IDF and Okapi BM25. The best performance is obtained by the TF_n vectorization matrix because all the information exists in the *vocabulary* index. TF*IDF and Okapi BM25 vectorizations are slower because they must be computed for each element during matrix construction.

492   C.-O. Truică et al.

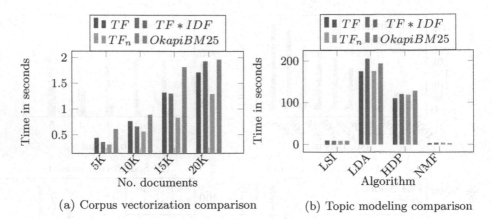

(a) Corpus vectorization comparison   (b) Topic modeling comparison

Fig. 6. Topic modeling comparison

The second set of tests presents the performance time of extracting topics from the entire corpus (Fig. 6b). LSI is faster then LDA and HDP by a factor of 21 and 13, respectively. NMF achieves the best performance.

5   Conclusion

In this paper, we present a new, complete architecture for text analysis that improves search performance, minimizes storage cost through efficient document-oriented storage, and scales up horizontally and vertically. Moreover, by exploiting MapReduce to parallelize index construction and by designing new structures for indexing and decreasing the number of records stored in the database, we minimize the number of CRUD operations and further enhance performance. Finally, the algorithm we propose for extracting top-k documents for a given search phrase also considerably improves query response time.

Our experimental results show that a document-oriented architecture is best-suited and improves performances when working with large volumes of text when adding documents into the database, cleaning text and constructing indexes. For all test cases, the mean time for populating the DODBMS is half that of the RDBMS. Cleaning texts and constructing inverted indexes is also faster when using a DODBMS. Although duplicates can be found inside a DODBMS, storage costs are significantly lower than with a RDBMS. A demo application that further shows the capabilities of this architecture is presented in [18].

In future work, we plan to add new features to our framework, such as automatic domain specific multiterm extraction, cross-language IR, word embedding and new topic models, e.g., dynamic topic modeling. From an architectural point of view, we also want to parallelize the algorithms and use a GPU for computations.

References

1. Arora, S., Ge, R., Halpern, Y., Mimno, D., Moitra, A., Sontag, D., Wu, Y., Zhu, M.: A practical algorithm for topic modeling with provable guarantees. In: International Conference on Machine Learning, pp. 939–947 (2013)
2. Ben Kraiem, M., Feki, J., Khrouf, K., Ravat, F., Teste, O.: OLAP of the tweets: from modeling toward exploitation. In: International Conference on Research Challenges in Information Science, pp. 1–10 (2014)
3. Blei, D.M., Ng, A.Y., Jordan, M.I.: Latent dirichlet allocation. J. Mach. Learn. Res. **3**, 993–1022 (2003)
4. Bringay, S., Béchet, N., Bouillot, F., Poncelet, P., Roche, M., Teisseire, M.: Towards an on-line analysis of tweets processing. In: Hameurlain, A., Liddle, S.W., Schewe, K.-D., Zhou, X. (eds.) DEXA 2011. LNCS, vol. 6861, pp. 154–161. Springer, Heidelberg (2011). doi:10.1007/978-3-642-23091-2_15
5. Cattell, R.: Scalable SQL and NoSQL data stores. ACM SIGMOD Rec. **39**(4), 12–27 (2011)
6. Deerwester, S., Dumais, S.T., Furnas, G.W., Landauer, T.K., Harshman, R.: Indexing by latent semantic analysis. J. Am. Soc. Inf. Sci. **41**(6), 391–407 (1990)
7. Ding, B., Zhao, B., Lin, C.X., Han, J., Zhai, C., Srivastava, A., Oza, N.C.: Efficient keyword-based search for top-k cells in text cube. Trans. Knowl. Data Eng. **23**(12), 1795–1810 (2011)
8. Han, J., Haihong, E., Le, G., Du, J.: Survey on NoSQL database. In: International Conference on Pervasive Computing and Applications, pp. 363–366 (2011)
9. Hecht, R., Jablonski, S.: NoSQL evaluation: a use case oriented survey. In: International Conference on Cloud and Service Computing, pp. 336–341 (2011)
10. Jivani, A.G.: A comparative study of stemming algorithms. Int. J. Comput. Technol. Appl. **2**, 1930–1938 (2011)
11. Kettunen, K., Kunttu, T., Järvelin, K.: To stem or lemmatize a highly inflectional language in a probabilistic IR environment? J. Documentation **61**(4), 476–496 (2005)
12. Lin, C.X., Ding, B., Han, J., Zhu, F., Zhao, B.: Text cube: computing IR measures for multidimensional text database analysis. In: International Conference on Data Mining, pp. 905–910 (2008)
13. Redmond, E., Wilson, J.R.: Seven Databases in Seven Weeks: A Guide to Modern Databases and the NoSQL Movement. The Pragmatic Bookshelf (2012)
14. Sharma, D.: Stemming algorithms: a comparative study and their analysis. Int. J. Appl. Inf. Syst. **4**, 7–12 (2012)
15. Tang, J., Wu, S., Sun, J., Su, H.: Cross-domain collaboration recommendation. In: ACM SIGKDD, pp. 1285–1293 (2012)
16. Teha, Y.W., Jordana, M.I., Beala, M.J., Bleia, D.M.: Hierarchical dirichlet processes. J. Am. Stat. Assoc. **101**(476), 1566–1581 (2012)
17. Truică, C.O., Boicea, A., Rădulescu, F., Bucur, I.: Performance evaluation for CRUD operations in asynchronously replicated document oriented database. In: International Conference on Control Systems and Computer Science, pp. 191–196 (2015)
18. Truică, C.O., Guille, A., Gauthier, M.: CATS: collection and analysis of tweets made simple. In: ACM Conference on Computer-Supported Cooperative Work and Social Computing, pp. 41–44 (2016)

19. Vishwakarma, S.K., Lakhtaria, K.I., Bhatnagar, D., Sharma, A.K.: An efficient approach for inverted index pruning based on document relevance. In: International Conference on Communication Systems and Network Technologies, pp. 487–490 (2014)
20. Zhang, D., Zhai, C.X., Han, J., Srivastava, A., Oza, N.: Topic cube: topic modeling for OLAP on multidimensional text databases. In: SIAM International Conference on Data Mining, pp. 1124–1135 (2009)

DAPPFC: Density-Based Affinity Propagation for Parameter Free Clustering

Hanning Yuan, Shuliang Wang$^{(\boxtimes)}$, Yang Yu, and Ming Zhong

School of Software, Beijing Institute of Technology,
100081 Beijing, China
slwang2011@bit.edu.cn

Abstract. In the clustering algorithms, it is a bottleneck to identify clusters with arbitrarily. In this paper, a new method DAPPFC (density-based affinity propagation for parameter free clustering) is proposed. Firstly, it obtains a group of normalized density from the unsupervised clustering results. Then, the density is used for density clustering for multiple times. Finally, the multiple-density clustering results undergo a two-stage synthesis to achieve the final clustering result. The experiment shows that the proposed method does not require the user's intervention, and it can also get an accurate clustering result in the presence of arbitrarily shaped clusters with a minimal additional computation cost.

Keywords: Density-based clustering · Affinity propagation · Parameter free · Clustering

1 Introduction

Clustering is a process of dividing observational dataset into a series of subsets, achieving the maximum intra-cluster similarities but the minimum inter-class similarities [1,2]. An effective clustering algorithm needs to address several challenges, including scalability to large data volume and dimensions, identifying clusters with arbitrarily shapes, and handling data with noises [3]. To meet these challenges, many clustering algorithms ask the users to manually adjust the parameters under their expert domain knowledge, e.g. DBSCAN (density-based spatial clustering of applications with noise) [4], the number of clusters or the density of clusters needs to be specified by the users, and thus making these algorithms ineffective against the practical scenarios, where it is impossible to choose the "best" set of parameters. The problem is exacerbated by factors such as human misinterpretation of data and errors.

AP (Affinity Propagation) is an unsupervised algorithm that identifies the clusters by imitating the message passing and feedback routine between the data objects [5]. Enjoying lower error rates than several traditional methods, AP is computationally efficient algorithm, making it suitable for many applications, such as artificial intelligence, biomedical science, image identification, image segmentation, social computing and facility location modeling [5–8]. However, AP is based on data centers, making it difficult to deal with arbitrary shaped clusters. There are two distinct drawbacks, the time-complexity and the inability to deal with non-convex clusters. Some researchers

© Springer International Publishing AG 2016
J. Li et al. (Eds.): ADMA 2016, LNAI 10086, pp. 495–506, 2016.
DOI: 10.1007/978-3-319-49586-6_34

attempted to improve AP's performance. For example, to address the low-efficiency in information spreading amongst the $N \times N$ dimension edges, Fujiwara, Irie, and Kitahara proposed to prune the unnecessary information exchange during the iteration process, and to calculate convergence information after clusters are created [9]. Under the scenarios where there ought to be a required number of clusters, AP may become slow.

To improve its efficiency under this setting, Wang and Zheng presented a fast algorithm on multi-grid that the calling times of AP was reduced by using multi-grid search, and the searching scope was reduced by improving the upper bound of preference parameter [10]. It could largely enhance the speed of AP under the fixed clusters numbers. To achieve the improvement over fixed damping factor which has negative impact on efficiency and convergence, Liu and Fu put forward a clustering algorithm F-AP that a constriction factor was given to speed up the convergence by dynamically adjusting the convergence coefficient along with the process of AP algorithm, and simultaneously ensured the same clustering results with original one from traditional AP [11]. Apart from the performance, the quality of clustering results is another fundamental issue. To overcome its inability of obtaining ideal clustering from datasets that are non-convex and uneven distributed, Feng and Yu firstly mapped datasets nonlinearly to a high dimensional space by using Kernel methods [12]. Secondly, according to the similarity magnanimity method shared by the nearest objects, a non-sensitive AP algorithm, DIS-AP was presented to improve the traditional AP in terms of addressing data which is non-convex and in high dimension.

In APSCAN (Affinity Propagation Spatial Clustering of Applications with Noise), the features were grouped by using double-density and non-parametric clustering based on AP algorithm, and then its density parameters of the clusters were normalized [13]. This dual density-clustering framework is applied to the multiple density clustering before the results are synthesized, and it is applicable to data with clusters of arbitrary shapes. However, APSCAN have several major drawbacks: Firstly, AP doesn't distinguish noise objects, it is no wise to directly utilize the capacity and radius of each AP result clusters as input parameters of density-based clustering. Secondly, APSCAN requires an intermediate process that the density of cluster center should be higher than the density of cluster edge, which is only suitable for clusters in specific density distribution. Finally, the clustering synthesize process of APSCAN may create a large number of tiny clusters, which cause the result hard to use.

In a word, AP had poor performance on non-convex datasets, whereas DBSCAN was sensitive to the input parameters. Trying to integrate AP and DBSCAN, APSCAN was only applicable to data with clusters in specific density distribution. To address the disadvantages, in this paper, AP and DBSCAN are integrated for a parameter free AP density-based clustering to further leverage the algorithm performance. Based on the unsupervised clustering results from AP, a group of monotonically increasing densities estimated parameters is firstly obtained. Secondly, several density-based clustering are performed using these parameters. Finally, the results of the clustering are combined together.

2 Principles

The whole process of the proposed algorithm is shown in Fig. 1. D is supposed to be a clustering dataset. In Fig. 1, three kinds of information on cluster structure are first obtained from the results of AP algorithm: (1) the number of objects, (2) the radius of cluster, and (3) the core region of cluster. Then, normalize the cluster information to get a number of minimum cluster objects and a set of non-monotonically decreasing radius sequence. Also take the normalized results as the density clustering parameters for multiple clustering. Finally, based on the synthesis rules, fuse the cluster core region from AP, and achieve the final result of density clustering.

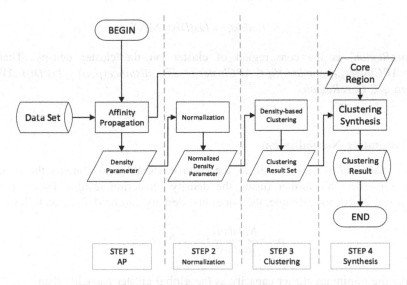

Fig. 1. The whole process of the proposed algorithm

2.1 Figure 1 the Whole Process of the Proposed Algorithm

Affinity Propagation (AP) is a distance-based algorithm for identifying exemplars in a dataset [14]. In AP, the measurements of the mutual similarity among N objects are recorded in an input matrix S that is $N \times N$ in dimension. The diagonal of S, i.e. $s(k, k)$, is treated as the reference P for the data object k to become the cluster center. The responsibility $r(i, k)$ that is sent from data object i to the candidate clustering center k, indicates how suitable an object k can be used as a cluster center for the object i. The availability $a(i, k)$ that is sent from the candidate cluster center k to the data object i, reflects how likely the object i chooses k as its cluster center. The larger the value of $r(i, k)$ and $a(i, k)$, the higher the probability that object k is to become the cluster center. Consequently, increase the chance that an object i belongs to a cluster with its center at object k. During this iterative process, AP keeps updating $r(i, k)$ and $a(i, k)$ between the data objects until the predefined convergence criteria is met.

The AP parameters adhere to what are used in its original settings, i.e. a maximum of iterations is 1000, the upper limit of steady times is 100, and the damping coefficient is 0.9. The reference of clustering center is chosen to be the median value of similarity matrix. Given a set of data objects, APClusters = {APCluster$_1$, APCluster$_2$, ..., APCluster$_k$}., there is three essential information associated with each *APCluster$_i$* $(i = 1, 2, \cdots, k)$, i.e. cluster capacity *Number$_i$*, cluster radius *Radius$_i$* and cluster core region *CoreRegion$_i$*.

A. *Number$_i$* is the cluster capacity on the cluster density. It is a scalar value that indicates the number of elements associated with a cluster *APCluster$_i$*.
B. *Radius$_i$* is the cluster radius on the cluster density. It documents the distance between every individual elements at cluster i and its respective cluster center. That is,

$$Radius_i = ListDist_i.MAX$$

C. *CoreRegion$_i$* is the core region of cluster on the cluster density. That is, $p : APCluster_i.examplar, \forall q \in APCluster_i$: If $distance(p, q) \leq ListDist_i.AVAG$, then $q \in CoreRegion_i$.

2.2   Parameter Normalization

Number$_i$ and *Radius$_i$*, are chosen as the control variables to conduct the multiple density clustering for further fusing the density clustering results. Taking a two-dimensional object for example, the horizontal density can be defined as follows:

$$\rho_i = \frac{Number_i}{Radius_i^2} i = 1, 2, \cdots, k \tag{1}$$

Take the minimum cluster capacity as the global cluster capacity, then

$$nNumber = \min(Number_i)i = 1, 2, \cdots, k \tag{2}$$

Combine Eq. (1) with Eq. (2), then

$$\rho_i = \frac{Number_i}{Radius_i^2} = \frac{nNumber}{\left(\sqrt{\frac{nNumber}{Number_i}} \times Radius_i\right)^2} \tag{3}$$

From Eq. (3), rearrange the terms, then

$$nRadius_i = \sqrt{\frac{nNumber}{Number_i}} \times Radius_i i = 1, 2, \cdots, k \tag{4}$$

From Eqs. (2) and (4), the parameters of cluster density information are normalized.

$$Nomalised\ Density\ List = \left\{ \begin{array}{l} (nNumber, nRadius_1), \\ (nNumber, nRadius_2), \cdots \\ , (nNumber, nRadius_k) \end{array} \right\} \tag{5}$$

Sort *Nomalised Density List* by *nRadius* in an ascending order, then

$$Nomalised\ Density\ Queue = \left\{ \begin{array}{l} (nNumber, nRadius_1), \\ (nNumber, nRadius_2), \cdots \\ , (nNumber, nRadius_m) \end{array} \right\} \tag{6}$$

Where, m is the number of clusters, an arbitrary object from $1 \leq j < k \leq m$ has the property, $\frac{nNumber_j}{nRadius_j^2} > \frac{nNumber_j}{nRadius_k^2}$. Thus, the *Nomalised Density Queue* is an ordered queue in a descending order by density.

2.3 Density Clustering

Nomalised Density Queue may be taken as the two parameters of DBSCAN, $MinPts = nNumber, \varepsilon = nRadius_i (i = 1, 2, \cdots, m)$. After the density clustering is implemented m times, the results are achieved:

$$Results = \{Result_1, Result_2, \cdots, Result_m\} \tag{7}$$

$$Result_i = \{Noises_i, Cluster_{i1}, Cluster_{i2}, \cdots, Cluster_{in}\} (1 \leq i \leq m) \tag{8}$$

2.4 Clustering Synthesis

The clustering synthesis has two procedures: (1) fusing the results of multiple density clustering, and (2) amalgamating the core region by connection.

2.4.1 Fusing the Results of Multiple Density Clustering

The method is performed as well as APSCAN algorithm. Here, DBSCAN [4] is used instead of DDBSCAN (Double-Density-Based SCAN) in APSCAN [13] Because the result from Eq. (7) is a group of clustering results in a descending queue by density, it is necessary to establish a set of rules to fuse the results of multiple density clustering in the context of Lemmas 1 and 2.

Lemma 1. AP normalized densities are in the *Nomalised Density Queue* from Eq. (2). They are clustered with DBSCAN, and the results are obtained as *Results* from Eqs. (7) and (8), if data object p is not marked as noise object in $Result_i$, then it will not be marked in $Result_j$ either.

Proof

- data object p is not marked as noise object in $Result_i$
- data object q in the input parameter of DBSCAN $MinPts = nNumber$, $\varepsilon = nRadius_i$ can be a core object and $distance(p, q) < nRadius_i$
- there exist at least $nNumber$ data objects in the $nRadius_i$ neighborhood of data object q
- $nRadius_j > nRadius_i$
- there exist at least $nNumber$ data objects in the $nRadius_j$ neighborhood of data object q
- data object q is core object in $Result_j$
- $distance(p, q) < nRadius_i < nRadius_j$
- data object p locates in the cluster consisted of core object q
- data object p is not marked as noise object in $Result_j$.

Lemma 2. AP normalized densities are in the Nomalised Density Queue from Eq. (2). They are clustered with DBSCAN, and the results are obtained as $Results$ Eqs. (7) and (8). If data object p and q are marked as the same cluster in $Result_i$, then it will also be marked the same cluster in $Result_j$.

Proof

- data object p and q are marked as the same cluster in $Result_i$
- in the input parameter $MinPts = nNumber$, $\varepsilon = nRadius_i$ of DBSCAN, data object p and q may be in 3 cases: p is directly density-reachable from q, p is density-reachable from q, and p is density-connected with q.

Case 1: p is directly density-reachable from q

- p is included in the ε neighborhood of q, and the integer part of the ε neighborhood is not less than $MinPts$
- p is directly density-reachable from q in the input parameter $MinPts = nNumber$, $\varepsilon = nRadius_i$ of DBSCAN.
- as for $nRadius_j > nRadius_i$, $nRadius_j$ neighborhood contains $nRadius_i$ neighborhood, of which the integer part is no less than $MinPts$.
- q is core object, and p is in the $nRadius_j$ neighborhood.
- data object p and q are marked as the same cluster in $Result_j$. Case 1 is symmetric if p, q are a pair of core data objects.

Case 2: p is density-reachable from q

- q is density-reachable from q in the input parameter $MinPts = nNumber$, $\varepsilon = nRadius_i$ of DBSCAN
- there is a chain of objects $p_1, p_2, \ldots, p_n$. In this chain, $p_1 = p$, $p_n = q$, and for $p_i \in D(1 \leq i \leq n)$, p_{i+1} is directly density-reachable from p_i about $nNumber$ and $nRadius_i$

- for $nRadius_j > nRadius_i$, $\forall p_i \in D (1 \leq i \leq n)$ in the object chain $p_2, ..., p_n$, $p_1 = p$, $p_n = q$ satisfies that p_{i+1} can be directly density-reachable from p_i about $nNumber$ and $nRadius_j$.
- data object p and q are marked as the same cluster in $Result_j$.

Case 3: p is density-connected with q

- p is density-connected with q about $nNumber$ and $nRadius_i$, $nRadius_i MinPts = nNumber$, $\varepsilon = nRadius_i$.
- there is core object $o \in D$, both p and q are density-reachable from o about $nNumber$ and $nRadius_i$.
- for $nRadius_j > nRadius_i$ both p and q are density reachable from o about $nNumber$ and $nRadius_j$.
- p is density-connected with q about $nNumber$ and $nRadius_j$
- data object p and q are marked as the same cluster in $Result_j$.

Based on the proved Lemmas 1 and 2, the rules to fuse the multiple density clusters are established. D is supposed to be a cluster dataset. Its AP normalized densities are in the *Nomalised Density Queue* from Eq. (2). They are clustered with DBSCAN, and the results are obtained as *Results* Eqs. (7) and (8). For arbitrary $1 \leq i < j \leq m$, it builds Relation Matrix [s][t]. s is the number of clusters in $Result_i$ and t is number of cluster in $Result_j$. RelationMatrix [u][v] is the object number of $Cluster_{jv}$ in $Cluster_{iu}$.

Rule 1: If object p is marked as noise object in $Result_i$ and $Result_j$, then it will also be marked as noise object in synthesized results.
Rule 2: If object p is marked as label $labelCluster_{iu}$ in $Result_i$ and label Clusterjv in $Result_j$, then it will be marked as $labelCluster_{iu}$ in synthesized results.
Rule 3: If all objects of $Cluster_{jv}$ are marked as noise objects in $Result_i$, then in synthesized results, all objects in $Cluster_{jv}$ are marked as a new cluster.
When rule 3 is executed, the RelationMatrix should be updated.
Rule 4: If object p of $Cluster_{jv}$ is marked as noise object in $Result_i$, but not all objects of $Cluster_{jv}$ are marked as noise objects in $Result_i$, then in synthesized results, object p will be marked as:

$$label = \arg MAX_{k=1,2,\cdots,s'} RelationMatrix[k][v]. \tag{9}$$

When Rule 4 is executed, the RelationMatrix should be updated. It executes from Rule 1 to Rule 4, in $Results = \{Result_1, Result_2, \cdots, Result_m\}$, $Result_{i+1}$ will be synthesized into $Result_i (1 \leq i \leq m - 1)$. Then it gets MergedClusters = {MNoises, MCluster1, MClusters2, ..., MClusterk}.

2.4.2 Core Region Connection Synthesis
The core region of D. from AP is CoreRegions = {CoreRegion1, CoreRegion2, ..., CoreRegionl}. The result of multiple density cluster is MergedClusters = {MNoises, MCluster1, MClusters2, ..., MClusterk}.

Rule 5: If $\exists p, q \in CoreRegion_u (1 \leq u \leq l)$, and p ∈ MClusteri, q ∈ MClusterj $(1 \leq i < j \leq k)$,then *MCluster$_i$* is supposed to synthesize with *MCluster$_j$*.

According to Rule 5, MergedClusters = {MNoises, MCluster$_1$, MClusters$_2$, ..., MClusterk} is synthesized into FinalClusters = {FNoises, FCluster$_1$, FCluster$_2$, ..., FCluster$_f$}.

3   Experiment Analysis

The experiments are performed on the environments that the hardware platform is CPU P8700, 8 G RAM, the software platform is Windows 8 Professional ×64, Framework 4.5, and the target program is written with C# 5.0. The test data are from University of Eastern Finland Clustering Datasets Shape Sets, i.e. Aggregation [15], Compound [16], R15 [17] and Spiral [18]. Taking optimal parameters of visualization from the original DBSCAN, the results of AP algorithm and APSCAN algorithm as reference, the proposed DAPPFC are experimented in the contrast on the test data.

3.1   The Clustering Quality

The clustering results on the four test data are shown in Figs. 2, 3, 4, 5, and are statistically concluded in Table 1.

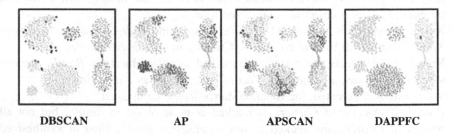

DBSCAN   AP   APSCAN   DAPPFC

Fig. 2. Clustering results on the test data of Aggregation

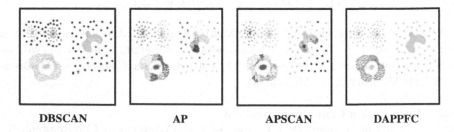

DBSCAN   AP   APSCAN   DAPPFC

Fig. 3. Clustering results on the test data of Compound

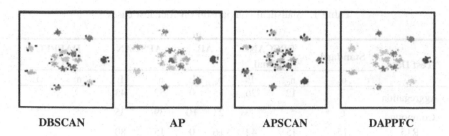

DBSCAN   AP   APSCAN   DAPPFC

Fig. 4. Clustering results on the test data of R15

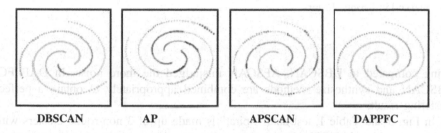

DBSCAN   AP   APSCAN   DAPPFC

Fig. 5. Clustering results on the test data of Spiral

As shown from Figs. 2, 3, 4, 5 and Table 1, the proposed DAPPFC has been close to or achieved even better clustering quality than parameter optimal DBSCAN, AP, and APSCAN.

In Fig. 2 and Table 1, test data "Aggregation" is made up of a few clusters in uniform density distribution. In DBSCAN test, after adjusting the parameters manually, generally satisfactory result can be achieved. In APSCAN test, due to use of DDBSCAN, which emphasizes the central density must be higher than the marginal density, dumbbell-shaped clusters on the right have been connected by mistake. In DAPPFC test, AP is used to estimate clusters' densities reasonably, and DBSCAN is used to avoid misunderstanding the data model. The clustering results are finally satisfactory.

In Fig. 3 and Table 1, test data "Compound" is made up of a few clusters with different densities in non-uniform density distribution. In DBSCAN test, the deficiency of single density parameter results in the mistakes that the objects in low distribution density are wronged noises. In APSCAN test, DDBSCAN is suited for two clusters in the upper left. However, because the densities are very uncommon among the test sample, too many tiny clusters are gotten finally, and the clusters are further synthesized erroneously. In DAPPFC test, the fragmentation of the results is avoided by its synthesized methods on multiple densities.

In Fig. 4 and Table 1, test data "R15" is made up of 15 round clusters, which may help the center-based cluster algorithm get excellent clustering results. In DBSCAN test, after adjusting the parameters manually, all clusters can be detected, and only misjudged a few noises. In APSCAN, DDBCAN algorithm did not show its advantage.

Table 1. Statistical comparison on four test cases

Test Data	Standard	DBSCAN optimal		AP		APSCAN		DAPPFC	
	n_c	n_c	E	n_c	E	n_c	E	n_c	E
Aggregation	7	12	26	35	0	52	43	9	0
Compound	6	5	39	23	10	39	35	5	72
R15	15	15	42	16	0	15	80	15	0
Spiral	3	3	0	34	0	37	0	5	0

Note: n_C is the number of clusters, and E means the number of objects that are marked in wrong clusters.

Being compared to DBSCAN, APSCAN misjudged the more noise. In DAPPFC, DBSCAN and synthesize methods are combined appropriately to obtain a perfect clustering result.

In Fig. 5 and Table 1, test data "Spiral" is made up of 3 non-round clusters with different densities. In DBSCAN test, due to distances between clusters, after manually adjusting the density parameter, correct clustering result can be got. In DBSCAN test, due to the significant differences in density in different areas among one cluster, too many insignificant clusters are finally gotten. In DAPPFC test, benefit from synthesizing methods, fragmentation is void.

3.2 The Extra Time

Furthermore, the extra time are compared between DAPPFC and APSCAN because these they both take use of AP results. The statistical comparison on the clustering elapsed time is concluded is in Table 2. Seen from Table 2, based on AP time, the extra time of DAPPFC are all lower than APSCAN on the four test data. That is, the saved additional rate is 0.01% (0.55%–0.54%) on "Aggregation", 0.06% (1.14%–1.08%) on

Table 2. Statistical comparison on the clustering elapsed time

Test data	AP	APSCAN		DAPPFC	
	Run Time	Extra Time	Addition Rate	Extra Time	Addition Rate
Aggregation	2196.55	+12.18	0.55%	+11.95	0.54%
Compound	183.08	+2.10	1.14%	+1.98	1.08%
R15	742.91	+3.28	0.44%	+3.04	0.41%
Spiral	92.84	+2.10	2.26%	+1.75	1.89%

"Compound", 0.03% (0.44%–0.41%) on "R15", and 0.37% (2.26%–1.89%) on "Spiral". Because APSCAN requires an intermediate process that the density of cluster center should be higher than the density of cluster edge, which is only suitable for clusters in specific density distribution. And the clustering synthesize process of APSCAN may create a large number of tiny clusters. Thus, DAPPFC has better performance than APSCAN, and only cost a little more extra time than AP.

4 Conclusions

In this paper, DAPPFC was proposed to improve the exiting APSCAN algorithm. The proposed DAPPFC may be is generic for multiple-density clustering instead of the division of the density core and density border. Without the user's intervention, it can also demonstrate an accurate clustering result in the presence of arbitrarily shaped clusters with a minimal additional computation cost. In the experiments, DAPPFC has been close to or achieved even better clustering quality than parameter optimal DBSCAN, AP, and APSCAN. And it also has better performance than APSCAN.

As with any other algorithms, there are also some limitations of the proposed algorithm. In terms of quality, the cluster results output by the proposed algorithm are still unable to identify a reliable hierarchy for data division. In the future, using some intrinsic properties of data, it is expected to achieve clustering hierarchy automatically. In terms of performance, the proposed algorithm is still limited by its efficiency when compared with AP algorithm.

Acknowledgements. This work was supported by National Natural Science Fund of China (61472039), National Key Research and Development Plan of China (2016YFC0803000, 2016YFB0502604), and Specialized Research Fund for the Doctoral Program of Higher Education (20121101110036).

References

1. Rodriguez, A., Laio, A.: Clustering by fast search and find of density peaks. Science **344** (6191), 1492–1496 (2014)
2. Seife, C.: Big data: the revolution is digitized. Nature **518**(7540), 480–481 (2015)
3. Aggarwal, C.C., Reddy, C.K. (eds.): Data Clustering: Algorithms and Applications. CRC Press, Boca Raton (2013)
4. Ester, M., Kriegel, H.P., Sander, J., Xu, X.: A density-based algorithm for discovering clusters in large spatial databases with noise. In: KDD, vol. 96, no. 34, pp. 226–231 (1996)
5. Dueck, D., Frey, B.J.: Non-metric affinity propagation for unsupervised image categorization. In: IEEE 11th International Conference on Computer Vision, pp. 1–8 (2007)
6. Dueck, D., Frey, Brendan, J., Jojic, N., Jojic, V., Giaever, G., Emili, A., Musso, G., Hegele, R.: Constructing treatment portfolios using affinity propagation. In: Vingron, M., Wong, L. (eds.) RECOMB 2008. LNCS, vol. 4955, pp. 360–371. Springer, Heidelberg (2008). doi:10. 1007/978-3-540-78839-3_31
7. Givoni, I.E., Frey, B.J.: Semi-supervised affinity propagation with instance-level constraints. In: AISTATS, pp. 161–168 (2009)

8. Xu, X.Z., Ding, S.F., Shi, Z.Z., Zhu, H.: Optimizing radial basis function neural network based on rough sets and affinity propagation clustering algorithm. J. Zhejiang University Science C **13**(2), 131–138 (2012)
9. Fujiwara, Y., Irie, G., Kitahara, T.: Fast algorithm for affinity propagation. In: IJCAI Proceedings-International Joint Conference on Artificial Intelligence, vol. 22, no. 3, p. 2238 (2011)
10. Wang, K., Zheng, J.: Fast algorithm of affinity propagation clustering under given number of clusters. Comput. Syst. Appl. **7**, 207–209 (2010)
11. Liu, X.Y., Fu, H.: A fast affinity propagation clustering algorithm. Shandong Daxue Xuebao (GongxueBan) **41**(4), 20–23 (2011)
12. Feng, X.L., Yu, H.T.: Research on density-insensitive affinity propagation clustering algorithm. Jisuanji Gongcheng/Computer Engineering, **38**(2) (2012)
13. Chen, X., Liu, W., Qiu, H., Lai, J.: APSCAN: a parameter free algorithm for clustering. Pattern Recogn. Lett. **32**, 973–986 (2011)
14. Frey, B.J., Dueck, D.: Clustering by passing message between data points. Science **315**, 972–976 (2007)
15. Gionis, A., Mannila, H., Tsaparas, P.: Clustering aggregation. ACM Trans. Knowl. Discov. Data **1**(1), 1–30 (2007)
16. Zahn, C.T.: Graph-theoretical methods for detecting and describing gestalt clusters. IEEE Trans. Comput. **100**(1), 68–86 (1971)
17. Veenman, C.J., Reinders, M.J.T., Backer, E.: A maximum variance cluster algorithm. IEEE Trans. Pattern Anal. Mach. Intell. **24**(9), 1273–1280 (2002)
18. Chang, H., Yeung, D.Y.: Robust path-based spectral clustering. Pattern Recogn. **41**(1), 191–203 (2008)

Effective Traffic Flow Forecasting Using Taxi and Weather Data

Xiujuan Xu[1,2], Benzhe Su[1,2], Xiaowei Zhao[1,2], Zhenzhen Xu[1,2(✉)],
and Quan Z. Sheng[3]

[1] School of Software, Dalian University of Technology, Dalian 116620, China
benzhe.su.123@foxmail.com, {xjxu,xiaowei.zhao,xzz}@dlut.edu.cn
[2] Key Laboratory for Ubiquitous Network and Service Software of Liaoning Province,
Dalian 116620, China
[3] School of Computer Science, The University of Adelaide,
Adelaide, SA 5005, Australia
michael.sheng@adelaide.edu.au

Abstract. Short-term traffic flow forecasting is an important component of intelligent transportation systems. The forecasting results can be used to support intelligent transportation systems to plan operation and manage revenue. In this paper, we aim to predict the daily floating population by presenting a novel model using taxi trajectory data and weather information. We study the problem of floating traffic flow prediction with weather-affected New York City, and a new methodology called *WTFPredict* is proposed to solve this problem. In particular, we target the busiest part of the city (i.e., the airports), and identify its boundary to compute the traffic flow around the area. The experimental results based on large scale, real-life taxi and weather data (12 million records) indicate that the proposed method performs well in forecasting the short-term traffic flows. Our study will provide some valuable insights to transport management, urban planning, and location-based services (LBS).

Keywords: Weather data · Prediction model · Big data · Intelligent transportation systems

1 Introduction

In a transportation network, the purpose of Intelligent Transportation Systems (ITS) is to provide dynamic traffic control and management real-time by forecasting traffic flows in the short-term future [6]. With the widespread use of GPS-equipped taxis, a huge amount of data about taxis' trajectories with location information is being generated by an increasing speed. Traffic flow prognostication [4] as an interesting problem draws an intensive attention from researchers for decades.

Traffic flow forecasting (*TFF*) has become important as the number of vehicles in big cities and freeway is continually increasing [2]. TFF could improve

© Springer International Publishing AG 2016
J. Li et al. (Eds.): ADMA 2016, LNAI 10086, pp. 507–519, 2016.
DOI: 10.1007/978-3-319-49586-6_35

the traffic and help people to do wiser decision, especially in big cities. Traffic flow is affected by many factors, for example, human behavior, vehicle trajectory, and transport lights and so on. In our work, we are interested to predict weather-affected traffic flow, and select the New York City as an example. As one of the biggest cities in the world, New York City has been the active target for scientists on transportation study (e.g., visualization of its urban taxi data [5,14]). Meanwhile, New York City Taxi data set was designed to build a wide range of spatio-temporal queries [5]. However, to our best knowledge, there are no research about the weather-affected traffic flow prediction. Generally speaking, New York City has a humid continental climate. It also experiences warm summers with long and cold winters. In the majority of winter seasons, a temperature of –25 °C or lower can be expected in the northern highlands (Northern Plateau) and 5 °C or colder in the southwestern and east-central highlands of the Southern Tier. For example, in January 2016, the winter storm resulted in a travel ban across the city and on Long Island, the shutdown of MTA buses, and the closure of above-ground subway lines through New York.

Undoubtedly, traffic could be affected significantly by the weather conditions. Traffic volume strongly depends on current and post volume of that location and its neighbours. However, to conduct an effective traffic flow prediction study, some important questions need to be answered, namely (i) which are the busiest area (e.g., an airport) of the New York City? (ii) how does the weather affect the traffic flow and the population of reaching the area?, and (iii) how should we design a model to predict the floating traffic flow?

There exist many challenges in the study. The first challenge lies on how we should identify the size of the area. The second challenge is how to deal with data sparseness and coverage. Finally, it is critical to develop consider information loss (e.g., taxi id) due to privacy issues. In this paper, we present our work for effective traffic flow forecasting, using New York City as an example. In a nutshell, the major contributions of our work are listed as the following:

- We study the problem of weather-affected traffic flow prediction in a traffic networks and we introduce a formal definition of short-term traffic flow prediction problem affected by weather.
- We develop an algorithm to extract bondaries of interested areas and also propose a novel approach called *WTFPredict* for traffic flow prediction.
- The results on the real taxis data demonstrate that by considering the weather and traffic flow in the traffic network collectively, the traffic flow prediction accuracy can be significantly improved.

The rest of this paper is organized as follows. Section 2 gives a brief overview on the related literature. Section 3 proposes some preliminaries about basic definitions and problem definitions. In Sect. 4, we first introduce the framework of our work and the algorithms, and then present detailed information about the data source used in our study. Section 5 reports the experimental results. Finally, we draw our conclusions in Sect. 6.

2   Related Work

In this section, we review the related efforts from two aspects: *short-term pre-diction techniques* and *application domains*.

2.1   Prediction Techniques

In recent years, traffic flow prediction (TFP) has received significant attention and many efficient algorithms have been proposed. The typical medium-long-term traffic flow forecasting approach includes four-stage prediction method (traffic generation, traffic distribution, traffic mode, and traffic assignment). However, this four-stage planning tool does not cover the dynamic properties of flow precisely [11]. These different prototypes can be generally divided into two categories: *model-based methods*, and *data-driven methods*.

Many model-based methods have been used for TFP, including support vec-tor machine (SVM) [4], Kalman Filter [13], neural network and some improved neural network. For example, Ding et al. proposed a new traffic flow time series prognostication by SVM in 2002 [4]. A divide-and-conquer method based on neural network and origin-destination (OD) matrix estimation was developed to forecast the short-term passenger flow in high-speed railway system [15]. In [17], the authors propose a Bayesian combined neural network approach to pre-dict short-term freeway traffic flow. Other researchers studied short-term traffic forecasting by tensor techniques. Based on dynamic tensor completion (DTC), a novel short-term traffic flow prediction approach was designed to use the mul-timode information to forecast traffic flow with a low-rank constraint [12]. In their model, the traffic data are represented as a dynamic tensor pattern so as to capture more information of traffic flow, namely, temporal variabilities, spatial characteristics, and multimode periodicity. Some researchers also have proposed a data-driven method for local traffic state estimation and prediction. A repre-sentative effort in this direction is the work in [1], which exploited available traffic and other information and developed data-driven computational approaches.

2.2   Applications

Traffic flow prediction has been applied in many application areas with a rich literature. The interesting applications can be grouped in the following areas: traffic flow prediction on high-speed railway, public bicycle traffic flow prediction, and population prediction on the beaches. Most of work about traffic flow pre-diction has been applied to the high-speed railway. For example, the researchers used a hybrid model combining clustering with support vector machine (SVM) to predict the uncertainty of traffic flow [16]. The authors in [9] designed a data fusion algorithm to fuse floating car data with stationary detector data on live traffic in order to eliminate individual source bias and alleviate source-specific limitations. However, there are a little work about traffic flow prediction by con-sidering the effect of weather conditions. In a recent effort, the authors in [10] designed a prediction model to predict the daily floating population based on weather factors, through a multiple liner regression analysis approach.

3   Problem Definition

This section defines the problem of traffic flow prediction with the effect of weather conditions. First, we present some basic definitions in Sect. 3.1. Then, we propose the formal problem definition in Sect. 3.2.

3.1   Basic Definitions

Assume a finite set of distinct taxis in a city, i.e., $Taxi = \{Taxi_1, \ldots, Taxi_n\}$. There are m passagers, i.e., $Passager = \{Passager_1, \ldots, Passager_m\}$. A recording database $D = Trip_1, Trip_2, \ldots, Trip_l$ is a set of trips, where each trip $Trip_r \in D$ is a subset of $Trip^*$ and has an unique identifier r, called $Trip_i d$. Every record of *GPS* data is denoted by a *GPS* data point. First, we present the definition of a GPS data point.

Definition 1. *GPS data point.* A piece of GPS data point is about a point in a map including a longitude i and a latitude j.

$$gps_{i,j} = \{longitude_i, latitude_j\} \tag{1}$$

A map consists of all GPS data point from taxis. For example, Fig. 1 shows a map of New York City including all GPS data from taxis in January, 2015 where a taxi is represented as a tiny dot in the figure. The formal definition of a map is shown in Definition 2.

Fig. 1. GPS data points of taxis of New York City in January, 2015

Definition 2. *Map.* A map of a city is composed by all GPS data points with k rows and l columns from $GPS_{1,1}$ to GPS_{kl}.

$$Map = \begin{pmatrix} GPS_{1,1} \cdots GPS_{1,l} \\ \cdots \quad \cdots \quad \cdots \\ GPS_{k,1} \cdots GPS_{k,l} \end{pmatrix} \tag{2}$$

Based on the definitions of GPS data point and the map, we present the definition of map segment.

Definition 3. *Map Segment.* A map segment MS is a boundary sub-map with a list of intermediate GPS points describing the size of the segment using a small block.

A map is divided into $u \times v$ map segments.

$$Map = \begin{pmatrix} MS_{1,1} \cdots MS_{1,v} \\ \cdots \quad \cdots \quad \cdots \\ MS_{u,1} \cdots MS_{u,v} \end{pmatrix} \tag{3}$$

Definition 4. *Trip.* A Trip record $Trip(P_o, taxi_t, time_p - time_q, GPS_{a,b} - GPS_{c,d})$ is about a passager P_o from a point $GPS_{a,b}$ of picking up by a taxi $taxi_t$ at time $time_p$ to another point $GPS_{c,d}$ of dropping off at time $time_q$.

Definition 5. *Traffic State.* A traffic state TS is about the number matrix of picking up in every map segment from time $time_p$ to time $time_q$.

Definition 6. *Traffic Flow.* A traffic flow TF is a sequence of traffic states $TS_1, TS_2, \ldots, TS_T$, i.e., $TF = \{TS_1, TS_2, \ldots, TS_T\}$, where TS_T is the latest matrix in the stream.

As Fig. 2 shows, the traffic flow is considered to be incrementally growing over time. TF_x is the latest matrix in the stream.

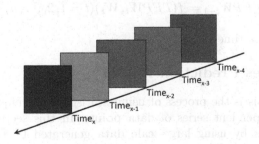

Fig. 2. A 4-order stream: the multimode data are growing incrementally over time

Definition 7. *Weather data.* A Weather data $Weather_u$ is a vector about the average of the weather in an hour, including temperature, humidity, visibility, wind-speed, rain, and so on at time $time_u$, i.e.,

$$Weather_u = \{temperature, humidity, visibility, wind - speed, rainfall\}$$

Similarly, we present the definition of the weather flow.

Definition 8. *Weather Flow.* A weather flow W_u is a sequence of weather data, i.e., $W = \{Weather_1, Weather_2, \ldots, Weather_u\}$.

3.2   Problem Definitions

In general, the floating traffic flow problem of this paper is a time predict problem affected by weather. The definition of traffic flow prediction (*TFP*) problem is introduced in Definition 9. The formal definition of weather-affected traffic flow prediction (*WTFP*) problem is given in Definition 10.

Definition 9. *Traffic flow prediction (TFP).* For any traffic flow dynamic window, the traffic data in the prediction horizon can be predicted on newly available data, so the prediction problem can be expressed as:

$$TF_{t+1} = f(TF_t) \qquad (t = 1, 2, \ldots, n) \tag{4}$$

where TF_x represents the traffic volume during a time section $time_x$.

Meanwhile, the floating traffic flow problem with the effect of weather is a predict problem, which is to predict the floating traffic flow depending on the weather at a certain time in a certain place.

Given large-scale GPS data and weather data of a city, the goal of this paper is to find the relationship function f between current/pass traffic data and the future traffic volume with the effect of the weather. In the following, the definition of the TFPW is introduced and the related notations is given.

Definition 10. *Weather-effected Traffic flow prediction (WTFP).* For any traffic flow dynamic window, the traffic data in the prediction horizon can be predicted on newly available data, so the prediction problem can be expressed as:

$$TFPW_{t+1} = f(TFPW_t, W_t)(t = 1, 2, \ldots, n) \tag{5}$$

where t represents a time.

4   Time Series Prediction Model

Time series analysis is the process of using statistical techniques to model and explain a time-dependent series of data points. In this section, we solve the following problems by using large-scale data generated by taxi GPS devices around the airport area. First we introduce our data source, including map data, taxi data and weather data. Then, we identify the boundary of an airport and compute the traffic flow of an airport. Finally, we design a novel algorithm to predict the short-term traffic flow in an airport.

4.1   Prediction Framework

Figure 3 depicts overall framework of the proposed prediction model including the analytical components. Algorithm 1 shows the pseudocode of our *WTFPrediction* model. In the following, we discuss different components of our model in detail.

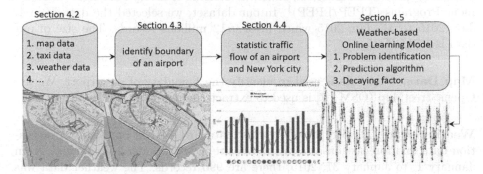

Fig. 3. WTFPrediction algorithm framework

The *WTFPredict* algorithm is detailed in Algorithm 1.

Algorithm 1. *WTFPredict* algorithm

Input: a data source *DB*, includes map data, weather data, traffic flow
Output: predicted traffic flow

---

1. Data preprocessing. All data are rescaled to a specific range for a time series forecasting problem.
2. The initial input data are the weather data and traffic flow.
3. Train a model according to Definition 10 to predict the results by a prediction algorithm.
4. Test the model.
5. Get the evaluation results.

4.2   Data Source

In this section, we present our data source about weather and traffic flow.

Taxi Data Source. Our dataset includes trip records from all trips completed in yellow and green taxis in NYC in 2014 and some selected months of 2015 [3]. Records include fields capturing pick-up and drop-off dates/times, pick-up and drop-off locations, trip distances, itemized fares, rate types, payment types, and driver-reported passenger counts. The data used in the attached datasets were collected and provided to the NYC Taxi and Limousine Commission (TLC) by technology providers authorized under the Taxicab & Livery Passenger Enhancement Programs (TPEP/LPEP)[1]. In our dataset, we selected the data from 1, Jan, 2015 to 31, Jan, 2015 which involve 12 millions records. The size of our data is 1.84 GB.

Map Data. Map data is used to identify the boundary of an airport. We use OpensStreetMap (OSM) [7] is used to extract the boundary of the airport.

Weather Data. Two types of weather data are used in our study: observation data and forecast data from the website (www.wunderground.com). From January 1, to January 31, 2015, there are 990 records. The weather data was observation data which recorded the basic information about the weather of a day: the highest temperature, the lowest temperature, average temperature, average humidity and so on.

4.3   Algorithm for Extracting Boundary

For detecting the change of traffic flow with weather-effect, we should identify the boundary of the area in order to compute the traffic flow in the area (e.g., the airport). In this section, we introduce a algorithm to extract the boundary, as shown in Algorithm 2. Figure 4 shows the boundary of the *JFK* airport.

Algorithm 2. Algorithm of Extracting Boundary

Input: a data source *DB* of map.ost.
Output: the boundary of the airport

---

1. Find root node from the file osm which includes the boundary of the airport;
2. Find a set of all nodes where (Name = the airport);
3. For each item in node
   (a) Create a new node;
   (b) Add an attribute of longitude;
   (c) Add an attribute of latitude;
   (d) Add the node into the file;
4. Save and output the new file.

---

[1] The trip data was not created by the TLC, and TLC makes no representations as to the accuracy of these data.

4.4   Number of Floating Taxis in Three Airports

After we identify the boundary of an airport, we compute the traffic flow by the real taxi data generated by yellow taxis in New York, USA.

Figure 5 shows the number of picking up and dropping off in the *JFK* airport in Jan, 2015. The tendency of the number of picking up marked by the purple line in the JFK airport is a little similar with the tendency of the temperature in the month in Fig. 5.

4.5   Prediction Algorithms

In this section, we select three different prediction algorithms to predict traffic flow, including linear regression (LR), multi-layer perception (MP), and SMOreg.

Prediction by Line Regression. Linear regression is a statistical technique for modeling the relationship between variables. We apply multiply linear regression to predict traffic flow.

Fig. 4. Boundary of JFK airports in New York City

Fig. 5. Distribution of picking up and dropping off in *JFK* airport by day (Jan-2015)

Prediction by Multi-layered Perception. Current weather status is usually affected by previous time. In fact, the weather change could be regarded even as a Markov chain. Therefore, we suppose that the weather change is correlated. Traffic flow is correlated, too. This is the theoretical foundation to adopt artificial neural network (abbreviated as *ANN*) to predict the traffic flow. ANN is a powerful tool to capture and represent complex input/output relationships. Backpropagation (*BP*) feed forward network is a common method of training *ANN* used in conjunction with an optimization method such as gradient descent. Therefore, we decide to use a multi-layered feedforward neural network and adopt the BP learning algorithm to predict the traffic flow.

Prediction by SMOreg Algorithm. The sequential minimal optimization algorithm for training a support vector classifier using RBF kernels (SMOreg), is used in this paper.

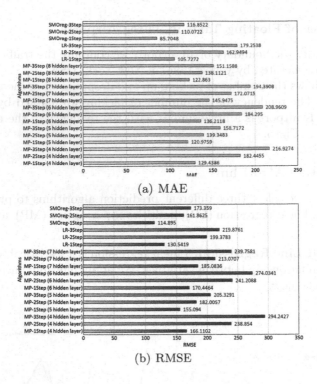

(a) MAE

(b) RMSE

Fig. 6. Comparison results of different algorithms

5   Experiment Study

In this section, we show that our proposed method can improve the weather-affected traffic flow accuracy compared with the baseline that only uses traffic flow information. We use weka [8] to test performance of the proposed prediction algorithm.

5.1   Prediction Performance for the *JFK* Airport

In this section, we consider the effect of the different algorithms in prediction of traffic flow. To compare the error of prediction, we select 70 % of the data to train a model and the rest 30 % to test the model. We use mean absolute error (*MAE*) and root mean squared error (*RMSE*) to measure the performance of our algorithm, which are calculated using:

$$MAE = sum(abs(predicted - actual))/N \tag{6}$$

$$RMSE = sqrt(sum((predicted - actual)^2)/N) \tag{7}$$

The test results of algorithms can be seen from Fig. 6(a) and (b). Smaller values of *MAE* and *RMSE* indicate better performance of the algorithms in prediction. From the figures, we can see that the better performance is obtained by the *SMOreg* algorithm in traffic flow prediction. It should be interesting to note that relatively good performance is also obtained by the *LR* algorithm.

5.2   Prediction Results for the *JFK* Airport

The prediction results of *LR* algorithm is shown in Fig. 7 and Fig. 8 shows the results of *SMOreg* algorithm. Compared Fig. 8 with Fig. 7, we can observe that the error is smaller in Fig. 8. Figure 9 presents the results of MP algorithm with 6 hidden layers. We can see that the SMOreg algorithm for regression performs the best in all algorithms. It should be particular mention that there was no taxi data on 27 January because of the weather condition. All algorithms do not consider that situation which leads to some error on that day.

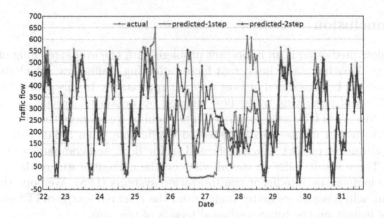

Fig. 7. Predict traffic flow (LR algorithm) for getting off in *JKF* airport by hour

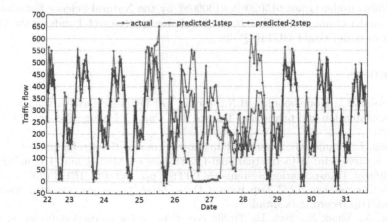

Fig. 8. Predict traffic flow (SMOreg algorithm) for getting off in *JKF* airport by hour

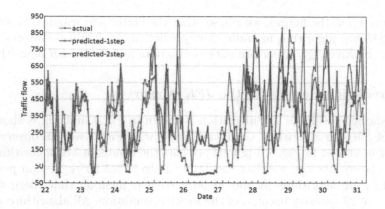

Fig. 9. Predict traffic flow (MP algorithm) for getting off in *JKF* airport by hour

6   Conclusion

Traffic flow prediction is an important problem with the ever stretching of cities nowadays. Weather is one important factor affecting traffic flow, which unfortunately has not be considered in most existing approaches. In this paper, we study the problem of predicting traffic flow affected by weather conditions, using airports of New York City as an example. In particular, we propose the *WTFPredict* model which first defines weather-affected traffic flow, then builds three different algorithms to predict weather-effected traffic flow, and compares those algorithms. These results offer an overall view on the current status of the research and development on intelligent transportation systems (ITS). We hope that our work can help researchers better understand the research status of ITS and gain valuable insight on the future technical trends of the area.

Acknowledgment. This work was supported in part by the Natural Science Foundation of China under Grant 61502069, 61300087 by the Natural Science Foundation of Liaoning under Grant 2015020003, by the Fundamental Research Funds for the Central Universities under Grant DUT15QY40.

References

1. Antoniou, C., Koutsopoulos, H.N., Yannis, G.: Dynamic data-driven local traffic state estimation and prediction. Transp. Res. Part C: Emerg. Technol. **34**, 89–107 (2013)
2. Barros, J., Araujo, M., Rossetti, R.J.: Short-term real-time traffic prediction methods: a survey. In: 2015 International Conference on Models and Technologies for Intelligent Transportation Systems (MT-ITS), pp. 132–139. IEEE (2015)
3. NTL Commission: TLC Trip Record Data. http://www.nyc.gov/html/tlc/html/about/trip_record_data.shtml
4. Ding, A., Zhao, X., Jiao, L.: Traffic flow time series prediction based on statistics learning theory. In: Proceedings of the IEEE 5th International Conference on Intelligent Transportation Systems, pp. 727–730. IEEE (2002)

5. Ferreira, N., Poco, J., Vo, H.T., Freire, J., Silva, C.T.: Visual exploration of big spatio-temporal urban data: a study of new york city taxi trips. IEEE Trans. Vis. Comput. Graph. **19**(12), 2149–2158 (2013)
6. Ghosh, B., Basu, B., O'Mahony, M.: Bayesian time-series model for short-term traffic flow forecasting. J. Transp. Eng. **133**(3), 180–189 (2007)
7. Haklay, M., Weber, P.: Openstreetmap: user-generated street maps. IEEE Pervasive Comput. **7**(4), 12–18 (2008)
8. Holmes, G., Donkin, A., Witten, I.H.: Weka: a machine learning workbench. In: Proceedings of the 1994 Second Australian and New Zealand Conference on Intelligent Information Systems, pp. 357–361. IEEE (1994)
9. Houbraken, M., Audenaert, P., Colle, D., Pickavet, M., Scheerlinck, K., Yperman, I., Logghe, S.: Real-time traffic monitoring by fusing floating car data with stationary detector data. In: 2015 International Conference on Models and Technologies for Intelligent Transportation Systems (MT-ITS), pp. 127–131. IEEE (2015)
10. Lee, K., Hong, B., Lee, J., Jang, Y.: A floating population prediction model in travel spots using weather big data. In: 2015 IEEE Fifth International Conference on Big Data and Cloud Computing (BDCloud), pp. 118–124. IEEE (2015)
11. Saw, K., Katti, B., Joshi, G.: Literature review of traffic assignment: static and dynamic. Int. J. Transp. Eng. **2**(4), 339–347 (2014)
12. Tan, H., Wu, Y., Shen, B., Jin, P.J., Ran, B.: Short-term traffic prediction based on dynamic tensor completion
13. Wang, Y., Papageorgiou, M., Messmer, A.: Real-time freeway traffic state estimation based on extended kalman filter: adaptive capabilities and real data testing. Transp. Res. Part A: Policy Prac. **42**(10), 1340–1358 (2008)
14. Wang, Z., Ye, T., Lu, M., Yuan, X., Qu, H., Yuan, J., Wu, Q.: Visual exploration of sparse traffic trajectory data. IEEE Trans. Vis. Comput. Graph. **20**(12), 1813–1822 (2014)
15. Xie, M.Q., Li, X.M., Zhou, W.L., Fu, Y.B.: Forecasting the short-term passenger flow on high-speed railway with neural networks. Comput. Intell. Neurosci. **2014**, 23 (2014)
16. Xu, H., Ying, J., Wu, H., Lin, F.: Public bicycle traffic flow prediction based on a hybrid model. Appl. Math. Inf. Sci **7**, 667–674 (2013)
17. Zheng, W., Lee, D.H., Shi, Q.: Short-term freeway traffic flow prediction: bayesian combined neural network approach. J. Transp. Eng. **132**(2), 114–121 (2006)

Understanding Behavioral Differences Between Short and Long-Term Drinking Abstainers from Social Media

Haripriya Harikumar[1(✉)], Thin Nguyen[1], Sunil Gupta[1], Santu Rana[1], Ramachandra Kaimal[2], and Svetha Venkatesh[1]

[1] Centre for Pattern Recognition and Data Analytics,
Deakin University, Geelong, Australia
{hharikum,thin.nguyen,sunil.gupta,santu.rana,
svetha.venkatesh}@deakin.edu.au
[2] Computer Science and Engineering Department,
Amrita University, Kollam, India
mrkaimal@am.amrita.edu

Abstract. Drinking alcohol has high cost on society. The journey from being a regular drinker to a successful quitter may be a long and hard journey, fraught with the risk to relapse. Research has shown that certain behavioral changes can be effective towards staying abstained. Traditional way to conduct research on drinking abstainers uses questionnaire based approach to collect data from a curated group of people. However, it is an expensive approach in both cost and time and often results in small data with less diversity. Recently, social media has emerged as a rich data source. Reddit is one such social media platform that has a community ('subreddit') with an interest to quit drinking. The discussions among the group dates back to year 2011 and contain more than 40,000 posts. This large scale data is generated by users themselves and without being limited by any survey questionnaires. The most predictive factors from the features (unigrams, topics and LIWC) associated with short-term and long-term abstinence are identified using Lasso. It is seen that many common patterns manifest in unigrams, topics and LIWC. Whilst topics provided much richer associations between a group of words and the outcome, unigrams and LIWC are found to be good at finding highly predictive solo and psycho linguistically important words. Combining them we have found that many interesting patterns that are associated with the successful attempt made by the long-term abstainer, at the same time finding many of the common issues faced during the initial period of abstinence.

Keywords: Feature selection · Health promotion · Reddit · Stop drinking · Abstinence

1 Introduction

Drinking alcohol excessively can be detrimental. According to Drug-Free World foundation, alcohol kills more teenagers than all other drugs combined. Alcohol

© Springer International Publishing AG 2016
J. Li et al. (Eds.): ADMA 2016, LNAI 10086, pp. 520–533, 2016.
DOI: 10.1007/978-3-319-49586-6_36

addiction not only harms the individual, but also damages relationships and society in terms of increased violence, crime and accidents resulting from drink driving. According to World Health Organization the harmful use of alcohol results in 3.3 million deaths every year which represent 5.9 % of all deaths and is the root cause of more than 200 disease and injury conditions [11]. The Center for Disease Control pegs the economic cost of excessive drinking for USA at over $220 billion[1]. Yet, a recent Time magazine article has reported that 1 in 10 adults has been able to quit drinking alcohol altogether[2]. The journey from being an excessive drinker to a successful quitter is onerous, and fraught with the risk of relapse. Effective formulation of successful strategies is required to provide proper guidance to the people on their journey to an alcohol-free life.

To formulate effective strategies it is necessary to understand the behavioral aspects of the people on the course of their journey to quitting. It is important to understand the issues faced by the people who has just started their journey and also learn the positive behavioral changes that are associated with successful people. Traditional approaches to collect useful data involves questionnaire based surveys [4, 19] from a curated set of people. This way of collection of data can be costly both in time and money. Additionally, such efforts are generally limited to a small group size due to the cost issue and thus generally lack variety in the sample. Although such methods have served specific purpose in the past, in today's web enabled world we have access to an alternative source of large data and almost at no cost. Recent emergence of social networking sites have enabled communication among a large group of people with shared interest. Such conversations can be used as an alternative to the questionnaire based surveys. Since, such data is generated spontaneously by users, not being limited by questionnaire design, they offer potential to discover yet unknown and interesting associations between different contributing factors and outcomes.

Recently, data collected from various social networking sites have fueled research studies into addiction related behavioral changes and health issues. Moreno et al. [9, 10] use data from publicly available Facebook profiles of college students in US to develop a screening tool for determining students at risk for problem drinking and its consequences. The study found strong correlation between Facebook posts mentioning intoxication and risk of alcohol overuse. In a separate study by Cook et al. [2] discussed the influence of peers, interconnectedness between individuals in a social network and its relation with alcohol abuse. The sample collected represents adults across US and provides useful insights on the relationship between social media data and substance abuse. Conversational data from Reddit have been analyzed in [16] to study the link between social media interaction and addiction related health outcomes.

In this work we use data collected from Reddit. It is a social media platform where registered users can share their thoughts and experiences with other people by submitting content such as text posts or web links. They can announce a

---

[1] http://www.cdc.gov/features/alcoholconsumption/.
[2] http://healthland.time.com/2012/03/07/quitting-drugs-or-alcohol-10-of-the-u-s-population-has-done-it/.

522   H. Harikumar et al.

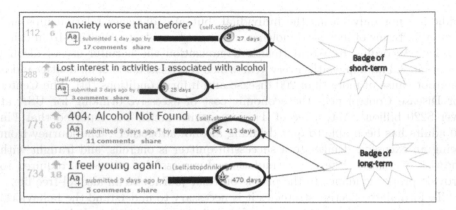

Fig. 1. The post submitted by users in "/r/stopdrinking" subreddit where badges which symbolize the abstinent days of users.

behavior they want to change, keep track of it and other users can provide social support. User posts are organized by areas of interest called 'subreddits'. The specific subreddit related to users with a shared interest in quitting alcohol addiction is 'stopdrinking'. In this forum they discuss issues and experiences on their journey to sobriety. We crawl all the posts from this subreddit from the year 2011 to April 2016. Total numbers of posts in this period is 42,337. In addition to the posts users can also keep a badge of honor which counts the number of days that the user has self-reported to be sober. The badge resets when a relapse is reported. We use the badge information shown in Fig. 1 to divide these posts into two groups. The first group is from users (referred to as 'Class 0') with the drinking abstinence period of at most 30 days while the second group (referred to as 'Class 1') is from users with the abstinence period of at least 365 days. The different time horizons are chosen to understand both the issues faced within the first month after deciding to quit and the characteristics associated with really long-term abstainer. This division results in a total of 10,551 posts where 6,910 posts belongs to 'Class 0' and the remaining 3,641 posts to 'Class 1'. A recent study has analyzed Reddit data [16] for alcohol drinkers but is limited to only long-term abstinence. We note that it is important to understand not only the factors behind success but also the factors behind failure so that users can avoid initial hurdles that are hard to overcome.

We pre-process the content of posts to extract a set of unigrams and represent each post as a vector of unigrams in a vector space model. Topics are also extracted from posts using Latent Dirichlet Allocation (LDA) model. Linguistic Inquiry and Word Count (LIWC) features are also used to extract the psycho linguistic features from the crawled posts. We use unigrams, topic proportion and LIWC vectors of each post to train three separate predictive models. We use a regularized linear classification method - Least Absolute Shrinkage and Selection Operator (Lasso). The most predictive factors (unigrams, LIWC and topics) related to both the short-term abstinence and long-term abstinence are identified and analyzed. It is seen that many common patterns are found in the analysis of unigrams, LIWC and topics. Many times topics provided a much richer association between a group of words and the outcome, although a few times a very strong and specific unigrams and LIWC have been found to be highly predictive. Combining them we have found that many interesting patterns that are associated with the successful attempt made by the long-term abstainer, whilst finding many of the common issues faced during the initial period of abstinence. For example, we have found that leaning towards spirituality can be transformational. Alternatively, we have also found that the unfortunate event of being in prison has also acted as a vehicle of change for regular drinkers. We also found the long-term abstainers as the agent of change who are very supportive to others on their journey. For beginners we have found that the issues they have to deal with are withdrawal symptoms such as sleeping disorder and anxiety, and circumstantial triggers such as Fridays and the feeling of loneliness. Some of these patterns have also been reported in previous psychological studies [5,6,8,15,18] on drinking behavior. The significance of our work is the demonstration that the low-cost social media data can be analyzed to look deeper into one of the greatest addiction problems faced by our society and possibly be used as an effective policy design tool.

2 Methods

2.1 Datasets

Reddit, is a fast growing[3] popular online community with a considerable number of registered users called 'redditors'. Redditors can freely interact, share their thoughts, experience, images, web links and comments with each other in the network. According to the area of interest the Reddit has subgroups called 'subreddits'. As an example, /r/AskReddit, /r/funny, /r/todayilearned, /r/pics, /r/science are some subreddits. Some categories of subreddits are educational, entertainment, technology, self-improvement and discussion-based. The reddit metrics from the year 2006 to 2014, shows that it has more than 340,000 subreddits[4]. Out of these subreddits, some self-improvement subreddits like /r/stopdrinking and /r/stopsmoking, the redditors were given badge as shown in Fig. 1 according to their request. The moderators of each subreddit have the right to give badge for those who request based on some terms and conditions. The user can re-request for badge, if he relapsed. It helps them to count the amount of days they have been sober since requesting this badge and thereby motivates them to remain sober. We focus on one of the self-motivating subreddits stopdrinking[5] for collecting addiction related self-reported data.

We used PRAW[6], a python package for collecting posts and its associated meta-data from Reddit. This Python Reddit API Wrapper allows simple access

---

[3] http://redditmetrics.com/history.
[4] http://redditmetrics.com/visuals/reddit-growth-gif.
[5] https://www.reddit.com/r/stopdrinking.
[6] https://praw.readthedocs.io/en/stable/.

to Reddit API. For each post, our crawler collected its title, content or self text submitted, author name and id, upvotes, downvotes, score got, number of comments, comment's content, badge information, post created time, post crawl time. A pre-processing had done on the collected posts by pruning away less significant variables.

We focus on the 'stopdrinking' subreddit which has more than 35,000 redditors and ranked 1,285 in Reddit[7]. It was created in the year 2010. This subreddit helps and motivates heavy drinkers to control or stop drinking. We use badge information which signifies redditors abstinent days, for grouping a total of 42,337 posts into two groups. But the badge value is automatically incrementing on each day and PRAW helps us to collect the badge information at the time of crawling. Even though a person submitted a post at the beginning phase of abstinence the Reddit API will give only latest badge while crawling. So for grouping the posts based on the badge value or capture the behavior of posts we computed the badge at the time of post submission. The newly computed badges are then used to categorize the posts in two separate groups, Class 0, one with badge days less than 30 days and Class 1, with badge days greater than 365 days. Finally we got 10,551 labeled posts where 6,910 belongs to less than 30 days category and 3,641 to greater than 365 days.

Most of the models generated by machine learning algorithms depends heavily on the quantity and quality of data. This unbalanced data may end up with biased results and generation of less predictive models. To avoid biased results we balanced the data by randomly selecting equal number of data, 3,500 posts from each group.

2.2   Feature Extraction

Pre-processing of text data influence feature extractors and affects model generation. Beautiful Soup, a python library is used for removing HTML and XML tags from our text data. We updated the stop word list by adding some common words (drink, sober, would) for getting efficient discriminating unigrams. As a second step stop word removal is applied on the raw post to erase most frequent and less useful features (and, the, his). We used three different types of feature sets for analyzing and understanding behavioral differences in both classes.

(1) Unigrams: This is a widely used, common and simple representation of documents [3] even though we cannot identify semantics and co-occurrence of words. Scikit-learn [12] is used for extracting unigrams from the posts. We had taken into consideration that the unigrams with less than a frequency of 10 contribute less for classification and the post vector with less than 5 non-zero values contains information less compared to the high dimensional space of post vectors. So after stop-word removal, from the cleaned posts, we extracted unigrams with a minimum threshold of occurrence 10 in whole posts and a length greater than 3. Finally we got a total of 4,344 unigrams. So given a post we converted

---

[7] http://redditmetrics.com/r/stopdrinking.

Table 1. Lasso co-efficients of unigrams with positive co-efficient for long-term group and negative co-efficient for short-term group.

Unigram	Co-efficient	Unigram	Co-efficient	Unigram	Co-efficient	Unigram	Co-efficient
theme	0.266	gets	0.063	weeks	−0.098	woke	−0.041
chat	0.148	summer	0.063	sleep	−0.079	ready	−0.041
anniversary	0.12	possible	0.062	beers	−0.077	advice	−0.04
gratitude	0.108	worth	0.06	feels	−0.073	booze	−0.039
join	0.101	struggles	0.06	sleeping	−0.07	strong	−0.038
answer	0.1	changed	0.057	supportive	−0.069	wine	−0.037
prison	0.098	dark	0.057	stopping	−0.069	boyfriend	−0.037
recovery	0.08	peace	0.056	reading	−0.062	super	−0.036
folks	0.079	struggling	0.055	slept	−0.06	withdrawal	−0.035
thread	0.078	topic	0.054	wait	−0.054	control	−0.034
date	0.078	powerful	0.054	friday	−0.053	guilt	−0.034
volunteer	0.071	idea	0.05	longest	−0.05	beer	−0.032
learned	0.068	rehab	0.05	problem	−0.049	anxiety	−0.031
message	0.066	drugs	0.048	girl	−0.048	forward	−0.031
matter	0.065	important	0.047	cravings	-0.044	late	−0.031

it into a vector of unigrams, where each value represents the frequency of that unigram within it and then removed the post vector with less than 5 unigrams. Finally we got a dictionary of size 6142 × 4344.

(2) Topics: Topic modeling is used for extracting clusters of words from large unstructured and unlabeled text data. It helps us to identify meaningful topics from a large corpus. We used Latent Dirichlet Allocation (LDA) [1] package for topic extraction. LDA is a probabilistic topic model and each document is viewed as a mixture of topics. The default dirichlet parameter over distribution over topics, alpha is 0.1 and over words beta is 0.01 is used in LDA. The feature space of each post is a vector of topic distributions, where each value is the proportion of the topic in the post. The size of feature space is reduced and the final dictionary size is 6142 × 100.

(3) Linguistic Inquiry and Word Count (LIWC): LIWC 2015 package [13,14] has a pre-defined set of features where each feature is a group of words categorized based on psychological similarity. We used this package for extracting LIWC feature set from the posts. The raw data after removing HTML and XML tags are given as input to the LIWC. LIWC reads this post and counts the percentage of words that reflect different emotions, thinking styles, social concerns, and even parts of speech. The output obtained from LIWC is a dictionary of size 6142 × 78.

2.3 Classification

A logistic regression model with L_1 norm regularizer, Lasso [17] is used for classification. This model has the ability to simultaneously classify, shrink and learn feature from the feature space. To evaluate the performance of three feature sets

Table 2. Lasso co-efficients of topics with positive co-efficient for long-term group and negative co-efficient for short-term group.

Topic	Co-efficient	Topic	Co-efficient	Topic	Co-efficient	Topic	Co-efficient
T93	1.447	T1	0.605	T68	−1.768	T44	−0.432
T55	1.279	T89	0.508	T62	−1.452	T9	−0.426
T56	1.182	T40	0.498	T3	−1.402	T86	−0.399
T13	1.181	T59	0.479	T88	−1.324	T87	−0.398
T78	1.143	T32	0.47	T36	−1.219	T66	−0.379
T67	1.116	T48	0.461	T0	−1.215	T27	−0.346
T23	1.101	T6	0.431	T46	−1.172	T82	−0.315
T81	1.073	T14	0.421	T64	−1.144	T92	−0.271
T90	0.948	T15	0.352	T75	−1.091	T12	−0.261
T61	0.927	T54	0.283	T79	−1.003	T4	−0.261
T49	0.926	T43	0.259	T5	−0.94	T98	−0.248
T80	0.853	T25	0.25	T95	−0.927	T94	−0.236
T28	0.85	T33	0.25	T47	−0.926	T22	−0.191
T72	0.848	T96	0.218	T20	−0.744	T85	−0.19
T38	0.828	T31	0.173	T91	−0.611	T18	−0.162
T57	0.797	T42	0.163	T2	−0.598	T37	−0.152
T16	0.78	T39	0.103	T17	−0.544	T65	−0.118
T76	0.766	T34	0.079	T41	−0.53	T10	−0.114
T99	0.734	T21	0.063	T97	−0.528	T74	−0.092
T30	0.722	T45	0.062	T58	−0.521	T51	−0.039
T83	0.703	T29	0.046	T52	−0.511	T7	−0.023
T53	0.648	T77	0.037	T11	−0.498		
T8	0.61	T60	−0.002	T63	−0.468		

in classifying the short-term and the long-term group we stratified our dataset into training and testing in a 6:4 ratio. The role of our classifier is to classify the data into short-term or long-term category by identifying the most discriminant features. Unigrams, topics and LIWC features are used as feature sets with label '0' for Class 0 and '1' for Class 1.

We used a 10 fold cross validation to fit 60 % (3,685) training data. Among 4,344 unigrams we got 208 most discriminant features (non-zero Lasso co-efficients). Lasso had taken more time for convergence and result generation because of the high dimensional data. The top 30 unigrams along with the Lasso co-efficients obtained for each class is shown in Table 1. The negative values are the co-efficients obtained for Class 0 and positive values are for Class 1.

Applying the Lasso classifier on the feature space of topics, resulted in a generation of 90 topics with non-zero Lasso co-efficients. The time taken for

Table 3. Lasso co-efficients of LIWC (long-term (left with positive co-efficient) short-term (right with negative co-efficient)).

LIWC	Co-efficient	LIWC	Co-efficient	LIWC	Co-efficient	LIWC	Co-efficient
wc	0.493	male	0.187	body	−0.881	article	−0.257
health	0.389	focuspast	0.162	ingest	−0.865	differ	−0.218
you	0.356	family	0.148	verb	−0.602	friend	−0.217
we	0.309	ipron	0.148	i	−0.430	focusfuture	−0.208
tentat	0.257	auxverb	0.120	feel	−0.406	shehe	−0.181
negate	0.230	discrep	0.109	wps	−0.406	netspeak	−0.180
reward	0.225	social	0.094	cogproc	−0.344	negemo	−0.165
interrog	0.225	hear	0.077	home	−0.329	quant	−0.156
they	0.225	relig	0.067	time	−0.325	adverb	−0.156
death	0.204	certain	0.065	prep	−0.278	leisure	−0.113

Table 4. Top 5 topics related to both the long-term abstinence and short-term abstinence, represented through word clouds.

Days of abstinence >365 (Class 1)		Days of abstinence < 30 (Class 0)	
Topic	Word Cloud	Topic	Word Cloud
93	prison	68	sleep
55	gratitude	62	anxiety
56	change	3	friday
13	spiritual	88	weeks
78	luck	36	stop

convergence and result generation is less compared to the unigram feature space. The topics along with their non-zero Lasso co-efficients is shown in Table 2 and the word clouds of the top 5 topics are shown in Table 4.

We got a total of 61 non-zero co-efficients after running Lasso over the LIWC feature space. The top 20 LIWC features from each class is shown in Table 3.

For evaluating the performance of classifications we used, F_1-score and accuracy. F_1-score uses precision and recall values. Accuracy is the percentage of mean of correct predictions.

2.4   Classification Results

The 60 % training data is used for generating models for prediction. The remaining 40 % data from our pre-processed dataset is used for evaluating the performance of the classifier. The F_1-score and accuracy obtained after running Lasso on the three different feature spaces of dataset is shown in Table 5. Even though the topics feature space has high F_1-score and accuracy compared to unigrams, LIWC outperforms both. The highest F_1-score and accuracy of LIWC feature helps to infer that the classifier performs well in psycho linguistically similar low dimensional feature space. Recalling that our data is class-balanced, all models have been able to perform reasonably well.

Table 5. The F_1-score and accuracy for different features.

Feature space	F_1-score	Accuracy (%)
Unigrams	0.63	69.76
Topics	0.67	70.57
LIWC	0.69	72.61

3   Discussion

Our analysis has captured important patterns that would be potentially useful in understanding the users of the two groups and identifying their strengths and challenges. We present and discuss these patterns in the remainder of the section.

3.1   Class 1 (Users with the Drinking Abstinence Period of at Least 365 days)

For the users in 'Class 1', our analysis captured several unigrams, topics, LIWC features that explain their **success**. Some of these are factors that are associated with the success (meaning that successful user group happened to have these behavioral patterns) and the factors that resulted from the success. We discuss them in the following.

Factors Associated to Success

– **Being in prison:** Our analysis captured a strongly predictive topic (Topic 93) related to *Prison*. Our unigram analysis too found it predictive. For many users, being in prison or interacting with people in prison happened to be a successful factor in quitting. For example, some posts associated with this topic are ''For many of them prison has been a place where they have transformed their lives, where they have grown emotionally and spiritually'' or ''I have been sober longer than I drank. I still go to 3 AA meetings a week including one on Wednesday evenings in a State Prison where I have met some incredible people, and where my growth in recovery has gone up a level''. Of course, visiting prison is not the recommended way for quitting, but it does reveal interesting insights. People who visit prison are forced to stay away from drinking for sometime. Further, they better appreciate the value of life, they realize what could they do if they were free. Such realizations help them both stop cravings and mental determination to quit. In addition, it also bring out the importance of continuing and possibly increasing funding for quit drinking effort [7] in prisons.

- *Being spiritually inclined:* Another factor that was found by our analysis was *Spiritual Inclination.* The LIWC features *'relig (altar, church)'* belongs to the long-term group support the importance of spiritual inclination in success. Topic 13 is also about spirituality, religion and is strongly present among the posts of users who have been long-term abstainers. Some of the posts that are associated with this topic are: ''I do all these things because they keep me spiritually fit, and when I am spiritually fit I don't even have to think about not drinking...''. This is an interesting association that we have found which can also be tried as a successful vehicle of change for addicted people.

- *Feeling of gratitude:* Both our topic analysis (Topic 55) and unigram analysis captured the pattern of *gratitude.* Users that have been long-term abstainers heavily used the words 'gratitude', 'happiness', 'world', 'amazing' and 'live'. Some of the posts associated with this pattern are: ''The gratitudes in particular have helped how I experience the world I live in, opened my eyes to recognise the beauty and wonder of the world around me...'' or ''There is a wonderful world out there that isn't based on drinking and isn't consumed with thoughts of sobriety. A world full of possibilities...''. One of the LIWC features *'negate (no, not, never)'* reveals the negative attitude of this group towards drinking alcohol.

- *Support from family:* The LIWC feature *'family (daughter, dad, aunt)'* which belongs to long-term group shows some interesting clues about the significance of family behind their success.

Factors Resulting from Success

- *Talking about the choices and change:* Our topic analysis (Topic 56) captured a pattern related to the *choices made and the change.* Many long-term abstainers talk about the change and the right choice they made in their life by not choosing alcohol and thus following a good path as mentioned in some posts ''If you do not change your behaviour and mindset, you will start to kill yourself with the anxiety and stress., For me, the

choice to quit wasn't about giving something up but about
gaining something back!''.

- *Wishing success to others:* 'Reward, certain' are some of the motivating
LIWC features extracted for long-term group. Top LIWC features 'we, social'
show the positive attitude of these group towards public. Our topic analysis
(Topic 78) also captured a pattern related to *wishing success to others*. The long-
term abstainers having experienced the path of quitting are useful motivators
for others in their quitting endeavors. This is evident from some of their posts:
''Good luck with your journey, inspire others, surprise yourself, live
life. :)'', ''Best of luck for all of you! :)'' and ''It can take months,
years even, but if you put in the work and remain sober your entire life
can and will change!''.

Together it shows that these people can act as the most useful agent of change
for the beginners. They are the people who would understand the beginners most
and who could provide the best support and care through their journey.

3.2 Class 0 (Users with the Drinking Abstinence Period of at Most 30 days)

For the users in 'Class 0', our analysis captured several unigrams, topics and
LIWC features that explain their **challenges**. Some of these challenges are where
clinical intervention may help while others are challenges related to environmental factors. We discuss them in the following.

Challenging Clinical Factors

- *Dealing with cravings:* LIWC features like 'feel, body, differ' describe the
issues related with the body and mind of short-term group. Our unigram
analysis also captured this withdrawal symptoms that are associated with
quitting. It is evident that initial days of abstinence may be helped by being
consulted with clinicians who can use medical drugs to combat the cravings.
- *Dealing with sleeplessness:* Both our unigram analysis and topic analysis
(Topic 68) captured a pattern related to *sleeplessness*. This topic is associated
with a large number of short-term abstainers whose posts indicate the problem
of sleeplessness, e.g., ''For the longest time I drank because I knew I
wouldn't have to worry about getting to sleep''. It may be useful to
get clinical help to solve the problem of sleeplessness and thus help them
quitting alcohol.
- *Dealing with anxiety:* Our topic analysis captured a pattern related to
anxiety. The user with posts associated to topic 62 are suffering from withdrawal symptoms such as anxiety, depression and mood disorders as evident
from an example post ''I'm pretty sure I drink like this to relieve
stress and as a crutch to deal with social anxiety''. These conditions are well known to be the problems of people with excessive drinking
habits [5]. Once again, clinical intervention to such users may be helpful in
quitting.

Challenging Environmental Factors

- *Fridays being the hardest:* Both our topic analysis and unigram analysis captured patterns related to Fridays being the hardest. Topic 3 and Topic 88 capture the drinking habits of people on Fridays for enjoying their weekends, getting relaxation from their stress after weekdays. It is evident from some posts, ``I think to some degree I feel it's justified as a reward for completing a hard week at work. But instead of being a'' healthy ``few beers, it turns into 6 or 8 for Friday night, then 6 or 8 Saturday night, and maybe a few sunday night.'', ``what do you guys suggest to stave off those cravings on those friday and saturday nights?''.
  'Friday' is also detected as an important predictive word of short-term abstainers by our unigram analysis. It may be useful for such users to indulge into other activities, e.g., go to a movie, attend live music.
- *Craving for supportive environment from loved ones:*   The LIWC feature *'shehe (she, her, him)'* supports that craving for supportive environment from loved ones is a challenging factor of short-term group. The feeling of loneliness can also be seen in the LIWC feature *'i'* in short-term group. From the unigrams we see one of the most predictive words for short-term is 'girl'. An example of one such post is: ``She's been away for the summer. Aside from missing her, I've been drinking a lot, in the evenings mostly''. It is evident that many of the posters are male and they are craving for love, support and care from the opposite gender.
- *Problems at the workplace:* Topic 36 from our topic analysis shows that people have plans to stop or moderate drinking but they cannot control it due to problems in their work place, work related stress.

4   Conclusion

In this paper we analyzed conversational data from stopdrinking subreddit to find both the factors associated with the successful abstainers and challenges faced by the beginners. Our data collection dates back to year 2011 and contain more than 40,000 posts. We analyze the posts by dividing them into two groups: one with abstinence shorter than 30 days and the other longer than 365 days. The most predictive factors (unigrams, topics, LIWC) related to both the short-term abstinence and long-term abstinence are identified using Lasso and then analyzed. It is seen that many common patterns manifest both in unigrams, topics and LIWC. Whilst topics provided much richer associations between a group of words and the outcome, unigrams and LIWC are found to be good at finding highly predictive solo and psycho linguistic words. Combining them we have found that many interesting patterns are associated with the successful attempt made by the long-term abstainer, at the same time finding many of the common issues faced during the initial period of abstinence. Our results shows the scope of using social network data mining for knowledge extraction. We identified some common results in our study and psychological studies as well. This ensures the possibility of using large scale online data for behavioral analysis for mental health and general well-being researches.

Acknowledgment. This work is partially supported by the Telstra-Deakin Centre of Excellence in Big Data and Machine Learning.

References

1. Blei, D.M., Ng, A.Y., Jordan, M.I.: Latent dirichlet allocation. J. Mach. Learn. Res. **3**, 993–1022 (2003)
2. Cook, S.H., Bauermeister, J.A., Gordon-Messer, D., Zimmerman, M.A.: Online network influences on emerging adults alcohol and drug use. J. Youth Adolesc. **42**(11), 1674–1686 (2013)
3. Feldman, R., Sanger, J.: The text mining handbook: Advanced approaches in analyzing unstructured data. Cambridge University Press (2007)
4. Gilpin, E.A., Pierce, J.P., Farkas, A.J.: Duration of smoking abstinence and success in quitting. J. Natl. Cancer Inst. **89**(8), 572 (1997)
5. Grant, B.F., Stinson, F.S., Dawson, D.A., Chou, S., Dufour, M., Compton, W., Kaplan, K.: Prevalence and co-occurrence of substance use disorders and independent mood and anxiety disorders. Alcohol. Res. Health **29**(2), 107–120 (2006)
6. Jonas, D.E., Garbutt, J.C., Amick, H.R., Brown, J.M., Brownley, K.A., Council, C.L., Viera, A.J., Wilkins, T.M., Schwartz, C.J., Richmond, E.M.: Behavioral counseling after screening for alcohol misuse in primary care: a systematic review and meta-analysis for the US Preventive Services Task Force. Ann. Intern. Med. **157**(9), 645–654 (2012)
7. Kaskutas, L.A., Bond, J., Humphreys, K.: Social networks as mediators of the effect of Alcoholics Anonymous. Addiction **97**(7), 891–900 (2002)
8. Miller, W.R., Walters, S.T., Bennett, M.E.: How effective is alcoholism treatment in the United States? J. Stud. Alcohol. **62**(2), 211–220 (2001)
9. Moreno, M.A., Christakis, D.A., Egan, K.G., Brockman, L.N., Becker, T.: Associations between displayed alcohol references on Facebook and problem drinking among college students. Arch. Pediatr. Adolesc. Med. **166**(2), 157–163 (2012)
10. Moreno, M.A., D'ngelo, J., Kacvinsky, L.E., Kerr, B., Zhang, C., Eickhoff, J.: Emergence and predictors of alcohol reference displays on Facebook during the first year of college. Comput. Hum. Behav. **30**, 87–94 (2014)
11. Organization, W.H.: Global status report on alcohol and health. World Health Organization (2014)
12. Pedregosa, F., Varoquaux, G., Gramfort, A., Michel, V., Thirion, B., Grisel, O., Blondel, M., Prettenhofer, P., Weiss, R., Dubourg, V.: Scikit-learn: machine learning in Python. J. Mach. Learn. Res. **12**, 2825–2830 (2011)
13. Pennebaker, J.W., Booth, R.J., Boyd, R.L., Francis, M.E.: Linguistic Inquiry and Word Count: LIWC 2015 [Computer software]. Pennebaker Conglomerates, Inc. (2015)
14. Pennebaker, J.W., Boyd, R.L., Jordan, K., Blackburn, K.: The development and psychometric properties of LIWC 2015. UT Faculty/Researcher Works (2015)

15. Regier, D.A., Farmer, M.E., Rae, D.S., Locke, B.Z., Keith, S.J., Judd, L.L., Goodwin, F.K.: Comorbidity of mental disorders with alcohol and other drug abuse: Results from the Epidemiologic Catchment Area (ECA) study. JAMA **264**(19), 2511–2518 (1990)
16. Tamersoy, A., De Choudhury, M., Chau, D.H.: Characterizing smoking and drinking abstinence from social media. In: Proceedings of the 26th ACM Conference on Hypertext & Social Media, pp. 139–148 (2015)
17. Tibshirani, R.: Regression shrinkage and selection via the lasso. J. Royal Stat. Soc. Ser. B (Methodological) **58**, 267–288 (1996)
18. Weitzman, E.R.: Poor mental health, depression, and associations with alcohol consumption, harm, and abuse in a national sample of young adults in college. J. Nerv. Ment. Dis. **192**(4), 269–277 (2004)
19. Witteman, J., Post, H., Tarvainen, M., Bruijn, A., Perna, E.D.S.F., Ramaekers, J.G., Wiers, R.W.: Cue reactivity and its relation to craving and relapse in alcohol dependence: a combined laboratory and field study. Psychopharmacology **232**(20), 3685–3696 (2015)

Discovering Trip Hot Routes Using Large Scale Taxi Trajectory Data

Linjiang Zheng[1,2(✉)], Qisen Feng[1,2], Weining Liu[1,2], and Xin Zhao[1,2]

[1] Key Laboratory of Dependable Service Computing in Cyber Physical Society,
Chongqing University, Ministry of Education, Chongqing 400030, China
zlj_cqu@cqu.edu.cn
[2] College of Computer Science, Chongqing University,
Chongqing 400030, China

Abstract. Discovering trip hot routes is very meaningful for drivers to pick up a passenger, as well as for managers to plan urban public transport. Riding by taxis is one of the important means of transportation. Large scale taxi trajectory data from taxi GPS device implicates residents' trip behavior. In this paper, we present a method to discover trip hot routes using large scale taxi trajectory data. Firstly, we measure taxi trajectory similarity with longest common subsequence (LCS). LCS-based DBSCAN trajectory clustering algorithm was proposed. Then hot routes were extracted using large scale taxi trajectory data. Our experiment shows that the trajectory clustering algorithm and hot route extraction method are effective.

Keywords: Taxi trajectory · LCS · Trajectory clustering · Hot routes

1 Introduction

With the rapid development of positioning equipment and technology such as Global Positioning System (GPS), Radio Frequency Identification (RFID), and the growing electronic map and emerging computing platform, trajectory of human activities is easier to obtain. People's work and life are completely and systematically recorded with trajectory data. Trajectory data produced by human activities contains human behavior. Mobile trajectory data is mainly from four kinds: mobile phone, public transport card, taxi and bank card [1]. Mobile trajectory data was generally divided into two kinds at existing research according to the data source. A kind of trajectory data is from public transportation such as buses and taxis. Another kind of trajectory data is from personal activities such as mobile phones and bank cards.

Taxi is one of the important means of resident's trip. Taxi passengers decided origin and destination (OD) at each trip. So taxi trajectory can well reflect the resident trip characteristics. In particular, resident trip hot routes are very meaningful for drivers to pick up a passenger, as well as for manager to plan urban public transport. Nowadays, Taxi in many cities have GPS device. Taxi trajectory data is easy to collect. Due to the characteristics of easy collection, wide distribution, and a large amount of data and so on, large scale taxi trajectory data is a good source of data for resident's trip behavior analysis.

J. Li et al. (Eds.): ADMA 2016, LNAI 10086, pp. 534–546, 2016.
DOI: 10.1007/978-3-319-49586-6_37

Resident's activities are restricted by time and space. For individual, his/her activities are periodic regular. For example, office worker is from home to the office in the morning and goes home from office after work. But for the community, this kind of regularity reflects a kind of group behavior, which can reflect the urban traffic situation and the resident's trip pattern directly. This rule can be explained by common sense. Resident trip hot routes are the embodiment of these group behavior patterns.

Taxis play important roles in modern urban transportation systems. Large scale taxi GPS trajectory data contain massive spatial and temporal information of urban human activity. The information derived from taxi trips benefits the city and transportation planning [1]. Taxi GPS data has been employed for a range of studies, particularly for travel time estimation [2–4], policy analysis [5–7] and travel flow patterns [8–14].

Understanding the travel patterns of residents by taxi GPS trajectory data is thus important for addressing many urban sustainability challenges [15]. Xiaolei Li et al. [16] introduced hot route as a general traffic flow pattern. They proposed a density-based algorithm named Flow Scan to find the set of hot routes by using a set of taxi trajectories in a road network. The method involves the determination of traffic statistics and road connectivity, focusing on the analysis of the group model of moving objects. However, despite the variety of researches and applications, taxi GPS trajectory data has so far not been explored for accessibility analysis in existing works.

Large scale taxi trajectory data from taxi GPS device implicates residents' trip behavior. In this paper, we presented a method to discover trip hot routes using large scale taxi trajectory data. Figure 1 shows the overall framework for discovering hot routes using large scale taxi GPS trajectory data. In the following section, we measured taxi trajectory similarity based on longest common subsequence (LCS). In Sect. 3, LCS-based DBSCAN trajectory clustering algorithm was proposed. In Sect. 4, hot routes were extracted using large scale taxi trajectory data. In Sect. 5, an experiment is carried out to validate the method. Conclusions are given in Sect. 6.

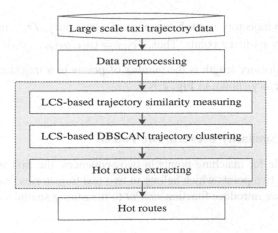

Fig. 1. Proposed framework for discovering hot routes using large scale taxi GPS trajectory data.

2 Measuring Trajectory Similarity Based on LCS

Euclidean distance is the simplest way to measure the similarity between the two tracks. As an alternative to traditional Euclidean distance to deal with time series trajectory with different lengths. Now longest common subsequence (LCS) has been widely used in time series trajectory clustering. The LCS model can match two sequences by allowing them to stretch, without rearranging the sequence of the elements. As a matter of fact, taxi trajectories are bound by the underlying road network. This suggests that trajectory data points can appear only in the space along the roads and this fact can reduce effort required to identify similar trajectories by focusing on their point sequence alignments.

In this section, each step is described to measure the trajectory similarity.

Definition 1. A taxi trajectory is a time-ordered sequence or time-series representing the x and y (longitude and latitude) coordinates of a taxi at time t. As shown in Fig. 2, $Tr = \{p_1, p_2, \ldots, p_7\}$ represents a trajectory, where $p_i(1 \leq i \leq 7)$ is the ith point of the sequence which contains longitude and latitude, time and speed. Given two taxi trajectories $Tr_1 = (p_{11}, p_{12}, \ldots, p_{1n})$, $Tr_2 = (p_{21}, p_{22}, \ldots, p_{2m})$ with length n and m, respectively.

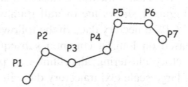

Fig. 2. A trajectory

Definition 2. Sub trajectory is a sub sequence of a trajectory. $Tr_{1(i)}$ represents the sub sequence of Tr_1 by its first i points. Thus, $Tr_{1(i)} = (p_{11}, p_{12}, \ldots, p_{1i})$.

Definition 3. Trajectory length is the number of points in a trajectory. For example, the length of trajectory shown in Fig. 2 is 7.

2.1 Measuring Similarity Between Two Points

To determine a rule for matching points in two sequences, the first step is measuring similarity between two points which belong to two taxi trajectories respectively. Given two points p_1, p_2, we introduce function $simPnt$ to measuring spatial similarity between two points:

$$simPnt(p_1, p_2) = \begin{cases} 0 & dist(p_1, p_2) > \delta \\ 1 - \frac{dist(p_1, p_2)}{\delta} & dist(p_1, p_2) \leq \delta \end{cases} \tag{1}$$

Where $dist(p_1, p_2) = \sqrt{(x_1 - x_2)^2 + (y_1 - y_2)^2}$ represents the spatial distance between p_1, p_2, δ is a parameter as distance threshold to specify the maximum allowable distance for two points. If distance between p_1, p_2 is too large, that is $dist(p_1, p_2) > \delta$, we regard the spatial similarity of two points as 0, that is $simPnt(p_1, p_2) = 0$. If $dist(p_1, p_2) \leq \delta$, that is $simPnt(p_1, p_2) > 0$. The two points of p_1, p_2 are considered to be matched. It is obvious that the value of $simPnt(p_1, p_2)$ is more closer to 1, the similarity of p_1, p_2 is higher. Using $simPnt(p_1, p_2)$, we can identify matched points in two trajectories sequences.

2.2 Measuring similarity between two sub trajectories

Given two trajectories $Tr_1 = (p_{11}, p_{12}, \ldots, p_{1n}), Tr_2 = (p_{21}, p_{22}, \ldots, p_{2m})$, we use a dynamic programing model to measure the LCS alignment between two sub trajectories by introducing a recursive function $simLCS(Tr_{1(i)}, Tr_{2(j)})$ as follows:

$$simLCS(Tr_{1(i)}, Tr_{2(j)}) = \begin{cases} 0 & i = 0 \text{ or } j = 0 \\ \max \begin{cases} simLCS(Tr_{1(i-1)}, Tr_{2(j-1)}) + simPnt(p_{1i}, p_{2j}) \\ simLCS(Tr_{1(i)}, Tr_{2(j-1)}) \\ simLCS(Tr_{1(i-1)}, Tr_{2(j)}) \end{cases} & \text{Otherwise} \end{cases}$$

(2)

Where $simPnt(p_{1i}, p_{2j})$ is calculated by Eq. (1). Recursively compute $simLCS(Tr_{1(i)}, Tr_{2(j)})$ from $i = 1, j = 1$ to $i = n, j = m$. If the value of

$$\max \begin{cases} simLCS(Tr_{1(i-1)}, Tr_{2(j-1)}) + simPnt(p_{1i}, p_{2j}) \\ simLCS(Tr_{1(i)}, Tr_{2(j-1)}) \\ simLCS(Tr_{1(i-1)}, Tr_{2(j)}) \end{cases}$$

is $simLCS(Tr_{1(i-1)}, Tr_{2(j-1)}) + simPnt(p_{1i}, p_{2j})$, that is to say, the similarity of p_{1i}, p_{2j} made contribution to the cumulative similarity. Finally, we can find the LCS alignment between full trajectories Tr_1 and Tr_2, which are denoted by $LCS(Tr_1)$ and $LCS(Tr_2)$. The value of $simLCS(Tr_{1(n)}, Tr_{2(m)})$ (also denote as $simLCS(Tr_1, Tr_2)$) reflects the maximum value of the spatial similarity between Tr_1 and Tr_2.

2.3 Measuring Similarity Between Two Trajectories

We denote the length of the longest common sub trajectory $LCS(Tr_1)$ and $LCS(Tr_2)$ as $|LCS(Tr_1, Tr_2)|$, the length of Tr_1 and Tr_2 as $|Tr_1|$ and $|Tr_2|$. We are aware that the proportion of the length of LCS and the length of trajectory is positively correlated with similarity between two trajectories, which denoted as $simTr(Tr_1, Tr_2)$.

The higher the value of proportion, the higher the similarity. If two trajectories overlap completely, we have $|LCS(Tr_1, Tr_2)| = |Tr_1| = |Tr_2|$. That is to say, the similarity between two trajectories is 1.

$$simTr(Tr_1, Tr_2) \propto \frac{2 \times |LCS(Tr_1, Tr_2)|}{|Tr_1| + |Tr_2|}$$

We also notice that, $simLCS(Tr_1, Tr_2)$ should be no more than $|Tr_1|$ or $|Tr_2|$ because maximum value of similarity between two points is 1. Therefore, we define the similarity between two trajectories as follows.

$$simTr(Tr_1, Tr_2) = \frac{2 \times |LCS(Tr_1, Tr_2)|}{|Tr_1| + |Tr_2|} \cdot \frac{simLCS(Tr_1, Tr_2)}{\max(|Tr_1|, |Tr_2|)} \quad (3)$$

Here an example is used to illustrate the calculation process of trajectory similarity. As is shown in Fig. 3, $Tr_1 = \{p_{11}, p_{12}, \ldots, p_{16}\}$, $Tr_2 = \{p_{21}, p_{22}, \ldots, p_{25}\}$ are two taxi trajectories. $|Tr_1| = 6, |Tr_2| = 5$. Assume that the similarity between two points was given. E.g. $simPnt(p_{11}, p_{21}) = 0.4$. If there is no lines between points, it means $simPnt(p_1, p_2) = 0$.

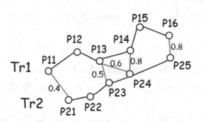

Fig. 3. Two taxi trajectories

Then calculate the LCS of Tr_1 and Tr_2.

$$simLCS(Tr_{1(1)}, Tr_{2(1)}) = 0.4, simLCS(Tr_{1(2)}, Tr_{2(1)}) = 0.4,$$

$$simLCS(Tr_{1(1)}, Tr_{2(2)}) = 0.4, simLCS(Tr_{1(2)}, Tr_{2(2)}) = 0.4,$$

$$simLCS(Tr_{1(3)}, Tr_{2(2)}) = 0.4, simLCS(Tr_{1(2)}, Tr_{2(3)}) = 0.4,$$

$$simLCS(Tr_{1(3)}, Tr_{2(3)}) = 0.9, simLCS(Tr_{1(4)}, Tr_{2(3)}) = 0.9,$$

$$simLCS(Tr_{1(3)}, Tr_{2(4)}) = 1.0, simLCS(Tr_{1(4)}, Tr_{2(4)}) = 1.7,$$

$$simLCS(Tr_{1(5)}, Tr_{2(4)}) = 1.7, simLCS(Tr_{1(4)}, Tr_{2(5)}) = 1.7,$$

$$simLCS(Tr_{1(5)}, Tr_{2(5)}) = 1.7, simLCS(Tr_{1(6)}, Tr_{2(5)}) = 2.5.$$

That is to say, $simLCS(Tr_1, Tr_2) = 2.5$, $LCS(Tr_1) = \{p_{11}, p_{13}, p_{14}, , p_{16}\}$, $LCS(Tr_2) = \{p_{21}, p_{23}, p_{24}, p_{25}\}$. The length of the longest common sub trajectory $|LCS(Tr_1, Tr_2)| = 4$.

Finally, according to Eq. (3), $simTr(Tr_1, Tr_2) = 0.33$. In other words, similarity between Tr_1 and Tr_2 is 0.33.

3 LCS-Based DBSCAN Trajectory Clustering

Clustering analysis in general requires two essential components, namely, similarity measure and clustering algorithm. Trajectory clustering approaches proposed in the literature also largely depend on the choice of the combination of these two. We have already discussed the similarity measure in the previous section. Now we propose the clustering algorithm based on DBSCAN (Density-based spatial clustering of applications with noise).

Given a trajectory Tr_s, the ε-neighborhood can be defined by similarity measure method mentioned before, denoted by $N_\varepsilon(Tr_s)$ as follows.

$$N_\varepsilon(Tr_s) = \{Tr_t \in Td | simTr(Tr_s, Tr_t) > \varepsilon, Tr_t \neq Tr_s\} \qquad (4)$$

The LCS-BASED DBSCAN trajectory clustering algorithm is as follows.

Algorithm 1. LCS-BASED DBSCAN

Input: $Td = \{Tr_1, Tr_2, ..., Tr_{|Td|}\}$, Parameters: ε (the maximum neighborhood distance) and σ(the minimum number of required trajectories in a cluster)

Output: $Clusters = \{C_1, C_2, ..., C_{num}\}$, where $C_i = \{Tr_{i1}, Tr_{i2}, ..., Tr_{i|C_i|}\}$

---

1. num= 0; visited = []; noise=[]; C=[]; Clusters=[]

2. for Tr_s in Td:

3.   if Tr_s not in visited:

4.   visited.append(Tr_s)

5.   if $|N_\varepsilon(Tr_s)| < \sigma$:

6.   noise.append(Tr_s)

7.   else:

8.   C= next cluster

9.   num=num+1

10.	ExpandCluster(Tr_s, $N_\varepsilon(Tr_s)$, C, σ)
11.	ExpandCluster(Tr_s, $N_\varepsilon(Tr_s)$, C, σ):
12.	C.append(Tr_s)
13.	for Tr_t in $N_\varepsilon(Tr_s)$:
14.	if Tr_t not in visited:
15.	visited.append(Tr_t)
16.	if $|N_\varepsilon(Tr_t)| \geq$ σ:
17.	$N_\varepsilon(Tr_s)=N_\varepsilon(Tr_s) \cup N_\varepsilon(Tr_t)$
18.	if Tr_t is not yet member of any cluster:
19.	C.append(Tr_t)
20.	Clusters.append(C)

4 Hot Routes Extracting

Kim et al. (2015) proposed the notion of cluster-representative subsequences (CRS), which can be viewed as the union of all pairwise LCSs of the trajectories within the cluster, to effectively represent a large number of trajectories in a given cluster [17]. Each CRS represents the typical trajectory of the cluster and describes a traffic flow pattern.

Given two trajectory Tr_1, Tr_2, treat trajectory as a set, LCS can be viewed as the intersection of Tr_1, Tr_2. CRS can be viewed as the union of Tr_1, Tr_2. Similar to this way, we regard CRS as a hot route. As we already obtained the longest common sub-trajectory(LCS) between any two trajectories above. The difference is how to obtain the CRS. We set a common sub-trajectory length threshold τ, if the length of the longest common subsequence trajectory exceeds the threshold, then we regard it as a hot route.

As shown in Fig. 4, there are three trajectories Tr_1, Tr_2, Tr_3, where the Longest common sub trajectory of Tr_1, Tr_2 is the red segment and $|LCS(Tr_1, Tr_2)| = 5$. The Longest common sub trajectory of Tr_2, Tr_3 is the blue segment and $|LCS(Tr_2, Tr_3)| = 2$. If we set τ = 3, then only the red segment can be regarded as a hot route.

The steps of LCS-BASED hot routes extraction method are as follows:

Step 1: Set the hot route HR to empty, randomly select a trajectory Tr_s from the first cluster Clusters[0], for each other trajectory Tr_t, calculate $|LCS(Tr_s, Tr_t)|$;

Step 2: if $|LCS(Tr_s, Tr_t)|$ is greater than the threshold τ, then the $LCS(Tr_s)$ and $LCS(Tr_t)$ will be added to the HR, otherwise go to the next step;

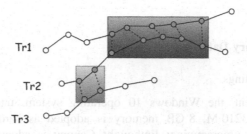

Fig. 4. Hot routes extraction method

Step 3: repeat Step 1 ~ Step 2 until all trajectories in Clusters[0] are processed;
Step 4: For other trajectories in Clusters[1... num], in accordance with the above steps until all the trajectories are processed.
Algorithm 2 shows a pseudo-code description.

Algorithm 2. LCS-BASED Hot Routes Extraction

Input: $Clusters = \{C_1, C_2, ..., C_{num}\}$, Parameters: τ(the length threshold of longest common sub trajectory)

Output: **Hot Routes HR[]**

LCS-BASED Hot Routes Extraction(Clusters[],τ):

1.   HR=[], visited = []

2.   for i in range(Clusters.length):

3.   for Tr_s in Clusters[i]:

4.   if Tr_s not in visted:

5.   visited.append(Tr_s)

6.   for Tr_t in Clusters[i]:

7.   if Tr_t not in visted:

8.   if $|LCS\,(Tr_s, Tr_t)| > \tau$:

9.   HR.append($LCS\,(Tr_s)$)

10.   HR.append($LCS\,(Tr_t)$)

11.   Return HR[]

---

5   Experiment

5.1   Taxi Trajectory Data

(1) Experiment settings

In the experiment, the Windows 10 operating system, Inter (R) Core (TM) 2.50 GHz CPU i5-3210 M, 8 GB memory is adopted as hardware environment. Python is language programming; Enthought Canopy is adopted as development environment.

(2) Data

In the experiment, Chongqing taxi GPS data is our experiment data. According to statistics, there is about 13000 taxis in Chongqing. Each taxi is equipped with GPS sensor which can record a time-stamped location. The real-time GPS data will be uploaded to the main server with a certain frequency of about 10 s. Each taxi GPS record consists of the following information (Table 1).

To simplify the experiment and get a clear result under the premise of no effect, 6717 taxi GPS records from 5 taxis at August 1, 2014 were used as experiment data.

5.2   Trajectory data preprocessing

In real world, trajectories are never perfectly accurate, due to sensor noise and other factors, such as receiving poor positioning signals in urban canyons. Those will inevitably lead to a variety of data quality problems. Before starting trajectory clustering, we dealt with a series of data preprocessing work on raw taxi GPS data, consisting of noise filtering, map matching, trajectory compression, and trajectory segmentation.

Some error is acceptable which can be fixed by map-matching algorithms. For example, some of the points that are clearly not on the road should be removed. But in

Table 1. Main fields of Chongqing taxi GPS data

No.	Fields	Type	Note
1	DEVICE_ID	VARCHAR2 (50)	the unique ID of each taxi
2	DATE_GPS	DATE	"YYYY-MM-DD HH:MM:SS"
3	LON	NUMBER (11,8)	the current longitude
4	LAT	NUMBER (11,8)	the current latitude
5	SPEED_GPS	NUMBER (6,1)	the current taxi speed in km/h
6	DIRECTION	NUMBER (*,0)	the direction the taxi, from $0\circ$ to $360\circ$ in clockwise (North is $0\circ$)
7	STATE	NUMBER	1: occupied, 0:unoccupied

other situations, we need to filter such noise points from trajectories by Kalman and other Filter algorithms before starting a mining task [18].

A period of raw taxi trajectory shown in Fig. 5 is from Chongqing Nan'an District Bureau Tongyuan subway station in the direction of Qixinggang. The sampling points in Chongqing Caiyuanba Yangtse River Bridge arranged very closely, indicates that the road is blocked, the vehicle speed is slow.

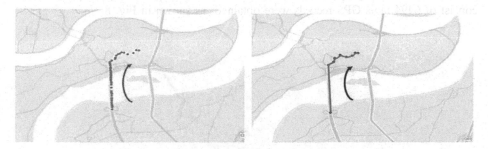

Fig. 5. A period of raw taxi trajectory   **Fig. 6.** A period of compressed taxi trajectory

Although the higher the trajectory sampling frequency, the higher the accuracy of the trajectory, there are a lot of battery power and the overhead for communication, computing, and data storage. In addition, we do not really need such a precision of location. So Douglas-Peucker algorithm was applied in this paper in order to reduce the size of a trajectory while not compromising much precision in its new data representation. The comparison result is shown in Fig. 6.

After noise filtering and trajectory compression from 6717 taxi GPS records, a total of 6693 taxi GPS records successfully retained, that is shown in Fig. 7.

Fig. 7. 6693 taxis GPS records at August 1, 2014 in Chongqing, China

The last step is to divide a trajectory into segments for further processes. The segmentation not only reduces the computational complexity but also enables us to mine richer knowledge, such as sub trajectory patterns, beyond what we can learn from an entire trajectory. In this paper, the trajectory is divided into two categories, which are occupied-trips trajectory and unoccupied-trips trajectory, according to the record of the STATE field. In other words, a record that changes in the STATE field is the first record of a trajectory. In addition, we only keep trajectory whose length is greater than 20, to ensure the integrity and effectiveness of the trajectory. Finally, 123 trajectories consist of 6494 taxis GPS records were obtained, as shown in Fig. 8.

Fig. 8. 123 taxi trajectories in Chongqing City

5.3   Results

There are three parameters in Algorithm 1: δ(maximum allowable distance for two points to match), ε(the maximum neighborhood distance) and σ(the minimum number of required trajectories in a cluster). The algorithm 2 has a length threshold τ. These parameters need to be specified by a user and there is no specific rule for finding optimal parameter values.

The number of matching point is affected by parameter δ and computation cost increases due to the increasing of δ; the number of clusters is affected by parameter ε and if ε is too large, all trajectories will be in one cluster; parameter the number of noise trajectory is affected by parameter σ and noise increases due to the increasing of σ. We conduct several experiments with different groups of parameters to observe its influence on the clustering effect. Finally, we set $\delta = 0.01, \varepsilon = 0.05, \sigma = 8$ in trajectory clustering. On this basis, we set the length threshold of longest common sub trajectory $\tau = 70$ and found hot path as shown in Fig. 9.

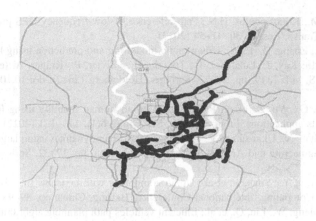

Fig. 9. Discovering hot routes

6 Conclusion

This paper focuses on identifying residents' trip hot routes in road networks using taxi trajectory data. We proposed a trajectory similarity measuring method based on the longest common subsequence. With combined DBSCAN algorithm, a taxi trajectory clustering algorithm called LCS-BASED DBSCAN was proposed. Furthermore, we extracted residents' trip hot routes based on LCS. Using real taxi trajectory data in the experiment, results verified the feasibility of the proposed method. The proposed clustering algorithm also has shortcomings that we only consider the distance factor. However, in real life, residents' trip will not only consider the distance factor, but also consider some impersonal and weather factors. By knowing the correlations between hot routes and other factors, one can enhance the usefulness of the discovered information.

Acknowledgments. This work was supported by the National High-tech R&D Program of China (2015AA015308), China Post-doctoral Science Foundation (2014T70852), Fundamental Research Funds for the Central Universities (106112014CDJZR188801), Chongqing Postdoctoral Science Foundation Project (Xm201305), and Key Projects of Chongqing Application Development (cstc2014yykfB30003).

References

1. Tang, J., Liu, F., Wang, Y., Wang, H.: Uncovering urban human mobility from large scale taxi GPS data. Phys. A: Stat. Mech. Appl. **438**(15), 140–153 (2015)
2. Mustary, N.R., Chander, R.P., Baig, M.N.A.: A performance evaluation of VANET for intelligent transportation system. World J. Sci. Technol. **2**(10), 89–93 (2012)
3. Feng, Y.H., Hourdos, J., Davis, G.A.: Probe vehicle based real-time traffic monitoring on urban roadways. Transp. Res. C **40**(40), 160–178 (2014)

4. Rahmani, M., Koutsopoulos, H.N.: Path inference of low-frequency GPS probes for urban networks. Transp. Res. C. **30**, 41–54 (2013)
5. Castro, P.S., Zhang, D., Li, S.: Urban traffic modelling and prediction using large scale taxi gps traces. In: Kay, J., Lukowicz, P., Tokuda, H., Olivier, P., Krüger, A. (eds.) Pervasive 2012. LNCS, vol. 7319, pp. 57–72. Springer, Heidelberg (2012). doi:10.1007/978-3-642-31205-2_4
6. Li, X., Pan, G., Wu, Z., et al.: Prediction of urban human mobility using large-scale taxi traces and its applications. Front. Comput. Sci. China **6**(1), 111–121 (2012)
7. Qian, X., Zhan, X., Ukkusuri, S.V.: Characterizing urban dynamics using large scale taxicab data. In: Engineering and Applied Sciences Optimization, pp. 17–32. Springer, Heidelberg (2015)
8. Zheng, Y., Liu, Y., Yuan, J., et al.: Urban computing with taxicabs. In: UBICOMP 2011: Ubiquitous Computing, International Conference, Beijing, China, pp. 89–98 (2011)
9. Zhang, J., Meng, W., Liu, Q., et al.: Efficient vehicles path planning algorithm based on taxi GPS big data. Optik Int. J. Light Electron Optics. **127**(5), 2579–2585 (2016)
10. Chu, V.W., Wong, R.K., Chen, F., et al.: Self-regularized causal structure discovery for trajectory-based networks. J. Comput. Syst. Sci. **82**(4), 594–609 (2015)
11. Cai, H., Jia, X., Chiu, A.S.F., et al.: Siting public electric vehicle charging stations in Beijing using big-data informed travel patterns of the taxi fleet. Transp. Res. Part D: Transport Environ. **33**, 39–46 (2014)
12. Kong, X., Xu, Z., Shen, G., et al.: Urban traffic congestion estimation and prediction based on floating car trajectory data. Future Gener. Comput. Syst. **61**, 97–107 (2016)
13. Xia, D., Wang, B., Li, H., et al.: A distributed spatial–temporal weighted model on MapReduce for short-term traffic flow forecasting. Neurocomputing **179**, 246–263 (2015)
14. Liu, X., Gong, L., Gong, Y., et al.: Revealing travel patterns and city structure with taxi trip data. J. Transport Geogr. **43**, 78–90 (2013)
15. Cai, H., Zhan, X., Zhu, J., et al.: Understanding taxi travel patterns. Phys. A: Stat. Mech. Appl. **457**, 590–597 (2016)
16. Li, X., Han, J., Lee, J.-G., Gonzalez, H.: Traffic density-based discovery of hot routes in road networks. In: Papadias, D., Zhang, D., Kollios, G. (eds.) SSTD 2007. LNCS, vol. 4605, pp. 441–459. Springer, Heidelberg (2007). doi:10.1007/978-3-540-73540-3_25
17. Kim, J., Mahmassani, H.S.: Spatial and temporal characterization of travel patterns in a traffic network using vehicle trajectories. Transp. Res. Part C Emerg. Technol. **9**, 164–184 (2015)
18. Zheng, Y.: Trajectory data mining: an overview. ACM Trans. Intell. Syst. Technol. **6**(3), 1–41 (2015)

Discovering Spatially Contiguous Clusters in Multivariate Geostatistical Data Through Spectral Clustering

Francky Fouedjio$^{(\boxtimes)}$

CSIRO Mineral Resources, Perth, WA, Australia
francky.fouedjiokameni@csiro.au

Abstract. Spectral clustering has recently become one of the most popular modern clustering algorithms for traditional data. However, the application of this clustering method on geostatistical data produces spatially scattered clusters, which is undesirable for many geoscience applications. In this work, we develop a spectral clustering method aimed to discover spatially contiguous and meaningful clusters in multivariate geostatistical data, in which spatial dependence plays an important role. The proposed spectral clustering method relies on a similarity measure built from a non-parametric kernel estimator of the multivariate spatial dependence structure of the data, emphasizing the spatial correlation among data locations. The capability of the proposed spectral clustering method to provide spatially contiguous and meaningful clusters is illustrated using the European Geological Surveys Geochemical database.

Keywords: Geostatistics · Spectral clustering · Spatial dependency · Spatial contiguity

1 Introduction

In recent years, spectral clustering has become one of the most popular modern clustering algorithms for classical data [11,19,22,23,27]. Spectral clustering is a class of partitional clustering algorithms that relies on the eigendecomposition of feature similarity matrices to partition the data points. Advantages of using spectral clustering include its flexibility in terms of incorporating diverse types of similarity measures, the superiority of its clustering solution compared to traditional clustering algorithms such as K-means algorithm, and its well-established theoretical properties [7,17,20,21,31].

However, applied to geostatistical data, spectral clustering method tends to produce spatially scattered clusters, which undesirable for many geoscience applications. This clustering method can not produce spatially contiguous and meaningful clusters because it makes the assumption that observations are independent. This fundamental assumption, however, does not hold in the realm of spatial data. Geostatistical data distinguish themselves from conventional data in that they often show properties of spatial dependency and heterogeneity over

© Springer International Publishing AG 2016
J. Li et al. (Eds.): ADMA 2016, LNAI 10086, pp. 547–557, 2016.
DOI: 10.1007/978-3-319-49586-6_38

the study domain. Observations located close to one another in the geographical space might have similar characteristics. Furthermore, the mean, the variance, and the spatial dependence structure can be different from one sub-domain to another.

Existing clustering approaches which take into account the specificities of geostatistical data can be classified into four groups: (1) non-spatial clustering with geographical coordinates as additional variables, (2) non-spatial clustering based on a spatial dissimilarity measure, (3) spatially constrained clustering, and (4) model-based clustering. The first group incorporates the spatial information by treating each observation as a point in a dimensional space formed by the geographical space and the attribute space, for a non-spatial clustering method. The second group uses existing non-spatial clustering methods by modifying the dissimilarity measure between two observations to take explicitly into account the spatial dependence [5,12,13,24]. The third group considers spatial contiguity constraints (rather than spatial dissimilarities) in a clustering process [25,26]. The latest group is not model-free. It relies on the assumption that observations are drawn from a particular distribution like a mixture of Gaussian or Markov random fields [1–4,10,14].

In this work, we propose a spectral clustering method designed for multivariate geostatistical data, in which spatial dependence plays an important role. The basic idea is to include the spatial information in the clustering procedure through a non-parametric kernel estimator of the multivariate spatial dependence structure of the data. This estimator is used to build a measure of similarity between two data locations, emphasizing the spatial correlation among data locations. The proposed clustering method is non-parametric, adapted to irregularly spaced data, and can produce spatially contiguous clusters without including any geometrical constraints. The proposed spectral clustering method is illustrated using the European Geological Surveys Geochemical database. The results derived from the proposed spectral clustering method are compared with those provided by two baseline clustering methods: K-means clustering and traditional spectral clustering.

The rest of the paper is organized as follows. Section 2 describes the proposed spectral clustering method through its basic ingredients. Section 3 illustrates using the European Geological Surveys Geochemical database, the capability of the proposed clustering method to providing spatially contiguous and meaningful clusters. Section 4 outlines concluding remarks.

2   Method

We consider a set of p standardized variables of interest $\{Z_1, \ldots, Z_p\}$ defined on a continuous domain of interest $G \subset \mathbb{R}^d (d \geq 1)$, and all measured at a set of distinct locations $\{\mathbf{x}_1, \ldots, \mathbf{x}_n\}$. The goal is to partition these data locations into spatially contiguous and meaningful clusters so that data locations belonging to the same cluster are more similar than those in different clusters. We describe in this section the different ingredients required to implement the proposed spectral clustering method.

2.1   Similarity Measure

One of the key tasks in spectral clustering as well as in other clustering methods is the choice of the similarity measure. The traditional spectral clustering usually calculates the similarity using the well-known Gaussian kernel based on the Euclidean distance in the attribute space. However, in the geostatistical setting, this type of similarity measure can not reflect the spatial dependence structure of the data, even if geographical coordinates are also considered as attributes. We propose a novel similarity measure that takes care of the spatial dependence between observations.

A non-parametric kernel estimator of the multivariate spatial dependence structure of the data described by the direct and cross variograms, at two locations $\mathbf{u} \in G$ and $\mathbf{v} \in G$ is given by:

$$\widehat{\gamma}_{ij}(\mathbf{u}, \mathbf{v}) = \frac{\sum_{l,l'=1}^{n} K_{\epsilon}^{\star}\left((\mathbf{u}, \mathbf{v}), (\mathbf{x}_l, \mathbf{x}_{l'})\right) \left(Z_i(\mathbf{x}_l) - Z_i(\mathbf{x}_{l'})\right)\left(Z_j(\mathbf{x}_l) - Z_j(\mathbf{x}_{l'})\right)}{2 \sum_{l,l'=1}^{n} K_{\epsilon}^{\star}\left((\mathbf{u}, \mathbf{v}), (\mathbf{x}_l, \mathbf{x}_{l'})\right)} \mathbb{1}_{\{\mathbf{u} \neq \mathbf{v}\}},$$

$$(1)$$

where $(i, j) \in \{1, \ldots, p\}^2$; $K_{\epsilon}^{\star}\left((\mathbf{u}, \mathbf{v}), (\mathbf{x}_l, \mathbf{x}_{l'})\right) = K_{\epsilon}(\|\mathbf{u} - \mathbf{x}_l\|)K_{\epsilon}(\|\mathbf{v} - \mathbf{x}_{l'}\|)$, with $K_{\epsilon}(\cdot)$ a non-negative kernel function with constant bandwidth parameter $\epsilon > 0$; $\mathbb{1}$ denotes the indicator function.

Given the set of estimated direct and cross variograms $\{\widehat{\gamma}_{ij}(\cdot, \cdot)\}_{i,j=1}^{p}$, the similarity between two sample locations $\mathbf{x}_t$ and $\mathbf{x}_{t'}$ $(t, t' = 1, \ldots, n)$ is defined by:

$$s(\mathbf{x}_t, \mathbf{x}_{t'}) = 1 - \frac{1}{\Gamma} \sum_{i,j=1}^{p} |\widehat{\gamma}_{ij}(\mathbf{x}_t, \mathbf{x}_{t'})|, \qquad (2)$$

with $\Gamma = \max_{(t,t') \in \{1,\ldots,n\}^2} \sum_{i,j=1}^{p} |\widehat{\gamma}_{ij}(\mathbf{x}_t, \mathbf{x}_{t'})|$. The resulting similarity matrix at all data locations is denoted $\mathbf{S} = [s(\mathbf{x}_t, \mathbf{x}_{t'})]_{t,t'=1,\ldots,n}$.

In Eq. (2), the term $\frac{1}{\Gamma} \sum_{i,j=1}^{p} |\widehat{\gamma}_{ij}(\mathbf{x}_t, \mathbf{x}_{t'})|$ represents the dissimilarity (normalized) between data locations $\mathbf{x}_t$ and $\mathbf{x}_{t'}$. Thus, the dissimilarity between two data locations is defined as the sum (normalized) of absolute values of all direct and cross variograms at these two data locations. Equation (2) well defines a measure of similarity [28].

2.2   Similarity Graph

Spectral clustering requires that the data are represented in the form of an undirected similarity graph $\mathcal{G} = (\mathcal{V}, \mathcal{E})$, where $\mathcal{V}$ is the set of vertices and $\mathcal{E}$ is the set of edges between pairs of vertices. We construct a graph $\mathcal{G}$ from the similarity measure defined in Eq. (2), where the vertices of the graph represent the data locations, and the edge weights represent similarities between data locations. The similarity graph resulting from this construction is a full connected (complete) graph. This construction is suited according to [19] since the similarity measure defined in Eq. (2) itself already encodes local neighbourhoods (through the kernel function $K_\epsilon(\cdot)$ in Eq. (1)). Moreover, this construction is coherent with the Tobler's first law of geography [29]: everything is related to everything else, but near things are more related than distant things.

Given the similarity graph, the next step in the spectral clustering is the computation of the graph Laplacian matrix. There are several versions of the graph Laplacian matrix [19,20]. We use the normalized graph Laplacian matrix defined as: $\mathbf{L} = \mathbf{D}^{-1/2}(\mathbf{D}-\mathbf{S})\mathbf{D}^{-1/2} = \mathbf{I} - \mathbf{D}^{-1/2}\mathbf{S}\mathbf{D}^{-1/2}$, where $\mathbf{S}$ is the affinity (similarity) matrix between every pair of the data locations built from the similarity measure defined in Eq. (2); $\mathbf{D}$ is a diagonal matrix whose elements are the degrees of the nodes of the graph $\mathcal{G}$ and corresponding to $d_{tt} = \sum_{t'=1}^{n} s(\mathbf{x}_t, \mathbf{x}_{t'})$; $\mathbf{I}$ denotes the identity matrix.

2.3 Spectral Clustering Algorithm

By representing data locations as a similarity graph, the clustering problem is equivalent to a graph partitioning problem, where we identify connected components with clusters. For a given number of clusters q, spectral clustering algorithm finds the top q eigenvectors. These q eigenvectors define a q-dimensional projection of the data. Then, a standard clustering algorithm such as K-means is applied to derive the final clusters of the data locations. The proposed spectral clustering algorithm for multivariate geostatistical data performs the following steps:

1. compute the similarity matrix of all data locations $\mathbf{S}$;
2. compute the degree matrix $\mathbf{D}$;
3. compute the graph Laplacian matrix $\mathbf{D}^{-1/2}\mathbf{S}\mathbf{D}^{-1/2}$;
4. compute the q largest eigenvalues of $\mathbf{D}^{-1/2}\mathbf{S}\mathbf{D}^{-1/2}$ and form the matrix $\mathbf{F} \in \mathbb{R}^{n \times q}$ whose columns are the associated q first eigenvectors of $\mathbf{D}^{-1/2}\mathbf{S}\mathbf{D}^{-1/2}$;
5. normalize the rows of $\mathbf{F}$ to norm 1;
6. cluster the rows of $\mathbf{F}$ with the K-means algorithm into clusters $C_1, \ldots, C_q$;
7. assign data location $\mathbf{x}_t$ to the same cluster the row t of $\mathbf{F}$ has been assigned.

2.4 Hyper-parameters Selection

The proposed spectral clustering method relies on the kernel function $K_\epsilon(\cdot)$ used in the estimation of the multivariate spatial dependence structure of the data (Eq. (1)). The choice of the kernel function $K_\epsilon(\cdot)$ is less important than the choice of its bandwidth parameter ϵ. We opt for the Epanechnikov kernel whose support is compact, showing optimality properties in density estimation [30]. To estimate the spatial dependence structure of the data reliably, the bandwidth parameter ϵ is chosen by using an empirical rule of thumb in geostatistics [9,15,16]: ϵ is chosen so that the support of the kernel function $K_\epsilon(\cdot)$ centered at each data location contains at least 35 observations. Thus, for each data location its distance to the 35th neighbour is computed; then, the maximum of resulting distances is taken as the value of the bandwidth parameter ϵ.

The optimal number of clusters is chosen so that it corresponds to the best clustering identified in terms of an internal clustering validation measure. A variety of internal cluster validation indexes has been proposed in the literature [8,28]. We choose the Caliński-Harabasz index [6] which is best suited for K-means clustering solutions with squared Euclidean distances. Given various number of clusters $q = 2, 3, \ldots$, the optimal number of clusters is the one that maximizes the Caliński-Harabasz index:

$$CH(q) = \frac{B(q)/(q-1)}{W(q)/(n-q)},$$ (3)

where $B(q) = \sum_{m=1}^{q} n_m \|\bar{\mathbf{y}}_m - \bar{\mathbf{y}}\|^2$ is the overall between-cluster variance, and $W(q) = \sum_{m=1}^{q} \sum_{t \in C_m} \|\mathbf{y}_t - \bar{\mathbf{y}}_m\|^2$ is the overall within-cluster variance; $\mathbf{y}_t \in \mathbb{R}^q$ is the vector corresponding to the t-th row of the matrix $\mathbf{F}$; $\bar{\mathbf{y}}_m = \frac{1}{n_m} \sum_{t \in C_m} \mathbf{y}_t$ is the average of points in cluster C_m, and $\bar{\mathbf{y}} = \frac{1}{n} \sum_{t=1}^{n} \mathbf{y}_t$ is the overall average; n_m is the number of points in cluster C_m.

3 Application

The proposed spectral clustering method is applied to the European Geological Surveys Geochemical database. The results provided by the proposed spectral clustering method are compared with those produced by two baseline clustering methods: K-means clustering and traditional spectral clustering.

3.1 Dataset

Data correspond to eight critical heavy metals in topsoils from the European Geological Surveys Geochemical database (26 European countries) [18]. Variables are: arsenic (As), cadmium (Cd), chromium (Cr), copper (Cu), mercury (Hg), nickel (Ni), lead (Pb), and zinc (Zn). On 1588 georeferenced available data, 1498 observations have been used in this application because there are some missing values for some variables. Prior to the clustering, all variables are logit-transformed and standardized. A representation of logit-transformed and standardized variables is given in Fig. 1. In the two baseline clustering methods, geographical coordinates are considered as attributes.

3.2 Results

Figure 2 shows the results provided by the baseline clustering methods and the proposed spectral clustering method, for different predefined number of clusters (from 2 to 4). As one can see, the baseline clustering (non-spatial clustering) methods fail to produce spatially contiguous clusters. The failure of these clustering methods is not surprising because they do not distinguish between the geographical space and the attribute space. It appears that the proposed spectral clustering method can produce spatially contiguous clusters. Moreover, the

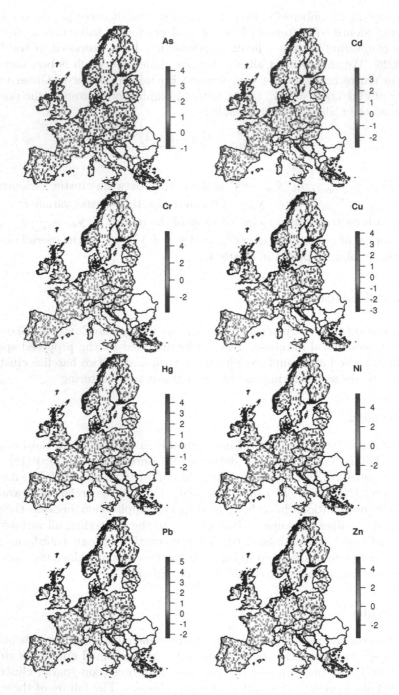

Fig. 1. Logit-transformed and standardized variables for clustering purpose. (Color figure online)

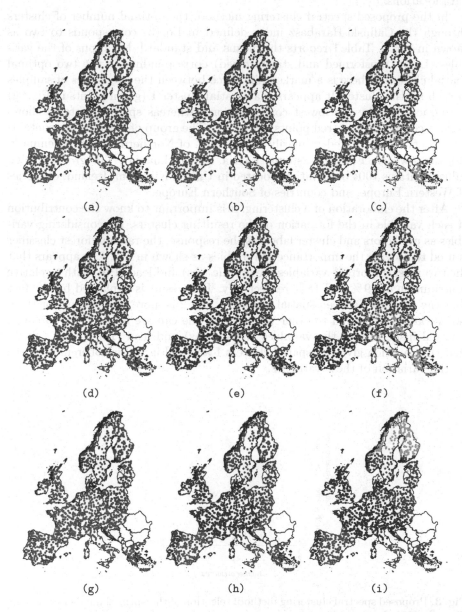

Fig. 2. (a, b, c) K-means clustering for 2, 3, and 4 clusters; (d, e, f) Traditional spectral clustering for 2, 3, and 4 clusters; (g, h, i) Proposed spectral clustering for 2, 3, and 4 clusters. The color of dots identifies the cluster membership. (Color figure online)

proposed spectral clustering method can produce disconnected clusters of similar data locations.

In the proposed spectral clustering method, the optimal number of clusters through the Caliński-Harabasz index defined in Eq. (3) corresponds to two as shown in Fig. 3. Table 1 reports the means and standard deviations of the variables (Logit-transformed and standardized) corresponding to the two optimal spatial clusters. There is a marked difference between the properties of samples in each spatial cluster. It appears that spatial cluster 1 (green points in Fig. 2g) is characterized by the lowest concentrations; whereas spatial cluster 2 shows highest concentrations (red points in Fig. 2g). The group of lower values contains 494 observations located primarily in countries of Northern Europe (Denmark, Norway, Sweden, Finland, Estonia, Latvia, and Lithuania). The group of high values contains 1004 observations located in United Kingdom, Ireland, countries of Western Europe, and countries of Southern Europe.

After the elaboration of a clustering, it is important to know the contribution of each variable in the formation of the resulting clusters. By considering variables as predictors and cluster labels as the response, the random forest classifier is used to provide the importance of variables as shown in Fig. 4. It appears that the two most important variables are arsenic (As) and lead (Pb), with a relative contribution of 19 % and 18 % respectively. This result is explained by the fact that the contrast between spatial clusters 1 and 2 is more pronounced for these two variables compared to other variables as one can see in Table 1. Moreover, a visual inspection of the variables arsenic (As) and lead (Pb) (Fig. 1) shows that the partition given by spatial clusters 1 and 2 (Fig. 2g) is coherent with the spatial variation of these variables.

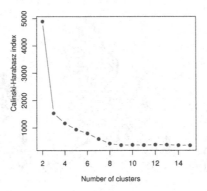

Fig. 3. Proposed spectral clustering method: selection of the optimal number of clusters through CH index.

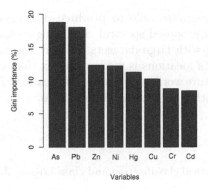

Fig. 4. Proposed spectral clustering method: contribution of each variable in the formation of the two optimal spatial clusters based on the Gini importance measure of the random forest classifier.

Table 1. Proposed spectral clustering method: means and standard deviations of the variables (Logit-transformed and standardized) corresponding to the two optimal spatial clusters.

	Spatial cluster 1 ($n_1 = 494$)		Spatial cluster 2 ($n_2 = 1004$)	
	Mean	Std.	Mean	Std.
As	-0.77	0.54	0.38	0.95
Cd	-0.50	0.79	0.25	1.00
Cr	-0.49	0.95	0.24	0.93
Cu	-0.65	0.94	0.32	0.86
Hg	-0.60	0.81	0.29	0.95
Ni	-0.65	0.77	0.32	0.94
Pb	-0.72	0.91	0.36	0.84
Zn	-0.62	0.88	0.31	0.91

4   Conclusion

In this work, a spectral clustering method aimed to discover spatially contiguous and meaningful clusters in multivariate geostatistical data has been developed. The proposed spectral clustering method relies on a similarity measure built from a non-parametric kernel estimator of the multivariate spatial dependence structure of the data, thereby reinforcing the spatial contiguity of the resulting clusters. The proposed spectral clustering approach is non-parametric; there is no distributional assumptions or spatial dependence structure assumptions. It is adapted to irregularly sampled data and can produce spatially contiguous and meaningful clusters without including any geometrical constraints. Applied to the European Geological Surveys Geochemical database, the proposed spectral clustering method highlights two spatially contiguous clusters with significant meaning. It is also able to produce disconnected clusters of similar data locations. The proposed spectral clustering method is computationally intensive when dealing with large datasets. Indeed, the calculation of the similarity matrix at all data locations is more complex than calculating the sum of squared deviations. Future work includes the application of the proposed spectral clustering method to other geostatistical databases.

References

1. Allard, D.: Geostatistical classification and class kriging. J. Geog. Inf. Decis. Anal. **2**, 87–101 (1998)
2. Allard, D., Guillot, G.: Clustering geostatistical data. In: Proceedings of the Sixth Geostatistical Conference (2000)
3. Allard, D., Monestiez, P.: Geostatistical segmentation of rainfall data. In: geoENV II: Geostatistics for Environmental Applications, pp. 139–150 (1999)
4. Ambroise, C., Dang, M., Govaert, G.: Clustering of spatial data by the EM algorithm. In: geoENV I: Geostatistics for Environmental Applications, pp. 493–504 (1995)
5. Bourgault, G., Marcotte, D., Legendre, P.: The multivariate (co)variogram as a spatial weighting function in classification methods. Math. Geol. **24**(5), 463–478 (1992)
6. Caliński, T., Harabasz, J.: A dendrite method for cluster analysis. Commun. Stat. **3**(1), 1–27 (1974)
7. Cao, Y., Chen, D.R.: Consistency of regularized spectral clustering. Appl. Comput. Harmonic Anal. **30**(3), 319–336 (2011)
8. Charu, C., Chandan, K.: Data Clustering: Algorithms and Applications. Chapman and Hall/CRC, Boca Raton (2013)
9. Chilès, J.P., Delfiner, P.: Geostatistics: Modeling Spatial Uncertainty. Wiley, Hoboken (2012)
10. Dempster, A.P., Laird, N.M., Rubin, D.B.: Maximum likelihood from incomplete data via EM algorithm (with discussion). J. Roy. Stat. Soc. Ser. **39**, 1–38 (1977)
11. Filippone, M., Camastra, F., Masulli, F., Rovetta, S.: A survey of kernel and spectral methods for clustering. Pattern Recogn. **41**(1), 176–190 (2008)
12. Fouedjio, F.: A clustering approach for discovering intrinsic clusters in multivariate geostatistical data. In: Perner, P. (ed.) MLDM 2016. LNCS, vol. 9729, pp. 491–500. Springer, Switzerland (2016)
13. Fouedjio, F.: A hierarchical clustering method for multivariate geostatistical data. Spatial Statistics (2016)
14. Guillot, G., Kan-King-Yu, D., Michelin, J., Huet, P.: Inference of a hidden spatial tessellation from multivariate data: application to the delineation of homogeneous regions in an agricultural field. J. Roy. Stat. Soc. Ser. C (Appl. Stat.) **55**(3), 407–430 (2006)
15. Haas, T.C.: Lognormal and moving window methods of estimating acid deposition. J. Am. Stat. Assoc. **85**(412), 950–963 (1990)
16. Journel, A., Huijbregts, C.: Mining Geostatistics. Blackburn Press, New York (2003)
17. Kannan, R., Vempala, S., Vetta, A.: On clusterings: good, bad and spectral. J. ACM **51**(3), 497–515 (2004)

18. Lado, L., Hengl, T., Reuter, I.: Heavy metals in European soils: a geostatistical analysis of the FOREGS geochemical database. Geoderma **148**(2), 189–199 (2008)
19. Luxburg, U.V.: A tutorial on spectral clustering. Stat. Comput. **17**(4), 395–416 (2007)
20. Luxburg, U.V., Belkin, M., Bousquet, O.: Consistency of spectral clustering. Ann. Stat. **36**(2), 555–586 (2008)
21. Luxburg, U.V., Bousquet, O., Belkin, M.: Limits of spectral clustering. In: Advances in Neural Information Processing Systems, pp. 857–864 (2004)
22. Nascimento, M.C., Carvalho, A.C.: Spectral methods for graph clustering – a survey. Eu. J. Oper. Res. **211**(2), 221–231 (2011)
23. Ng, A.Y., Jordan, M.I., Weiss, Y.: On spectral clustering: Analysis and an algorithm. In: Advances in Neural Information Processing Systems, pp. 849–856. MIT Press (2001)
24. Olivier, M., Webster, R.: A geostatistical basis for spatial weighting in multivariate classification. Math. Geol. **21**, 15–35 (1989)
25. Pawitan, Y., Huang, J.: Constrained clustering of irregularly sampled spatial data. J. Stat. Comput. Simul. **73**(12), 853–865 (2003)
26. Romary, T., Ors, F., Rivoirard, J., Deraisme, J.: Unsupervised classification of multivariate geostatistical data: two algorithms. Comput. Geosci. **85**, 96–103 (2015)
27. Schaeffer, S.E.: Graph clustering. Comput. Sci. Rev. **1**(1), 27–64 (2007)
28. Theodoridis, S., Koutroumbas, K.: Pattern Recognition, 4th edn. Academic Press, New York (2009)
29. Tobler, W.R.: A computer movie simulating urban growth in the Detroit region. Econ. Geogr. **46**, 234–240 (1970)
30. Wand, M., Jones, C.: Kernel Smoothing. Monographs on Statistics and Applied Probability. Chapman & Hall, Sanford (1995)
31. Zha, H., He, X., Ding, C., Gu, M., Simon, H.D.: Spectral relaxation for k-means clustering. In: Advances in Neural Information Processing Systems, pp. 1057–1064 (2001)

On Improving Random Forest
for Hard-to-Classify Records

Md Nasim Adnan[✉] and Md Zahidul Islam

School of Computing and Mathematics,
Charles Sturt University, Bathurst, NSW 2795, Australia
{madnan,zislam}@csu.edu.au

Abstract. Random Forest draws much interest from the research community because of its simplicity and excellent performance. The splitting attribute at each node of a decision tree for Random Forest is determined from a predefined number of randomly selected subset of attributes of the entire attribute set. The size of the subset is one of the most controversial points of Random Forest that encouraged many contributions. However, a little attention is given to improve Random Forest specifically for those records that are hard to classify. In this paper, we propose a novel technique of detecting hard-to-classify records and increase the weights of those records in a training data set. We then build Random Forest from the weighted training data set. The experimental results presented in this paper indicate that the ensemble accuracy of Random Forest can be improved when applied on weighted training data sets with more emphasis on hard-to-classify records.

Keywords: Decision forest · Random Forest · Ensemble accuracy

1 Introduction

In the arena of machine learning, supervised learning tasks such as classification play an important role in knowledge discovery and pattern understanding. Classification attempts to generate a model (commonly known as the classifier) that maps a set of non-class attributes $m = \{A_1, A_2, ..., A_m\}$ to a predefined class attribute C of a training data set D [39]. There are different types of classifiers including Artificial Neural Networks [25,43,44], Bayesian Classifiers [11,31], Decision Trees [14,33,34] and Support Vector Machines [15].

The use of ensembles in classification have been actively studied in recent years to find more accurate classification models [6,7,21,24,35]. Interestingly, an ensemble of classifiers is found to be more effective for unstable classifiers such as decision trees [39]. Decision trees are considered to be an unstable classifier because a slight change in a training data set can cause a significant dissimilarity between the decision trees obtained from the original and modified data sets [39]. A decision forest is an ensemble of decision trees where an individual decision tree acts as a base classifier. Classification is performed by taking a vote based on the predictions made by each decision tree of the decision forest [39].

© Springer International Publishing AG 2016
J. Li et al. (Eds.): ADMA 2016, LNAI 10086, pp. 558–566, 2016.
DOI: 10.1007/978-3-319-49586-6_39

Random Forest [13] is regarded as a state-of-the-art decision forest building algorithm [9,10] which is technically a combination of Bagging [12] and Random Subspace [23] algorithms. Bagging generates new training data set D_i for which the records are chosen randomly from the original training data set D. A new training data set D_i contains the same number of records as in D. Thus, some records of D may be chosen multiple times and some records may not be chosen at all. This approach of generating a new training data set is known as bootstrap sampling [22]. Generally, 63.2 % of the original records are selected in a bootstrap sample [22]. The number of the bootstrap samples is a user input that determines the number of trees to be generated for the forest. A decision tree building algorithm is then applied on each bootstrap sample D_i ($i = 1, 2, \ldots, |T|$) in order to generate $|T|$ number of trees for the forest. The Random Subspace algorithm ([23]) randomly draws a subset of attributes (subspace) f from the entire attribute space m and then selects the best attribute from f as the splitting attribute for each node of a decision tree. The size of f is commonly known to be a hyperparameter [10] and for Random Forest $|f|$ is chosen to be $int(\log_2 |m|) + 1$ [13].

It is worth to mention that the size of f ($int(\log_2 |m|) + 1$) does not increase at the same rate to the increase of m. For example, let us assume that we have a low dimensional data set consisting of 4 attributes. Thus a splitting attribute is determined from a randomly selected subspace of 3 attributes ($int(\log_2 4) + 1 = 3$) encompassing 75 % of the total attributes. As a result, the chance of appearing similar attributes in different subspaces becomes high, resulting in decreasing diversity among the trees. On the other hand, when $|m|$ is large say, 150 then f contains 8 randomly chosen attributes ($int(\log_2 150) + 1 = 8$) covering only 5 % of the total attributes. Hence, if the number of good attributes is not high enough in m then the chance of containing adequate number of good attributes in f becomes low, which is supposed to cause low individual accuracy of the trees.

Since its inception in 2001, Random Forest attains much interest from the research community and thereby numerous enhancements have been proposed in recent years [2–5,8,9,38,45]. In particular, the selection of more suitable subspace (f) invites much attention [1,10,17,21,30,36,40,41]. In [10], the authors proposed for random selection of $|f|$ between 1 to $|m|$ while [1] proposed for dynamic selection of $|f|$ based on the relative size of current data segment to the bootstrap samples at each node splitting event. The default size of the subspace drawn for Extremely Randomized Trees [21] is chosen to be $\sqrt{|m|}$ for classification problem. The Extremely Randomized Trees algorithm improvises more randomness for numerical attributes by selecting the cut-points fully at random while ensuring a minimum number of records in either sides of a cutpoint. Another algorithm [17] suggested setting the cut point midway between two training records that had been picked randomly.

In [41], the authors applied the stratified sampling of attributes for Random Forest to deal with high dimensional data set. The key idea behind the stratified sampling is to divide the attributes m into two groups. One group will contain the good attributes m_G and the other group will contain the bad attributes m_B. The attributes having the informativeness capacity higher than the average informativeness capacity are placed in the group of good attributes m_G and all other attributes are placed in the group of bad attributes m_B. Then $int(\log_2|m|)+1$ number of attributes are selected randomly from each group in proportion to the size of the groups. Unlike Random Forest, the stratified sampling method guarantees the presence of some good attributes in a subspace. However, in literature it is shown that dimensionality reduction/feature subset selection is more viable technique to deal with high dimensional data sets [16,26,42]. Yet, none of the above mentioned algorithms focused on improving Random Forest for those records that are hard to classify.

AdaBoost [18] that evolved as the first practical boosting algorithm adaptively changes the distribution of records in a training data set by focusing more on previously misclassified records. Initially, all records of the training data set are given the same weight so that each of them are equally likely to be selected in a sample. After the first classifier is built from the sample, it classifies all records from the training data set. Then, the weights of the misclassified records are increased and the weights correctly classified records are decreased to form a new weighted training data set. Any sample drawn from the new weighted training data set will have more misclassified records than correctly classified records and thus the next classifier will focus more on misclassified records from its immediate previous classifier. The process of assigning weights and building classifiers continues for a user defined number of iterations (called boosting rounds).

AdaBoost in its original form is designed assuming that the base classifiers are better than guessing random between two classes (accuracy level is better than 50 %) and thereby not suitable for multi-class classification problems without any modification [18,37]. One solution to overcome this problem was proposed to segregate the multi-class classification problem into two-class classification problem using available binarization techniques in literature [29]. However, it is also shown that binarization techniques help increasing accuracy of a wide range of classifiers individually [20] and also in the context of ensemble [5,19]. As a result, AdaBoost gains favour from binarization techniques in applicable circumstances. Nevertheless, the most serious shortcoming of AdaBoost is exposed when it is applied on noisy data (records with incorrect attribute values and/or class labels that are not possible to classify) as almost every classifier generated by AdaBoost will emphasise on noisy data to make the ensemble inaccurate. In addition, AdaBoost is not regarded as a parallel ensemble algorithm as only one classifier is generated at each boosting rounds as the generation process is dependent on the immediate previous base classifier.

In contrast, Random Forest is a parallel ensemble algorithm that can generate base classifiers very fast (due to use of subspace) and also robust to noise and overfitting [38]. As a result the application domain of Random Forest is larger. Thus, if Random Forest can be improved against hard-to-classify records the influence can be significant.

Recently, the authors conducted [28] an elaborated study to determine the minimum size of the ensemble required to obtain stable predictions assuming that majority voting [32] was used. Their analysis [28] showed that for most records only a small number of base classifiers are needed to reach the stable predictions. On the contrary, a small number of records require polling a large number of classifiers. Based on this remarkable finding, we reason that a small ensemble can correctly predict easily classifiable records and thus expose those records that are hard-to-classify. The main contribution of the paper is to employ a small-sized Random Forest to detect the hard-to-classify records at once and then increase their weight in training data set. Then Random Forest is generated from bootstrap samples drawn from the weighted training data set. We conduct an elaborate experimentation to evaluate the improvement of Random Forest in terms of ensemble accuracy when applied on the weighted training data set with more emphasis on hard-to-classify records. The experimental results indicate the effectiveness of the proposed technique.

The remainder of this paper is organized as follows: In Sect. 2 we explain the proposed technique. Section 3 discusses the experimental results. Finally, we offer some concluding remarks in Sect. 4.

2 Our Technique

The core of our technique is to generate a small-sized Random Forest from the original training data set to detect hard-to-classify records and then increase the weights of those records in such a way that does not drastically outbalance the proportion of records in bootstrap samples. The main steps of our proposed techniques are as follows:

Step 1: Generate a small-sized Random Forest (SRF) from the original training data set.

Step 2: Detect hard-to-classify records using SRF and then increase their weights in the training data set.

Step 3: Generate standard-sized Random Forest from the weighted training data set.

Step 1: **Generate a small-sized Random Forest (SRF) from the original training data set.** We generate the SRF from bootstrap samples drawn from the original training data set. We choose to generate 10 trees for SRF as the number is regarded to be quite small in ensemble standard [35].

Step 2: **Detect hard-to-classify records using SRF and then increase their weights in the training data set.** At first all records of the training data set are assigned the same weight (1.0). Then, SRF is employed to predict the class values of those records. Majority voting [32] is used to compile the ensemble prediction for SRF. Based on the prediction of SRF, the weight ω_i of a training record R_i is computed as follows:

$$\omega_i = \begin{cases} 1.0, & \text{if } R_i \text{ is predicted correctly.} \\ \frac{10-\alpha_i}{2.0}, & \text{if } R_i \text{ is predicted wrongly.} \\ 0.0, & \text{if } \alpha_i = 0. \end{cases}$$

Here, α_i is the number of trees in SRF that predict correctly for R_i. In accordance with [28], we consider the training records that were wrongly predicted by SRF to be hard-to-classify and their weights are increased relatively to the correctly predicted ones by the proposed technique. For example, let R_i be a training record that was wrongly predicted by SRF with $\alpha_i = 2$ (only two trees from the 10-tree SRF predict correctly for R_i). Accordingly, ω_i of R_i is calculated to be: $(10 - 2)/2.0 = 4.0$. The weight of any correctly predicted training record is retained as 1.0. In this way, a moderate increase in the weights for hard-to-classify records is ensured compared to AdaBoost. The weight increase for a misclassified record in AdaBoost can reach 99 times higher than a record that was classified correctly [39]. Compared to that, the highest weight increase according to the proposed technique occurs when α_i for R_i is 1 (i.e. only 1 tree from the 10-tree SRF predicts correctly for the record). At that time, ω_i of R_i will be: $(10 - 1)/2.0 = 4.5$. However, when $\alpha_i = 0$ meaning no tree from the 10-tree SRF predicts correctly for R_i, ω_i is set to 0.0. In this case, we treat R_i to be a noisy data.

Step 3: Generate standard-sized Random Forest from the weighted training data set. We generate 100-tree Random Forest from bootstrap samples drawn from the weighted training data set since the number is considered large enough to ensure convergence of the ensemble effect [8,10,21].

3   Experimental Results

We conducted an elaborated experimentation on 15 well known data sets that are publicly available from the UCI Machine Learning Repository [27]. The data sets used in the experimentation are described in Table 1. For example, the Libras Movement data set has 90 Non-Class Attributes, 360 Records with 15 Distinct Class Values.

For our experimentation, we remove records with missing values (Table 1 shows the number of records with no missing values) and identifier attributes such as *Transaction_ID* from each applicable data set. We generate 100 trees for both the original Random Forest (ORF) and the Random Forest generated from the weighted training data set (WRF). We use Gini Index [14] as a measure of classification capacity in accordance with the ORF [13]. The minimum Gini Index value is set to 0.01 for any attribute to qualify for splitting a node. Each leaf node of a tree requires at least two records and no further post-pruning is applied. We apply majority voting to aggregate results for the forests [13,32]. All the performance indicators reported in this paper are obtained using 10-fold-cross-validation (10-CV) [1] and the best results are stressed through **bold-face**.

Table 1. Description of the data sets

Data Set Name (DSN)	Non-Class Attributes	Records	Distinct Class Values
Balance Scale (BS)	04	625	3
Breast Cancer (BC)	33	602	2
Car Evaluation (CE)	06	1728	4
Chess (CHS)	36	3196	2
Credit Approval (CA)	15	653	2
Glass Identification (GI)	09	214	6
Hayes-Roth (HR)	04	132	3
Hepatitis (HEP)	19	80	2
Ionosphere (ION)	34	351	2
Iris (IRS)	04	150	3
Libras Movement (LM)	90	360	15
Liver Disorder (LD)	06	345	2
Statlog Heart (SH)	13	270	2
Statlog Vehicle (SV)	18	846	4
Wine (WNE)	13	178	3

Table 2. Ensemble Accuracy and tree related information for ORF and WRF

DSN	ORF			WRF		
	EA	TH	TGTS	EA	TH	TGTS
BS	80.50	**3.01**	0.15	**80.68**	**3.01**	**0.13**
BC	77.92	6.59	0.60	**78.88**	**6.09**	**0.46**
CE	91.19	**4.86**	**0.21**	**92.12**	5.03	0.23
CHS	95.22	**7.50**	**0.47**	**95.50**	7.79	0.50
CA	86.07	4.52	0.27	**86.68**	**4.21**	**0.24**
GI	74.12	8.66	0.40	**76.50**	**8.46**	**0.34**
HR	69.54	2.59	0.03	**75.49**	**2.48**	**0.02**
HEP	86.25	3.46	0.03	**87.50**	**3.45**	**0.02**
ION	**93.73**	**6.38**	1.09	**93.73**	6.42	**1.00**
IRS	**96.00**	3.81	**0.02**	**96.00**	3.75	**0.02**
LM	76.11	**11.01**	33.06	**76.67**	11.17	**29.60**
LD	**71.48**	10.11	0.20	**71.48**	**9.81**	**0.18**
SH	**82.96**	7.17	0.13	**82.96**	**7.09**	**0.12**
SV	74.14	14.37	2.50	**75.32**	**14.24**	**1.91**
WNE	97.25	**3.87**	**0.11**	**97.84**	3.90	**0.11**
Avg	83.50	6.53	2.62	**84.49**	**6.46**	**2.32**

Ensemble Accuracy (EA) is one of the most important performance indicators for any decision forest algorithm [1–3, 6]. Table 2 presents the comparison of EA (in percent) between ORF and WRF including some tree related information such as Tree Height (TH) and Tree Generation Time in Seconds (TGTS) for all data sets considered.

From Table 2, we observe that the EA of ORF can be improved through WRF. We also observe that the improvement is more significant for low-accuracy data sets (in this paper, data sets with EA < 80 % from ORF are termed as low-accuracy data sets). For low-accuracy data sets, the Average of EA from WRF is **75.72%** compared to 73.88 % from ORF. The rationale behind this is: low-accuracy data sets may have more hard-to-classify records compared to high-accuracy data sets. For high-accuracy data sets, the Average of EA from WRF is **90.33%** compared to 89.91 % from ORF. Besides, for 9 out of 15 data sets WRF has shorter TH compared to ORF (including one draw). Also, for 11 out of 15 data sets WRF generates trees faster than RF (including two draws).

4   Conclusion

In this paper, we propose a novel technique of detecting the hard-to-classify records and then increase their weights in the training data set and finally build Random Forest from the weighted training data set. We have gone through a rigorous empirical analysis to find the effectiveness of the proposed technique. The results of the experimentation show that the performance of Random Forest can be improved through the proposed technique. In future, we intend to extend our work by including data sets with more number of attributes and records. Further, we plan to apply the proposed technique on other decision forest algorithms.

References

1. Adnan, M.N.: On dynamic selection of subspace for random forest. In: Luo, X., Yu, J.X., Li, Z. (eds.) ADMA 2014. LNCS (LNAI), vol. 8933, pp. 370–379. Springer, Heidelberg (2014). doi:10.1007/978-3-319-14717-8_29
2. Adnan, M.N., Islam, M.Z.: A comprehensive method for attribute space extension for random forest. In: Proceedings of 17th International Conference on Computer and Information Technology, December 2014
3. Adnan, M.N., Islam, M.Z.: Complement random forest. In: Proceedings of the 13th Australasian Data Mining Conference (AusDM), pp. 89–97 (2015)
4. Adnan, M.N., Islam, M.Z.: Improving the random forest algorithm by randomly varying the size of the bootstrap samples for low dimensional data sets. In: Proceedings of the European Symposium on Artificial Neural Networks, Computational Intelligence and Machine Learning, pp. 391–396 (2015)
5. Adnan, M.N., Islam, M.Z.: One-vs-all binarization technique in the context of random forest. In: Proceedings of the European Symposium on Artificial Neural Networks, Computational Intelligence and Machine Learning, pp. 385–390 (2015)
6. Adnan, M.N., Islam, M.Z.: Forest CERN: a new decision forest building technique. In: In proceedings of the The 20th Pacific Asia Conference on Knowledge Discovery and Data Mining (PAKDD), pp. 304–315 (2016)

7. Ahmad, A., Brown, G.: Random projection random discretization ensembles - ensembles of linear multivariate decision trees. IEEE Trans. Knowl. Data Eng. **26**(5), 1225–1239 (2014)
8. Amasyali, M.F., Ersoy, O.K.: Classifier ensembles with the extended space forest. IEEE Trans. Knowl. Data Eng. **16**, 145–153 (2014)
9. Bernard, S., Adam, S., Heutte, L.: Dynamic random forests. Pattern Recogn. Lett. **33**, 1580–1586 (2012)
10. Bernard, S., Heutte, L., Adam, S.: Forest-RK: a new random forest induction method. In: Huang, D.-S., Wunsch, D.C., Levine, D.S., Jo, K.-H. (eds.) ICIC 2008. LNCS (LNAI), vol. 5227, pp. 430–437. Springer, Heidelberg (2008). doi:10.1007/978-3-540-85984-0_52
11. Bishop, C.M.: Pattern Recognition and Machine Learning. Springer, New York (2008)
12. Breiman, L.: Bagging predictors. Mach. Learn. **24**, 123–140 (1996)
13. Breiman, L.: Random forests. Mach. Learn. **45**, 5–32 (2001)
14. Breiman, L., Friedman, J., Olshen, R., Stone, C.: Classification and Regression Trees. Wadsworth International Group, CA (1985)
15. Burges, C.J.C.: A tutorial on support vector machines for pattern recognition. Data Min. Knowl. Disc. **2**, 121–167 (1998)
16. Chandrashekar, G., Sahin, F.: A survey on feature selection methods. Comput. Electr. Eng. **40**, 16–28 (2014)
17. Cutler, A., Zhao, G.: Pert: perfect random tree ensembles. Comput. Sci. Stat. **33**, 490–497 (2001)
18. Freund, Y., Schapire, R.E.: Experiments with a new boosting algorithm. In: Proceedings of the Thirteenth International Conference on Machine Learning, pp. 148–156 (1996)
19. Furnkranz, J.: Round robin classification. J. Mach. Learn. Res. **2**, 721–747 (2002)
20. Galar, M., Fernandez, A., Barrenechea, E., Bustince, H., Herrera, F.: An overview of ensemble methods for binary classifiers in multi-class problems: Experimental study on one-vs-one and one-vs-all schemes. Pattern Recogn. **44**, 1761–1776 (2011)
21. Geurts, P., Ernst, D., Wehenkel, L.: Extremely randomized trees. Mach. Learn. **63**, 3–42 (2006)
22. Han, J., Kamber, M.: Data Mining Concepts and Techniques. Morgan Kaufmann Publishers, San Francisco (2006)
23. Ho, T.K.: The random subspace method for constructing decision forests. IEEE Trans. Pattern Anal. Mach. Intell. **20**, 832–844 (1998)
24. Islam, M.Z., Giggins, H.: Knowledge discovery through SysFor - a systematically developed forest of multiple decision trees. In: Proceedings of the 9th Australian Data Mining Conference (2011)
25. Jain, A.K., Mao, J.: Artificial neural network: a tutorial. Computer **29**(3), 31–44 (1996)
26. Kwak, N., Choi, C.H.: Input feature selection for classification problems. IEEE Trans. Neural Netw. **13**(1), 143–159 (2012)
27. Lichman, M.: UCI machine learning repository. http://archive.ics.uci.edu/ml/datasets.html. Accessed 15 Mar. 2016
28. Lobato, D.H., Munoz, G.M., Suarez, A.: How large should ensembles of classifiers be? Pattern Recogn. **46**, 1323–1336 (2013)
29. Lorena, A.C., de Carvalho, A.C.P.L.F., Gama, J.M.P.: A review on the combination of binary classifiers in multiclass problems. Artif. Intell. Rev. **30**, 19–37 (2008)

30. Menze, B., Petrich, W., Hamprecht, F.: Multivariate feature selection and hierarchical classification for infrared spectroscopy: serum-based detection of bovine spongiform encephalopathy. Anal. Bioanal. Chem. **387**, 1801–1807 (2007)
31. Mitchell, T.M.: Machine Learning. McGraw-Hill, New York (1997)
32. Polikar, R.: Ensemble based systems in decision making. IEEE Circuits Syst. Mag. **6**, 21–45 (2006)
33. Quinlan, J.R.: C4.5: Programs for Machine Learning. Morgan Kaufmann Publishers, San Mateo (1993)
34. Quinlan, J.R.: Improved use of continuous attributes in c4.5. J. Artif. Intell. Res. **4**, 77–90 (1996)
35. Rodriguez, J.J., Kuncheva, L.I., Alonso, C.J.: Rotation forest: a new classifier ensemble method. IEEE Trans. Pattern Anal. Mach. Intell. **28**, 1619–1630 (2006)
36. Saeys, Y., Abeel, T., Peer, Y.: Robust feature selection using ensemble feature selection techniques. In: Daelemans, W., Goethals, B., Morik, K. (eds.) ECML PKDD 2008. LNCS (LNAI), vol. 5212, pp. 313–325. Springer, Heidelberg (2008). doi:10.1007/978-3-540-87481-2_21
37. Schapire, R.E.: Explaining AdaBoost. In: Schölkopf, B., Luo, Z., Vovk, V. (eds.) Empirical Inference, pp. 37–52. Springer, Heidelberg (2013)
38. Robnik-Šikonja, M.: Improving random forests. In: Boulicaut, J.-F., Esposito, F., Giannotti, F., Pedreschi, D. (eds.) ECML 2004. LNCS (LNAI), vol. 3201, pp. 359–370. Springer, Heidelberg (2004). doi:10.1007/978-3-540-30115-8_34
39. Tan, P.N., Steinbach, M., Kumar, V.: Introduction to Data Mining. Pearson Education, London (2006)
40. Tuv, E., Borisov, A., Runger, G., Torkkola, K.: Feature selection with ensembles, artificial variables, and redundancy elimination. J. Mach. Learn. Res. **10**, 1341–1366 (2009)
41. Ye, Y., Wu, Q., Huang, J.Z., Ng, M.K., Li, X.: Stratified sampling of feature subspace selection in random forests for high dimensional data. Pattern Recogn. **46**, 769–787 (2014)
42. Zhang, C., Masseglia, F., Zhang, X.: Discovering highly informative feature set over high dimensions. In: Proceedings of the IEEE 24th International Conference on Tools with Artificial Intelligence (ICTAI), pp. 1059–1064, November 2012
43. Zhang, G., Patuwo, B.E., Hu, M.Y.: Forecasting with artificial neural networks: The state of the art. Int. J. Forecast. **14**, 35–62 (1998)
44. Zhang, G.P.: Neural networks for classification: a survey. IEEE Trans. Syst. Man Cybern. **30**, 451–462 (2000)
45. Zhang, L., Suganthan, P.N.: Random forests with ensemble of feature spaces. Pattern Recogn. **47**, 3429–3437 (2014)

Scholarly Output Graph: A Graphical Article-Level Metric Indicating the Impact of a Scholar's Publications

Yu Liu[1,2(✉)], Dan Lin[1,2], Jing Li[1,2], and Shimin Shan[1,2]

[1] School of Software, Dalian University of Technology, Dalian 116620, China
{yuliu,ssm}@dlut.edu.cn, lindan0823@163.com, kobehz24@gmail.com
[2] Key Laboratory for Ubiquitous Network and Service Software of Liaoning Province, Dalian 116620, China

Abstract. Statistically, top scholars tend to accumulate a large number of publications during their tenure. While the patterns illustrating their scientific impact are monotonous and it is difficult to get a concrete comprehension to the academic development of the scholars' output. So we address the issue of graphically presenting and comparing the impact of individual scholars' publications. Besides, with the development of Web 2.0, more information about the social impact of a scholar's work is becoming increasingly available and relevant. Thus comes the challenge of how to quickly compare among a scholar's entire collection of publications, and pinpoint those with higher social popularity as well as academic influence. To this end, we propose a graphical article-level metric, namely Scholarly Output Graph (SOG). SOG captures three dimensions including journal impact factor (JIF), scientific impact and social popularity, and reflects not only the quality of the publications but also the immediate responses from social networks. With the visual cues of block length, width and color, users can intuitively locate articles of higher scientific impact, JIF and social popularity. Additionally, SOG proves to be widely applicable, practical and flexible as a navigation tool for filtering publications. To demonstrate the usability of SOG, we design a literature navigation homepage with a list of 50 researchers in computer science with their individual scholarly output graphs and the results can be found at http://impact.linkscholar.org/SOGExample.html.

Keywords: Graphical article-level metrics · Visualization · Scholarly Output Graph

1 Introduction

Distinguished researchers tend to publish a large number of articles nowadays. With the development of Web 2.0, an increasingly number of scholars being concerned on Internet as well as public media [2] and the sharing and discussion of these articles are moving away from traditional venues to the Web [1]. On the backdrop of science blossoming, bibliometric metrics of scholarly scientific

© Springer International Publishing AG 2016
J. Li et al. (Eds.): ADMA 2016, LNAI 10086, pp. 567–579, 2016.
DOI: 10.1007/978-3-319-49586-6_40

achievement are of particular significance. It requires an effective way to describe scholars' output comprehensively and systematically. Therefore, such situation presents a challenge to traditional metrics that are used to measure the impact of an individual's scientific works.

Traditionally, Journal impact factor (JIF) and citation counts have been used to measure the impact of a scholar's works in the scientific community. JIF is helpful while evaluating scholarly output impact as papers from high-IF journals are usually with high quality [4]. However, JIF is also much criticized because it shows the average citation score of all the articles published in a journal for a given period of time, instead of an individual article [5,6]. H-index is one of the most successful citation-based indicators proposed by Hirsch [7]. Citation is a good indicator in measuring an article's scientific impact, but it shows limited impact evaluation on who cite this paper. Besides, citation-based indexes take time to accumulate and therefore have the problem of time delay [12].

Nowadays, Article-Level Metrics (ALM) or alt-metrics are new, online-based indicators for evaluation of impact of publications [13]. They capture richer and timelier (daily or weekly) information derived from online open access archives, social bookmarking systems and social media [14,15]. They show us not only what impact looks like but also what makes impact [15]. Altmetrics have the potential to inform about social impact of scholarly output and be widely used and studied [16].

Although these studies of scholarly indicators do make some progress, there are still many challenges needed to be solved. First, it is not obvious how to intuitively display scholars' output and the dynamical changes of the scholars' output so that one can have a quick understanding about the scholars. Second, tools to comprehensively and intuitively hold the scholars' impact remain to be explored.

In order to help show and appraise individual scholar's publications in terms of both scientific impact and social popularity, we propose a graphical article level metric that represents publications as blocks in a planar histogram, named Scholarly Output Graph (SOG). We use journal impact factor (JIF) and citations to indicate scientific impact and readerships in Mendeley to indicate social popularity. We visualize these three dimensions by block width, block height and block color respectively. SOG provides an intuitive comparison among an individual scholar's entire collection of publications and is suitable for various kinds of application scenarios. As an example to prove feasibility and usability of SOG, we have released a literature navigation homepage with a list of 50 researchers in computer science, and the results indicate that SOG is flexible and convenient. SOG also contains interactive features that can be used to explore and filter important and relevant papers. In conclusion, SOG provides a comprehensive and graphical way to objectively describe a scholar's impact.

In the following sections, we introduce Scholarly Output Graph in detail and its test samples. The second section provides a review on related work, followed by the detail information section of Scholarly Output Graph. The section of Applications and Results shows the applications of SOG and their results. Finally, the last section gives a conclusion.

2 Related Work

In tradition, many objective scientometric indicators for scholars' output have been developed. At first, people qualify scholars' output by the impact of the journals where the articles get published. Journal Impact Factor (JIF) is an indicator to search the articles in journal [3] and Journal Impact Factor (JIF) shows the average cited times by other articles that a scientific received in a period of time [4]. Higher-JIF articles usually get more citations and attentions. And scholars prefer to get published by a higher-IF journal rather than a lower one as high-IF journals are stricter with article quality [5]. It is well recognized that JIF can be used to evaluate a scholar's works although there are doubt and criticize about that [5,6].

As citations indicator is developed, some new metrics are proposed. H-index is an index proposed to quantify the researchers' output, it is easy to compute and convenient to compare among different scholars [8,9]. Then, some variations of h-index are proposed such as G-index, hg-index, A-index, R-index and hw-index for web pages [10,11]. All of these indexes take into account some new factors that are not included in the h-index originally [10]. But these indexes are not comprehensive enough to show the scholars' output development visually.

With the development of web 2.0, New alternative metrics, shorted as 'alt-metrics', are organized to measure the impact of publications based online environment [20], such as social coverage [25], comments and blog post, social bookmarks [29], social citations [13]. Altmetrics assess a broader range of impact including papers quality, pre-publications, main argument or passage and so on [21]. Especially, the web-based social popularity of papers was intended to show the attention that articles received [16,17] in social media such as Twitter and Facebook, online reference manager such as Mendely and CiteULike, news media such as blogs and Wikipedia, and other popular medias [18,19].

Altmetrics get great attentions by the scientific community at the very beginning of proposal and large amount of researches are exceeded [32,33]. Important works have been carried out to construct alt-metrics as impact measures to filter important articles [22] and to study the correlation between alt-metrics and traditional measures. Several researches use social bookmarking systems for usage-based metrics of scientific evaluation of papers and journals [23,24]. Results from the studies of Zahedi et al. and Li et al. show that scholarly impact from social bookmarks significantly correlates with traditional citation-based impact [30,31]. A couple of other researches use 'citation' and 'references' within Twitter for scientific communication and impact measures [26,27]. Results from Bollen et al. show that it is possible to devise impact metrics based on usage information [28]. Bornmann provides an overview of research into three of the most important altmetrics: microblogging (Twitter), online reference managers and blogging [34,35].

3   Methods for the Metric

To build Scholarly Output Graph system, we first utilize and collect specific data for the three different dimensions discussed above, seen at Subsect. 3.1. Then in Subsect. 3.2, we introduce how the three metrics are calculated in SOG. Thus, the main idea of SOG is illustrated roundly.

3.1   Data Preparation

For the three different dimensions discussed above, we utilize and collect specific data. First, we use Scopus citations to indicate the dimension of scientific impact. We integrate Scopus citations from Scopus (www.scopus.com) via open API. Second, the dimension of JIFs of the articles are collected from ISI Web of Science (online resource). Third, the altmetrics for dimension of social popularity are choosed following.

We select four metrics including bookmarks in Mendeley, bookmarks in CiteULike, tweets in Twitter and readerships in Facebook. Those metrics are from four most popular and reliable resources including Twitter, Facebook, Mendeley and CiteULike. We collect 7,352 publications and their DOIs as our sample data from Scopus in subject of Computer science and based on DOIs of those publications. We extract these altmetrics data from altmetric.com (www.altmetric.com).

We compare the four metrics in aspects of publication coverage and correlation with citations in Scopus. The results are successfully, as is shown following in Figs. 1 and 2.

Figure 1 illustrates the coverage of different sources and we can visually see that Mendeley's coverage is over 80 %, far more than the other sources, followed

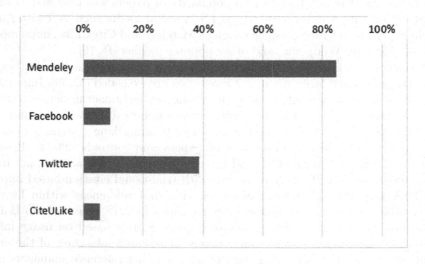

Fig. 1. The coverage of four altmetric sources.

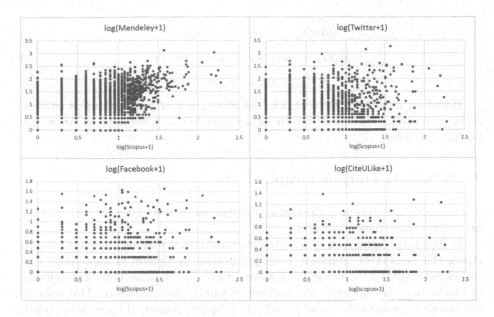

Fig. 2. The relationship between four metrics and Scopus citations processed by log-normalized functions.

by Twitter (no more than 40%). Facebook and CiteULike cover less than 20% of sampled publications. Figure 2 displays the relationship between Scopus citation count processed by log-normalized functions and the count from the other four data sources processed by log-normalized functions. The plot of Mendeley shows relationship tightens as the count of citations grow, which is more intuitional than other three plots. There is big noise in plot Facebook and CiteULike. The Figures show that Mendeley provides most valuable metrics for evaluating social popularity of scholar's output.

Based on the experiment above, we choose Mendeley bookmarks as the evaluation of social popularity dimension and Scopus citations to indicate the dimension of scientific impact. Thus, all the three-dimensional data is in order.

3.2 Three-Dimensional gALM

In order to visually enrich scholarly content with output impact, we utilize the graphical metrics to illustrate a scholar's publications in three dimensions including JIF, scientific impact and social popularity and integrate them in a coordinate system. The x axis represents the publication year, and the y axis represents citation count of paper. For each paper in SOG, we use $SOG_h(SOG - height)$ to show its scientific impact, $SOG_w(SOG - width)$ to show its JIF, and $SOG_c(SOG - color)$ to represent its social popularity:

$$SOG_h = scientific\ impact \tag{1}$$

$$SOG_w = journal\ impact\ factor \tag{2}$$

$$SOG_c = social\ popularity \tag{3}$$

Based on analyse in Subsect. 3.1, the SOG_h is counted by Scopus citation score, the SOG_w is counted by JIF score, and SOG_c is counted by Mendeley bookmark score.

4   Scholarly Output Graph

In this section, we propose the detailed definition of SOG for a scholar with all his publications, shown in Fig. 3.

In Fig. 3, the horizontal axis represents publication year and the vertical axis shows the number of citations. Each publication is visualized as a block and blocks belonging to the bar stand for the articles in the same year. The number of blocks in each bar shows the number of publications in years. The color of the blocks represents the social popularity property measured by social network citations represented by readership count extracted from Mendeley on November 31, 2015; the height indicates the scientific impact property measured by journal citations extracted from Google Scholar on November 31, 2015; the width of the blocks represents JIF citing from Journal Citation Reports (JCR) in 2015. Thus all the publications as well as the research development of a researcher are integrated in a single graph. The longer blocks reflect the publications of higher scientific importance; the wider blocks reflect the articles published in journals with higher JIF; the pinker blocks reflect more popular publications and the blue ones are publications with zero Mendeley readership count.

To efficiently compare publications among different years, we arrange the blocks in a way that the wider the blocks are the closer to the x axis. On the vertical aspect, for papers published in the same year, the height and width of a bar is respectively proportional to the number of citations and JIF. On the horizontal aspect, for papers published in different years, one can easily compare the impact of papers.

Compared to existing digital indicators, SOG comprehensively showcases a scholar's publications throughout the years. With a scholar's entire publications concentrated in SOG, users can quickly spot the changes in publication volumes as well as citations across years, intuitively understand the literature development of the scholar, and easily pinpoint the most popular and influential papers. The proportional width and height of blocks makes it more comparable among blocks in different years. Moreover, with different colors in the graph, one can intuitively spot the deep-colored block. The widest block is closest to the horizontal, thus one can easily get the papers from highest-IF journals. SOG gets results successfully and provides a graphical article level metric for scholars' output.

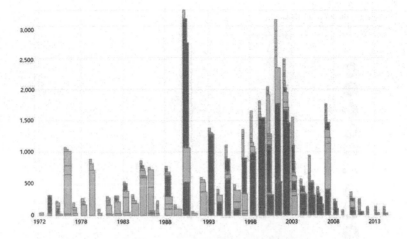

Fig. 3. The SOG for a scholar with all his publications.

5   Applications and Results

SOG has a wide range of practical applications. Areas where SOG can be used include expert system, personal homepage, scientific document navigation system, academic management system etc. To showcase the feasibility and usability of SOG, we design a literature navigation homepage with a list of 50 researchers in computer science, as is shown in Figs. 4 and 5. Upon clicking on any name in the list, a SOG page shows up that contains not only the researcher's output graph but also a detailed publication list showing titles, publication years, and citation counts of all of the researcher's publications. The navigation page can be found at http://impact.linkscholar.org/SOGExample.html.

For illustration purpose, we click on the top name in Figs. 4 and 5 (Tomaso Poggio) and are directed to the corresponding SOG page, as shown in the following gures.

Figure 6 shows the impact of Tomaso's whole publications visualized by SOG. Users can intuitively see that the scholar is productive and get high attraction and citation. The height of bars in recent years are low, indicating that citation metric needs time to accumulate. While the color of these bars are not influenced because social metrics timely snapshot the web-based impact. Moreover, users can easily spot the more popular papers, the higher papers as well as wider papers.

SOG also contains interactive features. Basic information and statistics about the publication are posted beneath the graph. The title, abstract and authors of the paper is shown to help users quickly learn about the main idea of the paper. The score of journal influence, citeby count, Mendeley readers, and CiteULike bookmarks (if any) listed on webpage gives users a general scan of scientific and social popularity. Besides, users can browse through the list of papers that cite this article and get detailed information. Upon clicking on any one block in SOG,

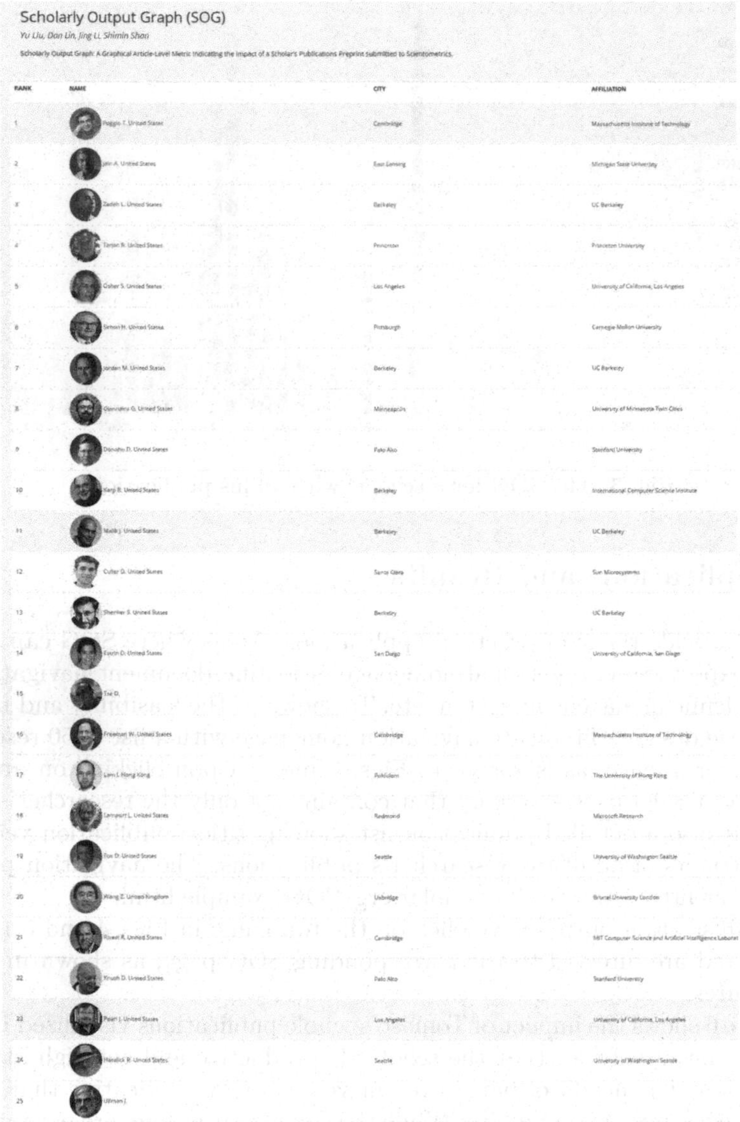

Scholarly Output Graph (SOG)

Yu Liu, Dan Lin, Jing Li, Shimin Shan

Scholarly Output Graph: A Graphical Article-Level Metric Indicating the Impact of a Scholar's Publications (Preprint submitted to Scientometrics).

RANK	NAME	CITY	AFFILIATION
1	Poggio T, United States	Cambridge	Massachusetts Institute of Technology
2	Jain A, United States	East Lansing	Michigan State University
3	Zadeh L, United States	Berkeley	UC Berkeley
4	Tarjan R, United States	Princeton	Princeton University
5	Osher S, United States	Los Angeles	University of California, Los Angeles
6	Simon H, United States	Pittsburgh	Carnegie Mellon University
7	Jordan M, United States	Berkeley	UC Berkeley
8	Giannakis G, United States	Minneapolis	University of Minnesota-Twin Cities
9	Donoho D, United States	Palo Alto	Stanford University
10	Karp R, United States	Berkeley	International Computer Science Institute
11	Malik J, United States	Berkeley	UC Berkeley
12	Culler D, United States	Santa Clara	Sun Microsystems
13	Sherber S, United States	Berkeley	UC Berkeley
14	Koch D, United States	San Diego	University of California, San Diego
15	Dai G,		
16	Freeman W, United States	Cambridge	Massachusetts Institute of Technology
17	Lam L, Hong Kong	Pokfulam	The University of Hong Kong
18	Lamport L, United States	Redmond	Microsoft Research
19	Fox D, United States	Seattle	University of Washington Seattle
20	Wang E, United Kingdom	Uxbridge	Brunel University London
21	Rivest R, United States	Cambridge	MIT Computer Science and Artificial Intelligence Laboratory
22	Knuth D, United States	Palo Alto	Stanford University
23	Perona P, United States	Los Angeles	University of California, Los Angeles
24	Hasslock R, United States	Seattle	University of Washington Seattle
25	Ullman J,		

Fig. 4. One application of SOG: literature navigation homepage of 50 scholars (part 1).

the block is highlighted and some interactive statistics are returned to the users beneath the graph including the respective paper's title, abstract, IF, citation count and Mendeley readership count as well as other publication statistics, seen in Fig. 7.

In Fig. 7, the first row of big font is the title, 'Networks for approximation and learning', followed by the name of the authors. The blue highlighted area tells

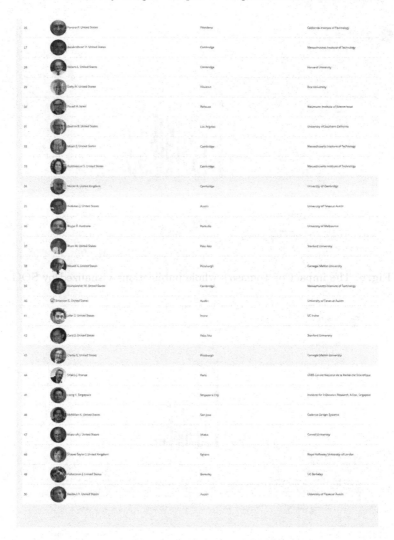

Fig. 5. One application of SOG: literature navigation homepage of 50 scholars (part 2).

you that the article was published by *Proceedings of the IEEE* in the year 1990. One can also find venue, volume number and page number in this section. Then comes the abstract of the paper followed by a statistic section that includes statistics on journal influence (5,466), citations (1,640), Mendeley readership (241) and CiteULike bookmark (3) counts. The last section is the list of papers that cite this article.

SOG provides a graphical metric to give a richer and more comprehensive demonstration for scholars' output. With the visual cues of block length, width and color, users can intuitively locate articles of high impact, JIF and popularity.

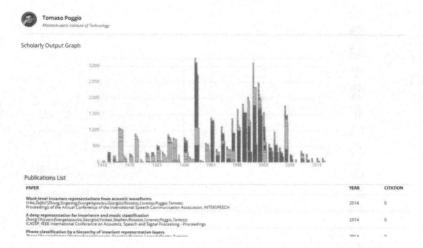

Fig. 6. The impact of Tomaso's whole publications visualized by SOG.

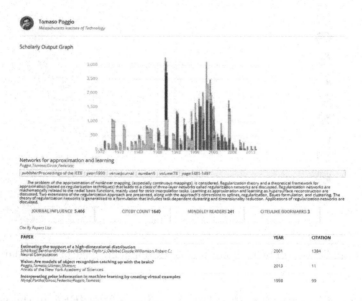

Fig. 7. The interactive statistic features of the clicked Tomaso's paper by SOG.

The detailed statistics provided on the webpage powerfully help users filter and find important papers.

6   Conclusion

In this paper, we propose an intuitive and effective way to appraise a scholar's scientific works, Scholarly Output Graph (SOG). SOG provides interactive guidance to explore and filter a scholar's publications and their impact in three dimensions: journal impact factor (JIF), scientific impact indicated by Scopus citations, and social popularity indicated by readership in Mendeley. With all of these factors visually represented in a histogram, SOG showcases the development of a scholar's publications and their impact over the years. SOG is comprehensive and flexible and can be extended to new platforms and fields in the era of Web 2.0. To prove the feasibility and usefulness, we design a literature navigation homepage with a list of 50 researchers in computer science. For illustration purpose, we click on the top name (Tomaso Poggio) and are directed to the corresponding SOG page. Therefore, SOG sets up in theory and can be used in practice.

SOG has several advantages. First, it visually and comprehensively presents the general development of a scholar's research. With all articles shown in one graph, users can directly see the changes of a researcher's publications throughout years. The three dimensions reflect not only the quality of the publications but also the immediate response from social networks. Second, SOG provides a more timely tool as compared to traditional metrics. To keep up with the latest data, SOG is updated constantly and persistently (every month or every year). Third, SOG provides the possibility of a more efficient way for online academic management and scientific navigation. SOG can effectively display a scholar's publications and its intuitive and interactive features make it potentially an effective tool for online academic management and scientific navigation systems.

In conclusion, SOG provides a comprehensive and intuitive way to objectively describe a scholar's impact in three dimensions including scientific impact, popularity and JIF. SOG captures not only traditional metrics but also the social popularity of a scholar's publications. Furthermore, SOG can be utilized in various scenarios and serve as an efficient navigation tool for exploring and filtering important articles of a scholar.

Acknowledgements. This work was supported in part by the Natural Science Foundation of China grant 61300087,61502069, 61672128; the Natural Science Foundation of Liaoning grant 2015020003; and by the Fundamental Research Funds for the Central Universities grant DUT15QY40, DUT16ZD(G)02.

References

1. Bar-Ilan, J., Haustein, S., Peters, I., Priem, J., Shema, H., Terliesner, J.: Beyond citations: Scholars' visibility on the social web. In: Proceedings of 17th International Conference on Science and Technology Indicators. 52900, pp. 98–109 (2012)
2. Haendel, M.A., Vasilevsky, N.A., Wirz, J.A.: Dealing with data: a case study on information and data management literacy. J. PLoS biol. **10**(5), e1001339 (2012)
3. Alexander, R., Sarah, D.R.: Accounting for Impact? the journal impact factor and the making of biomedical research in the Netherlands. J. Minerva **53**(2), 117–139 (2015)
4. Elliott, D.B.: The impact factor: a useful indicator of journal quality or fatally flawed? J. Ophthalmic Physiol. Optics **34**(1), 4–7 (2014)

5. Kamat, P.V., Schatz, G.C.: Journal impact factor and the real impact of your paper. J. Phys. Chem. Lett. **6**(15), 3074–3075 (2015)
6. Wenli, G.: Beyond journal impact and usage statistics: using citation analysis for collection development. Serials Libr. Printed Page Digital Age **70**(1–4), 121–127 (2016)
7. Hirsch, J.: An index to quantify an individual's scientific research output. Proc. Nat. Acad. Sci. **102**(46), 16569–16572 (2005)
8. Ball, P.: Index aims for fair ranking of scientists. Nature **436**, 900 (2005)
9. Glanzel, W.: On the opportunities and limitations of the h-index. Sci. Focus **1**, 10–11 (2006)
10. Alonso, S., Cabrerizo, F.J., Herrera-Viedma, E., et al.: h-Index: a review focused in its variants.computation and standardization for different scientific fields. J. Informetrics **3**(4), 273–289 (2009)
11. Bar-Ilan, J., Levene, M.: The hw-rank: An h-index variant for ranking web pages. Scientometrics **102**(3), 2247–2253 (2015)
12. Van Eck, N.J., Waltman, L., van Raan, A.F.J., et al.: Citation analysis may severely underestimate the impact of clinical research as compared to basic research. PLoS One **8**(4), e62395 (2013)
13. Das, A.K., Mishra, S.: Genesis of altmetrics or article-level metrics for measuring efficacy of scholarly communications: Current perspectives. Scientometrics **39**(2), 1–16 (2014)
14. Martin, F.: What can article-level metrics do for you? PLoS Biol. **10**(11), e1001687 (2013)
15. Neylon, C.: Article-level metrics and the evolution of scientific impact. PLoS Biol. **7**(11), e1000242 (2009)
16. Shema, H., Bar-Ilan, J., Thelwall, M.: Do blog citations correlate with a higher number of future citations? Research blogs as a potential source for alternative metrics. J. Assoc. Inf. Sci. Technol. **65**(5), 1018–1027 (2014)
17. Lu, Z.: PubMed and beyond: a survey of web tools for searching biomedical literature. Database **2011**(1), 56–65 (2011)
18. Priem, J., Hemminger, B.H.: Scientometrics 2.0: new metrics of scholarly impact on the social web. First Monday **15**(7) (2010)
19. Liu, Y., Huang, Z., Yan, Y., et al.: Science Navigation Map: an interactive data mining tool for literature analysis. In: Proceedings of the 24th International Conference on World Wide Web Companion, International World Wide Web Conferences Steering Committee, pp. 591–596 (2015)
20. Adie, E., Roe, W.: Altmetric: enriching scholarly content with article-level discussion and metrics. Learn. Publish. **26**(1), 11–17 (2013)
21. Thelwall, M., Haustein, S., Larivire, V., et al.: Do altmetrics work? Twitter and ten other social web services (2013)
22. Haustein, S., Siebenlist, T.: Applying social bookmarking dat to evaluate journal usage. J. Informetrics **5**(3), 446–457 (2011)
23. Gunn, W.: Social signals reflect academic impact: what it means when a scholar adds a paper to Mendeley. Inf. Stand. Q. **25**(2), 33–39 (2013)
24. Liu, Y., Huang, Z., Fang, J., Yan, Y.: An article level metric in the context of research community. In: Proceedings of the companion publication of the 23rd international conference on World Wide Web companion, pp. 1197–1202. International World Wide Web Conferences Steering Committee (2014)
25. Eysenbach, G.: Can tweets predict citations? Metrics of social impact based on twitter and correlation with traditional metrics of scientific impact. J. Med. Internet Res. **13**(4), e123 (2011)

26. Weller, K., Puschmann, C.: Twitter for scientific communication: how can citations/ references be identified and measured. In: Proceedings of the ACM WebSci 2011, pp. 1–4 (2011)
27. Thelwall, M., Tsou, A., Weingart, S., Holmberg, K., Haustein, S.: Tweeting links to academic articles. Cybermetrics: Int. J. Scientometrics Informetrics Bibliometrics 17, 1–8 (2013)
28. Bollen, J., Van de Sompel, H., Smith, J.A., Luce, R.: Toward alternative metrics of journal impact: a comparison of download and citation data. Inf. Process. Manage. 41(6), 1419–1440 (2005)
29. Zahedi, Z., Costas, R., Wouters, P.: How well developed are altmetrics? A cross-disciplinary analysis of the presence of alternative metrics in scientific publications. Scientometrics 101(2), 1491–1513 (2014)
30. Zahedi, Z., Fenner, M., Costas, R.: How consistent are altmetrics providers? Study of 1000 PLoS ONE publications using the PLOS ALM, Mendeley and Altmetric.com APIs. In altmetrics 14. Workshop at the Web Science Conference. Bloomington, USA (2014)
31. Li, X., Thelwall, M., Giustini, D.: Validating online reference managers for scholarly impact measurement. Scientometrics 91(2), 461–471 (2012)
32. Ling, X., Liu, Y., Huang, Z., Shah, P.K., Li, C.: A graphical article-level metric for intuitive comparison of large-scale literatures. Scientometrics 106(1), 41–50 (2015)
33. Alhoori, H., Kanan, T., Fox, E.A., Furuta, R., Giles, C.L., Pennsylvania, T.: On the relationship between open access and altmetrics. In: iConference 2015 Proceedings, pp. 1–8 (2015)
34. Bornmann, L.: What do altmetrics counts mean? A plea for content analyses. J. Assoc. Inf. Sci. Technol. 67(4), 1016–1017 (2016)
35. Bornmann, L.: Alternative metrics in scientometrics: a meta-analysis of research into three altmetrics. Scientometrics 103(3), 1123–1144 (2015)

Distributed Lazy Association Classification Algorithm Based on Spark

Xueming Li[1](✉), Chaoyang zhang[2], Guangwei Chen[2], Xiaoteng Sun[2],
Qi Zhang[2], and Haomin Yang[2]

[1] Key Laboratory of Dependable Service Computing in Cyber Physical Society,
Ministry of Education, Chongqing University,
Chongqing 400030, China
lixuemin@cqu.edu.cn
[2] College of Computer Science,
Chongqing University, Chongqing 400030, China

Abstract. The lazy association classification algorithms are inefficient when classifying multiple unclassified samples at the same time. The existing lazy association classification algorithms are sequential which can't deal with the big data problems. To solve these problems, we propose a distributed lazy association classification algorithm based on Spark, named as SDLAC. Firstly, it clusters the unclassified samples by K-Means algorithm. Secondly, it executes distributed projections according to clustered results, and mines classification association rules by a distributed mining algorithm based on spark. Then it constructs classifier to classify unclassified samples. The experiments are conducted on the 5 UCI datasets and a big dataset from the first national college competition on cloud computing(China). The results show that SDLAC algorithm is more accurate than the CBA algorithm. Besides, its efficiency is far more than the typical distributed lazy association classification algorithm. In other words, the SDLAC algorithm can adapt big data environment.

Keywords: Distributed lazy association classification · Big data · Spark

1 Introduction

Association classification algorithms(AC) [1] are the classification algorithms based on association rules which can be classified into the explicit association classification algorithms(EAC) and the lazy association classification algorithms(LAC) [2, 3] AC algorithms are more accurate than decision tree algorithm [4], bayesian algorithm [5], K-Means algorithm, support vector machines algorithm(SVM) [6] and have strong expandability. So AC algorithms are the important branch of classification algorithms.

Training and classification are independent in EAC algorithms, but LAC algorithms put off the training phase into the classification phase. LAC algorithms can solve the problems of small disjunction and oversize candidate rule set in EAC algorithms [3, 7, 8] but LAC algorithms will do projection operation and construct classifier for each training sample. So they are inefficient to classify multiple unclassified samples.

© Springer International Publishing AG 2016
J. Li et al. (Eds.): ADMA 2016, LNAI 10086, pp. 580–590, 2016.
DOI: 10.1007/978-3-319-49586-6_41

The existing LAC algorithms [3, 9, 10] are sequential which can't deal with the big data mining problem. It is necessary to develop the distributed lazy association classification algorithms.

To solve the problems mentioned above, we propose a distributed lazy association classification algorithm based on Spark, named as SDLAC. Firstly, SDLAC algorithm clusters the unclassified samples by K-Means algorithm. Secondly, it executes distributed projections according to the clustered results, and mines classification association rules by a distributed mining algorithm. Then it constructs classifier to classify unclassified samples. Our research shows that SDLAC algorithm is an efficient distributed LAC algorithm with no loss accuracy rate.

This paper is organized as follows. Section 2 is an introduction to the typical distributed lazy association classification algorithm (DLAC), Sect. 3 is the description of SDLAC algorithm, and Sect. 4 is the experimental results, the last section is conclusion.

2 Introduction of the Distributed Lazy Association Classification Algorithm(DLAC)

There is no a real distributed association classification algorithm at present. In order to verify our SDLC algorithm's accurate rate and efficiency, we need a comparing algorithm, so we combine the LAC algorithm and the DARM algorithm together to produce a distributed lazy association classification algorithm, named as DLAC. A typical DLAC algorithm is consisted of a classical LAC algorithm and C-DMA [10] (which is a distributed association rule mining algorithm).

2.1 The Procedure of DLAC Algorithm

The steps of general LAC algorithms are shown as follows.

Step 1: execute projection operation on training samples D or each unclassified sample z, and produce the dataset Dz;
Step 2: mine class association rules CARs from Dz by a DARM algorithm (for example C-DMA algorithm);
Step3: construct classifier C based on the CARs;
Step4: classify z by classifier C.

Usually, we can use kinds of association rule mining algorithms to realize step 2. The association rule mining algorithms can be serial or paralleled. The step 1, step 3 and step 4 are main parts of the lazy association rule classifier.

2.2 The Shortcomings of the DLAC Algorithm

The DLAC algorithm only adopt distributed association rule mining algorithms [11] in step 2. There still have two defects in DLAC algorithms:

(1) There is no distributed implementation in projection operation in DLAC algorithm. The DLAC algorithm only realizes distributed mining of association rules. It will do projection operation and mine class association rules for each unclassified sample. For example, if there are 10 unclassified samples to be classified, the DLAC algorithm will conduct 10 times projections, execute the DARM algorithm to mine the class association rules for 10 times, and build 10 classifiers, that is to say, the SDLAC is inefficient.

(2) Some DARM algorithms [11] based on MapReduce are proposed. But this kind of algorithms are not good enough to mine association rules because there are amounts of data interchanges completed through the file mode, which will obviously increase communication costs.

3 Distributed Lazy Association Classification Algorithm Based on Spark

3.1 The Procedure of the SDLAC Algorithm

The SDLAC algorithm is implemented based on spark framework. Firstly, it clusters unclassified samples by K-Means algorithm. Secondly, it executes distributed projections according to clustered results and uses center data mining algorithm(C-DMA) to mine association rules. Then it constructs classifier to classify unclassified samples.

The main challenges of SDLAC algorithm are shown as follows: (1) How to cluster unclassified samples; (2) How to execute distributed projection; (3) How to implement on the Spark.

Main steps of SDLAC algorithm are shown in Fig. 1. The innovative contents of SDLAC algorithm are in dotted box.

3.2 Clustering Unclassified Samples

Multiple similar unclassified samples only need once projection operation, and once mining operation for association rule mining, and only one classifier to be built, in other words, we can improve the efficiency of the LAC algorithms by clustering the unclassified samples.

There are two key problems with clustering unclassified samples. The first problem is to determine the number of clustered classes when using K-Means algorithm to cluster unclassified samples The second problem is how to judge the cluster whether it satisfied the clustered conditions.

Usually, there is inverse relationship between the number of clustered classes and particle size. The initial number of clustered classes is set to $KMeans\_ClassNum = 3N/c$ which is an empirical formula derived from our experiments. The N is the number of unclassified samples, and the c is the average value of options of each dimension of training data.

It will form several classes after clustering operation. In order to improve the projection efficiency, the attributes aggregation of the unclassified samples is necessary.

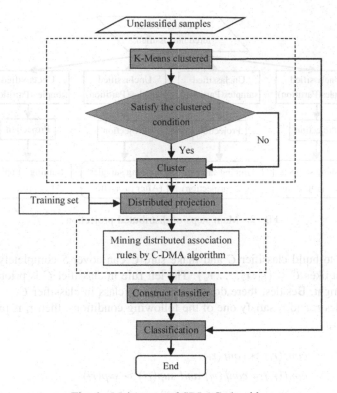

Fig. 1. Main steps of SDLAC algorithm

Suppose $\{S1, S2, \ldots, Sn\}$ represents unclassified samples of one class S. $LenS$ is the length of $S = S_1 \cup S_2 \cup \ldots \cup S_n$. $Len\overline{S}$ is the average length of $S1, S2, \ldots, Sn$. If a clustered class $\{S1, S2, \ldots, Sn\}$ meets the condition $LenS/Len\overline{S} < \ln n$, S represents the valid cluster center of the $\{S1, S2, \ldots, Sn\}$, and then S will be used in the next projection operation, otherwise S is invalid, $S1, S2, \ldots, Sn$ will be used respectively in the next projection operation.

3.3 Distributed Projection

DSLAC algorithm disseminates the results of clustered samples to each training partition. Then projection operation will be conducted independently in each partition. Main steps of distributed projection are shown in Fig. 2.

3.4 Classifier Construction

Suppose C represents the classifier built from the classification association rules CARs produced by C-DMA algorithm, S represents the unclassified samples, r represents one association rule in CARs. The basic idea of constructing classifier is to choose the high

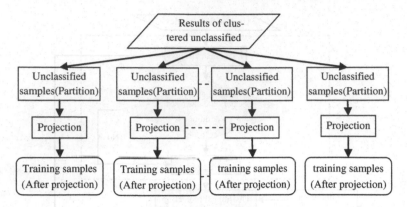

Fig. 2. Main steps of distributed projection

priority rule to build classifier C and make sure C can cover S completely. The final classifier just likes $C = \{r_1, r_2, \ldots, r_n\}$. The left rule in classifier C is prior than those rules on its right. Besides, there don't exist default class in classifier C.

If the rules r_i and r_j satisfy one of the following conditions, then r_i is prior than r_j, named $r_i > r_j$.

$$conf(r_i) > conf(r_j)$$
$$conf(r_i) = conf(r_j) \text{ and } sup(ri) > sup(rj)$$
$$conf(r_i) = conf(r_j), sup(ri) = sup(rj) \text{ and } ri \subset rj$$

The following are steps of constructing classifier:

Step 1: rank association rules in CARs according to their priority;

Step 2: choose the highest priority association rule r sequentially from CARs, and find the samples in S covered by r, if the number of found samples is at least greater than 1, then add the r into classifier C, delete the samples from S covered by r and delete the r from CARs at the same time, otherwise give up the rule r;

Step 3: repeat the step 2 until the S is empty or CARs is empty.

3.5   Implementation on the Spark Framework

The SDLAC algorithm uses the RDDs supported by Spark to represent data, and executes only once save operation in each cyclic, and doesn't execute sorting operations when SDLAC calls for combinBykey and reuceBykey functions during the shuffle procedure. The measurements mentioned above can reduce the unnecessary computation and improve the efficiency of SDLAC algorithm. The main steps of SDLAC algorithm are shown in Fig. 3.

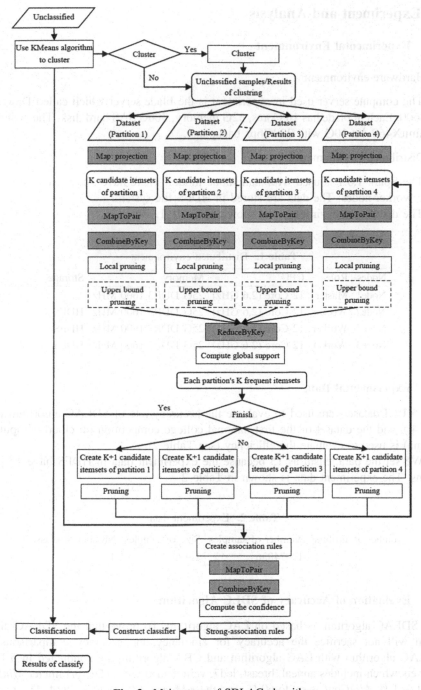

Fig. 3. Main steps of SDLAC algorithm

4 Experiment and Analysis

4.1 Experimental Environment

1. Hardware environment

The compute server used in experiment is the blade server which called Dawning TC4600. Each of blade has 16 Cores, 32G memory, 300G SAS hard disk. The switch is SummitX460-48t-10G with 320Gbps backplane.

2. Distributed environment

The Spark 1.0.1 is used with standalone model. There have one master node and three worker nodes. The data are stored in HDFS file system.

The distributed environment is shown in Table 1.

Table 1. Distributed environment

Nodes	Role	CPU	Memory	Storage
Node1	Master	12 Core (2.6 GHz)	25G DDR3 1600 MHz	
Node2	Worker	12 Core (2.6 GHz)	25G DDR3 1600 MHz	HDFS
Node3	Worker	12 Core (2.6 GHz)	25G DDR3 1600 MHz	HDFS
Node3	Worker	12 Core (2.6 GHz)	25G DDR3 1600 MHz	HDFS

4.2 Experimental Data

The 5 UCI datasets are used to evaluate the accuracy rate of SDLAC algorithm (see Fig. 4.), and the dataset of the first national college competition on cloud computing (China) is used to evaluate the efficiency (see Table 2).

We set the parallelism as 15 because the experiment data in HDFS have 15 partitions. The experiment data is shown in Table 2.

Table 2. Experiment data

Number of attribute	Number of samples	Size of samples	Number of class id
37	17974836	1.9 G	4

4.3 Evaluation of Accuracy of SDLC Algorithm

The SDLAC algorithm is based on LAC algorithm. To prove that the SDLAC algorithm will not sacrifice the accuracy for efficiency, the accuracy comparisons of SDLAC algorithm with LAC algorithm and CBA algorithm are conducted in 5 UCI datasets which includes anneal, breast, led7, vehicle and wave. The parameter $MinSup$ is set to 1 %, $MinConf$ is set to 60 %, and the $KMeans\_ClassNum = N/4$. The results of experiment are shown in Fig. 4.

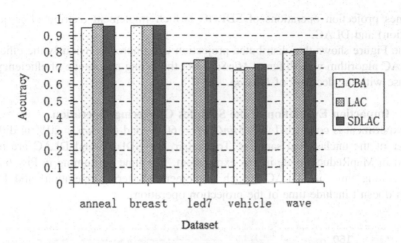

Fig. 4. The accuracy of CBA algorithm, LAC algorithm and SDLAC algorithm

In Fig. 4, the results show that the SDLAC algorithm is more accurate than CBA algorithm. The accuracy of SDLAC algorithm is almost same with the LAC algorithm.

4.4 Evaluation of the Efficiency of SDLC Algorithm

The distributed projection operation, clustering operation, and implementation method based on spark are the main factors which may impact on the efficiency of the SDLC algorithm.

4.4.1 Impaction Evaluation of the Distributed Projection Operation
This experiment is conducted with 50 unclassified samples and *MinConf* = 60 % at different *MinSup* (see Fig. 5). The time consuming in Fig. 5 stands for the average of

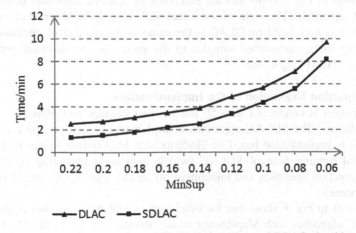

Fig. 5. The time consuming of SDLAC algorithm and DLAC algorithm

588   X. Li et al.

50 times projection operations of SDLAC (which implement distributed projection operation) and DLAC.

The Figure shows that distributed projection can improve obviously the efficiency of DLAC algorithm with different *MinSup*. But the lifting magnitude of efficiency will decrease with the decrease of *MinSup*.

4.4.2 Impaction Evaluation of the Samples Clustering Operation

This experiment is conducted with *MinConf* = 60 %, and *MinSup* = 20 % at different number of the unclassified samples. To be fair, the SDLAC and DLAC are implemented in MapReduce mode in this experiment. The time consuming in Fig. 6 is the total running time of SDLAC (which implement clustering operation) and DLAC (which doesn't include time of the projection operation).

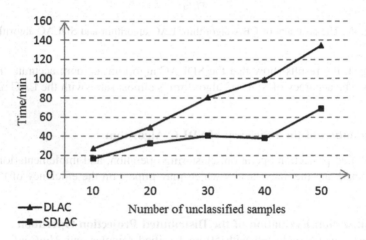

Fig. 6. The time consuming of DLAC algorithm and SDLAC

The results in Fig. 6 show that the efficiency of SDLAC algorithm is much better than DLAC algorithm. The main reason is that SDLAC only build serval classifiers, but number of classifiers build by DLAC is the same as the number of unclassified samples. Adding more unclassified samples to the model results in better performance when using SDLAC algorithm.

4.4.3 Impaction Evaluation of the Implementation

This experiment is conducted with 50 unclassified samples and *MinConf* = 60 % at different *MinSup*. The SDLAC algorithm is implemented based on Spark framework, and DLAC is implemented based on Hadoop with MapReduce mode. In Fig. 7, the time is total running time of SDLAC (which implement clustering operation, distributed projection operation and implement on Spark) and DLAC (which implement on MapReduce).

The results in Fig. 6 show that the efficiency of SDLAC algorithm is much better than DLAC algorithm with MapReduce mode. Besides, the lower the *MinSup* is, the

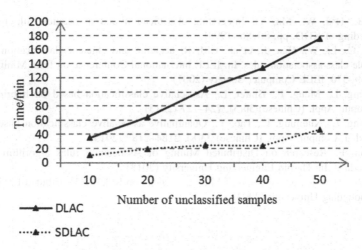

Fig. 7. The total running time of SDLAC and DLAC

higher the efficiency of SDLAC algorithm is. The first reason is that sorting operations are unnecessary and are ignored during the shuffle stage in SDLAC algorithm, but the sorting operations are necessary during the shuffle stage in DLAC algorithm with MapReduce. The second reason is that Spark is memory-based model and Hadoop MapReduce is file-based model for data interexchange.

5 Conclusions

The SDLAC algorithm proposed in this paper can improve obviously the efficiency of LAC algorithms with no loss of accuracy rate, when SDLAC algorithm is used to deal with the classification of the multiple unclassified samples and to handle the big data classification problems, while the classical LAC algorithm and general DLAC couldn't work well.

In other words, SDLAC algorithm can adapt big data environment, which can be used in practice.

References

1. Thabtah, F.: A review of associative classification mining. Knowl. Eng. Rev. **22**(1), 37–65 (2007)
2. Veloso, A., Meira, W., Zaki, M.J.: Lazy association classification. In: 6th International Conference on Data Mining, pp. 645–654. IEEE (2006)
3. Neapolitan, R.E.: Learning Bayesian Networks. Prentice Hall, Upper Saddle River (2004)
4. Quinlan, J.R.: Induction of decision trees. Mach. Learn. **1**(1), 81–106 (1986)
5. Cristianini, N., Shawe, J.: An introduction to Support Vector Machines. In: Cambridge University Press (2000)

6. Liu, B., Hsu, W., Ma, Y.: Integrating classification and association rule mining. In: Proceeding of KDD, pp. 80–86 (1998)
7. Rules, C., Li, W., Han, J., et al.: CMAR: accurate and efficient classification based on multiple class-association rules. In: IEEE International Conference on Data Mining, ICDM, pp. 369–376. IEEE Computer Society (2001)
8. Xueming, L., Meng, F., Binfei, L.: Associative classification based on hybrid strategy. J. Comput. Appl. (Chinese) **30**(3), 724–727 (2013)
9. Xueming, L., Xueming, L., Tao, Y.: Quantitative associative classification based on lazy method. J. Comput. Appl. (Chinese) **33**(8), 2184–2187 (2013)
10. Yanyan, F.: Research on Distributed Mining of Association rules algorithm based on MapReduce. In: Harbin Engineering University (2013)
11. Yue, W.: The Method Research of Mining Association Rules in Distributed Environments. In: Chongqing University (2003)

Event Evolution Model Based on Random Walk Model with Hot Topic Extraction

Chunzi Wu$^{(\boxtimes)}$, Bin Wu, and Bai Wang

Beijing Key Laboratory of Intelligent Telecommunication Software
and Multimedia, School of Computer Science, Beijing University of Posts
and Telecommunications, Beijing, China
wuchz0328@hotmail.com

Abstract. To identify the evolution relationship between events, this paper presents a new model which utilizes random walk model to weight the cosine similarity of events according to the chronological order. This model is proved to be effective and accurate in identifying relationship of events through comparisons with other models. This paper also puts forward an innovative method to detect hot topics which applies the concept of related events. This paper introduces parallel computing, including Spark and MapReduce, to significantly improve the efficiency of the event evolution calculation and hot topic detection.

Keywords: Event evolution · Hot topic · Random walk model · MapReduce · Spark

1 Introduction

People are able to obtain enormous amount of information everyday, which causes a problem called 'information overload'. Readers are struggling to keep up with the evolution paths of events and hot topics. For this reason, we figure out a novel model to process news articles to generate an event evolution graph and hot topics.

This paper proposes a novel model using random walk model and time stamps of event to calculate the correlation between events, which is capable of finding an acceptable result. Carlos Guestrin and Dafna Shahaf [1] also utilized random walk model to find the relationship between news articles. Our work differs from theirs as they used random walk model to evaluate the importance of words in two documents but we use it to calculate the importance of events which could happen in the next stage.

"Bag of words" method is used to construct vectors of documents and events, then evolution relationships are calculated through time relations, random walk model and cosine similarity. We also use parallelization platform to improve the time efficiency.

This work was supported by National Basic Research (973) Program of China (No. 2014CB329606) and Special Fund for Beijing Common Construction Project.

© Springer International Publishing AG 2016
J. Li et al. (Eds.): ADMA 2016, LNAI 10086, pp. 591–603, 2016.
DOI: 10.1007/978-3-319-49586-6_42

Section 1 in this paper introduces our work briefly. Some related work of the paper are included in Sect. 2. Details of calculating event evolution and detecting hot topics are discussed in Sects. 3 and 4. Section 5 shows the parallelization of the random walk model using Spark and MapReduce. Experiments and results are described in Sect. 6, and Sect. 7 summarizes the conclusion and future works of this model.

Our major contribution includes:

1. An innovative model of calculating the correlation of events, namely Event Evolution Model with Random Walk and Time (EEM_RW_T), is put forward. This model uses random walk model and time stamps of events to improve the accuracy of finding relationships among events.
2. The event evolution relationship is applied in the detection of hot topics for the first time. It is assumed that the more derived events a topic has, the hotter the topic is during that time.
3. The proposed model is executed on parallelization computing model (Spark and MapReduce) to improve the efficiency.

2 Relevant Work

2.1 Event Evolution Analysis

Various algorithms and models have been proposed in event evolution. In 2000, Russell Swan and David Jensen [2] used χ^2 statistics to justify term occurrence and co-occurrence. In 2003, Makkonen [3] pioneered in setting event evolution as a sub-objective of topic detection and tracking (TDT) research. In 2010, Carlos Guestrin and Dafna Shahaf [1] proposed an algorithm of calculating event correlation by using random walk model. Then Dafna Shahaf et al. [4] applied previous theory to the graph of event evolution instead of only to one single event chain. In the year of 2015, Dafna Shahaf et al. [5] introduced the concept of information cartography to multiple areas which promoted the development of event evolution. Hui Zhang et al. [6] proposed a modelling method for news event evolution. Five patterns of event evolution were introduced. Janani Kalyanam et al. [7] modeled the topic evolution through a non-negative matric factorization based model which considered both social context and text content.

Some researchers focused on large dynamic networks. Pei Lee et al. [8] tracked cluster evolution in highly dynamic networks by an incremental computation framework, they also introduced a "skeletal graph" to filter noise for a better result. Hongyun Cai et al. [9] proposed a novel model called MIL, which was based on upper bound pruning to search and update event in the dynamic database. Pei Lee et al. [8] also applied graph mining algorithm into event detection and evolution tracking.

2.2 Hot Topic Detection

Prominent achievement has been made in the hot topic detection area. In 2009, Xingxing Liu [10] put forward a formula of calculating hotness of topic through clustering documents into event. Zhifan Yang, et al. [11] combined bursty term identification and multi-dimension sentence modelling to automatically detect emerging hot topics. Guolong Liu et al. [12] utilized an improved Latent Dirichlet Allocation Model to extract hot topics.

2.3 Parallelization

Introduction of Apache Hadoop MapReduce. MapReduce, since it was introduced in 2004 by Jeffery Dean and Sanjay Ghemawat et al. [13], has become one of the most important programming models in data processing. It consists of two major steps of using "map" and "reduce" functions.

Introduction of Apache Spark. Apache Spark is a cluster computing framework developed by UC Berkeley AMP laboratory [14] which takes the advantage of MapReduce function and has better performance than MapReduce.

Spark has a feature of using an abstraction called resilient distributed datasets (RDD). During the computation of Spark, data is stored in memory instead of disk, which reduces the time cost on reading and writing data in iterative operation.

3 Event Evolution Analysis

Figure 1 demonstrates the flow chart of event evolution analysis.

Definition 1 (Event). *A clustered result extracted from large amount of news articles that is published during the time period of the dataset.*

Definition 2 (Event Evolution). *Event evolution contains association among events in a specific time period including the occurrence, development, evolution and termination of events.*

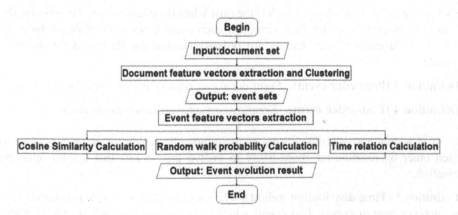

Fig. 1. Flow chart of event evolution model (EEM_RW_T)

3.1   Data Pre-processing

Original documents are tokenized by using HanLP. Term frequency–inverse document frequency (TF-IDF) is generated to extract feature vectors.

$$tf_{i,j} = \frac{n_{i,j}}{\sum_k n_{k,j}} \tag{1}$$

$$idf_i = \log \frac{|D|}{|\{j : t_i \in d_j\}|} \tag{2}$$

$$tfidf_{i,j} = tf_{i,j} \times idf_i \tag{3}$$

The formula (1) shows that the term frequency value of word i in document j is the times that word i is mentioned in document j, represented by $n_{i,j}$, divided by $\sum_k n_{k,j}$, which is the total number of words appear in document j. As the formula (2) describes, the logarithm of total number of documents divided by the number of documents which mentioned word i is the inverse document frequency value for word i.

After the TF-IDF values are calculated, the feature vector of each document can be obtained. All documents are clustered according to cosine similarity of feature vectors. Cosine similarity calculation is shown in formula (4).

$$\cos = \frac{\vec{A}_i \times \vec{A}_j}{|\vec{A}_i| \times |\vec{A}_j|} \tag{4}$$

A_i and A_j represent any two documents in the dataset. Feature vectors are processed by K-means algorithm to cluster all documents. The result of clustering are defined as events. Event evolution analysis will be performed on those events.

3.2   Time Relation

Each event has its time stamp, which is the time when the event occurs. To optimize the data processing process, the time stamp for each event is set to the time of the news report in clustering center. Events are ordered chronologically based on the time feature.

Definition 3 (Pre-order event). *Event occurs at relative earlier stage among events.*

Definition 4 (Post-order event). *Event occurs after a certain pre-order event.*

It is widely accepted that events evolve gradually, which means events close to each other on timeline are more likely to evolve than events that are far apart on timeline.

Definition 5 (Time distribution weight). *The time distribution weight is related to the occurrence time of events. Two events which occurred more closely on timeline will have higher time distribution weight, in other words, higher probability of evolution.*

3.3 Cosine Similarity

The term document-inverse event frequency (TF-IEF) [15], which is similar to TF-IDF, is used to evaluate the importance of a specific word in the event set. Then the feature vectors are used to calculate the cosine similarities between events. TF-IEF is calculated according to the formula (5) and (6).

$$\omega\left(f_j, E_i^{ti}\right) = \frac{\left[1 + \log_2 TF\left(f_j, E_i^{ti}\right)\right] \times IEF\left(f_j\right)}{\sqrt{\sum_{j=1}^{n} \left\{\left[1 + \log_2 TF\left(f_j, E_i^{ti}\right)\right] \times IEF\left(f_j\right)\right\}^2}} \tag{5}$$

$$IEF\left(f_j\right) = \log_2 \frac{|T| + 1}{|EF|\left(f_j\right) + 0.5} \tag{6}$$

$TF(f_j, Eiti)$ is the time of occurrence of feature f_j; the value of $|EF(f_j)|$ is the number of events which has the feature f_j and $|T|$ is the total number of events.

3.4 Random Walk Probability

Generate Association Graph. For each pre-order event, the set of all its post-order event and relevant words constitute a weighted directed graph $G = (V, E)$. In the graph, links are directed, including "event -> word" direction and "word -> event" direction. The weight of a link is the normalized TF-IEF values. Figure 2 introduces how to assign weight to links.

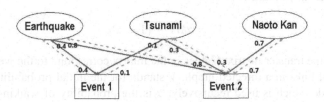

Fig. 2. An example of TF-IEF value of directional link

As shown in Fig. 2, words are shown in ellipse shapes while events are represented in rectangular shapes. For example, the TF-IEF value of earthquake for event 1 equals to 0.4 and 0.8 for event 2. Thus the weight of link "earthquake -> event 1" equals to *0.4/(0.4 + 0.8) = 0.33*. Similarly, the weight of link "event 1 -> earthquake" is *0.4/ (0.4 + 0.1) = 0.8*.

With TF-IEF values, a feature word set for events is generated. The feature word set is compared with the list of words in events. If a feature word is mentioned in both pre-order event and a certain post-order event of that pre-order event, this word can be recognized to connect the two events. Next, the association graph of events and words is plotted. The proposed algorithm need to calculate the probability of a pre-order event randomly walking to a post-order event. Thus each event has its specific directed association graph. The association graph of a pre-order event should be similar to the pattern of Fig. 3. It is assumed that post-order events are independent from each other to eliminate the influence between post-order events. Hence, links of "post-order event -> word" relationship should never appear in association graphs. Similarly, post-order events cannot walk back to pre-order event, which means no "word -> pre-order event" link in any association graphs.

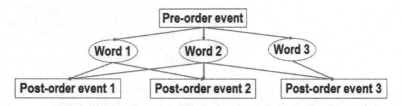

Fig. 3. An example of association graph

Random Walk Model. The probability of random walk between events is called random walk association. The calculation of random walk probability requires the random walk model which has been widely used in practice such as the PageRank algorithm developed by Google Inc. [16].

Definition 6 (Random walk association). *Random walk association is the probability that a walker walks randomly from a pre-order event to a post-order event. By introducing the idea of PageRank algorithm, the calculation of random walk probability is defined in formula (7)*

$$V' = \alpha M V + (1 - \alpha)e \tag{7}$$

M represents the transfer matrix. Values in the matrix correspond to the weighted value of all directed links in a directed graph. V stands for the initial probability of random walk of a node which is used recursively. α is the probability of walking to the next node and $(1 - \alpha)$ is the probability of restart. V' will replace parameter V in formula (7) in the recursive calculation. Recursion ends when result is stable.

4   Hot Topic Detection

Through the computation of event correlation, a set of events association can be derived. Such set can be used to depict the map of event evolution and can also facilitate the detection of hot topic in a period of time.

Definition 7 (Hot topic). *A topic that is widely discussed by plenty of people in a specific time period.*

A hot topic is an event which draws huge public attention. The dataset used in this paper is a collection of news reports. News agencies are highly concerned about hot topics to attract more readers. If they find a hot topic, more effort will be made to report following event of the topic. As such, the assumption that "events with a large number of post-order events in the event association set are likely to become hot topics" is valid. Hence, if multiple post-order events can be detected, the pre-order event of those post-order events is more likely to be a hot topic.

5 Parallelization of EEM_RW_T

Due to the news report dataset is huge, parallelized execution is introduced to improve the efficiency of processing data. Spark is used to implement the EEM_RW_T and we also use MapReduce to compare the computing efficiency of both model.

There are three major parts in the EEM_RW_T which goes cosine similarity calculation, time association computation and random walk association calculation. Especially the random walk association calculation, directed graph and transfer matrix for each pre-order event should be generated and the probability of each graph should be computed iteratively. Hence this part is the most time-consuming part in the model.

5.1 Implementation with Apache Spark

Apache Spark is developed under Scala, the steps taken to implement random walk association is illustrated in Table 1. Multiple recursive computations are introduced in the process, which is represented in "FOR" iteration in the table. These recursive computations can be processed by "reduceByKey()" function to enhance the efficiency.

Table 1. The calculation of random walk association on Spark

	Input: Graph G, where G is the "*event*-word-*event*" relations of all the preorder events.
	Output: V', which contains random walk probability of each node in every graph.
1:	*EventWordMap* ← w1, where w1 is a weighted value set of "*event*-word" pair
2:	*WordEventMap* ←w2, where w2 is a weighted value set of "word-*event*" pair
3:	**For** every g∈G do
4:	*Orderlist* ←o, where o is the meaning (topic or word) of each line (column) in transfer matrix
5:	**For** every o∈ *Orderlist* do
6:	Acquire weighted value from *EventWordMap* or *WordEventMap*
7:	M ← m, where m is transfer matrix
8:	V, e ← i, where i is initial value in random walk
9:	α← re-start probability
10:	*iterator* ← times of iteration
11:	**While**(iterator>0) do
12:	$V'=\alpha MV+(1-\alpha)e$

5.2   Implementation with MapReduce Model

Table 2 demonstrates the idea of implementing random walk association with MapReduce model. The idea is similar to the implementation with Spark.

Table 2. The calculation of random walk association on MapReduce

	Mapper class
	Input: Graph G, where G is the "topic-word-topic" relations of all the preorder events.
	Output: Transfer Matrix M.
1:	*EventWordMap* ← w1, where w1 is a weighted value set of "event-word" pair
2:	*WordEventMap* ←w2, where w2 is a weighted value set of "word-event" pair
3:	**For** every $g \in G$ do
4:	*Orderlist* ←o, where o is the meaning (topic or word) of each line (column) in transfer matrix
5:	**For** every o $\in$ *Orderlist* do
6:	Acquire weighted value from *EventWordMap* or *WordEventMap*
7:	Constructing Transfer Matrix M
	Reducer class
	Input: Transfer Matrix M.
	Output: V', which contains random walk probability of each node in every graph.
1	M ← m, where m is transfer matrix
2	V, e ← i, where i is initial value in random walk
3	α← re-start probability
4	*iterator* ← times of iteration
5	**While**(iterator>0) do
6	$V'=\alpha MV+(1-\alpha)e$

6   Valuation and Result

6.1   Evaluation Data

Two datasets are used to assess the system which is described in Table 3.

Table 3. Description of evaluation datasets

No.	Name	Number	Size
1	Japanese earthquake	6955	22.8 MB
2	Social news	382087	1.43 GB

Dataset 1 is about Japanese earthquake. After processing, two third of all events in the dataset are available to use. Rest events are discarded due to low association with other events. Next, a *threshold* is applied to filter the result. The process of setting *threshold* value is discussed later. This dataset is used to evaluate the result of EEM_RW_T.

Dataset 2 contains 382087 pieces of news which cover political, economic, military and social area. Dataset 2 is used to compare the efficiency improvement of parallelization pattern and evaluate the accuracy of hot topic detection.

6.2 Evaluation Methodology

Event Evolution Model. Once events are collected, several volunteers are invited to identify possible evolution among those events based on their background knowledge and common sense. The averaged result created by these volunteers is treated as the actual event evolution graph, which is treated as a standard to evaluate the accuracy and efficiency.

Two parameters, efficiency E and accuracy R are calculated to reflect the accuracy of the model. The system identifies N evolution, of which $N1$ is correct evolution. The actual evolution number is M.

$$E = \frac{N1}{N} \tag{8}$$

$$R = \frac{N1}{M} \tag{9}$$

Hot Topic Detection. Similar to the previous section, the same volunteers are asked to vote for hot topic and the voted hot topic is considered as correct answer. The topic detected by system will be compared with the correct answer.

Parallelization of EEM_RW_T. The EEM_RW_T model is implemented using Apache Spark and Apache Hadoop MapReduce separately. Various size of datasets are processed to calculate the random walk association of the EEM_RW_T model. The execution time of parallelized and serial versions is compared to evaluate performance improvement.

6.3 Evaluation Result and Analysis

Event Evolution Model. A threshold is applied to filter the link produced by the model. In Fig. 4, the vertical axis is accuracy and the horizontal axis is efficiency.

Dots in Fig. 4 represent the result of threshold value ranging from 0.00003 to 0.00010 accordingly. As the threshold increases, the accuracy decreases by less than 10 % yet the coverage ratio is increased with steady paces. It is obvious that after a point, the correct ratio drop rapidly, nevertheless, the improvement of coverage ratio is very limited. Thus the black dot, where threshold is set to 0.00007, is the optimum result.

Figure 5 contains the evolution graph generated by our method and actual evolution path. Broken lines represent the evolution paths of tsunami related events. Full lines stand for the evolution of events related to the leakage of radioactive material. Dot broken lines connect the events of the reaction to the catastrophe of Japan and

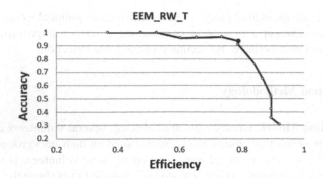

Fig. 4. The impact of threshold on the result

other countries. Black lines marks evolutions that is not identified by the model. Falsely identified evolutions are crossed out using cross.

The EEM_RW_T model is compared with Hui Zhang's [6] method of weighting cosine similarity using string fuzzy matching *(cs*ass)*. Non-weighted cosine similarity is also implemented to identify the impact of weighting. The result is shown in Fig. 6.

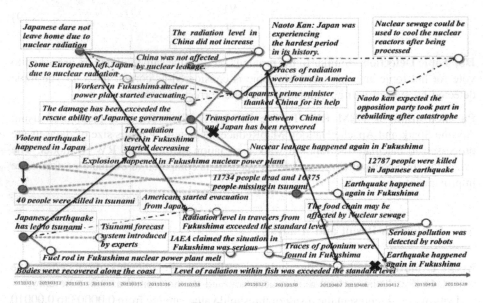

Fig. 5. Result graph

As Fig. 6 suggests, the EEM_RW_T has better performance in both efficiency and accuracy than the other two models. When the threshold decreases, the correct ratio of Hui Zhang's model is even worse than non-weighted model.

Detection of Hot Topic. The number of post-order event determines whether the pre-order event is a hot topic or not. For the dataset 1, if the number of post-order events is larger than 2, the event itself becomes a hot topic. In Fig. 5, detected hot topic

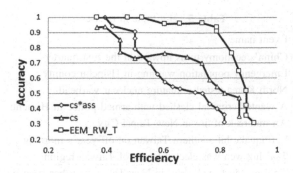

Fig. 6. The result of comparing different weighting strategy

Table 4. Detected and actual hot topics

Hot topics detected by system	Real hot topics
Violent earthquake happened in Japan	Violent earthquake happened in Japan
40 people were killed in tsunami	Japanese earthquake has led to tsunami
Japanese earthquake has led to tsunami	Japanese dare not leave home due to radiation
Japanese dare not leave home due to radiation	Traces of polonium were found in Fukushima
China was not affected by nuclear leakage	China was not affected by nuclear leakage
11734 people dead and 16375 people missing in tsunami	11734 people dead and 16375 people missing in tsunami
Transportation between China and Japan has been recovered	

is marked as solid point. The list of detected hot topic and actual hot topic is recorded in Table 4.

According to Table 4, the model identified 7 events as hot topics, 5 of them are correct. 2 false hot topics are picked up by the model and one actual hot topic is missed. The result indicates that the model is capable to detect most hot topics, which is an acceptable result.

Then we analyze the dataset of social news report. Ten hot topics in political, economic and livelihood areas are detected by the model, which is listed in Table 5.

Parallelization of EEW_RW_T. The hardware information of the platform, on which Spark and MapReduce are applied, is listed in Table 6. Spark cluster and MapReduce cluster on the platform owns 3 slave nodes and 1 master node.

In the comparison, time cost of serial version, MapReduce and Spark in calculating random walk association is plotted in Fig. 7.

The horizontal axis of Fig. 7 is the number of events processed, while the vertical axis is the elapsed time of execution. The dimension of each event's directed graph ranging from 300 to 1600. When the number of event is small, the advantage of

Table 5. Hot topic in dataset 2

ID	Event name
1	China's economy turns upward in the first quarter
2	Emergency population warning in Hsinchu county
3	North Korea tested miniature nucleear warhead
4	A wanted criminal clashed with armed police in Hsinchu
5	An outbreak of fire in New Taipei City
6	Earthquake hit southern Taiwan on off-year's morning
7	Tsai Ing-wen was elected leader of Taiwan region
8	Trial on 'dust explosion' in New Taipei City has been made
9	Terrorist attack raise alert in EU
10	People prefer to travel during Tomb-sweeping Day

Table 6. Hardware configuration of the platform

CPU Clock	2.10 GHz	Operation system	64-bit Centos6.5
Physical core	2 nodes (6 cores each)	JDK edition	JDK 1.7
Logical core	12 Cores	HADOOP edition	HADOOP 2.6.4
Disk I/O bandwidth	375.29 MB/s	Spark edition	Spark 1.5.1

parallelization with MapReduce is not obvious. As the number of input events increases, the superiority of using MapReduce pattern can be found. Spark is always the optimum option in the comparison. Thus, it is reasonable to claim that MapReduce is better than serial execution of program when the data set is large. Spark can benefit to a large extent in recursive or iterative computing due to the memory usage.

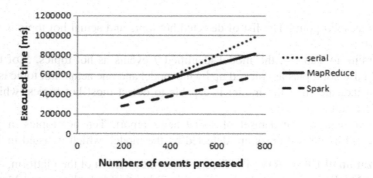

Fig. 7. Efficiency comparison

7 Conclusion

To accurately identify event evolution paths and detect hot topics, this paper proposed a novel model to investigate on event evolution by integrating cosine similarity, time distribution weight and random walk association. The model achieves acceptable result when being applied in detecting hot topic. It can be further used in the inspection of public opinions and the prediction of event evolution.

References

1. Shahaf, D., Guestrin, C.: Connecting the dots between news articles. In: Proceedings of the 16th ACM SIGKDD International Conference on Knowledge Discovery and Data Mining, 25-28 July, Washington, D.C. ACM Press, New York (2010)
2. Swan, R., Jensen, D.: TimeMines: constructing timelines with statistical models of word usage. In: Proceedings of the Sixth ACM SIGKDD International Conference on Knowledge Discovery and Data Mining, 20–23 August, Boston, MA. ACM Press, New York (2000)
3. Makkonen, J.: Investigations on event evolution in TDT. In: Proceedings of HLT NAACL 2003 Student Research Workshop, pp. 43–48 (2003)
4. Shahaf, D., Guestrin, C., Horvitz E.: Trains of thought: generating information maps. In: Proceedings of the 21st International Conference on the World Wide Web, 16–20 April, Lyon, France, pp. 899–908. ACM Press, New York (2012)
5. Shahaf, D., Guestrin, C., Horvitz, E., Leskovec, J.: Information cartography. Commun. ACM **58**(11), 62–73 (2015)
6. Hui, Z., Guohui, L., Li, J., et al.: A method of constructing news events evolution. J. Natl. Univ. Defense Technol. **2013**(04), 166–170 (2013)
7. Kalyanam, J., Mantrach, A., Saez-Trumper, D., Vahabi, H., Lanckriet, G.: Leveraging social context for modeling topic evolution. In: Knowledge Discovery and Data Mining, pp. 517–526 (2015)
8. Pei, L., Lakshmanan, L.V.S., et al.: Incremental cluster evolution tracking from highly dynamic network data. In: Proceedings of 2014 IEEE 30th International Conference on Data Engineering, 31 March – 4 April (2014)
9. Cai, H., Huang, Z., Srivastava, D., et al.: Indexing evolving events from tweet streams. In: Proceedings of 2016 IEEE 32nd International Conference on Data Engineering, 16-20 May 2016, pp. 1538–1539 (2016)
10. Xingxing, L.: Finding Trend and Automatic Topic Features Extraction. Central China Normal University (2009)
11. Zhifang, Y., Chao, W., Fan, Z., et al.: Emerging rumor identification for social media with hot topic detection. In: 2015 12th Web Information System and Application Conference (WISA), pp. 53–58 (2015)
12. Goulong, L., Xiaofei, X., Ying, Z., et al.: An improved latent Dirichlet allocation model for hot topic extraction. In: 2014 IEEE Fourth International Conference on Big Data and Cloud Computing (BdCloud), pp. 470–476 (2014)
13. Jeffrey, D., Sanjay, G.: MapReduce: simplified data processing on large clusters. Commun. ACM **51**, 107–113 (2008)
14. Matei, Z., Mosharaf, C., et al.: Spark: cluster computing with working sets. In: Proceedings of the 2nd USENIX Conference on Hot Topics in Cloud Computing (HotCloud 2010), p. 10 (2010)
15. Hui, Z., Guohui, L., Li, J., et al.: On-line news event detection based on TF·IEF model. J. Natl. Univ. Defense Technol. **35**(3), 9 (2013)
16. Brin, S., Page, L.: The anatomy of a large-scale hyper textual Web search engine. Comput. Netw. ISDN Syst. **30**, 107–117 (1998)

Knowledge-Guided Maximal Clique Enumeration

Steve Harenberg[1,2], Ramona G. Seay[1,2], Gonzalo A. Bello[1,2],
Rada Y. Chirkova[1], P. Murali Doraiswamy[3], and Nagiza F. Samatova[1,2(✉)]

[1] North Carolina State University, Raleigh, NC, USA
samatova@csc.ncsu.edu
[2] Oak Ridge National Laboratory, Oak Ridge, TN, USA
[3] Duke University, Durham, NC, USA

Abstract. Maximal clique enumeration is a long-standing problem in graph mining and knowledge discovery. Numerous classic algorithms exist for solving this problem. However, these algorithms focus on enumerating all maximal cliques, which may be computationally impractical and much of the output may be irrelevant to the user. To address this issue, we introduce the problem of knowledge-biased clique enumeration, a query-driven formulation that reduces output space, computation time, and memory usage. Moreover, we introduce a dynamic state space indexing strategy for efficiently processing multiple queries over the same graph. This strategy reduces redundant computations by dynamically indexing the constituent state space generated with each query. Experimental results over real-world networks demonstrate this strategy's effectiveness at reducing the cumulative query-response time. Although developed in the context of maximal cliques, our techniques could possibly be generalized to other constraint-based graph enumeration tasks.

1 Introduction

Mining for interesting patterns in large-scale graph data, or *graph pattern enumeration*, is at the core of graph data analytics in applications spanning many domains, including biomedical informatics, social network analysis, and climate science. These applications can often benefit from a *query-driven* enumeration of the patterns, where the output is constrained or influenced by given knowledge priors. For example, a user may only be interested in sets of vertices in a gene-gene functional association network that contain genes known to be associated with Alzheimer's disease. These genes could be viewed as the user's *knowledge bias* or *query vertices*. Such biased enumeration tasks are typically performed many times in an exploratory manner. Furthermore, the graphs to be analyzed can reach very large scales, containing millions of vertices, making queries non-trivial to answer. Hence, an efficient method to perform many queries over the same graph is critical.

To facilitate this type of analysis, two complementary strategies could be taken: (1) precompute and index some or all of the search space and answer

© Springer International Publishing AG 2016
J. Li et al. (Eds.): ADMA 2016, LNAI 10086, pp. 604–618, 2016.
DOI: 10.1007/978-3-319-49586-6_43

each query using this index, or (2) develop a new algorithm where states that are irrelevant to the query are filtered out during the search process. Some related work involving both these strategies can be found in Sect. 7. While the first strategy can yield a fast individual query response time, extensive computation time could be wasted generating solutions that never get queried, leading to a large cumulative query time. In contrast, the second strategy will not generate irrelevant solutions but will likely produce redundant computation between successive queries (e.g., a query for patterns containing u and v, and a query for patterns containing u and w). Most importantly, with both of these strategies, there is no synergy between successive queries despite the possibility of some computational overlap.

To address these limitations and improve the cumulative response time for users performing queries in an exploratory manner, we propose a dynamic state space indexing and querying strategy. The key idea is to store and index the constituent state space generated with each query. Subsequent queries can leverage this index to avoid recomputing any parts of the search space previously generated. Moreover, the index is incrementally updated as more queries are performed, thereby generating and storing exactly the query-relevant search space.

Specifically, we focus on the problem of maximal clique enumeration (MCE), a long-standing graph pattern enumeration task. MCE methods can be used for various applications, such as enumerating densely connected subnetworks in large protein-protein interaction graphs [26], discovering social hierarchy in email networks [6], and identifying regions of homogeneous long-term climate variability in climate networks [18]. Moreover, even though the scope of this paper is limited to the MCE problem, our proposed methodology could potentially be expanded to other graph pattern enumeration tasks, which is the subject of future work. In summary, we make the following contributions:

1. **Define a new problem of *biased maximal clique enumeration* (Sects. 2-3)**, a query-driven formulation of the MCE problem with specific applications in biomedical informatics and climate science.
2. **A dynamic state space index and querying strategy (Sect. 4)** that is able to store and leverage the constituent state space generated with each query to reduce cumulative query response time.
3. **Extensive Experiments (Sects. 5-6)** over real-world networks showing the applicability of our problem formulation and the efficacy of our dynamic indexing strategy at reducing cumulative query response time.

2   Problem Statement

Let $G = (V, E)$ be a simple, undirected graph. A set of vertices C is a *clique* iff every pair of vertices in C is connected by a single edge. A clique that is not contained in any larger clique is said to be *maximal*. We define a *query* as an ordered pair of the form (Q, η), where $Q \subseteq V$ and η is a non-negative integer. We then say that a clique C is *biased* with respect to a query (Q, η) if there are at least η query vertices from Q present in C; that is $|Q \cap C| \geq \eta$ (see Fig. 1).

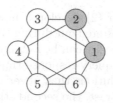

Fig. 1. An example graph with eight maximal cliques. For a query $(Q = \{1,2\}, \eta = 2)$, there are two biased maximal cliques, $\{1,2,3\}$ and $\{1,2,6\}$.

Problem Statement 1. *Given a simple, undirected graph $G = (V, E)$ and a query (Q, η), find all biased maximal cliques in G.*

There are several advantages of reducing generic maximal clique enumeration (MCE) to our query-driven problem formulation. First, the output is smaller and tailored towards specific query vertices of interest, since only biased cliques are returned rather than *all* maximal cliques. In addition, our problem can be solved more efficiently because less resources are wasted on computing cliques that are not relevant to the query. Both the output size and complexity are guided by the user-controlled bias parameter η. If η is large, then there will be fewer output cliques, but the cliques that are output will be more relevant to the query vertices.

3   Biased Clique Enumeration

The most naive way to mine biased cliques would be a *top-down* approach where all maximal cliques are computed using an existing method, stored (and indexed), and then queried. While having a fast query-response time (with an appropriate index), this approach may be impractical if not impossible. The MCE problem is NP-hard and the output size is worst-case exponential at $O(3^{|V|/3})$ [16], meaning the computation may be intractable and the resources insufficient to handle the full search space. Moreover, for users only interested in certain queries, the time to build and index over this full space will severely outweigh the savings in query-response time.

In contrast, a *bottom-up* approach could be used, where a novel algorithm is developed to mine only the maximal cliques relevant to the input query. Unfortunately, the bottom-up approach also has an issue: the search space is explored from scratch for every query. As a result, computations towards the beginning of the search are repeated across similar queries, producing redundant computations. However, we address this limitation via our dynamic state space indexing strategy (Sect. 4). Therefore, in this section, we focus on developing a bottom-up approach for mining biased cliques.

The problem can be characterized as a *search problem* consisting of (1) a start state, (2) a goal test, (3) a successor function, and (4) a validity test. In general, a start state is expanded to generate its children via the successor function.

Algorithm 1. Search	**Algorithm 2.** Successors

Input : Set of states $\mathcal{F}$
    A query (Q, η)
Output: Maximal biased cliques
	Input : A state $S = (R, P, X)$
	    A query (Q, η)
	Output: Set of child states

```
1 while F is not empty do           1 Let C ← ∅
2 │  S ← F.pop()                     2 for v ∈ P ∩ Q do
3 │  if is-goal(S) then              3 │  C ← C∪
4 │  │   output S                    │     (R ∪ {v}, P ∩ N(v), X ∩ N(v))
5 │  │   continue                    4 │  P ← P\{v};
6 │  for state C in                  5 │  X ← X ∪ {v};
   │  Successors(S, (Q, η)) do        6 Select a pivot u from P ∪ X
7 │  │   if is-valid((Q, η), C) then  7 for v ∈ P\N(u) do
8 │  │   │   F.push(S)                8 │  Repeat steps 3–6
                                      9 return C
```

This process is repeated on each of these states and so on, all while outputting the goal states, until no more states can be expanded. This search procedure is formalized in Algorithm 1. To help formulate the components of the search procedure in the context of biased cliques, we will be leveraging ideas from an existing well-known MCE algorithm by Bron and Kerbosch (BK algorithm) [5].

State. We define a state as the three-tuple $S = (R, P, X)$, where $R \subseteq V$ represents the current clique being constructed, $P \subseteq V$ represents the valid candidate vertices that may be added to R, and $X \subseteq V$ represents the vertices that must be excluded from R to prevent a redundant state from occurring.

Goal Test. A state $S = (R, P, X)$ is considered a goal state if $P = X = \emptyset$. In this case, the clique, represented by R, cannot get any larger ($P = \emptyset$) and a superset has not been previously generated ($X = \emptyset$).

Successor Function. The successor function, formalized in Algorithm 2, expands a state into a set of children states. Here, we leverage BK, which grows a clique (represented by R) by vertices in P. We further adapt this method to better incorporate query information by always creating a child for each query vertex (line 2) and doing so before the non-query vertices (line 7). This reduces the number of visited states by ensuring a query vertex does not continually remain in P, causing the state to appear valid. _Pivoting_ (line 6) is a technique in BK to reduce the number of child states; a good heuristic to use is $\mathrm{argmax}_v(|N(v) \cap P|)$.

Validity Test. The validity test determines whether a given state $S = (R, P, X)$ is valid with regards to the given query (Q, η). To do this, we simply check whether $|(R \cup P) \cap Q| \geq \eta$. If that is not true, then S can never be a biased maximal clique because R can never obtain enough vertices to be considered biased. Therefore, there is no reason to expand the state any further and it would be considered a _dead state_ with respect to the current query.

Start State. Assuming the query (Q, η), we create a start state for each vertex in Q; i.e., a state with $R = \{v\}$ for each $v \in Q$. Essentially, we run Successors

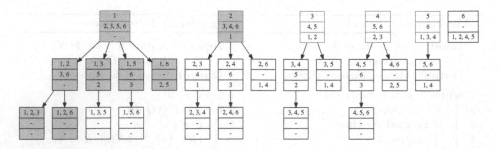

Fig. 2. Search space when mining all maximal cliques of the graph in Fig. 1. Gray nodes show the search space out of the total space (gray and white) that would be visited when mining biased cliques with query ($Q = \{1, 2\}, \eta = 2$). The rows in each node correspond to the sets R, P, and X, respectively. Note that this tree does not show pivoting.

with the input state as $S = (\emptyset, Q, \emptyset)$. Pivoting cannot be performed on Q itself because we are limiting the initial vertex set (as this improves efficiency since $|Q| \ll |V|$), which breaks the pivoting assumption.

Figure 2 depicts the search space generated when applying the algorithm on the example graph of Fig. 1. Assuming the query is ($Q = \{1, 2\}, \eta = 2$), the gray nodes show the search space that would be generated from the entire search space (white and gray). There is a start state for each vertex in Q, and the successor function is applied to each of these states. The successor function adds a vertex from P to R, growing the biased clique one vertex at a time with each call. Hence, the search tree is as deep as the largest clique. The validity test would prevent states from being expanded–*dead states*–if they weren't relevant to the given query, such as the state with $R = \{2\}$.

4   State Space Indexing and Querying Strategy

Here, we propose a strategy to facilitate the situation where a user is performing a sequence of queries to explore the graph. In this case, the main metric we are trying to optimize is cumulative query response time so that the user can explore the graph as fast as possible. The issue with our approach from Sect. 3, and solving constraint-based search problems in general, is the lack of synergy between repeated queries: you may perform redundant computations as overlapping portions of the search tree are explored for each subsequent query.

Our solution is to store and index the constituent state space that gets generated with each query that a user performs. Subsequent queries by the user can then leverage these existing states and use relevant components for their new query, thereby expanding less states. The main challenge is to keep the overhead low so that the time to search and update the index during new queries does not outweigh the time savings of having less states to expand.

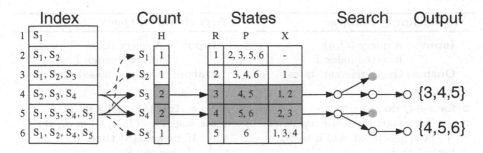

Fig. 3. Query process with query vertices $Q = \{4, 5\}$ and bias $\eta = 2$. Here, we assume that the red-outlined states of Fig. 2 are indexed. The states in the inverted Index of each query vertices have their Count incremented in H. Then, the States whose counts in H are at least η are fed to Search to produce Output.

4.1 Storing and Indexing the Search Space

One convenience of our problem formulation, is that we only need to store the states on the frontier, rather than all the states in the search space. As a consequence of our bias constraint and algorithm, we get the following proposition:

Proposition 1. *Given a state $S = (R, P, X)$ and one of its descendants $S' = (R', P', X')$, if R is a biased clique, then R' is also a biased clique. In other words, biased cliques are closed under the* Succesors *operation.*

This proposition follows from the fact that Succesors only produces supersets that are cliques, and if R is already biased (i.e., $R \cap Q \geq \eta$), then any superset of R will also be biased. Importantly, from this we can conclude that we only need to store the frontier and no ancestor states.

To store the frontier states, we simply need to keep a list of the (R, P, X) tuples. However, if only this is stored, then during query processing we must loop through *every* state to determine which ones are relevant to the given query. Since the number of states can be large, this can lead to an unacceptable overhead, particularly when Q is small, as the time to iterate through all the stored states will take longer than the time to simply generate the significantly fewer states associated with a small Q.

Therefore, to prevent looping over all states, we propose using an inverted index structure. We map each vertex in V to a list of state IDs where, for each state $S = (R, P, X)$ in the list, $v \in R \cup P$ (see Index in Fig. 3). In this case, we can easily find, for a given vertex v, all the states that *may* produce a biased clique containing v. Using the index in this way is critical for efficient query processing. Rather than checking every state to determine if it is biased, every query vertex can be checked to see which states occur at least η times.

Algorithm 3. Initialize	**Algorithm 4.** Query
Input : A query (Q, η) Inverted index $\mathcal{I}$ **Output**: Query-relevant states	**Input** : A query (Q, η) Inverted index $\mathcal{I}$ **Output**: Maximal biased cliques
1 $S_Q \leftarrow \emptyset$	1 $\mathcal{F} \leftarrow \emptyset$
2 **for** $v \in Q$ **do**	2 $S_Q \leftarrow$ Initialize$((Q, \eta), \mathcal{I})$
3 If the start state of v has never been generated, add it to S_Q	3 **for** state $S = (R, P, X)$ in S_Q **do**
4 **for** $v \in Q$ **do**	4 **if** is-goal(S) **then**
5 **for** state $S \in \mathcal{I}[v]$ **do**	5 output S
6 **if** S is in H **then**	6 **else**
7 $H[S] \leftarrow H[S] + 1$	7 **for** $v \in R \cup P$ **do**
8 **else**	8 $\mathcal{I}[v]$.remove(S)
9 $H[S] \leftarrow 1$	9 $\mathcal{F} \leftarrow \{S\} \cup \mathcal{F}$
10 **for** state S in H **do**	10 $D \leftarrow$ Search$(\mathcal{F}, (Q, \eta))$
11 **if** $H[S] \geq \eta$ **then**	11 **for** each goal or dead state
12 $S_Q \leftarrow \{S\} \cup S_Q$	$S = (R, P, X)$ from Search **do**
13 **return** S_Q	12 **for** $v \in R \cup P$ **do**
	13 $\mathcal{I}[v]$.add(S)

4.2 Query Processing

To process a query using our dynamic state space index, we first execute the Initialize procedure formalized in Algorithm 3. First, we have to check if every vertex in the given query has been a start state at some point; for those that haven't, we must add the start states to the frontier (lines 2–3). To check, we simply keep a global list of vertices that have been used as start states and check Q against that list.

Next, we must find all indexed states that are relevant to the given query via our inverted index (lines 4–12). This is accomplished by creating a table H that maps states to the number of query vertices in that state. This table is constructed by iterating through the inverted index for each query vertex, and, for every state listed, the corresponding value in H is incremented by one (lines 5–9). Typically most states will not contain a query vertex. Hence, initially creating H with all states would waste memory. Instead, we use a hash table to implement H and only insert a value once the first occurrence of a state is found. Once H is constructed, we can then iterate over H and collect only the state IDs with counts of at least η (lines 10–12).

After we have the relevant states from Initialize we can continue with the Query procedure as formalized in Algorithm 4. First, we check whether any of these states are goal states; if so, then we don't need to expand them (lines 4–5). More importantly, though, if they are not goal states, then we need to remove them from our inverted index (lines 6–8), as they will get expanded via the Successors function at least once, meaning we will be adding their descendants to the inverted index. During the search process, any state that does not pass the is-valid tests in the search process or any state that is a goal state must have its inverted index updated (lines 11–13). These new states are now part of the global frontier that has been pushed further by this latest query. An illustration of this complete pipeline on a simple example is depicted in Fig. 3.

5   Biased Clique Applications

Our biased MCE approach is applicable to a variety of problems in diverse fields, such as biomedical informatics and climate science.

Biomedical Informatics. Our biased MCE formulation can be applied to the discovery of gene-disease associations. A gene functional association network was constructed with approximately 7000 vertices and 25000 edges using the STRING database [10], where each vertex corresponds to a gene and each edge corresponds to a functional association between two genes (in our case, an interaction confidence over 950 of 1000). Genes known to be associated with a disease of interest (e.g., Alzheimer's disease), according to the Comparative Toxicogenomics Database (CTD) [9], were used as the query set. The genes from cliques biased towards this query set are likely to also be associated with the disease of interest. For example, the clique in Fig. 4, which is biased towards Alzheimer's-related genes ($\eta = 2$), includes the PSENEN gene, indicating a putative relationship between PSENEN and Alzheimer's disease. Although this relationship is not present in the CTD, the PSENEN gene is known to be associated with heart diseases, which in turn are correlated with Alzheimer's disease. Furthermore, a potential association between PSENEN and late onset Alzheimer's has been identified in the North Chinese population [13].

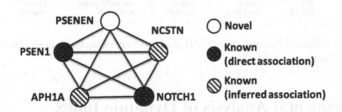

Fig. 4. A clique biased towards Alzheimer's-related genes, PSEN1 and NCSTN, including a novel gene, PSENEN, that is not yet known to be associated with Alzheimer's disease in the CTD.

Climate Science. The biased MCE formulation can also be applied to discover associations between global climate patterns and a weather event of interest (e.g., Atlantic hurricanes). A climate network of sea surface temperature (SST) was constructed with approximately 550 vertices and 19000 edges, where each vertex corresponds to a spatial point in a global grid and each edge represents a statistically significant correlation between the underlying SST time series at a pair of spatial points. Monthly SST data was obtained from the National Oceanic and Atmospheric Administration (NOAA) Extended Reconstructed SST (ERSST) dataset [1]. Vertices located in the Atlantic Ocean's Main Development Region (MDR), where the majority of Atlantic hurricanes form, were used as the query set. Cliques biased (with $\eta = 2$) towards the MDR are likely to be associated with Atlantic hurricane activity. These biased cliques include climate indices known to affect Atlantic hurricane variability (Fig. 5), such as the El Niño Southern Oscillation (ENSO) over the Pacific Ocean [11], the Madden-Julian Oscillation (MJO) over the Indian Ocean, Indonesia and the Pacific Ocean [14], and the Atlantic Multidecadal Oscillation (AMO), the Atlantic Meridional Mode (AMM) and the North Atlantic Oscillation (NAO) over the North Atlantic Ocean [21]. The analysis of these biased cliques could improve our understanding of the relationships between global climate and Atlantic hurricanes and help us identify previously undiscovered sources of Atlantic hurricane variability.

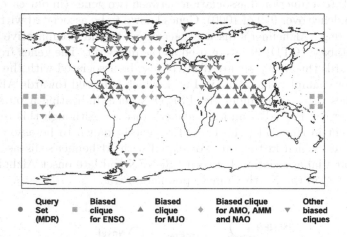

Fig. 5. Cliques biased towards the MDR and known climate indices associated with Atlantic hurricane variability.

6   Experimental Analysis of Dynamic Index

All of the experiments in this section were run on a machine with two hex-core Intel E5645 processors and 64GB DDR2 RAM running RHEL Server 6.4. Our algorithms were implemented in C++ and compiled with GCC 4.8.2 using optimization flag -O3.

To get reliable measurements, as query vertices were randomly selected when running our experiments, we ran ten trials for each data point and reported the average in the figures below. Moreover, we report CPU time and do not include the time to read the graph or write the output. Our analysis was performed on two real-world networks from the Stanford Large Network Dataset Collection [25]. The two graphs we use were social networks from the Epinions and Pokec websites. For both sets of graphs, any isolated vertices were removed and any directed edges were converted to undirected.

6.1   Benefits of Dynamic Indexing

To explore the runtime benefits of our dynamic state space index (referenced as *Dynamic*), we compare it to two baselines: full precomputation (referenced as *Full*) and no precomputation (referenced as *None*). For Full, we first find all maximal cliques in the graph, without concern for any query, effectively indexing the entire space. Once this index is built, we can then query it without ever having to perform any updates. For None, we do not build any index and do not store any states that were found during previous queries. Hence, there will be no large precomputation time and no need to update an index; however, there will likely be some redundant computations.

We attempt to simulate a realistic scenario where a user is searching for cliques with query vertices that are probably more closely related than a random pair of nodes (e.g., genes related to a certain disease). For a graph $G = (V, E)$, we generate a query with parameters n, k by first randomly selecting a set of n vertices from V. Then, for each chosen vertex v, an additional $min(k, |N(v)|)$ vertices are randomly selected from $N(v)$. To select n, we use $0 \leq \lambda \leq 1$ to denote the percentage of vertices to add to Q, and use $n = \lambda |V|/k$. In our experiments, we use $k = 4$ and $\lambda = 0.01$. Moreover, we allow overlap between successive query vertices; meaning, each successive query will contain a random fraction the query vertices from the previous query.

Figure 6 illustrates the benefits of our dynamic state space index strategy for varying degrees of query vertex overlap. In all cases, Dynamic has the lowest cumulative query response time: 17.4 % less time compared to None in the worst case (Fig. 6a) and 47.9 % in the best case (Fig. 6c). Moreover, these tests only spanned ten queries, so we could expect even higher total time savings as more queries are performed and the state space index fills out. Although Full has the fastest query response times, in this type of scenario, the time to compute and build the full index is overwhelming; hence, Full does not make sense to perform unless it is clear that the entire search space will be explored.

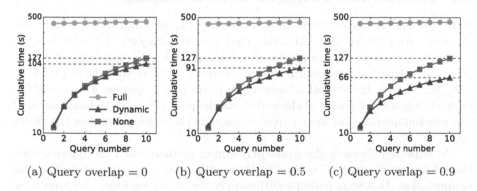

(a) Query overlap = 0   (b) Query overlap = 0.5   (c) Query overlap = 0.9

Fig. 6. Cumulative runtime efficiency of our dynamic state space index compared to the baseline methods (Full and None) when performing a number of queries (with bias value $\eta = 4$ and query size $|Q| = 0.01|V|$) on the Pokec graph.

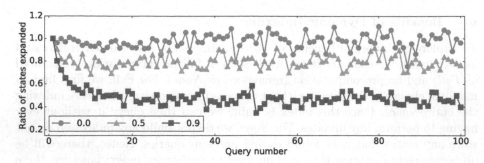

Fig. 7. Compared to no index (None), our dynamic state space index (Dynamic) allows for much fewer states to be expanded. This plot shows 100 queries (with $\eta = 2$ and query size $|Q| = 0.001|V|$) on the Pokec graph for three cases involving 0 %, 50 %, and 90 % overlap between successive queries.

Figure 7 shows how much computation our indexing strategy saves by tracking how many fewer states are expanded compared to not using an index (None). As expanding a state is the most computationally expensive part of the algorithm, this is the primary component driving up the cumulative query response time. For the first couple of queries, we can see Dynamic expands nearly the same number of states, but then it is consistently between 40 % and 60 % the number of states as None for queries with 90 % overlap. Over the 100 queries in this experiment, Dynamic expands 98 % ± 6 %, 80 % ± 6 %, and 48 % ± 8 % of the states that None expands for 0 %, 50 %, and 90 % query overlap, respectively. In the 0 % overlap case, some queries result in Dynamic expanding more states due to reusing dead states from previous queries; however, these expansions could be faster if they are deeper in the search tree, as implied by the 17.4 % query response time improvement in Fig. 6a.

6.2 Computational Overhead of Dynamic Indexing

Next, we analyze the overhead involved with performing our dynamic indexing strategy. We examine two situations that are advantageous for each baseline methods, but extremely adverse for our dynamic indexing strategy (see Fig. 8). In the first situation, we perform enough queries to fully expand the search space (Fig. 8a) and, in the second situation, we perform few queries whose searches expand only a small subset of the entire search space (Fig. 8b). In both cases, we use random queries that are completely disjoint (no overlap between the vertex sets).

As expected, None is the worst performing method when the queries cover the whole search space (Fig. 8a), as there will a large amount of overlapping computation. However, perhaps surprisingly, Dynamic performs with no overhead; in fact, it is marginally faster than Full. Intuitively, we would expect Full to perform the best since, in this case, there is no wasted computation by fully computing and indexing the entire space. Moreover, Full does not have to

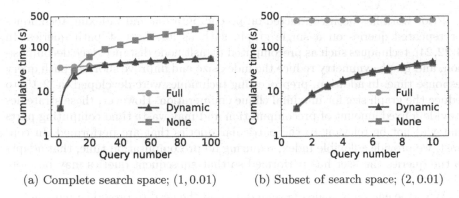

(a) Complete search space; $(1, 0.01)$ (b) Subset of search space; $(2, 0.01)$

Fig. 8. Time spent for our dynamic indexing strategy. For queries on Epinions covering a complete search space, there is no overhead compared to Full. For queries on Pokec covering only small subset of the search space, there is only a small overhead compared to None. Each plot is labeled with (bias value η, query size as fraction of $|V|$).

perform any updates to the index since it never changes. While Dynamic does have to pay an update cost, the index starts small and grows over time, so it has a faster `Initialize` procedure as there are less states to iterate through. In contrast, Full must search through a larger index to answer each query.

In the case where only small, mostly disjoint sections of the search space are explored (Fig. 8b), Full is the worst, as a large portion of the time is spent computing solutions that are never queried. Of course, None is the best as there is practically no redundant computation, effectively making the index useless. Despite this worst-case situation, None uses only 8.2 % less time than Dynamic. This difference could be erased by only a couple queries with overlapping computations. Therefore, this indicates that unless the user is exploring a very small subset of the search space with practically no overlapping computations, then it most likely makes sense to use the dynamic indexing strategy.

7 Related Work

A query-driven approach towards solving more complex graph problems has been explored from different contexts with the goal of reducing the search and output space. The community detection problem was reformulated to find a single community containing a given set of nodes [17] or to find the community or communities given a single query node [7,8]. Query vertices were also incorporated into quasi-clique discovery [19] in a similar manner. In addition, a constraint-based methodology for frequent subgraph mining has also been explored [27] as a means to efficiently incorporate users desired properties (e.g., density, minimum degree, size, etc.) and reduce the computation by filtering out states in the search space based on these properties. Complementary to these methods, our work focuses on building a synergy between *repeated* queries, via a state space index, so that a user can more efficiently explore multiple queries.

Prior work has been conducted on precomputation and indexing techniques for repeated queries on a single graph, such as for shortest path queries. In [4,22,24], techniques such as precomputed transit node distance, tree decomposition, and graph symmetry reduce the index size and improve shortest path query response time. In addition, preprocessing techniques were developed in [2,15] to reduce the graph size for maximal clique enumeration. However, these strategies provide a fixed amount of precomputation and may waste time computing parts that will not be relevant to the particular queries that are performed. In contrast, we provide a flexible index, requiring no precomputation time, that adapts to the queries the user has performed so that subsequent queries may be more efficient.

With the emerging scale of graph data and the need to provide a natural and efficient data model [20], graph databases have increased in popularity. Databases such as Neo4j, HyperGraphDB [12], and RDF databases (e.g., Jena [23]) support querying over large graphs; however, these databases generally support node/edge attribute-based, traversal-based, or exact subgraph matching queries [3], with little support for querying or indexing more complex structures.

8 Conclusion and Discussion

In this paper, we proposed a new query-driven problem formulation for maximal clique enumeration that provides a smaller, more relevant output to the user. We demonstrate concrete applications of this problem in biomedical informatics (discovering novel gene-disease relationship) and climate science (discovering associations between global climate patterns and Atlantic hurricanes). In addition, we proposed a dynamic state space indexing strategy that indexes the search space with each query to improve subsequent query response time. Experimental analysis on real-world networks showed that our strategy yields improved query-response time with minimal computational overhead, even for disjoint queries with little overlapping computation.

Although we developed our method in the context of biased maximal cliques, the underlying philosophy could possibly be generalized to other graph enumeration tasks, such as mining maximal quasi-cliques and graph pattern mining, including other constraints rather than just bias. Investigating the class of problems that are applicable, along with the effectiveness of our method, is the focus of future research efforts. For our strategy to be applicable, the enumeration task must be formulated with the search problem components specified in Sect. 3: states/start states, goal test, successor function, and validity test. This formulation along with the Search algorithm is what induces the search tree in a way that allows for the dynamic indexing.

In this work, our constraint (bias) allowed us to only store the frontier of our queried search space. If this weren't the case, then different queries may rely on the ancestors of the indexed states, requiring more than just the frontier to be stored. Although our method could arguably be extended to handle such a scenario, to ensure only the frontier needs to be stored, the following property should be observed: Given a state S, if the constraint is satisfied for S, then the constraint will also hold for all the children of S. This property limits the constraints that could be used with our strategy in its current form; for example, density would not be an applicable querying/indexing constraint because states may have a higher or lower density than their ancestors. However, there are still a number of constraints that would be applicable in the method's current form, including: bias (requiring certain nodes), structure (requiring a specific sub-pattern), size above some threshold, and maximum degree above some threshold.

Acknowledgments. The authors would like to thank David A. Boyuka II and Sriram Lakshminarasimhan for their initial discussions and feedback with this work. This material is based upon work supported in part by the Laboratory for Analytic Sciences and the U.S. DOE Office of Science ASCR. Any opinions or findings expressed in this material are those of the author(s) and do not necessarily reflect the views of any agent or entity of the US government.

References

1. Noaa earth system research laboratory. http://www.esrl.noaa.gov/psd/data/gridded/data.noaa.ersst.html
2. Abello, J., Resende, M.G.C., Sudarsky, S.: Massive Quasi-Clique detection. In: Rajsbaum, S. (ed.) LATIN 2002. LNCS, vol. 2286, pp. 598–612. Springer, Heidelberg (2002). doi:10.1007/3-540-45995-2_51
3. Angles, R.: A comparison of current graph database models. In: Workshops Proceedings of the IEEE 28th International Conference on Data Engineering, ICDE, pp. 171–177 (2012)
4. Bast, H., Funke, S., Matijevic, D., Sanders, P., Schultes, D.: In transit to constant time shortest-path queries in road networks. In: Proceedings of the Nine Workshop on Algorithm Engineering and Experiments, ALENEX (2007)
5. Bron, C., Kerbosch, J.: Finding all cliques of an undirected graph (algorithm 457). Commun. ACM **16**(9), 575–576 (1973)
6. Creamer, G., Rowe, R., Hershkop, S., Stolfo, S.J.: Segmentation and automated social hierarchy detection through email network analysis. In: Advances in Web Mining and Web Usage Analysis, 9th International Workshop on Knowledge Discovery on the Web, WebKDD 2007, pp. 40–58 (2007)
7. Cui, W., Xiao, Y., Wang, H., Lu, Y., Wang, W.: Online search of overlapping communities. In: Proceedings of the International Conference on Management of Data, SIGMOD 2013, pp. 277–288 (2013)
8. Cui, W., Xiao, Y., Wang, H., Wang, W.: Local search of communities in large graphs. In: International Conference on Management of Data, SIGMOD, pp. 991–1002 (2014)
9. Davis, A.P., Murphy, C.G., Johnson, R., Lay, J.M., Lennon-Hopkins, K., Saraceni-Richards, C.A., Sciaky, D., King, B.L., Rosenstein, M.C., Wiegers, T.C., Mattingly, C.J.: The comparative toxicogenomics database: update 2013. Nucleic Acids Res. **41**(D1), D1104–D1114 (2013)
10. Franceschini, A., Szklarczyk, D., Frankild, S., Kuhn, M., Simonovic, M., Roth, A., Lin, J., Minguez, P., Bork, P., von Mering, C., Jensen, L.J.: STRING v9.1: protein-protein interaction networks, with increased coverage and integration. Nucleic Acids Res. **41**(D1), D808–D815 (2013)

11. Goldenberg, S.B., Shapiro, L.J.: Physical mechanisms for the association of El Niño and West African rainfall with Atlantic major hurricane activity. J. Clim. **9**(6), 1169–1187 (1996)
12. Iordanov, B.: HyperGraphDB: a generalized graph database. In: Web-Age Information Management, pp. 25–36 (2010)
13. Jia, L., Ye, J., Haiyan, L.V., Wang, W., Zhou, C., Zhang, X., Xu, J., Wang, L., Jia, J.: Genetic association between polymorphisms of pen2 gene and late onset Alzheimer's disease in the north chinese population. Brain Res. **1141**, 10–14 (2007)
14. Klotzbach, P.J.: On the Madden-Julian oscillation-Atlantic hurricane relationship. J. Clim. **23**(2), 282–293 (2010)
15. Modani, N., Dey, K.: Large maximal cliques enumeration in large sparse graphs. In: Proceedings of the 15th International Conference on Management of Data (2009)
16. Moon, J.W., Moser, L.: On cliques in graphs. Israel J. Math. **3**(1), 23–28 (1965)
17. Sozio, M., Gionis, A.: The community-search problem and how to plan a successful cocktail party. In: Proceedings of the 16th International Conference on Knowledge Discovery and Data Mining, KDD. pp. 939–948 (2010)
18. Steinhaeuser, K., Chawla, N.V., Ganguly, A.R.: Complex networks as a unified framework for descriptive analysis and predictive modeling in climate science. Stat. Anal. Data Min. **4**(5), 497–511 (2011)
19. Tsourakakis, C.E., Bonchi, F., Gionis, A., Gullo, F., Tsiarli, M.A.: Denser than the densest subgraph: extracting optimal quasi-cliques with quality guarantees. In: The 19th International Conference on Knowledge Discovery and Data Mining, KDD, pp. 104–112 (2013)
20. Vicknair, C., Macias, M., Zhao, Z., Nan, X., Chen, Y., Wilkins, D.: A comparison of a graph database and a relational database: a data provenance perspective. In: Proceedings of the 48th Annual Southeast Regional Conference, p. 42 (2010)
21. Vimont, D.J., Kossin, J.P.: The Atlantic meridional mode and hurricane activity. Geophysical Research Letters **34**(7) (2007)
22. Wei, F.: TEDI: efficient shortest path query answering on graphs. In: Graph Data Management: Techniques and Applications., pp. 214–238 (2011)
23. Wilkinson, K., Sayers, C., Kuno, H.A., Reynolds, D.: Efficient RDF storage and retrieval in Jena2. In: Proceedings of The First International Workshop on Semantic Web and Databases, SWDB, pp. 131–150 (2003)
24. Xiao, Y., Wu, W., Pei, J., Wang, W., He, Z.: Efficiently indexing shortest paths by exploiting symmetry in graphs. In: 12th International Conference on Extending Database Technology, EDBT, pp. 493–504 (2009)
25. Yang, J., Leskovec, J.: Defining and evaluating network communities based on ground-truth. In: 12th International Conference on Data Mining, ICDM, pp. 745–754 (2012)
26. Zhang, B., Park, B.H., Karpinets, T., Samatova, N.F.: From pull-down data to protein interaction networks and complexes with biological relevance. Bioinformatics **24**(7), 979–986 (2008)
27. Zhu, F., Yan, X., Han, J., Yu, P.S.: gPrune: a constraint pushing framework for graph pattern mining. In: Zhou, Z.-H., Li, H., Yang, Q. (eds.) PAKDD 2007. LNCS (LNAI), vol. 4426, pp. 388–400. Springer, Heidelberg (2007). doi:10.1007/978-3-540-71701-0_38

Got a Complaint?- Keep Calm and Tweet It!

Nitish Mittal[1], Swati Agarwal[2], and Ashish Sureka[3(✉)]

[1] Netaji Subhas Institute of Technology (NSIT), New Delhi, India
nitishmittal94@gmail.com
[2] Indraprastha Institute of Information Technology, Delhi (IIITD), New Delhi, India
swatia@iiitd.ac.in
[3] ABB Corporate Research, Bangalore, India
ashish.sureka@in.abb.com

Abstract. Research shows that many public service agencies use Twitter to share information and reach out to the public. Recently, Twitter is also being used as a platform to collect complaints from citizens and resolve them in an efficient time and manner. However, due to the dynamic nature of the website and presence of free-form-text, manual identification of complaint posts is overwhelmingly impractical. We formulate the problem of complaint identification as an ensemble classification problem. We perform several text enrichment processes such as hashtag expansion, spell correction and slang conversion on raw tweets for identifying linguistic features. We implement a one-class SVM classification and evaluate the performance of various kernel functions for identifying complaint tweets. Our result shows that linear kernel SVM outperforms polynomial and RBF kernel functions and the proposed approach classifies the complaint tweets with an overall precision of 76 %. We boost the accuracy of our approach by performing an ensemble on all three kernels. Result shows that one-class parallel ensemble SVM classifier outperforms cascaded ensemble learning with a margin of approximately 20 %. By comparing the performance of each kernel against ensemble classifier, we provide an efficient method to classify complaint reports.

Keywords: Complaints and grievances · Government applications · Mining user generated content · Social media analytics · Text classification · Twitter

1 Introduction

Recently, there has been an increasing trend and adoption of social media by government organizations for not just disseminating information but also collect information such as complaints and grievances from citizens (a phenomenon referred to as *citizensourcing*) [5,7,8]. In particular, social media platforms like Facebook[1] and Twitter[2] are gaining popularity as social-media based grievance

---

[1] www.facebook.com.
[2] www.twitter.com.

© Springer International Publishing AG 2016
J. Li et al. (Eds.): ADMA 2016, LNAI 10086, pp. 619–635, 2016.
DOI: 10.1007/978-3-319-49586-6_44

@dtptraffic No one in Uttam Nagar follows traffic rule, no traffic police personnel is available. No enforcement, no fear. @AlokVermaCPDP

@IncomeTaxIndia My mom suppose to receive her #PANCard by 20June. Not received & No response from #FirstFlight.
@bookcomplaint #FFCLisCRAP

@DelhiPolice 10000₹ wrongly deducted frm my ac in Lakshmi Nagar.can u help me in refunding the amount.

@RailMinIndia @sureshpprabhu rain water is coming inside on seat through window sealing. Bedsheet n blanket both got wet. PNR NO.8248914739

Fig. 1. Concrete examples of citizen complaints reported to Government's Official Twitter Handlers- Tweets address various public issues such as traffic violation, inconvenience in train coach.

management system or platforms on which citizens can lodge complaints. Twitter is one of the most widely used micro-blogging website on Internet. Due to the wide reachability and connectivity among it's users, Twitter is being used by the Indian Government (such as IncomeTaxIndia) to reach out to the public and address their complaints[3,4]. Figure 1 shows examples of such complaints posted on various public services accounts of Indian government on Twitter. Statistics reveal that an active Government account on Twitter receives an approximate of 5 tweets per minute. Based on our analysis of several Indian government twitter account data we found that 50 % of the tweets an hour are complaints and grievances reported from various regions of India. The government bodies on Twitter forward these complaints and redirect authors to the concerned department for resolving these complaints efficiently. However, manual analysis and gaining insights from tweets is infeasible due to the high velocity and volume of the tweets posted. Further, in order to manually identify the type of the report and resolve it, many complaints remain unaddressed.

While there has been a lot of work in the area of mining twitter for extracting consumer insights, sentiments, preferences and issues by product and service companies; the field of analyzing twitter for citizen complaints and grievances is a relatively unexplored area. Heverin et al. [6] examine the use of Twitter by city police departments in large U.S. cities that have active Twitter accounts. Their analysis reveals that city police departments use Twitter to converse directly with the public and news media [6]. Anderson et al. [3] present a study on Twitter adoption across American municipal police departments serving populations over 100, 000. Meijer et al. [9] present an empirical analysis of Twitter usage by the Dutch police and their study reveals that Twitter is being used for external communication by the police officers. Edwards et al. [4] present a study on webcare, i.e. the act of engaging in online communication with citizens. They investigate 4 cases of webcare of Dutch public organizations by addressing the client feedback and related sentiments. Vanessa et al. [5] present a study for analyzing the behavioral similarities and differences of 3-1-1 phone service (formal)

---

[3] http://www.thehindu.com/business/Industry/government-to-introduce-twitter-seva-for-startups/article8483730.ece.

[4] https://blog.twitter.com/2016/modi-s-government-is-transformingindia-through-twitter-in.

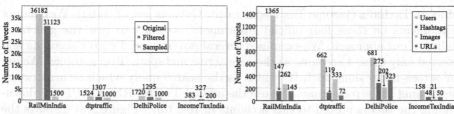

(a) Number of Tweets in Experimental Dataset   (b) Contextual Metadata Statistics

Fig. 2. Experimental Dataset Statistic- Illustrating the statistics of number of tweets collected, filtered and sampled for each account. Further, showing the variation in number of sampled tweets consisting of various entities (URL, hashtag, image, @user-mention)

and Twitter (informal) channels for reporting issues in the community. They present a supervised learning method to automatically classify the complaints and label them in order to show the comparison of types of reports posted on two channels.

The research work presented in this paper is motivated by the need to develop a solution to automatically resolve the challenges of manual inspection. Automated analysis of these complaints is a technically challenging problem due to the presence of free-form text consisting of multilingual scripts, spelling mistakes, short-forms and incorrect grammar. Further, filtering these complaint reports from non-complaint tweets is technically challenging due to the wide range of complaints. In particular, the research aim of the work presented in this paper is to investigate text classification based techniques for automatically identifying complaints tweets and assigning them to predefined labels based on the topic of the content. Furthermore, our aim is to create an annotated dataset and make it publicly available to the research community. In contrast to the existing work, our paper makes the following novel contributions:

1. We build a text analysis based ensemble classifier for identifying complaints and grievances reports from non-complaint tweets.
2. We apply core natural language processing techniques and address the challenge of presence of noisy content in the dataset.
3. We publish the first ever enhanced and enriched database of citizens' complaints tweets. We make our data publicly available for benchmarking, extension and comparison [1].

2 Experimental Setup and Characterization

We identify four Indian government related Twitter accounts for the purpose of our data collection and experimentation: @RailMinIndia (Railway Ministry of India), @dtpTraffic (Delhi Traffic Police), @DelhiPolice (Delhi Police) and

@IncomeTaxIndia (Income Tax Department, Government of India). We conduct experiments on these four official government accounts to test the generalizability of our approach. We examined several government Twitter accounts and selected only these 4 accounts as they are active and spans diverse topics. We use the screen-name along with @ symbol as the search query which returns tweets mentioning the screen-name. We use the official Twitter REST API[5] for downloading the tweets posted to this accounts in real time. Figure 2(a) displays the statistics of experimental dataset consisting of tweets (original tweets obtained, filtered and sampled) posted over a four week duration (11 April 2016 to 8 May 2016). Figure 2(a) reveals the total number of tweets collected for the four accounts were: @RailMinIndia- 36182, @dtpTraffic- 1524, @DelhiPolice- 1720 and @IncomeTaxIndia- 383. We conduct our study on English language tweets only. Therefore, we use "detected language" feature of Twitter API and filter the tweets identified as non-English or undefined ('und'). Undefined language tweets cover the posts which have no text and contains only image, video and URLs. Figure 2(a) displays the number of tweets remaining after applying the filter operation. In order to remove bias from the data, we perform a random sampling and select a sub-set of the dataset from the filtered tweets. The number of tweets after filtering and random sampling for the fours accounts (@RailMinIndia, @dtpTraffic, @DelhiPolice and @IncomeTaxIndia) are 1500, 1000, 1000 and 200 respectively.

Table 1. Frequently occurring Hashtags and Topics- Illustrating the examples of hashtags and related topics that are most discussed in the experimental dataset. The count shows the frequency of each tag in dataset.

Hashtag	Count	Topic	Hashtag	Count	Topic
@DelhiPolice			@dtpTraffic		
#OddEven	14	Vehicle Rule	#OddEven	53	Vehicle Rule
#IPSAKnowledgeSeries	13	IPS Discussion	#OddEvenDobara	44	Vehicle Rule
#sexualharassment	6	Harassment	#kotlamubarakpur	3	Car on Fire
@IncomeTaxIndia			@RailMinIndia		
#SovereignUnnathi	2	Construction Project	#RailDrishti	15	Initiative
#Theri	2	Raid	#Latur	10	Relief Operation
#Aadhaar	1	Unique ID	#RailwayZoneForVizag	8	Metro

Multi-media Content: In addition to downloading the textual content of the tweet, we use the API to extract contextual meta-data such as the type of tweet (re-tweet, reply or original), hashtag, URL, image, video and user mentions in the tweet. We also extract the general details of the blogger such as the @username and geo-location (if available). Figure 2(b) displays number of sampled tweets containing distinct users, distinct hashtags, image and URL. Figure 2(b) reveals that the number of tweets with media attachment vary for each account. For example, 50 % of sampled tweets in @dtpTraffic dataset contains external images.

---

[5] https://dev.twitter.com/rest/reference/get/search/tweets.

Table 2. Concrete examples of frequently occurring 7 and 8 character-gram strings in the experimental dataset of each public service account

Account	Hashtags
@DelhiPolice	traffic, missing, abusing, arrested, detained, criminal, communal
@dtpTraffic	flyover, oddeven, parking, pillion, crossing, redlight, hospital
@IncomeTaxIndia	website, efiling, pending, invoice, property, interest, marriage, passport
@RailMinIndia	toilets, sleeper, medical, delayed, cleaning, security, drinking, stoppage

Whereas, in @IncomeTaxIndia and @RailMinIndia datasets it varies from 15 % to 20 %. We observe a variety of topics being discussed in the tweets indicated by several distinct hashtags.

Popular User Mentions: We extract all the direct mentions in our dataset and compute their frequency. We observe that several tweets contain direct mentions to related official government Twitter handles. For example, out of 1000 tweets in our dataset for @DelhiPolice, we observe 131 direct mentions to @CPDelhi which is the official Twitter handle of the Commissioner of Police of Delhi. There are 97 mentions to @ArvindKejriwal (Chief Minister of Delhi), 44 mentions to @HMOIndia (Home Minister of India), 34 mentions to @PMOIndia (Office of the Prime Minister of India) and 26 mentions to @narendramodi (Prime Minister of India). Similarly, we observe 720 direct mentions to @sureshprabhu (Minister of Railways, Government of India) in 1500 tweets belong to the @RailMinIndia dataset. The top direct mention (total count of 26) in the dataset on @IncomeTaxIndia is @arunjaitley who is the Finance Minister and Minister of Corporate Affairs in Government of India. The Twitter handle of Chief Minister of Delhi (@ArvindKejriwal) is the top direct mention (total count 139) in our dataset for @dtpTraffic.

Topic Modeling: Table 1 displays frequently hashtags, their count and topic in our dataset. Table 1 reveals that the controversial odd-even traffic rule imposed by the Delhi government aimed at controlling the air pollution levels is one of the highly discussed topic on @dtpTraffic and @DelhiPolice. We also observe tweets on sexual harassment and a car on fire issue having direction mentions of @dtpTraffic and @DelhiPolice. Tweets on a government initiative (called as Rail Drishti) and bringing metro to a populated city were some of the topics on @RailMinIndia dataset. Topics on a unique identity card and an income tax raid were topics on @IncomeTaxIndia. We also compute frequently occurring character n-grams in our experimental dataset. Table 2 shows some of the frequently occurring 7-gram and 8-gram strings in the dataset of all 4 accounts. The frequently occurring character n-grams indicate the type of issues being discussed in the respective twitter accounts.

3   Data Enhancement and Enrichment

In this section, we discuss our proposed approach to address the challenge of noisy content in the tweets. Figure 3 shows the high-level block diagram of the proposed framework primarily consisting of 4 phases: Hashtag Expansion, Sentence Segmentation, Spell Error Correction and Acronyms & Slang Expansion. Table 3 shows examples of tweets before and after the execution of each phase.

Table 3. Examples of original and enriched complaint Tweets before and after performing the text pre-processing and enhancement of the content

	Before	After
Hashtag Expansion	@IncomeTaxIndia unearths Rs 52.5cr #blackmoney from Amritsar Rice miller via @timesofindia	unearths Rs 52.5cr black money from Amritsar Rice miller via.
	@ArvindKejriwal#OddEvenDobara being broken in broad daylight 2 men inside @abpnewstv @dtpTraffic	Odd Even Dobara being broken in broad daylight 2 men inside.
Spell Error	material snt by railways n 21/3. Current status:railways saying we haven't gt material.	Material sent by railways n 21/3. Current status:railways saying we haven't got material.
	my frnd geting call frm 8757969668 claiming to b from Naptol asking to dposit 12500 so that they will deliever Safari car	My friend getting call from 8757969668 claiming to b from naaptol asking to deposit 12500 so that they will deliver safari car.
Slang Expansion	Sir y is it so dat the vendor in S-9 of shramjeevi exp. Is taking more charge on a bottle of amul kool?	Sir why is it so that the vendor in S-9 of shramjeevi express. Is taking more charge on a bottle of amul cool?
	Pls tel n c JI dat failure of laws on cops led to dis in US learn from it	Please tel and see JI that failure of laws on cops led to this in US learn from it

3.1   Hashtag Expansion

Hashtags in a tweet are the key descriptive phrases written by simply adding a hash symbol # before the phrase. These hashtags are used to tag the tweets and make them easily searchable by other users. We observe that with the latest trend on Twitter, users create more descriptive hashtags by combining two or more strings (character sequences). For example, #GoodWorkRailwayPolice, #borivalitrainchaos, #WeRequestModiGovt and #ScamQueenOnRoad. As hashtags

Data Enhancement and Enrichment

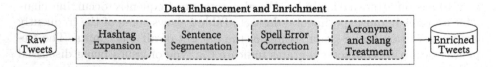

Fig. 3. High level framework design of Data Enhancement and Enrichment- The proposed framework addresses the challenge of noisy data in the tweets by expanding joint hashtags, normalizing text, correcting spelling errors and translating slang and abbreviations in a raw tweet.

are the user generated phrases, there is no one standard approach to create a joint hashtag. Therefore, expanding such hashtags is a technically challenging problem. We propose a four-step approach to split the strings in a joint hashtag and semantically enhance the tweets in our experimental dataset.

Common Separator: We expand the hashtag by simply splitting it around common separators ('_' and '-') used on Twitter. For example, #strong_action and #no-strong-action-by-police are converted into 'strong action' and 'no strong action by police' respectively.

Uppercase Letters: Unlike '_' and '-', due to the presence of acronyms splitting a hashtag only at uppercase letter can increases the noise in expansion. Therefore, we split a hashtag by keeping consecutive uppercase together until the last upper case in a string (if acronym is followed by lowercase letter). For example, #MarchForDemocracy is converted into 'March For Democracy', #CharchaOn-RWH is converted into 'Charcha On RWH' and #FANTomorrow is converted into 'FAN Tomorrow'. We however, do not expand a hashtag that contains only uppercase letters. For example, #CCTV and #FDDI.

Alphanumeric String: We split an alphanumeric phrase in a set of all numeric and character strings. If the expanded strings contain any uppercase letter, we expand it further by using Step 2. For example, #TheriJoins100crClub is converted into 'Theri Joins 100 cr Club'.

Porter Stemming: We use porter stemming algorithm[6] to identify the longest substring (a small character sequence within a large sequence) in a hashtag and split the string at that location. For example, #seriousissue and #have-somesenseofchecking are converted into 'serious issue' and 'have some sense of checking' respectively. If a hashtag is created by joining only Hindi language words then we do not perform any expansion on the hashtag. For example, #swacchbharat and #swaranshatabdi.

3.2   Sentence Segmentation

In sentence segmentation, we remove all the URLs and user mentions (@) from the tweets since they do not define the content or topic of the tweet. We remove the filler terms from the tweet and replace consecutive special character occurring together by single character. For example, '?????' is replaced with '?'. It has been seen that due to the character limit in a tweet, users do not write spaces after period or special characters. We expand our tweets by adding one space after each period, comma, question mark and exclamation mark. Later, we replace each joint hashtag with the expanded phrases obtained in previous phase. Further, if a hashtag occurs in between the text, we replace it as it was. Whereas, if the hashtag occurs at the end of the tweet, we treat each expanded hashtag as a new sentence.

---

[6] http://snowball.tartarus.org/algorithms/porter/stemmer.html

3.3 Spell Error Correction

Due to the presence of free-form and user generated text, a tweet is high likely to have spelling and grammar mistakes. In this phase, we address the challenge of spell errors in a tweet by correcting them using n-gram model. We split our tweet from each special character (_, -, ?, !) and create a set of n-consecutive word grams (3 in our approach). For example, for a tweet consisting of words 'a b c d' has a set of two 3-grams 'a b c' and 'b c d'. We use Bing Search Engine[7] and query each n-gram using GET method while the language of spell checking is set as English. We extract the translated n-gram resulted as "Including results for" in Bing Search and store the results for each n-gram separately and compute their extent of similarity with queried n-grams. For the first and last words in tweet, we replace them with the terms corrected by Bing Search. Whereas, for the terms appearing in multiple n-grams, we replace them by the term corrected in majority of n-grams. For example, for a given sentence, 'pleas answr my query asap', we create a set of three 3-grams [n1: 'pleas answr my', n2: 'answr my query', n3: 'my query asap']. Here, 'answr' is replaced by 'answer' only if both n1 and n2 are corrected as 'please answer my' and 'answer my query'.

3.4 Acronyms and Slang Treatment

In the last phase of data enhancement and enrichment, we expand the Internet Slangs and normalized text of a tweet written in 'sms' language. However, replacing all slang words with their standard phrases can increase the noise in the text. Therefore, we do a manual inspection on each account (@RailMinIndia, @dtpTraffic, @DelhiPolice and @IncomeTaxIndia) and create a list of domain specific Slangs and their respective definition. For example, in @RailMinIndia dataset, 'rly' is replaced by 'railway' rather than 'really'. If a term does not exist in domain specific slang list we replace them with their standard phrases by sending them through a POST request on noslang[8]- largest portal for Internet slang dictionary & translation. We however, do not replace a numeric character unless it occurs in an alphanumeric string. For example, 'B4' is converted into 'before' while '2' remains unconverted because semantically '2' can mean either of 'too', 'to' and 'two'.

4 Complaints and Grievance Tweets Classification

As discussed in Sect. 1, all public service accounts have open Twitter handlers and any one can mention them in their tweets. Therefore, not all tweets that are posted to these agencies are complaint reports. In order to classify complaints and grievances (C&G) tweets, we build a one-class classifier since the class of other tweets is unknown. We however observe that there are various features that are strong indicators of tweets to not be a complaint tweet. Despite having only

---

[7] https://www.bing.com.

[8] http://www.noslang.com/dictionary/.

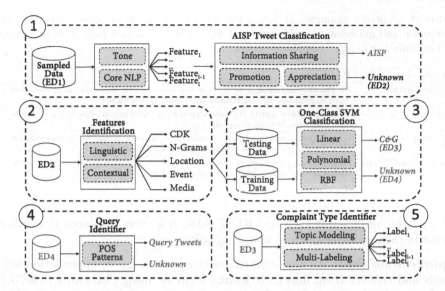

Fig. 4. A general research framework for complaints and grievances tweet classification

one target class, we build another independent one-class classifier that identifies the tweets that are certainly not complaints and grievance reports (non-C&G). Figure 4 shows the high level architecture of proposed framework primarily consisting of five phases. We discuss each phase in the following subsections.

4.1 Appreciation, Information Sharing and Promotion (AISP) Tweets Classifier

Based on our observation and analysis of non-complaint tweets, we divide them into four broad categories: appreciation, queries, information sharing and promotion tweets. Table 4 shows examples of tweets classified into each of these categories. Since, a query post can be ambiguous with a complaint tweet, we first classify other three categories of tweets and identify query tweets in later steps. We perform AISP tweet classification on the enhanced version of our experimental dataset (refer to Sect. 3).

News Update or Information Sharing: Tweets with the presence of an external URL (not image or video attached) are marked as non-complaint tweets. We identify such tweets by using the URL count feature of Twitter REST API. However, we do not classify tweets that has link to another tweet (referred as quote in Twitter).

Promotion: We observe that there are several tweets which are posted by other official accounts of same public agencies. For example, @sureshpprabhu (Minister of Railway Department of India) posting about a new policy and mentioning @RailMinIndia in his tweets. We classify such tweets by checking verified account

628   N. Mittal et al.

Table 4. Concrete examples of Non-C&G Tweets- presence of appreciation, query, promotion and information sharing content in a tweet are strong indicators of a tweet for certainly not being a complaint report.

Category	Tweet
Appreciation	Sir **thanks** for releasing all Income tax return forms and utilities in month of April as promised by you **Kudos to. Good Governance.**
Query	**Will Odd Even apply if** i take u turn from under rajokri flyover 2 go 2 ambience mall from IFFCO. Gurgaon.
Promotion	Income Declaration Scheme: Government assures complete confidentiality: http://goo.gl/jVU5qK @FinMinIndia **@IncomeTaxIndia**
Information Sharing	@RailMinIndia Unravelling the Central Railway?s hidden history. http://www.thehindu.com/todays-pape... @sureshpprabhu

value of the blogger using Twitter REST API. If a tweet is posted by verified accounts then we mark them as non-complaint posts.

Appreciation Post: We create an exhaustive lexicon L_K of appreciation keywords and convert them into their lemma form by using Stanford's CoreNLP API[9]. We create a bag of words L_W by converting each word of a tweet to their lemma form. If there exists an intersection between L_K and L_W, we compute the joy (emotion tone) feature of that tweet. If the confidence score of joy feature of a tweet is above certain threshold then we classify it as an appreciation post. In order to compute the threshold value, we take a sample of 50 posts annotated as appreciation post and compute their joy confidence score using Alchemy's Tone Analyzer API[10]. We compute the average of these confidence score and record it as the threshold for testing tweets.

4.2   Features Identification

In the second phase of our proposed approach, we identify various linguistic and contextual features that can be used to classify a complaint tweet.

Frequent N-Grams: As discussed in Sect. 2, we find several character n-grams that occur very frequently in complaint posts and defines the topic of the complaint. We create a triplet of such n-grams and defines a new feature vector to avoid keyword-based-flagging method. We create these triplets based on the frequent n-grams that are common for each account (to keep our approach generalized) and n-grams that are specific to an account and not seen in other complaints. Table 5 shows the triplets created for each account. We identify n-grams that are similar to each other to merge them into one item of triplet. For example, in a triplet <I1, I2, I3>, if we find a term I4 that is similar to I1 them we merge them together and expand our triplet as <I1/I4, I2, I3>. Table 5 shows the grouped triplets for each account. We further avoid keyword-spotting approach and improve the efficiency of our feature vector. We use WordNet lexical

---
[9] http://stanfordnlp.github.io/CoreNLP/.
[10] http://www.ibm.com/watson/developercloud/tone-analyzer.html.

Table 5. Grouped triplet of most frequent N-grams for each account in experimental dataset

Account	N-Grams	Grouped Triplets
@DelhiPolice	bribe, abuse, harrasment, FIR, phone, action, report, complaint	< {bribe, abuse, harass}, {FIR, phone}, {action, report, complaint}>
@dtpTraffic	bribe, challan, violation, abuse, harass-ment, jam, commotion, congestion, acci-dent	< {bribe, challan}, {violation, abuse, harassment}, {jam, commotion, conges-tion, accident}>
@RailMinIndia	train number, train name, coach, pnr number, bribe, corruption, report, com-plaint, action	< {train number, train name, coach, pnr number}, {bribe, corruption}, {report, complaint, action}>
@IncomeTaxIndia	pan number, ack, FIR, TIN, complaint, report, investigation, refund	< {pan number, ack, FIR, TIN}, {complaint, report, investigation}, {refund}>

Table 6. A sample of closed domain keywords identified for each account

Account	Key-Terms
@DelhiPolice	bribe, abuse, harassment, FIR, phone, action, report, complaint
@dtpTraffic	bribe, challan, violation, abuse, harassment, jam, commotion, congestion, acci-dent
@RailMinIndia	train number, train name, coach, pnr number, bribe, corruption, report, com-plaint, action
@IncomeTaxIndia	pan number, ack, FIR, TIN, complaint, report, investigation, refund

resource[11] to identify the synonyms of the terms present in the tweets and com-pare them with the words present in our lexicon. For example, if the triplet in our feature vector has the term "investigation" and an unknown tweet contains a term "enquiry", keyword spotting method will assign a value of 0. Whereas, "enquiry" is another term used for "investigation" and hence will be detected using a lexical database approach.

Closed Domain Key-Terms: Online government and public service agencies have several different departments and receives certain types of complaints that are specific to these departments. Based on our observation and manual inspec-tion, we create a lexicon of the keywords that are specific to these complaints without going into low-level details of the type of complaints and to keep our features generalized. Table 6 shows a sample of such keywords for each account.

Events and Substances: We find that not all complaints posted to these government accounts are related to the problems faced by the reporters. But these reports are also for common concern and bringing the attention of public service agencies to various issues. For example, violation of traffic rules and alcohol consumption. In order to find such activities in the tweets, we create two more feature vectors "events" and "substances". We use IBM Watson Concept and Relationship Extraction API[12] and extract various events and substances being performed and reported in the tweet.

---

[11] https://wordnet.princeton.edu.

[12] http://www.ibm.com/watson/developercloud/apis/relationship-extraction-apis.html.

Location: It has been seen that people reporting complaints about public issues often mention their locations in their tweets. For example, an accident happened at location L_1 or a train did not arrive at time at platform P_1. Therefore, we use location parameter as another feature vector for classifying complaint tweets. We use IBM Watson Relationship Extraction (IRE) API to identify the named entities in a given tweet. However, we find that due to the presence of free form text and people names in locations, IRE is not able to identify the location with 100 % accuracy and predict many locations as people and organization. For example, M. B. Road, Gandhi Nagar and AIIMS metro station. Therefore, we merge people, organization and location entities and further use Google geocoding API[13] to identify the terms that are locations. We apply IRE before performing geocoding as it groups the terms that are likely to be one entity. For example, Preet Vihar, Nehru Place and Hauz Khas Village.

Media Presence: Twitter allows users to attach multi-media files in their tweets such as video and pictures. Attaching a video and picture in the complaint tweets increases the credibility of their report as it provides an evidence and helps concerned authorities to understand the severity of problem. We identify the presence of media files (video or picture) in a tweet and create a Boolean feature vector in our model.

4.3   Classification

Complaint and Grievance (C&G) Tweet Classification: The third phase of our proposed framework is an ensemble learning based Support Vector Machine (SVM) classifier. We split our data (tweets classified as unknown in AISP classifier) into 1:3 ratio and use these 25 % and 75 % of the tweets as training and testing dataset. We train our model from feature vectors created in Phase 2 and perform one-class classification on each tweet in testing dataset. SVM classifier learns the features from the tweets verified and annotated as complaint tweets and identify whether a given tweet (in testing dataset) is a complaint tweet or not. If a tweet is not identified as C&G then it is classified as unknown. Research shows that the performance of an SVM classifier can be improved by modifying the kernels of the classifier [2]. Therefore, in order to investigate the efficacy of our proposed approach and identified features (linguistic and contextual), we run our classifiers for SVM by varying the kernel functions of SVM. We test our approach for linear, polynomial and RBF (Radial Basis Function) Kernels. Further, we ensemble all three kernels and execute them in cascaded and parallel manner. For each SVM classifier (linear, polynomial, RBF kernels and ensemble classifier), we get a set of tweets classified as complaints and grievances tweets and another set of unknown tweets.

Query Identification: Another indicator of a non-C&G tweet is a query post where users post their general queries about the things that are related to these public services. In order to perform a verification of unknown tweets classified by

---

[13] https://developers.google.com/maps/documentation/geocoding/intro.

C&G classifier, we identify the query posts (refer to Sect. 4.1) among unknown tweets. However, due to the presence of free-form text, it is hard to identify a query post just by spotting a 'Wh' word in the starting of a sentence or a question mark in the end of the sentence. We tag part-of-speeches in each tweets and defines 5 patterns that are strong indicators of a tweet to be query. "{}" denotes the optional part-of-speech.

1. Modal/VBP + PRP + VB + NN(P/S) or PRP
2. Modal/VBP + NN (P/S) + VB
3. WHADVP/WHNP/WP + Modal/VBZ + DT + VB + NNP/PRP
4. If + NNP/PRP + VBZ/Modal + NNP/PRP
5. WHADVP/WHNP/WP + VBZ/Modal + RB + NNP/PRP + JJ

We further avoid the pattern 'WP + DT + NN (P/S)' as it is commonly used pattern in complaint tweets.

Complaint Type Identification: In order to identify the type of complaint, we perform topic modeling on all tweets identified as complaint and grievances reports. Since, the type complaints posted on Twitter vary for each account and have a large dimension, for example, in @RailMinIndia, cleanliness, delay in train, refund-issue, waiting room, platform, berth allocation, poor service and assistance in coach and many more related complaints. Similarly, in @dtptraffic, various complaints on traffic rules violation, yellow line violation, bribery, illegal challans, riding motor-bikes without helmets and similar complaints with different issues can be there. Therefore, we use natural language processing technique and remove the dependency with keyword spotting methods. We use Alchemy concepts and taxonomy API[14] and label these complaints into the most likely topic and sub-topic defined in the taxonomy hierarchy. For example, riding without helmet or driving without a number plate both are traffic violation related complaints. Similarly, unhygienic food serving or low quality facilities to passengers are labeled as poor assistance in coaches.

5 Empirical Analysis and Experimental Results

In this Section, we present the accuracy results of each classifier and also discuss the influence of various kernel functions in SVM on the accuracy of proposed approach. We evaluate the accuracy of our classifier by comparing the observed results against actual labeled class. We conduct our experiments on the sampled tweets collected for @dtpTraffic (1000), @DelhiPolice (1000), @RailMinIndia (1500) and @IncomeTaxIndia (200) and report the accuracy of AISP classifier for each account. Proposed AISP classifier identifies 47 (A:12, IS:27, P:8), 132 (A:20, IS:99, P:13), 121 (A:41, IS:76, P:4) and 35 (A:4, IS:30, P:1) tweets as AISP for @dtpTraffic, @DelhiPolice, @RailMinIndia and @IncomeTaxIndia respectively. Based on our experimental results, we are able to correctly classify

---

[14] http://www.alchemyapi.com/products/alchemylanguage/taxonomy.

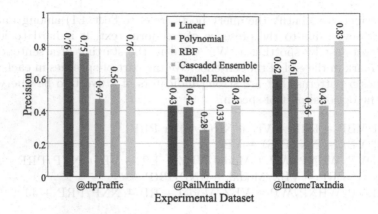

Fig. 5. Confusion matrix results for C&G Tweets classifiers (SVM with 3 different kernel functions and Ensemble Classifiers)- Column charts illustrate that linear kernel outperforms other kernels and by ensembling all kernels in cascaded or parallel boost the overall performance of each kernel.

124, 34, 32 and 103 tweets for @DelhiPolice, @dtpTraffic, @IncomeTaxIndia and @RailMinIndia respectively. While there is a misclassification of 8, 13, 3 and 18 tweets in similar order of accounts.

We execute our C&G classifier on unknown posts classified in previous phase (@dtpTraffic: 953, @DelhiPolice: 868, @RailMinIndia: 1379 and @IncomeTaxIndia: 165). We split these unknown tweets into training and testing dataset. Since, we implement a one-class classification algorithm, we train our model only for target class (complaints and grievance reports) tweets while test dataset contains tweets that belong to complaint or others categories. For @dtpTraffic, we use 239 and 714 tweets as training and testing tweets respectively. Similarly, for @RailMinIndia, @DelhiPolice and @IncomeTaxIndia, we use 345, 217 and 42 tweets as training dataset respectively. While, 1034, 651 and 123 tweets are used for testing dataset for these account. Since, the accuracy measures are biased towards the majority class in dataset, we evaluate the performance of our classifiers using the standard information retrieval metric i.e. precision. We however find that due to high imbalance of complaint reports in @DelhiPolice experimental dataset, our model does not find enough tweets for training the model (<50). Therefore, we discuss the complaint classification results for remaining three accounts.

Figure 5 shows that linear kernel in SVM outperforms RBF kernel with a reasonably high margin (varying from 20 % to 30 %). Figure 5 also shows that for @dtpTraffic (linear kernel), we are able to achieve the maximum precision rate i.e. 76 % (184/(184+58)). Whereas, for @IncomeTaxIndia and @RailMinIndia, we were able to achieve a precision rate of 62 % (31/(31+19)) and 43 % (188/(188+252)) respectively. Our result reveal that using linear kernel in one-class SVM classifier, we are able to achieve an overall accuracy up to 60 %.

Whereas, there is an overall misclassification of up to 10 % in identifying complaint tweets as unknown. The column charts in Fig. 5 shows that linear and polynomial kernels gives similar results with a difference 1 % to 2 % in precision rate. Using SVM polynomial kernel, in @dtpTraffic experimental dataset, we were able to identify complaint tweets with a precision of 75 % (170/(170+56)). While, for @RailMinIndia and @IncomeTaxIndia, we were able to identify complaints tweets with a precision rate of 42 %(139/(139+189)) and 61 % (22/(22+14)) respectively. In order to compute the efficacy of our approach for correct classification, we record an overall misclassification of 12% (complaint tweets wrongly classified as unknown) for all accounts for polynomial kernel SVM classifier.

Boosting of Base-line Approach: As discussed in the literature, the performance of SVM classifier can be boosted by modifying the kernels or combining more than one classifiers [2]. Therefore, in order to boost the efficiency of our proposed approach, we ensemble our SVM classifiers (3 different kernels) together and evaluate their performance while arranged in cascaded and parallel fashion. We compare the accuracy results of ensemble classifiers with three classifiers executed individually. Reveal that similar to individual classifiers, ensemble classifier also gives best results for @dtpTraffic (maximum precision). For a given testing dataset (dtptraffic: 714, RailMinIndia: 1034, IncomeTaxIndia: 123), parallel ensemble SVM classifier outperforms cascaded ensemble classifier. Using a combination of linear, polynomial and RBF kernels in parallel manner, we were able to achieve a precision of 75 %, 83 % and 39 % for @dtpTraffic, @IncomeTaxIndia and @RailMinIndia respectively. While, there is an overall misclassification of 18 % in identifying complaint tweets as unknown. Figure 5 reveals that by arranging these kernels in cascaded order, it decreases the performance of overall classification from 10 % to 20 %. For example, for @dtpTraffic and @IncomeTaxIndia datasets, we achieve a precision of 56 % and 43 % respectively which are approximately 20 % lesser than the individual precision of linear kernel SVM classifier. In comparison to cascaded ensembling, in parallel ensemble classification, we are able to boost the accuracy for @IncomeTaxIndia dataset by 21 % whereas for @dtpTraffic, the performance is maintained with a precision of 76 %.

We label complaint tweets of each account using taxonomy and concept feature. Our results reveal that for @dtpTraffic account, maximum complaints are posted regarding traffic light violation, illegal tax, bribe payment and license related issues. Similarly, in @RailMinIndia experimental dataset, maximum complaints belong to theft, food assistance, cleanliness and train delayed issues. Unlike @RailMinIndia and @dtpTraffic, @IncomeTaxIndia has a very small subset of tweets and therefore, we do not find a wide range of complaints. Maximum number of complaints in @IncomeTaxIndia belongs to PAN card related issues.

6   Conclusions and Future Work

Twitter has recently been used as a platform to make posts about complaints and grievances of citizens and to bring Government's attention to various public

issues. However, due to the presence of free-form social media text and high velocity of data, automatic identification of complaint tweets is a technically challenging problem. In this paper, we address the challenge of noisy and free-form data
by enhancing the hashtags, spell errors and Internet slang in a tweet. We publish the first ever database of complaint and grievances reports on Twitter and
make our enriched dataset publicly available for benchmarking and comparison.
We identify various linguistic and contextual features for identifying complaints
reports tweets. We also propose various features that are strong indicators of a
tweet to certainly not to be a complaint report. We propose a one-class ensemble
learning technique to classify complaint and grievances tweets from unknown or
non-complaint tweets. In order to evaluate the performance of our proposed approach, we execute our SVM classifier for three kernels (linear, polynomial and
RBF). Our results reveal that linear kernel one-class SVM outperforms RBF
with a margin of 20 % in precision rate while polynomial and linear kernels produce the similar results with a difference of 1 % to 2 % of precision. However,
the rate of misclassification in polynomial kernel is higher than the linear kernel
function. We further boost the accuracy of our proposed approach by combining
three kernels into a cascaded and parallel manner. Our result shows that parallel
ensemble classifier outperforms cascaded ensemble SVM. Using parallel ensemble
technique, we are able to improve the precision of complaint tweet classification
by 20 %.

Future work includes improving the accuracy of linguistic features and
addressing the limitations of present study. Expanding the list of domain specific
key-terms by using a lexical resource and improving the accuracy of proposed
approach over keyword-based flagging methods. Future work also includes creating a front-end data visualization tool and showing the status of complaints
and grievances in form of a dashboard.

References

1. Agarwal, S., Mittal, N., Sureka, A.: Enhanced dataset of citizen centric complaints
   and grievances on twitter mendeley data, v1 (2016). http://dx.doi.org/10.17632/
   w2cp7h53s5.1
2. Amari, S., Wu, S.: Improving support vector machine classifiers by modifying kernel
   functions. Neural Netw. **12**(6), 783–789 (1999)
3. Anderson, M., Lewis, K., Dedehayir, O.: Diffusion of innovation in the public sector:
   Twitter adoption by municipal police departments in the US. In: Portland International Conference on Management of Engineering and Technology (2015)
4. Edwards, A., Kool, D.: Webcare in public services: deliver better with less? In:
   Nepal, S., Paris, C., Georgakopoulos, D. (eds.) Social Media for Government Services, pp. 151–166. Springer, Cham (2015)
5. Frias-Martinez, V., Sae-Tang, A., Frias-Martinez, E.: To call, or to tweet?
   understanding 3-1-1 citizen complaint behaviors. In: ASE BigData/SocialCom/
   CyberSecurity Conference (2014)
6. Heverin, T., Zach, L.: Twitter for city police department information sharing. In:
   Proceedings of the American Society for Information Science and Technology (2010)

7. Khan, G.F., Swar, B., Lee, S.K.: Social media risks and benefits a public sector perspective. Soc. Sci. Comput. Rev. **32**(5), 606–627 (2014)
8. Loukis, E., Charalabidis, Y., Androutsopoulou, A.: Evaluating a Passive social media citizensourcing innovation. In: Tambouris, E., Janssen, M., Scholl, H.J., Wimmer, M.A., Tarabanis, K., Gascó, M., Klievink, B., Lindgren, I., Parycek, P. (eds.) EGOV 2015. LNCS, vol. 9248, pp. 305–320. Springer, Heidelberg (2015). doi:10.1007/978-3-319-22479-4_23
9. Meijer, A.J., Torenvlied, R.: Social media and the new organization of government communications an empirical analysis of twitter usage by the Dutch police. The American Review of Public Administration, p. 0275074014551381 (2014)

Query Classification by Leveraging Explicit Concept Information

Fang Wang[1](✉), Ze Yang[1], Zhoujun Li[1], and Jianshe Zhou[2]

[1] State Key Laboratory of Software Development Environment, Beihang University,
Beijing 100191, People's Republic of China
{fangwang,tobey,lizj}@buaa.edu.cn
[2] Beijing Advanced Innovation Center for Imaging Technology,
Capital Normal University, Beijing 100048, People's Republic of China
zhoujs@cnu.edu.cn

Abstract. A key task in query understanding is interpreting user intentions from the limited words that the user submitted to the search engines. Query classification (QC) has been widely studied for this purpose, which classifies queries into a set of target categories as user search intents. Query classification is an important as well as difficult problem in the field of information retrieval, since the queries are usually short in length, ambiguous and noisy. In this case, traditional "bag-of-words" based classification methods fail to achieve high accuracy in the task of QC. In this paper, we propose to mine explicit "Concept" information to help resolve this problem. Specifically, we first leverage existing knowledge bases to enrich the short query from the concept level. Then we discuss the usage of the mined concept information and propose a novel language model based query classification method which takes both words and concepts into consideration. Experimental results show that the mined concepts are very informative and effective to improve query classification.

1 Introduction

Query classification (QC) has been widely studied for query understanding [1–7], which essentially maps search queries into different intent categories. QC has many applications in information retrieval, Web search and online advertisement. Taking the Web search for example, it is important to organize the huge Web pages in the search results after the user issues a query, according to the potential categories of the results. Query classification can be used to effectively organize these results.

However, it is non-trivial to accurately classify user queries because the queries usually are short in length and query terms can be noisy [2]. Web search queries only contain less than three words on average [8]. At the same time, the vocabulary used in queries is vast. Thus, classifiers based on the "bag-of-words" text representation have to deal with a very sparse feature space, and often require large amounts of training data. Besides, many query terms are ambiguous. For example, "apple" means a kind of fruit in the query *"apple pie recipe"*,

J. Li et al. (Eds.): ADMA 2016, LNAI 10086, pp. 636–650, 2016.
DOI: 10.1007/978-3-319-49586-6_45

while in *"apple ipad"*, it means a computer company. Therefore, term ambiguity is another big issue which hinders the progress of query classification.

To address the above challenges, a variety of query classification approaches have been proposed in the literature. A major traditional approach is through query expansion using relevance feedback [9–11], typically, fetching search snippets from search engines. However, it is not an ideal solution for some applications, since it is very time consuming and heavily depending on the quality of search engines. Many other methods use query logs to expand the meanings of user queries [12–15]. However, in general, such logs are not available for the timely training and application of query classification models, as in the case of the KDDCUP2005 competition[1]. Another major method is query expansion by using a dictionary or thesaurus [1,5,16,17], whereby a set of explicit or implicit topics are discovered and used to enlarge the short queries. However, these expansions are still weak in semantics. For example, "Lady Gaga" and "Beyonce" are not synonymy, but they are very similar since both are famous singers, and tend to belong to the same category such as *Music*.

In this paper, we also mine explicit *Concept* information to help resolve this problem. The concepts used in existing methods are weak in semantics, which tends to be synonym sets. For example, Huang [18] represented "Obama", "President Obama", and "B. H. Obama" with the concept *Barack Obama* in Wikipedia. Differently, we are interested in high level concepts that cover a set or a class of terms or "things" within a domain. Also for the term "Obama", we mine its high level concepts such as *president, politician* and *American*. We argue that these concepts benefit query classification. For example, "Obama approval rating" and "Jinping Xi visit UK" are two different queries without any common words, but if we know that both "Obama" and "Jinping Xi" belong to the concept *politician*, then from the concept angle, the two queries are similar and tend to be classified to the same category *Politics*.

The contributions of this paper lies in: (1) We propose to mine concepts of different granularity from two knowledge bases. Compared with words, concepts are unambiguous [18]. However, an effective disambiguation method is needed to map an ambiguous term to its concepts. We explore different term disambiguation methods to mine concepts from two knowledge bases, and further study the effectiveness of concepts within different granularity in the task of query classification. (2) We discuss the usage of the mined concept information for the QC task and propose a novel language model based classification method which takes both words and concepts into consideration. (3) We conduct experiments on two real data sets to assess the effectiveness of concept-based query classification, and the experimental results show that concept information is very informative and effective in the task of query classification.

---

[1] http://www.kdd.org/kddcup/view/kddcup2005/Tasks.

2   Related Work

Mapping search queries to a predefined category with a reasonable degree of accuracy is an important but challenging task for many applications.

To overcome the problem of data sparsity, much work has been devoted to enrich query features. Web pages are the most used resources for query expansion [9–11]. However, how to seek related Web pages for a given query is one of the key problems for these approaches. Generally, existing methods leverage search engines to help retrieval related Web pages. Shen et al. [2,9] enriched the queries with its related Web pages together with their category information that were collected through the use of search engines. Broder et al. [11] used a blind feedback technique, whereby given a query, they determined its topic by classifying the web search results retrieved by the query. However, search engine based methods are unsatisfying because the search engines are not always reliable. For example, the Web page category information used in [2,9] is not available now. Besides, related Web pages searching is very time consuming, which is not applicable for many online applications such as search advertisement.

Another major method is to use query log to expand the meanings of user queries. Wen et al. [12] leveraged the co-clicked queries and most selected documents to enrich the query features. Beitzel et al. [13] used a very large web search engine query log as a source of unlabeled data to aid in automatic classification. Arguello et al. [15] mined query-log features from vertical query-logs and generic web query logs and built a unigram language model from each query-log for the task. There are limitations to leverage the query log information. A large proportion of search queries do not raise any click-through and two random queries rarely share the same click URLs [19]. The query log cannot help classify those queries that have no click information. Besides, the clicked URLs in query log data are noisy and may also be biased from some users with malicious intents [20].

Many other methods expand the short query by leveraging dictionaries, thesauruses or knowledge bases [1,5,16,17]. Shen et al. [1] leveraged the ODP directory as subcategory medium to help query classification and won in the KDD-CUP2005 competition. Huang [18] used the concepts in WordNet and Wikipedia to enrich the text representation. Gabrilovich et al. [16] proposed to enhance text categorization with encyclopedic knowledge, whereby they regarded each Wikipedia article a concept and enriched the text representation with relevant Wikipedia concepts. However, the concept information used above are still week in semantics. For instance, "Jeep" and "SUV" are two different terms but very similar to each other. Relying only on the surface concept information from WordNet and Wikipedia cannot capture this semantic information. Sepideh et al. [17] proposed to enrich short text by incorporating information about correlation between terms. Yang et al. [6] presented a topic-oriented word embedding approach to enrich the short and ambiguous queries with the learnt query embedding.

Our paper also mine explicit "Concept" information, but the concept granularity is different with existing concept based methods. For example, the term "Obama" can be linked to fine-grained concept *Barack Obama* in Wikipedia, while it can also belong to coarse-grained high level concepts such as *American president* and *politician*. We show the usage and effectiveness of concepts within different granularity in the QC task.

3   Concept-Based Query Representation

Much work has been devoted to building knowledge bases. Existing knowledge bases include WordNet [21], Wikipedia[2], Freebase [22], YAGO [23], Probase [24], etc. This paper takes Wikipedia and Probase as running examples to mine concepts from queries.

3.1   Identifying Wikipedia Concepts from Queries

The basic conceptual unit of Wikipedia is a Wikipedia article: a page dedicated to a specific concept that provides a detailed and well structured explanation of it. For instance, *agriculture, agricultural sector* and *agricultural* can be represented by the Wikipedia concept *Agriculture*. A mention is a phrase used to refer to something in the world such as Wikipedia concept. E.g., "Michael Jordan" may refer to the legendary American basketball player, or a famous researcher in machine learning and artificial intelligence. Generally, there mainly are two steps to map a mention to a Wikipedia concept.

Mention Detection. The first step is mention recognition and mention-candidate association. One of the most used approaches is dictionary-based method, which needs to construct a static dictionary with each entry as $\{m, (c_1, \ldots, c_k)\}$ using Wikipedia as the previous work [25] did. This paper uses four resources for building the dictionary: the title of c, the titles of pages redirecting to c, the titles of disambiguation pages containing c, and anchor texts with hyperlinks to c. To filter noise, we remove anchor texts composed by one character or numbers only, and anchor texts which appear less than 5 times. With the dictionary, we can identify mentions in queries using the longest prefix match algorithm[3], and associate the mentions with the entity candidates in the entries of the dictionaries. During the parsing, if one term is a substring of another term (e.g., New York and New York Times), we choose the longest term as the detected mention.

Mention Disambiguation. The second step, also the most important step, is disambiguation. Summarily, the following features have been proved to be useful:

- *The prior probability of a sense*: also known as sense probability. It captures the empirical rules of mapping a mention to a Wikipedia concept. There are many anchor texts in Wikipedia articles, with hyperlinks to the other concepts

---

[2] https://www.wikipedia.org/.
[3] http://en.wikipedia.org/wiki/Longest_prefix_match.

that the anchor texts refer to. Regarding the anchor texts as mentions, we can estimate the *sense popularity* with the following equation:

$$Probability(c_j|m_i) = \frac{count\,(c_j)}{\sum_{\forall c_k \in C(m_i)} count(c_k))} \qquad (1)$$

where $count(c_j)$ denotes the frequency of mention m_i linking to concept c_j in Wikipedia's articles and $C(m_i)$ indicates m_i's candidate concept set.

– *Text-based similarity*: Text-based features take the mention's context into consideration and define similarity measures based on the text information of the mention and the entity article in Wikipedia. Cosine similarity is one of the most typical textual features. Formally,

$$Similarity_{cs}(cxt(m_i), cxt(c_j)) = \frac{\sum_{k=1}^{n} V_k^{m_i} \times V_k^{c_j}}{\sqrt{\sum_{k=1}^{n}(V_k^{m_i})^2} \times \sqrt{\sum_{k=1}^{n}(V_k^{c_j})^2}} \qquad (2)$$

where V^{m_i} is the word vector of m_i and V^{c_i} for the vector of concept c_j.

– *Relatedness between Wikipedia concepts*: This kind of features assumes that the term's referent concept should be topical coherent with other concepts within the same context. Milne and Witten [26] first took account of the incoming link structure to calculate the relatedness of the entities. Formally,

$$Relatedness(c_i, c_j) = 1 - \frac{Log(Max(|L_i|, |L_j|)) - Log(L_i \cap L_j)}{Log(|W|) - Log(Min(|L_j, L_j|))} \qquad (3)$$

where W denotes the Wikipedia article collection, L_i and L_j are the sets of incoming links for entity c_i and c_j respectively.

Based on the above features, various models have been proposed to disambiguate a mention in the running text. Milne and Witten [26] proposed a supervised method using the above three features. Alhelbawy and Gaizauskas [27] built a graph with mention-concept pairs as nodes and associations of concepts as edges. A page rank algorithm run on the graph to get the ranking of candidate concepts for each mention. AIDA [28] and Wikifier [29] are two famous open source tools for mapping terms in running text to Wikipedia articles. These approaches have been proved to be effective on normal documents. However, their performances on query term disambiguation have not been investigated. This paper explores several methods for query term disambiguation and select the approach with best performance as our term-concept mapping method.

3.2   Identifying Probase Concepts from Queries

Probase [24] is a probabilistic semantic network that contains millions of concepts. It is rich enough to cover a large proportion of concepts about worldly facts. The version of Probase[4] we use contains almost 2.7 million concepts and 4.5 million *Is-A* relationships. For example, "robin" is-a *bird*, and "apple" is-a *computer company*.

---

[4] Probase data is publicly available at http://probase.msra.cn/dataset.aspx.

Probase Entity Detection. For mapping query terms to the Probase concepts, we first need to detect the Probase entities in the query so as to access concepts through entities. In this paper, Probase entities are regarded as the dictionary and *Backward Maximum Matching* is used for entity detection. Typicality measures the *Is-A* relations in a probabilistic way. It reflects how typical of c is among all concepts that contain instance e. Given an entity e and its concept c, *Typicality* $P(c|e)$ is given by Eq. 4,

$$P(c|e) = \frac{n(e, c)}{n(e)} \tag{4}$$

where $n(e, c)$ denotes the co-occur frequency of e and c. $n(e)$ and $n(c)$ are the frequencies of e and c occur during their extraction. In this paper, we leverage the *Typicality* to select typical concepts for the detected query entities.

Probase Concept Clustering. The concepts in Probase are fine-grained. On one hand, it can increase the capacity to distinguish between close classes such as "music star" to category *Music* and "movie star" to category *Movie*. On the other hand, many similar concepts exist, such as "country" and "nation", "music star" and "pop star", etc. We use a k-Medoids clustering method to cluster Probase concepts, such that the fine-grained concepts could be merged to general topics or senses. We implemented the k-Medoids clustering method proposed by Li et al. [30] and used the same parameter settings during the implementation. Finally, we obtained 4830 concept clusters. In this paper, we regard the concept cluster as **word sense** for Probase entity disambiguation.

Probase Entity Disambiguation. Entity ambiguity is a non-negligible problem in concept identification. Given the detected entity list E_q of a query q, we disambiguate vague entity by leveraging its unambiguous context entities. Our intuition is simple. The context has the ability to disambiguate a vague term. Taking the query *"apple ipad"* for example, "apple" is a vague term with two senses: *company* and *fruit*, but its context "ipad" is unambiguous. When they appear together, "apple" tends to be a *company*. In this paper, we regard concept clusters as term senses. For each sense (concept cluster), we weight them with Eq. 5.

$$P(s_i|e) = \sum_{e_j \neq e_i, e_j \in E_q, s_n \in S_{e_j}} P(s_i|e_i)P(s_n|e_j)CS(s_i, s_n) \tag{5}$$

where $P(s_i|e_j) = \sum_{IsA(e,c), c \in s_i} P(c|e_j)$ denotes the probability of e_j belonging to cluster s_i, which is the aggravation of the typicality scores for all its concepts belonging to s_i, $CS(s_i, s_n)$ denotes the concept cluster similarity, calculated by the co-entity jaccard similarity. We select the sense with the biggest weight as the disambiguation result for each vague entity. Then, given a query, we mine its concepts as the merged concept vectors of detected entity senses $\{\bigcup_j s_{e_j}\}$.

To this end, we have introduced the concept mining methods for enriching concept information in the task of query classification. For concept-based query representation, we can simply rely on the concept to represent the short query, or combine concepts with the original query words. We illustrate the usage of concept information for query classification in the following section.

4   Query Classification by Leveraging Concept Information

In this section, we discuss two typical usages of the mined explicit concept information in query classification.

4.1   Classification Using Concepts as Query Enrichment

Query text enrichment has been long studied in the past decades. For example, Shen et al. [2] used an ensemble of search engines to produce an enrichment of the queries from the contents of retrieved pages. Therefore, leveraging the mined explicit concepts to enrich the short query is a straightforward way for improving query classification.

Given a query q, we first extract its merged concept vector according to Sect. 3 and construct the concept feature $C(q) = (\alpha_1 c_1, \alpha_2 c_2, ..., \alpha_k c_k)$, where c_k is the kth concept mined from query q and α_k is the normalized weight of c_k in query q. For Wikipedia concept, we set $\alpha = \frac{1}{k}$, and for Probase concept, $\alpha_k = \frac{1}{Z} \times \sum_{e \in E_q} p(c_k|e)$, where Z is the normalization factor. Then we adapt a simple but powerful way to combine them with the original word vector $W(q) = (w_1 t_1, w_2 t_2, ..., w_n t_n)$ to form a new concept enriched feature for query q as follows:

$$F(q) = (w_1 t_1, w_2 t_2, ..., w_n t_n, \alpha_1 c_1, \alpha_2 c_2, ..., \alpha_k c_k) \qquad (6)$$

where w_n is the term weight for n^{th} term t_n in query q. In this paper, we weight each term in q using its normalized tf value. With the new features, we can train classifiers in traditional ways. We choose SVM as the classifier to investigate the classification performance of the concept enriched query representation.

4.2   Classification Incorporating Concepts

Instead of simply using concepts as query enrichment, other work studied new classification models based on concept information. For example, Fang et al. [31] proposed a concept based short text classification and ranking framework by leveraging Probase concept information. However, in their BocSTC model, only entities were considered. We argue that both concepts and words in a query can provide discrimination power. In this paper, we propose a language model based method that combines words and concepts for query classification.

Statistical language modeling has been successfully used for many tasks such as speech recognition, part-of-speech tagging, and information retrieval. Merkel et al. [3] first explored its usage in query classification. Formally, the classifier is defined by

$$\hat{CL} = argmax_{CL} P(q|CL)P(CL) \qquad (7)$$

where $P(q|CL)$ is the probability of a question q and a given semantic class CL. $P(CL)$ is the prior for that class. In most existing work, it is assumed to be uniform and does not affect the ranking. The probability of $P(q|CL)$ is a language model trained on the class CL. In case of unigram language model $P(q|CL) = \Pi_{w \in q} P(w|CL)$.

This paper, we estimate $P(q|CL)$ based on the assumption that the target class should generate the query with high probability from both word level and concept level. Formally,

$$P(q|CL) = (1 - \lambda)P_w(q|CL) + \lambda P_c(q|CL) \qquad (8)$$

where λ is a parameter for tuning the weight between word language model $P_w(q|CL)$ and concept language model $P_c(q|CL)$. The major advantage of the language modeling (LM) approach is that a huge amount of techniques are available to estimate and smooth probabilities even if there is just little training data available. In this paper, we use the Dirichlet smoothing method proposed by Zhai et al. [32] for estimating $P_w(w|CL)$.

$$P_w(w|CL) = \frac{N(w, CL) + \mu_1 P_{bg}(w|CL)}{\sum_w N(w, CL) + \mu_1} \qquad (9)$$

where $n(w, CL)$ are word frequencies on the training data, $P_{bg}(w|CL) = \frac{n(w,CL)}{\sum_{CL} n(w,CL)}$ is a background model used for smoothing, and μ_1 is a smoothing parameter to be tuned on the development data. For estimating $P_c(c|CL)$, the entity frequency and the typicality $P(c|e)$ are both considered, as Eq. 10 shows:

$$P_c(c|CL) = \frac{\sum_e P(c|e) \times n(e, CL) + \mu_2 P_{bg}(c|CL)}{\sum_{c,e} P(c|e) \times n(e, CL) + \mu_2} \qquad (10)$$

similarly, $n(e, CL)$ is the entity frequency, $P_{bg}(c|CL) = \frac{\sum_e P(c|e)n(e,CL)}{\sum_{CL,e} P(c|e)n(e,CL)}$, and μ_2 is the smoothing parameter for concept language model. Note that in case of using Wikipeida, there is no typicality $P(c|e)$ value to use, we can simply set it to 1.

5   Experiment

In this section, we first evaluate disambiguation performance during the concept mapping. Then, we assess the effectiveness of the mined two kinds of concept information for query enrichment. Finally, we evaluate the performance the proposed LM-based classification method through a comparison with QC models in current literature.

5.1   Experiment Setup

In the experiments, we used the following three date sets, one for disambiguation evaluation and the other two for testing query classification:

Questions with ambiguous entities: We selected 50 ambiguous entities such as "apple", "jorden" and "java" as our search queries and sought related questions from Yahoo! Answers. During the retrieval results, we selected questions belonging to different domains for each entity. For example, given the "apple" related questions, we chose questions within different domains such as *computer* and *food*. Finally, we obtained 500 questions and labeled them manually. For Wikipedia concept mapping, we labeled ambiguous entities with Wikipedia concepts, while for Probase, we labeled them with Probase clusters.

Queries from Bing query flow: We created a real query data set from Bing query flow. To alleviate the burden of manual annotation, we first filtered these queries with a pre-trained classifier, then annotated the pre-classified queries manually. We obtained 1000 labeled queries from 4 categories: *Music, Movie, TV, Money* and split them into a test set and validation set with a ration 4:1. For each query, we collected 20 top-ranked retrieval results from Bing as query expansion. For training data, we leveraged articles from the corresponding MSN categories and collected 5000 articles for each categories including the article title and body text.

Questions from Yahoo! Answers: We created a question data set for question classification. In fact, most CQA portals have a pre-defined category system for questions. By leveraging the categories of Yahoo! Answers, we crawled about 13,000 questions from 5 categories: *Computer, Food, Politics, Music*, and *Sports*. We also collected the question details and the 10 top-ranked answers as expansion for each question. For each category, we randomly selected 1500 questions for training, 300 for validation and 500 for testing from the data set.

5.2   Query Term Disambiguation Performance

In this subsection, we conduct an experiment to evaluate the query term disambiguation performance of different methods.

Disambiguation methods: For Wikipedia concept mapping, we explored the following four methods: *M&W* [26], *AIDA* [28], *Wikifier2.0* [33] and *GraphRanking* [27]. For M&W, we used the trained model downloaded from their open source data[5] and directly tested the disambiguation on our question data set. We utilized the open source codes for implementations of AIDA[6] and Wikifier2.0[7]. We implemented GraphRanking according to the best result in Alhelbawy's report [27], whereby we computed global entity popularity according to the number of incoming page links based on the Wikipedia dump on May 3,

---

[5] https://github.com/dnmilne/wikipediaminer.

[6] https://github.com/yago-naga/aida.

[7] http://cogcomp.cs.illinois.edu/page/download_view/Wikifier.

2013. For Probase concept mapping, we used the proposed unsupervised disambiguation method described in Sect. 3.2, marked as *ProbaseDis.*

Evaluation Metric: A term-concept pair $\langle t, c \rangle$ is judged as correct if and only if c is the correct concept for t. In this experiment, we used disambiguation accuracy as the evaluation metric, namely the proportion of mentions correctly disambiguated.

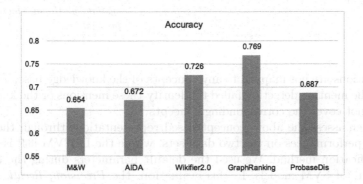

Fig. 1. Performance of different methods on query term disambiguation

Figure 1 shows the disambiguation results. We can see that for Wikipedia concept mapping, GraphRanking performs the best on question disambiguation, which is consistent with the conclusion that collective methods are better than individual methods drawn by the existing work. Therefore, we leverage GraphRanking for Wikipedia concept mapping in this paper.

As for Probase concept mapping, the accuracy is only 68.71 %. There is still much room for improvement. Our method only considered the unambiguous entities in the context. However, there are other useful information for helping disambiguation such as the verb. For instance, given the question "how to cook apple", we know that this "apple" means a *fruit* because it co-occurred with the verb "cook". We argue that the question disambiguation is not the key task of this paper, so we leave it for future work.

5.3 Concept-Based Query Representation

We used the proposed concept mining method in Sect. 3 to extract the concepts for our data sets. Table 1 shows some examples of our concept mapping results.

We draw several observations from this table. First, Wikipedia concepts are too fine grained, which can only map a mention to one Wikipedia concept. In contrast, we can extract more than one typical concepts for each detected Probase entity. Second, concept information is useful for query classification. For example, two queries "loves Eminem" and "Beyonce or Nicki Minaj" are kind of similar because they share the same concept *celebrity*. Third, there are

Table 1. Examples of query concept mapping

Query	Wikipedia concepts	Probase concepts
Loves Eminem	Eminem: *Eminem*	Eminem: *artist, rapper, celebrity*
Beyonce or Nicki Minaj	Beyonce: *Beyonc_Knowles*; Nick Minaj: *Nicki_Minaj*	Beyonce: *celebrity, artist, star*
Windows xp delete	Null	windows xp: *operating system, platform, electronic product*

some mentions are not mapped to any concepts of the knowledge base. This may because the mention detection failed to identify these mentions or the knowledge bases do not cover the corresponding concepts.

We then assess the above concept-based representations through their classification performances on the two data sets, where the LibSVM [34] is used as the classification method. We used the default parameters during the training and testing of SVM models. In this evaluation, the *Precision, Recall, F-Score* are employed as the performance metrics to evaluate the quality of Query Classification. The *F-Score* is the harmonic mean of precision and recall, where the precision and recall are evenly weighted.

Table 2 gives the experimental results. For the three metrics, we average the scores of all categories for each data set. From this table, we draw the following observations:

(1) Overall, the results show that concept-based representations are good enrichment for query representation. Query representations enriched by concepts successfully and consistently improve upon the word representations, and the improvements are significant. This indicates that concepts are very informative for conveying query topics which benefits the QC task a lot.
(2) Specially, the Probase concepts consistently performs better than Wikipedia concepts. This is not surprising. Wikipedia concepts are thematically dense descriptors, which are more fine grained concepts compared with Probase concepts. This makes it weak in semantics as illustrated by the example of "Obama" in Sect. 1.
(3) Furthermore, although the two pure concept representations are inferior to the "bag-of-words" model in terms of F-Score, their average precision scores are better than those of "bag-of-words" model. The low recall scores dramatically reduced their overall classification performances. This may because our dictionary-based mention detection performs not well on query mention recognition due to the query noise, which directly limits the concept coverage over the query data sets. We argue that better mention detection method could further improve the concept-based query classification. We leave it for future work.

Table 2. Performance of different representations on QC

Dataset	Scheme	Avg. recall	Avg. precision	Avg. F-score	Improvement
QueryData	Words	0.584	0.742	0.521	-
	Wikipedia	0.501	**0.803**	0.493	-
	Probase	0.511	**0.788**	0.517	-
	Words+wikipedia	0.688	0.804	**0.656**	+0.135
	Words+probase	0.673	0.891	**0.692**	+0.171
QuestionData	words	0.419	0.774	0.358	-
	Wikipedia	0.231	**0.849**	0.270	-
	Probase	0.384	**0.790**	0.329	-
	Words+wikipedia	0.534	0.772	**0.451**	+0.093
	Words+probase	0.562	0.760	**0.518**	+0.160

5.4   Query Classification Performance

In this experiment, we use Probase concepts as the concept information for query enrichment and our model, since Probase concepts performed better than Wikipedia concepts according to Table 2. In order to demonstrate the effectiveness of our method, we compare it with the following methods:

- SVM_{wc}: SVM is a state-of-the-art classification model. It was used as a base classifier for query categorization in the KDDCUP 2005 winning solution. In this paper, we use it as a baseline and enrich the query representation with Probase concepts.
- $BocSTC$: BocSTC [31] is a new concept based classification model, which relies only on query entities and their concepts for classifying search queries. We use it as a baseline to show the performance of our model combined words and concepts.
- $LMQC_w$: This is a simple LM-based QC model that only leverages word language models for this task. We take it as a baseline and investigate whether the concept language models can help the QC task.

We mark our LM-based QC model that combines words and concepts as $LMQC_{wc}$. During this evaluation, in training the SVM classifier, we use the LibSVM and verify the classifier with the verification datasets. For language models, we set $\mu_1 = 2000$, $\mu_2 = 50$ and $\lambda = 0.6$ after tuning. We implement the BocSTC model following the work of Fang et al. [31].

Table 3 shows the classification results. We can see that $LMQC_{wc}$ outperforms all the baseline methods on both query data and the question data. It performs much better than $LMQC_w$, indicating that the concept language models works well for the QC task. Compared with BocSTC, $LMQC_{wc}$ can significantly improve the classification results, with the average F-score increased by 4.4 % on query data and 4.3 % for question data. This is not surprising, since BocSTC only considered entities in queries while $LMQC_{wc}$ leveraged both query words

Table 3. Performance of different QC methods

Method	Query data				Question data			
	Avg. P	Avg. R	Avg. F	Impr.	Avg. P	Avg. R	Avg. F	Impr.
$LMQC_w$	80.6%	61.4%	65.6%	-	68.5%	40.2%	48.6%	-
BocSTC	90.3%	57.7%	66.5%	+0.9%	74.9%	38.4%	49.3%	+0.7%
SVM_{wc}	89.4%	67.3%	69.7%	+4.1%	74.2%	47.6%	52.5%	+3.9%
$LMQC_{wc}$	**90.5%**	**67.8%**	**70.9%**	**+5.3%**	**75.1%**	**47.2%**	**53.6%**	**+5.0%**

and concepts for this task. Although SVM_{wc} considered both words and concepts, it is inferior to $LMQC_{wc}$ in terms of average F-score for the QC task. It simply used concepts to expand the short queries. This does work well compared with word-based method $LMQC_w$ and concept-based method BocSTC. However, this expansion may further enlarge the feature space considering the large concept space, which may limits the improvement to some extend.

6   Conclusion

In this paper, we study the problem of using explicit concept information to enrich and classify short queries. We propose to mine explicit concept information from two knowledge bases. During the concept mining, we explore various methods to disambiguate query ambiguous terms and assess their performance on query term disambiguation. We discuss the usage of the mined concept information in the QC task, and propose a new LM-based classification model that leverages both query words and concepts to do the classification. Extensive experiments are conducted and the results show that the mined concept information are effective to improve query classification.

Acknowledgments. This work was supported by Beijing Advanced Innovation Center for Imaging Technology (No. BAICIT-2016001), the National Natural Science Foundation of China (Grand Nos. 61370126, 61672081), National High Technology Research and Development Program of China (No. 2015AA016004), the Fund of the State Key Laboratory of Software Development Environment (No. SKLSDE-2015ZX-16).

References

1. Shen, D., Sun, J.-T., Yang, Q., Chen, Z.: Building bridges for web query classification. In: SIGIR (2006)
2. Shen, D., Pan, R., Sun, J.-T., Pan, J.J., Wu, K., Yin, J., Yang, Q.: Query enrichment for web-query classification. ACM TOIS **24**(3), 320–352 (2006)
3. Merkel, A., Klakow, D.: Language model based query classification. In: Amati, G., Carpineto, C., Romano, G. (eds.) ECIR 2007. LNCS, vol. 4425, pp. 720–723. Springer, Heidelberg (2007). doi:10.1007/978-3-540-71496-5_77

4. Cao, H., Hu, D.H., Shen, D., Jiang, D., Sun, J.-T., Chen, E., Yang, Q.: Context-aware query classification. In: SIGIR, pp. 3–10. ACM (2009)
5. Hu, J., Wang, G., Lochovsky, F., Sun, J.-T., Chen, Z.: Understanding user's query intent with wikipedia. In: WWW, pp. 471–480. ACM (2009)
6. Yang, H., Hu, Q., He, L.: Learning topic-oriented word embedding for query classification. In: Cao, T., Lim, E.-P., Zhou, Z.-H., Ho, T.-B., Cheung, D., Motoda, H. (eds.) PAKDD 2015. LNCS (LNAI), vol. 9077, pp. 188–198. Springer, Heidelberg (2015). doi:10.1007/978-3-319-18038-0_15
7. KhudaBukhsh, A.R., Bennett, P.N., White, R.W.: Building effective query classifiers: a case study in self-harm intent detection. In: CIKM, pp. 1735–1738. ACM (2015)
8. Silverstein, C., Marais, H., Henzinger, M., Moricz, M.: Analysis of a very large web search engine query log. In: ACM SIGIR Forum, vol. 33, pp. 6–12. ACM (1999)
9. Shen, D., Pan, R., Sun, J.-T., Pan, J.J., Wu, K., Yin, J., Yang, Q.: Q2c@UST: our winning solution to query classification in KDDCUP 2005. SIGKDD 7(2), 100–110 (2005)
10. Dai, H.K., Zhao, L., Nie, Z., Wen, J.-R., Wang, L., Li, Y.: Detecting online commercial intention (OCI). In: WWW (2006)
11. Broder, A.Z., Fontoura, M., Gabrilovich, E., Joshi, A., Josifovski, V., Zhang, T.: Robust classification of rare queries using web knowledge. In: SIGIR, pp. 231–238. ACM (2007)
12. Wen, J.-R., Nie, J.-Y., Zhang, H.-J.: Query clustering using user logs. ACM Trans. Inf. Syst. 20(1), 59–81 (2002)
13. Beitzel, S.M., Jensen, E.C., Frieder, O., Lewis, D.D., Chowdhury, A., Kolcz, A.: Improving automatic query classification via semi-supervised learning. In: ICDM, pp. 42–49. IEEE (2005)
14. Beitzel, S.M., Jensen, E.C., Lewis, D.D., Chowdhury, A., Frieder, O.: Automatic classification of web queries using very large unlabeled query logs. ACM TOIS 25(2), 107–108 (2007)
15. Arguello, J., Diaz, F., Callan, J., Crespo, J.-F.: Sources of evidence for vertical selection. In: SIGIR, pp. 315–322. ACM (2009)
16. Gabrilovich, E., Markovitch, S.: Overcoming the brittleness bottleneck using wikipedia: enhancing text categorization with encyclopedic knowledge. In: AAAI (2006)
17. Seifzadeh, S., Farahat, A.K., Kamel, M.S., Karray, F.: Short-text clustering using statistical semantics. In: WWW, pp. 805–810. ACM (2015)
18. Huang, L.: Concept-based text clustering. Ph.D. thesis, The University of Waikato (2011)
19. Beeferman, D., Berger, A.: Agglomerative clustering of a search engine query log. In: SIGKDD, pp. 407–416. ACM (2000)
20. Craswell, N., Szummer, M.: Random walks on the click graph. In: SIGIR, pp. 239–246. ACM (2007)
21. Fellbaum, C.: WordNet. Wiley Online Library, New York (1998)
22. Bollacker, K., Evans, C., Paritosh, P., Sturge, T., Taylor, J.: Freebase: a collaboratively created graph database for structuring human knowledge. In: SIGMOD, pp. 1247–1250. ACM (2008)
23. Suchanek, F.M., Kasneci, G., Weikum, G.: Yago: a core of semantic knowledge. In: WWW, pp. 697–706. ACM (2007)
24. Wu, W., Li, H., Wang, H., Zhu, K.Q.: Probase: a probabilistic taxonomy for text understanding. In: SIGMOD, pp. 481–492. ACM (2012)

25. Cucerzan, S.: Large-scale named entity disambiguation based on wikipedia data. In: EMNLP-CoNLL, vol. 7, pp. 708–716 (2007)
26. Witten, I., Milne, D.: An effective, low-cost measure of semantic relatedness obtained from wikipedia links. In: Proceeding of AAAI Workshop on Wikipedia and Artificial Intelligence: An Evolving Synergy, pp. 25–30. AAAI Press, Chicago (2008)
27. Alhelbawy, A., Gaizauskas, R.: Graph ranking for collective named entity disambiguation. In: ACL, pp. 75–80. ACL (2014)
28. Hoffart, J., Yosef, M.A., Bordino, I., Fürstenau, H., Pinkal, M., Spaniol, M., Taneva, B., Thater, S., Weikum, G.: Robust disambiguation of named entities in text. In: EMNLP, pp. 782–792. ACL (2011)
29. Ratinov, L., Roth, D., Downey, D., Anderson, M.: Local and global algorithms for disambiguation to wikipedia. In: ACL-HLT, pp. 1375–1384. ACL (2011)
30. Li, P., Wang, H., Zhu, K.Q., Wang, Z., Wu, X.: Computing term similarity by large probabilistic ISA knowledge. In: CIKM, pp. 1401–1410. ACM (2013)
31. Wang, F., Wang, Z., Li, Z., Wen, J.-R.: Concept-based short text classification and ranking. In: CIKM, pp. 1069–1078. ACM (2014)
32. Zhai, C., Lafferty, J.: A study of smoothing methods for language models applied to ad hoc information retrieval. In: SIGIR, pp. 334–342. ACM (2001)
33. Cheng, X., Roth, D.: Relational inference for wikification. In: EMNLP 13. ACL (2013)
34. Chang, C.-C., Lin, C.-J.: Libsvm: a library for support vector machines. ACM Trans. Intell. Syst. Technol. (ACM TIST) 2(3), 27 (2011)

Stabilizing Linear Prediction Models
Using Autoencoder

Shivapratap Gopakumar$^{(\boxtimes)}$, Truyen Tran, Dinh Phung, and Svetha Venkatesh

Center for Pattern Recognition and Data Analytics,
Deakin University, Burwood, Australia
{sgopakum,truyen.tran,dinh.phung,svetha.venkatesh}@deakin.edu.au

Abstract. To date, the instability of prognostic predictors in a sparse high dimensional model, which hinders their clinical adoption, has received little attention. Stable prediction is often overlooked in favour of performance. Yet, stability prevails as key when adopting models in critical areas as healthcare. Our study proposes a stabilization scheme by detecting higher order feature correlations. Using a linear model as basis for prediction, we achieve feature stability by regularizing latent correlation in features. Latent higher order correlation among features is modelled using an autoencoder network. Stability is enhanced by combining a recent technique that uses a feature graph, and augmenting external unlabelled data for training the autoencoder network. Our experiments are conducted on a heart failure cohort from an Australian hospital. Stability was measured using Consistency index for feature subsets and signal-to-noise ratio for model parameters. Our methods demonstrated significant improvement in feature stability and model estimation stability when compared to baselines.

1 Introduction

Healthcare data is expected to increase by fifty-fold in the coming years [16]. While the direction of current machine learning research is to handle such data, clinical interpretability of models is often overlooked. In healthcare, interpretability is the ability of the model to explain the reason behind prognosis. Such models identify a small subset of strong features (predictors) from available data, and rank them according to their predictive power [28]. This act of feature selection and ranking need to be stable in the face of data re-sampling to ensure clinical adoption. Nonetheless, the nature of clinical data introduces several challenges.

For a particular condition, training data derived from electronic medical records (EMR) usually consist of small number of cases with a large number of events. Most of these events have high correlation with each other. As example, emergency admission events will be correlated with ward transfers, diagnosis of co-occurring diseases (heart failure and diabetes) will have high correlation, pathological measurements (amount of Sodium and Potassium in the body) will be related. To avoid over-fitting, such data require sparse methods in feature

© Springer International Publishing AG 2016
J. Li et al. (Eds.): ADMA 2016, LNAI 10086, pp. 651–663, 2016.
DOI: 10.1007/978-3-319-49586-6_46

selection and learning [24]. But sparsity in correlated features causes instability and results in non-reproducible models [2,12].

In the presence of correlated features, automatic feature selection using lasso has proven to be unstable for linear [27] and survival models [12]. Recent studies propose adapting lasso to acknowledge feature correlations. Such correlations or groupings can be identified using cluster analysis [1,13,15] or density estimation [25]. Group lasso and its variants have been introduced for scenarios where feature groupings are predefined [7,26]. When the features are ordered, or at least have a specification of the nearest neighbour of each feature, Tibshirani et al. proposed fused lasso to perform feature grouping [21]. In contrast, elastic net regularization forces sharing of statistical weights in correlated features without imposing any preconditions on data [29]. When applied to clinical prediction, elastic net regularization proved superior to lasso for prostate cancer dataset [19]. Another approach to stabilize a sparse model is by additional regularization using graphs, where the nodes are features and edges represent relationships [18]. This strategy has been successfully applied in bioinformatics, where feature interactions have been extensively documented and stored in online databases [5,11]. In clinical setting, recent studies have used covariance graph [9] and hand crafted feature graph using semantic relations in ICD-10[1] diagnosis codes and intervention codes[2] [6] as solutions to the instability problem. However, these studies did not consider higher order correlations, and lacked capability to automatically learn feature groupings.

In this paper, we propose a novel methodology to stabilize a sparse high dimensional linear model using recent advances in deep learning and self-taught learning [17]. We propose that the linear model parameter θ is a combination of a lower dimensional vector u, and a high dimensional matrix W, where W encapsulates the feature correlations. By modelling W as the encoding weights of an autoencoder network, we capture higher order feature correlations in data. The workflow diagram of our method is illustrated in Fig. 1.

To minimize variance in feature subsets and parameter estimation, we introduce three regularizers for our sparse linear model: (1) autoencoder derived from training cohort, (2) combination of autoencoder and feature graph derived from training cohort, (3) combination of feature graph derived from training cohort and autoencoder derived from augmenting an external cohort to training data. This process of augmenting external data to autoencoder training results in more robust estimation of higher order correlation matrix W.

We conducted our experiments on 1,885 heart failure admissions from an Australian hospital. The augmented external data consisted of 2,840 diabetic admissions. Feature stability was measured using consistency index [10]. Parameter estimation stability was measured using signal-to-noise ration (SNR). Our proposed stabilization methods demonstrated significantly higher stability when compared with the baselines. Our contribution is in understanding the need for stable prediction, when much research has been dedicated to improving performance. For critical

---

[1] http://apps.who.int/classifications/icd10.
[2] https://www.aihw.gov.au/procedures-data-cubes/.

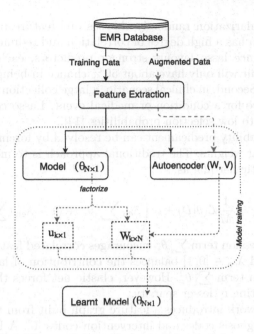

Fig. 1. The work-flow diagram of our framework for deriving autoencoder stabilized prediction model from EMR. The model parameter θ is factorized into a lower dimensional vector u and high dimensional matrix W. The W matrix is jointly modelled as encoding weights in an autoencoder network and is used to regularize the prediction model.

applications like healthcare, where data is sparse and redundant, stable features and estimates are necessary to lend credence to the model and its performance.

2 Framework

Sparse generalized linear models take the form $f(x) = \theta^T x$ subject to $\sum_{i=1}^{N} |\theta_i| \leq \alpha$, where $\theta \in \mathbb{R}^N$ is the model parameter derived from data: $x \in \mathbb{R}^N$. Here, α is the sparsity controlling parameter, typically enforced using lasso regularization [20]. More formally, let $\mathcal{D} = \{x_m, y_m\}_{m=1}^{M}$ denote the training data, where $x_m \in \mathcal{R}^N$ denotes the high dimensional feature vector of data instance m, and y_m is the outcome (for example, the occurrence of future readmission). If $\mathcal{L}(\theta|\mathcal{D})$ is a linear loss, we have:

$$\mathcal{L}_{\text{lasso}} = \frac{1}{M}\mathcal{L}(\theta|\mathcal{D}) + \alpha \sum_{i=1}^{N} |\theta_i| \qquad (1)$$

where $\alpha > 0$ controls the sparsity of the model parameters. The lasso regularization forces the weights of weak parameters towards zero. However, enforcing sparsity on high-dimensional, highly correlated and redundant data, as derived from EMR causes the following problems.

First, lasso regularization randomly chooses one feature from a correlated pair. The EMR data has a high degree of correlation and redundancy in hospital recorded events. Since lasso favours stronger predictors, each feature from a highly correlated pair will only have about 50 % chance in being selected during every training run. Second, in clinical scenario, a large collection of features could be weakly predictive for a condition or medical event. Lasso could ignore such feature groups due to low selection probabilities [14].

This sparsity-stability predicament can be resolved by forcing correlated features to have similar weights. The traditional approach is to modify lasso regularization with elastic net [29] as:

$$\mathcal{L}_{\text{elastic net}} = \frac{1}{M}\mathcal{L}(\theta|\mathcal{D}) + \alpha \left(\lambda_{\text{en}} \sum_i |\theta_i| + (1 - \lambda_{\text{en}}) \sum_i \theta_i^2 \right)$$

Here, the ridge regression term $\sum_i \theta_i^2$ encourages correlated feature pairs to have similar weights, and $\lambda_{\text{en}} \in [0,1]$ balances the contribution of lasso term $\sum_i |\theta_i|$ and ridge regression term $\sum_i \theta_i^2$. However, elastic net forces the weights to be equally small, resulting in lesser sparsity.

Another recent work introduces a feature graph built from hierarchical relations in ICD-10 diagnosis codes and intervention codes [6]. A Laplacian of this feature graph: $\mathbf{L}$, is used to ensure weight sharing between related features as:

$$\mathcal{L}_{\text{feature graph}} = \mathcal{L}_{\text{lasso}} + \frac{1}{2} \lambda_{\text{fg}}(\theta^T \mathbf{L} \theta) \tag{2}$$

We propose to automatically learn higher order correlations in data.

2.1   Correlation by Factorization in Linear Models

To model higher order correlations in data, we begin by decomposing model parameter θ into a lower order vector and a high dimensional matrix as: $\theta_{N \times 1} = W_{k \times N}^T u_{k \times 1}$, where $k \ll N$. This factorization offers several advantages. The lower dimensionality of u makes it more easier to learn and more stable to data variations. The W captures higher order correlations that be modelled using different auxiliary tasks. Greater number of tasks ensure better solution, since there are more constraints.

As a concrete example for generalized linear models, we work on binary prediction using logistic regression. The modified logistic loss function $\mathcal{L}(\theta|\mathcal{D})$ using u and W becomes:

$$\mathcal{L}_{\text{logit}}(u, W | \mathcal{D}) = \log(1 + \exp(-yu^T W \boldsymbol{x})) \tag{3}$$
$$= \log(1 + \exp(-yu^T z))$$

where $y \in \pm 1$ represents the data label[3]. Notice that $z = W\boldsymbol{x}$ is a data transformation from N dimensions to the smaller k dimension. To learn W, we need to choose a competent auxiliary task. We model W as the encoding weights of a classical autoencoder derived from the same data $\mathcal{D}$.

---

[3] We ignore the bias parameter for simplicity.

2.2 Learning Higher Order Correlations Using Autoencoder

An autoencoder is a neural network that learns by minimizing the reconstruction error using back-propagation [3]. The learning process is unsupervised, wherein the model learns the useful properties of the data. An autoencoder network consists of two components: (1) An *encoder function* that maps the input data $x \in \mathbb{R}^N$ as: $h(x) = \sigma(Wx + b_W)$, where σ can be any non-linear function (for e.g., the sigmoid function) and W, b_W are the weights and bias of the hidden layer (2) A *decoder function* that attempts to reconstruct the input data as: $\tilde{x} = Vh + b_V$, where V, b_V are the weights and bias of the output layer. The loss function is modelled as the reconstruction error:

$$\mathcal{L}_{\mathrm{AE}}(W, V, b_W, b_V | \mathcal{D}) = \frac{1}{2N} \| x - b_V - V\sigma(Wx + b_W) \|_2^2 \tag{4}$$

Once trained, evaluating a feed forward mapping using the encoder function gives a latent representation of the data. When the number of hidden units is significantly lesser than the input layer, W encapsulates the higher order correlations among features.

We propose to regularize our sparse linear model in (3) using the autoencoder framework in (4). The joint loss function becomes:

$$\begin{aligned}
\mathcal{L}_{\mathrm{model}}(u, W, V, b_W, b_V | \mathcal{D}) &= \mathcal{L}_{\mathrm{logit}}(u, W | \mathcal{D}) \\
&+ \alpha \Sigma_i | \Sigma_k W_{ik}^T u_k | \\
&+ \lambda_{\mathrm{AE}} \, \mathcal{L}_{\mathrm{AE}}(W, V, b_W, b_V | \mathcal{D}) \\
&+ \lambda_{\ell 2} \left(W^2 + V^2 + b_W^2 + b_V^2 \right)
\end{aligned} \tag{5}$$

where $\alpha > 0$ is the lasso regularization parameter which ensures weak $\theta_i = \Sigma_k W_{ik}^T u_k$ are driven to zero. While λ_{AE} controls the amount of regularization due to higher order correlation, $\lambda_{\ell 2}$ controls overfitting in autoencoder. The loss function in (5) is non-convex. We propose two extensions to our model.

Augmenting Feature Graph Regularization. While autoencoders can be used to find automatic feature grouping, we can also exploit the predefined associations in patient medical records. For example, diseases or conditions reoccurring over multiple time-horizons should be assigned similar importance [22]. Also, the ICD-10 diagnosis and procedure codes are hierarchical in nature [6,23]. We build a feature graph using these associations (as in (2)) and use it to further regularize our model in (5) as:

$$\begin{aligned}
\mathcal{L}_{\mathrm{model\text{-}fg}}(u, W, V, b_W, b_V | \mathcal{D}) &= \mathcal{L}_{\mathrm{model}}(u, W, V, b_W, b_V | \mathcal{D}) \\
&+ \frac{1}{2} \lambda_{\mathrm{fg}} \left[(u^T W) \, \mathbf{L} \, (W^T u) \right]
\end{aligned}$$

Augmenting External Data for Autoencoder Learning. The encoding weights W in (4) can be estimated from multiple sources. For example, in this paper, we propose to augment the current training data $\mathcal{D}$ (for example: heart failure cohort) with another cohort containing the same features (say, diabetic cohort). Training the autoencoder network on this augmented data will result in more robust estimation of W.

3 Experiments

The feature stability and model stability of our proposed framework is evaluated on heart failure (HF) cohort from Barwon Health[4], a regional hospital in Australia serving more than 350,000 residents. The Autoencoder learning was augmented with diabetes (DB) cohort from the same hospital. We mined the hospital EMR database for retrospective data for a period of 5 years (Jan 2007 to Dec 2011), focusing on emergency and unplanned admissions of all age groups. Inpatient deaths were excluded. All patients with at least one ICD-10 diagnosis code I50 were included in the HF cohort. The DB cohort contained all patients with at least one diagnosis code between E10-E14. This resulted in 1,885 heart failure admissions and 2,840 diabetic admissions. Table 1 shows the details of both cohorts.

Table 1. Characteristics of heart failure and diabetes cohort.

	Heart failure (HF)		Diabetes (DB)
	Derivation	Validation	(Augmented data)
Admissions	1,415	369	2,840
Unique patients	1,088	317	1,716
Gender:			
Male	541 (49.7%)	155 (48.9%)	908 (52.9%)
Female	547 (50.2%)	162 (51.1%)	808 (47.1%)
Mean age (years)	78.3	79.4	57.1
Mean length of stay	5.2 days	4.5 days	4.1 days
Total features	3,338		6,711
Common features	558		

The different features in EMR database (diagnosis, medications, treatments, procedures, lab results) were extracted using a one-sided convolutional filter bank introduced in [22]. The feature extraction process resulted in 3,338 features for HF and 6,711 features for DB cohort. A total of 558 features were common to both cohorts (Table 1).

---

[4] Ethics approval was obtained from the Hospital and Research Ethics Committee at Barwon Health (number 12/83) and Deakin University.

3.1 Models and Baselines

From this data, we derive a lasso regularized logistic regression model to predict heart failure readmissions in 6 months. We force lasso to consider higher order correlations in data by using the following three regularization schemes:

Lasso-Autoencoder. The linear model is regularized by an autoencoder derived from HF cohort.

Lasso-Autoencoder-Graph. We construct a feature graph from 3,338 features in HF cohort as in [6], and use it to further regularize the Lasso-Autoencoder model.

AG-Lasso-Autoencoder-Graph. AG denotes augmented data used to train the autoencoder. To estimate W, we used DB cohort augmented to the HF cohort. Training data consisted of 558 features common to both HF and DB. The sparse prediction model was built from common features in HF cohort, and regularized using a HF-based feature graph and autoencoder from augmented data.

We compare the stability of our proposed regularization methods with the following baselines: (1) pure lasso (2) elastic net and (3) recently introduced feature graph regularization (as in [6]).

3.2 Temporal Validation

The training and testing data were separated in time and patients. Patients discharged before September 2010 were included in training set. The validation set consisted of new admissions from September 2010 to December 2011. Model performance was measured using AUC (area under the ROC curve) based on Mann-Whitney statistic. A pre-defined threshold was used (chosen to maximize the F-score) to predict readmissions.

3.3 Measuring Stability

Variability in data resampling was simulated using bootstraps. We trained all models using 500 bootstraps of randomly sampled training data. At the end of each bootstrap, the features selected by a model were ranked based on importance. Feature importance was calculated as the product of mean feature weights across all bootstraps and feature standard deviation in the training data. The top k ranked features were collected to form a list of feature subsets: $S = \{S_1, S_2, \cdots, S_{500}\}$, where $|S_i| = k$.

We used consistency index [10] to measure pairwise feature subset stability. For a pair of subsets (S_i, S_j), with length k selected from a total of d features, consistency index becomes:

$$\mathrm{CI}(S_i, S_j) = \frac{rd - k^2}{k(d - k)}$$

where $r = |S_i \cap S_j|$. The overall stability score is the average of pairwise consistency index among all pairs. This stability score is bounded in $[-1, +1]$, with -1 for no overlap, 0 for independently drawn subsets, and $+1$ complete overlap of two subsets. The score calculation is monotonous and corrects overlap due to chance [10].

The stability of estimated model parameters was measured using signal-to-noise ratio (SNR). The variance in estimation of feature i can be calculated as:

$$\mathrm{SNR}(i) = w_i/\sigma_i$$

where w_i is the mean feature weight across bootstraps for feature i, and σ_i is its standard deviation.

4   Results

In this section, we demonstrate the effect of autoencoder regularization on model performance and stability, and compare with our baselines. The prediction models for heart failure readmission were derived from 3,338 features extracted from hospital database. The self taught learning stage during autoencoder training used an augmented 2,840 diabetic admissions with 558 features that were common in both cohorts. A grid search for the best hyper-parameter setting resulted in $\alpha = .001$, $\lambda_{\mathrm{en}} = .01$, $\lambda_{\mathrm{graph}} = .03$ for the baseline models, and $\alpha = .005$, $\lambda_{\mathrm{AE}} = 3000$, $\lambda_{\mathrm{graph}} = 0.3$ for our autoencoder regularized models.

4.1   Capturing Higher Order Correlations

The efficacy of autoencoder network to model higher order correlations was verified by comparing the correlation matrices of raw data and data from the encoding layer. The autoencoder derived correlation matrix was denser (matrix mean $= 0.19$) than the correlation matrix for raw data (matrix mean $= 0.05$).

Table 2. Effect of stabilization methods on model sparsity

Regularization	Features selected (%)
Lasso	550(16.5 %)
Elastic net	753(22.6 %)
Lasso-Graph	699(20.9 %)
Lasso-Autoencoder	513(15.4 %)
Lasso-Autoencoder-Graph	503(15.1 %)
AG-Lasso-Autoencoder-Graph	412(12.3 %)

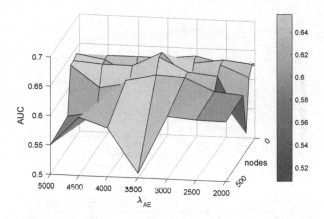

Fig. 2. Effect of number of hidden units (nodes) and autoencoder penalty (λ_{AE}) on AUC. Lasso parameter fixed at $\alpha = .005$

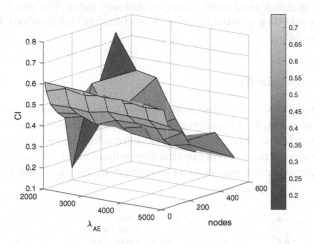

Fig. 3. Effect of number of hidden units (nodes) and autoencoder penalty (λ_{AE})feature stability measured by consistency of top 100 features. Lasso parameter fixed at $\alpha = .005$

4.2 Effect on Model Sparsity

Table 2 provides a summary of the effects of stabilization schemes on model sparsity. Autoencoder regularization resulted in sparser models with no loss in performance. Model performance was measured using area under the ROC curve (AUC). For autoencoder regularization, AUC critically depended on the choice of autoencoder penalty (λ_{AE}) and number of hidden units (see Fig. 2). A maximum AUC of 0.65 was obtained for AG-Lasso-Autoencoder-Graph model with 20 hidden units and hyperparameters as $\alpha = .005$, $\lambda_{en} = .03$, $\lambda_{graph} = .3$, $\lambda_{AE} = 3000$.

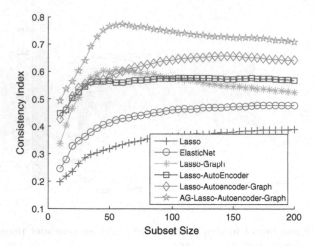

Fig. 4. Feature stability as measured using consistency index. The plot compares similarity in feature subsets generated by our proposed models and baselines. Higher values indicate more stability.

4.3   Effect on Stability

When compared to λ_{AE}, the choice of hidden units had more influence on feature stability (see Fig. 3). Consistency index measurements for feature selection stability is reported in Fig. 4. In general, capturing higher order correlations using autoencoder improved feature stability when compared to baselines. Even though pure autoencoder regularization proved to be more effective for larger

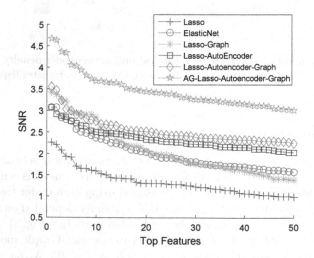

Fig. 5. Model stability as measured using signal-to-noise ratio (SNR) of feature weights. Higher values indicate more stability.

feature sets (>120), the combination of autoencoder and graph regularization consistently outperformed the baselines. Further, augmenting external cohort to autoencoder learning resulted in the most stable features. Similar observations were made when measuring model estimation stability. Figure 5 reports the signal-to-noise ratios of the top 50 individual features. At 95 % CI (approximately 1.96 std), lasso regularization identifies 3 features, elastic net identifies: 21, Graph regularization: 24, while the autoencoder regularized models identify all the 50 features. The variance in feature weight is greatly reduced by AG-Lasso-Autoencoder-Graph regularization.

5   Discussion and Conclusion

Sparsity and stability are two important characteristics of interpretable healthcare. Sparsity promotes interpretability and stability inspires confidence in the prediction model. Through our experiments, we have demonstrated that autoencoder regularization, when applied to high-dimensional clinical prediction results in a sparse model that is stable in features and estimation. Our proposed model is built from common clinical and administrative data recorded in the hospital database, and hence can be easily integrated into existing systems.

The predictive performance of our proposed model (as measured by AUC) is comparable with existing studies [4]. Our stabilization scheme did not improve classification performance. A similar observation was made by Kalousis et al. in their study on high dimensional feature selection stability [8]. However, as their study noted, stable models impart confidence on the features selected, and in turn lends credence to corresponding classification performance.

5.1   Conclusion

Traditionally, autoencoder variants are used to improve prediction/classification accuracy. In this paper, we demonstrate a novel use of autoencoders to stabilize high dimensional clinical prediction. Feature stable models are reproducible between model updates. Stable models encourage clinicians to further analyse the predictors in understanding the prognosis, thereby paving the way for interpretable healthcare. We have demonstrated that the encoding process of an autoencoder, though intrinsically unstable, can be applied to regularize sparse linear prediction resulting in more stable features. The encoding weights capture higher level correlations in EMR data. When collecting data becomes expensive, augmenting another cohort during autoencoder training resulted in a more robust estimation of encoding weights, translating to better stability. This approach belongs to the emerging learning paradigm of self-taught learning [17]. We believe this work presents interesting possibilities in the application of deep nets for model stability.

References

1. Au, W.H., Chan, K.C., Wong, A.K., Wang, Y.: Attribute clustering for grouping, selection, and classification of gene expression data. IEEE/ACM Trans. Comput. Biol. Bioinform. (TCBB) **2**(2), 83–101 (2005)
2. Austin, P.C., Tu, J.V.: Automated variable selection methods for logistic regression produced unstable models for predicting acute myocardial infarction mortality. J. Clin. Epidemiol. **57**(11), 1138–1146 (2004)
3. Bengio, Y.: Learning deep architectures for AI. Found. Trends Mach. Learn. **2**(1), 1–127 (2009)
4. Betihavas, V., Davidson, P.M., Newton, P.J., Frost, S.A., Macdonald, P.S., Stewart, S.: What are the factors in risk prediction models for rehospitalisation for adults with chronic heart failure? Aust. Crit. Care: Official J. Confederation Aust. Crit. Care Nurses **25**(1), 31–40 (2012). http://www.ncbi.nlm.nih.gov/pubmed/21889893
5. Cun, Y., Fröhlich, H.: Network and data integration for biomarker signature discovery via network smoothed t-statistics. PLoS One **8**(9), e73074 (2013)
6. Gopakumar, S., Tran, T., Nguyen, T.D., Phung, D., Venkatesh, S.: Stabilizing highdimensional prediction models using feature graphs. IEEE J. Biomed. Health Inform. **19**(3), 1044–1052 (2015)
7. Jacob, L., Obozinski, G., Vert, J.P.: Group lasso with overlap and graph lasso. In: Proceedings of the 26th Annual International Conference on Machine Learning, pp. 433–440. ACM (2009)
8. Kalousis, A., Prados, J., Hilario, M.: Stability of feature selection algorithms: a study on high-dimensional spaces. Knowl. Inf. Syst. **12**(1), 95–116 (2007)
9. Kamkar, I., Gupta, S.K., Phung, D., Venkatesh, S.: Exploiting feature relationships towards stable feature selection. In: IEEE International Conference on Data Science and Advanced Analytics (DSAA), 36678, pp. 1–10. IEEE (2015)
10. Kuncheva, L.I.: A stability index for feature selection. In: Artificial Intelligence and Applications, pp. 421–427 (2007)
11. Li, C., Li, H.: Network-constrained regularization and variable selection for analysis of genomic data. Bioinform. **24**(9), 1175–1182 (2008)
12. Lin, W., Lv, J.: High-dimensional sparse additive hazards regression. J. Am. Stat. Assoc. **108**(501), 247–264 (2013)
13. Ma, S., Song, X., Huang, J.: Supervised group lasso with applications to microarray data analysis. BMC Bioinform. **8**(1), 1–17 (2007)
14. Meinshausen, N., Bühlmann, P.: Stability selection. J. Roy. Stat. Soc. B (Stat. Methodol.) **72**(4), 417–473 (2010)
15. Park, M.Y., Hastie, T., Tibshirani, R.: Averaged gene expressions for regression. Biostatistics **8**(2), 212–227 (2007)
16. Raghupathi, W., Raghupathi, V.: Big data analytics in healthcare: promise and potential. Health Inf. Sci. Syst. **2**(1), 1–10 (2014)
17. Raina, R., Battle, A., Lee, H., Packer, B., Ng, A.Y.: Self-taught learning: transfer learning from unlabeled data. In: Proceedings of the 24th International Conference on Machine Learning, pp. 759–766. ACM (2007)
18. Sandler, T., Blitzer, J., Talukdar, P.P., Ungar, L.H.: Regularized learning with networks of features. In: Advances in Neural Information Processing Systems, vol. 21, pp. 1401–1408. Curran Associates, Inc. (2009)
19. Simon, N., Friedman, J., Hastie, T., Tibshirani, R., et al.: Regularization paths for cox's proportional hazards model via coordinate descent. J. Stat. Softw. **39**(5), 1–13 (2011)

20. Tibshirani, R.: Regression shrinkage and selection via the lasso. J. Roy. Stat. Soc.: Ser. B (Methodol.) **58**(1), 267–288 (1996)
21. Tibshirani, R., Saunders, M., Rosset, S., Zhu, J., Knight, K.: Sparsity and smoothness via the fused lasso. J. Roy. Stat. Soc. Ser. B (Stat. Methodol.) **67**(1), 91–108 (2005)
22. Tran, T., Phung, D., Luo, W., Harvey, R., Berk, M., Venkatesh, S.: An integrated framework for suicide risk prediction. In: 19th ACM SIGKDD International Conference on Knowledge Discovery and Data Mining, pp. 1410–1418. ACM (2013)
23. Tran, T., Phung, D., Luo, W., Venkatesh, S.: Stabilized sparse ordinal regression for medical risk stratification. Knowl. Inf. Syst., 1–28 (2014)
24. Ye, J., Liu, J.: Sparse methods for biomedical data. ACM SIGKDD Explor. Newsl. **14**(1), 4–15 (2012)
25. Yu, L., Ding, C., Loscalzo, S.: Stable feature selection via dense feature groups. In: Proceedings of the 14th ACM SIGKDD International Conference on Knowledge Discovery and Data Mining, pp. 803–811. ACM (2008)
26. Yuan, M., Lin, Y.: Model selection and estimation in regression with grouped variables. J. Roy. Stat. Soc. B (Stat. Methodol.) **68**(1), 49–67 (2006)
27. Zhao, P., Yu, B.: On model selection consistency of lasso. J. Mach. Learn. Res. **7**, 2541–2563 (2006)
28. Zhou, J., Sun, J., Liu, Y., Hu, J., Ye, J.: Patient risk prediction model via top-k stability selection. In: Proceedings of the 13th SIAM International Conference on Data Mining. SIAM (2013)
29. Zou, H., Hastie, T.: Regularization and variable selection via the elastic net. J. Roy. Stat. Soc. B **67**, 301–320 (2005)

Mining Source Code Topics Through Topic Model and Words Embedding

Wei Emma Zhang[1]($^{(\boxtimes)}$), Quan Z. Sheng[1], Ermyas Abebe[2],
M. Ali Babar[1], and Andi Zhou[2]

[1] School of Computer Science, The University of Adelaide, Adelaide, Australia
{wei.zhang01,michael.sheng,ali.babar}@adelaide.edu.au
[2] IBM Research, Melbourne, Australia
etabebe@au1.ibm.com, aus.azh@gmail.com

Abstract. Developers nowadays can leverage existing systems to build their own applications. However, a lack of documentation hinders the process of software system reuse. We examine the problem of mining topics (i.e., topic extraction) from source code, which can facilitate the comprehension of the software systems. We propose a topic extraction method, Embedded Topic Extraction (EmbTE), that considers word semantics, which are never considered in mining topics from source code, by leveraging word embedding techniques. We also adopt Latent Dirichlet Allocation (LDA) and Non-negative Matrix Factorization (NMF) to extract topics from source code. Moreover, an automated term selection algorithm is proposed to identify the most contributory terms from source code for the topic extraction task. The empirical studies on Github (https://github.com/) Java projects show that EmbTE outperforms other methods in terms of providing more coherent topics. The results also indicate that method name, method comments, class names and class comments are the most contributory types of terms to source code topic extraction.

Keywords: Source code mining · Topic model · Word embedding

1 Introduction

Source code repositories and open source systems are becoming popular resources for software development because developers are able to reuse the shared source code of existing systems to build their own software [7]. However, many such systems have poor or no documentation. Many works apply text summarization techniques to summarize source code [8,9,16] or parts of source code [14,19] aiming to assist the comprehension. In this way, the source code can be comprehended by reading generated summaries. These approaches follow a common process that first extracts top-N keywords from source code and then builds summaries using these keywords. Thus, the quality of summarization depends on the extracted keywords, which represent the topics (we use *keywords* and *topics* interchangeably in this paper hereafter). However, these works fail to discover latent topics by only considering term frequencies [9,17].

© Springer International Publishing AG 2016
J. Li et al. (Eds.): ADMA 2016, LNAI 10086, pp. 664–676, 2016.
DOI: 10.1007/978-3-319-49586-6_47

Word embedding is a group of feature learning techniques in Natural Language Processing (NLP) that maps the words into vectors which are regarded as the underlying representation of the original words. Neural networks is one of the methods for learning word embedding. It provides simple and efficient neural architecture to learn word vectors and has shown desirable performance in NLP tasks such as measuring word-to-word relationships [12]. The learned representations of two words reflect the latent relationship of the words, which can be used to discover the latent topics of documents. However, these techniques haven't been adopted in mining source code topics. In this paper, we propose the first attempt to apply word embedding techniques to mining source code topics. For this purpose, we propose an extraction method, Embedded Topic Extraction (EmbTE), that leverages the underlying relationships among terms from source code files via learned term vectors. Specifically, we first obtain vectors for terms in source code files using a neural embedding method proposed in [13]. Thus each document is mapped to a group of term vectors. Then we build a K-dimensional tree (K-d tree) [3] for term vectors belonging to one document. The centroid of the term vectors in this document is computed. Top-N nearest neighborhoods to the centroid are obtained by searching the K-d tree and are regarded as the top keywords (i.e., topics).

For comparison, we examine the effectiveness of the most popular topic model Latent Dirichlet Allocation (LDA) [4] in mining topics from source codes. LDA is a topic model that considers the term distribution in documents. It has shown effectiveness in finding latent topics from natural language text corpus. But limited works apply LDA to extract topics from source code. A classic dimension reduction method Non-negative Matrix Factorization (NMF) [10] is also evaluated for our topic extraction task. NMF factorizes a non-negative matrix to two matrices that are in lower dimension space. NMF can be adopted to topic extraction with the document-term matrix as the original matrix. The dimension of the two factored matrices is the number of topics. One factored matrix is document-topic matrix and the other one is topic-term matrix.

The second problem we address in this paper is the evaluation metric of source code topic extraction methods. Existing works involve human expertise to perform the evaluation. Except for the works use pure human judgement [9,19], other works compare the extracted keywords with human picked ones by adopting distance measurements on top-N ranked lists [17]. However, all these methods highly depend on human efforts, which have threats to validity. Instead, we adopt the evaluation metric for topic models in our work. Specifically, we use coherence measurement, which measures the coherence among extracted topics. A list of extracted topics with high coherence are considered as good extraction. An algorithm that combines the indirect cosine measure with Normalized Pointwise Mutual Information (NPMI) and the boolean sliding window [18] is used.

The third problem we consider in this paper is to identify the most contributory types of terms for source code topic extraction. Existing works consider comments [16,17], method names [8], class names and identifier names [9] in source code topic extraction. But limited explanation is given on why these types of terms are chosen and how the terms affect the source code topic extraction performance. Therefore, we develop an automated process that heuristically select the types of terms that increase the coherence of extracted topics. The ones who reduce the coherence will be ignored. The proposed selection approach draws ideas from the wrapper family in feature selection algorithms [6] in the machine learning community.

In a nutshell, the main contributions of this work are summarized as follows:

- We propose a source code topic extraction method EmbTE by leveraging the word embedding techniques. We compare EmbTE with LDA and NMF on extraction topics from source code. The coherence measurement is used as the performance metric which has not been adopted in existing works on mining source code.
- We develop a selection process that automatically identifies the most contributory terms of source code for topic extraction. This automated process can also be applied to other software engineering applications (e.g., link recovery, bug location and component categorization etc.) by modifying the performance assessment metric.
- We evaluate our methods on Github Java projects. Although only Java projects are considered in this work, our approach is language-independent.

The remainder of this paper is structured as follows. We describe our extraction methods in detail in Sect. 2. Existing related research efforts are discussed in Sect. 3. Section 4 reports the experimental results. Finally, we conclude the paper in Sect. 5.

2   Methodology

To extract topics from source code, we first pre-process the source code files (Sect. 2.1), then apply topic extraction methods (Sect. 2.2). An automated terms selection algorithm is proposed to select the most contributory terms (Sect. 2.3) and we use coherence measurement as metric to evaluate topic extraction performance (Sect. 2.4).

2.1   Data Pre-processing

The pre-processing step focuses on extracting feature terms from source code which are used as input to topic extraction methods. Unlike text document, the source code consider much programming language-specific terms, so our goal is to prune the invalid and meaningless ones and construct clean input to further steps. The details of pre-processing are as follows:

- *Feature terms extraction.* We extract feature terms such as method names, class names from source code. Each type of feature terms form a document. For example, all the method names of one project are in one document.

- *Identifier splitting.* Unlike in natural language text, where each word can be found in the dictionary, source code identifier names are generally a combination of several words and delimiters following certain naming convention, e.g., "loanInterests". We split this kind of identifiers into a set of terms, e.g., "loanInterests" to "loan" and "Interests".
- *Stop words and language keywords pruning.* Programming language keywords such as "implements" in Java do not contribute to comprehend the projects. Hence we prune the language keywords. We also remove the stop words.
- *Stemming and lemmatization.* Like natural language text, terms are used in various forms such as singular and plural, comparatives and superlatives. In our analysis, we do not consider them as different terms. Thus we unify them by using stemming and lemmatization techniques.
- *Filtering.* Keywords such as "set" are very generic and not contributory to the overall comprehension. Thus we filter out such words in our analysis.

2.2 Topic Extraction

2.2.1 Topic Extraction via Words Embedding

The embedding methods [12,13] proposed to reveal the semantic information embedded in documents, topics and terms. These methods represent words by learning essential concepts and representations and show effectiveness in many NLP and machine learning tasks. Mikolov et al. [13] propose an efficient method using neural networks to learn distributed representation of words from large data sets. The method includes both Continuous Bag of Words (CBOW) and Skip-gram models. Given a word sequence $S = \{w_1, ..., w_q\}$, the objective functions for learning based on CBOW and Skip-gram are as follows:

$$F_{CBOW}(S) = \frac{1}{q} \sum_{i=1}^{q} \log p(w_i|w_c), \tag{1}$$

$$F_{Skip-gram}(S) = \frac{1}{q} \sum_{i=1}^{q} \sum_{-l \leqslant \delta \leqslant l, \delta \neq 0, j=i+\delta} \log p(w_j|w_i), \tag{2}$$

where w_c in Eq. (1) denotes the context of the word w_i and δ in Eq. (2) is the window size of context. Both equations aim to maximize the corresponding log-likelihoods.

We adopt the learned vector representation of words in EmbTE. Figure 1 depicts how EbmTE works. We first extract terms from source code files of systems and construct one document for each system. The document is preprocessed according to Sect. 2.1. Then EmbTE learns words vectors from a corpus of documents $\{d_1, ..., d_n\}$, which represents a corpus of systems that need to extract topics. Each document d_i can be represented by a group of vectors $\{\mathbf{w}_1, \mathbf{w}_2, ..., \mathbf{w}_m\}$, where m is the number of words in d_i. EmbTE computes the centroid $\mathbf{w}_{ct}$ of $\{\mathbf{w}_1, \mathbf{w}_2, ..., \mathbf{w}_m\}$ using k-medoids clustering with only one cluster and cosine distance as the distance metric. $\mathbf{w}_{ct}$ is considered as the key topic

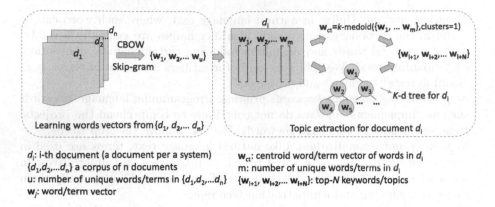

d_i: i-th document (a document per a system) $\quad$ $\mathbf{w}_{ct}$: centroid word/term vector of words in d_i
$\{d_1,d_2,...d_n\}$ a corpus of n documents $\quad$ m: number of unique words/terms in d_i
u: number of unique words/terms in $\{d_1,d_2,...d_n\}$ $\quad$ $\{\mathbf{w}_{l+1}, \mathbf{w}_{l+2},... \mathbf{w}_{l+N}\}$: top-N keywords/topics
$\mathbf{w}_j$: word/term vector

Fig. 1. Topic extraction via words embedding (EmbTE)

of d_i. Then EmbTE searches $\mathbf{w}_{ct}$'s N nearest neighbors and considers them as the top-N keywords/topics of d_i. If $\mathbf{w}_{ct}$ represents a word in d_i, then only N-1 nearest neighbors are returned.

2.2.2 Topic Extraction via Latent Dirichlet Allocation

LDA is a widely applied topic model which has successfully extract topics from natural language texts. We adopt this model to extract source code topics. For completeness, we briefly introduce how LDA extracts topics. Given a corpus of document $D = \{d_1, ..., d_n\}$ where each document $d_i, i = 1, ...n$ is a sequence of m words denoted by $d_i = \{w_1, ..., w_m\}, w_j \in W, j = 1, ...m$. W is the vocabulary set. Each document d_i can be modelled as a multinomial distribution $\theta^{(d_i)}$ over t topics and each topic $z_k, k = 1, ...t$ is modelled as a multinomial distribution $\phi^{(k)}$ over the set of words W. LDA assumes a prior Dirichlet distribution on θ thus allowing the estimation of ϕ without requiring the estimation of θ. LDA is based on the following document generation process [4]:

– Choose $N \sim Poission(\xi)$: Select the number of words m.
– $\theta \sim Dir(\alpha)$: Select θ from the Dirichlet distribution parameterized by α.
– For each $w_j \in W$ do

• Choose a topic $z_k \sim Multinomial(\theta)$
• Choose a word w_j from $p(w_j|z_k, \beta)$, a multinomial probability ϕ^{z_k}.

LDA assumes each document is a word mixture of multiple topics. Thus the topic extraction is a reverse process of document generation.

To apply LDA for source code topics extraction, we construct one document per system's source code. As mentioned previously, pre-processed terms of one system construct a document. We assume the document contains terms/words from different topics as in natural language documents. Then we apply LDA on multiple documents. We use a variational Bayes approximation [4] of the posterior distribution in LDA. We first train LDA model on these document and obtain the terms distribution. Then top keywords for each document are returned. Thus we obtain topics from each document at the same time.

2.2.3 Topic Extraction via Non-negative Matrix Factorization

NMF considers the problem of factorizing a non-negative matrix into two smaller non-negative matrices. The problem is formulated as: Given $\mathbf{X} \in \mathbb{R}^{n \times m}$ and a positive integer $r < min(n, m)$, find non-negative matrices $\mathbf{U} \in \mathbb{R}^{n \times r}$ and $\mathbf{V} \in \mathbb{R}^{r \times m}$ which satisfy:

$$\min_{\mathbf{U}, \mathbf{V}} \; (\|\mathbf{X} - \mathbf{UV}\|_F^2), \tag{3}$$

where F is the Frobenius Norm. By multiplying $\mathbf{U}$ and $\mathbf{V}$ together, the product approximates the original matrix $\mathbf{X}$. NMF aims to reveal the meaningful latent features that are hidden in the input matrix $\mathbf{X}$, where each entry can be considered as the combination of these latent features with different weights. It can perform dimension reduction and clustering simultaneously. Thus it has enjoyed much success in many areas such as text mining, image processing and recommendation systems.

To extract topics using NMF, the input matrix $\mathbf{X}$ is the document-term matrix, where each entry is the weight of each term in each document. We use Term Frequency/Inverse Document Frequency (TF/IDF) as the weight. The dimension r is the number of topics. $\mathbf{U}$ represents the association between documents and topics while the $\mathbf{V}$ indicates the association between topics and terms. By ordering the values in a column of $\mathbf{U}$ and selecting the top ones, we can get the topics of the document. By ordering the values of rows of $\mathbf{V}$, we get the top terms/keywords for a topic. Thus we are able to associate document with keywords, which are considered as topics.

2.3 Automated Terms Selection for Topic Extraction

We have discussed the three extraction methods in previous section. Except for the extraction method, the terms/words used for extraction also heavily impacts the topic extraction performance. Existing works give limited explanation on why they use terms like comments, method names and identifier names for topic extraction. To this end, we propose a term selection algorithm that automatically selects the most contributory terms. We divide the terms into several types, e.g., method names, method comments, and identifier names etc. Each type of terms are recorded in a separate document. When a type is selected, the content of selected document is merged with the document that represents the system. We select the types of terms based on the performance of topic extraction (the metric will be discussed in Sect. 2.4) and use this metric to guide a heuristic search. The algorithm performs a best-first search in the feature space. It starts with constructing the terms set using one type of terms. Here we start from method names. Then it obtains top-N keywords by applying one of the extraction models discussed from Sects. 2.2.1 to 2.2.2. After obtaining keywords, the algorithm measures the contribution of the new selected type of terms and decide if it will be kept/removed. If it improves the topic extraction performance, the type of terms is selected. Otherwise, it is removed from the terms set. The algorithm then moves forward to construct a larger terms set by including more types of terms. In our work, we form a new document by merging the new selected term document with the term set document. Then the selection process is repeated until all types of terms are traversed.

The algorithm adopts idea from feature selection in machine learning community, but it differs with it. Because in feature selection, the metric evaluation is directly performed on the set of features, while in this algorithm, the metric evaluation is based on the topics extracted from the extraction method, and the selected terms are the input of extraction model. This is the reason why we do not apply feature selection algorithms in our work.

2.4 The Coherence Measurement

We adopt coherence measurement for evaluating the topics extraction performance because does not require human effort to judge the extracted topics. Coherence measurement measures the coherence within the keywords list. High coherence score indicates that the extracted key words are closely coherent which further indicates that the extraction of topics is good. Röder et al. [18] empirically show that the combination of indirect cosine measure with NPMI and the boolean sliding window gives best coherence measurement through their extensive studies on natural language corpus. We apply this measurement to the source code. Given a list of keywords $\{w_i\}, i \le K$, where K is the number of keywords, four dimensions are required to measure the coherence on $\{w_i\}$: the segmentations of $\{w_i\}$, the probability P, the confirmation measurement m and the aggregation method. The segmentation of $\{w_i\}$ is defined as:

$$S_{set}^{one} = \{(W', W^*)\|W' = \{w_i\}; w_i \in W; W^* = W\}, \tag{4}$$

where W is the total word set. The probability method used is boolean sliding window, which determines word counts using a sliding window. Confirmation measure takes a single pair $S_i = (W'; W^*)$ of word sublists as well as the corresponding probabilities to compute how strong the conditioning word set W' supports W^*. Indirect confirmation is defined as:

$$m_{s_{cos}(m,\gamma)}^*(W', W^*) = s_{cos}(\mathbf{u}, \mathbf{v}),$$
$$= \frac{\sum_{i=1}^{|W|} u_i \cdot v_i}{||\mathbf{u}||_2 \cdot ||\mathbf{v}||_2}, \tag{5}$$

where vector $\mathbf{u}$ is generated by:

$$\mathbf{u} = \left\{ \sum_{w_i \in W'} m(w_i, w_j)^\gamma \right\}, \tag{6}$$

where m, which is the direct confirmation measure using NPMI, is:

$$m_{nmpi}(S_i) = \frac{log\dfrac{P(W',W^*)+\epsilon}{P(W')*P(W^*)}}{-log(P(W',W^*)+\epsilon)}. \tag{7}$$

Similarly, vector $\mathbf{v} = \mathbf{v}_{m,\gamma}(W^*)$ is calculated using Eq. 6 by replacing W' with W^*. Finally, all confirmations of all subset pairs S_i are aggregated to a single coherence score using the arithmetic mean.

3   Related Works

Much research effort has been devoted to tackle different tasks of source code mining. We discuss some of the works that are closely related to our work.

Source code summarization works apply text summarization techniques to generate summary for source code [8,9,16] or parts of source code [19]. Haiduc et al. [9] proposed an automated text summarization method using Vector Space Model (VSM). They extracted method comments, method names and class names as the terms and formed document-term matrix with TF/IDF weighing scheme. The terms with top-weighting are considered as the top keywords of the document. Finally, they used the keywords to construct the summary for the software system. The evaluation includes two parts: judgement from experienced developers and feedback from follow-up questionnaires. Rodeghero et al. [17] improved the work in [9] by introducing eye-tracking techniques. They also used method comments and VSM. Their method assigns weights according to the code position obtained from eye-tracking experiments. It uses *Kendall's* τ [5] distance to evaluate the keywords extraction by computing the distance between extracted keywords and the ones picked by experienced developers. Sridhara et al. contributed several works on generating comments and summaries for Java methods [19]. They designed heuristics to choose method statements and then transformed these statements to natural language comments and summaries. Their approach requires scanning the content of a method code.

Rama et al. [16] proposed to use topic model LDA to extract business concepts from source code. This work assumes a software system is a text corpus, in which the files are constructed using function names, comments and data structures. However, using this approach, each project requires an independent LDA model which is not applicable to processing large volume of projects at the same time. Many following works adopted LDA for solving various software engineer problems. For example, Asuncion et al. [2] applied LDA to enhance software traceability. They recorded traceability links during the software development process and learned a LDA model over artifacts. The learned model categorized artifacts according to their topics. Lukins et al. [11] used LDA to localize bugs in software systems. They learned topic model from source codes. Queries generated from bug reports were performed against the learned model to localize the code that might produce the bug.

4 Experiments

We perform the experiments on Java projects randomly chosen from the Java project corpus provided by [1]. This corpus consists 14,807 Java projects in Github. All the .java files are kept and other files are removed. We compare EmbTE, LDA and NMF on extracting topics from these source codes. Both CBOW and Skip-gram models in EmbTE are considered. Firstly, we examine the impact of different weighting scheme on LDA and NMF methods (Sect. 4.2.1). Second, we compare the performance of the three methods on various types of terms in the source codes (Sect. 4.2.2). Third, we train the models on different number of systems and see the difference when the number changes (in both Sects. 4.2.2 and 4.2.3). The coherence is computed via Eq. (5).

4.1 Implementation

We extract source code terms from Java projects using JavaParser[1]. Stop words are pruned using Snowball English stop words lists[2]. We leverage StanfordNLP[3] for the stemming and lemmatization. For the vector representation of words, we use gensim[4] implementation of Word2Vec[5]. The implementation of coherence measurement adopts Palmetto[6] tool. The scikit-learn[7] tool is used to implement LDA/NMF and build K-d tree for EmbTE. Other functions are implemented by us using Java. We perform all the experiments on a PC with Intel Core i7 2.40 GHZ processor and 8 GB RAM. All the source codes are available per request.

4.2 Results

The performance given in Sect. 4.2.1 is based on the evaluation of source codes from 100 systems. Sections 4.2.2 and 4.2.3 report the performances on the source codes from a small size of 10 and a medium size of 100 systems.

4.2.1 Impact of Different Weighting Schemes

In this experiment, we compare the performances of LDA and NMF on TF/IDF and TF weighting schemes when adding different types of terms. The bars from left to right show the coherence when adding new types of terms. We start from *MeNM* (method names), then the second bar of each group is the result based on merging *MeParNM* (method parameter names) to *MeNM*. Thus, until the last bar, the result on *VarType*, all the types of terms are added. We can see the coherence fluctuates when adding new types of terms. If the coherence

---

[1] https://github.com/javaparser/javaparser.
[2] http://snowball.tartarus.org/algorithms/english/stop.txt.
[3] http://nlp.stanford.edu/.
[4] https://radimrehurek.com/gensim/.
[5] https://code.google.com/archive/p/word2vec/.
[6] https://github.com/AKSW/Palmetto.
[7] http://scikit-learn.org/.

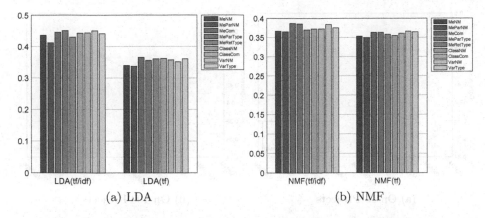

Fig. 2. Impact of different weights on topic extraction methods using LDA and NMF. The meaning of legends are as following: MeNM (method name), MeParNM (method parameter name), MeCom (method comments), MeParType (method parameter type), MeRetType (method return type), ClassNM (class name), ClassCom (class comments), VarNM (variable name) and VarType (variable type).

depicted by one bar is higher than the value of previous bar, it means this type of terms increases the coherence (i.e., contributory). If the value is lower than the one of previous bar, this type of terms decreases the coherence (i.e., not contributory). Both Fig. 2(a) and (b) show the same trends (i.e., increase or decrease the coherence) when adding new types of terms. We will discuss in more details of the impact on different types of terms in Sect. 4.2.2 as we compare more methods there. In Fig. 2, we can also find that the TF/IDF weight gives better performance than only using TF as weight because TF considers only the frequency of the term in a document, while the TF/IDF also considers the number of documents that contain the term (using a logarithmically scaled inverse fraction of the documents that contain the term). Thus TF/IDF will lower the weight of terms that with high frequency in a document and appears in many documents. When we mine topics from a document, the ones appear in many documents will not well represent the topic of this document. That is why the extraction performances on TF/IDF weights are better than on TF weights.

4.2.2 Impact of Different Types of Terms

Figure 3 shows the topic extraction performance when adding terms. Figure 3(a) depicts the average value performance on 10 projects, while Fig. 3(b) depicts the average performance on 100 projects. The X-axis shows the results from adding different types of terms. Correspondingly, the Y-axis gives the coherence when the terms pointed by X-axis are added. For example, in Fig. 3(b) the coherence on *MeCom* (method comments) is higher than its previous coherence on *MeParNM*. It indicates that method names positively contribute to the topics coherence (i.e.,

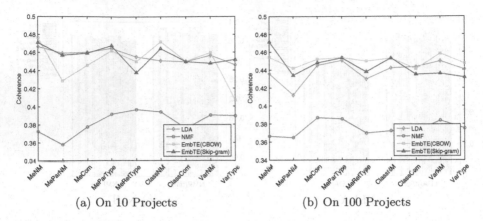

(a) On 10 Projects   (b) On 100 Projects

Fig. 3. Impact of different types of terms. The meaning of X-axis ticks are the same with legend meanings in Fig. 2

improve performance). The coherence on *MeRetType* (method return types) is lower than its previous coherence which means method return types negatively impact the topic coherence (worsen performance). Figure 3(a) shows unstable trends among four considered models, but in Fig. 3(b), the four models give similar performance on different types of terms. It indicates that the more documents are considered, the better topic extraction performance is achieved. When adding *MeNM* (method names), *MeCom* (method comments), *ClassNM* (class names) and *ClassCom* (class comments), the coherence increases, which means they are contributory to the topic extraction. The types of terms which decrease the coherence give negative impact on the topic extraction performance. We can also observe from the figures that EmbTE(CBOW) performs consistently the best and EmbTE(Skip-gram) has close performance with LDA. NMF shows the worst performance. EmbTE gives better performance than LDA is because that although the probability distribution obtained from LDA describes the statistical relationship among documents, topics and terms, sometimes the terms with higher probability can not represent topics well [15]. Instead, EmbTE considers semantic relationships between terms and is able to identify the central semantic of the documents. NMF performs worst is because it sets fixed values for the probability of multinomial distribution of words and topics in documents, however, this is often unlikely in real documents. EmbTE with CBOW model performs better than with Skip-gram model. It indicates that the frequency of terms gives more impact than the context of terms in extracting topics from source codes.

4.2.3   Impact of Different Number of Keywords

According to the results in Sect. 4.2.2, we use the contributory types of terms, namely method name, method comments, class names and class comments from source codes in this experiment. Figure 4 (a) and (b) gives the results when

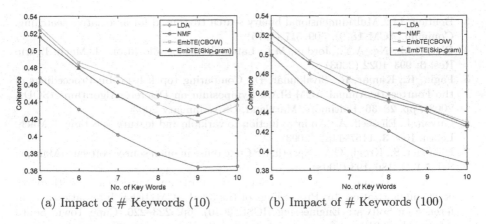

(a) Impact of # Keywords (10)   (b) Impact of # Keywords (100)

Fig. 4. Impact of number of keywords

varying the number of keywords/topics. Both Fig. 4(a) and (b) show similar trends for four methods. When the number of keywords is increasing, most methods produce lower coherent topics. EmbTE(CBOW) consistently performs the best while EmbTE(Skip-gram) and LDA performs the less best and NMF is the worst. The results show that more extracted topics will reduce the coherence of topics. This is because the coherence is normalized, when the new added topic is not 100 % coherent with existing topics, the overall coherence will decrease.

5 Conclusion

In this paper, we develop a method EmbTE, for source code topic extraction, based on word embedding techniques. We also adopt LDA and NMF to extract topics from source code. The empirical comparisons show that EmbTE outperforms LDA and NMF in terms of providing more coherent topics. EmbTE with CBOW model performs better than Skip-gram model. We also identify the most contributory terms from source code via our proposed term selection algorithm. We find that the method name, method comments, class names and class comments are the most contributory term types. In the future, we will apply our methods on larger scale projects and evaluate on source codes of other languages.

References

1. Allamanis, M., Sutton, C.A.: Mining source code repositories at massive scale using language modeling. In Proceedings of the 10th Working Conference on Mining Software Repositories (MSR 2013), pp. 207–216, San Francisco, CA, USA, May 2013
2. Asuncion, H.U., Asuncion, A.U., Taylor, R.N.: Software traceability with topic modeling. In: Proceedings of the 32nd ACM/IEEE International Conference on Software Engineering (ICSE 2010), pp. 95–104, Cape Town, South Africa, May 2010

3. Bentley, J.L.: Multidimensional binary search trees used for associative searching. Commun. ACM **18**(9), 509–517 (1975)
4. Blei, D.M., Ng, A.Y., Jordan, M.I.: Latent Dirichlet allocation. J. Mach. Learn. Res. **3**, 993–1022 (2003)
5. Fagin, R., Kumar, R., Sivakumar, D.: Comparing top k lists. In: Proceedings of the Fourteenth Annual ACM-SIAM Symposium on Discrete Algorithms (SODA 2003), pp. 28–36, Baltimore, Maryland, USA, January 2003
6. Guyon, I., Elisseeff, A.: An introduction to variable and feature selection. J. Mach. Learn. Res. **3**, 1157–1182 (2003)
7. Haefliger, S., Krogh, G.V., Spaeth, S.: Code reuse in open source software. Manage. Sci. **54**(1), 180–193 (2008)
8. Haiduc, S., Aponte, J., Marcus, A.: Supporting program comprehension with source code summarization. In: Proceedings of the 32nd ACM/IEEE International Conference on Software Engineering (ICSE 2010), pp. 223–226, Cape Town, South Africa, May 2010
9. Haiduc, S., Aponte, J., Moreno, L., Marcus, A.: On the use of automated text summarization techniques for summarizing source code. In: Proceedings of the 17th Working Conference on Reverse Engineering (WCRE 2010), pp. 35–44, Beverly, MA, USA, October 2010
10. Lee, D.D., Seung, H.S.: Learning the parts of objects by non-negative matrix factorization. Nature **401**(6755), 788–791 (1999)
11. Lukins, S.K., Kraft, N.A., Etzkorn, L.H.: Bug localization using latent Dirichlet allocation. Inf. Softw. Technol. **52**(9), 972–990 (2010)
12. Mikolov, T., Chen, K., Corrado, G., Dean, J.: Efficient estimation of word representations in vector space. CoRR, abs/1301.3781 (2013)
13. Mikolov, T., Sutskever, I., Chen, K., Corrado, G.S., Dean, J.: Distributed representations of words and phrases and their compositionality. In: Proceedings of the 27th Annual Conference on Neural Information Processing Systems (NIPS 2013), pp. 3111–3119, Lake Tahoe, United States, December 2013
14. Moreno, L., Aponte, J., Sridhara, G., Marcus, A., Pollock, L.L., Vijay-Shanker, K.: Automatic generation of natural language summaries for java classes. In: Proceedings of the 21st IEEE International Conference on Program Comprehension (ICPC 2013), pp. 23–32, San Francisco, NC, USA, May 2013
15. Niu, L., Dai, X., Zhang, J., Chen, J.: Topic2Vec: learning distributed representations of topics. In: Proceedings of the International Conference on Asian Language Processing 2015 (IALP 2015), pp. 193–196, Suzhou, China, October 2015
16. Rama, G.M., Sarkar, S., Heafield, K.: Mining business topics in source code using latent Dirichlet allocation. In: Proceedings of the 1st Annual India Software Engineering Conference (ISEC 2008), pp. 113–120, Hyderabad, India, February 2008
17. Rodeghero, P., McMillan, C., McBurney, P.W., Bosch, N., D'Mello, S.K.: Improving automated source code summarization via an eye-tracking study of programmers. In: Proceedings of the 36th International Conference on Software Engineering (ICSE 2014), pp. 390–401, Hyderabad, India, June 2014
18. Röder, M., Both, A., Hinneburg, A.: Exploring the space of topic coherence measures. In: Proceedings of the Eighth ACM International Conference on Web Search and Data Mining (WSDM 2015), pp. 399–408, Shanghai, China, February 2015
19. Sridhara, G., Pollock, L.L., Vijay-Shanker, K.: Automatically detecting and describing high level actions within methods. In: Proceedings of the 33rd International Conference on Software Engineering (ICSE 2011), pp. 101–110, Waikiki, Honolulu, HI, USA, May 2011

IPC Multi-label Classification Based on the Field Functionality of Patent Documents

Sora Lim(✉) and YongJin Kwon

Department of Information and Telecommunication Engineering,
Korea Aerospace University, Goyang, Korea
{ebbunsora, yjkwon}@kau.ac.kr

Abstract. The International Patent Classification (IPC) is used for the classification of patents according to their technological area. Research on the IPC automatic classification system has focused on applying various existing machine learning methods rather than considering the data characteristics or the field structure of the patent documents. This paper proposes a new method for IPC automatic classification using two structural fields, the technical field and the background field selected by applying the characteristics of patent documents. The effects of the structural fields of the patent document classification are examined using a multi-label model and 564,793 registered patents of Korea at the IPC subclass level. An 87.2% precision rate is obtained when using titles, abstracts, claims, technical fields and backgrounds. From this sequence, it is verified that the technical field and background field play an important role in improving the precision of IPC multi-label classification at the IPC subclass level.

Keywords: Patent classification · IPC classification · Patent document fields · Field function · Multi-label classification

1 Introduction

In knowledge-based societies, knowledge and information are key sources of competitive advantage. As such, protecting the intellectual property that results from knowledge-related activities has become more important. Globally, therefore, vast numbers of patents, a representative form of intellectual property, have already been filed, and the current trend is for continued upward growth. In Korea, there were approximately 210,000 patents filed just in 2014 [1]. The number of cumulated patents is expected to explosively increase in the near future. Accordingly, it is important to classify patents according to their technical and industrial fields so that patent investigations can be performed effectively. In general, patent classification is used to (1) investigate existing patents to effectively evaluate novelty and non-obviousness, (2) arrange patent documents to facilitate access to the technological and legal information contained therein, (3) create statistics related to technological advancements in various areas, and (4) improve the correctness, consistency and quality of patent-related work [2].

According to the Guidelines for Examination [3] offered by the Korean Intellectual Property Office (KIPO), the granting of patent classifications should follow a process. This process is shown in Fig. 1. First, for submitted patent documents, KIPO requests a

© Springer International Publishing AG 2016
J. Li et al. (Eds.): ADMA 2016, LNAI 10086, pp. 677–691, 2016.
DOI: 10.1007/978-3-319-49586-6_48

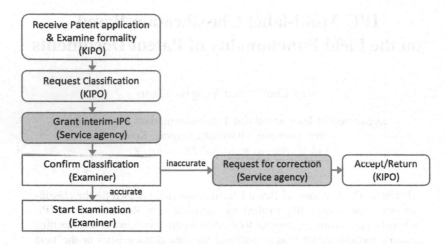

Fig. 1. IPC assignment flowchart

service agency to label the classifications for the patents only if no problems are identified in the submission process. The service agency then manually performs an interim-classification for the patent applications. This is done by experts according to the technical subject matter. In addition, a KIPO examiner checks whether the pre-classifications of the patent applications are correctly labeled based on their technical content. In this process, the International Patent Classification (IPC) is used to categorize patents. Further details regarding the IPC system are provided in Sect. 3. To replace current human-based processes, developing an efficient learning model and constructing a practical system are important research topics in data mining and machine learning areas. There is also a need for substantive research to be applied to the real world.

Second, patents have to meet novelty, non-obviousness and utility requirements to take effect. Here, novelty means that an invention can be patented only if it is new, non-obviousness means that an invention should be sufficiently inventive to be patented, and utility means that an invention is patentable only if the invention is considered to industrially applicable. These three conditions are asserted by patent document claims. That is, claims show the value of an idea in a new invention. When patent documents are classified, putting emphasis on claims that contain newly invented features may cause negative consequences with respect to the classification. However, claims need to be assessed if the classification categories are narrowly defined with only small variances between each category, such as the categories in IPC main group or subgroup level. This is because claims must show how they can be differentiated from potentially competing claims.

Third, patent documents can simultaneously possess multiple IPC codes. In statistics and machine learning, classification problems are widely divided using binary, multi-class and multi-label classification problems. IPC classification is a multi-label classification problem. From a practical point of view, it would be more appropriate to use a multi-label classification method to assign IPCs to real patent documents rather than a multi-class classification method, which would eliminate the possibility of labeling many IPCs to a patent at the same time.

These patent document characteristics are also put to good use when humans categorize documents according to multiple IPCs. In that process, particular fields such as title or abstract are first applied to broadly define the technical subject of the submitted patents. This is done based on an understanding of the roles of patents and the structure of the documents. Ultimately, the claims are analyzed in depth to determine their more specific categories. Therefore, this manual classification process, which is based on data characteristics, needs to be applied to the learning process used in a patent auto-classification process. However, previous research on IPC-based patent classification has tended to focus on the selection of a machine learning model or discriminative words in patent documents rather than to consider data characteristics and their potential benefits.

Therefore, this paper aims to construct an IPC multi-label classification system based on the multinomial naive Bayes method to replace the IPC classification process carried out by humans (the service agency) and to assist the examiner's decision by recommending relevant IPCs at the IPC subclass level (which has 630 categories). This paper also investigates how the structural fields in the patent documents affect the result of the auto-classification.

The rest of the paper is organized as follows. The next section introduces prior re-search on IPC classification, and Sect. 3 presents the IPC auto-classification process using the technical field and background, which are considered to be more effective structural fields for the present patent classification. Section 4 shows the experiments and the results of the IPC auto-classification system using the technical field and background. Section 5 states the conclusion and discusses avenues of future research.

2 Related Work

With recent technical advances in big data and data mining areas, much of the literature on patent classification has focused on using various machine learning models. Fall et al. [4] studied IPC auto-classification with several machine learning models – naive Bayes (NB), k-nearest neighbor (KNN), support vector machines (SVM) and sparse network of winnows (SNoW). They performed an IPC classification for the IPC class and subclass level and made use of titles, abstracts, claims and the first 300 words of the patent documents in the WIPO-alpha collection. According to their results, the first 300 words of the titles, inventors, applicants, abstracts and descriptions facilitated the best classification results, and the SVM model showed the best performance with a 41% top-prediction precision rate and a 48% all-category precision rate when they evaluated the classification performance according to three measures – top-predictions, three-guesses and all-category. The KNN model showed a 62% precision rate with three guesses measure.

Larkey [5] developed a classification system for US patents in accordance with the United States Patent Classification (USPC) scheme based on KNN. The results indicated that classification accuracy could be increased by using the first 20 lines of the background and summary as well as all of the claims together. The study achieved classification accuracy ranging from 25% to 32% when evaluating for signal processing subclasses under subclass 2.09 of class 395.

Fall et al. and Larkey investigated the effects of particular fields in patent documents on classification performance. However, these studies did not seem to consciously base their choice of fields on the characteristics of the patent documents as shown in the use of the first 300 words or first 20 lines.

Tikk et al. [6] proposed a hierarchical classification model called HITEC. This model classified patents at the IPC main group level by combining several machine learning algorithms. Their results showed 36% to 56% classification accuracy when using inventors, titles, abstracts and claims. These accuracy rates went up by approximately 6% to 10% at the IPC class level and 12% to 14% at the IPC subclass level compared to the results provided by Fall et al. Y. Chen et al. [7] presented a three-phase classification model constructed with SVMs and a KNN hierarchically. They classified patent documents at the IPC subgroup level and obtained 36.07% accuracy. These studies by Tikk et al. and Y. Chen et al. improved IPC classification performance through hierarchical methods but paid little attention to the meaning and function of the structural fields in patent documents.

Seneviratne et al. [8] suggested an efficient method using document signatures in connection to search times for patent classification and scalability. This was proposed as an alternative method to bag-of-words modeling, which is mainly used for vector space modeling. They generated the document signatures by using the titles, abstracts and the first 300 words from patent documents. Their work was heavily influenced by the aforementioned research by Fall et al.

In addition, several studies were conducted to extract discriminative features from patent documents to improve classification accuracy. Park et al. [9] proposed a new feature extraction method named dominant information and compared the effects of their method and existing feature extraction methods. This research employed the KNN model to classify Korean patent documents at the IPC section level, and achieved 43% accuracy with k = 2 and 10% of the feature selections. However, there are still significant practicality issues to overcome because their target level, the IPC section, consisted of only eight categories.

Kim et al. [10] proposed using specific fields of patent documents to search for similar patent documents with respect to their structural characteristics. They determined similarity among Japanese patent documents by carrying out a cross comparison of several fields, i.e. technical field, purpose, method, claims, explanation and examples of the patent documents. According to their experiment, technical field, purpose and claims were more effective when searching for similar patent documents. Their study did not treat IPC classification directly, but it consciously made use of each field of patent documents in the searching process.

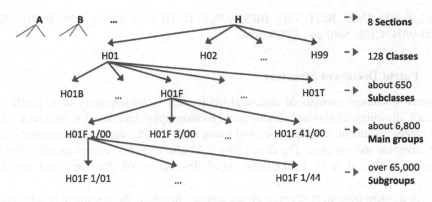

Fig. 2. IPC structure

3   Proposed Method

In this section, the structure of the IPC and patent documents are investigated, and the classification method is described. In particular, this method uses technical field and background field, taking into account the meaning of the structural fields of patent documents using these two fields.

3.1   IPC Structure

The IPC system assigns its own classification codes to patent documents. The intention here is to create an internationally uniform classification system that classifies patents according to their technological areas.

The IPC system has a hierarchical structure of five levels consisting of eight sections, 128 classes, about 650 subclasses, about 6,800 main groups and more than 65,000 subgroups. At the highest level – the section level – the eight sections are denoted with letters A to H. They are (A) Human necessities, (B) Performing operations and transporting, (C) Chemistry and metallurgy, (D) Textiles and paper, (E) Fixed constructions, (F) Mechanical engineering, lighting, heating, weapons and blasting, (G) Physics and (H) Electricity. As an example, the IPC code 'H01F 1/01' falls under Section H (Electricity), Class H01 (Basic electric elements), Subclass H01F (Magnets, inductances, transformers and the selection of materials for their magnetic properties), Main Group H01F 1/00 (Magnets or magnetic bodies characterized by the magnetic materials therefor, and the selection of materials for their magnetic properties) and Subgroup H01F 1/01 (Of inorganic materials). Therefore, the technical subject scopes are more specified as they descent to the lowest IPC level. Figure 2 shows the hierarchical structure of IPC.

Moreover, each patent document could have more than one IPC code. For instance, the patent 'Portable terminal device comprising bended display and method for controlling thereof' has two IPC codes – G06F 3/048 and H04B 1/40. The patent 'Integration of sample storage and sample management for life science' has four IPCs – B01L 3/00, G06 K 19/07, G01S 13/00 and C12Q 1/68. The patent 'Manufacturing apparatus and method of solid fuel using food waste' has 11 IPCs – C10L 5/46, A47 J

37/12, B01D 47/00, B01F 7/02, B02C 23/08, B04B 5/10, B07B 1/28, B09B 3/00, B30B 9/04, C10L 5/40 and F26B 3/04.

3.2 Patent Document Structure

A patent document consists of structural fields such as bibliography data, a title, an abstract, drawings, claims and descriptions. Bibliography data, the first structural field of a patent document, contains specified fields such as IPC, application number, field date, inventor and assignee. The description field also includes several specified fields, including technical field, background, brief description of drawings and detailed description. Figure 3 depicts the structure of a patent document. The order of these fields in a patent document depends on the country in which the document is submitted. In this paper, images in the drawings field are not used; focus is only placed on patent document text.

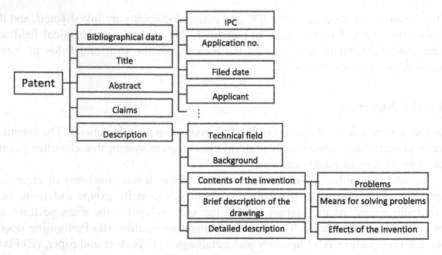

Fig. 3. Patent structure

3.3 Construction of Proposed Patent Classification Model

In this paper, patent classification in the IPC subclass level is performed via data collecting, preprocessing, feature selection, vector space modeling, classifier construction and patent classification. Figure 4 shows these processes.

In the first step, patent data collecting, Open API (Kipris Plus [11]) is used to collect sufficient data to classify the patents. Kipris Plus provides two types of services – Bulk Data and Open API – to make it possible to use all of the patent information that KIPO opens to the public in real-time. Using Open API, the patent documents registered in Korea from Jan 1 to Dec 31, 2015, are collected. These documents are then parsed so that the representative fields, title, abstract, claims, technical field and background that have an effect on patent classification are extracted from each of the collected documents. At this point, patent documents are removed if they do not

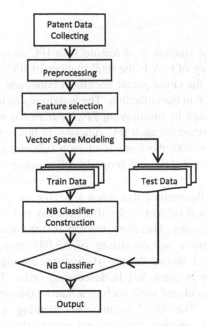

Fig. 4. IPC Multi-label Classification Process

include technical field and background. Finally, a patent collection is constructed with 564,793 patents categorized into 630 IPC subclasses.

Preprocessing is carried out next. This is a crucial step, particularly because there are 630 target classes and the quality of the preprocessed data could have a significant effect on classification performance. In this step, nouns are extracted as the minimum unit with meaning from each patent document in the collection. In this task, a Korean language morphological analyzer, KLT2000 [12] is used. In the patent document collection, there are a number of terminologies that have no influence on the meaning of the patent documents, including 'invention', 'claim', 'application', 'patent', 'description' and 'drawing'. These words are generally called stopwords. To remove stopwords effectively, a stopwordlist is constructed and the words are then removed from each document in the patent collection. The stopwordlist is based on the index terms in the glossary of the intellectual property terms and patent examination guidelines. There are 1,860 stopwords.

The next step is feature selection. In this step, deterministic words (features) are selected for the 630 IPC subclasses by using a TF-ICF scheme [7], which is a variation of the well-known TF-IDF weighting scheme for multi-classifications. Word weights are calculated for each IPC subclass using TF-ICF, as follows.

$$TF_{ij} = avg(freq_{ij}), ICF_i = \log_2(N/n_i)$$

$$W_{ij} = TF_{ij} * ICF_i$$

where TF_{ij} is the average frequency of feature f_i in IPC subclass c_j, and ICF_i is the inverse category frequency of f_i. N is the total number of IPC subclasses used to label patent documents across the entire patent document collection and n_i is the number of IPC subclasses including f_i in the collection. The weight w_{ij} of feature f_i in IPC subclass c_j is subsequently calculated by multiplying TF_{ij} by ICF_i. To construct the feature set, the top k features are selected for each IPC subclass. In this paper, it is experimentally determined that when k = 100, the features of each subclass aptly represent the technical subject of the IPC subclass and properly differentiate the subclass from other subclasses. In addition, by analyzing the rankings of features weighted by TF-ICF for the IPC subclasses, it is determined that when k = 100, the features of each subclass aptly represent the technical subject of the IPC subclass and properly differentiate the subclass from other subclasses. Therefore, the top 100 features are extracted for each IPC subclass, and a feature set consisting of 33,047 words is constructed after removing overlap. Table 1 shows the top 10 features among the extracted top 100 features of the subclasses in class A01 in descending order. The words in the A01B column are related to agricultural tools such as a harrow (sseorae)[1], tractor, implement (jageopki) and rotavator. The A01C column lists planting terms and includes seed (jongja), sowing (pajong), seedling (mojong) and manure (biryo). The A01P column is related to agricultural chemicals such as herbicides (jechoje) and pesticides (sal-chunje). Compared to the IPC definition [13] as shown in Table 2, the features selected by the TF-ICF weighting scheme represent each IPC subclass.

In the vector space modeling step, patent documents are transformed into the fea-ture vectors based on the feature set of the previous step. The vectors with raw fre-quencies of features are weighted using the TF-IDF scheme.

The classification model is constructed to classify patent documents with multi-labels for 630 target classes. For example, the patent 'Refrigerator for wireless power transmission to sensor for detecting condition of stored food' is labeled with two different IPC subclasses, F25D (Refrigerators; Cold rooms; Ice-boxes; Cooling or freezing apparatus not covered by any other subclass) and A23B (Preserving, e.g. by canning, meat, fish, eggs, fruit, vegetables, edible seeds; chemical ripening of fruit or vegetables; The preserved, ripened, or canned products). This is to be expected from the title. Likewise, the patent 'A method and apparatus for transmitting a media transport packet in a multimedia transport system' is categorized into H04 N (Pictorial communication, e.g. television) and H04L (Transmission of digital information) at the same time. To classify patent documents that could have multi-labels, the multinomial naive Bayes model is used. This is one of the most representative machine learning models.

Finally, in the classification step, the IPC subclasses are predicted for the patent document based on the multinomial naive Bayes model [14]. This model is used in both the classifier construction and classification steps, as follows.

---

[1] This is Korean pronunciation of the corresponding original feature.

Table 1. Extracted features for each IPC subclass using TF-ICF (Class A01) *(This is translated from the Korean features.)*

Rank	A01B	A01C	A01D	A01F	A01G	A01H
1	Harrow	Seed	Brush cutter	Baler	Cultivation	Gene
2	Tractor	Seeder	Cutting	Bale	Flowerpot	Transgenic
3	Harrow plate	Sowing	Combine	Threshing	Greenhouse	Plant body
4	Ridge	Seedling	Mowing	Cutting	Plants	Promoter
5	For tractor	Fertilizer	Cutting part	Bale wrapper	Nutrient solution	Plants
6	Implement	Rice planting machine	Harvest	Forage crops	Mushroom	Arabidopsis
7	Rotavator	Seed	For brush cutter	Rice straw	Planting	Protein
8	Ridger	Germination	Harvester	Combine	Tree	Seed
9	Plow part	Seedling box	Grass	Grain	Crop	Rice
10	Rice field	Manure	Weeder	Thresher	Vegetation mat	Recombination

Rank	A01 J	A01 K	A01 M	A01 N	A01P
1	Dairy cattle	Artificial reef	Pest	Prevention	Prevention
2	Teat cup	Fishing	Agricultural chemicals	Strain	Herbicide
3	Teat cup	Fishing rod	Eradication	Herbicide	Pesticide
4	Cheese	Aquarium	Prevention	Agricultural chemicals	Composition
5	Raw milk	Fishing line	Mosquito	Herbicide	Plant disease
6	Milking	Fishing rod	Pest control machine	Plant disease	Pesticidal
7	Brest pump	Sea cucumber	Scattering	Plants	Plants
8	Yogurt	Sinker	Liquid chemical	Ylamino	Pest
9	Teat cup	Pet	Algae	Composition	Agricultural chemicals
10	Teat	Fish	Wild animal	Hexa hydro phthalazin	Control agent

$$P(c|f_1, f_2, \ldots, f_n) \propto P(c) \prod_{i=1}^{n} P(f_i|c)$$

$$\hat{c} = argmax_c P(c) \prod_{i=1}^{n} P(f_i|c) = argmax_c P(c) \prod_{i=1}^{n} \frac{N_{ci} + \alpha}{N_c + an}$$

Table 2. Subclasses in Class A01 (2015.01 ver.)

Section	A	Human Necessities
Class	A01	Agriculture; Forestry; Animal Husbandry; Hunting; Trapping; Fishing
Subclass	A01B	Soil Working In Agriculture or Forestry; Parts, Details, or Accessories of Agricultural Machines or Implements In General
	A01C	Planting; Sowing; Fertilizing
	A01D	Harvesting; Mowing
	A01F	Threshing; Baling of Straw, Hay or The Like; Stationary Apparatus or Hand Tools for Forming or Binding Straw, Hay or The Like into Bundles; Cutting of Straw, Hay or The Like; Storing Agricultural or Horticultural Produce
	A01G	Horticulture; Cultivation of Vegetables, Flowers, Rice, Fruit, Vines, Hops, or Seaweed; Forestry; Watering
	A01H	New Plants or Processes for Obtaining Them; Plant Reproduction by Tissue Culture Techniques
	A01 J	Manufacture of Dairy Products
	A01 K	Animal Husbandry; Care of Birds, Fishes, Insects; Fishing; Rearing or Breeding Animals, Not Otherwise Provided For; New Breeds of Animals
	A01L	Shoeing of Animals
	A01 M	Catching, Trapping or Scaring of; Apparatus for The Destruction of Noxious Animals or Noxious Plants
	A01 N	Preservation of Bodies of Humans or Animals or Plants or Parts Thereof; Biocides, E.G. As Disinfectants, As Pesticides or As Herbicides; Pest Repellants or Attractants; Plant Growth Regulators
	A01P	Biocidal, Pest Repellant, Pest Attractant or Plant Growth Regulatory Activity of Chemical Compounds or Preparations

where N_{ci} is the number of occurrences of the feature f_i in category c, and N_c is the total number of occurrences of every feature in category c. This paper uses Laplace smoothing in which α is 1, and the top five predictions are provided at the IPC subclass level.

In addition, when classification model is constructed and classification is carried out, the patent document collection is divided into a training set and test set by using the 10-fold cross validation method. Next, a classifier is constructed using the training set, and multi-label classification is performed for the test set at the IPC subclass level.

4   Experiments and Results

In this section, IPC classification performance is evaluated for the patent documents based on technical field and background. The results are then discussed in comparison with previous research efforts that relied on title, abstract and claims.

4.1 Precision Measurements

The goal of this paper is to provide a practical patent classifier that is able to support human decision-making processes. Therefore, two precision measures – single match precision and all match precision – are used to evaluate the classification performance for each field of the patent documents. Single match precision considers the classification successful if one of the predicted IPC subclasses for a patent document corresponds to one of the true IPC subclasses of that patent document. All match precision considers the classification successful if all of the predicted results for a patent document correspond to the true subclasses.

$$Precision_{single} = \frac{1}{n} \sum_{i=1}^{n} I(|T_i \cap P_i|)$$

$$Precision_{all} = \frac{1}{n} \sum_{i=1}^{n} I(|T_i \subseteq P_i|)$$

where I is the indicator function, T_i is the true IPC subclass categories of patent document i, P_i is the predicted IPC subclass categories, and n is the number of patent documents in the test patent document set.

4.2 Evaluation

To examine the effects of the technical field and background, comparative experiments are designed with: (1) titles, (2) abstracts, (3) claims, (4) technical fields, backgrounds, (5) titles, abstracts, (6) titles, abstracts, claims, (7) titles, abstracts, technical fields, backgrounds, (8) titles, abstracts, claims, technical fields and backgrounds, according to the used fields for indexing. The experiments are performed with an Intel Xeon 8-Core CPU, 128 GB RAM and 64-bit Linux OS. The multinomial naive Bayes classifier is implemented in Python using the scikit-learn library.

Before beginning the classification process, the average length of the eight different fields shows that (1) titles have four words, (2) abstracts have 48.19 words, (4) technical fields, backgrounds have 141.60 words, (3) claims have 212.72 words and (8) titles, abstracts, claims, technical fields, backgrounds have 333.52 words on average. Table 3 shows these results.

IPC classification is performed for the patent document sets consisting of these eight different combinations of fields. Table 4 lists the classification results at the IPC subclass level. The results show the highest precision (87.2 %) for (7) titles, abstracts, technical fields, backgrounds, which means that classification is successful for 492,513 of the total 564,793 test documents. 86.60% precision is achieved for (4) technical fields, backgrounds, and 85.67% precision is achieved for (8) titles, abstracts, claims, technical fields, backgrounds.

Table 3. Average length of patent documents by field

Indexing fields	Average length
(1) Titles	4.00
(2) Abstracts	48.19
(3) Claims	212.72
(4) Technical Fields, Backgrounds	141.60
(5) Titles, Abstracts	51.62
(6) Titles, Abstracts, Claims	243.47
(7) Titles, Abstracts, Technical Fields, Backgrounds	182.60
(8) Titles, Abstracts, Claims, Technical Fields, Backgrounds	333.52

Table 4. Classification precision at IPC subclass level

Indexing fields	Precision (single)	Precision (all)
(1) Titles	78.24%	60.78%
(2) Abstracts	77.57%	59.40%
(3) Claims	76.66%	58.65%
(4) Technical fields, Backgrounds	86.60%	69.00%
(5) Titles, Abstracts	79.59%	61.62%
(6) Titles, Abstracts, Claims	79.13%	61.18%
(7) Titles, Abstracts, Technical Fields, Backgrounds	87.20%	70.00%
(8) Titles, Abstracts, Claims, Technical Fields, Backgrounds	85.67%	68.31%

However, the classification performance shows the lowest rate of precision (76.66%) for (3) claims, 77.57% precision for (2) abstracts, and 78.24% precision for (1) titles.

For the second precision measure – all match precision – (7) titles, abstracts, technical fields, backgrounds is determined to be 70%, and (4) technical fields, backgrounds and (8) titles, abstracts, claims, technical fields, backgrounds are determined to be 69% and 68.31%, respectively.

Looking at the results closely, the patent 'Battery separator, and battery separator manufacturing method' has five IPC subclasses of H01 M (Processes or means, e.g. batteries, for the direct conversion of chemical energy into electrical energy), B32B (Layered products, i.e. products built-up of strata of flat or non-flat, e.g. cellular or honeycomb, form), C08 J (Working-up; General processes of compounding; After-treatment not covered by subclasses C08B, C08C, C08F, C08G or C08H), C08 K (Use of inorganic or non-macromolecular organic substances as compounding ingredients) and C08L (Compositions of macromolecular compounds). When this patent document is classified using (4) technical fields, backgrounds and (7) titles, abstracts, technical fields, backgrounds, the results are correctly predicted for the five IPC subclasses. When using (2) abstracts, (5) titles, abstracts and (8) titles, abstracts, claims, technical fields, backgrounds, four of the five IPC subclasses are correctly predicted, with the exception being C08 K. When using (1) titles, (3) claims and

(6) titles, abstracts, claims, three IPC subclasses are correctly predicted, with the exceptions being C08 K and C08L.

The patent 'Treatment apparatus and method of leachate using animal and vegeta-ble oil' is labeled with seven IPC subclasses. They are C02F (Treatment of water, waste water, sewage, or sludge), A47 J (Kitchen equipment; Coffee mills; Spice mills; Apparatus for making beverages), A61L(Methods or apparatus for sterilising materials or objects in general; Disinfection, sterilisation, or deodorisation of air; Chemical aspects of bandages, dressings, absorbent pads, or surgical articles; Materials for bandages, dressings, absorbent pads, or surgical articles), B01D (Separation), B04B (Centrifuges), B09B (Disposal of solid waste) and F26B (Drying solid materials or objects by removing liquid therefrom). In this case, five IPC subclasses are predicted correctly when using (7) titles, abstracts, technical fields, backgrounds, four IPC subclasses are predicted properly when using (2) abstracts, (3) claims, (4) technical fields, backgrounds, (5) titles, abstracts, (6) titles, abstracts, claims and (8) titles, abstracts, claims, technical fields, backgrounds, and three IPC subclasses are predicted properly when using only (1) titles.

It can be concluded that technical field and background play important roles in patent classification at the subclass level of the IPC, while claims decreases the precision of patent classification. Furthermore, the use of the claims field is regarded as inefficient in terms of processing time when carrying out classification. The time necessary for claims is quadruple that of abstract, and double that of technical field, background.

In addition, cases exist in which every subclass for a patent document is misclassified. In the patent 'Sensor', even though there are four IPC subclasses – G01L (Measuring force, stress, torque, work, mechanical power, mechanical efficiency, or fluid pressure), G01H (Measurement of mechanical vibrations or ultrasonic, sonic or infrasonic waves), B82Y (Specific uses or applications of nano-structures; Measurement or analysis of nano-structures; Manufacture or treatment of nano-structures) and G01B (Measuring length, thickness or similar linear dimensions; Measuring angles; Measuring areas; Measuring irregularities of surfaces or contours), the subclasses including G01R (Measuring electric variables; Measuring magnetic variables), G06F (Electric digital data processing), H04L (Transmission of digital information, e.g. telegraphic communication) and H04 W (Wireless communication networks) are predicted. A close examination of the results in this case reveal that when fields have a variety of applicable areas, titles such as 'Sensor' cause an inability to determine specified subclasses. Sometimes, patent categories are determined not by the title, abstract, technical field, background and claims, but by almost the entire patent document, including the detailed descriptions. Therefore, the selection of patent document fields needs to be understood in detail to improve classification performance and to distinguish further subdivided technological areas at the lowest IPC level.

5 Conclusions

In order to classify patent documents effectively based on the IPC system, this paper focused on the data structure of the patent documents themselves rather than on the classification models. It was suggested that the technical fields and background of the patent documents were significant in the classification task at the IPC subclass level. This suggestion was derived from an analysis of the characteristics of the patent documents and the IPC structure. An experiment was conducted to determine the effects of the structural fields on the classification performance for 564,793 Korean patent documents at the IPC subclass level. As a result, the use of the title, abstract, technical field, and background achieved the highest classification precision. In every case that included two significant fields, technical field and background showed better performance than their counterparts did. However, the use of claims caused a decrease in classification precision at the IPC subclass level. It was confirmed that the classification of Korean patent documents could be applied in the real world through multi-label classification at the IPC subclass level, which is divided into 630 categories.

However, further research is needed to construct a complete multi-label IPC auto-classification system that targets the lowest IPC level, which contains over 65,000 subgroups. To accomplish this, future research could employ a hierarchical algorithm that uses each patent document field selectively in each classification process IPC based on the characteristics of the patent documents. For example, title, abstract, technical field and background could be used first to fix the upper levels of the IPC, and claims could subsequently be used to decide the final IPC categories.

Acknowledgments. This research was supported by Gyeonggi Province's GRRC Program [(GRRC-B01), Development of Ambient Mobile Broadcasting Service System].

References

1. Choi, D.K.: Intellectual Property Statistics for 2014, Korean Intellectual Property Office (2015)
2. International Patent Classification Guide. http://www.wipo.int/export/sites/www/classifi cations/ipc/en/guide/guide_ipc.pdf
3. Kim, Y.M.: Guidelines for Examination, Korean Intellectual Property Office (2015)
4. Fall, C.J., Törcsvári, A., Benzineb, K., Karetka, G.: Automated categorization in the international patent classification. ACM SIGIR Forum **37**(1), 10–25 (2003). ACM
5. Larkey, L.S.: A patent search and classification system. In: The 4th ACM Conference on Digital Libraries, pp. 119–187. ACM (1999)
6. Tikk, D., Biró, G., Törcsvári, A.: A Hierarchical online classifier for patent categorization. In: Emerging Technologies of Text Mining: Techniques and Applications, pp. 244–267 (2007)
7. Chen, Y.-L., Chang, Y.-C.: A three-phase method for patent classification. Inf. Process. Manage. **48**(6), 1017–1030 (2012)
8. Seneviratne, D., Geva, S., Zuccon, G., Ferraro, G., Chappell, T., Meireles, M.: A signature approach to patent classification. In: Zuccon, G., Geva, S., Joho, H., Scholer, F., Sun, A., Zhang, P. (eds.) AIRS 2015. LNCS, vol. 9460, pp. 413–419. Springer, Heidelberg (2015). doi:10.1007/978-3-319-28940-3_35

9. Park, C., Kim, K., Seong, D.: Automatic IPC classification for patent documents of convergence technology using KNN. J. KIIT. **12**(3), 175–185 (2014)

10. Kim, J.-H., Choi, K.-S.: Patent document categorization based on semantic structural information. Inf. Process. Manage. **43**(5), 1200–1215 (2007)

11. KIPRIS (Korea Intellectual Property Rights Information Service) plus. http://plus.kipris.or.kr/

12. KLT2000, Korean Morphological Analyzer. http://nlp.kookmin.ac.kr/

13. International Patent Classification Official Publication. http://web2.wipo.int/classifications/ipc/ipcpub/#refresh=page

14. Kibriya, A.M., Frank, E., Pfahringer, B., Holmes, G.: Multinomial naive Bayes for text categorization revisited. In: Webb, G.I., Yu, X. (eds.) AI 2004. LNCS (LNAI), vol. 3339, pp. 488–499. Springer, Heidelberg (2004)

Unsupervised Component-Wise EM Learning for Finite Mixtures of Skew t-distributions

Sharon X. Lee$^{(\boxtimes)}$ and Geoffrey J. McLachlan

University of Queensland, St Lucia, Brisbane, QLD 4017, Australia
{s.lee11,g.mclachlan}@uq.edu.au

Abstract. In recent years, finite mixtures of skew distributions are gaining popularity as a flexible tool for modelling data with asymmetric distributional features. Parameter estimation for these mixture models via the traditional EM algorithm requires the number of components to be specified *a priori*. In this paper, we consider unsupervised learning of skew mixture models where the optimal number of components is estimated during the parameter estimation process. We adopt a component-wise EM algorithm and use the minimum message length (MML) criterion. For illustrative purposes, we focus on the case of a finite mixture of multivariate skew t distributions. The performance of the approach is demonstrated on a real dataset from flow cytometry, where our mixture model was used to provide an automated segmentation of cell populations.

Keywords: Mixture models · EM algorithm · Skew distributions · Minimum message length

1 Introduction

Finite mixture models provide a powerful tool for the modelling and analysis of heterogeneous data. In recent times, mixture models with skew component distributions have received increasing attention due to their increased flexibility over the traditional normal mixture model. These models adopt densities that can take more flexible distributional shapes such as asymmetry and heavy-tailedness, rendering them suitable for a wider range of applications. In particular, the skew normal and skew t-mixture models have found many interesting applications in a variety fields, including biology, finance, imaging, medicine, pharmacy, and social sciences [1,3,5,10,12,13,16–18,21,22,25–29,31].

The standard method for parameter estimation for finite mixture model is by maximum likelihood (ML) via the Expectation-Maximization (EM) algorithm. These have been developed for various characterizations of skew component distributions. Some notable examples include the models and EM algorithms considered in [6,9,19,20,33] and Lee and McLachlan [11,14,15].

An important issue with the fitting of mixture models is the choice of the number of mixture components g. The common approach to tackle this is to

© Springer International Publishing AG 2016
J. Li et al. (Eds.): ADMA 2016, LNAI 10086, pp. 692–699, 2016.
DOI: 10.1007/978-3-319-49586-6_49

perform model selection, where a set of candidate models for a range of g must first be obtained. One or more model selection criteria are then calculated for each candidate model and the model that performs best according to the criteria is chosen. Some commonly used criterion include the Bayesian information criterion (BIC), Akaike's information criterion (AIC), and Integrated Completed Likelihood (ICL).

In this paper, we consider a more recent approach that uses the minimum message length (MML) criterion [8,32]. We adopt a component-wise EM (CEM) algorithm, which updates the components incrementally [7]. We start with an overestimation of the number of components, then trim out a component when its support is below a minimum threshold. When a component is trimmed, its probability mass is redistributed to other components. This approach was considered for the case of the normal mixture model in Figueiredo and Jain [8]. Here, we adapt the CEM-MML approach for the case of finite mixtures of multivariate skew t (MST) distributions (FM-MST).

It is interesting to note that there are various characterizations of the MST distribution. For our purpose here, we shall adopt the characterization used in Lee and McLachlan [15] and proposed by Arellano-Valle and Genton [2], known as the canonical fundamental skew t (CFUST) distribution. This represents a fairly general characterization that encompasses some of the most commonly used characterizations such as that by Azzalini and Capitanio [4] and Sahu et al. [30]. Further details and discussions on the this topic can be found in Lee and McLachlan [15] and McLachlan and Lee [23,24].

2   Finite Mixtures of Skew t-distributions

We begin by giving the definition of the CFUST distribution, hereafter referred as the MST distribution. The density of the MST distribution can be expressed as a product of a multivariate t-density and the cdf of a t-distribution. Its density is given by

$$f\left(\boldsymbol{y};\boldsymbol{\mu},\boldsymbol{\Sigma},\boldsymbol{\Delta},\nu\right) = 2^q\, t_p\left(\boldsymbol{y};\boldsymbol{\mu},\boldsymbol{\Omega},\nu\right) T_q\left(\mathbf{c}(\boldsymbol{y})\sqrt{\frac{\nu+p}{\nu+d(\boldsymbol{y})}};\mathbf{0},\boldsymbol{\Lambda},\nu+p\right), \quad (1)$$

where $t_p\left(\boldsymbol{y};\boldsymbol{\mu},\boldsymbol{\Omega},\nu\right)$ denote the p-dimensional t-distribution with location parameter $\boldsymbol{\mu}$, scale matrix $\boldsymbol{\Omega}$, and degrees of freedom ν, and $T_p(.;\boldsymbol{\mu},\boldsymbol{\Omega},\nu)$ denote its corresponding cdf.

A finite mixture model is a convex combination of component densities. The density of g-component finite mixture model takes the form

$$f(\boldsymbol{y};\boldsymbol{\Psi}) = \sum_{h=1}^{g} \pi_h\, f_h(\boldsymbol{y};\boldsymbol{\theta}_h), \quad (2)$$

where π_h $(h = 1,\dots,g)$ are the mixing proportions and $f_h(\cdot)$ denotes the density of the hth mixture component of the mixture model. The mixing proportions are nonnegative and sum to one, that is, they satisfy $\pi_h \geq 0$ and $\sum_{h=1}^{g} \pi_h = 1$.

The vector $\boldsymbol{\Psi} = (\pi_1, \ldots, \pi_{g-1}, \boldsymbol{\theta}_1^T, \ldots, \boldsymbol{\theta}_g^T)$ contains all the unknown parameters of the mixture model, with $\boldsymbol{\theta}_h$ containing the unknown parameters of the hth component. In this paper, we shall adopt the MST distribution (1) as component densities of our finite mixture model. We refer to this as the finite mixture of MST (FM-MST) model.

3 Existing Method for Parameter Estimation

Before we present the CEM-MML approach for the FM-MST model, we briefly describe the traditional EM algorithm for this model. The EM algorithm starts with an initial estimate of the model parameters, then alternates between the E- and M-steps until a stopping criterion is met.

The E-step of the EM algorithm requires the calculation of the so-called Q-function, $Q(\boldsymbol{\Psi}; \boldsymbol{\Psi}^{(k)})$, which is the conditional expectation of the complete-data log likelihood given the observed data $\boldsymbol{y}$, using the current estimate of $\boldsymbol{\Psi}$, which is denoted by $\boldsymbol{\theta}^{(k)}$ after the kth iteration. It follows that on the $(k+1)$th iteration, the E-step requires the following five conditional expectations to be calculated,

$$z_{hj}^{(k)} = E_{\boldsymbol{\Psi}^{(k)}} \left[z_{hj} = 1 \mid \boldsymbol{y}_j \right], \tag{3}$$

$$w_{hj}^{(k)} = E_{\boldsymbol{\Psi}^{(k)}} \left[w_{hj} \mid \boldsymbol{y}_j, z_{hj} = 1 \right], \tag{4}$$

$$e_{1hj}^{(k)} = E_{\boldsymbol{\Psi}^{(k)}} \left[\log(w_{hj}) \mid \boldsymbol{y}_j, z_{hj} = 1 \right], \tag{5}$$

$$e_{2hj}^{(k)} = E_{\boldsymbol{\Psi}^{(k)}} \left[w_{hj} \boldsymbol{u}_{hj} \mid \boldsymbol{y}_j, z_{hj} = 1 \right], \tag{6}$$

$$e_{3hj}^{(k)} = E_{\boldsymbol{\Psi}^{(k)}} \left[w_{hj} \boldsymbol{u}_{hj} \boldsymbol{u}_{hj}^\top \mid \boldsymbol{y}_j, z_{hj} = 1 \right]. \tag{7}$$

On the $(k+1)$th iteration of the M-step, the current estimate of $\boldsymbol{\Psi}$, $\boldsymbol{\Psi}^{(k)}$, is updated to $\boldsymbol{\Psi}^{(k+1)}$, which is chosen to globally maximize $Q(\boldsymbol{\Psi}; \boldsymbol{\Psi}^{(k)})$ over $\boldsymbol{\Psi}$:

$$\pi_h^{(k+1)} = \frac{1}{n} \sum_{j=1}^{n} z_{hj}^{(k)}, \tag{8}$$

$$\boldsymbol{\mu}_h^{(k+1)} = \frac{\sum_{j=1}^{n} z_{hj} w_{hj}^{(k)} \boldsymbol{y}_j - \boldsymbol{\Delta}_h^{(k)} \sum_{j=1}^{n} z_{hj}^{(k)} e_{2hj}^{(k)}}{\sum_{j=1^n} z_{hj}^{(k)} w_{hj}^{(k)}}, \tag{9}$$

$$\boldsymbol{\Delta}_h^{(k+1)} = \left[\sum_{j=1}^{n} z_{hj}^{(k)} \left(\boldsymbol{y}_j - \boldsymbol{\mu}_h^{(k+1)} \right) e_{2hj}^{(k)^\top} \right] \left[\sum_{j=1}^{n} z_{hj}^{(k)} e_{3hj}^{(k)} \right]^{-1}, \tag{10}$$

$$\boldsymbol{\Sigma}_h^{(k+1)} = \left\{ \sum_{j=1}^{n} z_{hj}^{(k)} \left[w_{hj}^{(k)} \left(\boldsymbol{y}_j - \boldsymbol{\mu}_h^{(k+1)} \right) \left(\boldsymbol{y}_j - \boldsymbol{\mu}_h^{(k+1)} \right)^T \right.\right.$$
$$\left. - \boldsymbol{\Delta}_h^{(k+1)} e_{2hj}^{(k)} \left(\boldsymbol{y}_j - \boldsymbol{\mu}_h^{(k+1)} \right)^\top - \left(\boldsymbol{y}_j - \boldsymbol{\mu}_h^{(k+1)} \right) e_{2hj}^{(k)^\top} \boldsymbol{\Delta}_h^{(k+1)^\top} \right.$$
$$\left.\left. + \boldsymbol{\Delta}_h^{(k+1)} e_{3hj}^{(k)^\top} \boldsymbol{\Delta}_h^{(k+1)^\top} \right] \right\} \left[\sum_{j=1}^{n} z_{hj}^{(k)} \right]^{-1}. \tag{11}$$

An update of the degrees of freedom ν_h is obtained by solving the following equation for $\nu_h^{(k+1)}$,

$$0 = \left(\sum_{h=1}^{n} z_{hj}^{(k)} \right) \left[\log \left(\frac{\nu_h^{(k+1)}}{2} \right) - \psi \left(\frac{\nu_h^{(k+1)}}{2} \right) + 1 \right] - \sum_{j=1}^{n} z_{hj}^{(k)} \left(e_{1hj}^{(k)} - w_{hj}^{(k)} \right), \quad (12)$$

where $\psi(\cdot)$ denotes the digamma function.

Further technical details of this EM algorithm, including the calculation of (3) to (7), for the FM-MST model can be found in the recent work of Lee and McLachlan [15].

4   Minimum Message Length (MML) Approach for FM-MST

In ML estimation, one seeks to maximize the log likelihood function of the model. With the MML approach, the objective is to minimize a modified likelihood function that includes additional penalty terms. Let $\ell(\Psi)$ denote the usual log likelihood function and $\ell^{(k)}(\Psi)$ its value using the current estimates of the parameters at the kth iteration. It follows that the objective function under the MML framework is given by

$$\ell^{*(k)} = \frac{m}{2} \sum_{\alpha_h > 0} \log \left(\frac{n\alpha_h^{(k)}}{12} \right) + \frac{g^*}{2} \log \left(\frac{n}{12} \right) + \frac{g^*}{2}(m+1) - \ell^{(k)}(\Psi), \quad (13)$$

where n is the number of observations, m is the number of free parameters in the component distribution, and g^* is (current) number of components with non-zero mixing proportions. The quantity α_h is a modified version of the mixing proportion, given by

$$\alpha_h = \frac{\max \left(0, \left(\sum_{j=1}^{n} z_{hj}^{(k)} \right) - \frac{m}{2} \right)}{\sum_{h=1}^{g} \max \left(0, \left(\sum_{j=1}^{n} z_{hj}^{(k)} \right) - \frac{m}{2} \right)}. \quad (14)$$

In effect, we are truncating a component h if its support lower than $\frac{m}{2}$.

The CEM-MML algorithm proceeds as follows:

1. Choose an overestimate of g.
2. Perform initialization as described in Lee and McLachlan [15].
3. Start with component $h = 1$
4. Compute $z_{hj}^{(k)}$ using (3) for component h.
5. Compute α_h using (14) for component h.
6. If $\alpha_h > 0$, perform the standard E- and M-steps for component h. Then normalize the mixing proportions using

$$\pi_h^{(k+1)} = \frac{\pi_h^{*(k+1)}}{\sum_{h=1}^{g} \pi_h^{*(k+1)}}, \quad (15)$$

for $h = 1, \ldots, g$, where $\pi_h^{*(k+1)}$ refers to the value obtained using (8).

7. Repeat steps 4 to 6 for the remaining components.
8. Calculate the objective function value using (13) and check if its increase over the previous iteration is larger than a specified threshold ϵ. If yes, stop the algorithm; otherwise go to Step 4.

5   Cell Population Segmentation from a DLBCL Sample

As an illustration, we consider the application of automated segmentation of cell populations in a flow cytometric dataset. This dataset was collected by the British Columbia Cancer Agency, consisting of measurements of single cells in a blood sample obtained from a patient diagnosed with Diffuse Large B-Cell Lymphoma (DLBCL). The three variables correspond to three different protein markers used by experts to discriminate between two cell populations in the sample. There were over 5,000 cells in the sample, each belonging to one of the two distinct cell types. A 3D plot of the markers for this dataset is displayed in Fig. 1, where the colours of the cells correspond to the labels given by the expert analyst. We applied the CEM-MML algorithm to this dataset with the initial g

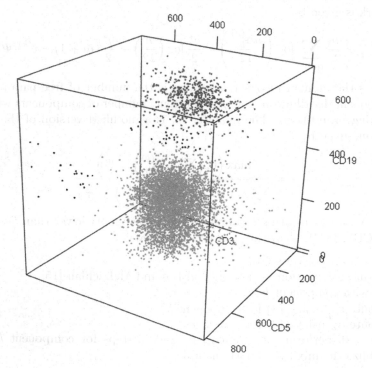

Fig. 1. Scatter plot of the DLBCL dataset with cells color-coded according to cell population label. (Color figure online)

set to 5. It correctly identified the true number of populations and provided a segmentation with high accuracy. It obtained an error rate of 0.003, suggesting the clustering given by the FM-MST model is in close agreement with the expert analyst.

6 Conclusions

We have presented a modified EM algorithm for the fitting of finite mixtures of multivariate skew t-distributions where the optimal number of components can be inferred from the data simultaneously during parameter estimation. We adopt a minimum message length (MML) approach that trims out mixture components with low support from the data. It proceeds similarly to the component-wise EM algorithm but with a modified expression for the updates of the mixing proportions. Our approach is illustrated on a flexible characterization of the skew t-mixture model. The approach is applied to cell population segmentation of a real flow cytometric data.

References

1. Abanto-Valle, C.A., Lachos, V.H., Dey, D.K.: Bayesian estimation of a skew-student-t stochastic volatility model. Methodol. Comput. Appl. Probab. **17**, 721–738 (2015)
2. Arellano-Valle, R.B., Genton, M.G.: On fundamental skew distributions. J. Multivar. Anal. **96**, 93–116 (2005)
3. Asparouhov, T., Muthén, B.: Structural equation models and mixture models with continuous non-normal skewed distributions. Structural Equation Modeling (2015)
4. Azzalini, A., Capitanio, A.: Distributions generated by perturbation of symmetry with emphasis on a multivariate skew t-distribution. J. Roy. Stat. Soc. B **65**, 367–389 (2003)
5. Bernardi, M.: Risk measures for skew normal mixtures. Stat. Probab. Lett. **83**, 1819–1824 (2013)
6. Cabral, C.R.B., Lachos, V.H., Prates, M.O.: Multivariate mixture modeling using skew-normal independent distributions. Comput. Stat. Data Anal. **56**, 126–142 (2012)
7. Celeux, G., Chrétien, S., Forbes, F., MkhadrA.: A component-wise EM algorithm for mixtures. Journal of Computational and Graphical Statistics **10**(4) (2001)
8. Figueiredo, M.A.T., Jain, A.K.: Unsupervised learning of finite mixture models. IEEE Trans. Pattern Anal. Mach. Intell. **24**, 3813 (2002)
9. Frühwirth-Schnatter, S., Pyne, S.: Bayesian inference for finite mixtures of univariate and multivariate skew-normal and skew-t distributions. Biostatistics **11**, 317–336 (2010)
10. Hu, X., Kim, H., Brennan, P.J., Han, B., Baecher-Allan, C.M., Jager, P.L., Brenner, M.B., Raychaudhuri, S.: Application of user-guided automated cytometric data analysis to large-scale immunoprofiling of invariant natural killer t cells. In: Proceedings of the National Academy of Sciences USA, vol. 110, pp. 19030–19035 (2013)

11. Lee, S., McLachlan, G.J.: Finite mixtures of multivariate skew t-distributions: Some recent and new results. Stat. Comput. **24**, 181–202 (2014)
12. Lee, S.X., McLachlan, G.J.: Model-based clustering and classification with non-normal mixture distributions. Stat. Methods Appl. **22**, 427–454 (2013)
13. Lee, S.X., McLachlan, G.J.: Modelling asset return using multivariate asym- metric mixture nodels with applications to wstimation of value-at-risk. In: MODSIM 2013, 20th International Congress on Modelling and Simulation, pp. 1228–1234, Adelaide, Australia (2013)
14. Lee, S.X., McLachlan, G.J.: On mixtures of skew-normal and skew t-distributions. Adv. Data Anal. Classif. **7**, 241–266 (2013)
15. Lee, S.X., McLachlan, G.J.: Finite mixtures of canonical fundamental skew t-distributions: the unification of the restricted and unrestricted skew t-mixture models. Stat. Comput. **26**, 573–589 (2016)
16. Lee, S.X., McLachlan, G.J.: Risk measures based on multivariate skew normal and skew t-mixture models. In: Alcock, J., Satchell, S. (eds.) Asymmetric Dependence in Finance. Wiley, Hoboken, New Jersey (2016, to appear)
17. Lee, S.X., McLachlan, G.J., Pyne, S.: Supervised classification of flow cytometric samples via the Joint Clustering and Matching (JCM) procedure. arXiv:1411.2820 [q-bio.QM] (2014)
18. Lee, S.X., McLachlan, G.J., Pyne, S.: Modelling of inter-sample variation in flow cytometric data with the Joint Clustering and Matching (JCM) procedure. Cytometry A (2016)
19. Lin, T.I.: Robust mixture modeling using multivariate skew-t distribution. Stat. Comput. **20**, 343–356 (2010)
20. Lin, T.I., Ho, H.J., Lee, C.R.: Flexible mixture modelling using the multivariate skew-t-normal distribution. Stat. Comput. **24**, 531–546 (2014)
21. Lin, T.I., McLachlan, G.J., Lee, S.X.: Extending mixtures of factor models using the restricted multivariate skew-normal distribution. J. Multivar. Anal. **143**, 398–413 (2016)
22. Lin, T.I., Wu, P.H., McLachlan, G.J., Lee, S.X.: A robust factor analysis model using the restricted skew t-distribution. TEST **24**, 510–531 (2015)
23. McLachlan, G.J., Lee, S.X.: Comment on "Comparing Two Formulations of Skew Distributions with Special Reference to Model-Based Clustering" by A. Azzalini, R. Browne, M. Genton, and P. McNicholas. arXiv:1404.1733 (2014)
24. McLachlan, G.J., Lee, S.X.: Comment on "On nomenclature for, and the relative merits of, two formulations of skew distributions" by A. Azzalini, R. Browne, M. Genton, and P. McNicholas. Statistics and Probaility Letters **116**, 1–5 (2016)
25. Muthén, B., Asparouhov, T.: Growth mixture modeling with non-normal distrib- utions. Stat. Med. **34**, 1041–1058 (2014)
26. Pyne, S., Hu, X., Wang, K., Rossin, E., Lin, T.I., Maier, L.M., Baecher-Allan, C., McLachlan, G.J., Tamayo, P., Hafler, D.A., Jager, P.L., Mesirow, J.P.: Automated high-dimensional flow cytometric data analysis. In: Proceedings of the National Academy of Sciences USA, vol. 106, pp. 8519–8524 (2009)
27. Pyne, S., Lee, S.X., Wang, K., Irish, J., Tamayo, P., Nazaire, M.D., Duong, T., Ng, S.K., Hafler, D., Levy, R., Nolan, G.P., Mesirov, J., McLachlan, G.: Joint modeling and registration of cell populations in cohorts of high-dimensional flow cytometric data. PLOS ONE **9**, e100334 (2014)
28. Pyne, S., Lee, S., McLachlan, G.: Nature and man: The goal of bio-security in the course of rapid and inevitable human development. J. Indian Soc. Agric. Stat. **69**, 117–125 (2015)

29. Riggi, S., Ingrassia, S.: A model-based clustering approach for mass composition analysis of high energy cosmic rays. Astropart. Phys. **48**, 86–96 (2013)
30. Sahu, S.K., Dey, D.K., Branco, M.D.: A new class of multivariate skew distributions with applications to Bayesian regression models. Can. J. Stat. **31**, 129–150 (2003)
31. Schaarschmidt, F., Hofmann, M., Jaki, T., Grün, B., Hothorn, L.A.: Statistical approaches for the determination of cut points in anti-drug antibody bioassays. J. Immunol. Methods **25**, 295–306 (2015)
32. Wallace, C.S., Boulton, D.M.: An information measure for classification. Comput. J. **11**, 185–189 (1968)
33. Wang, K., Ng, S.K., McLachlan, G.J.: Multivariate skew t mixture models: applications to fluorescence-activated cell sorting data. In: Proceedings of Conference of Digital Image Computing: Techniques and Applications, pp. 526–531, Los Alamitos, California (2009)

Supervised Feature Selection by Robust Sparse Reduced-Rank Regression

Rongyao Hu[1,2], Xiaofeng Zhu[1,2]($\boxtimes$), Wei He[1,2], Jilian Zhang[3],
and Shichao Zhang[1,2]($\boxtimes$)

[1] Guangxi Key Lab of Multi-source Information Mining and Security,
Guangxi Normal University, Guilin 541004, Guangxi, People's Republic of China
seanzhuxf@gmail.com
[2] College of CS and IT, Guangxi Normal University,
Guilin 541004, Guangxi, People's Republic of China
[3] Guangxi University of Finance and Economics,
Nanning 530003, Guangxi, People's Republic of China

Abstract. Feature selection keeping discriminative features (*i.e.,* removing noisy and irrelevant features) from high-dimensional data has been becoming a vital important technique in machine learning since noisy/irrelevant features could deteriorate the performance of classification and regression. Moreover, feature selection has also been applied in all kinds of real applications due to its interpretable ability. Motivated by the successful use of sparse learning in machine learning and reduced-rank regression in statics, we put forward a novel feature selection pattern with supervised learning by using a reduced-rank regression model and a sparsity inducing regularizer during this article. Distinguished from those state-of-the-art attribute selection methods, the present method have described below: (1) built upon an $\ell_{2,p}$-norm loss function and an $\ell_{2,p}$-norm regularizer by simultaneously considering subspace learning and attribute selection structure into a unite framework; (2) select the more discriminating features in flexible, furthermore, in respect that it may be capable of dominate the degree of sparseness and robust to outlier samples; and (3) also interpretable and stable because it embeds subspace learning (*i.e.,* enabling to output stable models) into the feature selection framework (*i.e.,* enabling to output interpretable results). The relevant results of experiment on eight multi-output data sets indicated the effectiveness of our model compared to the state-of-the-art methods act on regression tasks.

Keywords: Subspace learning · Reduced-rank regression · sparse learning · Supervised feature selection

1   Introduction

Feature selection methods directly selecting a subset of features lead to interpretable results and thus enabling to explain the relationship of the corresponding internal data [20,21,27]. A number of feature selection methods thus have

© Springer International Publishing AG 2016
J. Li et al. (Eds.): ADMA 2016, LNAI 10086, pp. 700–713, 2016.
DOI: 10.1007/978-3-319-49586-6_50

been proposed in pattern recognition and machine learning research community. For example, Fisher Score (FS) method conducts feature selection by evaluating the relevance of features based on the ground truth distribution to the samples. Currently, the more and more models based on sparsity representation have been successfully applied for feature selection [18,19]. For example, Nie *et al.* proposed to leverage a joint $\ell_{2,1}$-norm minimization on the two aspects, *i.e.*, loss function and regularizer [9], while Zhu *et al.* proposed to use a least square loss function plus a graph Laplacian regularizer and an $\ell_{2,1}$-norm regularizer [25]. The motivation of the sparse feature selection methods is that a lot of real life data are usually get sparse attribute, so that searching for the data with sparse representation structure and corresponding results in conducting feature selection.

In real applications, the available sparse feature selection methods already have been proved to be effective but they still existing limitations [17,22]. For example, the widely used both ℓ_1-norm and $\ell_{2,1}$-norm are too rigid to control the corresponding sparseness of relevant feature representation matrix and thus not enabling to output stable results, *e.g.*, inconsequential noisy/irrelevant features may be selected and significant informative features may be neglected. On the contrary, subspace learning methods [15,16] have been established to map all high-dimensional features into a low-dimensional space so that removing noisy/irrelevant features in data. In this way, the high-dimensional problem can be solved to achieve stable performance. However, subspace learning methods transferring the original high-dimensional space to a new low-dimensional space prohibit for interpretation. Therefore, it will be interesting to integrate feature selection with subspace learning in a supervised feature selection framework for achieving interpretable ability and stable performance [3,26].

Motivated by the successful application of sparse learning and reduced-rank regression [1,7], during this article, we put forward a new supervised feature selection model by conducting a robust and flexible sparse reduced-rank regression, which embeds subspace learning structure into the constructed attribute selection combine with sparsity framework. Specifically, the presented reduced-rank sparse regression pushes the reduced rank constraint application for the regression coefficient matrix to conduct subspace learning by taking into account the relations among label information, and also pushes a sparsity inducing regularizer to conduct feature selection by considering the relations among the features [10,13]. Furthermore, we apply the $\ell_{2,p}$-norm to the two aspects that include the loss function and the regularizer to flexible control the influence of outlier samples and the sparseness of the feature selection (via tuning the vale of p), and thus resulting a data-driven feature selection model. By considering the relations among both the samples and the features, the proposed method also enables to results in a stable and small capacity subset from the initial feature set [28].

Compare to those state-of-the-art available feature selection methods, the contributions of our presented method with supervised learning are two-fold:

– The presented feature selection method outputs stable and interpretable results due to taking subspace learning method combine with the feature selection in a unite framework, rather than the previous methods [3,9,25] only output either stable performance (via subspace learning) or interpretable results (via feature selection). Moreover, this paper employs reduced-rank regression to conduct subspace learning, which has been widely used in statistics but seldom for machine learning and data mining.

– The proposed method utilizes the $\ell_{2,p}$-norm on the two aspects which include the loss function term and the regularizer term, and aim to output stable performance by avoiding the affect of noisy/redundant features and outlier samples. Specifically, by adjusting the value of p, the $\ell_{2,p}$-norm loss function can relief the influence of outliers while the $\ell_{2,p}$-norm regularizer is capable of selecting discriminative features.

The rest parts of this article are revealed and organized as follows: Sect. 2 gives the details of our proposed feature selection model. Sections 3 and 4, respectively, analyzes experimental results and summarizes our paper.

2   Approach

In this part, we first give some notations definition and used in the whole paper and then to describe the detail of the proposed method, in Sects. 2.1 and 2.2, respectively. We further report the proposed optimization method to the resulting objection function in Sect. 2.3.

2.1   Notations

During the whole article, we take boldface uppercase letters to express the corresponding all matrices, and then let boldface lowercase letters to represent the corresponding all vectors.

For a matrix $\mathbf{X} = [x_{ij}]$, its i-th row and j-th column are denoted as $\mathbf{x}_i$ and $\mathbf{x}_j$, respectively. Also we denote the Frobenius norm, $\ell_{2,p}$-norm of a matrix $\mathbf{X}$ respectively as $||\mathbf{X}||_F = \sqrt{\sum_i ||\mathbf{x}_i||_2^2} = \sqrt{\sum_j ||\mathbf{x}_j||_2^2}$ and $||\mathbf{X}||_{2,p} = (\sum_i \sqrt{\sum_j x_{ij}^2}^{\frac{p}{2}})^{\frac{1}{p}}$. We further denote the transpose operator and the inverse of a matrix $\mathbf{X}$ as $\mathbf{X}^T$ and $\mathbf{X}^{-1}$, respectively.

2.2   Objective Function

We denote $\mathbf{X} \in \mathbb{R}^{m \times n}$ and $\mathbf{Y} \in \mathbb{R}^{n \times k}$, respectively, as the feature matrix and the label matrix, of training samples, where m, n, and k, respectively, are the number of features, training samples, and classes of features, and $\mathbf{x}_i \in \mathbb{R}^n$ stands for the i-th feature vector. In general, we use the following formulation to construct a linear relationship between the feature matrix and the label matrix:

$$\min_{\mathbf{Z}} g(\mathbf{Z}) = f(\mathbf{Z}) + \lambda \phi(\mathbf{Z}) \tag{1}$$

where $\mathbf{Z} \in \mathbb{R}^{m \times k}$ denotes the reconstruction coefficient matrix, $f(\mathbf{Z})$ denotes the loss term imposed on $\mathbf{Z}$, $\phi(\mathbf{Z})$ denotes the regularization term, and λ denotes a positive constant. According to the previous literature [11], $f(\mathbf{Z})$ is defined as $f(\mathbf{Y} - \mathbf{X}^T\mathbf{Z})$ that its purpose is obtaining the relevant minimum regression error which is revealed between the label matrix $\mathbf{Y}$ and their prediction $\mathbf{X}^T\mathbf{Z}$, thus we could define the least square loss function between the feature matrix and the label matrix as follows:

$$\min_{\mathbf{Z}} ||\mathbf{Y} - \mathbf{X}^T\mathbf{Z}||_F^2 + \lambda||\mathbf{Z}||_F^2 \qquad (2)$$

The objective function in Eq. (2) is convex and smooth, so the optimal coefficient matrix $\mathbf{Z}$ is obtained as $\mathbf{Z}^* = (\mathbf{X}\mathbf{X}^T + \lambda\mathbf{I})^{-1}\mathbf{X}\mathbf{Y}$, where $\mathbf{I} \in \mathbb{R}^{n \times n}$ is an identity matrix. However, the optimal $\mathbf{Z}$ in Eq. (2) does not make use of relations among the label matrix. Actually, multiple responses (*i.e.*, labels) are likely correlated. To do this, one way of taking advantages of possible correlation between label information may impose a constraint on the rank of $\mathbf{Z}$, such as:

$$rank(\mathbf{Z}) = r \leq \min(m, k) \qquad (3)$$

A direct intuition on Eq. (3) is that there is a number of linear constraints on regression coefficient $\mathbf{Z}$, and hence the estimation efficiency is improved and the effective number of parameters is reduced.

In real applications, there is an underlying correlations between samples/features. The reduced rank constraint on Eq. (3) has been shown to be a reasonable way to explore such hidden structures in high-dimensional data to improve the learning model [1]. In this way, a reduced rank constrain structure introduced on $\mathbf{Z}$ can naturally be expressed as a product of two r-rank matrices as follows:

$$\mathbf{Z} = \mathbf{AB} \qquad (4)$$

where $\mathbf{A} \in \mathbb{R}^{m \times r}$ and $\mathbf{B} \in \mathbb{R}^{r \times k}$. Then, for the fixed r, a reduced rank regression can naturally be formulated as the following problem about the optimization:

$$\min_{\mathbf{A},\mathbf{B}} ||\mathbf{Y} - \mathbf{X}^T\mathbf{AB}||_F^2 \qquad (5)$$

where the matrix $\mathbf{A}$ and $\mathbf{B}$ have the reduced rank constraint simultaneously.

In the applications for data of high-dimensional structure, there are a large number of features or samples, some of them may not be related to the learning tasks. The noisy/irrelevant features/samples may affect the extraction of both r latent factors of $\mathbf{X}$ (*i.e.*, conducting subspace learning) and the selection of the discriminative features (*i.e.*, conducting feature selection). This motivates us to conduct feature selection and control the influence of outlier samples via structured sparsity constraints on both $\mathbf{AB}$ and $(\mathbf{Y} - \mathbf{X}^T\mathbf{AB})$ while conducting subspace learning via the reduced rank constraint. The rationale of using a structured sparsity constraint is that it effectively selects highly predictive features/samples (*i.e.*, discarding the unimportant features/samples from the model) by considering the correlations among the features/samples as well as effectively relieves the influences of outlier samples by assigning small magnitude on their loss term. Thus, in this paper, we impose the $\ell_{2,p}$-norm (where $0 < p < 2$)on both loss function term and regularization term, and then change Eq. (2) or (5) to our final objective function as follows:

$$\min_{\mathbf{A},\mathbf{B}} ||\mathbf{Y} - \mathbf{X}^T\mathbf{AB}||_{2,p} + \lambda||\mathbf{AB}||_{2,p} \qquad (6)$$

where $\mathbf{X} \in \mathbb{R}^{m \times n}$ and $\mathbf{Y} \in \mathbb{R}^{n \times k}$, respectively, denote the feature matrix and the response matrix, both $\mathbf{A} \in \mathbb{R}^{m \times r}$ and $\mathbf{B} \in \mathbb{R}^{r \times k}$ are regression coefficient matrices, and λ and p $(0 < p < 2)$ are tuning parameters.

Different from the previous sparse feature selection models [3, 9, 25] using an ℓ_1-norm or an $\ell_{2,1}$-norm to yield sparse results, Eq. (6) applies the $\ell_{2,p}$-norm to the two aspects that include the loss function and the regularizer. A direct benefit of using $\ell_{2,p}$-norm, and compared to utilizing ℓ_1-norm or $\ell_{2,1}$-norm, is that one may better adjust the value of p to control the degree of the sparseness of regression coefficient (for the regularizer term) and the influences of outliers (for the loss function term). More specifically, the $\ell_{2,p}$-norm regularizer on $\mathbf{AB}$ penalizes $\mathbf{AB}$ in a way that encourages joint row sparsity, i.e., each row of $\mathbf{AB}$ is either all zeros or all non-zeros, to select or un-select the corresponding features in $\mathbf{X}$. Moreover, if the value of p is lower, the more rows of $\mathbf{AB}$, i.e., the corresponding coefficient would further shrink to zeros, in other words, the more sparse $\mathbf{AB}$ is. On the other hand, instead of using conventional Frobenious norm on the loss function (e.g., Eq. (2)), we use the $\ell_{2,p}$-norm loss function to make our formulation robust against outliers. Specifically, each row of $(\mathbf{Y} - \mathbf{X}^T\mathbf{AB})$ in Eq. (6) is corresponding to the prediction residuals of one sample. Under $\ell_{2,p}$-norm operation, the residual values of each row (sample) are combined via ℓ_p-norm, and thus is less affected by the outliers, if compared with Frobenius norm, where the residual values of each row are combined via the square of ℓ_2-norm. Moreover, the lower p is, the less magnitude the $(\mathbf{Y} - \mathbf{X}^T\mathbf{AB})$ is, or in other words, the less influence the outlier sample is, compared to the ℓ_1-norm loss function or the $\ell_{2,1}$-norm loss function.

It is noteworthy the varied p value makes the resulting objective function in Eq. (6) data-driven. That is, Eq. (6) selects different p values for different distribution of the data. In this way, Eq. (6) uses a reduced-rank constraint and a sparsity inducing regularizer to simultaneously conduct subspace learning (i.e., Linear Discriminative Analysis (LDA) [8], please see the detail in Eq. (10)) and feature selection, as well as uses an $\ell_{2,p}$-norm on the two aspects which include the loss function term and the regularizer term, and aim to achieve data-driven learning model. This enables the proposed feature selection method to output interpretable and stable results.

2.3   Optimization

In Eq. (6), both the loss function and the regularizer are convex but the $\ell_{2,1}$-norm is non-smooth, so Eq. (6) is thus convex and non-smooth so that taking a global solution. In this section, we employ the framework of Iteratively Reweighted Least Square (IRLS) [5] to solve Eq. 6) by listing its pseudo in Algorithm 1.

Based on the framework of the IRLS methods, we define two diagonal matrices $\mathbf{R}$ and $\mathbf{D}$ and their corresponding diagonal elements are defined as follows:

$$r_{ii} = \frac{1}{(2/p)||(\mathbf{Y} - \mathbf{X}^T \mathbf{AB})^i||_2^{2-p}} \quad s.t. \quad i = 1, 2, \ldots, n, \quad 0 < p < 2 \quad (7)$$

$$d_{ii} = \frac{1}{(2/p)||(\mathbf{AB})^i||_2^{2-p}} \quad s.t. \quad i = 1, 2, \ldots, m, \quad 0 < p < 2 \quad (8)$$

where $(\mathbf{Y} - \mathbf{X}^T \mathbf{AB})^i$ and $(\mathbf{AB})^i$, respectively, denote the i-th row of the matrix $(\mathbf{Y} - \mathbf{X}^T \mathbf{AB})$ and $\mathbf{AB}$. Therefore, the objective function in Eq. (6) is equivalent to:

$$\min_{\mathbf{A},\mathbf{B}} Tr((\mathbf{Y} - \mathbf{X}^T \mathbf{AB})^T \mathbf{R}(\mathbf{Y} - \mathbf{X}^T \mathbf{AB})) + \lambda Tr(\mathbf{B}^T \mathbf{A}^T \mathbf{DAB}) \quad (9)$$

Then by setting the derivative of Eq. (9) with respect to $\mathbf{B}$ to zero, we have:

$$\mathbf{B} = \mathbf{A}^T(\mathbf{XRX}^T + \lambda \mathbf{D})\mathbf{A}^{-1}\mathbf{A}^T \mathbf{XRY} \quad (10)$$

Obviously, the optimization of $\mathbf{B}$ actually conducts LDA. However, in Eq. (10), we do not know the matrices $\mathbf{R}$, $\mathbf{D}$, and $\mathbf{A}$ in advance so that cannot optimize $\mathbf{B}$. In this paper, according to the IRLS framework, we iteratively optimize each of them by fixing the others: (i) update $\mathbf{R}$ and $\mathbf{D}$ by fixing $\mathbf{B}$ and $\mathbf{A}$, $i.e.$, Eq. (7) and (8); (ii) update $\mathbf{B}$ by fixing $\mathbf{R}$, $\mathbf{D}$ and $\mathbf{A}$, $i.e.$, Eq. (10); and iii) update $\mathbf{A}$ by fixing $\mathbf{R}$, $\mathbf{D}$ and $\mathbf{B}$. We iteratively conduct these steps until one of the predefined stopping criteria is satisfied.

Next, we optimize $\mathbf{A}$ via bring the expression of $\mathbf{B}$ into the Eq. (9), and then we can obtain as follows.

$$\max_{\mathbf{A}} Tr((\mathbf{A}^T(\mathbf{XRX}^T + \lambda \mathbf{D})\mathbf{A})^{-1}\mathbf{A}^T \mathbf{XRYY}^T \mathbf{R}^T \mathbf{X}^T \mathbf{A}) \quad (11)$$

In the process, we need to pay attention to that.

$$\mathbf{S}_t = \mathbf{XRX}^T + \lambda \mathbf{D}, \quad \mathbf{S}_b = \mathbf{XRYY}^T \mathbf{R}^T \mathbf{X}^T \quad (12)$$

where the between-class and total-scatter matrices of data in the LDA subspace are indicated as $\mathbf{S}_b$ and $\mathbf{S}_t$, respectively. In this way, the ultimately solution of Eq. (9) that we can be re-written as:

$$\mathbf{A} = \arg\max_{\mathbf{A}}\{Tr((\mathbf{A}^T \mathbf{S}_t \mathbf{A})^{-1}\mathbf{A}^T \mathbf{S}_b \mathbf{A})\} \quad (13)$$

The above mentioned is generally called the problem of LDA, meanwhile, in order to achieve global optimization solution to Eq. (13) is that, through the top r eigenvectors of $\mathbf{S}_t^{-1}\mathbf{S}_b$ with respect to the nonzero eigenvalues.

We iteratively optimize the parameters, such as $\mathbf{A}$, $\mathbf{B}$, $\mathbf{R}$ and $\mathbf{D}$ until the distance between two sequential objective function values is less than 10^{-5}.

From Algorithm 1, we get the optimal solution $\mathbf{A}$, $\mathbf{B}$, $\mathbf{R}$, and $\mathbf{D}$. In the following, we will prove that the proposed algorithm converges.

Proposition 1. Algorithm (1) monotonically decreases the objective function in Eq. (6) until convergence.

Proof. Assuming that, in the t-th iteration, we have

$$< \mathbf{A}(t+1), \mathbf{B}(t+1) > =$$
$$\underset{\mathbf{A},\mathbf{B}}{\arg\min} Tr[(\mathbf{Y} - \mathbf{X}^T \mathbf{AB})^T \mathbf{R}(t)(\mathbf{Y} - \mathbf{X}^T \mathbf{AB})] + \lambda Tr(\mathbf{B}^T \mathbf{A}^T \mathbf{D}(t)\mathbf{AB}) \quad (14)$$

In other words, in the $(t+1)$-th iteration step, we have the following inequality:

$$Tr[(\mathbf{Y} - \mathbf{X}^T \mathbf{A}(t+1)\mathbf{B}(t+1))^T \mathbf{R}(t)(\mathbf{Y} - \mathbf{X}^T \mathbf{A}(t+1)\mathbf{B}(t+1))]$$
$$+ \lambda\, Tr(\mathbf{B}(t+1)^T \mathbf{A}(t+1)^T \mathbf{D}(t)\mathbf{A}(t+1)\mathbf{B}(t+1))$$
$$\leq Tr[(\mathbf{Y} - \mathbf{X}^T \mathbf{A}(t)\mathbf{B}(t))^T \mathbf{R}(t)(\mathbf{Y} - \mathbf{X}^T \mathbf{A}(t)\mathbf{B}(t))]$$
$$+ \lambda\, Tr(\mathbf{B}(t)^T \mathbf{A}(t)^T \mathbf{D}(t)\mathbf{A}(t)\mathbf{B}(t)) \quad (15)$$

Denote $\mathbf{Z}(t) = \mathbf{A}(t)\mathbf{B}(t)$ and $\mathbf{R}(t) = \mathbf{Y} - \mathbf{X}^T \mathbf{A}(t)\mathbf{B}(t)$, respectively, we have $\mathbf{Z}^{(t+1)} = \mathbf{A}(t+1)\mathbf{B}(t+1)$ and $\mathbf{R}(t+1) = \mathbf{Y} - \mathbf{X}^T \mathbf{A}(t+1)\mathbf{B}(t+1)$. We further plug the matrix $\mathbf{D}$ defined in Eq. (8) into above inequality to yield the following:

$$\sum_{i=1}^{k} \frac{\|\mathbf{r}^i(t+1)\|_2^{2(2-p)}}{(2/p)\|\mathbf{r}^i(t+1)\|_2^{2-p}} + \lambda \sum_{i=1}^{m} \frac{\|\mathbf{z}^i(t+1)\|_2^{2(2-p)}}{(2/p)\|\mathbf{z}^i(t+1)\|_2^{2-p}}$$
$$\leq \sum_{i=1}^{k} \frac{\|\mathbf{r}^i(t)\|_2^{2(2-p)}}{(2/p)\|\mathbf{r}^i(t)\|_2^{2-p}} + \lambda \sum_{i=1}^{m} \frac{\|\mathbf{z}^i(t)\|_2^{2(2-p)}}{(2/p)\|\mathbf{z}^i(t)\|_2^{2-p}} \quad (16)$$

Algorithm 1. The Pseudo code of solving Eq. (6).

Input: $\mathbf{X} \in \mathbb{R}^{m \times n}$, $\mathbf{Y} \in \mathbb{R}^{n \times k}$, p, λ;
Output: $\mathbf{A} \in \mathbb{R}^{m \times r}$, $\mathbf{B} \in \mathbb{R}^{r \times k}$;
.   Initialize $t = 0$;
.   Initialize $\mathbf{A}(0)$ and $\mathbf{B}(0)$ as two random matrices;
.   Initialize $\mathbf{D}^{(t)} = \mathbf{I} \in \mathbb{R}^{m \times m}$
.   Initialize $\mathbf{R}^{(t)} = \mathbf{I} \in \mathbb{R}^{n \times n}$
.   **Repeat**
.   Update $\mathbf{A}(t+1)$ via Eq. (13);
.   Update $\mathbf{B}(t+1)$ via Eq. (10);
.   Update the diagonal matrix $\mathbf{R}(t+1)$ via Eq. (7);
.   Update the diagonal matrix $\mathbf{D}(t+1)$ via Eq. (8);
.   **Until** The difference between the objective function value of Eq. (6) in two sequential iterations less than 10^{-5}.

---

where $\mathbf{z}^i(t)$ and $\mathbf{z}^i(t+1)$ are the i-th row of the matrix $\mathbf{Z}(t)$ and $\mathbf{Z}(t+1)$ respectively. For each i, we get

$$\|\mathbf{z}^i(t+1)\|_2^{2-p} - \frac{\|\mathbf{z}^i(t+1)\|_2^{2(2-p)}}{(2/p)\|\mathbf{z}^i(t)\|_2^{2-p}} \leq \|\mathbf{z}^i(t)\|_2^{2-p} - \frac{\|\mathbf{z}^i(t)\|_2^{2(2-p)}}{(2/p)\|\mathbf{z}^i(t)\|_2^{2-p}} \qquad (17)$$

Thus, summing up m inequalities and multiplying the summation with the regularization parameter λ, we obtain:

$$\lambda \sum_{i=1}^{m}(\|\mathbf{z}^i(t+1)\|_2^{2-p} - \frac{\|\mathbf{z}^i(t+1)\|_2^{2(2-p)}}{(2/p)\|\mathbf{z}^i(t)\|_2^{2-p}})$$

$$\leq \lambda \sum_{i=1}^{m}(\|\mathbf{z}^i(t)\|_2^{2-p} - \frac{\|\mathbf{z}^i(t)\|_2^{2(2-p)}}{(2/p)\|\mathbf{z}^i(t)\|_2^{2-p}}) \qquad (18)$$

Based on the same principle from Eqs. (16) to (18), we have:

$$\sum_{i=1}^{n}(\|\mathbf{r}^i(t+1)\|_2^{2-p} - \frac{\|\mathbf{r}^i(t+1)\|_2^{2(2-p)}}{(2/p)\|\mathbf{r}^i(t)\|_2^{2-p}})$$

$$\leq \sum_{i=1}^{n}(\|\mathbf{r}^i(t)\|_2^{2-p} - \frac{\|\mathbf{r}^i(t)\|_2^{2(2-p)}}{(2/p)\|\mathbf{r}^i(t)\|_2^{2-p}}) \qquad (19)$$

According to Eqs. (18) and (19), we can yield:

$$\|\mathbf{Y} - \mathbf{X}^T\mathbf{A}(t+1)\mathbf{B}(t+1)\|_{2,p} + \lambda\|\mathbf{A}(t+1)\mathbf{B}(t+1)\|_{2,p}$$

$$\leq \|\mathbf{Y} - \mathbf{X}^T\mathbf{A}(t)\mathbf{B}(t)\|_{2,p} + \lambda\|\mathbf{A}(t)\mathbf{B}(t)\|_{2,p} \qquad (20)$$

Equation (20) indicates that the proposed optimization algorithm (*i.e.*, Algorithm 1) will monotonically decrease the objective function in each iteration, so the proposed optimization algorithm could converge to its global optimal solution.

3 Experiments

In this section, we used twenty data sets for classification and regression to compare the proposed method (namely STR_SFS) and the comparison methods. Specifically, Sect. 3.1 listed the details of the used data sets and the settings of experiments. In the Sect. 3.2 that we reported the performance of regression tasks.

3.1 Experiments Setup

We used eight publicly available data sets for regression tasks. We downloaded the data sets (such as SF1 and SF2) from UCI Machine Learning Repository[1], the EDM data set from the literature [6], and the WQ data set from [2], and the data sets (*e.g.*, ATP1d, ATP7d, OES97 and OES10) from [12], for regression tasks. We listed more details of all data sets in Table 1.

---

[1] http://archive.ics.uci.edu/ml/.

Table 1. The details of datasets for regression

Data sets	Instances	Features	Targets	Domain
EDM	154	16	2	Manufacture
SF1	323	10	3	Environment
SF2	1066	10	3	Environment
WQ	1060	16	14	Environment
ATP1d	337	411	6	Price prediction
ATP7d	296	411	6	Price prediction
OES97	334	263	16	Artificial
OES10	403	298	16	Artificial

We listed the details of the comparison methods as follows:

- Non Feature Selection (NFS) uses original variable to conduct classification by the SVM classifier directly. Here we use NFS to indicate if dimensionality reduction methods make sense in real applications.
- Robust feature selection (RFS) [9] is a classical supervised feature selection that it integrates the supervised learning with sparse learning into a unite framework, then to choose the better dependent features with sparsity.
- Linear Discriminant Analysis (LDA) [8], as one of subspace learning methods, preserves the neighborhood relation of each class sample to conduct subspace learning.
- Logistic Regression (LGR) [1] is a biased estimator regression model for linear data analysis.
- Robust Self-Representation (RSR) [23] chooses a representative response matrix through the self-representation method. Then embedded into the sparse learning model for feature selection.
- Least Absolute Shrinkage and Selection Operator (Lasso) [14] uses the absolute value function of model coefficient as a penalty term, and compress the regression coefficient that it can shrink some absolute values of the coefficients to zero.
- Elastic Net (NE) [4] selects representative features by integrating the lasso method and ridge method into a unite model.

In the following experimental setting in [24], we compared all methods with a 10-fold cross-validation. In detail, we first divided the entire of the original data set into ten parts by random. Moreover, we toke nine parts for training data and choose the rest of part for testing. Then we took the whole process repeat with 10 times, so that one can avoid the reveal bias in possible during the initial data set. At last, the ultimate results usually were computed by the average of ten results plus the standard deviation results for all experiments.

In all experiments, we tune the regularization parameters (*e.g.,* λ among the values $\lambda \in \{10^{-5}, \ldots 10^5\}$ and p among the values $p \in \{0.5, 1, 1.5\}$) to select the best parameters automatically, and we set $[c, g] \in \{2^{-5}, \dots 2^5\}$ in the SVM (*i.e.*, via the LIBSVM toolbox[2]) by a 5-fold inner cross-validation to distinguish different types of samples. Moreover, to select the best classification of a group and to return its coefficient as the coefficient final result with the 10-fold cross-validation of the model. For fair comparison, we used both 5-fold inner cross-validation and 10-fold outer cross-validation on all comparison methods, which parameters' setting was followed their corresponding literatures. We also repeated each of the comparison methods ten times to report the averaged performance.

Among the regression task, we used average Correlation Coefficient (aCC) and average Root Mean Squared Error (aRMSE) as evaluation metric to evaluate the regression performance of all multi-target regression data sets and comparison methods.

We defined average Correlation Coefficient (aCC) as follows:

$$aCC = \frac{1}{d} \sum_{i=1}^{d} CC = \frac{1}{d} \sum_{i=1}^{d} \frac{\sum_{l}^{N_{test}} (y_i^{(l)} - \bar{y}_i)(\hat{y}_i^{(l)} - \bar{\hat{y}}_i)}{\sqrt{\sum_{l=1}^{N_{test}} (y_i^{(l)} - \bar{y}_i)^2 \sum_{l=1}^{N_{test}} (\hat{y}_i^{(l)} - \bar{\hat{y}}_i)^2}} \tag{21}$$

Then, we defined average root mean squared error (aRMSE) as follows:

$$aRMSE = \frac{1}{d} \sum_{i=1}^{d} RMSE = \frac{1}{d} \sum_{i=1}^{d} \sqrt{\frac{\sum_{l}^{N_{test}} (y_i^{(l)} - \hat{y}_i^{(l)})^2}{N_{test}}} \tag{22}$$

where N_{test} means the size of test data set, then $\hat{y}^{(l)}$ and $y^{(l)}$ be the vectors of the predicted and actual targets for $x^{(l)}$, respectively. Besides, $\bar{\hat{y}}$ and $\bar{y}$ be the vectors of averages of the predicted and actual targets, respectively. A larger aCC shows better correlation coefficient results, while a smaller aRMSE means better robust.

3.2   Regression Results and Analysis

We reported the regression performance of average correlation coefficient and average root mean squared error in Tables 2 and 3, respectively. And the result of computational cost showed in Table 4.

From Table 2, we found that the proposed STR_SFS method achieved the minimal correlation coefficient, in the most of regression tasks. For example, our method increased on average by 4.3 % and 1.52 %, respectively, than the worst method NFS and RSR which achieved the best performance among the comparison methods. Our proposed method achieved the second best performance among WQ and OES10 data set, and also had similar performance to LR. Furthermore, our proposed method achieved the minimum standard deviation in all of regression data sets, compared to all the comparison methods. This indicated

---

[2] http://www.csie.ntu.edu.tw/~cjlin/libsvm/.

Table 2. average correlation coefficient (aCC) for all of multi-output data sets. The best performance are emphasized by boldface in each row.

	NFS	Lasso	NE	LR	LGR	RFS	RSR	STR_SFS
EDM	0.6035	0.8355	**0.8697**	0.7446	0.6800	0.6797	0.7785	0.8040
SF1	0.5273	0.5562	0.5740	0.5316	0.5259	0.5540	0.5540	**0.5840**
SF2	0.5441	0.5211	0.5430	0.5412	0.5285	0.5447	0.5430	**0.5497**
WQ	0.3603	0.3849	**0.4210**	0.3746	0.3482	0.3595	0.3603	0.3663
ATP1d	0.9251	0.8840	0.9120	0.9272	0.9190	0.9347	0.9344	**0.9428**
ATP7d	0.8694	0.8221	0.8342	0.8669	0.8517	0.8674	0.8713	**0.8769**
OES97	0.8592	0.8442	0.8466	0.8811	0.8671	0.8615	0.8613	**0.8976**
OES10	0.9352	0.8705	0.9101	**0.9473**	0.9256	0.9364	0.9406	0.9468
Average Value	0.7030	0.7148	0.7388	0.7268	0.7058	0.7172	0.7308	**0.7460**

Table 3. average root mean squared error (aRMSE ± STD) for all of multi-output data sets. The least aRMSE are emphasized by boldface in each row.

	NFS	Lasso	NE	LR	LGR	RFS	RSR	STR_SFS
EDM	0.0483	0.0850	0.0791	0.0506	0.0479	0.0503	0.0493	**0.0126**
SF1	0.0230	0.0600	0.0801	0.0205	0.0204	0.0222	0.0202	**0.0200**
SF2	0.0108	0.0351	0.0223	0.0108	0.0109	0.0108	0.0108	**0.0107**
WQ	0.0257	0.0541	0.0380	0.0245	0.0249	0.0247	0.0247	**0.0246**
ATP1d	0.0067	0.0281	0.0113	0.0068	0.0067	0.0067	0.0067	**0.0066**
ATP7d	0.0073	0.0225	0.0211	0.0072	0.0074	0.0073	0.0073	**0.0071**
OES97	0.0165	0.0421	0.0310	**0.0161**	0.0163	0.0164	0.0164	0.0164
OES10	0.0129	0.0258	0.0330	0.0128	0.0125	0.0125	0.0126	**0.0124**
Average Value	0.0189	0.0441	0.039	0.0187	0.0184	0.0189	0.0185	**0.0139**

that the proposed method has the best stability than all comparison methods for regression tasks.

Meanwhile, our proposed method had the minimum root mean squared error results in the most of data sets in Table 3, compared to all the comparison methods. This is similar to the conclusion of the evaluation of correlation coefficient. For example, our proposed STR_SFS method decreased on average by 0.5 %, 0.48 %, 0.45 %, 0.5 % and 4.6 %, respectively, compared to NFS, LR, LGR, RFS and RSR, in terms of average root mean squared error. Furthermore, we can more intuitive observe the Table 2 and 3 with three-dimensional histogram in the Fig. 1.

At last, we have compared computational cost with all methods. Except NFS method that the reason is it only directly use the LIBSVM to select features without construct model. The details revealed in the Table 4.

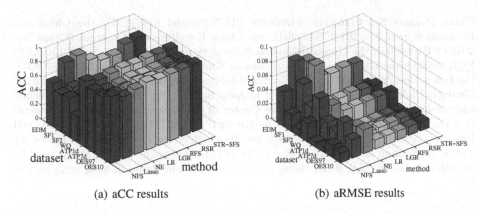

(a) aCC results   (b) aRMSE results

Fig. 1. all methods with aCC and aRMSE evaluate index for all multi-output data sets.

Table 4. The corresponding computational cost in the each iteration.

	Lasso	NE	LR	LGR	RFS	RSR	STR_SFS
EDM	0.0064	0.0060	0.0012	0.0008	0.0210	0.0216	0.0096
SF1	0.0078	0.0075	0.0015	0.0099	0.0336	0.0333	0.0207
SF2	0.0105	0.0105	0.0024	0.0023	0.0358	0.0570	0.1248
WQ	0.0476	0.0504	0.0126	0.0098	0.3094	0.3080	1.0612
ATP1d	0.8700	0.8496	0.6888	0.5106	3.2124	3.1878	0.6324
ATP7d	0.8664	0.8454	0.7230	0.5082	3.1314	3.0546	0.6120
OES97	1.2000	1.1600	0.9584	0.6192	4.4000	4.1616	1.0736
OES10	2.6400	2.5616	1.8688	1.5520	8.7200	8.0016	1.7376

4   Conclusion

In this paper, we put forward a novel supervised feature selection methods using a robust framework of reduced-rank regression. The proposed method utilized the $\ell_{2,p}$-norm on the two aspects which include the loss function term and the relevant regularizer term, and aim to flexible conduct subspace learning and feature selection in a framework, to output interpretable and stable results. The relevant results of experiment on the real life data set of multi-output indicated that the presented model outperformed the comparison methods, in terms of binary and multi-class classification tasks and regression tasks.

Due to the future task, we would like improve and extend the presented framework to measure the nonlinear relationship between the feature matrix and the label matrix via kernel methods.

Acknowledgment. This work was supported in part by the China "1000-Plan" National Distinguished Professorship; the Nation Natural Science Foundation of

China (Grants No: 61263035, 61363009, 61573270 and 61672177), the China 973 Program (Grant No: 2013CB329404); the China Key Research Program (Grant No: 2016YFB1000905); the Guangxi Natural Science Foundation (Grant No: 2015GXNS-FCB139011); the China Postdoctoral Science Foundation (Grant No: 2015M570837); the Innovation Project of Guangxi Graduate Education under grant YCSZ2016046; the Guangxi High Institutions' Program of Introducing 100 High-Level Overseas Talents; the Guangxi Collaborative Innovation Center of Multi-Source Information Integration and Intelligent Processing; and the Guangxi "Bagui" Teams for Innovation and Research, and the project "Application and Research of Big Data Fusion in Inter-City Traffic Integration of The Xijiang River - Pearl River Economic Belt(da shu jv rong he zai xijiang zhujiang jing ji dai cheng ji jiao tong yi ti hua zhong de ying yong yu yan jiu)".

References

1. Cai, X., Ding, C., Nie, F., Huang, H.: On the equivalent of low-rank linear regressions and linear discriminant analysis based regressions. In: Proceedings of the 19th ACM SIGKDD international conference on Knowledge discovery and data mining, pp. 1124–1132. ACM (2013)
2. Dzeroski, S., Demsar, D., Grbovic, J.: Predicting chemical parameters of river water quality from bioindicator data. Appl. Intell. **13**(1), 7–17 (2000)
3. Gu, Q., Li, Z., Han, J.: Joint feature selection and subspace learning. In: IJCAI 2011, Proceedings of the International Joint Conference on Artificial Intelligence, Barcelona, Catalonia, Spain, July, pp. 1294–1299 (2011)
4. Hui, Z., Trevor, H.: Regularization and variable selection via the elastic net. J. Roy. Stat. Soc. **67**(2), 301–320 (2005)
5. Jorgensen, M.: Iteratively reweighted least squares. Encyclopedia of Environmetrics (2006)
6. Karalic, A., Bratko, I.: First order regression. Mach. Learn. **26**(26), 147–176 (1997)
7. Liu, X., Guo, T., He, L., Yang, X.: A low-rank approximation-based transductive support tensor machine for semisupervised classification. IEEE Trans. Image Process. **24**(6), 1825–1838 (2015)
8. Luo, D., Ding, C.H.Q., Huang, H.: Linear discriminant analysis: New formulations and overfit analysis. In: Proceedings of the Twenty-Fifth AAAI Conference on Artificial Intelligence, AAAI 2011, San Francisco, California, USA, 7–11., August 2011
9. Nie, F., Cai, X., Huang, H., Ding, C.: Efficient and robust feature selection via joint $\ell_{2,1}$-norms minimization. In: Advances in Neural Information Processing Systems, pp. 1813–1821 (2010)
10. Qin, Y., Zhang, S., Zhu, X., Zhang, J., Zhang, C.: Semi-parametric optimization for missing data imputation. Appl. Intell. **27**(1), 79–88 (2007)
11. Shi, X., Guo, Z., Lai, Z., Yang, Y., Bao, Z., Zhang, D.: A framework of joint graph embedding and sparse regression for dimensionality reduction. IEEE Trans. Image Process. **24**(4), 1341–1355 (2015)
12. Spyromitros-Xioufis, E., Tsoumakas, G., Groves, W., Vlahavas, I.: Multi-target regression via input space expansion: treating targets as inputs. Mach. Learn. **26**, 1–44 (2016)
13. Tang, Z., Zhang, X., Li, X., Zhang, S.: Robust image hashing with ring partition and invariant vector distance. IEEE Trans. Inf. Forensics Secur. **11**(1), 200–214 (2016)

14. Tibshirani, R.: Regression shrinkage and selection via the lasso. J. Royal Stat. Soc. **58**(1), 267–288 (1996)
15. Wang, T., Qin, Z., Zhang, S., Zhang, C.: Cost-sensitive classification with inadequate labeled data. Inf. Syst. **37**(5), 508–516 (2012)
16. Wu, X., Zhang, C., Zhang, S.: Efficient mining of both positive and negative association rules. ACM Trans. Inf. Syst. **22**(3), 381–405 (2004)
17. Wu, X., Zhang, S.: Synthesizing high-frequency rules from different data sources. IEEE Trans. Knowl. Data Eng. **15**(2), 353–367 (2003)
18. Zhang, C., Qin, Y., Zhu, X., Zhang, J., Zhang, S.: Clustering-based missing value imputation for data preprocessing. In: IEEE International Conference on Industrial Informatics, pp. 1081–1086 (2006)
19. Zhang, S.: Decision tree classifiers sensitive to heterogeneous costs. J. Syst. Softw. **85**(4), 771–779 (2012)
20. Zhang, S., Cheng, D., Zong, M., Gao, L.: Self-representation nearest neighbor search for classification. Neurocomputing **195**, 137–142 (2016)
21. Zhang, S., Li, X., Zong, M., Cheng, D., Gao, L.: Learning k for knn classification. ACM Transactions on Intelligent Systems and Technology (2016, accepted)
22. Zhang, S., Wu, X., Zhang, C.: Multi-database mining. IEEE Comput. Intell. Bull. **2**(1), 5–13 (2003)
23. Zhu, P., Zuo, W., Zhang, L., Hu, Q., Shiu, S.C.K.: Unsupervised feature selection by regularized self-representation. Pattern Recogn. **48**(2), 438–446 (2015)
24. Zhu, X., Suk, H.I., Shen, D.: A novel matrix-similarity based loss function for joint regression and classification in ad diagnosis. NeuroImage **100**, 91–105 (2014)
25. Zhu, X., Zhang, L., Huang, Z.: A sparse embedding and least variance encoding approach to hashing. IEEE Trans. Image Process. **23**(9), 3737–3750 (2014)
26. Zhu, X., Zhang, S., Jin, Z., Zhang, Z., Xu, Z.: Missing value estimation for mixed-attribute data sets. IEEE Trans. Knowl. Data Eng. **23**(1), 110–121 (2011)
27. Zhu, X., Zhang, S., Zhang, J., Zhang, C.: Cost-sensitive imputing missing values with ordering. AAAI Press **2**, 1922–1923 (2007)
28. Zhu, Y., Lucey, S.: Convolutional sparse coding for trajectory reconstruction. IEEE Trans. Pattern Anal. Mach. Intell. **37**(3), 529–540 (2015)

PUEPro: A Computational Pipeline
for Prediction of Urine Excretory Proteins

Yan Wang[1], Wei Du[1,4(✉)], Yanchun Liang[1], Xin Chen[1,4],
Chi Zhang[4], Wei Pang[3], and Ying Xu[1,2,4(✉)]

[1] College of Computer Science and Technology, Changchun, China
[2] College of Public Health, Jilin University, Changchun, China
[3] School of Natural and Computing Sciences,
University of Aberdeen, Aberdeen AB24 3UE, UK
[4] Computational Systems Biology Lab, Department of Biochemistry
and Molecular Biology and Institute of Bioinformatics,
University of Georgia, Athens, GA, USA
weidu@jlu.edu.cn, xyn@uga.edu

Abstract. A computational pipeline is developed to accurately predict urine excretory proteins and the possible origins of the proteins. The novel contributions of this study include: (i) a new method for predicting if a cellular protein is urine excretory based on unique features of proteins known to be urine excretory; and (ii) a novel method for identifying urinary proteins originating from the urinary system. By integrating these tools, our computational pipeline is capable of predicting the origin of a detected urinary protein, hence offering a novel tool for predicting potential biomarkers of a specific disease, which may have some of their proteins urine excreted. One application is presented for this prediction pipeline to demonstrate the effectiveness of its prediction. The pipeline and supplementary materials can be accessed at the following URL: http://csbl.bmb.uga.edu/PUEPro/.

Keywords: Urine excretory proteins · Support vector machine recursive feature elimination · Biomarkers of disease

1 Introduction

Early detection is essential for disease control and possible prevention [1]. Among the existing techniques, detection of biomarkers in body fluids such as blood, urine or saliva represents the least invasive and most efficient approaches, which can offer an initial indication of diseases in specific organs. A key to accomplishing this lies in our ability to accurately identify informative biomarkers. Technical challenges involve (1) accurate identification of overly produced biomolecules in targeted disease tissues, which are specific to the disease, and (2) reliable prediction of which of such biomolecules can enter a specific type of body fluid.

Compared to blood, urine is probably equally information rich in term of the types of biomolecules from different origins. This makes urinary biomarkers more desirable, considering that (i) urine tends to have a simpler composition, which simplifies the

© Springer International Publishing AG 2016
J. Li et al. (Eds.): ADMA 2016, LNAI 10086, pp. 714–725, 2016.
DOI: 10.1007/978-3-319-49586-6_51

detection problem compared to blood; (ii) the dynamic range across different proteins is substantially smaller in urine than in blood; and (iii) collecting urine is substantially less invasive and easier to do than blood collection.

Proteins in urine originate mainly from glomerular filtration of serum proteins [2] and from the urinary system through secretion and membrane shedding. Therefore, it is necessary to identify and remove proteins that are from the urinary system among proteins found in urine, in order to identify biomarkers for diseases in distal organs [3]. Currently, the most useful disease markers in urine have been largely for urogenital diseases, such as urothelial cancer [4], renal cell carcinoma [5], prostate cancer [6], and bladder cancer [7]. A few recent studies have demonstrated the feasibility in using urinary proteins as disease markers in distal organs, such as ovarian carcinoma [8], lung cancer [9], hepatocellular carcinoma [10], and gastric cancer [11].

Only a few studies have been published on the prediction of urinary proteins, ours being one of them [11]. The present study extends the previous study by including novel capabilities for identification of origins of detected proteins in urine in addition to an improved prediction tool for proteins that are urine excretory. Our study utilizes a few data sources of urinary proteins to build a predictor for such proteins, including those given in [12–16]. The current knowledge is: 70 % of the urinary proteins originate from the kidney and the urinary tracts, and the remaining 30 % are filtered from blood circulation by the glomerulus [17]. Specifically, the origins of urinary proteins are: (i) glomerular filtration of blood proteins; (ii) proteins from renal tubular epithelial cells and other urinary cells, including those secreted from these cells or shed from their plasma membranes; (iii) membrane shedding proteins from renal tubular epithelial cells and other urinary system; (iv) exosome secretion; and (v) the whole cell shed from urinary tracts [2, 18].

A few studies have been published on the identification of the origins of detected urinary proteins, such as the work presented in [19], which identified urinary proteins which originated from kidney using an isolated rat kidney model, and studies that identified urinary proteins as being from the urinary tracts [3, 20]. These data are used to train our computational predictor for the origins of detected urinary proteins. Overall, the current study has made the following novel contributions: (i) a novel approach to predicting excretory proteins in urine; and (ii) a novel method for predicting the origins of detected urinary proteins. A server called PUEPro (Prediction of Urine Excretory Proteins) has been developed based on these novel methodological developments, and it can be accessed at: http://csbl.bmb.uga.edu/PUEPro/, from which the supplementary files of this paper can be downloaded.

2 Materials and Methods

2.1 Data Collection

Collecting urinary proteins and generating negative training data. Several datasets of proteins have been identified in human urine, including those in the Sys-BodyFluid database [12] and the Human Proteome Project (HPP) database [13]. The Sys-BodyFluid database consists of 1,941 distinct human proteins that have been experimentally identified in nine urinary proteomic studies. Over 2,000 experimentally verified urinary proteins are available and retrieved from the HPP database. In addition, we have also gathered urinary proteins identified by other urinary proteomic studies [14–16]. Overall, a total of 3,133 unique human urinary proteins were collected. To rule out the possibility of false identification of urinary proteins, we have used 1,495 out of the 3,133 proteins that have been detected by more than one study as the positive data in our study. Among the 1,495 proteins, 1,000 are used as training data and the remaining 495 as the test data.

Since we do not have a very clear understanding about which cellular proteins cannot be excreted to urine, generating a negative dataset is a challenge. In this study, we applied a selection process similar to the one presented by Cui et al. [21] through choosing proteins from the Pfam protein families [22] that do not contain any proteins that have been detected in urine. For each Pfam family (with at least ten members), ten members are randomly selected as part of the negative data. As a result, 1,821 proteins are selected as the negative data, of which 1,000 are used as training data and 821 proteins as the test data.

Collecting urological proteins in urine and generating negative training data. A few studies have been published regarding identified urinary proteins with originate from the urinary system. For example, 990 human proteins are predicted to be homologs of rat kidney proteins [19]. In addition, other studies have identified more urinary proteins with origins in the urinary system [3, 20]. We have compared these proteins with the above 1,495 urinary proteins, and found that only 430 of them originates from the urinary system. To predict which urinary proteins do not originate from the urinary system, we used a similar procedure discussed earlier, i.e., to select proteins from the Pfam families which do not contain any of the 430 proteins detected above. This gives us 365 urinary proteins which do not originate from the urinary system.

2.2   Model Construction

Feature construction. We aim to identify sequence or structure-based features that can distinguish between a specified positive set and a negative set as discussed in the previous section. We have examined features of the following types: (1) general sequence features; (2) physicochemical properties; (3) specific domains/motifs; (4) structural properties. The general sequence features include sequence length, amino acid composition, auto-correlation and quasi-sequence-order of each protein. The physicochemical properties include hydrophobicity, polarity, charge, secondary structure, and molecular weight. Specific domains/motifs include transmembrane score, signal peptide, and the number of glycosylation sites. Structural properties include secondary structure composition, radius of gyration among a few others. Overall, 39 features, represented by 1,537 feature elements, are considered and are shown in Table S1 of the supplementary material.

Distinguishing feature selection. For these features elements, there are four major categories: relevant features, redundant features, irrelevant features, and noisy features. For the feature set containing many features, the relevant features are only very small part of the whole feature set, and most of the features are irrelevant features. So, many feature selection methods for expression data analysis remove the irrelevant features firstly. In this research, we have employed a two-stage feature-selection procedure to distinguish the positive datasets from the negative ones on the training dataset. A t-test, which is a simple and effective filter feature selection method, was used to determine and eliminate the features without discerning power for our problem. Based on the calculated p-value, a q-value for each feature was calculated to control the False Discovery Rate [23]. Q-value = 0.005 was used as the threshold of q-value for removing non-contributing features. In the second step, a support vector machine (SVM)-recursive feature elimination (RFE) procedure [24], which is one of the best embedded feature selection methods, was applied to rank the remaining features, and to remove the lowly ranked and non-contributing features by the backward elimination technology, which selects relevant features by iteratively removing the most irrelevant feature at one time until the predefined size of the final features subset is reached. In each loop, the feature ranking of the remaining features can be possibly modified. At the end, 87 feature elements were selected and used in our analysis. The method eliminates irrelevant features and selects relevant features according to a criterion related to their support to a discrimination function DJ, which is measured by training SVM at each step. The discrimination function DJ is defined as follows:

$$\begin{cases} H = y_i y_j K(x_i, x_j) \\ DJ(i) = (1/2)a^T Ha - (1/2)a^T H(-i)a \end{cases} \tag{1}$$

where y_i and y_j are the class labels of samples x_i and x_j. $K(x_i, x_j)$ is the kernel function that measures the similarity between x_i and x_j. and α is obtained by training the classifier of SVM in the algorithm of SVM-train. The algorithm of SVM-RFE [24] is defined as follows:

SVM-RFE Algorithm:

Input:

    Training examples: $X_0 = [x_1, x_2, \dots x_i, \dots x_n]^T$

    Class label: $y = [y_1, y_2, \dots y_i, \dots y_n]^T$

Initialize:

    Subset of surviving features: $s = [1, 2, \dots m]$

    Feature ranked list: r=[]

    Repeat until $s = []$

        Restrict training examples to good feature indices: $X = X_0(:, s)$

        Train the classifier by SVM : $\alpha = SVM\text{-}train(X, y)$

        Compute the matrix H: $H = y_i y_j K(x_i, x_j)$

        Compute the ranking criteria: $DJ(i) = (1/2)\alpha^T H\alpha - (1/2)\alpha^T H(-i)\alpha$

        Find the feature with the smallest ranking criterion: $f = argmin(DJ)$

        Update feature ranking list: r = [s(f), r]

        Eliminate the feature with smallest ranking criterion: s = s(1:f-1, f+1:length(s))

    End

Output:

    Feature ranking list r.

Classification and assessment. For determining the class labels for the new proteins correctly, a classifier needs to be constructed. In this research, the Support Vector Machine (SVM) is used as the classifier with several evaluation criteria, which are used to guide the choice of parameters. In SVM, the hyperplane of a high dimensional space, which is called feature space, is constructed to separate two classes. A good separation of one hyperplane needs to have the largest distance to the nearest training data of any class. The kernel functions, the wide coefficient of kernel functions, and the penalty coefficient C are the main parameters of SVM. Gaussian kernel with a single parameter q is a common choice for classification. Then, we can select the combination of C and q by grid search to improve the effectiveness. Using different parameters for the classifier, we can derive the distance d between the positions of the prediction data in the feature space and the optimal separating hyperplane. A larger distance d means more reliable prediction results. The SVM-based classifier was trained on the training data using the selected features to predict if a protein is urinary or not. Similarly, a second classifier is trained to predict whether a urinary protein originates from the urinary system.

The following measures are used to evaluate the prediction performance: specificity $= \frac{TN}{TN+FP}$, precision $= \frac{TP}{TP+FP}$, accuracy $= \frac{TP+TN}{N_{total}}$, MCC $= \frac{(TP \times TN - FP \times FN)}{\sqrt{(TP+FN)(TP+FP)(TN+FP)(TN+FN)}}$ and AUC (the area under the curve) of the sensitivity-specificity curve [25], where TP is the number of true positives, FP refers to the number of false positives, TN means the number of true negatives, FN for the number of false negatives, and N_{total} is the total number of proteins for prediction in a given test set.

2.3 Identification of Differentially Expressed Genes

We have applied our prediction method developed above to the gene-expression data of lung cancer versus control samples. The dataset consists of RNA-seq data of 101 paired samples of lung cancer and control samples, which are downloaded from the TCGA database [26]. The following formula is used to estimate the fold-change of each gene:

$$FC_{ij} = \begin{cases} fc_{ij} - 1 & (fc_{ij} \geq 1) \\ 1 - \frac{1}{fc_{ij}} & (fc_{ij} < 1) \end{cases}, \ where \ fc_{ij} = C_{ij}/N_{ij}, \tag{2}$$

$$FC_i = \frac{1}{m} \sum_{j=1}^{m} FC_{ij}, \tag{3}$$

where m is the number of samples, and C_{ij} and N_{ij} are the expression levels of gene i in the jth pair of cancer and normal control. If FC_i is greater than zero, the relevant gene is considered as up-regulated in cancer; otherwise down-regulated or no change. 0.5 (-0.5) was used as the threshold for defining differentially expressed genes. In addition, a Wilcoxon test was used to assess the statistical significance of the observed differential expression in cancer vs. normal samples, and the statistical significance cutoff value is set at 0.05.

3 Results

3.1 Features of Urine-Excretory Proteins

39 features, represented as a vector of 1,537 elements (see Table S1), are used to distinguish between the positive and the negative training data by training an SVM-based classifier with an RBF kernel. 10-fold cross-validation was done to assess the performance of the trained classifier, and the classifier has the average sensitivity, specificity, precision and accuracy of 84.40 %, 82.87 %, 83.20 %, and 83.63 %, respectively.

A feature selection process was then conducted to select the most discerning parameters among the 1,537. At the end, 87 parameters were selected, which achieve comparable performance to the above. Among the selected parameters, transmembrane domains and signal peptides have been found to be useful for predicting protein secretion to blood circulation [11, 21]. The radius of gyration is an interesting one, which has been suggested to play a role in proteins passing through the GBM (glomerular basement membrane). Published studies have observed that proteins with a radius smaller than 1.8 nm can pass through the GBM-slit diaphragm barrier, whereas proteins with a radius larger than 4.0 nm are retained [27].

3.2 Performance of Urine-Excretory Proteins

Urine-excretory proteins prediction. We then retrained an SVM-based classifier based on the 87 selected parameters, using both the linear and RBF kernel. The performance assessment of the classifier was done using 10-fold cross-validation by repeating the prediction 100 times on the training set containing 1,000 positive and 1,000 negative samples. It is found that the classification accuracy ranges from 81.00 % to 97.00 % for the positive data and from 74.51 % to 94.12 % for the negative data. The average performance based on the linear and RBF kernel is shown in Table 1. In addition, the ROC curve is given in Fig. 1 (left).

Table 1. Average performance of urine protein prediction by 10-fold cross validation on training set

SVM Kernel	Sensitivity	Specificity	Precision	Accuracy	MCC	AUC
Linear	88.77 %	88.70 %	88.61 %	88.74 %	0.775	0.947
RBF	88.05 %	87.57 %	87.52 %	87.81 %	0.756	0.937

We then assessed the trained models on an independent test set composed of 495 urine-excretory proteins and 821 non-urine excretory proteins, with the detailed protein names given in Table S2 of the supplementary material. The prediction performance is presented in Table 2 along with the ROC curve in Fig. 1(right). At the end, we have selected the classifier using the RBF kernel as it performs better than the linear model on the test set.

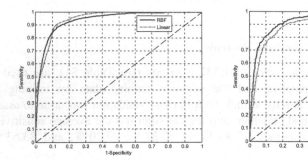

Fig. 1. The ROC curve for 10-fold cross validation on the training set (left) and on an independent testing set (right).

Predicting and ranking the known excretory proteins in urine. We define the D-value of a protein in the UniProt database [28] as follows:

$$D = d * p, \tag{4}$$

where p = 1 if the protein is predicted to be urine-excretory and −1 otherwise; and d is the distance between the position of the protein in the feature space and the separating hyper-plane defined by the trained SVM classifier. 228 (22.8 %) of the positive training data (1,000) are ranked among the top 1,000 proteins. Among these 1,000 proteins, 110 (22.2 %) are in the positive test dataset (495).

We also ranked the urine-excretory proteins that have been detected to be associated with human diseases in the literature and do not overlap with our training data. To accomplish this, we have collected such proteins from the Urinary Protein Biomarker Database [3]. 261 proteins are found to be in both this database and the UniProt, with the detailed protein names given in Table S3. 56 (21.46 %) of these 261 proteins ranked among the top 1,000, 91 (34.87 %) among the top 2,000 and 123 (47.13 %) among the top 3,000, as detailed in Table 3 along with the p-values being 6.45e-21, 4.22e-28, and 1.83e-35, respectively. A comparison was also included in Table 3 between the results by our model and by a previous study [11], which is the only relevant study in the literature. In our model, we have employed a two-stage feature-selection procedure to distinguish the positive datasets from the negative ones on the training dataset and applied SVM with different kernel functions.

Table 2. Average **performance** of our classifier on the independent testing set

SVM Kernel	Sensitivity	Specificity	Precision	Accuracy	MCC	AUC
Linear	83.84 %	83.19 %	75.05 %	83.43 %	0.658	0.906
RBF	83.84 %	87.82 %	80.58 %	86.32 %	0.712	0.931

We have also conducted a function enrichment analysis of the top 1,000 D-value ranked proteins, using DAVID [29] against the Gene Ontology, KEGG, BBID and BIOCARTA databases, and using the whole set of UniProt as the background set.

The goal is to check the subcellular locations as well as the biological processes enriched by these proteins. For understanding the cellular functions and subcellular locations of these predicted excretory proteins in urine, we noted that the most significantly enriched biological processes and cellular components were cell adhesion and extracellular region. In addition, the most significantly enriched pathways are cell adhesion molecules, ECM-receptor interaction, and complement and coagulation cascades (see Table S4), which are all closely involved in the urine excretory process.

3.3   The Prediction of Origins of Urinary Proteins

The prediction of urological origins of the predicted excretory proteins. We have developed a classifier for predicting the urological origin of excretory proteins. The training of the classifier was done on a set of 430 proteins known to be of urological origin and 365 proteins known to be not of urological origin. An SVM-based classifier was trained along with a feature selection procedure based on the same set of 39 features totaling 1,537 dimensions (see Sect. 3.1), which gives rise to 111 final parameters.

Table 3. A comparison among the ranking results of known urinary biomarkers for diseases by our classifier *versus* a published classifier [11]

Top Ranked Proteins	The number of urinary biomarkers included[a]	P-value[a]	The number of urinary biomarkers included[b]	P-value[b]
500	12	0.0045	29	1.50e-11
1,000	21	0.0012	56	6.45e-21
1,500	34	1.15e-05	74	1.60e-24
2,000	48	2.60e-08	91	4.22e-28
2,500	68	2.83e-14	112	3.88e-35
3,000	86	4.24e-20	123	1.83e-35
3,500	106	3.74e-28	140	1.13e-40
4,000	113	6.83e-28	146	3.28e-38
4,500	116	3.01e-25	154	1.21e-37
5,000	120	1.34e-23	164	8.01e-39
5,500	126	2.06e-23	176	8.37e-42
6,000	129	1.43e-21	182	1.18e-40

[a]By using the classifier in a previous study [11].
[b]By using our classifier.

Then 10-fold cross-validation was applied to the training set to evaluate the prediction performance of excretory proteins of urological origin. The performance by the trained classifier using a linear and RBF kernel, respectively, is shown in Table 4. Figure 2 shows the ROC curves.

Table 4. The average performance of the prediction of urological origins for excretory proteins assessed by 10-fold cross validation

SVM Kernel	Sensitivity	Specificity	Precision	Accuracy	MCC	AUC
Linear	83.10 %	77.90 %	81.72 %	80.72 %	0.611	0.873
RBF	83.39 %	77.90 %	81.77 %	80.88 %	0.614	0.875

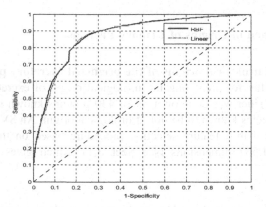

Fig. 2. The ROC curves for predicted urological origins of excretory proteins.

For the top 5,000 human proteins that have been predicted as excretory proteins in urine, we predicted that 2,357 are of urological origin and 2,643 are not. The function enrichment analysis of Gene Ontology and Pathway is used to understand the cellular functions and subcellular locations of the 2,357 urological origin proteins. The most significantly enriched biological processes, cellular components, and molecular function are biological adhesion, extracellular region, and GTP binding. Meanwhile, the most significantly enriched pathways are Glycolysis/Gluconeogenesis and Pyruvate metabolism (see Table S5). The function enrichment analysis of Gene Ontology and Pathway is used to understand the cellular functions and subcellular locations of the 2,643 non-urological origin proteins. The most significantly enriched biological processes, cellular components, and molecular function are immune response, intrinsic to plasma membrane, and hormone activity. Meanwhile, the most significantly enriched pathways are Cytokine-cytokine receptor interaction and Complement and coagulation cascades (see Table S5). We can see that the significantly enriched pathways of urological origin proteins are related to metabolic pathways, and the significantly enriched pathways of non-urological origin proteins are related to immune systems and tissue repair (Table 5).

Table 5. Proteins predicted as urinary biomarkers of two types of lung cancer.

Not included in the training dataset

Accession	Protein Name	Ratio (cancer/normal)	D-value
Q6ZMP0	Thrombospondin type-1 domain-containing protein 4	−1.74	1.466
P28908	Tumor necrosis factor receptor superfamily member 8	−1.92	1.431
P08833	Insulin-like growth factor-binding protein 1	1.45	1.591
O43240	Kallikrein-10	−15.05	1.110
P01127	Platelet-derived growth factor subunit B	−2.13	1.324
P39900	Macrophage metalloelastase	87.24	1.119

Included in the training dataset

Accession.	Protein Name	Ratio (cancer/normal)	D-value
P13688	Carcinoembryonic antigen-related cell adhesion molecule 1	2.23	2.401
P01033	Metalloproteinase inhibitor 1	1.62	1.479
P39060	Collagen alpha-1(XVIII) chain	2.23	2.401
P39059	Collagen alpha-1(XV) chain	2.34	1.238
P01024	Complement C3	−5.20	2.131
P10909	Clusterin	−4.19	1.791
P04085	Platelet-derived growth factor subunit A	−1.00	1.000

3.4 Identification of Urinary Biomarkers for Lung Cancer

We have applied the methods presented above to the gene-expression data of lung adenocarcinoma and squamous cell carcinoma, with an aim to predict urinary markers for the disease. By examining 102 lung cancer tissue versus 102 matching control tissues in the TCGA database [26], 5,491 genes are found to be differentially expressed in the cancer versus the control tissues. Using the prediction method given in Sect. 2.2, 587 of these genes are predicted to be urine excretory. Out of these proteins, 116 have been identified in human urines, including 13 that have been reported as potential urine biomarkers for non-small-cell lung carcinoma [30].

4 Discussions and Conclusion

Early diagnosis plays a vital role in controlling diseases. Identifying disease-informing biomarkers represents an effective way for early diagnosis of a disease. The key is to identify the most useful biomarkers for disease detection. With the rapid development of omic technologies, a variety of disease tissue omic data are being generated and stored into publicly available databases. These data provided unprecedented opportunities to computational data analysts to develop effective methods to discover the most effective biomarkers for specific diseases.

Comparable to the existing biomarker prediction methods, our study has two novel aspects: (i) a new method for predicting if a cellular protein is urine excretory based on unique features of proteins known to be urine excretory; and (ii) a novel method for identifying urinary proteins originated from the urinary system. We anticipate that these ideas and methods will ultimately lead to substantially improved abilities for reliable identification of urinary biomarkers.

Acknowledgments. This work is supported by the National Natural Science Foundation of China (Grant Nos. 81320108025, 61402194, 61572227), Development Project of Jilin Province of China (20140101180JC) and China Postdoctoral Science Foundation (2014T70291).

References

1. Lee, S., Huang, H., Zelen, M.: Early detection of disease and scheduling of screening examinations. Stat. Methods Med. Res. **13**, 443–456 (2004)
2. Thongboonkerd, V.: Practical points in urinary proteomics. J. Proteome Res. **6**, 3881–3890 (2007)
3. Shao, C., Li, M., Li, X., Wei, L., Zhu, L., Yang, F., Jia, L., Mu, Y., Wang, J., Guo, Z., Zhang, D., Yin, J., Wang, Z., Sun, W., Zhang, Z., Gao, Y.: A tool for biomarker discovery in the urinary proteome: a manually curated human and animal urine protein biomarker database. Mol. Cell. Proteomics **10**(M111), 010975 (2011)
4. Abogunrin, F., O'Kane, H.F., Ruddock, M.W., Stevenson, M., Reid, C.N., O'Sullivan, J.M., Anderson, N.H., O'Rourke, D., Duggan, B., Lamont, J.V., Boyd, R.E., Hamilton, P., Nambirajan, T., Williamson, K.E.: The impact of biomarkers in multivariate algorithms for bladder cancer diagnosis in patients with hematuria. Cancer **118**, 2641–2650 (2012)
5. Raimondo, F., Morosi, L., Corbetta, S., Chinello, C., Brambilla, P., Della Mina, P., Villa, A., Albo, G., Battaglia, C., Bosari, S., Magni, F., Pitto, M.: Differential protein profiling of renal cell carcinoma urinary exosomes. Mol. BioSyst. **9**, 1220–1233 (2013)
6. Malavaud, B., Salama, G., Miedouge, M., Vincent, C., Rischmann, P., Sarramon, J.P., Serre, G.: Influence of digital rectal massage on urinary prostate-specific antigen: interest for the detection of local recurrence after radical prostatectomy. Prostate **34**, 23–28 (1998)
7. Ghoniem, G., Faruqui, N., Elmissiry, M., Mahdy, A., Abdelwahab, H., Oommen, M., Abdel-Mageed, A.B.: Differential profile analysis of urinary cytokines in patients with overactive bladder. Int. Urogynecol. J. **22**, 953–961 (2011)
8. Abdullah-Soheimi, S.S., Lim, B.K., Hashim, O.H., Shuib, A.S.: Patients with ovarian carcinoma excrete different altered levels of urine CD59, kininogen-1 and fragments of inter-alpha-trypsin inhibitor heavy chain H4 and albumin. Proteome Sci. **8**, 58 (2010)
9. Li, Y., Zhang, Y., Qiu, F., Qiu, Z.: Proteomic identification of exosomal LRG1: a potential urinary biomarker for detecting NSCLC. Electrophoresis **32**, 1976–1983 (2011)
10. Abdalla, M.A., Haj-Ahmad, Y.: Promising urinary protein biomarkers for the early detection of hepatocellular carcinoma among high-risk hepatitis C virus Egyptian patients. J Cancer. **3**, 390–403 (2012)
11. Hong, C.S., Cui, J., Ni, Z., Su, Y., Puett, D., Li, F., Xu, Y.: A computational method for prediction of excretory proteins and application to identification of gastric cancer markers in urine. PLoS ONE **6**, e16875 (2011)
12. Li, S.J., Peng, M., Li, H., Liu, B.S., Wang, C., Wu, J.R., Li, Y.X., Zeng, R.: Sys-BodyFluid: a systematical database for human body fluid proteome research. Nucleic Acids Res. **37**, D907–D912 (2009)

13. Legrain, P., Aebersold, R., Archakov, A., Bairoch, A., Bala, K., Beretta, L., Bergeron, J., Borchers, C.H., Corthals, G.L., Costello, C.E., Deutsch, E.W., Domon, B., Hancock, W., He, F., Hochstrasser, D., Marko-Varga, G., Salekdeh, G.H., Sechi, S., Snyder, M., Srivastava, S., Uhlen, M., Wu, C.H., Yamamoto, T., Paik, Y.K., Omenn, G.S.: The human proteome project: current state and future direction. Mol. Cell. Proteomics 10(M111), 009993 (2011)

14. Adachi, J., Kumar, C., Zhang, Y., Olsen, J.V., Mann, M.: The human urinary proteome contains more than 1500 proteins, including a large proportion of membrane proteins. Genome Biol. 7, R80 (2006)

15. Li, Q.R., Fan, K.X., Li, R.X., Dai, J., Wu, C.C., Zhao, S.L., Wu, J.R., Shieh, C.H., Zeng, R.: A comprehensive and non-prefractionation on the protein level approach for the human urinary proteome: touching phosphorylation in urine. Rapid Commun. Mass Spectrom. 24, 823–832 (2010)

16. Marimuthu, A., O'Meally, R.N., Chaerkady, R., Subbannayya, Y., Nanjappa, V., Kumar, P., Kelkar, D.S., Pinto, S.M., Sharma, R., Renuse, S., Goel, R., Christopher, R., Delanghe, B., Cole, R.N., Harsha, H.C., Pandey, A.: A comprehensive map of the human urinary proteome. J. Proteome Res. 10, 2734–2743 (2011)

17. Decramer, S., Gonzalez de Peredo, A., Breuil, B., Mischak, H., Monsarrat, B., Bascands, J. L., Schanstra, J.P.: Urine in clinical proteomics. Mol. Cell. Proteomics 7, 1850–1862 (2008)

18. Hoorn, E.J., Pisitkun, T., Zietse, R., Gross, P., Frokiaer, J., Wang, N.S., Gonzales, P.A., Star, R.A., Knepper, M.A.: Prospects for urinary proteomics: exosomes as a source of urinary biomarkers. Nephrology (Carlton) 10, 283–290 (2005)

19. Jia, L., Li, X., Shao, C., Wei, L., Li, M., Guo, Z., Liu, Z., Gao, Y.: Using an isolated rat kidney model to identify kidney origin proteins in urine. PLoS ONE 8, e66911 (2013)

20. Wood, S.L., Knowles, M.A., Thompson, D., Selby, P.J., Banks, R.E.: Proteomic studies of urinary biomarkers for prostate, bladder and kidney cancers. Nat. Rev. Urol. 10, 206–218 (2013)

21. Cui, J., Liu, Q., Puett, D., Xu, Y.: Computational prediction of human proteins that can be secreted into the bloodstream. Bioinformatics 24, 2370–2375 (2008)

22. Finn, R.D., Bateman, A., Clements, J., Coggill, P., Eberhardt, R.Y., Eddy, S.R., Heger, A., Hetherington, K., Holm, L., Mistry, J., Sonnhammer, E.L., Tate, J., Punta, M.: Pfam: the protein families database. Nucleic Acids Res. 42, D222–D230 (2014)

23. Storey, J.D., Tibshirani, R.: Statistical significance for genomewide studies. Proc. Natl. Acad. Sci. U.S.A. 100, 9440–9445 (2003)

24. Guyon, I., Weston, J., Barnhill, S., Vapnik, V.: Gene selection for cancer classification using support vector machines. Mach. Learn. 46, 389–422 (2002)

25. Scholkopf, B., Platt, J.C., Shawe-Taylor, J., Smola, A.J., Williamson, R.C.: Estimating the support of a high-dimensional distribution. Neural Comput. 13, 1443–1471 (2001)

26. Hampton, T.: Cancer genome atlas. JAMA 296, 1958 (2006)

27. Harris, N.S., Winter, W.E.: Multiple myeloma and related serum protein disorders: an electrophoretic guide, Demos Medical Publishing (2012)

28. Consortium, U: Reorganizing the protein space at the Universal Protein Resource (UniProt). Nucleic Acids Res. 40, D71–D75 (2012)

29. Huang, D.W., Sherman, B.T., Lempicki, R.A.: Systematic and integrative analysis of large gene lists using DAVID Bioinformatics Resources. Nature Protoc. 4, 44–57 (2009)

30. Nolen, B.M., Lomakin, A., Marrangoni, A., Velikokhatnaya, L., Prosser, D., Lokshin, A.E.: Urinary protein biomarkers in the early detection of lung cancer. Cancer Prev. Res. (Phila) (2014)

Partitioning Clustering Based on Support Vector Ranking

Qing Peng[1], Yan Wang[1(✉)], Ge Ou[1], Yuan Tian[1], Lan Huang[1], and Wei Pang[2(✉)]

[1] College of Computer Science and Technology,
Jilin University, Changchun, China
wy6868@jlu.edu.cn
[2] School of Nature and Computing Sciences,
University of Aberdeen, Aberdeen, Scotland, UK
pang.wei@abdn.ac.uk

Abstract. Support Vector Clustering (SVC) has become a significant boundary-based clustering algorithm. In this paper we propose a novel SVC algorithm named "Partitioning Clustering Based on Support Vector Ranking (PC-SVR)", which is aimed at improving the traditional SVC, which suffers the drawback of high computational cost during the process of cluster partition. PC-SVR is divided into two parts. For the first part, we sort the support vectors (SVs) based on their geometrical properties in the feature space. Based on this, the second part is to partition the samples by utilizing the clustering algorithm of similarity segmentation based point sorting (CASS-PS) and thus produce the clustering. Theoretically, PC-SVR inherits the advantages of both SVC and CASS-PS while avoids the downsides of these two algorithms at the same time. According to the experimental results, PC-SVR demonstrates good performance in clustering, and it outperforms several existing approaches in terms of Rand index, adjust Rand index, and accuracy index.

Keywords: Support vector clustering · Support vector ranking · Partitioning clustering

1 Introduction

Data Clustering has been an important task in data mining, and existing clustering algorithms can be classified into five categories [1]: partitioning methods [2–4], hierarchical methods [2, 5–7], density-based methods [8–10], grid-based methods [11, 12], and model-based methods [13, 14].

Among many clustering algorithms, support vector clustering (SVC) [15, 16] has become a significant boundary-based clustering algorithm in several applications such as community discovery, speech recognition and bioinformatics analysis [17]. SVC has the following features: first, it can be applied to various shapes of the clusters; second,

---

The original version of this chapter was revised: An acknowledgement has been added. The erratum to this chapter is available at DOI: 10.1007/978-3-319-49586-6_61

J. Li et al. (Eds.): ADMA 2016, LNAI 10086, pp. 726–737, 2016.
DOI: 10.1007/978-3-319-49586-6_52

the number of clusters is not needed in advance; third, it can deal with structured data by using kernel functions; fourth, it can reduce the impact of noise on the cluster partition.

However, there is still room for improvement for SVC. The algorithm is still inadequate due to two bottlenecks: expensive computational cost and poor labeling piece, and this degrades the popularity of SVC [17]. To address these limitations, some work has been done: Ben-Hur *et al.* [15] improved the original algorithm and proposed a method called support vector graph (SVG). The main idea of this method was that support vectors (SVs) were used to construct the adjacency matrix and derive connected component with an aim to reduce time complexity; Yang et al. [18] proposed the proximity graph (PG), and its time complexity was reduced to $O(N \log N)$ or $O(N)$; Lee et al. [19] devised gradient descent (GD) by looking for the stable equilibrium point (SEP); Jung et al. [20] proposed the fast support vector clustering (FSVC), which improved the speed of the algorithm as well as the quality of clustering; Sei-Hyung Lee [21] designed a cone-based cluster partition method to avoid random operations, and it was called Cone Cluster Labeling (CCL), which improved the quality of clustering but increased operation cost; Convex decomposition based cluster labeling (CDCL) [22] was proposed to improve both the efficiency and accuracy of clustering based on convex decomposition; L-CRITICAL was a novel SVC cluster labeling algorithm, and it solved the labeling phase of SVC within competitive processing time [23]; Proximity Multi-sphere Support Vector Clustering (PMS-SVC) was developed based on the multi-sphere approach to support vector data description [24]; Rough–Fuzzy Support Vector Clustering (RFSVC) can obtain rough fuzzy clusters using the support vectors as cluster representatives [25].

The clustering algorithm of similarity segmentation based point sorting (CASS-PS) [26] has a faster speed in clustering. However, the similarity measure of the algorithm is based on distance, which is likely to cause staggered sorting issue between different cluster elements, and this will reduce the accuracy of clustering results.

In this paper, we propose an improved SVC algorithm called partitioning clustering based on support vector ranking (PC-SVR). The algorithm's crucial components are (1) SV's sorting based on their geometric properties in the feature space and (2) cluster partition that uses the clustering algorithm of similarity segmentation based point sorting (CASS-PS). The proposed algorithm guarantees the quality of the clustering and improves the speed of clustering at the same time.

2 Partitioning Clustering Based on Support Vector Ranking

Our PC-SVR algorithm combines the first stage of SVC and CASS-PS, and the algorithm is composed of two stages: first, sort the support vectors (SVs) into an array; second, split the sorted array.

2.1 Support Vector Sorting

In the feature space, data are mapped to the minimal sphere. Assume this sphere is S, and the center is a. According to $K(x, x) = \exp\left(-q \cdot \|x - x\|^2\right) = 1$, we can get

$K(x, x) = <\Phi(x) \cdot \Phi(x)> = \|\Phi(x)\|^2 = 1$, which means all the data points are located on the surface of the unit ball. Assume this ball is B, and the center of B is O. So, the covering is the intersection, whose shape is like a cap. The center of this cap is denoted as a', as shown in Fig. 1. Since SVs are on the surface of S, they are also on the intersection hyper line of S and $B.\Phi(v_i)$ and $\Phi(v_j)$ are SVs in the feature space, and θ is the angle between the support vectors and two sphere center. The transverse section of the cap is illustrated in Fig. 2.

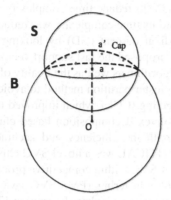

Fig. 1 Intersection between ball and sphere

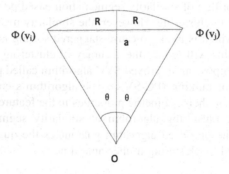

Fig. 2 Transverse section of the cap

Given a dataset containing N data points $\{x_i | x_i \subseteq x, 1 \le i \le N\}$, let $V = \{v_i | v_i$ is a SV, $1 \le i \le N_{SV}\}$. In this research, we will use the geometric properties of samples in the feature space as follows [21]:

Lemma 1. $\angle(\Phi(v_i oa')) = \angle(\Phi(v_j oa')) \ \forall v_i, v_j \in V$

Lemma 2. $\forall x \in X, v \in V, \angle(\Phi(v)o\Phi(x)) < \theta \Leftrightarrow \|v - x\| < \|v - \Phi^{-1}(a')\|$

Lemma 3. $x \in X, v \in V, \|v - x\| < \|v - \Phi^{-1}(a')\| \Leftrightarrow x, v$ belongs to the same cluster.

The following corollary can be proven by the above three properties, as in [17]:

Corollary. In the feature space of Gaussian Kernel, SVs are collected in terms of clusters on the intersection hyper line of S and B. This is illustrated in Fig. 3.

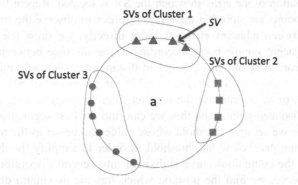

Fig. 3 The distribution of SVs in three clusters

From the above properties, for any $v_i, v_j \in V$, in the feature space, they have the same angle as Oa' in Fig. 1, and if the angle between the sample point and the SV is less than θ, the point and the SV belong to the same cluster. In the data space, the point to which the distance from the SV is less than $||v - \Phi^{-1}(a')||$ has the same cluster label as the SV, and thus the computation of the feature space can be converted to the computation of input space.

Therefore, we can use the angle between two SVs (v_i and v_j) and O to measure the distance between two SVs. The relation between the angle and the Gaussian kernel function is as follows: $\cos(\angle\Phi(v_i)O\Phi(v_j)) = <\Phi(v_i) \cdot \Phi(v_j)> = K(v_i \cdot v_j)$. Namely, the comparison of the distance of two SVs is transformed into the comparison of the kernel function, and the greater the distance between the two SVs is, the larger the angle is, and the smaller the kernel value is.

The similarity matrix for SVs is constructed according to the values of the kernel function. Then, according to the matrix, SVs are sorted as follows: first, the two SVs whose $\Phi(v_i)$, $\Phi(v_j)$ have the minimum distance are selected as the head and tail of an ordered array; Second, find the SV whose $\Phi(v_k)$ has the minimum distance from the head (or tail) of the sorted array as the head (or tail) of the array. Repeat this step until all data points are stored in an array and we can get the sequence of SVs in the feature space on the circumference, as shown in Fig. 4.

$$\Phi(v_k) \quad \Phi(v_i) \quad \Phi(v_j)$$

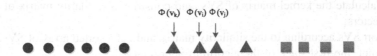

Fig. 4 The Sequence of SVs

2.2   Partition Clustering

This part mainly uses the CASS-PS algorithm to partition the cluster.

In the data space, we firstly calculate the distance between each sample point (except for SVs) and each SV, and find the nearest SV. Then we insert the point into the adjacent position of the array in which the SV is located. Repeat this process until all the sample points are stored in the array. In order to observe the transformation of the distance between adjacent elements more directly, we draw the distance curve between the adjacent sample points according to the distance between elements. The distance curve can show obvious changes in distance between adjacent elements, and especially it has a great wave between adjacent sample points of SVs.

At the same time, we can use the wavelet filter function to reduce the impact of noise points or isolated points, and thus we can find the best segmentation point more accurately. Then we set up a threshold whose value can be set as the mean amplitude, and we ignore the part below the threshold in order to simplify the determination of split points. So the continuous curve is divided into several discontinuous curve segments. Furthermore, we find the position which has the maximum distance between adjacent elements as the splitting point in each curve. Then we sort these splitting points after finding the splitting points of all (the whole) curve, and the position which has the maximum distance between adjacent splitting points is selected as the first splitting place. According to this procedure, it can be decided that the next step is re-segmentation or termination. After algorithm terminates, we output the number of clusters.

2.3   The Implementation of PC-SVR Algorithm

In this research we mainly use the geometric properties of sample points in the feature space and CASS-PS to improve the cluster partitioning, which is the second stage of the SVC algorithm. In the feature space, SVs are collected based on the clusters on the intersection hyper line of the minimal sphere and the unit ball. Sorting SVs is based on the similarity between two SVs, which is based on the value of the kernel function. Since SVs are already sorted, it is useful to avoid the limitation of the CASS-PS algorithm, that is, the sample points of different clusters tend to overlap.

The detailed steps of our algorithm are finally given as follows:

(1) Given a sample set $S = \{x_i | x_i \subseteq X\}$, its sample size is N, and set parameters q and C. Initialize a one-dimensional array based on the sample size;

(2) Calculate the kernel matrix of the sample set;

(3) Calculate the radius R of the minimal sphere and SVs according to Lagrange polynomial;

(4) Calculate the kernel matrix of SVs, and construct a similarity matrix of support vectors;

(5) Sort SVs according to the similarity matrix and get a sorted array of SVs. At this point, the first stage of the algorithm is completed;

(6) Calculate the distance from other sample points to all SVs, and find the closest SV to the sample point to be sorted. Insert the sample point into the back of the SV;

(7) Repeat Step 6, until all other sample points are completed with the interpolation, and we get a new sample point array;

(8) Draw the curve of the distance between adjacent sample points. Apply the wavelet filter function to the sample point array to reduce noise. Set a certain threshold and retain the portion above the threshold as the split segment;

(9) Find the points that have the maximum distance (the peak of the distance curve) in various segments, and sort these points. According to the number of clusters, select the corresponding points as the splitting points to split the array of sample points;

(10) Label cluster labels on the sample points.

3 Experiment Analysis

3.1 Evaluation Criteria of Experimental Results

In this paper, Rand index [27], Adjust Rand index [28] and Accuracy index [29] are used to evaluate the clustering results.

The Rand index is an external evaluation metric and it evaluates the effectiveness of clustering by comparing the actual results and the results obtained by the clustering algorithms. Given a dataset that contains n elements and its known partition result P, we run the algorithm to be evaluated to get another partition result Q. Suppose r is the number of data which belong to the same cluster in P and Q, s is the number of data which belong to different clusters in P and Q, t is the number of data belong to the same cluster in P but belong to the different cluster in Q, and v is the number of data belong to the same cluster in Q but belong to a different cluster in P. On the base of the above, r and s can determine the similarity of clustering results, while t and v can describe the inconsistency of the results. Rand index is given as follows:

$$RI = \frac{r+s}{r+s+t+v} \tag{1}$$

The values of Rand index range in [0, 1], and the greater the value of RI is, the better the clustering results are.

The adjust Rand index will standardize the clustering results in addition to the comparison of the known clustering results and the results obtained by an algorithm. The formula is as follows:

$$s_1 = \sum_{i=1}^{K_P} C_{N_i}^2, \quad s_2 = \sum_{j=1}^{K_Q} C_{N_j}^2, \quad s_3 = \frac{2 s_1 s_2}{N(N-1)}, \quad ARI = \frac{\sum_{i=1}^{K_P}\sum_{j=1}^{K_Q} C_{N_{ij}}^2 - s_3}{(s_1 + s_2)/2 - s_3} \tag{2}$$

In the above, P and Q represent the two clustering results of a sample set consisting of n elements, and K_P and K_Q are the numbers of clusters in P and Q, respectively. N_i and N_j represent the numbers of elements in clusters i and j in P and Q, respectively, and N_{ij} represents the number of elements in both cluster i in P and cluster j in Q. Adjust Rand index ranges in [-1, 1], and the greater the index value is, the more similar the results of the two clustering results are. Adjust Rand index can also be used as a method for determining whether the algorithm is applicable to certain datasets.

Accuracy is one of the most commonly used external evaluation indices. The formula is as follows:

$$AC = \frac{\sum_{i=1}^{m} c_i}{N} \qquad (3)$$

In the above, m represents the number of clusters, and N represents the number of elements in the sample set. The above formula is based on the principle of similarity comparison between the correct results and the results obtained by the clustering algorithm.

3.2   Experimental datasets

In this research, the experiments are carried out by both artificial data and real data. All the datasets are described in Table 1. The two artificial datasets: Example 1 and Example 2.

Example 1 is a set of two-dimensional datasets with the size of 150×2. In order to verify the feasibility of the algorithm, the dataset is relatively easy to separate.

Table 1. Description of datasets

Dataset	Dims	Size	Clusters
Example 1	2	150	3
Example 2	2	250	2
Iris	4	150	3
Wine	13	178	3
Wisconsin	9	683	2
Balance Scale	4	625	3

Example 2 is a set of concentric ring datasets with the size of ... This type of dataset is difficult to cluster, and the purpose is to verify whether the algorithm can deal with the linearly inseparable situations.

The four real datasets used in this research are taken from UCI [30], and they are frequently used in clustering analysis: Iris dataset, Wine dataset, Wisconsin dataset and Balance Scale dataset.

3.3   Experimental Results and Analysis

(1) We make a comparison between the experimental results on two artificial datasets based on the PC-SVR algorithm and the original clustering algorithm of similarity segmentation based point sorting algorithm (CASS-PS). The two artificial datasets are shown in Fig. 5.

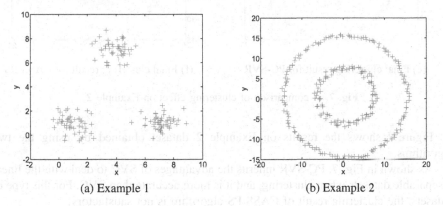

(a) Example 1   (b) Example 2

Fig. 5 Two artificial datasets

As shown in Fig. 6, the clustering results of the two algorithms on Example 1 are both satisfactory. But the PC-SVR algorithm is more accurate than the CASS-PS algorithm, and it does not have wrong clustering points. Figure 6 shows that PC-SVR which uses the sorting and clustering algorithm after sorting SVs makes the sorting process more accurate, and it does not tend to assign the data points of the same cluster to the wrong ones.

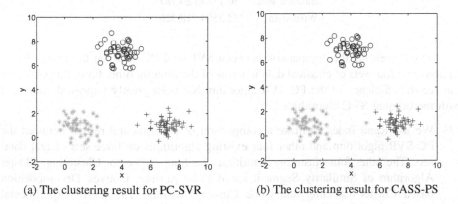

(a) The clustering result for PC-SVR   (b) The clustering result for CASS-PS

Fig. 6 A comparison of Example 1 clustering effect

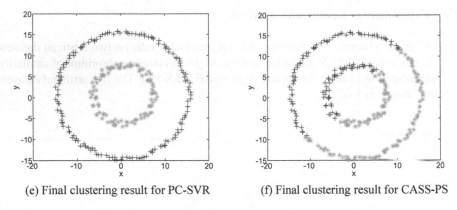

(e) Final clustering result for PC-SVR   (f) Final clustering result for CASS-PS

Fig. 7 A comparison of clustering effect on Example 2

Figure 7 shows the results on Example 2 dataset obtained by using the two algorithms.

As shown in Fig. 7, PC-SVR inherits the advantages of SVC to deal with the linear inseparable datasets when clustering, and it is more accurate than SVC. For this type of datasets, the clustering result of CASS-PS algorithm is not satisfactory.

(2)   The comparison of time cost between PC-SVR and SVC is presented in Table 2.

Table 2. Time comparison between SVC and PC-SVR (in second)

Datasets	SVC	PC-SVR
Example 1	215.0942	5.6316
Example 2	47.1435	9.3601
Iris	39.8895	4.9608
Wine	637.2953	5.6940
Balance scale	50.1387	20.787
Wisconsin	37.3594	18.326

Table 2 presents the comparison between SVC and PC-SVR on the two artificial datasets and four sets of classical data in terms of the running time. From this table, we can see the efficiency of the PC-SVR algorithm has been greatly improved compared with the original SVC algorithm.

(3)   We use Rand Index to make a comparison of experimental results between the PC-SVR algorithm and other four existing algorithms on three sets of real datasets. The other four clustering algorithms are Support Vector Clustering, Cluster Algorithm of Similarity Segment based Point Sorting, Convex Decomposition based Cluster Labeling and Cone Cluster Labeling. The above experimental results about CDCL and CCL is from reference [22]. The results are shown in Fig. 8.

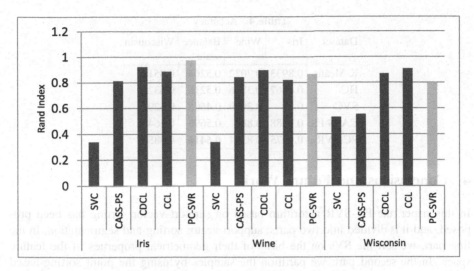

Fig. 8 A comparison of Rand Index for all five algorithms

Figure 8 reports the results for the Rand Index on the three datasets with five algorithms. We can see that PC-SVR performs the best on Iris dataset. Although does not getting the highest Rand Index values on other two datasets, PC-SVR is only 3.46 % lower than CDCL (the best one on Wine dataset) and 12.15 % lower than CCL (the best one on Wisconsin dataset).

In addition, in order to fully verify the clustering performance of the PC-SVR algorithm, we use the other two indices of clustering results to evaluate and compare the clustering performance of PC-SVR algorithm and the other four classical algorithms which are K-means, Hierarchical Clustering, Support Vector Clustering and Cluster Algorithm of Similarity Segment based Point Sorting on the four real datasets, that is, Adjust Rand Index and Accuracy. The results are listed in Tables 3 and 4.

The above results show that the PC-SVR algorithm can ensure the quality of the clustering and improve the speed of clustering, and the clustering performance is excellent.

Table 3. Ajust Rand Index

Dataset	Iris	Wine	Balance scale	Wisconsin
K-Means	0.7302	0.3711	**0.1335**	0.4914
HC	0.5621	−0.0054	0.0854	0.0073
SVC	0.00018143	0	0	0.0024
CASS-PS	0.5621	**0.6569**	0.131	0.0073
PC-SVR	**0.941**	0.6162	0.1298	**0.5921**

Table 4. Accuracy

Dataset	Iris	Wine	Balance scale	Wisconsin
K-Means	0.8933	0.7022	0.5264	0.8514
HC	0.6867	0.3876	0.5216	0.6327
SVC	0.34	0.3989	0.4608	0.6292
CASS-PS	0.7688	**0.882**	0.5696	0.6643
PC-SVR	**0.9975**	0.8384	**0.6416**	**0.8858**

4 Conclusions and Future Work

In this paper, the PC-SVR algorithm based on support vector sorting has been proposed, and it is divided into two parts: support vector sorting and segmentation. In the first part, we sort the SVs on the basis of their geometrical properties of the feature space. In the second part, we partition the samples by using the point sorting-based partition cluster algorithm and generate the clustering. Experimental results demonstrate the effect of PC-SVR for improving the performance of SVC, and better clustering performance has been achieved compared with existing approaches.

In the future work, we would explore the potential application fields of our approach, for instance, in the field of bioinformatics and social media analysis.

Acknowledgement. This work is supported by the National Natural Science Foundation of China (Grant Nos. 61472159, 61572227), Development Project of Jilin Province of China (20140101180JC,20160204022GX).

References

1. Tung, A.K., Hou, J., Han, J.: Spatial clustering in the presence of obstacles. In: 17th IEEE International Conference on Data Engineering, pp. 359–367. IEEE, Heidelberg (2001)
2. Kaufman, L.R., Rousseeuw, P.: Finding Groups in Data: An Introduction to Cluster Analysis. Hoboken NJ John Wiley & Sons Inc., New York (1990)
3. Ng, R.T., Han, J.: Efficient and Effective Clustering Methods for Spatial Data Mining. University of British Columbia, Vancouver (1994)
4. Bradley, P.S., Fayyad, U.M., Reina., C.: Scaling Clustering Algorithms to Large Databases. In: KDD, New York, pp. 9–15 (1998)
5. Zhang, T., Ramakrishnan, R., Livny, M.: BIRCH: an efficient data clustering method for very large databases. In: ACM SIGMOD International Conference on Management of Data, pp. 103–114. ACM, New York (1996)
6. Guha, S., Rastogi, R., Shim, K.: CURE: an efficient clustering algorithm for large databases. In: ACM SIGMOD Record, vol. 27, No. 2, pp. 73–84. ACM, Seattle (1998)
7. Karypis, G., Han, E.-H., Kumar, V.: Chameleon: Hierarchical clustering using dynamic modeling. Computer **32**(8), 68–75 (1999)
8. Ester, M., et al.: A density-based algorithm for discovering clusters in large spatial databases with noise. In: KDD, vol. 96, No. 34, Portland, pp. 226–231 (1996)

9. Ankerst, M., et al.: OPTICS: ordering points to identify the clustering structure. In: ACM Sigmod Record. vol. 28, pp. 49–60. ACM, Philadelphia (1999)
10. Hinneburg, A., Keim, D.A.: An efficient approach to clustering in large multimedia databases with noise. In: KDD, vol. 98, New York, pp. 58–65 (1998)
11. Wang, W., Yang, J., Muntz, R.: STING: A statistical information grid approach to spatial data mining. In: VLDB, vol. 97, Athens, pp. 186–195 (1997)
12. Sheikholeslami, G., Chatterjee, S., Zhang, A.: Wavecluster: a multi-resolution clustering approach for very large spatial databases. In: VLDB, vol. 98, New York, pp. 428–439 (1998)
13. Shavlik, J.W., Dietterich, T.G.: Readings in machine learning. Morgan Kaufmann (1990)
14. Kohonen, T.: Self-organized formation of topologically correct feature maps. Biol. Cybern. 43(1), 59–69 (1982)
15. Ben-Hur, A., et al.: Support vector clustering. J. Mach. Learn. Res. 2(12), 125–137 (2001)
16. Schölkopf, B., et al.: Estimating the support of a high-dimensional distribution. Neural Comput. 13(7), 1443–1471 (2001)
17. Ping, L., Chun-Guang, Z., Xu, Z.: Improved support vector clustering. Eng. Appl. Artif. Intell. 23(4), 552–559 (2010)
18. Yang, J., Estivill-Castro, V., Chalup, S.K.: Support vector clustering through proximity graph modelling. In: 9th International Conference on Neural Information Processing, pp. 898–903. IEEE, Singapore (2002)
19. Lee, J., Lee, D.: An improved cluster labeling method for support vector clustering. IEEE Trans. Pattern Anal. Mach. Intell. 27(3), 461–464 (2005)
20. Jung, K.H., Lee, D., Lee, J.: Fast support-based clustering method for large-scale problems. Pattern Recogn. 43(5), 1975–1983 (2010)
21. Lee, S.H., Daniels, K.M.: Gaussian kernel width exploration and cone cluster labeling for support vector clustering. Pattern Anal. Appl. 15(3), 327–344 (2012)
22. Ping, Y., et al.: Convex decomposition based cluster labeling method for support vector clustering. J. Comput. Sci. Technol. 27(2), 428–442 (2012)
23. D'Orangeville, V., et al.: Efficient cluster labeling for support vector clustering. IEEE Trans. Knowl. Data Eng. 25(11), 2494–2506 (2013)
24. Le, T., et al.: Proximity multi-sphere support vector clustering. Neural Comput. Appl. 22(7–8), 1309–1319 (2013)
25. Saltos, R., Weber, R.: A rough-fuzzy approach for support vector clustering. Inf. Sci. 339, 353–368 (2016)
26. Li, H.B., Wang, Y., Huang, L., et al.: Clustering algorithm of similarity segmentation based on point sorting. In: The International Conference on Logistics Engineering, Management and Computer Science (2015)
27. Rand, W.M.: Objective criteria for the evaluation of clustering methods. J. Am. Stat. Assoc. 66(336), 846–850 (1971)
28. Naganathan, G.K., et al.: A prototype on-line AOTF hyperspectral image acquisition system for tenderness assessment of beef carcasses. J. Food Eng. 154, 1–9 (2015)
29. Shao-Hong, Z., Yang, L., Dong-Qing, X.: Unsupervised evaluation of cluster ensemble solutions. In: 7th International Conference on Advanced Computational Intelligence, pp. 101–106. IEEE, Wuyi (2015)
30. UCI machine learning repository. http://archive.ics.uci.edu/ml

Global Recursive Based Node Importance Evaluation

Lu Zhao[1], Li Xiong[1], and Shan Xue[1,2(✉)]

[1] School of Management, Shanghai University, Shanghai 200444, China
{lulukeerle,xiongli8,xueshan}@shu.edu.cn
[2] Lab of Decision Systems and e-Service Intelligence, Faculty of Engineering
and Information Technology, School of Software,
University of Technology Sydney, Sydney 2007, Australia

Abstract. The world city network (WCN) research has been promoting the evaluation algorithms in complex networks. The study of urban network focuses on the measurements of city position in the WCN. In previous study, a set of algorithms such as centricity evaluation, power evaluation and their recursive power are proposed to support the WCN research. In this paper, we propose a novel global recursive based node importance evaluation (GRNIE) algorithm for WCN, which improves the performance in evaluating the network centricity and power by being applied to the Friedmann Basics Network. The results of the experiment show that the proposed GRNIE outperforms the previous algorithms (i.e., degree, recursive centricity, recursive power) in task of classification with the improvements of 72 %, 32 % and 20 % respectively in accuracy.

1 Introduction

In recent decade, the researchers have made great contributions to the hot topics of complex systems and urban strategies and developed a new research area of the world city network (WCN). The algorithm developments of world city have six stages, i.e., attribute analysis, relationship analysis, rank evaluation, simple WCN evaluation, WCN evaluation based on statistical analysis, and importance evaluation of nodes and links in WCN. The first stage of attribute analysis studied the attribute information of cities [19], e.g., concentration of multinational company headquarters and financial capitals. The algorithms of attribute analysis cannot well describe the relationship between cities and their resource flows status. To this end, Taylor et al. [20] proposed the WCN for world city research that was represented by the information from multinational enterprises. The global city system is denoted as a WCN in which consists a set of internal and external interactions [1]. Cities and their interactions are the main concepts in WCN. Neal [2] firstly proposed the centrality and power theory for WCN and made the foundations for other researchers in the following aspects: (1) the concept of centricity has been clearly defined and extended for WCN, (2) the concept of power has been distinguished from the concept of centricity with its own definition since then, and (3) the measurements of the centricity and the power in WCN are proposed respectively based on recursive. These three definitions of the main concepts in WCN make an important contribution to evaluating the key nodes in WCN.

© Springer International Publishing AG 2016
J. Li et al. (Eds.): ADMA 2016, LNAI 10086, pp. 738–750, 2016.
DOI: 10.1007/978-3-319-49586-6_53

The recursive algorithm provides a new approach of WCN measurement. The previous study of global urban networks only considered the direct links between cities to evaluate the locations and statuses of world cities in WCN, but ignored the influence of their neighbors' links (indirect links). The recursive algorithm indicates that the status of a node in WCN is not only influenced by the links of one node, but also by the links of all connected nodes. However, the recursive algorithm cannot completely achieve its effects and the connection strength between nodes has been ignored in these studies. To this end, in this paper, we comprehensively consider the direct links and the indirect links which are reflected by different values between nodes from one to another, and propose a novel algorithm to evaluate the centricity and power based on global recursive. The global recursive centricity algorithm can manage the links between one node to its neighbors, but the links between its neighbors as well. The process of the global recursive centricity starts from one node and walks through the network. It maps the node connections and forms a hierarchical network structure. Similarly, the global recursive power algorithm performs on all nodes and their connections in the network. Its process starts from one node along the connections between nodes through the network, and also forms a hierarchical network structure. In such a way, connections in this paper are given two meanings, i.e., the connectivity (a link) and the connection strength (a weight). The experiment results in our paper show that the proposed algorithm can better evaluate the node importance of WCN in classification accuracy.

The main contributions of this paper are as follows, (1) the effect of indirect links is involved in WCN evaluation, so that the links of a set of connected nodes are measured for the node in the set; (2) a global recursive algorithm is designed for real-world applications such as urban network; (3) parameters (e.g., weights on global links) are given by a learning algorithm of global recursive to represent the connection strength between city, which improve the algorithm of Friedmann Basics Network that calculated the weight only based on the degree of the city and avoided using learning mechanisms for the parameter training.

The rest of this paper is organized as follows. We briefly review the recursive centricity and the recursive power in Sect. 2. In Sect. 3, we propose a novel algorithm which consists of global recursive centricity and global recursive power. Finally, we conduct the experimental analysis in Sect. 4 and show our conclusions and future work in Sect. 5.

2 Related Work

Although the world city researches have been made great progress through the six stages, the WCN still faces challenges in methodologies and applications. Since the concepts of centricity and power were separated in recent research, there are a set of gaps between the concepts and the evaluation algorithms. To this end, we review the main evaluation algorithms of node importance for WCN.

In 1986, Friedmann [6] pointed out that the world city is the centre of the global economic system and the organization node, which focuses on the control and command on the strategic function of the world economy. Friedmann proposed a set of indicators of the world city: major financial centres, the concentration degree of the headquarters of multinational companies, international institutions, business services, fast growth, major manufacturing center, the main transport hub, population scale, and the world city system of 25 basis points in the classification. The top rank of the city is the world's economic control and command center, such as New York, London and Tokyo. Secondary cities are within powerful economy countries and the connections of the world economy such as Paris, Madrid and Sao Paulo [21]. In the previous studies, Friedmann [6] and Sassen [5] only used one aspect of centricity to measure world cities, i.e., resource accumulations. In such way, cities like New York, Chicago and Los Angeles, which are diffusion centers in aviation and logistics of WCN [12], should be centricities but are missing. In fact, the centricity of a WCN forms in the process of resource concentrations and diffusions [3]. The resource concentration indicates a quick process of resource gathering; and the resource diffusion represents the process that world city resources are rapidly spread in the network [16]. The definition of state-of-arts on centricity re-defined the WCN structure.

The power on each city evaluates the level of resource concentrations and diffusions which determined by the physical position and real-world function of the city in WCN [4]. Hesse [13] studied of the global supply chains and found that the cities with higher power had strong dominations than other cities. The concepts of centricity and power haven't been distinguished in the previous world city researches (stages 1 to 5). Most researchers regarded power as centricity and they used the same definition and measurements to centricity and power [10, 11]. It is the social interaction theory that first suggested WCN to separate these two concepts [7, 8]. The power in WCN is determined by the degree of dependence between cities when the resource demands are equal, but this mechanism is not suitable for some special cases which is common in real-world application, e.g., one city strongly controls another by rare resources. Cook et al. [14] found that the link strength between resource exchangers in the economic exchange network directly influenced the centricity. Those resource exchangers sharing high reliance in their relationships always show stronger power of control on their neighbors (i.e., high power) than the resource exchangers have high value of degree (i.e., high centricity). Therefore, centricity and power are not same [9].

In Neal's theory, which firstly defined centricity and power respectively, the centricity and the power are determined by degree of nodes and also by the degree of the neighbor nodes by considering the effects of direct links and indirect links between nodes.

As shown in Fig. 1, we assume there are two networks A and B, the city s illustrated by the red dot in networks A and B are with the same degree value of 3. However, as we can see, the city s is connected with three clusters in network A but the city s is connected with three single nodes in network B. If we use the evaluation algorithms proposed before Neal's, the city s gets same centricity score in two network structure; but if we use Neal's theory, the city s can be better evaluated based on the different network structure and the results will be closer to the fact.

Above example indicates the evaluation of direct links over a network is not sufficient for network analysis and involving the evaluation of indirect links is reasonable. To deal with the kind of problem in WCN node importance evaluation that cities are in a same value of direct links but in different values of indirect links, recursive centricity and recursive power were suggested by Neal.

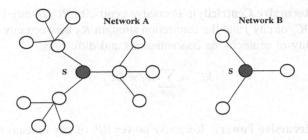

Fig. 1. An illustration of two WCNs. (Color figure online)

However, Neal's recursive algorithms on centricity and power have the following limitations.

The evaluation algorithms are poorly applied in real-work problems. We conclude this shortage into two reasons: (1) Neal proposed the framework, but did not indicate how to use the algorithm on applications; and (2) recursive algorithms work well on global network evaluations, while Neal focused on the cities in clusters and ignored the influence of the cities that belonged to other clusters [18].

The weights on links were not considered in WCN. Neal exampled a basic evaluation algorithm by detecting the connectives over WCN and set a same value of connection strength for all links. However, to determine the weights on links are important for evaluating the relationship (connection strength) which can be affected by multiple factors. The weight learning always contributes the evaluation results on accuracy.

To do the improvements on node importance evaluation in WCN, we employ the global recursive and apply it on the Friedmann Basics Network.

3 Global Recursive Based WCN Node Importance Evaluation

3.1 Preliminaries

For evaluating the nodes (cities) in WCN, the algorithms such as degree centrality, recursive centrality and recursive power have been used for node importance evaluation and link importance evaluation [10].

Definition 1 (Degree): Degree D_i of city i denotes the number of edges that nodes directly connected. It reflects direct influence of a city in the network.

$$D_i = \sum_{j \in V} a_{ij}, \tag{1}$$

where a_{ij} denotes the connection strength between city i and city j, $a_{ij} \in \{0, 1\}$.

Definition 2 (Recursive Centricity): Recursive centricity RC_i of city i is the product of the degree DC_j of city j and the connection strength R_{ij} between city i and city j. It reflects the ability of centricity on concentrations and diffusions.

$$RC_i = \sum_{j \in V} R_{ij} \times DC_j \qquad (2)$$

Definition 3 (Recursive Power): Recursive power RP_i of city i means the quotient of the degree DC_j of city j and the connection strength R_{ij} between city i and city j. It reflects the ability to control the resource flows in the network.

$$RP_i = \frac{\sum_{j \in V} R_{ij}}{DC_j} \qquad (3)$$

Based on Neal's assumption that the connection strengths of WCN are the same, we evaluate the networks in Fig. 1. As seen in Table 1, we have the following observations: (1) in network A, the recursive centricity of city s is 9, and the recursive power is 0.75; (2) in network B, the recursive centricity and recursive power are both 3.

Table 1. The evaluation results of different networks in Fig. 1.

	Degree	RC_i	RP_i
Network A	3	9	0.75
Network B	3	3	3

3.2 Algorithm of Global Recursive Based Node Importance Evaluation

In this paper, the node importance evaluation algorithm based on the global recursive is as follows. We first define the global recursive centricity and the global recursive power. Then, we design a flow chart for the evaluation framework. Finally, we propose the evaluation algorithm.

Definition 4 (Global Recursive Centricity): Global recursive centricity GRC_i is determined by the degree D_i of city i, the connection strength ω_{ij} between city i and city j, and the recursive value r_q in the qth layer. It reflects the ability of centricity on concentrations and diffusions over the world.

$$GRC_i = \sum_{q=1}^{m} \sum_{j \in V} r_q \times \omega_{ij} \times D_j, \qquad (4)$$

where m is the number of total layers, $i \neq j$, $j = 1, \dots, n$.

In the recursive process, the relationship between the nodes in each layer has different values, so we set r_q to represent the weight of the qth layers.

Definition 5 (Global Recursive Power): Global recursive power GRP_i is also determined by the degree D_i of city i, the connection strength ω_{ij} between city i and city j, and the recursive value r_q in the qth layer. It reflects the ability to control the resource flows in the network over the world.

$$GRP_i = \sum_{q=1}^{m} \sum_{j \in V} \frac{r_q \times \omega_{ij}}{D_j}, \qquad (5)$$

where $i \neq j$, $j = 1, \ldots, n$.

Based on above definitions, we propose a novel algorithm of global recursive based node importance evaluation (GRNIE) and design a workflow of the algorithm (see in Fig. 2). The algorithm of GRINE is conducted on an array model. The array model also initializes the recursive GRC and GRP. A new array is created to record the current layer by a layer recorder. In the iterations, null values are pre-processed, the layer structure is recorded, and the two-dimensional array, value of GRC and GRP are updated.

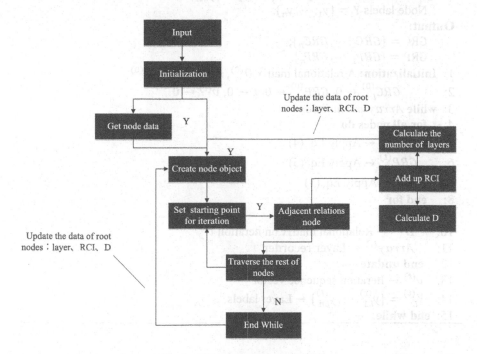

Fig. 2. The workflow of GRNIE

Step 1. A feature space $F = \{I_1, I_2, \cdots I_i, \cdots, I_n\}$ is designed for each node i, where n is the number of nodes in the WCN. The instance I_i in feature space consists of the attributes of the node $\{f_1, f_2, \cdots, f_m\}$, GRC of the node GRC_i, GRP of the node GRP_i, the layer No. of the node $l_i \in L$ and the label of the node $y_i \in Y$.

Step 2. Construct a relational matrix D in upper or lower triangular that denotes the relationships between nodes. Input it to the algorithm.

Step 3. Make initializations on relational matrix D and feature space F (exclude the attributes of the nodes).

Step 4. Design iterations on nodes $1 \sim n$ (1) Through mining the WCN, an iteration sequence vector $v^{(t)}$ in iteration t is returned. $v^{(t)}$ labels the nodes in sequence by adjacent nodes based on the global node connectives. (2) Based on the sequence of $v^{(t)}$, calculate $GRC_i^{(t)}$ and $GRP_i^{(t)}$, $i = 1, \cdots, n$. Update $F^{(t)}$. (3) Update $D^{(t)}$ and layer labels $L^{(t)}$.

Step 5. Return feature space $F = F^{(t)}$.

Algorithm 1. Global Recursive based Node Importance Evaluation (GRNIE)

Input:

        A set of city nodes $N = \{N_1, \cdots, N_n\}$;

        Node feature space $F = \{I_1, \cdots I_i, \cdots, I_n\}$;

        Attributes of feature spaces $I_i = \{f_1, \cdots, f_m\}$;

        Node labels $Y = \{y_1, \cdots, y_n\}$.

Output:

        GRC $= \{GRC_1, \cdots, GRC_n\}$;

        GRP $= \{GRP_1, \cdots, GRP_n\}$.

1: **Initialization:** A relational matrix $D^{(0)}$, $Array^{(0)} \leftarrow D^{(0)}$,

2:       $GRC_i^{(0)} \leftarrow 0, GRP_i^{(0)} \leftarrow 0, t \leftarrow 0, v^{(t)} \leftarrow \mathbf{0}$.

3: **while** $Array$ *not null* **do**

4:   **for all** nodes **do**

5:     $GRC_i^{(t)} \leftarrow$ Apply Eq. (4)

6:     $GRP_i^{(t)} \leftarrow$ Apply Eq. (5)

7:     $D_i^{(t)} \leftarrow$ Apply Eq. (1)

8:   **end for**

9:   **update**

10:     $D^{(t)} \leftarrow$ Relational matrix on iteration t

11:     $Array^{(t)} \leftarrow$ Layer recorder

12:   **end update**

13:   $v^{(t)} \leftarrow$ Iteration sequence vector

14:   $Y_L^{(t)} = \{y_{L1}^{(t)}, \cdots, y_{Ln}^{(t)}\} \leftarrow$ Layer labels

15: **end while**

---

4 Experiment

4.1 Experimental Conditions

We implement the proposed algorithm using Java data-mining tool and validate its performance on Friedmann Basics Network. We obtain the matrix data in the Excel through Java Excel API, and store them in a two-dimensional array. For each node to build a storage space, storage of every node attributes. The parameters D_i, ω_{ij}, and r_q, are initialized respectively. Moreover, all experiments are conducted on a Linux cluster node with an Interl(R) Xeon(R) @3.33 GHZ CPU and 3 GB fixed memory size.

The experiment is programed by jdk1.7.0 on eclipse_mars using three main classes, i.e., CreateTest, Port, CountRCI.

Class 1: Port. Class Port abstractions leaf nodes in the tree structure model and their attributes of name, level No. RCI information and so on. This class well maps the feature space of all nodes.

Class 2: CreateTest. Class CreateTest is used to access to the distance matrix stored in Excel which is achieved by Java Excel API. The two-dimensional array is created and updated by this class. It is form-transferred from the distance matrix. In this way, the relationship between nodes can be easily learned. We calculate the degree of the array in the class. For example, if the first row in the distance matrix has four none-zero values, the degree of the first array is 4.

Class 3: CountRCI. Class CountRCI is the main class in this experiment. It creates an object for each leaf node of the tree structure model first, which includes all attributes of the nodes; creates the two-dimensional array, which describes the relationships between the nodes based on the distance matrix, to record the key components in each iterations. In every iteration, the learning process starts from leaf nodes and follows the path recorded in the array. This class output all leaf nodes in final tree structure and the RCI node values.

The summary of code is as follows:

```
Input: Matrix"Rworldcity.xls" // java excel api
Output:
    (1) Each node of the RCI values
    (2) Each node of the Dj values: the first j points de-
grees, label each line in the matrix have how much a non-
empty value;
    (3) Rq: The qth layers, the number of iterations.
Step 1: Access data from matrix, abstract into two-
dimensional arrays, each field has a value of type int
Step 2: Traverse the two-dimensional array
for (int i=1;i<sheet[0].getRows();i++){
        for (int j=1;j<sheet[0].getColumns();j++){
            Wij[i][j] = CreateTest.get_W(i,j,workbook);
        }
    }
Step 3: Loop for each element of the array and print
    for (int i = 1; i < Wij.length; i++) {
        for (int j = 1; j < Wij[i].length; j++) {
            if (j!=Wij[i].length-1) {
                System.out.print(Wij[i][j]+",");
    }
}
```

```
        else{
          System.out.print(Wij[i][j]);
        }
      }
      System.out.println();
    }
```
Step 4: Create the node object, and for its loop initial-
ization
```
Port  pts=new Port[Wij[0].length-1];
      for (int i = 0; i < pts.length; i++) {
      pts[i]=new Port();
    pts[i].putname(i+1);
    pts[i].putLevelnums(0);
    //System.out.println(pts[i].getname());
    }
```
Step 5: Simulation as its starting from each node calcu-
lation RCI and so on.
end.

4.2 Baseline Algorithms

For comparison purposes, we use the following baseline algorithms in our experiments.

1. Degree centrality [17]: In this paper, Class CreateTest is used to access to the distance matrix stored in Excel. Node degree is the total number that each line is not empty in the matrix. For example, if the first row in the distance matrix has four none-zero values, the degree of the first array is 4;
2. Recursive centricity (RC) [22]: DC_j is calculated by degree centricity. And then, set the number of iterations of CouneRCI in two. This class output all leaf nodes in final tree structure and the RC node values;
3. Recursive power (RP) [22]: Similar with the method to get parameter of RC, into the formula to calculate of the RP.

4.3 Friedmann Dataset

We use the WCN data of Friedmann's, and establish the WCN model as shown in Fig. 3.

The results are shown in Table 2, that our evaluation algorithm of GRNIE s more efficient to evaluate urban cities in the WCN.

(1) **GRC vs. RC vs. degree:** The evaluation results on degree centricity in 25 cities only contains seven different degrees, so the accuracy is 28 %; the evaluation results of recursive centricity contains 17 different recursive central values, so the accuracy is 68 %; however, the evaluation results of GRC contains 25 different global recursive central values, so the accuracy of node recognition is 100 %.

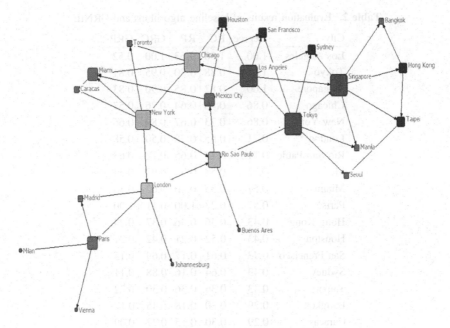

Fig. 3. Friedmann dataset sample of WCN

(2) **GRP vs. RP vs. degree:** The accuracy of degree centricity is 28 %; the evaluation results of recursive power contains 20 different recursive power values, so the accuracy is 80 % [15]; however, the evaluation results of GRC contains 25 different global recursive power values, so the accuracy of node recognition is 100 %.

In our algorithm, the results of GRC and GRP have no duplicate values. It means that the 25 cities have their individual scores. Therefore, GRNIE can accurately distinguish the cities from one another that it is more effective in WCN evaluation than other algorithms.

On the other hand, GRNIE is more accurate than other algorithms in locating the city in WCN. Those algorithms in baselines cannot apart the characteristics of centricity from power. They only can form a single indicator evaluation system in one dimension. The centricity and power can be used as two different dimensions to assess the status of city in the WCN, so that building a two-dimensional evaluation system is necessary to sub-divide the WCN. Compared with the WCN evaluation algorithms in local network areas, the global evaluation algorithm that we propose in this paper is more accurate in the process of recognition.

4.4 Learning Curves

In this section, we analysis the fitting curves of RC vs. GRC, and RP vs. GRP, respectively.

Table 2. Evaluation results of baseline algorithms and GRNIE

City	Degree	RC	RP	GRC	GRP
Los Angeles	1.00	1.00	0.63	1.00	0.52
Tokyo	1.00	0.88	0.80	0.95	0.81
Singapore	1.00	0.82	0.85	0.90	0.83
Chicago	0.86	0.76	0.64	0.76	0.57
New York	0.86	0.73	0.67	0.89	0.66
London	0.71	0.55	0.78	0.50	0.58
Rio Sao Paulo	0.71	0.70	0.65	0.78	0.63
Mexico City	0.57	0.58	0.34	0.92	0.34
Miami	0.57	0.55	0.40	0.62	0.38
Paris	0.57	0.27	1.00	0.11	1.00
Hong Kong	0.43	0.36	0.36	0.37	0.24
Houston	0.43	0.52	0.21	0.42	0.21
San Francisco	0.43	0.61	0.17	0.69	0.15
Sydney	0.43	0.64	0.16	0.88	0.11
Taipei	0.43	0.36	0.36	0.30	0.22
Bangkok	0.29	0.30	0.18	0.45	0.17
Caracas	0.29	0.30	0.15	0.27	0.09
Madrid	0.29	0.27	0.17	0.15	0.18
Manila	0.29	0.42	0.11	0.46	0.06
Seoul	0.29	0.30	0.18	0.19	0.19
Toronto	0.29	0.36	0.12	0.35	0.07
Buenos Aires	0.14	0.15	0.07	0.09	0.02
Johannesburg	0.14	0.15	0.07	0.07	0.01
Milan	0.14	0.12	0.09	0.06	0.05
Vienna	0.14	0.12	0.09	0.03	0.04

In Figs. 4 and 5, the curve fitting results show that the GRC and GRP of GRNIE have better performance in city rankings than the local recursive algorithms on RC and RP. The fitting results of accuracy indicate that GRNIE is more accurate and more suitable to identify different types of nodes. Therefore, the GRNIE proposed in this paper is effective and feasible.

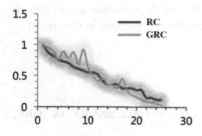

Fig. 4. Curve fittings of RC and GRC

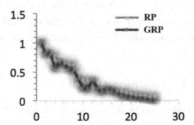

Fig. 5. Curve fittings of RP and GRP

5 Conclusions

In this paper, we focus on how to identify cities in the world, analyze their physical locations in WCN and their characteristics of centricity and power. To improve the recursive algorithms on node importance evaluations in WCN, we propose a novel GRNIE algorithm with a workflow. The GRC and GRP are defined with equations. In the experiment on the Friedmann dataset, we conclude the following findings.

(1) *Centricity and power should be distinguished.* Since a city in the WCN often have bi-level functions, a two-dimensional description for a city is more suitable than the single description on a city. To this end, to evaluate the two characteristics of centricity and power are reasonable for real-world applications.

(2) *Global recursive is well performed on node importance evaluation in WCN.* The evaluation algorithm of node importance based on global recursive is effective. The algorithm can accurately rank the city in WCN.

(3) *GRNIE is accurate in WCN.* The position of a city in WCN is not only influenced by its own functions and statues, but also the functions and statues of its neighbor too. The direct links and the indirect links of a city node contribute to node importance evaluation in global weights. Therefore, the proposed GRNIE can accurately evaluate the positions of the city in the WCN.

Acknowledgments. This work is supported by the Shanghai Education Commission (No. 14ZS085).

References

1. Beckfield, J., Alderson, A.S.: Reply: whither the parallel paths? The future of scholarship on the world city system. Am. J. Sociol. **112**(3), 895–904 (2006)
2. Neal, Z.: Differentiating centrality and power in the world city network. Urban Stud. **48**, 2733–2748 (2011)
3. Castells, M.: The Informational City: Information Technology, Economic Restructuring, and the Urban Regional Process. Blackwell, Oxford, Cambridge (1989)
4. Sassen, S.: The Global City: New York, London, Tokyo, Princeton. Princeton University Press, Princeton (1991)
5. Sassen, S.: Global intercity networks and commodity chains: any intersections. Glob. Netw. **10**, 150–163 (2010)
6. Friedmann, J.R.: The world city hypothesis: development and change. Urban Stud. **23**(2), 59–137 (1986)
7. Homans, G.C.: Social behavior as exchange. Am. J. Sociol. **63**(6), 597–606 (1958)
8. Liu, X., Derudder, B.: Two-mode networks and the interlocking world city network model: a reply to Neal. Geogr. Anal. **44**, 171–173 (2012)
9. Sanderson, M.R., Derudder, B., Timberlake, M., Witlox, F.: Are world cities also world immigrant cities? An international, cross-city analysis of global centrality and immigration. Int. J. Comp. Sociol. **56**(3–4), 173–197 (2015)
10. Hu, P., Fan, W., Mei, S.: Identifying node importance in complex networks. Phys. A **429**, 169–176 (2015)

11. Wheeler, J.O., Mitchelson, R.L.: Information flows among major metropolitan areas in the United States. Ann. Assoc. Am. Geogr. **79**(4), 523–543 (1989)
12. Hesse, M.: Cities, material flows and the geography of spatial interaction: urban places in the system of chains. Glob. Netw. **10**(1), 75–91 (2010)
13. Cook, K.S., Emerson, R.M., Gillmore, M.R., Yamagishi, T.: The distribution of power in exchange networks: theory and experimental results. Am. J. Sociol. **89**(2), 275–305 (1983)
14. Boyd, J.P., Mahutga, M.C., Smith, D.A.: Measuring centrality and power recursively in the world city network: a reply to Neal. Urban Stud. **50**(8), 1641–1647 (2013)
15. Taylor, P.J., Derudder, B., Faulconbridge, J., Hoyler, M., Ni, P.: Advanced producer service firms as strategic networks, global cities as strategic places. Econ. Geogr. **90**(3), 267–291 (2014)
16. Alderson, A.S., Beckfield, J., Sprague-Jones, J.: Intercity relations and globalisation: the evolution of the global urban hierarchy, 1981–2007. Urban Stud. **47**(9), 1899–1923 (2010)
17. Peng, G., Wu, J.: Optimal network topology for structural robustness based on natural connectivity. Phys. A **443**, 212–220 (2016)
18. Hall, P.: The World Cities. Heinemann, London (1966)
19. Taylor, P.J., Ni, P., Derudder, B., et al.: Measuring the world city network: new developments and results (2009). http://www.lboro.ac.uk/gawc/rb/rb300html
20. Friedmann, J.: Where we stand: a decade of world city research. In: World Cities in a World-System, pp. 21–47 (1995)
21. John, P.B., Matthew, C.M., David, A.S.: Measuring centrality and power recursively in the world city network: a reply to neal. Urban Stud. **50**(8), 1641–1647 (2013)
22. Neal, Z.: Does world city network research need eigenvectors? Urban Stud. **50**(8), 1648–1659 (2013)

Extreme User and Political Rumor Detection on Twitter

Cheng Chang, Yihong Zhang[✉], Claudia Szabo, and Quan Z. Sheng

School of Computer Science, The University of Adelaide, Adelaide, SA 5005, Australia
cheng.chang@student.adelaide.edu.au,
{yihong.zhang,claudia.szabo,michael.sheng}@adelaide.edu.au

Abstract. Twitter, as a popular social networking tool that allows its users to conveniently propagate information, has been widely used by politicians and political campaigners worldwide. In the past years, Twitter has come under scrutiny due to its lack of filtering mechanisms, which lead to the propagation of trolling, bullying, and other unsocial behaviors. Rumors can also be easily created on Twitter, e.g., by extreme political campaigners, and widely spread by readers who cannot judge their truthfulness. Current work on Twitter message assessment, however, focuses on credibility, which is subjective and can be affected by assessor's bias. In this paper, we focus on the actual message truthfulness, and propose a rule-based method for detecting political rumors on Twitter based on identifying *extreme users*. We employ clustering methods to identify news tweets. In contrast with other methods that focus on the content of tweets, our unsupervised classification method employs five structural and timeline features for the detection of extreme users. We show with extensive experiments that certain rules in our rule set provide accurate rumor detection with precision and recall both above 80 %, while some other rules provide 100 % precision, although with lower recalls.

1 Introduction

In recent years, Twitter as a popular social networking tool has gained increasing popularity. In addition to daily conversations among friends and observations of newsworthy events created by ordinary users, politicians and political campaigners also use Twitter to create and publicise campaigns and advance their political agendas. For example, the current US president Barack Obama[1] and the former State Secretary and current Democrat Party presidential candidate hopeful Hillary Clinton[2] are known for their frequent use of Twitter as a publicity tool. However, Twitter does not filter its content, leading to rumors, trolling, and bullying becoming widespread in some topics, such as GamerGate[3] and discussions about the current refugee crises[4]. In politics, rumors can be created by

---

[1] https://twitter.com/BarackObama.
[2] https://twitter.com/HillaryClinton.
[3] https://en.wikipedia.org/wiki/Gamergate_controversy.
[4] https://en.wikipedia.org/wiki/European_migrant_crisis.

© Springer International Publishing AG 2016
J. Li et al. (Eds.): ADMA 2016, LNAI 10086, pp. 751–763, 2016.
DOI: 10.1007/978-3-319-49586-6_54

extreme political campaigners, and then widely spread by readers. In this paper, we propose an analysis method to automatically detect political rumors.

Extensive research has been done to evaluate the credibility of tweets on Twitter [2,4,5]. However, studies have illustrated the gap between perceived credibility and the actual truthfulness [11]. Rumors are very difficult to model. Semantically, there is no explicit border between rumor and non-rumor statements, and the difference between them is highly conceptual. To address this issue, we develop a method based on modeling tweet authors and rumor propagators, rather than the rumor content. Our method employs structural information about the tweet feed and does not rely on content analysis.

We propose a rule-based method to detect rumor tweets, based on the observation that particular kinds of malicious users, which we call *"extreme users"*, tend to post false news constantly on Twitter. We hypothesize that the truthfulness of news tweets could be judged according to the number of the extreme users involved. In our method, the retrieved tweets are first grouped into clusters so that each cluster represents the propagation of a piece of news. Then the extreme users involved in each cluster are identified. Finally the truthfulness of each cluster is determined by the proportion of extreme users. As the main contribution of our work, we identify two kinds of rumors: (i) news rumors, where the tweets have URLs pointing to external news articles, and (ii) picture rumors, where tweets contain pictures with rumor statements.

Experiments have been performed to prove the effectiveness of our news rumor detection and picture rumor detection. We collected and manually labeled two datasets involving Hillary Clinton and Barack Obama. The Hillary Clinton dataset contains 45 rumors and 151 non-rumors and the Barack Obama dataset contains 87 rumors and 107 non-rumors. We tuned the parameters using exhaustive search on possible values and determined the best rules. The results show that the picture rumor detection task was performed satisfactorily with high precision and recall in both datasets, while the news rumor detection only achieves good results in the Obama dataset.

The remainder of this paper is organized as the following: in Sect. 2, we discuss related works. Section 3 presents our rule-based rumor detection approach. In Sect. 4, we present our experiments on detecting news and picture rumors in real Twitter datasets. Finally, Sect. 5 concludes this paper.

2   Related Work

Rumor detection-related researches include emerging topic detection, topic classification, credibility analysis, and information propagation. We discuss these in the following. The first step to detect rumors on Twitter in previous studies is usually to identify emerging events and group tweets together based on the event they describe. TwitterMonitor is a trend detection system over Twitter developed by Mathioudakis et al. in 2010 [8]. It could identify emerging topics on Twitter in real time and also provide meaningful analytics. Unankard et al. proposed another method to detect emerging events on Twitter, by clustering the tweets based on text similarity and location correlation [14]. The clustering algorithm is also useful for rumor detection, since the first step to detect rumors is that of grouping similar tweets so that more information about the topic could be discovered.

Another group of work focuses on detecting special contents such as controversial topics and spams. Popescu and Pennacchiotti developed a method to detect controversial events on Twitter [12]. A controversial event is defined as a public discussion with opposite opinions. Three detection models based on sentiment and controversy dictionaries are proposed. Lee et al. analyzed the collective attention spam activities and trained a classifier to detect such spam tweets [7]. They found that the spam behavior has great correlation with the short longevity of the account, user features such as a small number of following/follower/tweets, and content features such as containing same URLs and adult keywords. There are other applications sharing similar methods to rumor detection. Imran et al. analyzed disaster related tweets on Twitter and separated disaster informative messages from non-informative ones by applying supervised machine learning techniques [1]. Zhang et al. studied perspective classification for Twitter messages [15] and employed supervised machine learning methods to classify the tweets into observations, affection messages, and speculations.

With respect to misinformation on social media, however, current works focus more on message credibility, rather than its truthfulness, and tend to be supervised rather than unsupervised approaches, thus further introducing assessor bias [2,11]. Moreover, supervised methods requires manual effort in preparing training examples, which is expensive and does not have guaranteed quality. Castillo et al. developed supervised learning methods to discover news events and determine their credibility [2]. Tweets are first clustered into topics and then newsworthy topics and their credibility are determined by a model trained in advance. It achieves 89 % accuracy in the news discovery task and 70 % in the credibility determination task. Alethiometer is a framework to assess tweet credibility [5] that utilizes supervised machine learning and more complex features, including the author's reputation measured by the number of comments or likes, the author's history measured by the frequency of posts, and the message popularity measured by the number of retweets. TweetCred is a credibility assessment tool [4] that is based on semi-supervised learning and is installed as web browser plugin in the front-end. Another event credibility assessment tool utilizes a PageRank-like algorithm to model the credibility network and using global optimization to propagate the credibility among related users and events [3]. The authors found that users often forward the tweets they read without verification because they trust the users they are following. However, as discussed by Morris et al. [11], credibility could be affected by many characteristics such as the assessor's bias, and does not always reflect the truthfulness of the message. Another study by Kakol et al. [6] showed that strong bias toward positive values exists when users read web content although the subjectivity factors are not significant, and also revealed some of the bias in evaluating credibility or detecting rumors in previous works.

Studies that model the propagation of rumor on Twitter are also of interest. Metaxas et al. developed a tool called TwitterTrails[5] to investigate the origin and propagation of rumors and analyze their truthfulness by visualizing burst activities, propagation timeline, propagation networks, level of visibility and skepticism [9]. Miyabe et al. proposed a rumor tweets extraction system based on rumor-disaffirmation tweets [10]. They found that rumor tweets have stronger behavior facilitation and let readers feel more negative, rather than positive as normal tweets. They also found that the rumors could be divided into two categories: obvious rumors and unapparent rumors, with different propagation behaviors. Qazvinian et al. developed a rumor tweet extraction algorithm given the rumor events, and they found that user history is a good indicator of posting rumors [13]. These papers focused on extracting rumors after they were disaffirmed and studying their characteristics. In contrast, in our paper we focus on identifying rumor tweets without any prior knowledge.

3   Political Rumor Detection

When employing only tweet texts, our analysis shows that rumors are indistinguishable from real news. As such, we focus our approach on the users who post the rumors. The main insight of our method is based on the finding that a kind of users, which we call *extreme users*, tend to post false news or statements on Twitter. If most of the tweets about a news or statement are propagated by these extreme users, it can be determined that the news or statement is a rumor. Our analysis found that the extreme users share many common features. They might have a large number of followers, a high tweeting frequency, and show passion about the target topic. We define a set rules for these features to determine whether a user is extreme. A sample rule is "IF users following count ≥ 2000 AND tweeting frequency ≤ 5 min AND the percentage of the users recent tweets containing extreme keywords $\geq 10\%$ THEN the user is extreme ELSE the user is not extreme". These kind of extreme users are most common in political discussions, and therefore we currently only focus on political news tweets, and leave other types of tweets for future work.

Our approach mainly involves two steps. First, we cluster messages to find news and pictures that are potentially rumor. Second, we identify extreme users, who we consider likely to create and propagate rumors. Then based on the proportion of extreme users in the cluster, we identify rumor clusters. In this section, we will present our approach in detail.

3.1   Determining News Propagation by Clustering

We can broadly distinguish political tweets into two broad categories, namely, tweets that contain a URL pointing to an article on a news website, and tweets that contain a picture. In both cases, a URL link is associated with the tweet. The

---

[5] http://twittertrails.com/.

news are usually generated by news agencies such as news websites, or some news Twitter accounts. Most of Twitter users who post message containing the news URL are propagating the information or discussing the content of the news. Since we do not judge the news by its content but rather by the users who propagate it, first we generate the state of propagation of a news, by clustering the tweets that contain the same URL link, including the original tweet and retweets. We also merge similar clusters that discuss the same news into one cluster based on cosine similarities. The clustering process is shown in Algorithm 1.

Algorithm 1. Determining News Propagation by Clustering

1: **for** each tweet **do**
2:    Process the URLs in the tweet (resolve shortened URLs &eliminate non-news links)
3:    Add tweet into existing clusters by matching the news URLs.
4:    Create new clusters for tweet with first time appearing news URLs
5: **end for**

The pre-processing of the URL (line 2) first filters out URLs that are unlikely to be news articles. By observation, most web pages reporting news have the URL format like: "http://www.theguardian.com/us-news/2015/may/13/bill-clinton-hopes-move-back-white-house-2016-if-invited". We can see that the path is very long, has many levels of sub-directories and contains a news title in the last sub-directory. Therefore we filter out URLs that are too short (≤ 50 characters), have few sub-directories (≤ 3), and do not contain dash in the last directory. Also shortened URLs are resolved using LongURL.orgs shortend URL expanding service[6]. Clusters are expressed as news links pointing to the full story, propagating tweets for the news, and the involved user information.

3.2   Rumor Detection Based on Extreme Users

In the next step, authors of clustered tweets are analyzed, and, if the percentage of extreme users in a cluster is above a given threshold θ, the cluster and the news is considered a rumor. The overall process is shown in Algorithm 2.

We do not count the number of unique users in a cluster. Instead, we aggregate the author of a retweet with the author of the original tweet. If any of them is judged as an extreme user, the tweet is considered as written by an extreme user (**Lines 4, 8**). At the same time, duplicate users who post more than one tweet in a cluster are also counted. This avoids the case whereby the rumor is posted by an extreme user and widely believed and retweeted by normal users.

We aggregate the author of a retweet with the original author because a tweet reader will see both authors' names above the tweet. This is equivalent to that both authors are responsible for the credibility of tweet. Besides, the retweet

---

[6] http://longurl.org.

Algorithm 2. Judge rumor by proportion of extreme users

INPUT: c: the cluster to be judge
OUTPUT: : whether c is a rumor or not
1: $cnt \leftarrow 0$, $cnt_{ex} \leftarrow 0$
2: **for** each tweet t in c **do**
3:     $u \leftarrow$ the author of t
4:     **if** $JudgeExtremeUser(u)$ **then**
5:         $cnt_{ex} \leftarrow cnt_{ex} + 1$
6:     **else if** t is retweet **then**
7:         $ou \leftarrow$ the original author of t
8:         **if** $JudgeExtremeUser(ou)$ **then**
9:             $cnt_{ex} \leftarrow cnt_{ex} + 1$
10:         **end if**
11:     **end if**
12:     $cnt \leftarrow cnt + 1$
13: **end for**
14: **if** $(cnt_{ex}/cnt) \geq \theta$ **then**
15:     **return** *true*
16: **else**
17:     **return** *false*
18: **end if**

---

depth is always one on Twitter. For example, if B retweeted a tweet written by A, and C saw B's retweet and decided to forward it, it would be shown that "C retweeted the tweet from A" instead of "C retweeted the retweet from B". Thus, the original author will appear in all of the retweets.

There are three common cases of the propagation of rumors we observed on Twitter. First, a rumor tweet is created or first propagated by an extreme user, and then widely retweeted by other extreme users and normal users. In this case, if the original extreme user is identified, the rumor would be identified. Second, the rumor is created outside Twitter such as from a news website, and then it is posted on Twitter by the websites official account and retweeted by many users. Third, the rumor is created from external website and posted on Twitter by many extreme users and normal users. In the last two cases, if enough extreme users are determined, the rumor can be identified.

3.3 Judging Extreme Users

Extreme users are a kind of users who have great passion about a topic and tend to post messages that are not credible with some unknown intentions, regardless of whether they are driven by beliefs or money. An example of an extreme user is shown in Fig. 1.

It can be seen that the author posts a status about "The uranium deal makes you forget about the $6 billion" in the picture and "Can you think of anyone in real world losing 6 billion", which are rumors. A photo of Hillary Clinton is shown on the side of the words highlighting that this event is related

Fig. 1. An extreme user example

to her. Three abnormal hashtag keywords are presented at the end of the tweets: #Hillary4Prison2016, #StopHillary2016 and #WakeUpAmerica. From the hashtags, we can infer that the intention of this tweet is to fight against Hillary Clintons 2016 presidential candidacy.

We observe that extreme users have some of the following common characteristics:

- a large number of followers (e.g., 23 K, which significantly exceeds normal users)
- high tweeting frequency (e.g., several minutes between tweets)
- topic concentration (e.g., 80 % recent tweets are about the same topic)
- use of extreme keywords in description or tweets

Since extreme users could be motivated by some strong intentions, they constantly post tweets supporting their opinions about certain topics. The frequency of their tweeting behavior is very high and they employ extreme keywords as labels. At the same time, these users desire to enlarge their influence and tend to follow a large number of users, expecting them to follow back. Based on these observations, we define and measure the key features of the users and set up rules to identify them. Table 1 lists our six selected features.

We compare the user characteristics to a list of thresholds, each corresponding to a feature in Table 1. Given thresholds $\Lambda = \{\lambda_{TF}, \lambda_{FC}, \lambda_{RT}, \lambda_{RP}, \lambda_{RE}\}$, the process of judging an extreme user is shown in Algorithm 3.

4   Experimental Analysis

We set up experiments to validate our approach. More specifically, we aim to find out that whether the best parameters perform satisfactorily (i.e., precision ≥ 0.80 and recall ≥ 0.80). We test various combinations of parameters, separately for news rumors and picture rumors, as shown in Tables 2 and 3, where "NU" means the parameter and the associated rule were *not used*.

Table 1. Extreme user features

Code	Description
DE	Whether the description contains extreme keywords
FC	The following count of the user
TF	Tweeting frequency, measured by the median interval of the user's recent 100 tweets in minutes
RT	The percentage of the user's recent 100 tweets that contain topic-related keywords
RP	The percentage of the user's recent 100 tweets that contain pictures
RE	The percentage of the user's recent 100 tweets that contain extreme keywords

---

Algorithm 3. Judge Extreme User

INPUT: the feature values of a user
OUTPUT: whether the user is extreme or not
1: **if** DE **then return** *true*
2: **else if** $(TF \leq \lambda_{TF}) \wedge (FC \geq \lambda_{FC}) \wedge (RT \geq \lambda_{RT}) \wedge (RP \geq \lambda_{RP}) \wedge (RE \geq \lambda_{RE})$ **then return** *true*
3: **else**
4:     **return** *false*
5: **end if**

Using collected tweets regarding two prominent US political actors, Hillary Clinton and Barack Obama, we tested these rules and measured the detection accuracy. In the end we found that the best performing rules achieved satisfying accuracy, and in both news rumor and picture rumor detection, we found rules that reached ≥ 0.80 precision and ≥ 0.80 recall.

4.1 News Rumor Detection

We collected tweets about Hillary Clinton in August 2015 and tweets about Barack Obama in September 2015. Hillary Clinton announced that she would campaign for the 2016 U.S. presidency election in April, but at the time of tweet collection she was addressing the e-mail server controversy. Obama is the current

Table 2. The parameters of the best performing rules in news rumor detection

Code	θ	DE	λ_{FC}	λ_{TF}	λ_{RT}	λ_{RP}	λ_{RE}
Rule 1	60 %	Use	NU	5 min	10 %	70 %	NU
Rule 2	10 %	NU	NU	NU	70 %	30 %	5 %
Rule 3	50 %	Use	NU	5 min	70 %	10 %	5 %
Rule 4	10 %	NU	1000	NU	50 %	30 %	5 %
Rule 5	10 %	NU	NU	5 min	70 %	30 %	5 %

Table 3. The parameters of the best performing rules in picture rumor detection

Code	θ	DE	λ_{FC}	λ_{TF}	λ_{RT}	λ_{RP}	λ_{RE}
Rule 11	50 %	NU	NU	NU	30 %	30 %	10 %
Rule 12	60 %	Use	NU	NU	50 %	30 %	10 %
Rule 13	40 %	NU	NU	NU	10 %	70 %	5 %
Rule 14	60 %	NU	5000	NU	30 %	70 %	10 %
Rule 15	40 %	NU	5000	15 min	70 %	50 %	10 %

U.S. president and will end his presidency in 2016. The Hillary dataset contains 60,511 tweets posted by 32,957 users, and the Obama dataset contains 84,827 tweets posted by 38,753 tweets. Shortened URLs were resolved and tweets with invalid links, such as those containing only a domain name, were removed, before we clustered the tweets using our news clustering algorithm.

In the Clinton dataset, our program identifies 6,221 news clusters. The largest news cluster contains 1,908 tweets focused on the news "*Hillary Clinton's email firm had its servers in the bathroom of a Denver apartment*". The largest picture cluster contains 619 tweets saying "*now that the Supreme Court has once again re-affirmed the ACA as the law of the land. It's time for Republican attacks to end - Hillary Clinton*". In the Obama dataset, our program identified 7,078 news clusters. The largest news cluster contains 920 tweets reporting "*Obama staffers party, smoke Cuban cigars on anniversary of Benghazi, 9–11 terror attacks*". The largest picture cluster containing 14,567 tweets reports "*Obama debt forgiveness law passed*", which is an old false rumor.

In the Clinton dataset, we manually labeled 22 large rumor clusters containing 3,812 tweets and 75 non-rumors containing 10,406 tweets. In the Obama dataset, we manually labeled 13 rumors containing 1,655 tweets and 22 non-rumors containing 2,644 tweets.

The precision, recall and F1Score results for detecting news rumors using the selected rules are listed in Table 4. The highest precision, recall and F1Score in each dataset are highlighted in bold. According to the results, rule 1 is the best rule in the news rumor detection task, which obtained the highest F1score (0.85) in the Obama dataset and relatively high F1score (0.70) in the Clinton dataset. Rule 2 is the second best rule in the Obama dataset, which has a higher precision (0.91) and lower recall (0.77) than Rule 1 in the Obama dataset and lower precision (0.63) and higher recall (0.88) in the Clinton dataset. Rule 3 has the highest precision in the Clinton dataset (1.00) and rule 4 has the highest precision (1.00) in the Obama dataset. Rule 5 has the highest F1score in the Clinton dataset (0.77) and similar score in the Obama dataset (0.78).

We note that many rules behave oppositely in the Clinton dataset and the Obama dataset. Rule 1 achieves 0.85 precision and 0.85 recall in the Obama dataset, which means the news rumor detection performs very well in this dataset, while it performs not well in the Clinton dataset, with only 0.65 precision and 0.76 recall. Other rules, such as rule 2, 4 and 5, having high precision

Table 4. The results of the best performing rules in news rumor detection

The Clinton dataset

Rule	Detected	TP	Precision	Recall	F1Score
Rule 1	20	13	0.65	0.76	0.70
Rule 2	24	15	0.63	0.88	0.73
Rule 3	7	7	**1.00**	0.41	0.58
Rule 4	21	14	0.67	0.82	0.74
Rule 5	22	15	0.68	0.88	**0.77**

The Obama dataset

Rule	Detected	TP	Precision	Recall	F1Score
Rule 1	13	11	0.85	0.85	**0.85**
Rule 2	11	10	0.91	0.77	0.83
Rule 3	5	5	**1.00**	0.38	0.56
Rule 4	9	9	**1.00**	0.69	0.82
Rule 5	10	9	0.90	0.69	0.78

in the Obama dataset gain high recall in Clinton dataset (e.g. Rule 2 obtains 0.63 precision and 0.88 recall in Clinton dataset but gets 0.91 precision and 0.77 recall in Obama dataset). Thus, these rules are considered as problematic and should be discarded.

In addition, most of the top rules did not utilize the parameter λFC, which means it is not a good indicator in the judgment of extreme users. Probably because the following count is always increasing along with the time using Twitter. There is not an explicit border between the following counts of extreme users and non-extreme users. This point brings an insight for us to include the proportion of following count over Twitter age or the proportion of following count over follower count as features in the future work.

4.2 Picture Rumor Detection

In the Clinton dataset, our program identifies 2,773 picture clusters, and we manually labeled 23 rumors containing 956 tweets and 76 non-rumors containing 3,797 tweets. In the Obama dataset, our program identified 3,028 picture clusters, and we manually labeled 74 rumors containing 2,956 tweets and 85 non-rumors containing 3,355 tweets. Examples of labeled picture are shown in Fig. 2.

The precision, recall and F1Score results for detecting picture rumors using selected rules are list in Table 5. According to the results, Rule 11 is considered as the best rule in both datasets with high precision and recall (0.80 precision and 0.87 recall in the Clinton dataset and 0.83 precision and 0.81 recall in the Obama dataset). Rule 12 performs well in the Clinton dataset (0.83 precision and 0.83 recall) but not well in the Obama dataset (0.76 precision and 0.76 recall), while rule 13 performs the best in the Obama dataset (0.83 precision and 0.89

Fig. 2. Examples of picture rumors

recall) but not well in the Clinton dataset (0.71 precision and 0.87 recall). Rule 14 and rule 15 have the highest precisions in the two datasets.

Table 5. The results of the best performing rules in picture rumor detection

The Clinton dataset					
Rule	Detected	TP	Precision	Recall	F1Score
Rule 11	25	20	0.80	0.87	**0.83**
Rule 12	23	19	0.83	0.83	**0.83**
Rule 13	28	20	0.71	0.87	0.78
Rule 14	13	11	0.85	0.48	0.61
Rule 15	13	13	**1.00**	0.57	0.72
The Obama dataset					
Rule	Detected	TP	Precision	Recall	F1Score
Rule 11	72	60	0.83	0.81	0.82
Rule 12	74	56	0.76	0.76	0.76
Rule 13	80	600	0.83	0.89	**0.86**
Rule 14	34	33	**0.97**	0.45	0.61
Rule 15	35	30	0.86	0.41	0.55

From the results we can see that the picture rumor detection performs very well in both datasets (Rule 11 obtains 0.80 precision and 0.87 recall in Clinton dataset and 0.83 precision and 0.81 recall in Obama dataset). Also, higher precision could be reached (Rule 14 and rule 15: precision ≥0.85) but the recall rates are much lower (0.4 0.6). These rules are suitable for the cases that the rumor detection method must have lowest false positive rate or lowest false negative rate. In addition, most of the rules in Clinton dataset and Obama dataset behave consistently, i.e. the rules having high precision and low recall (or low precision and high recall) in the Obama dataset also achieve high precision and low recall (or low precision and high recall) in the Clinton dataset. Thus, the rules in the picture rumor detection are more stable than the rules in the previous news rumor detection.

5   Discussion

The limitation in our approach mainly lies in the use of parameters. As our experiments have shown, the best rules for different dataset and different type of rumors are not the same. This implies that given a new dataset, we may not know which rules are the best without first run some tests. Also the use of extreme keywords for identifying extreme user is a drawback. Here we show that the extreme users who participate in discussing two political actors often have extreme keywords in their description. However, in other types of political discussions, extreme keywords may not present. For example, when we examined rumors about European Refugee Crisis, we have not found users who helped spreading rumors have extreme keywords in their user description. The applicability of our method therefore requires further investigation.

However, where our method do apply, most often in political debates that involve extreme users attacking political actors, it can achieve satisfactory accuracy in detecting rumors, as our experiments have shown. The method we present here can therefore be used as an additional analysis for online information filtering that involves political debates, as well as preventing rumor propagation in early stages.

6   Conclusion

As Twitter emerged as an important information exchange platform, identifying rumors and false information on Twitter has become a critical and challenging research topic. In this paper, we propose a method for detecting political rumors on Twitter based on identifying *extreme users*. We show with extensive experiments that, with the right choice of parameters, news and picture rumors can be detected with very high accuracy. In the experiment, we observe that certain rules provide accurate detection with precision and recall both above 80 %, while some other rules provide 100 % precision, although with lower recalls.

We consider our findings as an important first step to build an automatic political rumor detection system on Twitter. Our studies of effects of various parameters on rumor detection allow them to be effectively incorporated into learning models and classifiers. In the future, we plan to investigate methods that automatically generate topic and extreme keywords, as well as tune the parameters and provide realtime rumor detection.

References

1. Al-Ali, A., Zualkernan, I., Aloul, F.: A mobile GPRS-sensors array for air pollution monitoring. IEEE Sensors J. **10**(10), 1666–1671 (2010)
2. Castillo, C., Mendoza, M., Poblete, B.: Information credibility on Twitter. In: Proceedings of the 20th International World Wide Web Conference, pp. 675–684 (2011)
3. Gupta, A., Kumaraguru, P.: Credibility ranking of tweets during high impact events. In: Proceedings of the 1st Workshop on Privacy and Security in Online Social Media, p. 2. ACM (2012)
4. Gupta, A., Kumaraguru, P., Castillo, C., Meier, P.: TweetCred: real-time credibility assessment of content on twitter. In: Aiello, L.M., McFarland, D. (eds.) SocInfo 2014. LNCS, vol. 8851, pp. 228–243. Springer, Heidelberg (2014). doi:10. 1007/978-3-319-13734-6_16
5. Jaho, E., Tzoannos, E., Papadopoulos, A., Sarris, N.: Alethiometer: a framework for assessing trustworthiness and content validity in social media. In: Proceedings of the 23rd International Conference on World Wide Web Companion, pp. 749–752 (2014)
6. Kakol, M., Jankowski-Lorek, M., Abramczuk, K., Wierzbicki, A., Catasta, M.: On the subjectivity and bias of web content credibility evaluations. In: Proceedings of the 22nd International Conference on World Wide Web Companion, pp. 1131–1136 (2013)
7. Lee, K., Caverlee, J., Kamath, K.Y., Cheng, Z.: Detecting collective attention spam. In: Proceedings of the 2nd Joint WICOW/AIRWeb Workshop on Web Quality, pp. 48–55. ACM (2012)
8. Mathioudakis, M., Koudas, N.: TwitterMonitor: trend detection over the twitter stream. In: Proceedings of the ACM SIGMOD International Conference on Management of data, pp. 1155–1158 (2010)
9. Metaxas, P.T., Finn, S., Mustafaraj, E.: Using twittertrails.com to investigate rumor propagation. In: Proceedings of the 18th ACM Conference Companion on Computer Supported Cooperative Work & Social Computing, pp. 69–72. ACM (2015)
10. Miyabe, M., Nadamoto, A., Aramaki, E.: How do rumors spread during a crisis? Analysis of rumor expansion and disaffirmation on twitter after 3.11 in Japan. Int. J. Web Inf. Syst. **10**(4), 394–412 (2014)
11. Morris, M.R., Counts, S., Roseway, A., Hoff, A., Schwarz, J.: Tweeting is believing?: understanding microblog credibility perceptions. In: Proceedings of the ACM Conference on Computer Supported Cooperative Work, pp. 441–450 (2012)
12. Popescu, A.-M., Pennacchiotti, M.: Detecting controversial events from twitter. In: Proceedings of the 19th ACM International Conference on Information and Knowledge Management, pp. 1873–1876 (2010)
13. Qazvinian, V., Rosengren, E., Radev, D.R., Mei, Q.: Rumor has it: identifying misinformation in microblogs. In: Proceedings of the Conference on Empirical Methods in Natural Language Processing, pp. 1589–1599. Association for Computational Linguistics (2011)
14. Unankard, S., Li, X., Sharaf, M.A.: Emerging event detection in social networks with location sensitivity. World Wide Web **18**(5), 1393–1417 (2015)
15. Zhang, Y., Szabo, C., Sheng, Q.Z., Fang, X.S.: Classifying perspectives on twitter: immediate observation, affection, and speculation. In: Proceedings of the 16th International Conference on Web Information Systems Engineering, Part I, pp. 493–507 (2015)

Deriving Public Sector Workforce Insights: A Case Study Using Australian Public Sector Employment Profiles

Shameek Ghosh[1(✉)], Yi Zheng[1], Thorsten Lammers[1], Ying Ying Chen[2], Carolyn Fitzmaurice[2], Scott Johnston[2], and Jinyan Li[1]

[1] University of Technology Sydney (UTS), Ultimo, Australia
{Shameek.Ghosh,Yi.Zheng-8}@student.uts.edu,
{Thorsten.Lammers,Jinyan.Li}@uts.edu.au
[2] NSW Public Service Commission, Sydney, Australia
{Lucy.Chen,Carolyn.Fitzmaurice,Scott.Johnston}@psc.nsw.gov.au

Abstract. Effective approaches for measurement of human capital in public sector and government agencies is essential for robust workforce planning against changing economic conditions. To this purpose, adopting innovative hypotheses driven workforce data analysis can help discover hidden patterns and trends about the workforce. These trends are useful for decision making and support the development of policies to reach desired employment outcomes. In this study, the data challenges and approaches to a real life workforce analytics scenario are described. Statistical results from numerous workforce data experiments are combined to derive three hypotheses that are useful to public sector organisations for human resources management and decision making.

Keywords: Workforce analytics · Public sector · Data mining

1 Introduction

Human Resource Management in public administration plays a critical role in targeting and executing long term employment and workforce retention objectives. From a government perspective, it turns out to be a function correlating the employees performance to align with the government's strategic goals and priorities like sustainable budgeting for example [1,2]. Additionally, with the increased use of digital technologies in the workforce, traditional jobs are at risk of being displaced and replaced by transformed job requirements and descriptions creating new types of jobs [6]. Thus the modernisation of existing workforce requires a robust workforce planning process that integrates future workforce needs so that capabilities are developed earlier. Typically, workforce planning processes involve the identification of strengths, weaknesses, opportunities and threats to develop appropriate interventions to mitigate future risks

---

S. Ghosh and Y. Zheng—Contributed equally.

© Springer International Publishing AG 2016
J. Li et al. (Eds.): ADMA 2016, LNAI 10086, pp. 764–774, 2016.
DOI: 10.1007/978-3-319-49586-6_55

[3–5]. Previous applications of advanced data analysis techniques on workforce data have provided actionable insights on the effectiveness, efficiency, and impact of the workforce initiatives in the private sector related to retention and talent management [7–9,16]. Data mining techniques like clustering and sequence mining have been used for resource management, staffing allocations and predicting employee attritions [13–15,17]. In contrast, public sector organisations tend to have massive amounts of fragmented, disparate and unstructured datasets posing initial data processing challenges. Recent advances in large-scale data fusion [10], efficient service provisioning, and disaster mitigation [11] have been explored for public sector organisations. In the context of large-scale public sector workforce data, data-driven analysis drives a larger objective of conceptualizing and development of policy interventions that are relevant to the economic problems associated with maintaining an efficient and diverse national or state level public sector workforce.

The important contributions of this study are described as follows.

- A real life data exploration study for a subset of the Australian public sector workforce is reported.
- Existing data mining approaches are combined with summary statistics to derive actionable insights that can be used for workforce planning.
- Three data driven hypotheses are derived for future workforce data investigations.

2 Methods

2.1 Problem Description

Exploratory data mining questions were posed to derive workforce in-sights in two broad categories namely (1) Workforce Diversity, and (2) Ageing Workforce. In workforce diversity, the data analysis required finding emergent trends among employees for specific subpopulations viz. ethnic employees, employees with a disability and Aboriginal people. For an ageing workforce, data analysis was required to derive workforce implications due to population ageing.

2.2 Approach for Addressing Workforce Diversity

For deriving relevant insights about workforce diversity, a number of subpopulations were extracted from the original datasets. Table 1 lists the variable names and valid values which were used to select three types of subpopulation datasets based on (1) Ethnicity, (2) Disability, and (3) Aboriginality.

Table 2 shows the various subpopulation combinations that were used to compare the population results for Aboriginal, Ethnic and mainstream groups of employees. Analogously, for comparisons with the Disability subpopulation, we use the combinations groups listed in Table 3.

To understand the temporal trends of a certain subpopulation, two different methods are utilized. Taking the leadership as an example, in the first method,

Table 1. Variable names used in diversity analysis and their valid values

Variable name (ID)	Valid values
Aboriginal person or torres strait islander (2a)	1 = Aboriginal, 2 = Torres Strait Islander, 3 = Aboriginal and Torres Strait Islander, 4 = Not Aboriginal or Torres Strait Islander, −7777/−9999 = Missing/Withdrawn
Person with a disability (2b)	1 = Disability requiring adjustment at work, 2 = Disability not requiring adjustment at work, 3 = Person with a disability, 4 = No disability, −7777/−9999 = Missing/Withdrawn
Ethnicity (2c)	1 = Person from an Ethnic Minority Group, 2 = Person not from an Ethnic Minority Group, −7777/−9999 = Missing/Withdrawn

Table 2. Selecting subpopulations for aboriginal and ethnic minority

Name of the subpopulation	Subpopulation selection criteria
Aboriginal	2a = 1 or 2 or 3
Ethnic	2c = 1 and 2a = 4
Mainstream$_1$	2a = 4 and 2c = 2

Table 3. Selecting subpopulations for disability

Name of the subpopulation	Subpopulation selection criteria
Disability	2b = 1 or 2 or 3
Mainstream$_2$	2b = 4

a temporal trend was derived using the percentage of leadership in a subpopulation - $Perc_L$, as given in Eq. 1. Leaders in the workforce data were defined as employees with an annual base salary greater than a specific salary band threshold (band 6 for senior executives encoded as GSE6, corresponding to ≥ \$130929 for the year 2015).

$$Perc_L^S = \frac{(Employee\,Count\,of\,sub-population \geq Band6)}{(Total\,Number\,of\,Employees\,in\,sub-population)} \tag{1}$$

For the second method, the percentage of employees (for each subpopulation) within the total leadership was calculated, as shown in Eq. 2.

$$Perc_L^T = \frac{(Employee\,Count\,of\,sub-population \geq Band6)}{(Total\,Number\,of\,Employees \geq GSE6band)} \tag{2}$$

To acquire the distribution of the Aboriginal, ethnic, disability and mainstream populations in different occupations, the ANZSCO (Australian and New

Zealand Standard Classification of Occupations) coding standard was employed [12]. ANZSCO is a classification system developed to collect, and analyse occupation statistics across government agencies in Australia. The x (x = 1, 2, 3, 4, 5, 6) digits of an ANZSCO (Australian and New Zealand Standard Classification of Occupations) code follow a hierarchical tree structure, such that any digit at a given position is a child node of the prefix ANZSCO sequence. This means each position (or a digit) in the ANZSCO code corresponds to the suboccupation within the prefix code.

2.3 Approach for Addressing Workforce Ageing Issues

For data analysis purposes, the calculation of age was carried out using the date-of-birth (DOB) variable, as shown in Eq. 3.

$$Age = CensusDate_{Year} - DOB \qquad (3)$$

Within workforce ageing, for the identification of critical roles, specific occupations of interest were verified. Specifically, the occupations used for the ageing analyses are given in Table 4.

Table 4. Mapping of ANZSCO codes to selected occupations

Occupation	ANZSCO
Accountants, Auditors and Company Secretaries	221
Information and Organisation Professionals	224
Engineering Professionals	233
Natural and Physical Science Professionals	234
Education Professionals	24
ICT Professionals	26
Office Managers and Program Administrators	51
General Clerical Workers	53
Numerical Clerks	55
Clerical and Office Support Workers	56

Additionally, to determine correlation scores with age, each of the numeric variables in the workforce dataset, were initially processed to remove missing values. Later, normalization of the data was carried out, with mean as 0 and variance as 1. Following this, the Pearson product-moment correlation coefficient (PCC) was computed using the corresponding workforce numeric variable and age. Typically, the PCC is a measure of the linear correlation between two variables X and Y, giving a value between +1 and 1. Here, 1 is maximum positive correlation, 0 is no correlation, and 1 is maximum negative correlation. The PCC can be calculated as given in Eq. 4.

$$PCC = \frac{\sum (X_i - Mean(X))(Y_i - Mean(Y))}{\sqrt{\sum (X_i - Mean(X))^2}\sqrt{\sum (Y_i - Mean(Y))^2}} \qquad (4)$$

Here, X and Y are considered to be the two variables for which PCC is being calculated.

3   Results

3.1   Dataset Description

For performing data exploration experiments, we obtained public sector workforce datasets from 1999 to 2015. The total number of variables in these datasets ranged between 40 to 83, for a given year. The number of variables for year varied due to the addition of new variables in later years. Summary statistics related to the Australian workforce data are reported next.

3.2   On Workforce Diversity

The ethnic sub-population demonstrated a three times higher likelihood of moving into a leadership position (defined by a salary band of 6 and above), in comparison to the aboriginal population (Fig. 1).

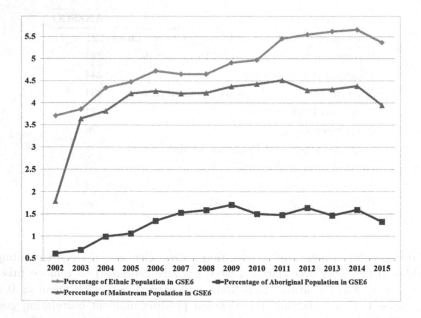

Fig. 1. Percentage of employees in each subpopulation with a GSE6 salary

The majority of the aboriginal population is working in the Community and Personal Service Profession (ANZSCO 4). Three of the top five ranked

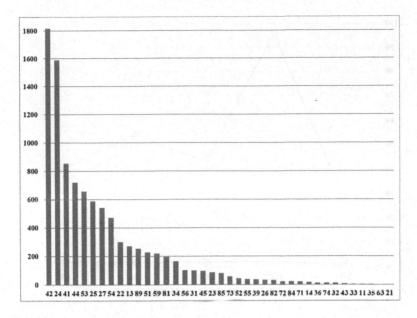

Fig. 2. Percentage of employees in each subpopulation with a GSE6 salary

occupations belonged to ANZSCO code 4. However, there are only two aboriginal leaders of a total of 3493 aboriginal employees in ANZSCO code 4. In contrast, for every 956 mainstream employees in ANZSCO code 4, there is one mainstream leader (Figs. 2 and 3).

In the ANZSCO code 4 occupations, which have been the third most popular occupation for the ethnic population and have almost 14 % of the ethnic population, there is only one ethnic leader over GSE6 out of 5870 ethnic employees (Fig. 3). Each subpopulation in the workforce exhibits an ageing trend (average age of 41.4 1999, compared to the average age of 45.2 in 2015). However, the disability population is ageing twice as fast as the main-stream group (7 years older on average for the disability population from 1999 to 2015, while only 2.7 years older for the mainstream population from 1999 to 2015, see Figs. 4 and 5).

The disability population in the lower salary bands of 2, 3, 4 and 5 are the main part leading to the population decrease of the disability group (Fig. 6).

Typically, in the leadership population, age has the highest negative correlation with unpaid sick leave (Table 5). Generally, in the leadership population leave variables tend to negative correlated with age.

3.3 On Workforce Ageing

Accountants had a median age of 43 years in 2015, which is two years younger than the median age of accountants in 2008. The median age for Clerical and Office Support Workers was 57 years in 2015, three years older than the median age of the workers of this profession in 2008.

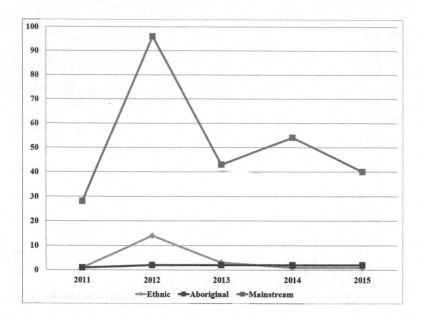

Fig. 3. Percentage of employees in each subpopulation with a GSE6 salary

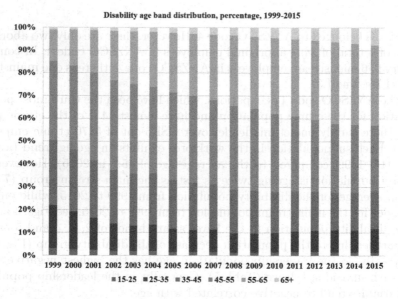

Fig. 4. Percentage of employees in each subpopulation with a GSE6 salary

In the leadership population, age has the highest negative correlation with unpaid sick leave (Table 5). Generally, in the leadership population leave variables tend to negative correlated with age. Evidence from PCC demonstrates that within the leadership cohort, age has a negative correlation to the unpaid

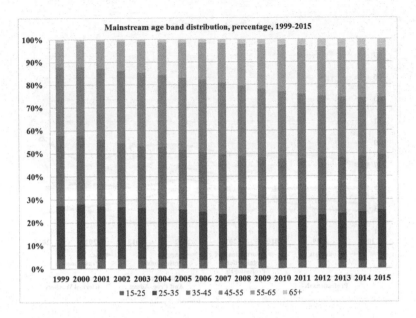

Fig. 5. Percentage of employees in each subpopulation with a GSE6 salary

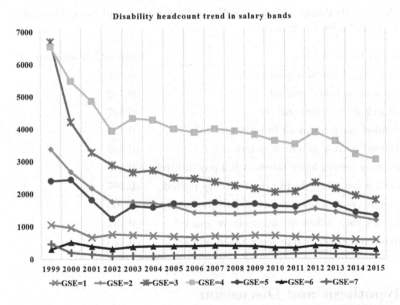

Fig. 6. Percentage of employees in each subpopulation with a GSE6 salary

and paid sick leave taken, when compared to the general population. This means that as age increases, less sick leaves had been taken in the senior leadership cohort, in 2015 (Table 5).

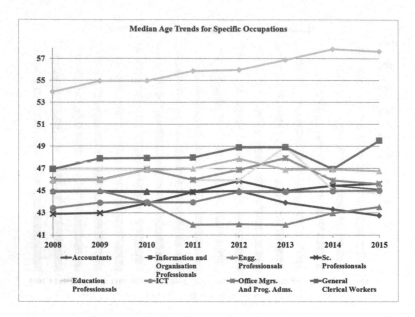

Fig. 7. Percentage of employees in each subpopulation with a GSE6 salary

Table 5. Top 10 Ranked Workforce Variables for the Leadership Population using Pearsons Correlation Coefficient.

Name of the variable	Correlation with age (PCC)
Unpaid sick leave taken ann ref	−0.446
Unpaid leave taken ann ref	−0.442
Paid sick leave taken ann ref	−0.437
Extended leave taken on full pay ann ref	−0.427
Extended leave taken on half-pay ann ref	−0.352
Mat pater parent leave taken at full pay ann ref	−0.198
Special leave taken during the reference period	−0.187
Mat pater parent leave taken at half pay ann ref	−0.165
Paid sick leave taken as careers leave ann ref	−0.152
Unpaid mat pater parent leave taken ann ref	−0.125

4   Hypothesis and Discussion

As part of the workforce study, the results lead us to determine specific insights about the public sector workforce which could be framed as hypotheses that may be further investigated. These are described further.

Hypothesis 1: Given that bands 6 and 7 workers are older and ageing faster than the middle-band workers in the disability population, there might be a succession lag and gap in the leadership by the executives with a disability.

Our results showed that the headcount of the workforce with a disability has been gradually decreasing from the year 1999 to 2015, which is mainly attributed to the decreasing subpopulation in salary bands 2, 3, 4 and 5. However, the headcounts in the salary bands 6 and 7 are maintained over recent years (Fig. 7).

Hypothesis 2: Within the leadership cohort, there exists a comparatively strong negative correlation between leave variables and age. This negative correlation does not exist for the general population.

The senior leaders in the workforce have taken much less sick leave (of different types including paid and unpaid) than the other senior staff.

Hypothesis 3: There does not exist adequate Aboriginal representation in leadership cohort, which both includes and affects the profession of "Community and Personal Service Workers".

The Community and Personal Service Workers (ANZSCO major 4) occupation ranks among the top 5 occupations for both ethnic and Aboriginal subpopulations in terms of the number of employees working in it. Each year, more employees of ethnic and aboriginal populations choose a profession within Community and Personal Service Workers as their occupation. However, there exist significantly less aboriginal leaders in Community and Personal Service Workers, when compared to the whole population.

5   Conclusion

In the current study, a real life public sector workforce dataset was employed to perform trend analysis from 1999 to 2015. In recent years, there have been few research efforts that employed a combination of data analytic methods to derive insights about the Australian public sector workforce. Specifically, our study emphasises the importance of statistical analysis and the need to combine findings with knowledge discovery techniques for generating actionable outcomes that can be usable for workforce interventions.

References

1. Adams, C.A., Muir, S., Hoque, Z.: Measurement of sustainability performance in the public sector. Sustain. Account. Manage. Policy J. **5**(1), 46–67 (2014)
2. Christensen, T., Lgreid, P.: The whole of government approach to public sector reform. Publ. Admin. Rev. **67**(6), 1059–1066 (2007)
3. MacCrory, F., Westerman, G., Alhammadi, Y., Brynjolfsson, E.: Racing with and against the machine: changes in occupational skill composition in an era of rapid technological advance. In: ICIS, May 2014
4. Bessen, J.E.: How Computer Automation Aects Occupations: Technology, Jobs, and Skills. Boston Univ. School of Law, Law and Economics Research Paper (15–49) (2015)

5. Woon, W.L., Aung, Z., AlKhader, W., Svetinovic, D., Omar, M.A.: Changes in occupational skills - a case study using non-negative matrix factorization. In: Arik, S., Huang, T., Lai, W.K., Liu, Q. (eds.) ICONIP 2015. LNCS, vol. 9491, pp. 627–634. Springer, Heidelberg (2015). doi:10.1007/978-3-319-26555-1_71
6. Caselli, M.: Trade, skill-biased technical change and wages in Mexican manufacturing. Appl. Econ. **46**(3), 336–348 (2014)
7. Wei, D., Varshney, K.R., Wagman, M.: Optigrow: people analytics for job transfers. In: IEEE International Congress on Big Data, pp. 535–542. IEEE, June 2015
8. Ramamurthy, K.N., Singh, M., Davis, M., Kevern, J.A., Klein, U., Peran, M.: Identifying employees for re-skilling using an analytics-based approach. In: 2015 IEEE International Conference on Data Mining Workshop (ICDMW), pp. 345–354. IEEE, November 2015
9. Varshney, K.R., Chenthamarakshan, V., Fancher, S.W., Wang, J., Fang, D., Mojsilovi, A.: Predicting employee expertise for talent management in the enterprise. In: Proceedings of the 20th ACM SIGKDD International Conference on Knowledge Discovery and Data Mining, pp. 1729–1738. ACM, August 2014
10. Amorim, J.A., Andler, S.F., Gustavsson, P.M., Agostinho, O.L.: Big data analytics in the public sector: improving the strategic planning in world class universities. In: International Conference on Cyber-Enabled Distributed Computing and Knowledge Discovery (CyberC), pp. 155–162. IEEE, October 2013
11. Morabito, V.: Big data and analytics for government innovation. In: Morabito, V. (ed.) Big Data and Analytics: Strategic and Organizational Impacts, pp. 23–45. Springer, Heidelberg (2015)
12. Trewin, D., Trewin, D.J., Pink, B.N.: Australian and New Zealand standard classification of occupations. Australian Bureau of Statistics/Statistics New Zealand (2006)
13. Datta, R., Hu, J., Ray, B.: Sequence mining for business analytics: building project taxonomies for resource demand forecasting. In: Frontiers in Artificial Intelligence and Applications, vol. 133 (2008)
14. Hu, J., Ray, B.K., Singh, M.: Statistical methods for automated generation of service engagement staffing plans. IBM J. Res. Dev. **51**(3/4), 281–293 (2007)
15. Mojsilović, A., Connors, D.: Workforce analytics for the services economy. In: Maglio, P.P., Kieliszewski, C.A., Spohrer, J.C. (eds.) Handbook of Service Science, pp. 437–460. Springer, Heidelberg (2010)
16. Richter, Y., Naveh, Y., Gresh, D.L., Connors, D.P.: Optimatch: applying constraint programming to workforce management of highly skilled employees. Int. J. Serv. Oper. Inf. **3**(3–4), 258–270 (2008)
17. Singh, M., Varshney, K.R., Wang, J., Mojsilovic, A., Gill, A.R., Faur, P.I., Ezry, R.: An analytics approach for proactively combating voluntary attrition of employees. In: 2012 IEEE 12th International Conference on Data Mining Workshops, pp. 317–323. IEEE, December 2012

Real-Time Stream Mining Electric Power Consumption Data Using Hoeffding Tree with Shadow Features

Simon Fong[1], Meng Yuen[1], Raymond K. Wong[2(⊠)], Wei Song[3], and Kyungeun Cho[4]

[1] Department of Computer Information Science,
University of Macau, Macau SAR, China
ccfong@umac.mo, 250625774@qq.com
[2] School of Computer Science and Engineering,
University of New South Wales, Kensington, Australia
wong@cse.unsw.edu.au
[3] College of Information Engineering,
North China University of Technology, Beijing, China
sw@ncut.edu.cn
[4] Department of Multimedia Engineering,
Dongguk University, Seoul, South Korea
cke@dongguk.edu

Abstract. Many energy load forecasting models have been established from batch-based supervised learning models where the whole data must be loaded to learn. Due to the sheer volumes of the accumulated consumption data which arrive in the form of continuous data streams, such batch-mode learning requires a very long time to rebuild the model. Incremental learning, on the other hand, is an alternative for online learning and prediction which learns the data stream in segments. However, it is known that its prediction performance falls short when compared to batch learning. In this paper, we propose a novel approach called Shadow Features (SF) which offer extra dimensions of information about the data streams. SF are relatively easy to compute, suitable for lightweight online stream mining.

Keywords: Electric power consumption prediction · Data stream mining · Shadow features

1 Introduction

Supply and demand forecasting of power load consumption has a long history of resource planning in power industry. Collecting the utility data, analyzing them in real time and harvesting timely knowledge are useful for tuning for optimal production capacity and planning for the best distribution network, as well as preventing for under/over-supply and power black-outs. In the past, such power utilities data, from households or commercial buildings, are aggregated into an archive from which descriptive statistics are derived as references for resource planning. Rigorous studies

© Springer International Publishing AG 2016
J. Li et al. (Eds.): ADMA 2016, LNAI 10086, pp. 775–787, 2016.
DOI: 10.1007/978-3-319-49586-6_56

along this direction began in early 70's resulting in a wide variety of forecasting models in pre-millennium. These methods are based on time-series forecasting with means of removing the cyclic components, fluctuations and sporadic outliers in an effort of ensuring the best possible fitting curve by the model onto the actual data. Further sophisticated models in the name of machine learning have emerged in the post-millennium. There are fancy tools such as fuzzy logic, genetic algorithms, metaheuristic optimization, support vector machine, artificial neural networks, even deep learning; they have been applied to solve the problems of finding the best configuration of power network or predicting the future demand loads, most of them claimed to produce superior results.

In this paper, we take a completely different angle of data analytics perspective. While for all the forecasting from the 70's to 90's and most of the advanced machine learning models in the years beyond, the power utility data as input to the models is assumed static. A full set of data must be loaded for the models to learn the varying patterns, trying to reverse engineer the relations between the input attributes (as some unique usage patterns) and the outcomes in some forms of consumption conditions, e.g. very low, low, average, high and overly high usage. It has no doubt of the inference power and the predictive power by these machine learning algorithms. The very non-linear nature of the complex relations between the input data and the outcomes can be accurately model, the accuracy is established based on the stationarity of the incoming data. In practice, however, we know that the power consumption data stream is anything but static or stationary. They vary irregularly and unexpectedly due to many unforeseen factors, e.g. tourist influx, extreme weather conditions, recreation events etc.

Instead of using traditional data mining algorithms, we turn to data stream mining algorithms which build a prediction model incrementally without the need of seeing the whole dataset. A prediction model updates itself quickly when fresh data streams in. This lightweight prediction mechanism is favored by the motivation of the current sensing infrastructure as advocated for smart city, Internet of things and big data analytics. These new campaigns advocated real-time information and insights from our daily activities with the help of big data for improving the residents' quality of life. Better power management is one of them. Despite of conventional power grid resource planning and pricing by the electricity industry, electricity consumption prediction is lately welcomed by smart home users. From the composition of electricity usage patterns, the use of electric appliances is monitored and scheduled for the sake of saving energy, thereby saving money at the consumer level using innovative power plugs[1].

While traditional data mining which embraces the long-term knowledge from a complete set of usage data, data stream mining is able to reflect the short-term insights from a moving flow of data streams. The knowledge extraction is adaptive to the changes of the usage data.

Many energy load forecasting models that have been developed in the past are established from batch-based supervised learning models. The whole data must be loaded to learn. Due to the sheer volumes of the accumulated consumption data which arrive in the form of continuous data streams, such batch-mode learning requires a very

---

[1] http://www.innolinks.cn.

long time to rebuild the model and the knowledge extraction is not adaptive in real time. Incremental learning, on the other hand, is an alternative for online learning and prediction which learns the data stream in segments. However, it is known that its prediction performance falls somewhat behind that of the batch learning. This is owing to nature of the power usage data – the data is long temporal sequence but usually characterized with few attributes. In light of this, we propose a novel approach called Shadow Features (SF) which offer extra dimensions of information about the data streams for improving the supervised learning in incremental mining. Since only a window of data can be seen at a time in incremental learning, more relevant information about the data stream given at each window will help boost the learning efficacy. Given the real-time speed constraints, SF must be easy to compute, so to fulfill the lightweight requirements of online stream mining. The new data stream mining model using SF is tested with empirical electricity consumption datasets from three countries.

The remainder of this paper is structured as follows. Section 2 provides some basis and background about shadow features. Section 3 presents our proposed classification model incorporating the new shadow features method into Hoeffding Tree. Experiments comparing the original and new methods using datasets from electricity markets are reported in Sect. 4. Section 5 concludes the paper.

2   Background

Real-time prediction of electric load demand is centred on the collected measurement data of electricity usage from households. The analytic module tries to infer the fluctuation patterns for prediction and classification. The usage data arrive in the form of continuous and sequential values, similar to time series. Time series, such as those collected from power meters and other sensors, are usually multi-variate, typically about the voltage in use, current intensity, sub-metering power consumption from different rooms or electric appliances. These consumption features plus the time-stamp information in the time series represent the temporal consumption by different households and appliances in motion. More complicated time series may contain extra information, such as important aspects of the home environment. They include but not limited to average household electricity usage per second. If it is a smart home, they may include usage at individual circuit or socket, that covers every plug load. The same data could be referenced with the electricity generation data from other sustainable generation sites that supply the electricity to the house, such as wind turbines and solar panels. These generation instruments are in turn influenced by outdoor weather data, that cover outdoor and indoor temperature and humidity etc. Even though attribute information from related accessories could be collected, the ultimate power consumption usage data has few features ranging from a singular power load in voltage to approximately an extra dozens of relevant features from accessories and weather conditions. The relatively small number of features makes it difficult for data stream mining algorithms to sufficiently learn about the underlying mapping between the input values and the target outcomes, especially when the information of the full dataset is unavailable.

In this paper, a new type of feature, which we call a 'shadow feature' is presented. The concept of shadow features is inspired by Etienne-Jules Marey (1830-1904), who instigated chronophotography or cinematographic movement recording [1] as a fundamental technique of capturing motion pictures. His lifetime passion 'to see the invisible' motivated him towards a single goal: recording all existing movement, produced by humans or anything else that moves, on a recordable surface. In alignment with Etienne-Jules' motivation, shadow features are derived from the dynamics of moving objects. The shadow feature values are subtly approximated by the underlying motion resulting from performing activities. For example, in Fig. 1, the motion of a marching solider can be seen in a chronophotograph as longitudinal curves in a time series.

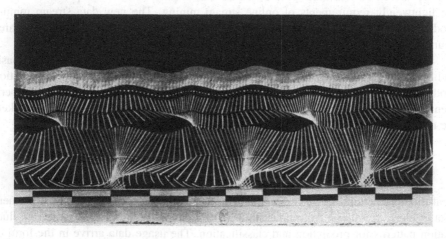

Fig. 1. Cinematographic movement of marching solider. *(Image courtesy of Motion Studies predating Taylorism, Images produced by Etienne-Jules Marey).*

The shadow feature approximation is carried out quickly by a time-series smoothing technique coupled with the Hurst factor for fast-processing (described in detail in Sect. 3). Unlike other feature transformation techniques that have been previously reported in the literature, the shadow features method does not transform and replace the original features; instead, it creates a new set of motion dynamic data that augments the original feature space of the data. Hence the term 'shadow'.

Shadow features offer extra dimensions of information for characterising the moving dynamics in time series prediction process. The extra feature information gives insight into the motion of the fluctuating item (e.g. stock price or power load usage) in supervised learning. The classification algorithm tries to learn and induce relationship mapping between the input feature values and the target classes.

Not only is the shadow features method designed to improve classification accuracy, its advantage over peer techniques is its incremental nature. Simple and lightweight in computation, this method can generate extra feature values on the fly, making it suitable for data stream mining time series data. This quality is important in power load measurement environment, where the incoming data are fast-moving in nature and the sensing data can potentially amount to infinity. Shadow features can be incrementally extended such that the classification model learns incrementally as data streams in. In contrast, feature transformation techniques such as wavelet transformation [2] and statistical transformation techniques [3] require all the data to be available to calculate the new feature values. This induces heavy latency in real-time prediction because the full dataset must be reloaded into the classification model to rebuild (instead of refresh) the whole classifier all over again. Another advantage of shadow feature perhaps is the ability to smooth out the data, thereby subsiding any sporadic outliers, or "spikes" that may occur as white-noise in the data transmission.

3 Proposed Shadow Feature Prediction Model

A new classification model is described in this section. It extends the traditional power load prediction model in which the raw power usage data is pre-processed prior to either training or testing the classifier in the batch learning mode and data stream mining mode.

3.1 New Prediction Model

The new prediction model comprises a power meter recording module, data sampling and parameter calibration modules, the shadow feature generation module and a classification module. The recording module and classification module are generic. They can be implemented by any suitable software/hardware and algorithms. The recorder is supposed to produce data in time-stamped ordered sequences. The training dataset has a labelled column that already contains the types of outcomes (e.g. increase/decrease, greater than mean, heavy/light usage, etc.) corresponding to each particular row of instance. The training dataset is used to train or build a classifier after passing through the shadow feature generation process that extends the original feature set of the training data with shadow features. In the testing phase, an unlabelled dataset is subject to a shadow feature generation process configured by the same parameters that controlled the shadow feature generation process for the training dataset. Then the extended testing dataset with the corresponding shadow features is tested against the built classifier, for activity recognition and the predicted result is outputted from the classifier.

In incremental learning, which is also known as data stream mining, the training dataset streams continuously into the classifier construction process. The dataset passes through the shadow feature generation process as through a pipeline, with new features continuously being generated using a sliding window of size m, over the input data stream. Likewise, the testing dataset streams into the classifier at any time (though preferably after some warm-up period at the initial stage of classifier training), the corresponding shadow features are added on the fly, and the results are predicted continuously at the end of the pipeline as streaming progresses.

The sampling and calibration process is an important module in the proposed model. It has two parts. First, small training data samples are taken to evaluate several key parameters of the time series of the training data. The module tests whether the time series is stationary or non-stationary, whether a trend exists and whether the time series is merely random-walk or it can be forecasted. The testing method is detailed in [4]. Consequently, a suitable time series forecasting method is deduced from the salient characteristic of the time series obtained from the initial samples. The chosen forecasting method will be used as a curve-fitting method for generating shadow features in the subsequent step. Thus, the model is adaptive and generalised for different shapes of time series in most types of electric load forecasting scenarios.

3.2   Shadow Feature Generation

Shadow features are defined as supplementary information that models the smoothed movements of a time series. In the case of the electricity usage data to be used in our experiments, although the time series fluctuate greatly in the time domain, they are stationary and no obvious trend is observed.

The training/testing dataset takes form of a data matrix, D_{train} / D_{test}, with dimensions $(M + 1) \times N$, where there are $M + 1$ columns covering M features and 1 target class, and N rows of data instances. Each column has a time series corresponding to a specific feature. Given a time series Y_j for each original feature a_j, there is a complimentary time series Y_j^{shadow} for the corresponding shadow feature a_j^{shadow}, where $j \in [1, M]$. Let ri be a data record holding data of M features at the ith order of D. The shadow time series is derived from its counterpart original time series that has the following form:

$Y_j = \{(r_1, t_1), ..., (r_n, t_n)\}$, $n \in N$, and $Y_j^{shadow} = \{(r_1^{shadow}, t_1), ..., (r_n^{shadow}, t_n)\}$, both time series are synchronized by the same timestamps, where r_i and r_i^{shadow} share the common data tuple format $(x_{i,1}, x_{i,2}, ..., x_{i,M} \mid class_i)$ at the ith position of the time series sequence, where $x_{i,j}$ is a real-number extracted from the raw sensing data with respect to the jth feature. Each data tuple r in D_{train} would carry a target class label; the class is empty in the data records in D_{test}.

At the beginning of the model building process, a certain amount of initial samples are used for calibration. From there, salient statistics are calculated in the start-up. Hurst factors are computed. The time series in D_{train} /D_{test} are determined whether they are stationary or otherwise, as to decide which curve fitting algorithm were to. For whichever curve fitting algorithm that is chosen to use, Hurst factors play a part in the shadow feature values generation. Two Hurst exponentials are calculated, one is H^{start} calculated during calibration from a subset of Y ($Y^{subset} \subseteq Y$), the other is $H^{dynamic}$ calculated during the curve-fitting process in shadow feature generation. H^{start} is the standard Hurst exponent defined by the asymptotic behaviour of the standard rescaled range as a function of time period of Y^{subset} that contains certain amount of data points in the initial part of the time series:

$$\varepsilon \left[\frac{\mathcal{R}\left(Y^{subset}\right)}{\sigma\left(Y^{subset}\right)} \right] = C(Y^{subset})^{H^{start}} \tag{1}$$

where $\varepsilon[\cdot]$ is the expected value pertaining to the jth feature in the bounded time-series Y^{subset}, $\mathcal{R}$ is the range of the values in the Y^{subset}, σ is their standard deviation and C is an arbitrarily chosen constant. The scaled range is calculated as:

$$\frac{\mathcal{R}\left(Y^{subset}\right)}{\sigma\left(Y^{subset}\right)} = \frac{max\left(\delta_1,\delta_2\ldots\delta_{|Y^{subset}|}\right) - min\left(\delta_1,\delta_2\ldots\delta_{|Y^{subset}|}\right)}{\sqrt{\frac{1}{|Y^{subset}|}\sum_{i=1}^{|Y^{subset}|}\left(x_i - \mu\right)^2}} \qquad (2)$$

where μ is the mean, $\mu = \frac{1}{|Y^{subset}|}\sum_{i=1}^{|Y^{subset}|} r_i$; and δ_t is an element of the cumulative deviate series δ, so $\delta_t = \sum_{i=1}^{t}(r_i - \mu)$, for $t = 1,2,\ldots,|Y^{subset}|$. H^{start} is calculated by fitting the power law in Eq. (1) to the initially sampled time series. Repeat the estimation of H^{start} for each feature j.

At the start, H^{start} is used as one of the salient indicators [4] in deciding a stochastic model for fitting the time series for shadow feature generation. H^{start} is also being referred to for quickly finding an appropriate sliding window size. The length of the sliding window is pegged at the long-term memory dependence (LRMD) which is approximated by Hurst factor.

Instead of dealing with power-like exponential decay and auto-covariance function which are heavy in computation, an alternative lightweight approach using Hurst factor is adopted here. Assuming the power usage time series is stationary (by which the consumer is using about the same levels of energy on regular basis), there should be some minor fluctuations within an overall fluctuation, and they do repeat in similar manner along the way. E.g. seasonality. Quantitative techniques are available to estimate the appropriate window length [5] though they incur substantial computing overheads. A quick alternative is to manually prepare a group of various predefined window lengths according to the durations and extents of the fluctuations. The value of H^{start} is then used to pick a size by referring to the continuity intensity Table 1 [6]. In general, the stronger the LRMD is, the longer the window size would give good result. Calculating H^{start} once at calibration enables us to choose a curve fitting method and estimating the sliding window length.

Table 1. Performance results of experiments of different m at Australian electricity data

HT (current mode)				HT (mean mode)				C4.5			
m =	Accuracy	Kappa	Times (s)	m =	Accuracy	Kappa	Times (s)	m =	Accuracy	Kappa	Times (s)
0	83.6	67.66	1.84	0	79.23	56.04	0.86	0	90.2016	0.8003	2.27
5	84.2	68.12	1.47	5	81.33	60.36	0.62	5	88.208	0.7592	3.2
100	92.4	84.7	2.22	100	87.18	73.22	0.85	100	92.813	0.8531	3.03
200	85.4	70.66	2.31	200	83.11	63.98	0.93	200	92.2834	0.8424	3.3
300	86.8	73.46	1.97	300	82.57	62.53	0.77	300	91.739	0.8308	3.14
400	83.2	66.4	1.75	400	82.6	62.9	0.72	400	90.7827	0.8117	3.13
500	86	71.72	1.77	500	82.54	62.93	0.69	500	91.3197	0.8225	3.42
1000	86	71.33	1.78	1000	81.9	61.4	0.71	1000	91.2976	0.8222	3.45
PartMb	91.4	82.7	301.34	PartMb	90.28	79.8	111.32	PartMb	out of memory		

During the shadow feature generation process, another type of Hurst factor called $H^{dynamic}$, is calculated along with the curve fitting process. In moving average smoothing, $H^{dynamic}$ is a variable whose value is derived from the range of elements within the current sliding window. Then it is updated when the sliding window advances to the next data record. At the same time the window size (tailing position) is dynamically updated according to the latest $H^{dynamic}$ value.

The $H^{dynamic}$ is calculated in the same manner as in Eqs. (1) and (2), except that the data range for $H^{dynamic}$ is bounded by the data length in the current window w along the time series Y, instead of Y^{subset}. This dynamic Hurst factor $H_i^{dynamic}$ is calculated iteratively whenever the window advances to the next item till the end. By considering Hurst exponent can be drawn on a $Log(R/S)$ and $Log(n)$ plot; the slope of the regression model is approximated as follow.

$$Log(R/S)_{w_i} = Log(C) + H_i^{dynamic} \cdot Log(|w_i|) \ \forall i \in [1, N] \qquad (3)$$

$$H_i^{dynamic} = \frac{Log(R/S)_{w_i} - Log(C)}{Log(|w_i|)} \qquad (4)$$

Assume the values of items in the time series characterized by the shadow feature Y^{shadow} take the same format as Y, the item value can be computed rapidly by calculating the mean of the previous w data including the current one, as follow.

$$r_i^{shadow} = (\varphi_i + 1) \cdot \frac{r_i + r_{i-1} + \ldots + r_{i-(w_i-1)}}{w_i} \forall i \in [1..n] \qquad (5)$$

where φ_i is a scaling factor signifying the importance or relevance of the data points in the current i^{th} window to the shadow feature. The factor can be estimated by normalizing the dynamic Hurst factor, such that

$$\varphi_i = \frac{\left| H_i^{dynamic} \right| - 0.5}{0.5} \qquad (6)$$

There is an option to assign more weights to the data near the front of the window by multiplying the window positions. So that the time series becomes a convolution of data points with different weights. For example, the current movement data has the maximum weight ω, and the second latest has ω -1 etc., down to one.

$$r_i^{shadow} = (\varphi_i + 1) \cdot \frac{w_i \cdot r_i + (w_i - 1) \cdot r_{i-1} + \ldots + 2r_{(i-w_i+2)} + r_{(i-w_i+1)}}{w_i + (w_i - 1) + \ldots + 2 + 1} \forall i \in [1..n] \quad (7)$$

Other curve fitting methods do optionally exist, such as exponential moving average, and exponentially weighted moving average etc. However, they may not be suitable for fast processing in incremental learning environment because recursive and power operators are involved, that incur heavy computing time.

4 Experiment

The objective of the experiment is to validate the efficacy of the proposed shadow feature prediction model, as well as to investigate how data stream mining perform over some empirical electricity dataset in real time. The performance measure thus come in three aspects: accuracy, kappa and time cost. Considering the hit counts of how many instances are predicted in positive or negative (P/N) classes, and how many are truly there or otherwise (True/False = T/F), the accuracy is defined as: (TP + TN)/(TP + FP + FN + TN). Accuracy is in percentage, [0, 100]. In data stream mining environment, there are two types of accuracy possible: one is the mean accuracy which is the sum of all the individual accuracy measured at the end of each movement of sliding window, over the total number of instances; the second is the current accuracy which is recorded as the individual accuracy at the final movement of the sliding window. This represents also the accuracy of the decision tree after the final update at where the sliding window stopped. Kappa is rough estimation about how well the prediction model can generalize itself when other datasets than the one used for trained would be used. In a simple term, it is often perceived as an index of [0, 1]; Kappa index is normalized to percentage for easy comparison in this paper. The time cost is measured by considering the amount of time incurred during each decision tree update (or refresh) upon testing new instances of data streaming in. This is proportional to the complexity of the decision tree, which is anticipated to grow when further instances have been used to train the classifier. As the classifier gets more sophisticated thereby it is needing longer time to refresh.

The simulation is conducted in the benchmarking software program called Massive Online Analysis [7] which is a free Java based data stream mining platform by University of Waikato. The computing environment is Intel Core i7 with 2.20 GHz CPU, 8 Gb RAM and x64-based processor and operating system.

The dataset put under test in the experimentation is a real-life electricity dataset called *elecNormNew*. The electricity data was obtained from the electricity market of New South Wales state of Australia. The prices in the dataset are not constant, rather they change adaptively according to the dynamic demand and supply in the market of the NSW state. Therefore, it is a classification problem of identifying and predicting whether the price should go up or down (Increase or Decrease as in the target label) given 8 attributes that describe the current condition of the consumption and supply. The features are the current (changing) prices of electricity in Victoria state and NSW state, the demands of the two states, and how much is transferred from one state to another. The instances are sampled and recorded one per five minutes. The dataset contains recording of the 26.2 days with labels that indicate the change of the price relative to the moving average of the past 24 h.

Two data mining environments are used for comparison in the experimentation. First the whole dataset is subject to traditional data mining, it is loaded in full to induce a C4.5 decision tree. It is considered as a full tree because all the instances have been seen in the supervised training. It is used as a benchmark for comparing with the Hoeffding tree (HT) which is the counterpart of C4.5. HT in contrast only reads the data instances in one pass with a default window size of 1000. Each time only 1000 instances are taken from the data stream for testing and check if the HT needs to be updated upon the testing with the new 1000 samples. The sliding window rolls from the first 1000 instances to the end of the data stream. For comparison the pre-processing scheme, a powerful feature transformation method called *PartitionMembership* (PM) [8] is used here. This is available as a pre-processing filter that uses the function *PartitionGenerator* to create membership of many partitions. The instances are transformed from the original feature space to partitions which the function would automatically find by clustering the feature values and the target classes. It transforms the feature space into a sparse matrix of instances, attributed by partitions. The dataset that is transformed by PM is tested in both traditional batch learning environment and data stream mining environment. In the traditional batch learning environment where all the data that are in a very high (sparse) feature dimensions caused the simulator out of memory, because too many tree branches are to be induced from such sparse matrix. When the PM-ed dataset is applied in data stream mining environment, nevertheless, a HT with very high performance is produced. This PM results are serving as comparison benchmark referencing that this probably could be the highest performance which a classifier can get using feature transformation technique. Such high performance however is only hypothetical because in real-life, this transformation is costing too high the time, making it unsuitable for data stream mining environment. This comparison is shown in the following Figs. 2, 3 and 4.

It is apparent that PM outperforms the results, but the time cost is too high even in data stream mining environment. Due to the extremely large decision tree it becomes using sparse feature, the time required for refreshing a model scales up to several hundred seconds. On the other hand, the time cost for the shadow feature method goes up linearly in the scale of seconds in a single digit. The accuracies and kappa fluctuate along the instances as shown in Figs. 4 and 5. At some times, the accuracies peak over 90 % and they fluctuate around 80 %. It can be observed that at around 17,000 instances shadow feature method has an edge over the original version. It prevents the drop of accuracy too deep at the curve valleys. At instances 37,000 and near the end of the data streams, the accuracies by shadow feature method have increased significantly over the original version. In this case, the shadow feature method is empowered by window size $m = 5$. (Please note that there two sliding windows, one is for incremental learning, set at default 1000. The other one, m, is for generating the shadow features during pre-processing).

In the next set of experiment, the size of m is varied so to observe the effect of sliding window size in shadow feature generation on the ultimate prediction performance. Three performance indicators, accuracy, kappa and time are charted in Figs. 2, 3 and 4 respectively. A small window ($m = 5$), a large window ($m = 100$), and an intermediate range of window sizes ($m = 100, 200.. 500$) are used in this test. The PM method is not shown in the Figures because sliding window size is irrelevant to this technique. However, the corresponding results of PM are shown in the tabulated results of this experiment in Table 1.

It is observed from Figs. 5, 6 and 7 that the size of the shadow sliding window has effects on both C4.5 and HT, traditional data mining and data stream mining environments respectively. However, it is more obvious the effects on data stream mining environment than on C4.5. They follow about the same patterns. The performance

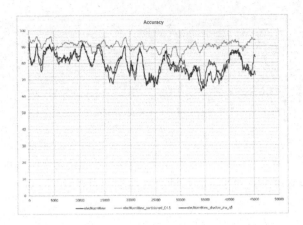

Fig. 2. Accuracy of data stream mining Australian electricity data.

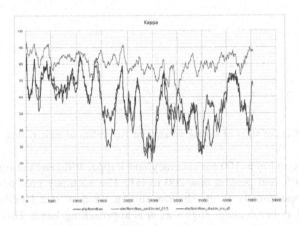

Fig. 3. Kappa of data stream mining Australian electricity data.

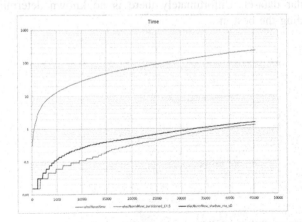

Fig. 4. Time consumption of data stream mining Australian electricity data.

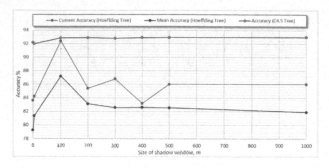

Fig. 5. Effects of sizes of sliding window at Australian electricity data on Accuracy

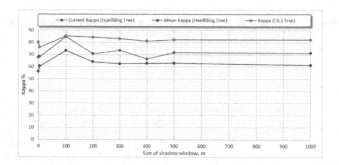

Fig. 6. Effects of sizes of sliding window at Australian electricity data on Kappa

gains peak at around $m = 100$ for accuracy and kappa. Whereas the time cost peak at elsewhere in C4.5, and peak at $m = 200$ in HT. The accuracy gain nevertheless does not increase proportionally as the shadow sliding window size lengthens. The mean accuracy for HT rather declines slights as m is approaching large. This may be due to the prolong memory effect by the shadows becomes somehow irrelevant to the prediction power. The best shadow window length is found to be at $m = 100$ in this case for this particular dataset. Unfortunately there is no known deterministic equation available in finding the best m.

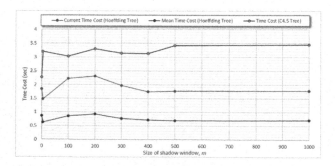

Fig. 7. Effects of sizes of sliding window at Australian electricity data on Time

5 Conclusions

Accurate prediction on the electricity demand is crucial for electricity pricing, resource allocations, power grid settings etc. It also benefits individual households in the era of smart home and smart cities, where real-time information about electricity consumption is made available by sensing technologies. However, reliable prediction plays a central role optimizing costs and usage. In the past decades many prediction models have been studied including those from machine learning to econometrics techniques. Most of these data mining model assume batch-based supervised learning models where the whole data must be loaded to learn. Due to the continuous streaming nature of electricity data and the real-time information requirements, we opt for an alternative online, in-memory, one-pass, incremental learning which learns and forgets. One disadvantage of incremental learning is the lack of long term memories when it comes to inducing a useful prediction model. In this paper, we looked into a new feature generation scheme called shadow feature which provides additional information about the long-range-memory to complement the original features. Experiments have been conducted and the results show that shadow feature is feasible when coupled with Hoeffding Tree which is a popular data stream mining algorithm.

Acknowledgement. The authors are thankful for the financial support from the Research Grant Temporal Data Stream Mining by Using Incrementally Optimized Very Fast Decision Forest (iOVFDF), Grant no. MYRG2015-00128-FST, offered by the University of Macau, FST, and RDAO.

References

1. Getty Museum, J.P.: Photography: Discovery and Invention. ISBN 0-89236-177-8 (1990)
2. Vishwakarma, D.K., Rawat, P., Kapoor, R.: Human activity recognition using gabor wavelet transform and ridgelet transform. In: 3rd International Conference on Recent Trends in Computing 2015 (ICRTC-2015), vol. 57, pp. 630–636 (2015)
3. Zhang, M., Sawchuk, A.A.: A feature selection-based framework for human activity recognition using wearable multimodal sensors. In: Proceedings of the 6th International Conference on Body Area Networks, pp. 92–98 (2011)
4. Fong, S.: Adaptive forecasting of earthquake time series by incremental decision tree algorithm. Inf. J. **16**(12), 8387–8395 (2013). International Information Institute (Tokyo)
5. Witt, A., Malamud, B.D.: Quantification of long-range persistence in geophysical time series: conventional and benchmark-based improvement techniques. Surv. Geophys. (Springer) **34** (5), 541–651 (2013)
6. Zhou, N.: Earthquake Forecasting Using Dynamic Hurst Coefficiency, MSc thesis, Department of Computer and Information Science, University of Macau, Macau SAR (2013)
7. Holmes, B.A.: Bernhard Pfahringer, Philipp Kranen, Hardy Kremer, Timm Jansen, Thomas Seidl. In: MOA: Massive Online Analysis, a Framework for Stream Classification and Clustering. Workshop and Conference Proceedings. vol. 11: Workshop on Applications of Pattern Analysis, pp. 1–14 (2010)
8. Frank, E., Pfahringer, B.: Propositionalisation of multi-instance data using random forests. In: Cranefield, S., Nayak, A. (eds.) AI 2013. LNCS (LNAI), vol. 8272, pp. 362–373. Springer, Heidelberg (2013). doi:10.1007/978-3-319-03680-9_37

Real-Time Investigation of Flight Delays Based on the Internet of Things Data

Abdulwahab Aljubairy, Ali Shemshadi$^{(\boxtimes)}$, and Quan Z. Sheng

School of Computer Science, The University of Adelaide, Adelaide, SA 5005, Australia
abdulwahab.aljubairy@student.adelaide.edu.au,
{ali.shemshadi,michael.sheng}@adelaide.edu.au

Abstract. Flight delay is a very important problem resulting in the wasting of billions of dollars each year. Other researchers have studied this problem using historical records of flights. With the emerging paradigm of Internet of things (IoT), it is now possible to analyze sensors data in real-time. We investigate flight delays using real-time data from the IoT. We crawl IoT data and collect the data from various resources including flights, weather and air quality sensors. Our goal is to improve our understanding of the roots and signs of flight delays in order to be able to classify a given flight based on the features from flights and other data sources. We extend the existing works by adding new data sources and considering new factors in the analysis of flight delay. Through the use of real-time data, our goal is to establish a novel service to predict delays in real-time.

Keywords: Internet of Things · Flight delay analysis · Data mining · Machine learning · Prediction

1 Introduction

Flight delay is a longstanding problem with the aviation industry, which massively affects the productivity of airlines and airports around the world. Direct and indirect losses of flight delays are mind-blowing in terms of cost and span. A study by the National Center of Excellence for Aviation Operations Research (NEXTOR) estimates that the annual cost of air transportation delays only in the US surpass $32.9 billion in the year 2007 [1]. This number includes $8.3 billion airline component (consisting of increased expenses for crew, fuel, and maintenance, among others), $16.7 billion passenger component (based on the passenger time lost due to schedule buffer, delayed flights, flight cancellations, and missed connections) and $3.9 billion cost from lost demand. The indirect costs of flight delays can also be much higher in terms of the number and the span. However, flight delays are often subjected to be caused by a number of sources of irregularity. In particular, weather is responsible for nearly 75 % of delays [12]. Moreover, due to the recent changes in weather patterns as an effect of global warming, we expect to see it a rise in those numbers as a result of increased harsh conditions.

© Springer International Publishing AG 2016
J. Li et al. (Eds.): ADMA 2016, LNAI 10086, pp. 788–800, 2016.
DOI: 10.1007/978-3-319-49586-6_57

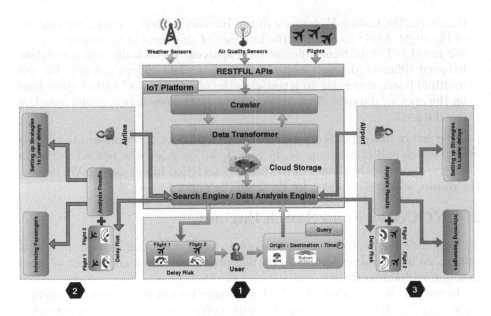

Fig. 1. Motivation scenario

Prediction and the analysis of flight delays is useful to reduce the direct and/or indirect associated costs. However, due to the highly dynamic environment, relying on a single historical dataset of flight delays in previous works [6,13] may not be sufficient. For instance, the users of a flight delay prediction system would be interested to find out the chance of the delay for a scheduled flight rather than a flight in the past.

The emerging paradigm of the Internet of Things (IoT) aims at establishing a worldwide pool of sensors to interconnect physical devices [3,7]. Thus, sensors will become the main generator of data on the Internet and enable a ubiquitous sensing of the environment. Based on the IoT data, Context-Aware Computing [8] can increase the effectiveness of the flight delay analysis. We use the scenario in Fig. 1 to illustrate this idea.

In this study, we tackle a number of technical challenges to enable real-time flight delay analysis based on the IoT data. To the best of our knowledge, due to privacy issues, the access to the real-world IoT data remains very limited. In addition, none of the previous works has investigated the connection between contextual IoT data and flight schedules. In this paper, we crawl and use real-world datasets to identify the correlation of the different data sources which consist of flight, weather and air quality data sources. We summarize our contributions as follows:

- We create an IoT search engine to crawl the data from publicly available websites. In our crawler, we identify and standardize a set of steps to facilitate the Extract, Transform and Load processes in acquiring IoT data. In the context of IoT, users would normally be less interested in finding the pages of

things (unlike finding Web pages in the Internet). Thus, we add the analysis of the flight delays to enhance the interests of the users in the result.
- We crawl IoT data from different data sources. We examine the correlation between different datasets and the projected flight delays dataset. We use multiple linear regression to investigate the effectiveness of each feature base on the crawled datasets. This helps us to prepare a prediction model based on our datasets.

The rest of this paper is organized as follows: In Sect. 2, we present the details of our search and crawler engine. In Sect. 3, we overview the related works. In Sect. 4, we present the details of the features included in our model. We present the results in Sect. 5. Finally, we conclude the paper in Sect. 6.

2   IoT Crawler and Search Engine

To minimize the required amount of work when collecting data from a new source, we have broken down the crawling procedure into a certain set of steps in a unified framework.

In the first step of crawling, a URL generator initializes the queue of queries. Each entry in the queue is supplied with certain parameters to construct a query to a page or a specific location. The parameters can be the time window, the boundaries of the querying region and/or other parameters. Then for each entity in the queue, a reader function reads the selected part of the page, and the contents are converted to a set of vectors and refined using a refiner. The refiner basically bind all read data from the previous step into subsets. The data for each subset is separately held until all subsets are refined where we merge all of the subsets of the resource's data. In this step, a specific enricher can be possibly used to collect the missing information, if any, from other sources. This can, for example, fill the incomplete fields such as IP address by acquiring them from Shodan. Finally, the collected data from different sources are integrated and stored on a distributed back-end.

Due to the size and dynamics of the sensor-generated data, IoT data sources often provide a subset of their data with a call to their API. Thus, pagination techniques such as location-based queries are deployed to present the data. We use the same mechanism through implementing the URL generator. The URL generator plays a key role in adjusting the workload on the data source. It converts a set of spatial segments to a sequence of queries which can be submitted via the API of the data source. Thus, a highly populated area can be placed multiple times in the processing queue while an empty area may appear only once (or not appear) in the queue. For example, through a URL generator, URL b will be repeated three times for others during a scan as it contains more dynamic objects than others:

We have developed our crawler using a set of tools to collect, process and visualize the dataset. Some of the tools we used are as follows: R programming language, SparkR, Apache Spark 1.4.1 and Rails framework. We initialized the crawler with around 3 data sources for air quality, weather watch and aircraft tracking.

3   Related Works

In this section, we overview related works to our paper. Flight delay is not a new problem, and it has been considered by many researchers. Here it is the most relevant work to our work. In [9] the authors analyzed the time factor influence of the flight delay in twenty airports in the US. They observed the changes of the delay rate using historical data. They used ANOVA and k-means clustering model in order to demonstrate the periodic of the delay rate. Although their model was able to predict accurately for the first airport they were studying, they found out that their model should be improved in order to be applied to the other 19 airports. However, they did not consider the airline influence.

Liu and Yang studied in [5] the flight delay propagation in the flight chain. So they proposed a new algorithm that could estimate the delay from the beginning in order to determine how much time the flights in chain could be delayed. Authors of [5] did not focus on the potential causes of the delay. They only modelled the problem utilizing the Bayesian Network. Liu and Ma (2009) [4] analyzed how flight delay is influenced by delay propagation using Bayesian Network. First, they investigated the correlation between the departure delay and the arrival delay at a particular airport. They found that the majority of delays happens in the period between 8 am and 9 pm. They measured the delays as light, medium, or heavy. They proposed that canceling flights when there is a heavy delay in the chain will relief the problem. Even though canceling the flights will definitely help other subsequent flights in the chain to be on-time, other factors that may cause the flight delay should be taken in account.

In the study in [13], authors studied the major factors that contribute to flight delay. They developed a model to predict the flight delay using historical records of Denver International Airport. Basically, their model considers two types of delays. First is daily propagation patterns that might be caused by crew connection problems, propagated delay from previous flights, or other factors. Second is seasonal trend where weather or seasonal demand have impact on it. However, as in [9] predicting the status of the flight in the future would require additional dynamic resources that could enrich the model. In [3] the authors looked at how the arrival delay could be propagated and impact the other subsequent flights in the stream. They believe all these types of delay only happen in busy hub-airports. They created three models. This study claims that the arrival delay is the source that mainly cause the departure delay.

Geng in his paper [2] provided statistical analysis of the flight delay. He listed all potential factors that may cause the flight delay. Some of these factors are airports, airlines, passengers, public safety, weather, fuel, departure control system, and air force. All these factors are actually play a role on the flight delay. Then he discussed some countermeasures in order to deal with the flight delay. Another study [11] focused on study the flight delay problem based on the random flight point delays. They used time series analysis on airline data and presented an influence factor model of the random flight points.

As the best of our knowledge, there is no study has considered the real-time data to investigate the flight delay. In [9] the authors recommend for the future work to combine the analysis of historical data with real time data. That would predict the on-time performance of any airport. Our work will consider the real time data to predict the performance of individual flights. Rebollo and Balakrishnan [10] presented a new model to predict the flight delay. They consider the temporal and the spatial delay states as explanatory variables. Their approach is to predict the delay sometime in the future between 2 to 24 h. They use the Random Forest algorithm to do so. Although this model predicts the flight status in the future, the aforementioned interval seems too short because people require time more than that when they book their flights.

4   Model Features

Since all the previous studies only considered the historical data of flights and weather, our data model will be based on real-time and new data. As a result, we use real-time data sources from sensors publicly distributed in order to explore and analyse the flight delay phenomenon. Our main method to find such sources is by using set of keywords such as real-time map [of application], live map [of application], and tracker map of [application]. The term application in this context refers to any IoT data source.

Our main goals for this research are to identify the most important factors that contribute to the flight delay, to create a model that predict the possibility if an individual flight would be on-time or delayed, and finally to estimate the magnitude of this phenomenon. In the next section, we describe a subset of real-time data sources types with some examples that we use in this research. Then, we discuss all features from the data sources in order to provide some understanding of each one of them.

We come up with list of features. Figure 2 shows some of the features for some IoT data sources. We get these features when read the data sources. We collect them in order to combine them with other features from other data sources. In the next section we will explain each feature in the chart. Table 1 describes features we get from IoT sources. These features will be used later in our analysis while studying the flight delay problem and creating the predictive model.

5   Results

5.1   Data Sets

There are several data sources which provide live data of flights every day. They leverage data from several resources such as air traffic control systems. More importantly, they utilize the network ADS-B ground stations. We can realize the flight number, the origin, the destination, the scheduled and actual departure time, the scheduled and actual arrival time, the aircraft type, and many other flight details.

Many weather websites incorporate real-time weather data obtained from various weather and climate agencies. These sources offer wide range of relevant

Table 1. Feature List in Real-Time sensors

Source	Feature	Description
Flight	Time of day	the time of the flight during the day
	Day of week	the day of the flight during the week
	Departure/Arrival Delay	the departure delay and the arrival delay of the flight in minutes
	Origin/Destination	the origin airport, city, country of the flight
	Airport	the airport where the flight de-parts or arrives
	Airline	the airline that operates the flight
	Scheduled/Actual Departure	the scheduled/actual departure time of the flight
	Scheduled/Actual Arrival	the scheduled/actual arrival time of the flight
	Aircraft Type	the airplane type of the flight
	Flight Number	the flight number
	International/Domestic	if the flight domestic or international
Weather	Temperature	the current temperature
	Dew	the dew at the airport
	Humidity	the Humidity at the airport
	Wind Direction	the Wind Direction at the airport
	Wind Speed	the Wind Speed at the airport
	Wind Gust	the Wind Gust at the airport
	Wind Chill	the Wind Chill at the airport
	Raining	the Raining at the airport
	Snowing	the Snowing at the airport
	Visibility	the Visibility at the airport
	Pressure	the Pressure at the airport
	Heat Index	the Heat Index at the airport
Air Quality	aqi	the air quality index at the airport
	PM2.5	particulate matter $2.5\,\mu m$
	PM10	particulate matter $10\,\mu m$
	NO2	the chemical compound Nitrogen dioxide
	SO2	the chemical compound Sulfur dioxide
	O3	the Ozone
	CO	the Carbon monoxide

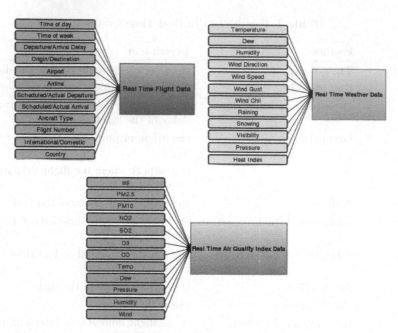

Fig. 2. Features from each source

weather information. They deliver their data in various format such as XML or map format. Weather Underground is one of the well-known data sources that provide live weather data. Its large network includes more than 180,000 weather stations.

An air quality index (AQI) is an index that indicates the quality of air in a place. This number is measured by monitoring the air data, and this index reflects the air quality standards. It tells how good or bad the air quality is.

5.2 Data Collection

We select 7 airports in 7 big cities in China. The cities are: Beijing (PEK), Shanghai (SHA), Guangzhou (CAN), Wuhan (WUH), Chengdu (CTU), Harbin (HRB), and Dalian (DLC). We collect all flights among these cities. The number of flights is 800. The distribution of the number of flights varies according to the city size and the population number. The airlines operate the flight between these cities are: China Air (CA), Shanghai Airlines (FM), China Eastern Airlines (MU), China Southern Airlines (CZ), Juneyao Airlines (HO), Hainan Airlines (HU), Xiamen Airlines (MF), Sichuan Airlines (3U), Shandong Airlines (SC), Chongqing Airlines(OQ), Grand China Air (CN), Shenzhen Airlines (ZH), Spring Airlines (9C), Tibet Airlines (TV), Beijing Capital Airlines (JD), Chengdu Airlines (EU).

identification.row	identification.number.default	identification.number.alternative	status.live	status.text
2938392811	CA1855	CA1855	FALSE	Scheduled
2934625405	CA1855	CA1855	FALSE	Scheduled
2930890219	CA1855	CA1855	FALSE	Scheduled
2930890220	CA1855	CA1855	FALSE	Scheduled
2920607116	CA1855	CA1855	FALSE	Scheduled
2916225693	CA1855	CA1855	FALSE	Scheduled
2912409039	CA1855	CA1855	FALSE	Scheduled
2908630624	CA1855	CA1855	FALSE	Scheduled
2904833455	CA1855	CA1855	FALSE	Scheduled

Fig. 3. Example of the flight records we get when we run the crawler

We run the crawler in order to collect data for the China case study. We get around 14,000 records for all flights among the selected cities mentioned above (Fig. 3). The data requires some cleaning as we will do in the next the step.

The process of collecting the weather data for a country differs from the process we do when we collect flights data. To collect the flight data, we need to prepare the flights data set in order to allow the crawler to check them online and bring all the required records. However, to collect the weather data for a particular country such as China, we need to write a function to instruct the crawler to fetch the data from the stations that are in the range of the search. Otherwise, the crawler ignores reading data from the other stations. We do that for two reasons. First we only need weather data of China. Second, we build this method in order to generalized it for any other countries. As a result, we get 2091 records. To collect the air quality index data, we use the same idea of collecting weather data. So we write a function that sets the required location parameters of our case. We get 3084 records.

5.3 Data Cleaning

After we read the data from all targeted IoT data sources, we should clean the data of each source. We use some criteria to clean. For the flight data, we summarize the data set by only selecting the data that we are interested in. In addition, we change the columns names and the order of the columns to make meaningful as you can see in the Fig. 4. We do the same process for the weather data and the air quality data. As you can see from Fig. 4, the flight number is duplicated. Each record of that flight is in a different date, so we will use them in our study.

Furthermore, we create some extra columns by transforming the content of some columns and by doing some calculation such as the delay at departure and the delay at arrival in minutes.

5.4 Data Integration

Performing exploration and analysis of the data to predict the flight delay requires us to combine all data sources in one container. We find the contents

Flight ID	Status with Time	Status	Status Type	Airline	Airline IATA Code	Origin Airport	Origin Airport IATA	Orig Airport Latitude	Orig Airport Longitude
CA1855	Scheduled	scheduled	departure	Air China	CA	Beijing Capital International Airport	PEK	40.080109	116.584503
CA1855	Scheduled	scheduled	departure	Air China	CA	Beijing Capital International Airport	PEK	40.080109	116.584503
CA1855	Scheduled	scheduled	departure	Air China	CA	Beijing Capital International Airport	PEK	40.080109	116.584503
CA1855	Scheduled	scheduled	departure	Air China	CA	Beijing Capital International Airport	PEK	40.080109	116.584503
CA1855	Scheduled	scheduled	departure	Air China	CA	Beijing Capital International Airport	PEK	40.080109	116.584503
CA1855	Scheduled	scheduled	departure	Air China	CA	Beijing Capital International Airport	PEK	40.080109	116.584503
CA1855	Scheduled	scheduled	departure	Air China	CA	Beijing Capital International Airport	PEK	40.080109	116.584503

Fig. 4. Flight data after cleaning and changing the columns names

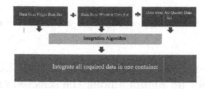

Fig. 5. Integration structure

of all data sources share the longitude and latitude columns in common. The longitude and the latitude represent the location of a particular object such as an airport, a weather station, etc. That means the integration process should be done in a particular way. Therefore, we combine the flight data and weather data based on the location of objects. We integrate data sources together by using this algorithm (Fig. 5).

1. Get the location of the airport from the flight data.
2. For each weather record, check the distance between the airport and the weather station.
3. If the distance is 5 Km or less then take the weather information from that station
   - We set the distance to be 5 Km because we tried to use smaller distances, but we did not find enough information. So 5 Km seems a realistic distance to get the data from the nearest station.
   - The calculation of the distance is based on the Haversine formula.

5.5 Data Exploration and Visualization

Airlines Performance. We analyse the collected airline performance data by exploring them using one of the statistical methods. We use boxplot in order to find out the overall performance of each individual airline. We get interesting plot that gives us an indication about the airline impact on the flight delay problem. We believe that airline factor plays a significant role on the flight delay. As we can see from the Fig. 6, some airlines do not operate the majority of their flight on-time. For example, as shown in Fig. 6, the majority of the Tibet Airlines (TV) flights are delayed likewise the Spring Airlines (9C).

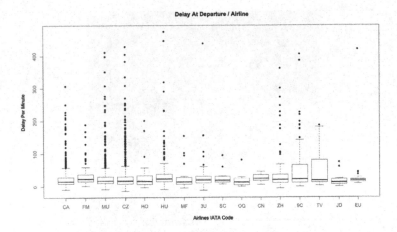

Fig. 6. Delay at departure performance for airlines

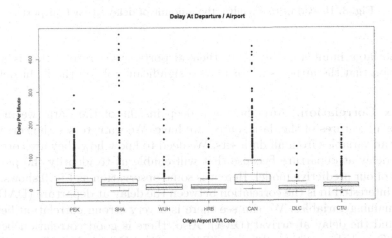

Fig. 7. Delay at departure performance for airports

Airports Performance. When looking at the individual airports, we also use boxplot to find the performance of each one of them. At this stage, we do not consider the capacity of the airport and how busy it is. We just want to see how much delay each airport have. Interestingly, we find that Guangzhou airport (CAN) does not perform well since the majority of flights are delayed. We know that Beijing airport (PEK) is as big as Guangzhou (CAN), but the performance of (PEK) airport seems normal. As a result, this indicates that the airport factor should be taken in account in the future (Fig. 7).

Heat Map. We create a heat map in order to visualize the delay size of each airport. As the Fig. 8 shows, most of the flight delays happen in large cities Shanghai, Beijing, and Guangzhou. This is due to the large size of these cities

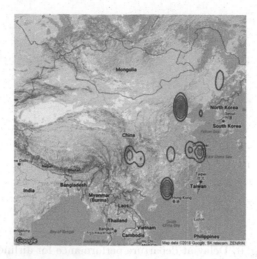

Fig. 8. HeatMap to visualize the amount of delay at each airport

and the large number of flights in their airports. As a result, this is a good indication that the airports factor plays a significant role on the flight delay.

Matrix Correlation. After having a deep insight of the data, we move to analyse all features of the data sources we have. We want to see the correlation among all variables from all data sets. We need to know how they are correlated to the delay at departure because that will enable us to identify the potential factors of our predictive model. Here it is some observation as Fig. 9 shows: what we are interested in is the correlation between the delay at departure (DAD) and the remaining variables. We can see there is a very strong correlation between DAD and the delay at arrival (DAA). Also, there is good correlation between the DAD and the weather elements (Temperature, Heat index, Dew, visibility, and elevation). When we look at the correlation among the weather data, we can observe that some of them have almost perfect correlation. So, we will continue to analyse the data more with having various study cases where the weather plays a significant role. We will also see the correlation again when we add more extra data sets.

5.6   Predictive Model

The purpose of studying the correlation among the features in the IoT data sources is to determine their impact on the flight delay. When we identify how each contributes to this problem, we will be able to create a predictive model using some machine learning methods. The predictive model will predict the flight delay. As a result, when we pass a given flight ID along with the time of the flight, our model should classify the status of the flight whether delayed or

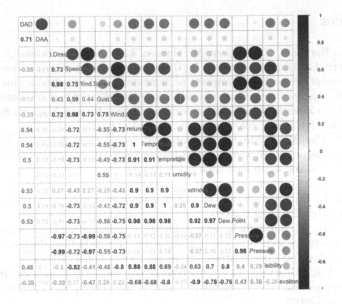

Fig. 9. Results of the correlation analysis

on-time. Then we will let the model determine the period of delay because that will increase the awareness of the user if she wants to accept that delay.

6 Conclusion

Previous studies have addressed the flight delay problem in terms of historical data that were collected by the Bureau of Transportation and the Federal Aviation Administration. These studies were helpful to determine some of the major factors that cause the flight delay. However, studying this phenomenon requires to consider other data rather than depending on the data related to the Air traffic. In order to tackle this problem, we need to widen our vision and the context of the problem by incorporating some extra data sources that provide real-time data. Therefore, we utilize the real-time data provided from the various resources as indicated above. This research will look for the contextual data which is not considered before. That means this study would provide a new step toward investigating this issue. We build a crawler to collect data. We analyze the data to see how the data from different data sets are correlated. After we determine the key factors that cause the delay, we will come up with a mathematical model in order to predict the flight delay in advance. That will enable all stakeholders to make the right decisions. This study will provide a novel method to discuss the flight delay, and it will contribute to allow further investigations by utilizing the contextual data. Furthermore, this is study will be significant in both academic and industry. With the emergence of the IoT paradigm, huge amount of data is there. This research is significant because it will contribute to put the first brick in order to fill the gap since there is a lack in this field.

References

1. Ball, M., Barnhart, C., Dresner, M., Hansen, M., Neels, K., Odoni, A., Peterson, E., Sherry, L., Trani, A.A., Zou, B.: Total delay impact study: a comprehensive assessment of the costs and impacts of flight delay in the united states (2010)
2. Geng, X.: Analysis and countermeasures to flight delay based on statistical data. In: 2013 5th International Conference on Intelligent Human-Machine Systems and Cybernetics (IHMSC), vol. 2, pp. 535–537. IEEE (2013)
3. Liu, Y.J., Cao, W.D., Ma, S.: Estimation of arrival flight delay and delay propagation in a busy hub-airport. In: 2008 Fourth International Conference on Natural Computation, vol. 4, pp. 500–505. IEEE (2008)
4. Liu, Y.J., Ma, S.: Flight delay and delay propagation analysis based on Bayesian network. In: Intl. Symposium on. Knowledge Acquisition and Modeling, KAM 2008, pp. 318–322. IEEE (2008)
5. Liu, Y., Yang, F.: Initial flight delay modeling and estimating based on an improved Bayesian network structure learning algorithm. In: 2009 Fifth International Conference on Natural Computation, vol. 6, pp. 72–76. IEEE (2009)
6. Mueller, E.R., Chatterji, G.B.: Analysis of aircraft arrival and departure delay characteristics. In: AIAA aircraft technology, integration and operations (ATIO) Conference (2002)
7. Perera, C., Liu, C.H., Jayawardena, S., Chen, M.: A survey on internet of things from industrial market perspective. IEEE Access 2, 1660–1679 (2014)
8. Perera, C., Zaslavsky, A., Christen, P., Georgakopoulos, D.: Context aware computing for the internet of things: a survey. IEEE Commun. Surv. Tutorials 16(1), 414–454 (2014)
9. Qin, Q., Yu, H.: A statistical analysis on the periodicity of flight delay rate of the airports in the US. Advances in Transportation Studies (2014)
10. Rebollo, J.J., Balakrishnan, H.: Characterization and prediction of air traffic delays. Transp. Res. Part C Emerg. Technol. 44, 231–241 (2014)
11. Rong, F., Qianya, L., Bo, H., Jing, Z., Dongdong, Y.: The prediction of flight delays based the analysis of random flight points. In: 34th Chinese Control Conference (CCC), 2015, pp. 3992–3997. IEEE (2015)
12. Rosenberger, J.M., Schaefer, A.J., Goldsman, D., Johnson, E.L., Kleywegt, A.J., Nemhauser, G.L.: A stochastic model of airline operations. Transp. Sci. 36(4), 357–377 (2002)
13. Tu, Y., Ball, M.O., Jank, W.S.: Estimating flight departure delay distributions a statistical approach with long-term trend and short-term pattern. J. Am. Stat. Assoc 103(481), 112–125 (2008)

Demo Papers

IRS-HD: An Intelligent Personalized Recommender System for Heart Disease Patients in a Tele-Health Environment

Raid Lafta[1]($\boxtimes$), Ji Zhang[1], Xiaohui Tao[1], Yan Li[1], and Vincent S. Tseng[2]

[1] Faculty of Health, Engineering and Sciences, University of Southern Queensland,
Toowoomba, Australia
{RaidLuaibi.Lafta,ji.zhang,xtao,yan.li}@usq.edu.au
[2] Department of Computer Science, National Chiao Tung University,
Hsinchu, Taiwan
vtseng@cs.nctu.edu.tw

Abstract. The use of intelligent technologies in clinical decision making support may play a promising role in improving the quality of heart disease patients' life and helping to reduce cost and workload involved in their daily health care in a tele-health environment. The objective of this demo proposal is to demonstrate an intelligent prediction system we developed, called IRS-HD, that accurately advises patients with heart diseases concerning whether they need to take the body test today or not based on the analysis of their medical data during the past a few days. Easy-to-use user friendly interfaces are developed for users to supply necessary inputs to the system and receive recommendations from the system. IRS-HD yields satisfactory recommendation accuracy, offers a promising way for reducing the risk of incorrect recommendations, as well saves the workload for patients to conduct body tests every day.

1 Introduction

The chronical diseases such as heart disease have become the main public health issue worldwide accounting for 50 % of global mortality burden [1]. Today, the survival rates have been increased partially due to technological advancements in disease prediction. Extensive research work has been carried out on data mining and analytic in various medical domains [2].

One of the important problems in medical science is accurate prediction of disease based on analysing historical data of patients. The data mining techniques and statistical analysis have been extensively used to reduce various healthcare and medical issues. They have provided major assistance to experts in disease prediction [5], which can help in minimizing medical errors and providing more detailed data analysis in a shorter time. There are various predictive data mining techniques such as classification by decision tree induction, Support Vector Machine (SVM), Neural Networks, Bayesian classification, and classification based on association that are accepted for disease risk assessment and prediction

© Springer International Publishing AG 2016
J. Li et al. (Eds.): ADMA 2016, LNAI 10086, pp. 803–806, 2016.
DOI: 10.1007/978-3-319-49586-6_58

in medical domain [3–7]. Intelligent technologies can particularly be developed to greatly facilitate the development and deployment of tele-health systems for patients including those who suffer from chronical diseases such as heart disease and require continuous monitoring of their heart-related medical measurements.

When performing remote continuous monitoring of patients' key measurement readings, an abundance of time series data are generated. Most of the existing predictive analytic methods on medical time series data are used to predict the long-term risk (e.g., the chance of survival) or the diagnosis of diseases. Nevertheless, it turns out that short-term prediction is more difficult than long-term projection due to a higher level of short-term uncertainty existing in the readings of various medical measurements. In addition, short-term recommendations are equally useful for patients as they provide guidance as to what the patients need to do within a short timeframe.

Motivated by the need of a strong intelligent prediction system, this demo proposal presents an intelligent recommender system we developed, called IRS-HD (stands for Intelligent Recommender System for Heart Disease patients). The system is supported by several predictive algorithms for short-term risk assessment on patients in telehealth environment based on analysis of a patient's historical medical data. On the basis of assessment results, the system provides personalized recommendations to patients suffering from heart diseases in relation to the necessity of medical tests taken on a daily basis.

2 Proposed Recommendation System

In this demo proposal, we present an intelligent recommendation system equipped with an array of predictive algorithms to analyze the medical data of heart disease patients, assess their risk and provide them with appropriate recommendations regarding the necessity of taking a medical test in the following day based on the outcome of the prediction.

Easy-to-use user friendly interfaces are developed for users to supply necessary inputs to the system and receive recommendations from the system. IRS-HD involves human computer interaction to receive input from human users concerning the values of the parameters that are used in the algorithm of the system. The recommendations generated by IRS-HD will be returned back to users through different channels and platforms including desktops, laptops and even tablets to embrace the latest technological advancement for quick information dissemination. Besides returned back to the patients, the results can also be sent remotely to medical practitioners such as doctors and nurses so that they can be informed and keep track of the physical checkups and overall health conditions of the patients.

The demonstration system utilizes heterogenous prediction algorithms, including Basic Heuristic Algorithm (BHA), Regression-Based Algorithm (BRA), Hybrid Algorithm (HA), Neural network (NN), Least Square-Support Vector Machine (LS-SVM), Naive Bayes (NB) and K-Nearest Neighbor (KNN), for conducting predictive analysis and produce for ensemble-based recommendations to patients.

Specifically, there are three parameters in the recommendation system requiring configuration before use. The first parameter is the medical measurement (e.g., heart rate, blood pressure, etc.) that the prediction is working on. The minimum (min) and maximum (max) normal values for the selected medical measurement are also specified to establish its normal range. The second parameter is the length of the sliding time window k which determines the historical data to be utilized for prediction and recommendation. The final parameter is p for determining the minimum percentage of days when physical test is conducted by the patient for the measurement in the past k days.

All phases of IRS-HD were implemented using Matlab. The user must enter the three parameters to the system and then the predictive analytic methods for prediction. Users can choose one or multiple different analytical methods and, if multiple methods are selected, IRS-HD utilizes ensemble-based method to produce the prediction and recommendation. Through this function, IRS-HD can assess the patient's status and then generate recommendations based on the analysis of his/her measurement readings for the past k days and decide whether a given patient needs to take a medical measurement such as the heart rate test today or not.

Here, if the patient does not need to take the test on the following day for a selected medical measurement, a recommendation of "no test needed" will be generated and presented on the interface, and stored as well into the backend database as a part of the patient's historical records, as shown in Fig. 1. Otherwise, a recommendation of "test required" will be generated and the patient is suggested to take the medical test on the following day. Then, the recommendation system will ask the user to enter the his/her test value for today which will be stored into the system as a historical record, as shown in Fig. 2.

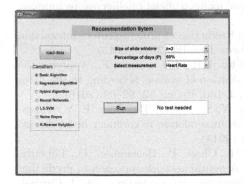 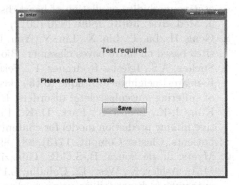

Fig. 1. Generation of the recommendation "No test needed" in IRS-HD

Fig. 2. Generation of the recommendation "Test required" in IRS-HD

3   Demonstration Plan

This section presents the detailed plan for our demonstration, which consists of the following parts:

First, we will describe to the audience the limitation in the existing predictive algorithms used to various healthcare and medical issues which motivates the development of our intelligent recommender system. We demonstrate the prototype of an intelligent recommendation system that is supported internally by innovative prediction and recommendation algorithms and feature intuitive user friendly system interface;

Second, we will display the system architecture of IRS-HD. The technical detail of the system will be introduced for its components and how connect each other in the recommendation system. We will also present to the audience how the ensemble-based prediction is performed within the system;

Finally, we will present the interactive interfaces developed for our recommendation system and how they significantly facilitate receiving input from human users concerning the values of parameters that are used in the algorithm of our system. Experimental evaluation results will be presented to the audience to show the performance of IRS-HD in terms of recommendation accuracy and workload saving (That is the percentage of days when the physical medical test is not required based on the recommendation).

References

1. Kuh, D., Shlomo, Y.B.: A Life Course Approach to Chronic Disease Epidemiology. Oxford University Press, Oxford (2004)
2. Hsieh, N.-C., Hung, L.-P., Shih, C.-C., Keh, H.-C., Chan, C.-H.: Intelligent postoperative morbidity prediction of heart disease using artificial intelligence techniques. J. Med. Syst. **36**(3), 1809–1820 (2012)
3. Geng, H., Lu, T., Lin, X., Liu, Y., Yan, F.: Prediction of protein-protein interaction sites based on naive bayes classifier. Biochem. Res. Int. **2015** (2015)
4. Snchez, A.S., Iglesias-Rodrguez, F., Fernndez, P.R., de Cos Juez, F.: Applying the K-nearest neighbor technique to the classification of workers according to their risk of suffering musculoskeletal disorders. Int. J. Ind. Ergon. **52**, 92–99 (2016)
5. Kim, J.-K., Lee, J.-S., Park, D.-K., Lim, Y.-S., Lee, Y.-H., Jung, E.-Y.: Adaptive mining prediction model for content recommendation to coronary heart disease patients. Cluster Comput. **17**(3), 881–891 (2014)
6. Myers, J., de Souza, B.-S.C.R., Guazzi, M., Chase, P., Bensimhon, D., Peberdy, M.A., Ashley, E., West, E., Cahalin, L.P.: A neural network approach to predicting outcomes in heart failure using cardiopulmonary exercise testing. Int. J. Cardiol. **171**(2), 265–269 (2014)
7. Bashir, S., Qamar, U., Khan, F.H.: BagMOOV: A novel ensemble for heart disease prediction bootstrap aggregation with multi-objective optimized voting. Australas. Phys. Eng. Sci. Med. **38**(2), 305–323 (2015)

Sentiment Analysis for Depression Detection on Social Networks

Xiaohui Tao[✉], Xujuan Zhou, Ji Zhang, and Jianming Yong

University of Southern Queensland, Toowoomba, Australia
{xtao,susan.zhou,ji.zhang,jianming.yong}@usq.edu.au

Abstract. As a response to the urgent demand of methods that help detect depression at early stage, the work presented in this paper has adopted sentiment analysis techniques to analyse users' contributions of social network to detect potential depression. A prototype has been developed, aiming at demonstrating the mechanism of the approach and potential social effect that may be delivered. The contributions include a depressive sentiment knowledge base and an algorithm to analyse textual data for depression detection.

1 Introduction

In this era of global digital connectivity and consequent increasing human disconnection, depression numbers are rising at an alarming rate. Depression can lead to suicide [5]. The World Health Organisation (WHO) states that suicide is the second leading cause of death among 15 to 29-year-olds [3]. A number of researches have been done to understand if social media include Twitter can be used to determine depression and how Twitter can be used to detect depression accurately by using machine learning and nature language processing for sentiment analysis techniques [4, 6–8].

The "Sentiment Analysis for Depression" tool in the demonstration is designed to provide alerts to social workers or a designated care person or parent, in the event that a user shows signs of depression. Whilst the work will not claim to replace or represent any type of professional mental health care, it will be useful as a tool to monitor a specific threshold of decline in the perspective of the user. The sentiment result is based on the analysis of text from the user's Twitter account. The concept of the work is to rate the user's level of depression by using sentiment analysis techniques. The data consists of user's contribution to social media such as tweets and is measured for the occurrence of depressive words over a period of time. When the result presents either an increase in the use of such words, or the incidence of depressive words reaches a given threshold, an alert is triggered. Providing such a tool can help social workers to detect and assist potential depressive people at early stage, as well for parents to monitor any decline in the state of mind of vulnerable children and adolescents. Similarly, adults who are prone to depression, without

---

The original version of this chapter was revised: An acknowledgement has been added. The erratum to this chapter is available at DOI: 10.1007/978-3-319-49586-6_61

© Springer International Publishing AG 2016
J. Li et al. (Eds.): ADMA 2016, LNAI 10086, pp. 807–810, 2016.
DOI: 10.1007/978-3-319-49586-6_59

recognising it, may benefit from an automatic alert to a designated mentor. Early intervention, in this case, is key to prevention against further decline, and thus, motivated us in the work presented in this demonstration. An approach to provide availability of such early intervention is then the main contribution of our work to both knowledge advancement and social effect.

2   Framework

The architecture of proposed system is illustrated in Fig. 1. The system will first collect data (e.g., tweets posted by social network players) from social media. The textual data will then be processed and analysed by the system communicating with a knowledge base storing depressive sentiment vocabulary. The analysis result, for instance, any issues regarding potential depression will then be alerted to the monitor (e.g., social workers, the parents of children) via a friendly user interface. The research challenges remained in two core components of the system; the *knowledge base* with depressive sentiment vocabulary and the *sentiment analysis engine*.

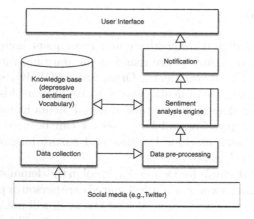

Fig. 1. The flow of proposed system architecture

Depressive sentiment vocabulary is the most important tool of this project. To the best of our knowledge, there is not a comprehensive vocabulary available in digital format. Aiming at generating the vocabulary, many web sources have been explored such as Synonyms.com and vocabulary.com with explanation of clinical depression and depressive disorder. Myvocabulary.com [2] and thesaurus.com [1] also provided some meaningful words of depression and can be helpful to build depression vocabulary. The knowledge base with depressive sentiment vocabulary in the work is a collection from all these sources manually and a result confirmed by multiple psychologists. Definition 1 defines the knowledge base used in the work for sentiment analysis, where 1 indicates the least level of depression and 3 the most serious level.

Definition 1. *The knowledge base is a feature vector* $\bar{V} := \{\langle t, w \rangle\}$, *where* $t \in \mathcal{V}$, *the depressive sentiment vocabulary, and* $w \in \{1, 2, 3\}$.

The collected social media data (textual documents like tweets) will first go through the text pre-processing phase including tasks such as stopword removal and word stemming, aiming at reducing dimensionality and improving accuracy in analysis. The documents are then represented as a feature vector after counting the statistic of terms, considering all as just "bag-of-words". Definition 2 formally defines the representation of documents in our work.

Definition 2. *A document is a feature vector* $\bar{d} := \{\langle t, w \rangle\}$, *where* $t \in \mathbb{T}$, *the universal term set, and* $w \in \{1, \ldots, +\infty\}$, *the frequency of* t *being observed in* d.

On the basis of Definitions 1 and 2, one may see that $\mathcal{V} \subset \mathbb{T}$ as $\mathbb{T}$ is the universal term set covering all possible terms including those collected in $\mathcal{V}$. The depression level of a social media document, denoted by $\text{dep}(d)$, can then be evaluated by calculating its semantic distance to the depressive sentiment vocabulary, because they are all represented as feature vectors. In this work, cosine similarity is used in calculation for the distance. Equation (1) presents the calculating the depression level revealed in a document. Apparently, a higher $\text{dep}(d)$ value suggests more severe depression revealed from document d.

$$\text{dep}(d) = \cos(\bar{d}, \bar{V}) = \frac{\bar{d} \cdot \bar{V}}{|\bar{d}| \, |\bar{V}|}. \tag{1}$$

The potential depression level of a user can then be evaluated by accumulating the $\text{dep}(d)$ value for all documents (e.g., tweets) contributed by him (her) over a period of time. Equation (2) presents the normalised form of $\text{dep}(u)$, where u denotes the user, D_u^τ denotes the collected social media documents contributed by u over a period of time τ, and $n = |D|$:

$$\text{dep}(u^\tau) = \frac{\sum\limits_{i=1}^{n} \text{dep}(d_i)}{|D_u^\tau|}. \tag{2}$$

The user's activities (contributions to social network like Twitter) will be observed and the $\text{dep}(u)$ will be evaluated in daily basis. An alert will be issued to the monitor if (i) the depression level $\text{dep}(u)$ is observed over a threshold on any day; or (ii) the depression level $\text{dep}(u)$ keeps of increasing for three continuing time units (e.g., days).

3   Demonstration Plan

The research work and the prototype will be demonstrated in two different versions. One is for offline analysis and will be demonstrated using a laptop. Visitors can press the buttons, check out the ranking list and alert list by comparing with the raw data (tweets) showing on the same screen. Such kind of tools will be valuable to social workers who may like to detect and help potential depressive people at early stage. The other version will be on mobile devices (e.g., iPhones) demonstrating online analysis. Visitors make tweet on an designated account on one device, and alerts are issued on the other mobile device. Such kind of tools will be in demand for close, one-to-one monitoring, for example, parents for their beloved child. The application can offer them a tool to monitor their children's psychological status with preservation of the children's privacy.

4   Conclusions

The work presented in this paper has successfully analysed the urgent demand of methods to help detect potential depression at early stage. As a response to the call, an approach has been developed in the work by analysing users' contribution to social networks such as Twitter using sentiment analysis techniques. A prototype is developed, which will serve to monitor user's tweets on Twitter and alert notifications to a nominated person in the event that depression is observed. Such an approach and the tool offer social workers the ability to access potential depressive people who deserve warm help at early stage, as well parents of teenagers, and concerned friends alike, the ability to be attentive to their beloved one's psychological status, without invasion of privacy, to ensure the wellbeing of their friends and relatives.

Acknowledgement. Special thanks go to the "Helping Minds" team from University of Southern Queensland, Australia, specifically, Heather Wallace, Declan Keyes-Bevan, Jason Alexander, and Jodie Coles, for implementation of the demo system.

References

1. Dictionary.com, "Depression" (2016). http://www.thesaurus.com/browse/depression
2. myvocabulary.com, "Depression Vocabulary Word List" (2016). https://myvocabulary.com/word-list/depression-vocabulary/
3. WHO, "Mental health: suicide data" (2016). http://www.who.int/mental_health/prevention/suicide/suicideprevent/en/
4. Tsugawa, S., Kikuchi, Y., Kishino, F., Nakajima, K., Itoh, Y., Ohsaki, H.: Recognizing depression from twitter activity. In: Proceedings of the 33rd Annual ACM Conference on Human Factors in Computing Systems, CHI 2015, New York, NY, USA, pp. 3187–3196. ACM (2015)
5. O'Dea, B., Wan, S., Batterham, P.J., Calear, A.L., Paris, C., Christensen, H.: Detecting suicidality on twitter. Internet Interventions 2(2), 183–188 (2015)
6. De Choudhury, M., Counts, S., Horvitz, E.: Social media as a measurement tool of depression in populations. In: Proceedings of the 5th Annual ACM Web Science Conference, WebSci 2013, New York, NY, USA, pp. 47–56. ACM (2013)
7. Resnik, P., Armstrong, W., Claudino, L., Nguyen, T., Nguyen, V.-A., Boyd-Graber, J.: Beyond lda: exploring supervised topic modeling for depression-related language in twitter. In: NAACL HLT 2015, p. 99 (2015)
8. Nambisan, P., Luo, Z., Kapoor, A., Patrick, T.B., Cisler, R.A.: Social media, big data, public health informatics: ruminating behavior of depression revealed through twitter. In: 2015 48th Hawaii International Conference on System Sciences (HICSS), pp. 2906–2913, January 2015

Traffic Flow Visualization Using Taxi GPS Data

Xiujuan Xu[1,2], Zhenzhen Xu[1,2], and Xiaowei Zhao[1,2](✉)

[1] School of Software, Dalian University of Technology, Dalian 116620, China
{xjxu,xzz,xiaowei.zhao}@dlut.edu.cn
[2] Key Laboratory for Ubiquitous Network and Service Software of Liaoning Province,
Dalian 116620, China

Abstract. Intelligent transportation systems (ITSs) became an essential tool for a broad range of transportation applications. Traffic flow visualization is an important problem in ITS. The visualized results can be used to support ITSs to plan operation and manage revenue. In this paper, we aim to visualize the daily floating taxis by presenting a novel figure using taxi trajectory data and weather information. Many visualization platforms feature a online-offline phase, in which taxi GPS trajectory data is processed by two phases. This approach incurs high costs though, since trajectory data is huge generated by taxis every second continually. To support the frequent trajectories, we present an analysis tool for mining frequent trajectories of taxis (FTMTool). It allows us to find the driver's routes by collecting input on the most frequent roads, thereby achieving a set of high quality routes. The tool also supports the task statistic in selecting the specific roads. We demonstrate the usefulness of our tool using real data from New York city.

Keywords: Taxi GPS data · Frequent trajectories · Big data · Intelligent transportation systems

1 Introduction

With the blooming development of positioning technology, the emerging GPS location can accurately record taxis' location and moving information, which provide the opportunity to gain insight on taxis' driving behavior at an unprecedented level. A taxi trajectory is consisted by GPS point data. Taxi drivers, however, have different backgrounds and wide-ranging levels of expertise and motivation in a city, so that the taxis trajectories shows the smart driving directions. Involving an intelligent expert on selecting roads, the historical GPS trajectories of a large number of taxis provides the status of roads in a city.

Some researchers have conducted on spatio-temporal analysis and visualization of mobility data in this field. Pablo Samuel Castro et al. provided exhaustive survey on mining trajectories [1]. Wei Ling-Yin et al. presented a Route Inference framework based on collective knowledge to construct the popular routes from uncertain trajectories [8]. Ding Chu et al. transformed the geographic coordinates (latitude and longitude) to street names so as to reflect contextual semantic information [2]. Ferreira Nivan et al. [4] proposed a new model that allowed

© Springer International Publishing AG 2016
J. Li et al. (Eds.): ADMA 2016, LNAI 10086, pp. 811–814, 2016.
DOI: 10.1007/978-3-319-49586-6_60

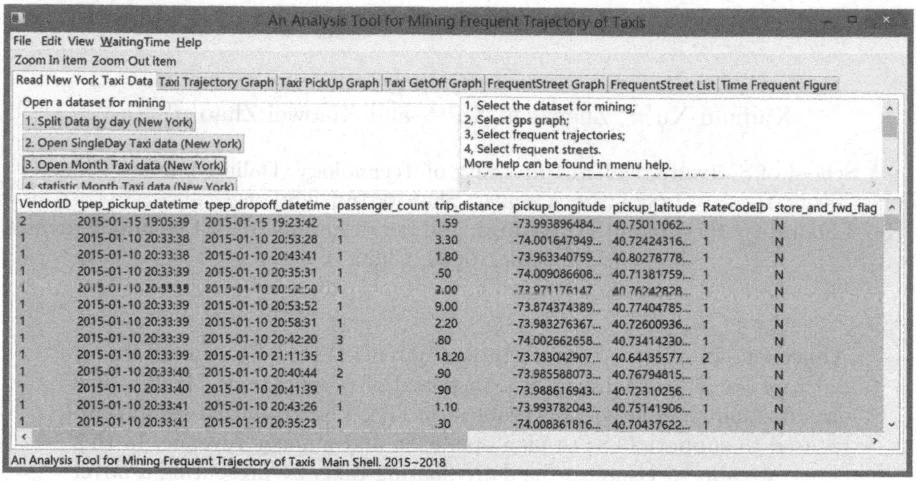

812   X. Xu et al.

Fig. 1. Frame traffic flow

users to visually query taxi trips. Wang Zuchao et al. presented a visual analysis system to explore sparse traffic trajectory data recorded by transportation cells [7]. Fei Wang et al. proposed a novel road-based query model for interactively conducting evaluation tasks [6]. Besides, Masahiko Itoh et al. [5] proposed a visual interaction method for exploring caution spots from large-scale vehicle recorder data. Different from the above work, we focus on visualizing the taxi trajectories for taxi drivers and passengers.

In this paper, we aim at mining taxis' historical riding patterns, to show the most frequent streets in a city. We hope FTMTool will help taxi drivers and passengers explore and analyze transportation status of a city.

2 Implementation

This section discusses how FTMTool handles large amounts of taxi trajectories and implements the above functionalities. The demo presents FTMTool: a tool for frequent trajectories mining for taxis in Fig. 1. Its functionalities are summarized as follows:

(1) Loading the taxi trajectories data in a list: We load the first 1,000 taxi trajectories in the FTMTool. In the dataset, every record in use contains the latitude, longitude of the picking-up point and getting-off point, and time and so on [3].
(2) Showing the GPS data map in a month: GPS data generated by taxis, can form a map of a city. By exploiting the depth of color in every GPS point, it is darker when the GPS points are denser.
(3) Mining frequent trajectories: User of transportation system, such as taxi drivers or passengers, can check the status of the whole city. Be computing

the map formed by taxis' gps trajectories, FTMTool could compute the frequent streets and frequent zone. FTMTool enables to obtain the top-k frequent streets on the map. Meanwhile, FTMTool enables to obtain the top-k frequent zones on the map.

(4) Showing the frequent by 24 hours of a day: FTMTool could show the frequent by 24 hours of a day.

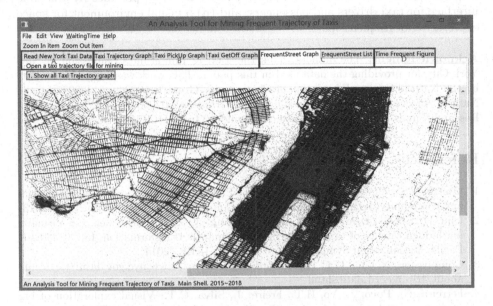

Fig. 2. Gpsgraph traffic flow

3   Demonstration

FTMTool is a taxis trajectories exploratory tool. By selecting a dataset to start with, the user tries to visualize the taxi trajectories of a city. With this basic information the user navigates and finds taxi trajectories according to the selected dimension. Figure 1 shows the overview of a taxi gps dataset of New York City. In the section A of Fig. 2, the users can observe the attributions of a dataset. The section B of Fig. 2 shows the map formed by taxi trajectories of all points, picking-up points and getting-off points. As can be observed, the darker points when more persons select the points as their pick-up points. The section C of Fig. 2 shows the map formed by taxi trajectories of all points. The section D of Fig. 2 presents the number of frequent trajectories by hours.

4   Summary

It is vital for the transportation systems which performs most of their work by automobiles to reduce its number of traffic accidents [5]. Traffic flow prediction is an important problem with the ever stretching of cities nowadays. In this paper, we present FTMTool that supports visual exploration of big space data. A key componet of this demo is a visual frequent trajectories model that allows users to quickly select a dataset and explore it. Our tool provides (i) a flexible interface for taxi drivers and passengers, and (ii) a visual environment for space exploration of frequent spots.

Acknowledgment. The authors thank the Taxi & Limousine Commission of New York City for providing the data used in this paper. This work was supported in part by the Natural Science Foundation of China under Grant 61502069, 61300087 by the Natural Science Foundation of Liaoning under Grant 2015020003, by the Fundamental Research Funds for the Central Universities under Grant DUT15QY40.

References

1. Castro, P.S., Zhang, D., Chen, C., Li, S., Pan, G.: From taxi gps traces to social and community dynamics: A survey. ACM Comput. Surv. (CSUR) **46**(2), 17 (2013)
2. Chu, D., Sheets, D.A., Zhao, Y., Wu, Y., Yang, J., Zheng, M., Chen, G.: Visualizing hidden themes of taxi movement with semantic transformation. In: 2014 IEEE Pacific Visualization Symposium, pp. 137–144. IEEE (2014)
3. Commission, N.T.L.: TLC Trip Record Data. http://www.nyc.gov/html/tlc/html/about/trip_record_data.shtml
4. Ferreira, N., Poco, J., Vo, H.T., Freire, J., Silva, C.T.: Visual exploration of big spatio-temporal urban data: A study of new york city taxi trips. IEEE Trans. Vis. Comput. Graph. **19**(12), 2149–2158 (2013)
5. Itoh, M., Yokoyama, D., Toyoda, M., Kitsuregawa, M.: Visual interface for exploring caution spots from vehicle recorder big data. In: 2015 IEEE International Conference on Big Data (Big Data), pp. 776–784. IEEE (2015)
6. Wang, F., Chen, W., Wu, F., Zhao, Y., Hong, H., Gu, T., Wang, L., Liang, R., Bao, H.: A visual reasoning approach for data-driven transport assessment on urban roads. In: 2014 IEEE Conference on Visual Analytics Science and Technology (VAST), pp. 103–112. IEEE (2014)
7. Wang, Z., Ye, T., Lu, M., Yuan, X., Qu, H., Yuan, J., Wu, Q.: Visual exploration of sparse traffic trajectory data. IEEE Trans. Vis. Comput. Graph. **20**(12), 1813–1822 (2014)
8. Wei, L.Y., Zheng, Y., Peng, W.C.: Constructing popular routes from uncertain trajectories. In: Proceedings of the 18th ACM SIGKDD International Conference on Knowledge Discovery and Data Mining, pp. 195–203. ACM (2012)

Erratum to: Advanced Data Mining and Applications

Jinyan Li[1]([⊠]), Xue Li[2], Shuliang Wang[3], Jianxin Li[4],
and Quan Z. Sheng[5]

[1] University of Technology Sydney, Ultimo, NSW, Australia
`jinyan.li@uts.edu.au`
[2] University of Queensland, Brisbane, QLD, Australia
[3] Beijing Institute of Technology, Beijing, China
[4] University of Western Australia, Crawley, WA, Australia
[5] University of Adelaide, Adelaide, SA, Australia

Erratum to:
Chapter 52
Partitioning Clustering Based on Support Vector Ranking
DOI: 10.1007/978-3-319-49586-6_52

The original version of the paper starting on p. 726 was revised. An acknowledgement has been added. The original chapter was corrected.

Erratum to:
Chapter 59
Sentiment Analysis for Depression Detection on Social Networks
DOI: 10.1007/978-3-319-49586-6_59

The original version of the paper starting on p. 807 was revised. An acknowledgement has been added. The original chapter was corrected.

The updated original online version for this chapters can be found at
DOI: 10.1007/978-3-319-49586-6_52
DOI: 10.1007/978-3-319-49586-6_59

Author Index

Printed in the United States
By Bookmasters

Printed in the United States
By Bookmasters